Probability and Statistics with Reliability, Queuing, and Computer Science Applications

Probability and Statistics with Reliability, Queuing, and Computer Science Applications

KISHOR SHRIDHARBHAI TRIVEDI

Duke University
Durham, North Carolina

PRENTICE-HALL, INC., ENGLEWOOD CLIFFS, NJ 07632

Library of Congress Cataloging in Publication Data

Trivedi, Kishor Shridharbhai (1946–   )
Probability and statistics with reliability, queuing,
and computer science applications.

Bibliography
Includes index.
1. Probabilities—Data processing. 2. Mathematical
statistics—Data processing.. 3. Algorithms
1. Title
QA273.19.E4T74   519.2   81-20960
ISBN 0-13-711564-4   AACR2

Editorial/production supervision and interior design: Nancy Milnamow
Cover design: Tony Ferrara Studio
Manufacturing buyer: Gordon Osbourne

ISBN 0-13-711564-4

Printed in the United States of America

10 9 8 7 6 5 4 3 2

Prentice-Hall International, Inc., *London*
Prentice-Hall of Australia Pty. Limited, *Sydney*
Prentice-Hall of Canada, Ltd., *Toronto*
Prentice-Hall of India Private Limited, *New Delhi*
Prentice-Hall of Japan, Inc., *Tokyo*
Prentice-Hall of Southeast Asia Pte. Ltd., *Singapore*
Whitehall Books Limited, *Wellington, New Zealand*

Contents

PREFACE ix

Chapter 1
INTRODUCTION 1

1.1 Motivation *1*
1.2 Probability Models *2*
1.3 Sample Space *3*
1.4 Events *6*
1.5 Algebra of Events *7*
1.6 Graphical Methods of Representing Events *11*
1.7 Probability Axioms *14*
1.8 Combinatorial Problems *20*
1.9 Conditional Probability *23*
1.10 Independence of Events *26*
1.11 Bayes' Rule *33*
1.12 Bernoulli Trials *41*

Chapter 2
DISCRETE RANDOM VARIABLES 54

2.1 Introduction *54*
2.2 Random Variables and Their Event Spaces *55*
2.3 The Probability Mass Function *57*
2.4 Distribution Functions *58*
2.5 Special Discrete Distributions *62*

2.6 Analysis of Program MAX *84*
2.7 The Probability Generating Function *88*
2.8 Discrete Random Vectors *92*
2.9 Independent Random Variables *98*

Chapter 3
CONTINUOUS RANDOM VARIABLES **109**

3.1 Introduction *109*
3.2 The Exponential Distribution *114*
3.3 The Reliability, Failure Density, and Hazard Function *118*
3.4 Some Important Distributions *125*
3.5 Functions of a Random Variable *139*
3.6 Jointly Distributed Random Variables *144*
3.7 Order Statistics *148*
3.8 Distribution of Sums *154*
3.9 Functions of Normal Random Variables *170*

Chapter 4
EXPECTATION **181**

4.1 Introduction *181*
4.2 Moments *185*
4.3 Expectation of Functions of More Than One Random Variable *188*
4.4 Transform Methods *196*
4.5 Moments and Transforms of Some Important Distributions *205*
4.6 Computation of Mean Time to Failure *215*
4.7 Inequalities and Limit Theorems *223*

Chapter 5
CONDITIONAL DISTRIBUTION AND CONDITIONAL EXPECTATION **232**

5.1 Introduction *232*
5.2 Mixture Distributions *239*
5.3 Conditional Expectation *245*
5.4 Imperfect Fault Coverage and Reliability *251*
5.5 Random Sums *260*

Chapter 6
STOCHASTIC PROCESSES **268**

6.1 Introduction *268*
6.2 Classification of Stochastic Processes *274*
6.3 The Bernoulli Process *280*

6.4 The Poisson Process *283*
6.5 Renewal Processes *292*
6.6 Availability Analysis *297*
6.7 Random Incidence *302*
6.8 Renewal Model of Program Behavior *305*

Chapter 7
DISCRETE-PARAMETER MARKOV CHAINS **309**

7.1 Introduction *309*
7.2 Computation of n-step Transition Probabilities *311*
7.3 State Classification and Limiting Distributions *317*
7.4 Distribution of Times Between State Changes *325*
7.5 Irreducible Finite Chains with Aperiodic States *326*
7.6 The $M/G/1$ Queuing System *336*
7.7 Discrete-Parameter Birth-Death Processes *344*
7.8 Finite Markov Chains with Absorbing States: Analysis of Program Execution Time *351*

Chapter 8
CONTINUOUS-PARAMETER MARKOV CHAINS **360**

8.1 Introduction *360*
8.2 The Birth and Death Process *365*
8.3 Other Special Cases of the Birth-Death Model *388*
8.4 Non-Birth-Death Processes *393*
8.5 Markov Chains with Absorbing States *400*

Chapter 9
NETWORKS OF QUEUES **411**

9.1 Introduction *411*
9.2 Open Queuing Networks *416*
9.3 Closed Queuing Networks *423*
9.4 Nonexponential Service-Time Distributions and Multiple Job Types *446*
9.5 Non-Product-Form Networks *454*
9.6 Summary *464*

Chapter 10
STATISTICAL INFERENCE **469**

10.1 Introduction *469*
10.2 Parameter Estimation *471*
10.3 Hypothesis Testing *507*

Chapter 11
REGRESSION, CORRELATION, AND ANALYSIS OF VARIANCE **538**

11.1 Introduction *538*
11.2 Least-Squares Curve Fitting *543*
11.3 The Coefficient of Determination *546*
11.4 Confidence Intervals in Linear Regression *549*
11.5 Correlation Analysis *552*
11.6 Simple Nonlinear Regression *556*
11.7 Higher-Dimensional Least-Squares Fit *557*
11.8 Analysis of Variance *559*

Appendix A
BIBLIOGRAPHY **573**

Appendix B
PROPERTIES OF DISTRIBUTIONS **579**

Appendix C
STATISTICAL TABLES **582**

Appendix D
LAPLACE TRANSFORMS **602**

Appendix E
PROGRAM ANALYSIS **608**

AUTHOR INDEX **611**

SUBJECT INDEX **617**

Preface

The aim of this book is to provide an introduction to probability, stochastic processes, and statistics for students of computer science, electrical/computer engineering, reliability engineering, and applied mathematics. The prerequisites are two semesters of calculus, a course on introduction to computer programming, and preferably, a course on computer organization.

I have found that the material in the book can be covered in a two-semester or three-quarter course. However, through a choice of topics, shorter courses can also be organized. I have taught the material in this book to seniors and first-year graduate students but with the text in printed form, it could be given to juniors as well. The book is also suitable for self-study by computer professionals and mathematicians interested in applications.

With the specified audience in mind, I have attempted to provide examples and problems, with which the student can identify, as motivation for the probability concepts. The majority of applications are drawn from reliability analysis and performance analysis of computer systems and from probabilistic analysis of algorithms. Although there are many good texts on each of these application areas, I felt the need for a text that treats them in a balanced fashion.

The first five chapters provide an introduction to probability theory. These five chapters provide the core for a one-semester course on introduction to applied probability. Chapters 6 through 9 deal with stochastic processes and their applications. These four chapters form the core of a second course with a title such as systems modeling. I have included an entire chapter on networks of queues. The last two chapters cover statistical inference and regression, respectively. I have placed the material on sampling distributions in Chapter 3, dealing with continuous random variables. Portions of the chapters on statistics can be taught with the first course and other portions in the second course.

The appendices contain a bibliography classified by topics, properties of important distributions, statistical tables, a primer on Laplace transforms, and a table of formulae for program performance analysis.

In addition to more than two hundred worked examples, each section concludes with a number of exercises. Difficult exercises are indicated by an asterisk. A solution manual for instructors is available from the publisher.

I am indebted to the Department of Computer Science, Duke University and to Merrell Patrick for their encouragement and support during this project. The efficient typing skills of Patricia Land helped make the job of writing the book much easier than it might have been otherwise.

Many of my friends, colleagues, and students carefully read several drafts and suggested many changes improving the readability and the correctness of this text. Many thanks to Robert Geist, Narayan Bhat, Satish Tripathi, John Meyer, Frank Harrell, Veena Adlakha, and Jack Stiffler for their suggestions. Joey de la Cruz and Nelson Strothers helped in the editing and the typing process very early in the project. The help by the staff of Prentice-Hall in preparation of the book is also appreciated.

I would like to thank my wife, Kalpana, and my daughters, Kavita and Smita, for enduring my preoccupation with this work for so long. The book is dedicated to my mother, Jayaben, and to my father, Shridharbhai.

Kishor S. Trivedi

Probability and Statistics with Reliability, Queuing, and Computer Science Applications

Chapter 1

Introduction

1.1 MOTIVATION

Computer scientists need powerful analytic tools to analyze algorithms and computer systems. Many of the tools necessary for these analyses have their foundations in probability theory. For example, in the analysis of algorithm execution times, it is common to draw a distinction between the *worst-case* and the *average-case* behavior of an algorithm. The distinction is based on the fact that for certain problems, while an algorithm may require an inordinately long time to solve the least favorable instance of the problem, the average solution time is considerably shorter. When many instances of a problem have to be solved, the probabilistic (or average-case) analysis of the algorithm is likely to be more useful. Such an analysis accounts for the fact that the performance of an algorithm depends upon the distributions of input data items. Of course, we have to specify the relevant probability distributions before the analysis can be carried out. Thus, for instance, while analyzing a sorting algorithm, a common assumption is that every permutation of the input sequence is equally likely to occur.

Similarly, if the storage is dynamically allocated, a probabilistic analysis of the storage requirement is more appropriate than a worst-case analysis. In a like fashion, a worst-case analysis of the accumulation of round-off errors in a numerical algorithm tends to be rather pessimistic; a probabilistic analysis, though harder, is more useful.

When we consider the analysis of a computer system serving a large number of users, several types of random phenomena need to be accounted for. First, owing to a large population of diverse users, the arrival pattern of jobs is subject to randomness. Second, the resource requirements of jobs will likely fluctuate from job to job as well as during the execution of a single job. Finally, the resources of the computer system are subject to random failures due to environmental conditions and aging phenomena. The theory of stochastic processes is very useful in evaluating various measures of system effectiveness, such as throughput, response time, reliability, and availability.

Before an algorithm or a system can be analyzed, various probability distributions have to be specified. Where do the distributions come from? We may collect data during the actual operation of the system (or the algorithm). These measurements can be performed by hardware monitors, software monitors, or both. Such data need to be analyzed and compressed to obtain the necessary distributions that drive the analytical models discussed above. Mathematical statistics provides us with tools for this purpose, such as the **design of experiments, hypothesis testing, estimation, analysis of variance, linear and nonlinear regression**.

1.2 PROBABILITY MODELS

Probability theory is concerned with the study of random (or chance) phenomena. Such phenomena are characterized by the fact that their future behavior is not predictable in a deterministic fashion. Nevertheless, owing to certain statistical regularities, such phenomena are usually capable of mathematical descriptions. This can be accomplished by constructing an idealized probabilistic model of the real-world situation. Such a model consists of a list of all possible outcomes and an assignment of their respective probabilities. The theory of probability then allows us to predict or deduce patterns of future outcomes.

Since a model is an abstraction of the real-world problem, predictions based on the model must be validated against actual measurements collected from the real phenomena. A poor validation may suggest modifications to the original model. The theory of statistics facilitates the process of validation. Statistics is concerned with the inductive process of drawing inferences about the model and its parameters based on the limited information contained in real data.

The role of probability theory is to analyze the behavior of a system or algorithm assuming the given probability assignments and distributions. The results of this analysis are as good as the underlying assumptions. Statistics helps us in choosing these probability assignments and in the process of validating model assumptions. The behavior of the system (or the algorithm) is observed, and an attempt is made to draw inferences about the underlying unknown distributions of random variables that describe system activity. Methods of statistics, in turn, make heavy use of probability theory.

Consider the problem of predicting the number of job arrivals to a computer center in a fixed time interval $(0, t)$. A common model of this situation is to assume that the number of job arrivals in this period has a particular distribution, such as the Poisson distribution (see Chapter 2). Thus we have replaced a complex physical situation by a simple model with a single unknown parameter, namely the average job arrival rate λ. With the help of probability theory we can then deduce the pattern of future job-arrivals. On the other hand, statistical techniques help us estimate the unknown parameter λ based on actual observations of past arrival patterns. Statistical techniques also allow us to test the validity of the Poisson model.

As another example, consider a computer system with automatic error-recovery capability. Model this situation as follows: probability of successful recovery is c and probability of an abortive error is $1 - c$. The uncertainty of the physical situation is once again reduced to a simple probability model with a single unknown parameter c. In order to estimate parameter c in this model, we observe N errors out of which n are successfully recovered. A reasonable estimate of c is the relative frequency n / N, since we expect this ratio to converge to c in the limit $N \to \infty$. Note that this limit is a limit in a probabilistic sense; that is:

$$\lim_{N \to \infty} P(|\frac{n}{N} - c| > \epsilon) = 0,$$

and not the usual mathematical limit:

$$\lim_{N \to \infty} \frac{n}{N}.$$

In other words, given any small $\epsilon > 0$, it is not possible to find a value M such that

$$|\frac{n}{N} - c| < \epsilon \qquad \text{for all } N > M,$$

as would be required for the mathematical limit. Axiomatic approaches to probability allow us to define such limits in a mathematically consistent fashion (e.g., see the law of large numbers in Chapter 4) and hence allow us to use relative frequencies as estimates of probabilities.

1.3 SAMPLE SPACE

Probability theory is rooted in the real-life situation where a person performs an experiment the outcome of which may not be certain. Such an experiment is called a **random experiment**. Thus, an experiment may consist of the simple process of noting whether a component is functioning properly or has failed; it may consist of determining the execution time of a program; or it may consist of determining the response time of a terminal request. The result of any such observations, whether they be simple "yes" or "no" answers, meter readings, or whatever, are called **outcomes** of the experiment.

Definition (Sample Space). The totality of the possible outcomes of a random experiment is called the **sample space** of the experiment and it will be denoted by the letter S.

The sample space is not determined completely by the experiment. It is partially determined by the purpose for which the experiment is carried out. If the status of two components is observed, for some purposes it is sufficient to consider only three possible outcomes: two functioning, two malfunctioning, and one functioning and one malfunctioning. These three outcomes constitute the sample space S. On the other hand, we might be interested in exactly which of the components has failed, if any has failed. In this case the sample space S must be considered as four possible outcomes, where the earlier single outcome of one failed, one functioning is split into two outcomes: first failed, second functioning and first functioning, second failed. Many other sample spaces can be defined if we take into account such things as type of failure and so on.

Frequently, we use a larger sample space than is strictly necessary because it is easier to use; specifically, it is always easier to discard excess information than to recover lost information. For instance, in the preceding illustration, the first sample space might be denoted $S_1 = \{0, 1, 2\}$ (where each number indicates how many components are functioning) and the second sample space might be denoted $S_2 = \{(0, 0), (0, 1), (1, 0), (1, 1)\}$ (where $0 =$ failed, $1 =$ functioning). Given a selection from S_2, we can always add the two components to determine the corresponding choice from S_1; but, given a choice from S_1 (in particular one), we cannot necessarily recover the corresponding choice from S_2.

It is useful to think of the outcomes of an experiment, the **elements** of the sample space, as points in a space of one or more dimensions. For example, if an experiment consists of examining the state of a single component, it may be functioning properly (denoted by the number 1), or it may have failed (denoted by the number 0). The sample space is one-dimensional, as shown in Figure 1.1. If a system consists of two components there are four possible outcomes, as shown in the two-dimensional sample space of Figure 1.2. Here each coordinate is 1 or 0 depending on whether the corresponding component is functioning properly or has failed. In general, if a system has n components there are 2^n possible outcomes, each of which can be regarded as a point in an n-dimensional sample space. It should be noted that the sample space used here in connection with the observation of the status of components could also serve to describe the results of other experiments; for example, the experiment of observing n successive executions of an **if** statement, with 1 denoting the execution of the **then** clause and 0 denoting the execution of the **else** clause.

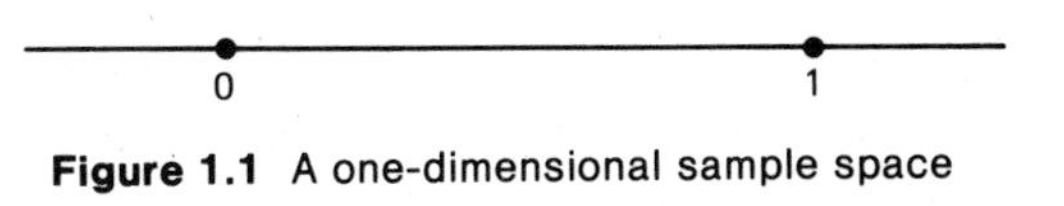

Figure 1.1 A one-dimensional sample space

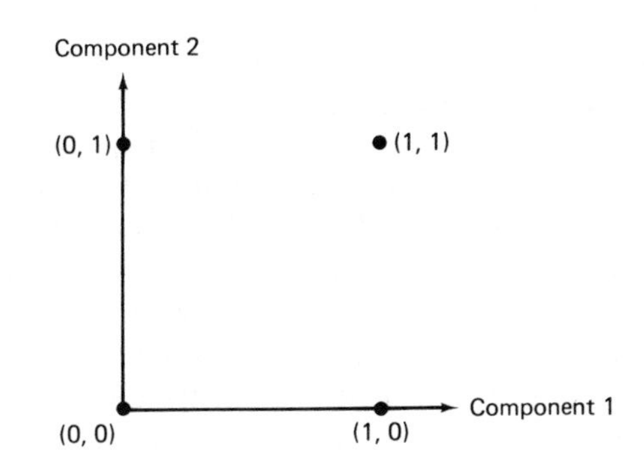

Figure 1.2 A two-dimensional sample space

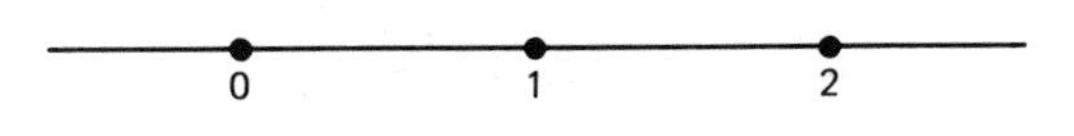

Figure 1.3 A one-dimensional sample space

The geometrical configuration that is used to represent the outcomes of an experiment (e.g., Figure 1.2) is not necessarily unique. For example, we could have regarded the outcomes of the experiment of observing the two-component system to be the total number functioning, and the outcomes would be 0, 1, 2, as depicted in the one-dimensional sample space of Figure 1.3. Note that the point 1 in Figure 1.3 corresponds to the two points $(0, 1)$ and $(1, 0)$ in Figure 1.2. It is often easier to use sample spaces whose elements cannot be further "subdivided"; that is, the individual elements of a sample space should not represent two or more outcomes that are distinguishable in some fashion. Thus, sample spaces like those of Figures 1.1 and 1.2 should be used in preference to sample spaces like the one in Figure 1.3.

It is convenient to classify sample spaces according to the number of elements they contain. If the set of all possible outcomes of the experiment is finite, then the associated sample space is a **finite sample space**. Thus, the sample spaces of Figures 1.1, 1.2, and 1.3 are finite sample spaces.

To consider an example where a finite sample space does not suffice, suppose we inspect components coming out of an assembly line and that we are interested in the number inspected before we observe the first defective component. It could be the first, the second, . . ., the hundredth, . . ., and, for all we know, we might have to inspect a billion or more before we find a defective component. Since the number of components to be inspected before the first defective one is found is not known in advance, it is appropriate to take the sample space to be the set of natural numbers. The same sample space results for the experiment of tossing a coin until a head is observed. A sample space such as this, where the set of all outcomes can be put into a one-to-one correspondence with the natural numbers, is said to be **countably infinite**. Usually it is not necessary to distinguish between finite and countably infinite sample spaces. Therefore, if a sample space is either finite or countably infinite, we say that it is a **countable** or a **discrete sample space**.

Measurement of the time until failure of a component would have an entire interval of real numbers as possible values. Since the interval of real numbers cannot be enumerated—that is, they cannot be put into one-to-one correspondence with natural numbers—such a sample space is said to be **uncountable** or **nondenumerable**. If the elements (points) of a sample space constitute a continuum, such as all the points on a line, all the points on a line segment, all the points in a plane, the sample space is said to be **continuous**. Certainly, no real experiment conducted using real measuring devices can ever yield such a continuum of outcomes, since there is a limit to the fineness to which any instrument can measure. However, such a sample space can often be taken as an idealization of, an approximation to, or a model of a real-world situation, which may be easier to analyze than a more exact model.

Problems

1. Describe a possible sample space for each of the following experiments:
   (a) A large lot of RAM (random access memory) chips is known to contain a small number of ROM (read only memory) chips. Three chips are chosen at random from this lot and each is checked to see whether it is a ROM or a RAM.
   (b) A box of ten chips is known to contain one defective and nine good chips. Three chips are chosen at random from the box and tested.
   (c) An **if** . . . **then** . . . **else** . . . statement is executed four times.

1.4 EVENTS

An **event** is simply a collection of certain sample points, that is, a subset of the sample space. Equivalently, any statement of conditions that defines this subset is called an event. Intuitively, an event is defined as a statement whose truth or falsity is determined after the experiment. The set of all experimental outcomes (sample points) for which the statement is true defines the subset of the sample space corresponding to the event. A single performance of the experiment is known as a **trial**. Let E be an event defined on a sample space S; that is, E is a subset of S. Let the outcome of a specific trial be denoted by s, an element of S. If s is an element of E, then we say that the event E has occurred. Only one outcome s in S can occur on any trial. However, every event that includes s will occur.

Consider the experiment of observing a two-component system and the corresponding sample space of Figure 1.2. Let event A_1 be described by the statement "Exactly one machine has failed." Then it corresponds to the subset $\{(0,1),(1,0)\}$ of the sample space. We will use the term **event** interchangeably to describe the subset or the statement. There are sixteen different subsets of this sample space with four elements, and each of these subsets defines an event. In particular, the entire sample space $S = \{(0,0),(0,1),(1,0),(1,1)\}$ is an event (called the **universal event**), and so is the null set $\varnothing$ (called the **null** or **impossible event**). The event $\{s\}$ consisting of a single sample point will be called an **elementary event**.

Consider the experiment of observing the time to failure of a component. The sample space, in this case, may be thought of as the set of all nonnegative real numbers, or the interval $[0, \infty) = \{t|0 \leqslant t < \infty\}$. Note that this is an example of a continuous sample space. Now if this component is part of a system that is required to carry out a mission of certain duration t, then an event of interest is "The component does not fail before time t." This event may also be denoted by the set $\{x|x \geqslant t\}$, or by the interval $[t, \infty)$.

1.5 ALGEBRA OF EVENTS

Consider an example of a computer system with five identical tape drives. One possible random experiment consists of checking the system to see how many tape drives are currently available. Each tape drive is in one of two states: busy (labeled 0) and available (labeled 1). An outcome of the experiment (a point in the sample space) can be denoted by a 5-tuple of 0's and 1's. A 0 in position i of the 5-tuple indicates that tape drive i is busy and a 1 indicates that it is available. The sample space S has $2^5 = 32$ sample points, as shown in the table on page 8.

The event E_1 described by the statement "At least four tape drives are available" is given by:

$$E_1 = \{(0,1,1,1,1),(1,0,1,1,1),(1,1,0,1,1),\\ (1,1,1,0,1),(1,1,1,1,0),(1,1,1,1,1)\}\\ = \{s_{15}, s_{23}, s_{27}, s_{29}, s_{30}, s_{31}\}.$$

The **complement** of this event, denoted by $\overline{E}_1$, is defined to be $S - E_1$ and contains all the sample points not contained in E_1. That is, $\overline{E}_1 = \{s \in S | s \notin E_1\}$. In our example, $\overline{E}_1 = \{s_0$ through s_{14}, s_{16} through s_{22}, s_{24} through s_{26}, $s_{28}\}$. $\overline{E}_1$ may also be described by the statement "At most three tape drives are available." Let E_2 be the event "At most four tape drives are avail-

$s_0 = (0, 0, 0, 0, 0)$	$s_{16} = (1, 0, 0, 0, 0)$
$s_1 = (0, 0, 0, 0, 1)$	$s_{17} = (1, 0, 0, 0, 1)$
$s_2 = (0, 0, 0, 1, 0)$	$s_{18} = (1, 0, 0, 1, 0)$
$s_3 = (0, 0, 0, 1, 1)$	$s_{19} = (1, 0, 0, 1, 1)$
$s_4 = (0, 0, 1, 0, 0)$	$s_{20} = (1, 0, 1, 0, 0)$
$s_5 = (0, 0, 1, 0, 1)$	$s_{21} = (1, 0, 1, 0, 1)$
$s_6 = (0, 0, 1, 1, 0)$	$s_{22} = (1, 0, 1, 1, 0)$
$s_7 = (0, 0, 1, 1, 1)$	$s_{23} = (1, 0, 1, 1, 1)$
$s_8 = (0, 1, 0, 0, 0)$	$s_{24} = (1, 1, 0, 0, 0)$
$s_9 = (0, 1, 0, 0, 1)$	$s_{25} = (1, 1, 0, 0, 1)$
$s_{10} = (0, 1, 0, 1, 0)$	$s_{26} = (1, 1, 0, 1, 0)$
$s_{11} = (0, 1, 0, 1, 1)$	$s_{27} = (1, 1, 0, 1, 1)$
$s_{12} = (0, 1, 1, 0, 0)$	$s_{28} = (1, 1, 1, 0, 0)$
$s_{13} = (0, 1, 1, 0, 1)$	$s_{29} = (1, 1, 1, 0, 1)$
$s_{14} = (0, 1, 1, 1, 0)$	$s_{30} = (1, 1, 1, 1, 0)$
$s_{15} = (0, 1, 1, 1, 1)$	$s_{31} = (1, 1, 1, 1, 1)$

able.'' Then $E_2 = \{s_0 \text{ through } s_{30}\}$. The **intersection** E_3 of the two events E_1 and E_2 is denoted by $E_1 \cap E_2$ and is given by:

$$\begin{aligned} E_3 &= E_1 \cap E_2 \\ &= \{s \in S \mid s \text{ is an element of both } E_1 \text{ and } E_2\} \\ &= \{s \in S \mid s \in E_1 \text{ and } s \in E_2\} \\ &= \{s_{15},\ s_{23},\ s_{27},\ s_{29},\ s_{30}\}. \end{aligned}$$

Let E_4 be the event ''Tape drive 1 is available.'' Then $E_4 = \{s_{16} \text{ through } s_{31}\}$. The **union** E_5 of the two events E_1 and E_4 is denoted by $E_1 \cup E_4$ and is given by:

$$\begin{aligned} E_5 &= E_1 \cup E_4 \\ &= \{s \in S \mid \text{either } s \in E_1 \text{ or } s \in E_4 \text{ or both}\} \\ &= \{s_{15} \text{ through } s_{31}\}. \end{aligned}$$

Note that E_1 has six points, E_4 has sixteen points, and E_5 has seventeen points. In general:

$$\begin{aligned} |E_5| &= |E_1 \cup E_4| \\ &\leqslant |E_1| + |E_4|. \end{aligned}$$

Here, the notation $|A|$ is used to denote the number of elements in the set A (also known as the **cardinality** of A).

Two events A and B are said to be **mutually exclusive events** or **disjoint events** provided $A \cap B$ is the null set. If A and B are mutually exclusive, then it is not possible for both events to occur on the same trial. For example, let E_6 be the event "Tape drive 1 is busy." Then E_4 and E_6 are mutually exclusive events, since $E_4 \cap E_6 = \emptyset$.

Although the definitions of union and intersection are given for two events, we observe that they extend to any finite number of sets. However, it is customary to use a more compact notation. Thus we define:

$$\bigcup_{i=1}^{n} E_i = E_1 \cup E_2 \cup E_3 \cup \cdots \cup E_n$$
$$= \{s \text{ element of } S \,|\, s \text{ element of } E_1 \text{ or } s \text{ element of } E_2 \text{ or} \ldots s \text{ element of } E_n\}.$$

$$\bigcap_{i=1}^{n} E_i = E_1 \cap E_2 \cap E_3 \cap \cdots \cap E_n$$
$$= \{s \text{ element of } S \,|\, s \text{ element of } E_1 \text{ and } s \text{ element of } E_2 \text{ and} \ldots s \text{ element of } E_n\}.$$

These definitions can also be extended to the union and intersection of a countably infinite number of sets.

The algebra of events may be fully defined by the following five laws or axioms, where A, B, and C are arbitrary sets (or events) and S is the universal set (or event):

(E1) ***Commutative laws:***

$$A \cup B = B \cup A, \qquad A \cap B = B \cap A.$$

(E2) ***Associative laws:***

$$A \cup (B \cup C) = (A \cup B) \cup C,$$
$$A \cap (B \cap C) = (A \cap B) \cap C.$$

(E3) ***Distributive laws:***

$$A \cup (B \cap C) = (A \cup B) \cap (A \cup C),$$
$$A \cap (B \cup C) = (A \cap B) \cup (A \cap C).$$

(E4) ***Identity laws:***

$$A \cup \emptyset = A, \qquad A \cap S = A.$$

(E5) ***Complementation laws:***

$$A \cup \bar{A} = S, \qquad A \cap \bar{A} = \emptyset.$$

Any relation that is valid in the algebra of events can be proven by using these axioms [(E1)–(E5)]. Some of the other useful relations are:

(R1) ***Idempotent laws:***

$$A \cup A = A, \qquad A \cap A = A.$$

(R2) ***Domination laws:***

$$A \cup S = S, \qquad A \cap \varnothing = \varnothing.$$

(R3) ***Absorption laws:***

$$A \cap (A \cup B) = A, \qquad A \cup (A \cap B) = A.$$

(R4) ***De Morgan's laws:***

$$\overline{(A \cup B)} = \bar{A} \cap \bar{B}, \qquad \overline{(A \cap B)} = \bar{A} \cup \bar{B}.$$

(R5)

$$\overline{(\bar{A})} = A.$$

(R6)

$$A \cup (\bar{A} \cap B) = A \cup B.$$

From the complementation laws, we note that A and $\bar{A}$ are mutually exclusive, since $A \cap \bar{A} = \varnothing$. In addition, A and $\bar{A}$ are collectively exhaustive, since any point s (an element of S) is either in $\bar{A}$ or in A. These two notions can be generalized to a list of events.

A list of events $A_1, A_2, \ldots, A_n$ is said to be composed of **mutually exclusive** events if and only if:

$$A_i \cap A_j = \begin{cases} A_i & \text{if } i=j, \\ \varnothing & \text{otherwise.} \end{cases}$$

Intuitively, a list of events is composed of mutually exclusive events if no point in the sample space is included in more than one event in the list.

A list of events $A_1, A_2, \ldots, A_n$ is said to be **collectively exhaustive** if and only if:

$$A_1 \cup A_2 \cup \cdots \cup A_n = S.$$

Given a list of events that is collectively exhaustive, each point in the sample space is included in at least one event in the list. An arbitrary list of events may be mutually exclusive, collectively exhaustive, both, or neither. For each point s in the sample space S, we may define an event $A_s = \{s\}$. The resulting list of events is mutually exclusive and collectively exhaustive (such a list of events is also called a **partition** of the sample space S). Thus, a sample space may be defined as the mutually exclusive and collectively exhaustive listing of all possible outcomes of an experiment.

Problems

1. Four components are inspected and three events are defined as follows:

   A = "All four components are found defective."

   B = "Exactly two components are found to be in proper working order."

   C = "At most three components are found to be defective."

   Interpret the following events:
   (a) $B \cup C$.
   (b) $B \cap C$.
   (c) $A \cup C$.
   (d) $A \cap C$.

2. Use axioms of the algebra of events to prove the relations:
   (a) $A \cup A = A$.
   (b) $A \cup S = S$.
   (c) $A \cap \emptyset = \emptyset$.
   (d) $A \cap (A \cup B) = A$.

1.6 GRAPHICAL METHODS OF REPRESENTING EVENTS

Venn diagrams often provide a convenient means of ascertaining relations between events of interest. Thus, for a given sample space S and the two events A and B, we have the Venn diagram shown in Figure 1.4. In this figure, the set of all points in the sample space is symbolically denoted by the ones within the rectangle. The events A and B are represented by certain regions in S.

The union of two events A and B is represented by the set of points lying in either A or B. The union of two mutually exclusive events A and B is represented by the shaded region in Figure 1.5. On the other hand, if A and B are not mutually exclusive, they might be represented by a Venn diagram like Figure 1.6. $A \cup B$ is represented by the shaded region; a portion of this shaded region is $A \cap B$ and is so labeled.

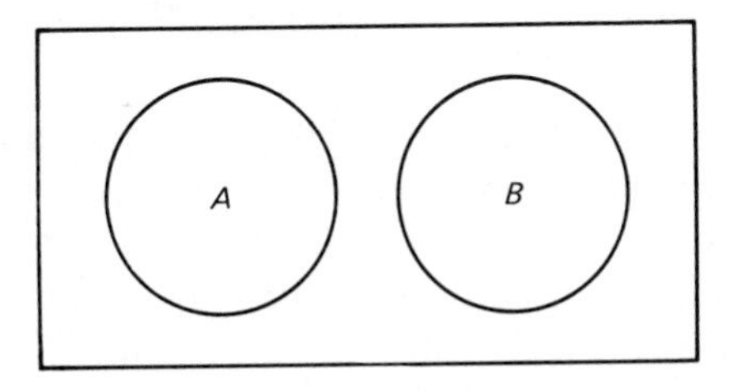

Figure 1.4 Venn diagram for sample space S and events A and B

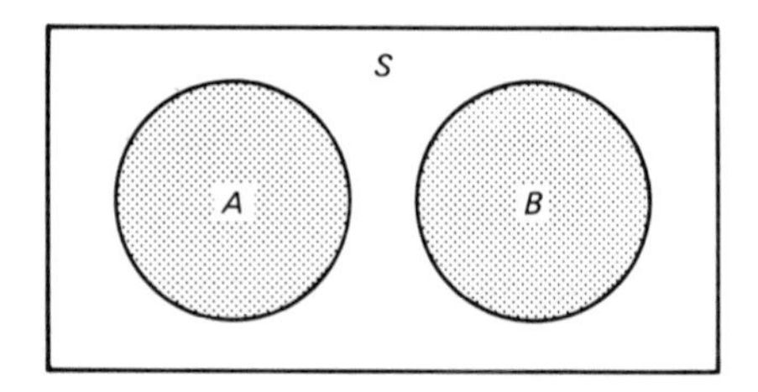

Figure 1.5 Venn Diagram of Disjoint Events A and B

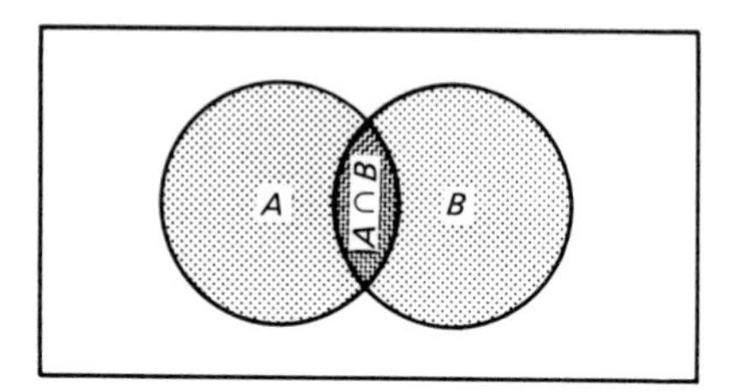

Figure 1.6 Venn Diagram for two Intersecting Events A and B

For an event A, the complement $\bar{A}$ consists of all points in S that do not belong to A; thus $\bar{A}$ is represented by the unshaded region in Figure 1.7. The usefulness of Venn diagrams becomes apparent when we see that the following laws of event algebra, discussed in the last section, are easily seen to hold true by reference to Figures 1.6 and 1.7:

$$A \cap S = A,$$

$$A \cup S = S,$$

$$\overline{(\bar{A})} = A,$$

$$\overline{(A \cup B)} = \bar{A} \cap \bar{B},$$

$$\overline{(A \cap B)} = \bar{A} \cup \bar{B}.$$

Another useful graphical device is the **tree diagram**. As an example, consider the experiment of observing two successive executions of an **if** statement in a certain program. The outcome of the first execution of the **if** statement may be the execution of the **then** clause (denoted by T_1) or the execution of the **else** clause (denoted by E_1). Similarly the outcome of the second execution is T_2 or E_2. This is an example of a **sequential sample space** and leads to the tree diagram of Figure 1.8. We picture the experiment proceeding sequentially downward from the root. The set of all leaves of the tree is the sample space of interest. Each sample point represents the event

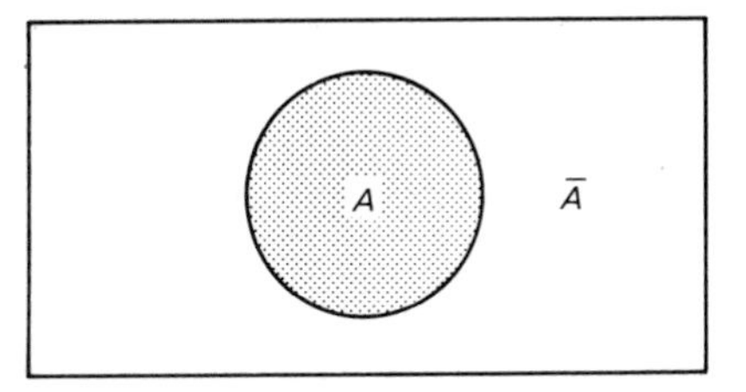

Figure 1.7 Venn diagram of A and its complement

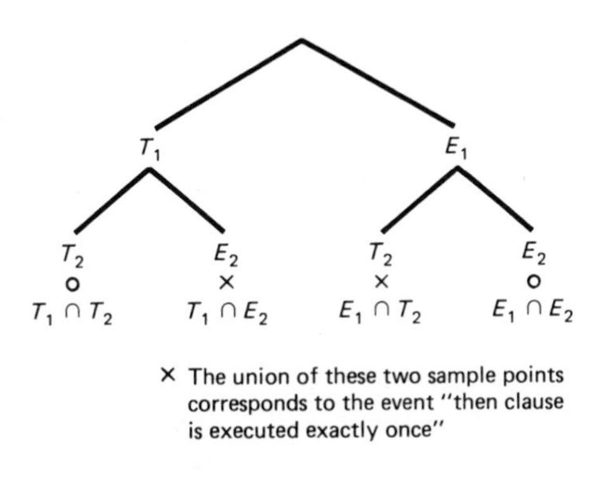

Figure 1.8 Tree diagram of a sequential sample space

corresponding to the intersection of all events encountered in tracing a path from the root to the leaf corresponding to the sample point. Note that the four sample points (the leaves of the tree) and their labels constitute the sample space of the experiment. However, when we deal with a sequential sample space, we normally picture the entire generating tree as well as the resulting sample space.

When the outcomes of the experiment may be expressed numerically, yet another graphical device is a coordinate system. As an example, consider a system consisting of two subsystems. The first subsystem consists of four components, and the second subsystem of three. Assuming that we are concerned only with the total number of defective components in each subsystem (not with what particular components have failed), the cardinality of the sample space is $5 \cdot 4 = 20$, and the corresponding two-dimensional sample space is illustrated in Figure 1.9. The three events identified in Figure 1.9 are easily seen to be:

A = "The system has exactly one defective component."

B = "The system has exactly three defective components."

C = "The first subsystem has more defective components than the second subsystem."

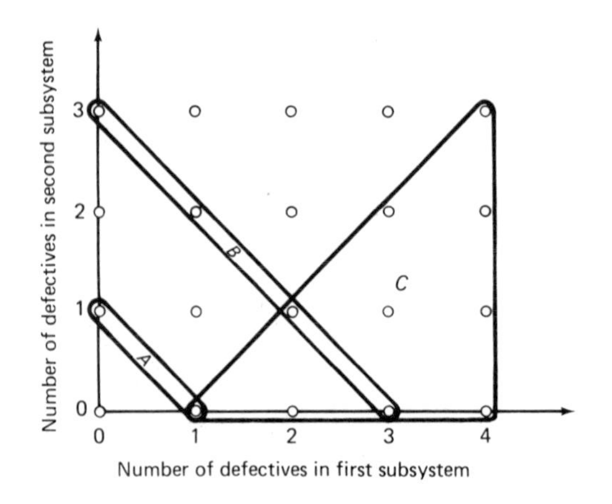

Figure 1.9 A two-dimensional sample space

1.7 PROBABILITY AXIOMS

We have seen that the physical behavior of random experiments can be modeled naturally using the concepts of events in a suitably defined sample space. To complete our specification of the model, we shall assign **probabilities** to the events in the sample space. The probability of an event is meant to represent the "relative likelihood" that a performance of the experiment will result in the occurrence of that event. $P(A)$ will denote the probability of the event A in the sample space S.

In many engineering applications and in games of chance, the so-called relative-frequency interpretation of the probability is used. For many applications, however, such an approach is inadequate. We would like the mathematical construction of the probability measure to be independent of the intended application. This leads to an *axiomatic* treatment of the theory of probability. The theory of probability starts with the assumption that probabilities can be assigned so as to satisfy the following three basic **axioms of probability**. The assignment of probabilities is perhaps the most difficult aspect of constructing probabilistic models. Assignments are commonly based on intuition, experience, or experimentation. The theory of probability is neutral; it will make predictions regardless of these assignments. However, the results will be strongly affected by the choice of a particular assignment. Therefore if the assignments are inaccurate, the predictions of the model will be misleading and will not reflect the behavior of the real-world problem being modeled.

Let S be a sample space of a random experiment. We use the notation $P(A)$ for the probability measure associated with event A. If the event A consists of a single sample point s, then $P(A) = P(\{s\})$ will be written as $P(s)$. The probability function $P(.)$ must satisfy the following axioms:

(A1) For any event A, $P(A) \geqslant 0$.

(A2) $P(S) = 1$.

(A3) $P(A \cup B) = P(A) + P(B)$, whenever A and B are mutually exclusive events—that is, when $A \cap B = \varnothing$.

The first axiom states that all probabilities are nonnegative real numbers. The second axiom attributes a probability of unity to the universal event S, thus providing a normalization of the probability measure (the probability of a certain event, an event that *must* happen, is equal to 1). The third axiom states that the probability function must be additive. These three axioms are consistent with our intuitive ideas of how probabilities behave.

The **principle of mathematical induction** can be used to show [using axiom (A3) as the basis of induction] that for any positive integer n the probability of the union of n mutually exclusive events $A_1, A_2, \ldots, A_n$ is equal to the sum of their probabilities:

$$P(A_1 \cup A_2 \cup \cdots \cup A_n) = \sum_{i=1}^{n} P(A_i).$$

The three axioms (A1) through (A3) are adequate if the sample space is finite, but to deal with problems having infinite sample spaces, we need to modify axiom (A3):

(A3′) For any countable sequence of events $A_1, A_2, \ldots, A_n, \ldots$, that are mutually exclusive (that is, $A_j \cap A_k = \varnothing$ whenever $j \neq k$):

$$P\left(\bigcup_{n=1}^{\infty} A_n\right) = \sum_{n=1}^{\infty} P(A_n).$$

All of conventional probability theory follows from the three axioms [(A1) through (A3′)] of probability measure and the five axioms [(E1) through (E5)] of the algebra of events discussed earlier. These eight axioms can be used to show several useful relations:

(Ra) For any event A, $P(\overline{A}) = 1 - P(A)$.

Proof: A and $\overline{A}$ are mutually exclusive, and $S = A \cup \overline{A}$. Then by axioms (A2) and (A3), $1 = P(S) = P(A) + P(\overline{A})$, from which the assertion follows.

(Rb) If $\varnothing$ is the impossible event, then $P(\varnothing) = 0$.

Proof: Observe that $\varnothing = \overline{S}$, so that the result follows from from relation (Ra) and axiom (A2).

(Rc) If A and B are any events, not necessarily mutually exclusive, then

$$P(A \cup B) = P(A) + P(B) - P(A \cap B).$$

Proof: From the Venn diagram of Figure 1.6, we note that $A \cup B = A \cup (\bar{A} \cap B)$ and $B = (A \cap B) \cup (\bar{A} \cap B)$, where in each equation the events on the right-hand side are mutually exclusive. By axiom (A3):

$$P(A \cup B) = P(A) + P(\bar{A} \cap B),$$
$$P(B) = P(A \cap B) + P(\bar{A} \cap B).$$

The second equation implies $P(\bar{A} \cap B) = P(B) - P(A \cap B)$, which, after substitution in the first equation, yields the desired assertion.

The relation (Rc) can be generalized to a formula similar to the principle of inclusion and exclusion of combinatorial mathematics [LIU 1968]:

(Rd) If $A_1, A_2, \ldots, A_n$ are any events, then:

$$\begin{aligned} P(\bigcup_{i=1}^{n} A_i) &= P(A_1 \cup A_2 \cup \cdots \cup A_n) \\ &= \sum_i P(A_i) - \sum_{1 \leqslant i < j \leqslant n} P(A_i \cap A_j) \\ &\quad + \sum_{1 \leqslant i < j < k \leqslant n} P(A_i \cap A_j \cap A_k) + \cdots \\ &\quad + (-1)^{n-1} P(A_1 \cap A_2 \cap \cdots \cap A_n), \end{aligned}$$

where the successive sums are over all possible events, pairs of events, triples of events, and so on.

Proof: We prove this result by induction on the number of events n. The result (Rc) above can serve as the basis of induction. Assume inductively that (Rd) holds for a union of $n - 1$ events. Define the event $B = A_1 \cup A_2 \cup \cdots \cup A_{n-1}$. Then:

$$\bigcup_{i=1}^{n} A_i = B \cup A_n.$$

Using the result (Rc) above, we get:

$$\begin{aligned} P(\bigcup_{i=1}^{n} A_i) &= P(B \cup A_n) \\ &= P(B) + P(A_n) - P(B \cap A_n). \end{aligned} \tag{1.2}$$

Now

$$B \cap A_n = (A_1 \cap A_n) \cup (A_2 \cap A_n) \cup \cdots \cup (A_{n-1} \cap A_n)$$

is a union of $n - 1$ events and hence, using the inductive hypothesis, we get:

$$
\begin{aligned}
P(B \cap A_n) = {} & P(A_1 \cap A_n) + P(A_2 \cap A_n) + \cdots + P(A_{n-1} \cap A_n) \\
& - P[(A_1 \cap A_n) \cap (A_2 \cap A_n)] \\
& - P[(A_1 \cap A_n) \cap (A_3 \cap A_n)] \\
& - \cdots \\
& + P[(A_1 \cap A_n) \cap (A_2 \cap A_n) \cap (A_3 \cap A_n)] \\
& + \cdots - \cdots \\
& + (-1)^{n-2} P[(A_1 \cap A_n) \cap (A_2 \cap A_n) \cap \cdots \cap (A_{n-1} \cap A_n)] \\
= {} & P(A_1 \cap A_n) + P(A_2 \cap A_n) + \cdots + P(A_{n-1} \cap A_n) \\
& - P(A_1 \cap A_2 \cap A_n) - P(A_1 \cap A_3 \cap A_n) - \cdots \\
& + P(A_1 \cap A_2 \cap A_3 \cap A_n) + \cdots \\
& - \cdots \qquad (1.3) \\
& + (-1)^{n-2} P(A_1 \cap A_2 \cap A_3 \cap \cdots \cap A_{n-1} \cap A_n).
\end{aligned}
$$

Also, since $B = A_1 \cup A_2 \cup \cdots \cup A_{n-1}$ is a union of $n - 1$ events, the inductive hypothesis gives:

$$
\begin{aligned}
P(B) = {} & P(A_1) + P(A_2) + \cdots + P(A_{n-1}) \\
& - P(A_1 \cap A_2) - P(A_1 \cap A_3) - \cdots \qquad (1.4) \\
& + \cdots \\
& + (-1)^{n-2} P(A_1 \cap A_2 \cap \cdots \cap A_{n-1}).
\end{aligned}
$$

Substituting (1.3) and (1.4) into (1.2), we obtain the required result.

To avoid certain mathematical difficulties, we must place restrictions on which subsets of the sample space may be termed events and to which probabilities can be assigned. In a given problem there will be a particular class of subsets of S that is "measurable" and will be called the "class of events" $\mathfrak{F}$. Since we would like to perform the standard set operations on events, it is reasonable to demand that $\mathfrak{F}$ be closed under countable unions as well as under complementation. A collection of subsets of a given set S that is closed under countable unions and complementation is called a σ-field of subsets of S. Now a **probability space** or **probability system** may be defined as a triple $(S, \mathfrak{F}, P)$, where S is a set, $\mathfrak{F}$ is a σ-field of subsets of S which includes S, and P is a probability measure on $\mathfrak{F}$, assumed to satisfy axioms (A1) through (A3′).

If the sample space is discrete (finite or countable) then every subset of S

can be an event belonging to $\mathcal{F}$. However, in the case that S is uncountable, this is no longer true. For example, let S be the interval $[0, 1]$ and assume the probability assignment $P(a \leqslant x \leqslant b) = b - a$ for $0 \leqslant a \leqslant b \leqslant 1$. Then it can be shown that not all possible subsets of S can be assigned a probability in a manner consistent with the three axioms of P. In such cases, the smallest σ-field of subsets of S containing all open and closed intervals is usually adopted as the class of events $\mathcal{F}$.

In summary, P is a function with domain $\mathcal{F}$ and range $[0, 1]$, which satisfies the three axioms (A1), (A2), and (A3′). P assigns a number between 0 and 1 to any event in $\mathcal{F}$. In general, $\mathcal{F}$ does not include all possible subsets of S, and the subsets (events) included in $\mathcal{F}$ are called measurable. However, for our purposes, every subset of a sample space constructed here can be considered an event having a probability.

We now outline the steps of a basic procedure to be followed in solving problems [GOOD 1977]:

1. *Identify the sample space S.* The sample space S must be chosen so that all its elements are mutually exclusive and collectively exhaustive; that is, no two elements can occur simultaneously and one element must occur on any trial. Many of the "trick" probability problems are based on some ambiguity in the problem statement or an inexact formulation of the model of a physical situation. The choice of an appropriate sample space resulting from a detailed description of the model will do much to resolve common difficulties. Since many choices for the sample space are possible, it is advisable to use a sample space whose elements cannot be further "subdivided"—that is, all possible distinguishable outcomes of the experiment should be listed separately.

2. *Assign probabilities.* Assign probabilities to the elements in S. This assignment must be consistent with the axioms (A1), (A2), and (A3′). In practice, the assignment of probabilities is based either on estimates obtained from past experience, or on a careful analysis of conditions underlying the random experiment, or on assumptions, such as the common assumption that various outcomes in a finite sample space are equiprobable (equally likely).

3. *Identify the events of interest.* The events of interest, in a practical situation, will be described by statements. These need to be recast as subsets of the sample space. The laws of event algebra [(E1)—(E5) and (R1)—(R6)] may be used for any simplification. Pictorial devices such as Venn diagrams, tree diagrams, or coordinate-system plots may also be used to advantage.

4. *Compute desired probabilities.* Calculate the probabilities of the events of interest using the axioms (A1), (A2), and (A3′) and any derived laws such as (Ra), (Rb), (Rc), and (Rd). It is usually helpful to express the event of interest as a union of mutually exclusive points in the sample space and summing the probabilities of all points included in the union.

As a simple illustration of this procedure, consider the example of the computer system with five tape drives:

Example 1.1

Step 1. An appropriate sample space consists of 32 points (see p. 8), each represented by a 5-tuple of 0's and 1's. A 0 in position i indicates that tape drive i is busy and a 1 indicates that it is available.

Step 2. In the absence of detailed knowledge about the system, we assume that each sample point is equally likely. Since there are 32 sample points, we assign a probability of $1/32$ to each; that is, $P(s_0) = P(s_1) = \cdots = P(s_{31}) = 1/32$. It is easily seen that this assignment is consistent with the three probability axioms.

Step 3. Assume that we are required to determine the probability that a process is scheduled immediately for execution, given that the process needs at least three tape drives for its execution. The event E of interest, then, is "Three or more tape drives are available." From the definition of the sample points, we see that:

$$\begin{aligned} E &= \{s_7, s_{11}, s_{13}, s_{14}, s_{15}, s_{19}, s_{21}, s_{22}, s_{23}, s_{25} \text{ through } s_{31}\} \\ &= \{s_7\} \cup \{s_{11}\} \cup \{s_{13}\} \cup \{s_{14}\} \cup \{s_{15}\} \cup \{s_{19}\} \cup \{s_{21}\} \cup \{s_{22}\} \\ &\quad \cup \{s_{23}\} \cup \{s_{25}\} \cup \{s_{26}\} \cup \{s_{27}\} \cup \{s_{28}\} \cup \{s_{29}\} \cup \{s_{30}\} \\ &\quad \cup \{s_{31}\}. \end{aligned}$$

Step 4. We have already simplified E so that it is expressed as a union of mutually exclusive events. The probability of each of these elementary events is $1/32$. Thus, a repeated application of axiom (A3) gives us:

$$\begin{aligned} P(E) &= \sum_{s_i \in E} P(s_i) \\ &= 1/32 + 1/32 + 1/32 + 1/32 + 1/32 + 1/32 + 1/32 + 1/32 + 1/32 + 1/32 \\ &\quad + 1/32 + 1/32 + 1/32 + 1/32 + 1/32 + 1/32 \\ &= 1/2. \end{aligned}$$

Alternatively, we could have noted that E consists of sixteen sample points and, since each of the 32 sample points is equally likely, $P(E) = 16/32$. #

Problems

1. Consider a pool of six I/O buffers. Assume that any buffer is just as likely to be available (or occupied) as any other. Compute the probabilities associated with the following events:

   A = "At least two but no more than five buffers occupied."

   B = "At least three but no more than five buffers occupied."

   C = "All buffers available or an even number of buffers occupied."

   Also determine the probability that at least one of the events A, B, and C occurs.

2. Show that if event B is contained in event A, then $P(B) \leqslant P(A)$.

1.8 COMBINATORIAL PROBLEMS

If the sample space of an experiment consists of only a finite number n of sample points, or **elementary events**, then the computation of probabilities is often simple. Assume that assignment of probabilities is made such that for s_i (an element of S), $P(s_i) = p_i$ and:

$$\sum_{i=1}^{n} p_i = 1.$$

Since any event E consists of a certain collection of these sample points, $P(E)$ can be found, using axiom (A3), by adding up the probabilities of the separate sample points that make up E (recall the computer example of the last section).

Example 1.2

Consider the following **if** statement in a program:

if B **then** s_1 **else** s_2.

The random experiment consists of "observing" two successive executions of the **if** statement. The sample space consists of the four possible outcomes:

$$\begin{aligned} S &= \{(s_1, s_1), (s_1, s_2), (s_2, s_1), (s_2, s_2)\} \\ &= \{t_1, t_2, t_3, t_4\}. \end{aligned}$$

Assume that on the basis of strong experimental evidence the following probability assignment is justified:

$$P(t_1) = 0.34, \quad P(t_2) = 0.26, = P(t_3) = 0.26, \quad P(t_4) = 0.14.$$

The events of interest are given as E_1 = "At least one execution of the statement s_1" and E_2 = "Statement s_2 is executed the first time." It is easy to see that:

$$\begin{aligned} E_1 &= \{(s_1, s_1), (s_1, s_2), (s_2, s_1)\} \\ &= \{t_1, t_2, t_3\} \\ E_2 &= \{(s_2, s_1), (s_2, s_2)\} \\ &= \{t_3, t_4\} \\ P(E_1) &= P(t_1) + P(t_2) + P(t_3) = 0.86 \\ P(E_2) &= P(t_3) + P(t_4) = 0.4. \end{aligned}$$

#

In the special case when $S = \{s_1, \ldots, s_n\}$ and $P(s_i) = p_i = 1/n$ (equally likely sample points), the situation is even simpler. Calculation of probabilities is then reduced to simply counting the number of sample points in the event of interest. If the event E consists of k sample points, then:

$$P(E) = \frac{\text{number of points in } E}{\text{number of points in } S}$$
$$= \frac{\text{favorable outcomes}}{\text{total outcomes}} \quad (1.5)$$
$$= \frac{k}{n}.$$

Example 1.3

A group of four integrated-circuit (IC) chips consists of two good chips, labeled g_1 and g_2, and two defective chips, labeled d_1 and d_2. If three chips are selected at random from this group, what is the probability of the event E = "Two of the three selected chips are defective"?

A natural sample space for this problem consists of all possible three-chip selections from the group of four chips: $S = \{g_1g_2d_1,\ g_1g_2d_2,\ g_1d_1d_2,\ g_2d_1d_2\}$. It is customary to interpret the phrase "selected at random" as implying equiprobable sample points. Since the two sample points $g_1d_1d_2$ and $g_2d_1d_2$ are favorable to the event E, and since the sample space has four points, we conclude that $P(E) = 2/4 = 1/2$. #

We have seen that under the equiprobability assumption, finding $P(E)$ simply involves counting the number of outcomes favorable to E. However, counting by hand may not be feasible when the sample space is large. Standard counting methods of combinatorial analysis can often be used to avoid writing down the list of favorable outcomes explicitly.

1.8.1 Ordered Samples of Size *k*, with Replacement

Here we are interested in counting the number of ways we can select k objects from among n objects where order is important and when the same object is allowed to be repeated any number of times (**permutations with replacement**). Alternatively, we are interested in the number of ordered sequences $(s_{i_1}, s_{i_2}, \ldots, s_{i_k})$, where each s_{i_r} belongs to $\{s_1, \ldots, s_n\}$. It is not difficult to see that the required number is $(n \cdot n \cdot \cdots \cdot n\ (k \text{ times}))$, or n^k.

Example 1.4

Assume that we are interested in finding the probability that some randomly chosen k-digit decimal number is a valid k-digit octal number. The sample space, in this case, is:

$$S = \{(x_1, x_2, \ldots, x_k) | x_i \in \{0, 1, 2, \ldots, 9\}\},$$

and the event of interest is:

$$E = \{(x_1, x_2, \ldots, x_k) | x_i \in \{0, 1, 2, \ldots, 7\}\}.$$

By the above counting principle, $|S| = 10^k$ and $|E| = 8^k$. Now, if we assume that all the sample points are equally likely, then the required answer is:

$$P(E) = \frac{|E|}{|S|} = \frac{8^k}{10^k} = \frac{4^k}{5^k}.$$

#

1.8.2 Ordered Samples of Size *k*, Without Replacement

The number of ordered sequences $(s_{i_1}, s_{i_2}, \ldots, s_{i_k})$, where s_{i_r} belongs to $\{ s_1, \ldots, s_n \}$, but repetition is not allowed (i.e., no s_i can appear more than once in the sequence), is given by:

$$n(n-1)\cdots(n-k+1) = \frac{n!}{(n-k)!} \qquad \text{for } k = 1, 2, \ldots, n.$$

This number is also known as the number of permutations of n distinct objects taken k at a time, and denoted by $P(n, k)$.

Example 1.5

Suppose we wish to find the probability that a randomly chosen three-letter sequence will not have any repeated letters.

Let $I = \{a, b, \ldots, z\}$ be the alphabet of 26 letters. Then the sample space is given by:

$$S = \{(\alpha, \beta, \gamma) \mid \alpha \in I, \ \beta \in I, \ \gamma \in I\}$$

and the event of interest is:

$$E = \{(\alpha, \beta, \gamma) \mid \alpha \in I, \ \beta \in I, \ \gamma \in I, \ \alpha \neq \beta, \ \beta \neq \gamma, \ \alpha \neq \gamma\}.$$

By the above counting principle, $|E|$ is simply $P(26, 3) = 15,600$. Furthermore, $|S| = 26^3 = 17,576$. Therefore, the required answer is:

$$P(E) = \frac{15,600}{17,576} = 0.8875739.$$

#

1.8.3 Unordered Samples of Size *k*, Without Replacement

The number of unordered sets $\{s_{i_1}, s_{i_2}, \ldots, s_{i_k}\}$, where the s_{i_r} $(r = 1, 2, \ldots, k)$ are distinct elements of $\{ s_1, \ldots, s_n \}$ is:

$$\frac{n!}{k!(n-k)!}.$$

This is also known as the number of combinations of n distinct objects taken k at a time, and is denoted by $\binom{n}{k}$.

Example 1.6

If a box contains 75 good IC chips and 25 defective chips, and 12 chips are selected at random, find the probability that at least one chip is defective.

By the above counting principle, the number of unordered samples without replacement is $\binom{100}{12}$ and hence the size of the sample space is $|S| = \binom{100}{12}$. The event of interest is E = "At least one chip is defective." Here we find it easier to work with the complementary event $\bar{E}$ = "No chip is defective." Since there are 75 good chips, the above counting principle yields $|\bar{E}| = \binom{75}{12}$. Then:

$$
\begin{aligned}
P(\bar{E}) &= \frac{|\bar{E}|}{|S|} \\
&= \frac{\binom{75}{12}}{\binom{100}{12}} \\
&= \frac{75! \cdot 12! \cdot 88!}{12! \cdot 63! \cdot 100!} \\
&= \frac{75! \cdot 88!}{63! \cdot 100!}.
\end{aligned}
$$

Now since $P(E) = 1 - P(\bar{E})$, the required probability is easily obtained. #

Problems

1. How many even two-digit numbers can be constructed out of the digits 3, 4, 5, 6 and 7? Assume first that you may use the same digit again; then repeat the question, assuming that you may not use a digit more than once.
2. If a three-digit decimal number is chosen at random, find the probability that exactly k digits are $\geqslant 5$, for $0 \leqslant k \leqslant 3$.
3. A box with fifteen integrated circuit chips contains five defectives. If a random sample of three chips is drawn, what is the probability that all three are defective?
4. In a party of five persons, compute the probability that at least two have the same birthday (month/day), assuming a 365-day year.

*5. A series of n jobs arrive at a computing center with n processors. Assume that each of the n^n possible assignment vectors (processor for job 1, . . . , processor for job n) is equally likely. Find the probability that exactly one processor will be idle.

1.9 CONDITIONAL PROBABILITY

So far, we have assumed that the only information about the outcome of a trial of a given experiment, available before the trial, is that the outcome will correspond to some point in the sample space S. With this assumption, we can compute the probability of some event A. Suppose we are given the added information that the outcome s of a trial is contained in a subset B of the sample space, with $P(B) \neq 0$. Knowledge of the occurrence of the event B may change the probability of the occurrence of the event A. We wish to define the **conditional probability** of the event A given that the event B occurs, or the **conditional probability of A given B,** symbolically as:

$$P(A|B).$$

Given that event B has occurred, the sample point corresponding to the outcome of the trial must be in B and cannot be in $\bar{B}$. To reflect this partial information, we define the conditional probability of a sample point s (an element of S) by:

$$P(s|B) = \begin{cases} \dfrac{P(s)}{P(B)} & \text{if } s \in B, \\ 0 & \text{if } s \in \bar{B}. \end{cases}$$

Thus the original probability assigned to a sample point in B is scaled up by $1/P(B)$, so that the probabilities of the sample points in B will add up to 1. Now the conditional probability of any other event, such as A, can be obtained by summing over the conditional probabilities of the sample points included in A (noting that $A = (A \cap B) \cup (A \cap \bar{B})$):

$$\begin{aligned} P(A|B) &= \sum_{s \in A} P(s|B) \\ &= \sum_{s \in A \cap \bar{B}} P(s|B) + \sum_{s \in A \cap B} P(s|B) \\ &= \sum_{s \in A \cap B} P(s|B) \\ &= \sum_{s \in A \cap B} \frac{P(s)}{P(B)} \\ &= \frac{P(A \cap B)}{P(B)}, \quad P(B) \neq 0 \end{aligned}$$

This leads us to the following definition:

Definition (Conditional Probability). The conditional probability of A given B is

$$P(A|B) = \frac{P(A \cap B)}{P(B)}$$

if $P(B) \neq 0$ and it is undefined otherwise.

A rearrangement of the above definition yields the following:

MR (Multiplication Rule).

$$P(A \cap B) = \begin{cases} P(B)P(A|B) & \text{if } P(B) \neq 0, \\ P(A)P(B|A) & \text{if } P(A) \neq 0, \\ 0 & \text{otherwise.} \end{cases}$$

Example 1.7

We are given a box containing 5,000 IC chips, of which 1,000 are manufactured by company X and the rest by company Y. Ten percent of the chips made by company X and 5 percent of the chips made by company Y are defective. If a randomly chosen chip is found to be defective, find the probability that it came from company X.

Define the events A = "Chip is made by company X" and B = "Chip is defective." Since out of 5,000 chips, 1,000 are made by company X, we conclude that $P(A) = 1{,}000/5{,}000 = 0.2$. Also, out of a total of 5,000 chips, 300 are defective. Therefore, $P(B) = 300/5{,}000 = 0.06$. Now the event $A \cap B$ = "Chip is made by company X and is defective." Out of 5,000 chips, 100 chips qualify for this statement. Thus $P(A \cap B) = 100/5{,}000 = 0.02$. Now the quantity of interest is:

$$P(A|B) = \frac{P(A \cap B)}{P(B)} = \frac{0.02}{0.06} = \frac{1}{3}.$$

Thus the knowledge of the occurrence of event B has increased the probability of the occurrence of event A. Similarly we find that the knowledge of the occurrence of A has increased the chances for the occurrence of the event B, since $P(B|A) = 0.1$. In fact, note that:

$$\frac{P(A|B)}{P(B|A)} = \frac{1/3}{0.1} = \frac{0.2}{0.06} = \frac{P(A)}{P(B)}.$$

This interesting property of conditional probabilities is easily shown to hold in general:

$$\frac{P(A|B)}{P(B|A)} = \frac{P(A \cap B)/P(B)}{P(A \cap B)/P(A)} = \frac{P(A)}{P(B)}.$$

#

Problems

1. Consider four computer firms A, B, C, D bidding for a certain contract. A survey of past bidding success of these firms on similar contracts shows the following probabilities of winning:

$$P(A) = 0.35, \quad P(B) = 0.15, \quad P(C) = 0.3, \quad P(D) = 0.2.$$

Before the decision is made to award the contract, firm B withdraws its bid. Find the new probabilities of winning the bid for A, C, and D.

1.10 INDEPENDENCE OF EVENTS

It is possible for the probability of an event A to decrease, remain the same, or increase given that event B has occurred. If the probability of the occurrence of an event A does not change whether or not event B has occurred, we are likely to conclude that the two events are independent. Thus we define two events A and B to be independent if and only if:

$$P(A \mid B) = P(A).$$

From the definition of conditional probability, we have [provided $P(A) \neq 0$ and $P(B) \neq 0$]:

$$P(A \cap B) = P(A)P(B \mid A) = P(B)P(A \mid B).$$

From this we conclude that the condition for the independence of A and B can also be given either as $P(A \mid B) = P(A)$ or as $P(A \cap B) = P(A)P(B)$. Note that $P(A \cap B) = P(A)P(B \mid A)$ [if $P(A) \neq 0$] holds whether or not A and B are independent, but $P(A \cap B) = P(A)P(B)$ holds only when A and B are independent. In fact this latter condition is the usual definition of independence:

Definition (Independent Events). Events A and B are said to be independent if

$$P(A \cap B) = P(A)P(B).$$

Some authors use the phrases "stochastically independent events" or "statistically independent events" in place of just "independent events." Note that if A and B are not independent, then $P(A \cap B)$ is computed using the multiplication rule of the last section.

The above condition for independence can be derived in another way by first noting that the event A is a disjoint union of events $A \cap B$ and $A \cap \bar{B}$. Now the conditional probability [given that B has occurred] of all the sample points in the latter event is zero, while the conditional probability of all the sample points in the former event is increased by the factor $1/P(B)$. Therefore, for $P(A \mid B) = P(A)$ to hold, the decrease in probability due to points in $A \cap \bar{B}$ must be balanced by the increase in probability due to points in $A \cap B$. In other words,

$$\frac{P(A \cap B)}{P(B)} - P(A \cap B) = P(A \cap \bar{B}) - 0$$

or,

$$\frac{P(A \cap B)}{P(B)} = P(A \cap \bar{B}) + P(A \cap B)$$

$$= P(A).$$

Example 1.8

A microcomputer system consists of a microprocessor CPU chip and a random access main memory chip. The CPU is selected from a lot of 100, of which 10 are defective, and the memory chip is selected from a lot of 300, of which 15 are defective. Define A to be the event "The selected CPU is defective," and let B be the event "The selected memory chip is defective." Then $P(A) = 10/100 = 0.1$, and $P(B) = 15/300 = 0.05$. Since the two chips are selected from different lots, we may expect the events A and B to be independent. This can be checked, since there are $10 \cdot 15$ ways of choosing both defective chips and there are $100 \cdot 300$ ways of choosing two chips. Thus:

$$P(A \cap B) = \frac{10 \cdot 15}{100 \cdot 300}$$

$$= 0.005$$

$$= 0.10 \cdot 0.05$$

$$= P(A)P(B).$$

#

Several important points are worth noting about the concept of independence:

1. If A and B are two mutually exclusive events, then $A \cap B = \emptyset$, which implies $P(A \cap B) = 0$. Now if they are independent as well, then either $P(A) = 0$ or $P(B) = 0$.

2. If an event A is independent of itself—that is, if A and A are independent—then $P(A) = 0$ or $P(A) = 1$, since the assumption of independence yields $P(A \cap A) = P(A)P(A)$ or $P(A) = P(A)^2$.

3. If the events A and B are independent and the events B and C are independent, then events A and C need not be independent. In other words, the relation of independence is not a transitive relation.

4. If the events A and B are independent, then so are events $\bar{A}$ and B, events A and $\bar{B}$, and events $\bar{A}$ and $\bar{B}$. To show the independence of events $\bar{A}$ and B, note that $A \cap B$ and $\bar{A} \cap B$ are mutually exclusive events whose union is B. Therefore:

$$P(B) = P(A \cap B) + P(\bar{A} \cap B)$$

$$= P(A)P(B) + P(\bar{A} \cap B).$$

since A and B are independent. This implies that $P(\overline{A} \cap B) = P(B) - P(A)P(B) = P(B)[1 - P(A)] = P(B)P(\overline{A})$, which establishes the independence of $\overline{A}$ and B. Independence of A and $\overline{B}$, and $\overline{A}$ and $\overline{B}$ can be shown similarly.

The concept of independence of two events can be naturally extended to a list of n events.

Definition (Independence of a Set of Events). A list of n events $A_1, A_2, \ldots, A_n$ is defined to be mutually independent if and only if for each set of k $(2 \leqslant k \leqslant n)$ distinct indices $i_1, i_2, \ldots, i_k$ that are elements of $\{1, 2, \ldots, n\}$, we have:

$$P(A_{i_1} \cap A_{i_2} \cap \cdots \cap A_{i_k}) = P(A_{i_1})P(A_{i_2}) \cdots P(A_{i_k}).$$

Given that a list of events $A_1, A_2, \ldots, A_n$ is mutually independent, it is straightforward to show that for each set of distinct indices $i_1, i_2, \ldots, i_k$ that are elements of $\{1, 2, \ldots, n\}$:

$$P(B_{i_1} \cap B_{i_2} \cap \cdots \cap B_{i_k}) = P(B_{i_1})P(B_{i_2}) \cdots P(B_{i_k}) \qquad (1.6)$$

where each B_{i_k} may be either A_{i_k} or $\overline{A}_{i_k}$. In other words, if the A_i are independent and we replace any event by its complement, we still have independence.

By the probability axiom (A3), if a list of events is mutually exclusive, the probability of their union is the sum of their probabilities. On the other hand, if a list of events is mutually independent, the probability of their intersection is the product of their probabilities. The additive and multiplicative natures, respectively, of two event lists should be noted.

Note that it is possible to have $P(A \cap B \cap C) = P(A)P(B)P(C)$ with $P(A \cap B) \neq P(A)P(B)$, $P(A \cap C) \neq P(A)P(C)$, and $P(B \cap C) \neq P(B)\,P(C)$. Under these conditions, events A, B, and C are not mutually independent. Similarly, the condition $P(A_1 \cap A_2 \cap \cdots \cap A_n) = P(A_1)P(A_2) \cdots P(A_n)$ does not imply a similar condition for any smaller family of events, and therefore this condition does not imply that events $A_1, A_2, \ldots, A_n$ are mutually independent.

Example 1.9 [ASH 1970]

Consider the experiment of tossing two dice. Let the sample space $S = \{(i, j) | 1 \leqslant i, j \leqslant 6\}$. Also assume that each sample point is assigned a probability $1/36$. Define the events A, B, and C so that:

A = "First die results in a 1, 2, or 3."

B = "First die results in a 3, 4, or 5."

C = "The sum of the two faces is 9."

Then $A \cap B = \{(3, 1), (3, 2), (3, 3), (3, 4), (3, 5), (3, 6)\}$, $A \cap C = \{(3, 6)\}$, $B \cap C = \{(3, 6), (4, 5), (5, 4)\}$, and $A \cap B \cap C = \{(3, 6)\}$. Therefore:

$$P(A \cap B) = 1/6 \neq P(A)P(B) = 1/4,$$

$$P(A \cap C) = 1/36 \neq P(A)P(C) = 1/18,$$

$$P(B \cap C) = 1/12 \neq P(B)P(C) = 1/18,$$

but:

$$P(A \cap B \cap C) = 1/36 = P(A)P(B)P(C).$$

#

If the events $A_1, A_2, \ldots, A_n$ are such that every pair is independent, then they are called **pairwise independent**. It does not follow that the list of events is **mutually independent**.

Example 1.10 [ASH 1970]

Consider the above experiment of tossing two dice. Let

A = "First die results in a 1, 2, or 3."

B = "Second die results in a 4, 5, or 6."

C = "The sum of the two faces is 7."

Then:

$A \cap B = \{(1, 4), (1, 5), (1, 6), (2, 4), (2, 5), (2, 6), (3, 4), (3, 5), (3, 6)\}$

and:

$$\begin{aligned} A \cap C = B \cap C \\ &= A \cap B \cap C \\ &= \{(1, 6), (2, 5), (3, 4)\}. \end{aligned}$$

Therefore:

$$P(A \cap B) = 1/4 = P(A)P(B),$$

$$P(A \cap C) = 1/12 = P(A)P(C),$$

$$P(B \cap C) = 1/12 = P(B)P(C),$$

but:

$$P(A \cap B \cap C) = 1/12 \neq P(A)P(B)P(C) = 1/24.$$

In this example, events A, B, and C are **pairwise independent** but not **mutually independent**.

#

We illustrate the idea of independence by considering the problem of computing reliability of so-called series-parallel systems. A **series system** is one in which all components are so interrelated that the entire system will fail if any one of its components fails. On the other hand, a **parallel system** is one that will fail only if all its components fail. We will assume that failure events of components in a system are mutually independent.

First consider a series system of n components. For $i = 1, 2, \ldots, n$, define events A_i = "Component i is functioning properly." Let the **reliability**, R_i, of component i be defined as the probability that the component is functioning properly; then $R_i = P(A_i)$. By the assumption of series connections, the system reliability:

$$
\begin{aligned}
R_s &= P(\text{"The system is functioning properly."}) \\
&= P(A_1 \cap A_2 \cap \cdots \cap A_n) \\
&= P(A_1)P(A_2) \cdots P(A_n) \qquad (1.7) \\
&= \prod_{i=1}^{n} R_i .
\end{aligned}
$$

This simple **product law of reliabilities**, applicable to series systems of independent components, demonstrates how quickly system reliability degrades with an increase in complexity. For example, if a system consists of five components in series, each having a reliability of 0.970, then the system reliability is $0.970^5 = 0.859$. Now if the system complexity is increased so that it contains ten similar components, its reliability is reduced to $0.970^{10} = 0.738$. Consider what happens to system reliability when a large system such as a computer system consists of tens to hundreds of thousands of components!

One way to increase the reliability of a system is to use **redundancy**. The first scheme that comes to mind is to replicate components with small reliabilities (**parallel redundancy**). First consider a system consisting of n independent components in parallel, so that it will fail to function only if all n components have failed. Define event A_i = "The component i is functioning properly" and A_p = "The parallel system of n components is functioning properly." Also let $R_i = P(A_i)$ and $R_p = P(A_p)$. To establish a relation between A_p and the A_i's, it is easier to consider the complementary events. Thus:

$$
\begin{aligned}
\overline{A}_p &= \text{"The parallel system has failed."} \\
&= \text{"All } n \text{ components have failed."} \\
&= \overline{A}_1 \cap \overline{A}_2 \cap \cdots \cap \overline{A}_n .
\end{aligned}
$$

Therefore:

$$
\begin{aligned}
P(\overline{A}_p) &= P(\overline{A}_1 \cap \overline{A}_2 \cap \cdots \cap \overline{A}_n) \\
&= P(\overline{A}_1)P(\overline{A}_2) \cdots P(\overline{A}_n)
\end{aligned}
$$

by independence. Now let $F_p = 1 - R_p$ be the **unreliability** of the parallel system and similarly let $F_i = 1 - R_i$ be the unreliability of component i. Then, since A_i and $\bar{A}_i$ are mutually exclusive and collectively exhaustive events, we have:

$$\begin{aligned} 1 &= P(S) \\ &= P(A_i) + P(\bar{A}_i) \end{aligned}$$

and:

$$\begin{aligned} F_i &= P(\bar{A}_i) \\ &= 1 - P(A_i). \end{aligned}$$

Then:

$$\begin{aligned} F_p &= P(\bar{A}_p) \\ &= \prod_{i=1}^{n} F_i \end{aligned}$$

and:

$$\begin{aligned} R_p &= 1 - F_p \\ &= 1 - \prod_{i=1}^{n} (1 - R_i). \end{aligned} \tag{1.8}$$

Thus, for parallel systems of n independent components, we have a **product law of unreliabilities** analogous to the product law of reliabilities of series systems. If we have a parallel system of five components, each with a reliability of 0.970, then the system reliability is increased to:

$$\begin{aligned} 1 - (1 - 0.970)^5 &= 1 - (0.03)^5 \\ &= 1 - 0.0000000243 \\ &= 0.9999999757. \end{aligned}$$

However, we should be aware of a **law of diminishing returns**: the rate of increase in reliability with each additional component decreases rapidly as n increases. This is illustrated in Figure 1.10, where we have plotted R_p as a function of n. [This remark is easily formalized by noting that R_p is a concave function of n since $R'_p(n) = -(1 - R)^n \ln(1 - R) > 0$, and $R''_p(n) = -(1 - R)^n (\ln(1 - R))^2 < 0$.]

The basic formulas (1.7) and (1.8) for the reliability computation of series and parallel systems can be used in combination to compute the reliability of a system having both series and parallel parts (**series-parallel systems**). Consider a series-parallel system of n serial stages where stage i consists of n_i identical components in parallel. Let the reliability of each component at

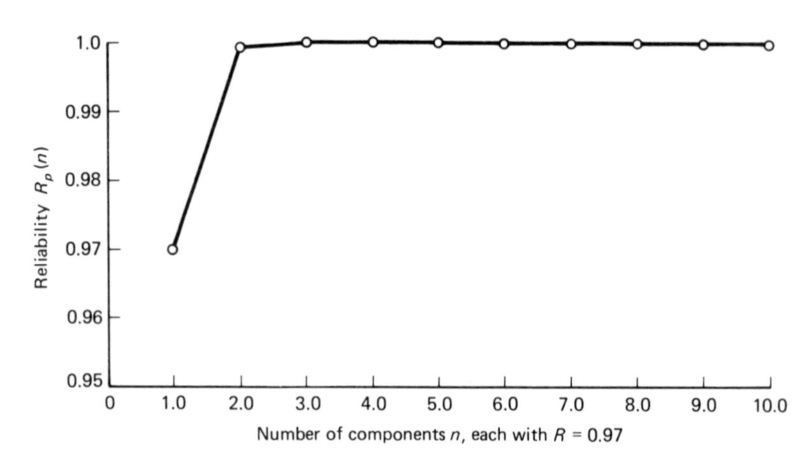

Figure 1.10 Reliability curve of a parallel redundant system

stage i be R_i. Assuming that all components are independent, system reliability R_{sp} can be computed from the formula:

$$R_{sp} = \prod_{i=1}^{n} [1 - (1 - R_i)^{n_i}]. \tag{1.9}$$

Example 1.11

Consider the system shown in Figure 1.11, consisting of five stages, with $n_1 = n_2 = n_5 = 1$, and $n_3 = 3$ and $n_4 = 2$. Also:

$$R_1 = 0.95, \quad R_2 = 0.99, \quad R_3 = 0.70, \quad R_4 = 0.75, \quad R_5 = 0.9.$$

Then:

$$\begin{aligned} R_{sp} &= 0.95 \cdot 0.99 \cdot (1 - 0.3^3) \cdot (1 - 0.25^2) \cdot 0.9 \\ &= 0.772. \end{aligned}$$

#

Reliability of systems with more general interconnections cannot be computed with the above formula. In such a case, we may make use of Bayes' rule (to be discussed in the next section). Reliability of systems with standby redundancy cannot be computed using methods discussed in this chapter, but techniques to be discussed later will enable us to do so.

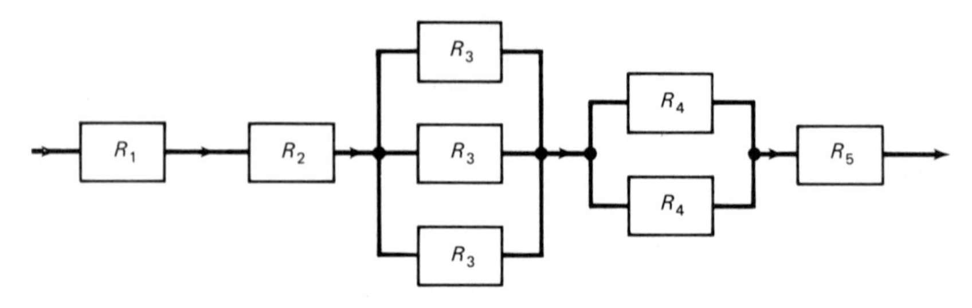

Figure 1.11 A series-parallel system

Problems

1. Two towns are connected by a network of communication channels. The probability of a channel's failure-free operation is R, and channel failures are independent. Minimal level of communication between towns can be guaranteed provided at least one path containing properly functioning channels exists. Given the network of Figure 1.P.1, determine the probability that the two towns will be able to communicate. Here $-||-$ denotes a communication channel.

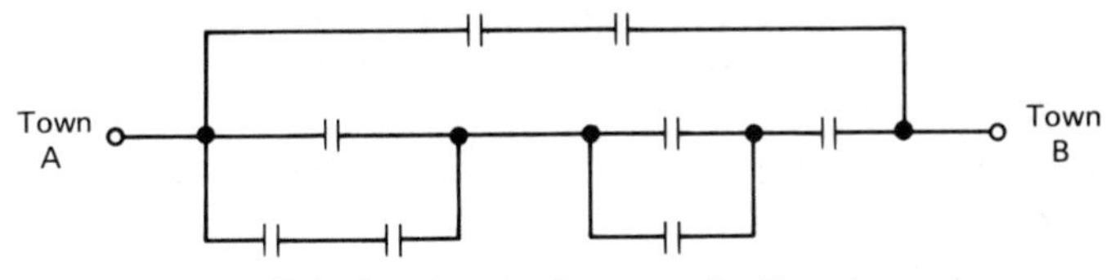

Figure 1.P.1 A network of communication channels

2. Given three components with respective reliabilities $R_1 = 0.8$, $R_2 = 0.75$, and $R_3 = 0.98$, compute the reliabilities of the three systems shown in Figure 1.P.2.

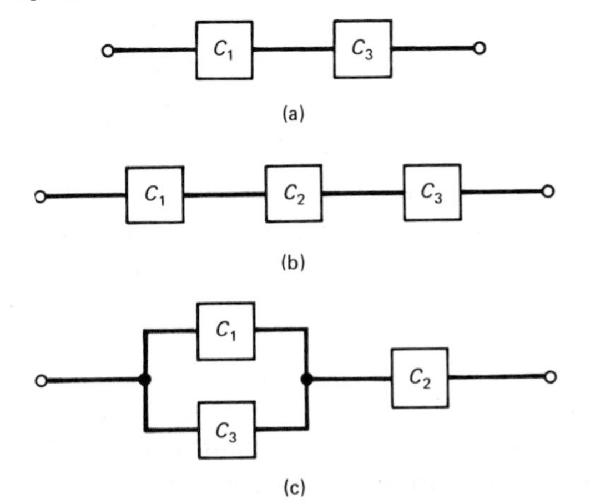

Figure 1.P.2 Reliability block diagrams

3. Determine the conditions under which an event A is independent of its subset B.

4. *General multiplication rule (GMR).* Given a list of events $A_1, A_2, \ldots, A_n$ (not necessarily independent), show that

$$
\begin{aligned}
P(A_1 \cap A_2 \cap \cdots \cap A_n) = {} & P[A_1|(A_2 \cap A_3 \cap \cdots \cap A_n)] \\
& \cdot P[A_2|(A_3 \cap \cdots \cap A_n)] \\
& \cdot P[A_3|(A_4 \cap \cdots \cap A_n)] \\
& \cdots \\
& \cdot P(A_{n-1}|A_n)P(A_n),
\end{aligned}
$$

provided all the conditional probabilities on the right-hand side are defined.

1.11 BAYES' RULE

A given event B of probability $P(B)$ partitions the sample space S into two disjoint subsets B and $\bar{B}$. If we consider $S' = \{B, \bar{B}\}$ and associate the probabilities $P(B)$ and $P(\bar{B})$ to the respective points in S', then S' is very similar to a sample space, except that there is a many-to-one correspondence between the outcomes of the experiment and the elements of S'. A space such as S' is often called an **event space**. In general, a list of n events $B_1, B_2, \ldots, B_n$ that are collectively exhaustive and mutually exclusive form an **event space**, $S' = \{B_1, B_2, \ldots, B_n\}$.

Returning to the event space $S' = \{B, \bar{B}\}$, note that an event A is partitioned into two disjoint subsets:

$$A = (A \cap B) \cup (A \cap \bar{B}).$$

Then by axiom (A3):

$$\begin{aligned} P(A) &= P(A \cap B) + P(A \cap \bar{B}) \\ &= P(A|B)P(B) + P(A|\bar{B})P(\bar{B}) \end{aligned}$$

by definition of conditional probability.

This relation is analogous to Shannon's Theorem in switching theory and can be generalized with respect to the event space $S' = \{B_1, B_2, \ldots, B_n\}$:

$$P(A) = \sum_{i=1}^{n} P(A|B_i)P(B_i). \tag{1.10}$$

This relation is also known as the **theorem of total probability** and is sometimes called the **rule of elimination**. This situation can be visualized by constructing a **tree diagram** (or a **probability tree**) as shown in Figure 1.12,

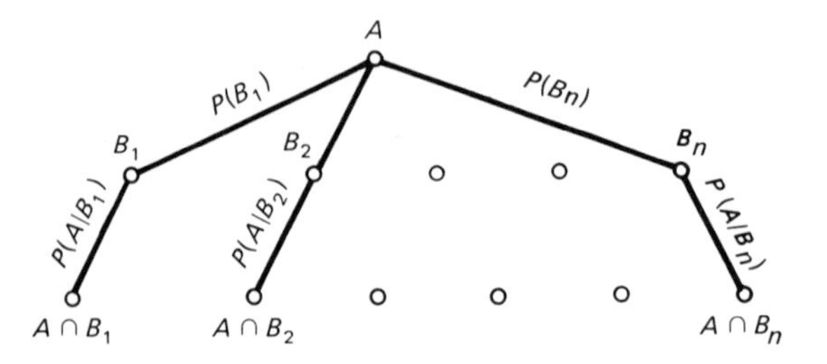

Figure 1.12 The theorem of total probability

where each branch is so labeled that the product of all branch probabilities from the root of the tree to any node equals the probability of the event represented by that node. Now $P(A)$ can be computed by summing probabilities associated with all the leaf nodes of the tree.

In practice, after the experiment, a situation often arises in which the event A is known to have occurred, but it is not known directly which of the mutually exclusive and collectively exhaustive events $B_1, B_2, \ldots, B_n$ has occurred. In this situation, we may be interested in evaluating $P(B_j|A)$, the conditional probability that one of these events B_j occurs, given that A occurs. By applying the definition of conditional probability followed by the use of theorem of total probability, we find that:

$$P(B_j|A) = \frac{P(B_j \cap A)}{P(A)} \tag{1.11}$$

$$= \frac{P(A|B_j)P(B_j)}{\sum_i P(A|B_i)P(B_i)}.$$

This relation is known as **Bayes' rule** and is useful in many applications. This rule also forms the basis of a statistical method called **bayesian procedure**. $P(B_j|A)$ is sometimes called an **a posteriori probability**.

Example 1.12

Measurements at the Triangle Universities Computation Center (TUCC) on a certain day indicated that the source of incoming jobs is 15 percent from Duke, 35 percent from North Carolina, and 50 percent from North Carolina State. Suppose that the probabilities that a job initiated from these universities requires (operator intervention for tape) set-up are 0.01, 0.05, and 0.02, respectively. Find the probability that a job chosen at random at TUCC is a set-up job. Also find the probability that a randomly chosen job comes from the University of North Carolina, given that it is a set-up job.

Define the events B_i = "Job is from university i (i= 1, 2, 3 for Duke, UNC, and NC State, respectively), and A= "Job requires set-up." Then by the theorem of total probability:

$$P(A) = P(A|B_1)P(B_1) + P(A|B_2)P(B_2) + P(A|B_3)P(B_3)$$

$$= (0.01) \cdot (0.15) + (0.05) \cdot (0.35) + (0.02) \cdot (0.5)$$

$$= 0.029.$$

Now the second event of interest is $[B_2|A]$, and from Bayes' rule:

$$P(B_2|A) = \frac{P(A|B_2)P(B_2)}{P(A)}$$

$$= \frac{0.05 \cdot 0.35}{0.029}$$

$$= 0.603.$$

Note that the knowledge that the job requires set-up increases the chance that it came from UNC from 35 percent to about 60 percent. #

Example 1.13

A binary communication channel carries data as one of two types of signals denoted by 0 and 1. Owing to noise, a transmitted 0 is sometimes received as a 1 and a transmitted 1 is sometimes received as a 0. For a given channel, assume a probability of 0.94 that a transmitted 0 is correctly received as a 0 and a probability of 0.91 that a transmitted 1 is received as a 1. Further assume a probability of 0.45 of transmitting a 0. If a signal is sent, determine:

1. Probability that a 1 is received.
2. Probability that a 0 is received.
3. Probability that a 1 was transmitted, given that a 1 was received.
4. Probability that a 0 was transmitted, given that a 0 was received.
5. Probability of an error.

Define events T_0 = "A 0 is transmitted" and event R_0 = "A 0 is received." Then let $T_1 = \bar{T}_0$ = "A 1 is transmitted" and $R_1 = \bar{R}_0$ = "A 1 is received." Then the events of interest under questions 1, 2, 3, and 4 are respectively given by R_1, R_0, $[T_1|R_1]$, and $[T_0|R_0]$. An error in the transmitted signal is the union of the two disjoint events $[T_1 \cap R_0]$ and $[T_0 \cap R_1]$. The operation of a binary communication channel may be visualized by a **channel diagram** in Figure 1.13. In the given problem, we have $P(R_0|T_0) = 0.94$, $P(R_1|T_1) = 0.91$, and $P(T_0) = 0.45$. From these we get:

$$P(R_1|T_0) = P(\bar{R}_0|T_0) = 1 - P(R_0|T_0) = 0.06,$$

$$P(R_0|T_1) = P(\bar{R}_1|T_1) = 1 - P(R_1|T_1) = 0.09,$$

$$P(T_1) = P(\bar{T}_0) = 1 - P(T_0) = 0.55.$$

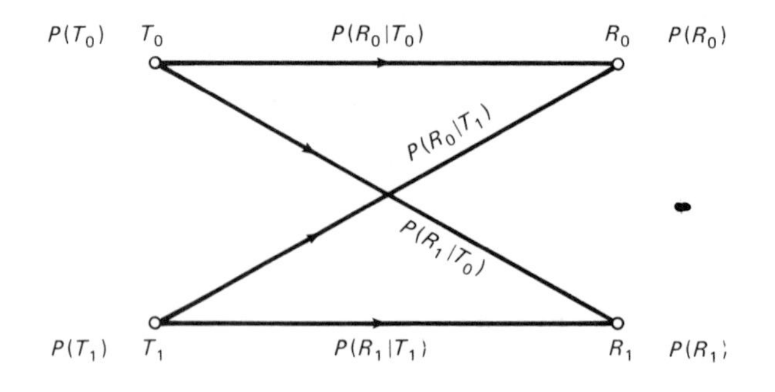

Figure 1.13 A channel diagram

Now from the theorem of total probability:

$$\begin{aligned} P(R_0) &= P(R_0|T_0)P(T_0) + P(R_0|T_1)P(T_1) \\ &= (0.94) \cdot (0.45) + (0.09) \cdot (0.55) \\ &= 0.423 + 0.0495 \\ &= 0.4725. \\ P(R_1) &= P(\bar{R}_0) \\ &= 1 - P(R_0) \\ &= 1 - 0.4725 \\ &= 0.5275. \end{aligned}$$

Using Bayes' rule, we have:

$$\begin{aligned} P(T_1|R_1) &= \frac{P(R_1|T_1)P(T_1)}{P(R_1)} \\ &= \frac{0.91 \cdot 0.55}{0.5275} \\ &= 0.9488, \\ P(T_0|R_0) &= \frac{P(R_0|T_0)P(T_0)}{P(R_0)} \\ &= \frac{0.94 \cdot 0.45}{0.4725} \\ &= 0.8952. \end{aligned}$$

Now:

$$\begin{aligned} P(T_1|R_0) &= P(\bar{T}_0|R_0) \\ &= 1 - P(T_0|R_0) \\ &= 0.1048, \\ P(T_0|R_1) &= 1 - P(T_1|R_1) \\ &= 0.0512, \end{aligned}$$

and

$$\begin{aligned} P(\text{``error''}) &= P(T_1 \cap R_0) + P(T_0 \cap R_1) \\ &= P(T_1|R_0)P(R_0) + P(T_0|R_1)P(R_1) \\ &= 0.1048 \cdot 0.4725 + 0.0512 \cdot 0.5275 \\ &= 0.0765. \end{aligned}$$

Alternatively, the error probability can be evaluated by:

$$P(\text{"error"}) = P(T_1 \cap R_0) + P(T_0 \cap R_1)$$
$$= P(R_0|T_1)\,P(T_1) + P(R_1|T_0)\,P(T_0)$$
$$= 0.09 \cdot 0.55 + 0.06 \cdot 0.45 = 0.0765.$$

[Quiz: Construct an appropriate sample space for this problem.] #

Example 1.14

A given lot of IC chips contains 2 percent defective chips. Each chip is tested before delivery. The tester itself is not totally reliable so that:

$$P(\text{"Tester says chip is good"}|\text{"Chip is actually good"}) = 0.95,$$
$$P(\text{"Tester says chip is defective"}|\text{"Chip is actually defective"}) = 0.94.$$

If a tested device is indicated to be defective, what is the probability that it is actually defective?

By Bayes' rule, we have:

$$P(\text{"Chip is defective"}|\text{"Tester says it is defective"})$$
$$= \frac{P(\text{"Tester says it is defective"}|\text{"Chip is defective"})\ P(\text{"Chip is defective"})}{P(\text{"Tester says defective"}|\text{"Chip defective"})\ P(\text{"Chip defective"}) + P(\text{"Tester says defective"}|\text{"Chip is good"})\ P(\text{"Chip is good"})}$$
$$= \frac{0.94 \cdot 0.02}{0.94 \cdot 0.02 + 0.05 \cdot 0.98}$$
$$= \frac{0.0188}{0.0188 + 0.049}$$
$$= \frac{0.0188}{0.0678}$$
$$= 0.2772861.$$ #

Example 1.15

We have seen earlier how to evaluate the reliability of series-parallel systems. However, many systems in practice do not conform to a series-parallel structure. As an example, consider evaluating the reliability R of the five-component system shown in Figure 1.14. The system is said to be functioning properly only if all the components on at least one path from point A to point B are functioning properly.

Define for $i = 1, 2, \ldots, 5$ event X_i = "Component i is functioning properly," and let R_i = reliability of component $i = P(X_i)$. Let X = "System is functioning properly" and let R = system reliability $= P(X)$. It is clear that X is union of four events:

$$X = (X_1 \cap X_4) \cup (X_2 \cap X_4) \cup (X_2 \cap X_5) \cup (X_3 \cap X_5). \quad (1.12)$$

These four events are not mutually exclusive. Therefore, we cannot directly use axiom (A3). Note, however, that we could use relation (Rd), which does apply to a un-

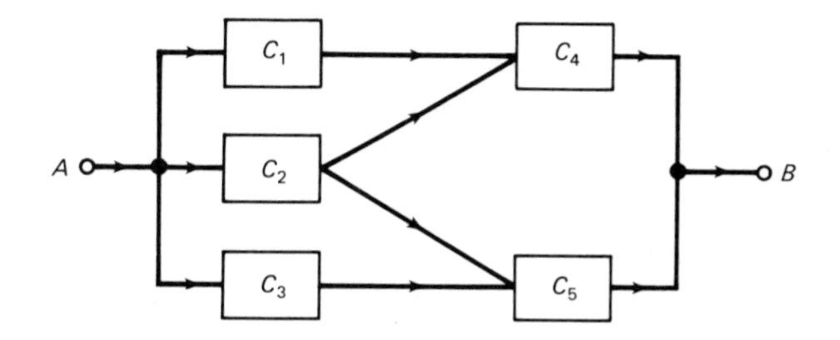

Figure 1.14 A non-series-parallel system

ion of intersecting events. But this method is computationally tedious for a relatively long list of events. Instead, using the theorem of total probability, we have:

$$P(X) = P(X|X_2)P(X_2) + P(X|\bar{X}_2)P(\bar{X}_2)$$
$$= P(X|X_2)R_2 + P(X|\bar{X}_2)(1 - R_2). \tag{1.13}$$

Now to compute $P(X|X_2)$, we observe that since component C_2 is functioning, the status of components C_1 and C_3 is irrelevant. Thus, under this condition, the system is equivalent to two components C_4 and C_5 in parallel. Therefore using formula (1.8), we get:

$$P(X|X_2) = 1 - (1 - R_4)(1 - R_5). \tag{1.14}$$

To compute $P(X|\bar{X}_2)$, we observe that, since component C_2 is known to have malfunctioned, the resulting equivalent system is a series-parallel one whose reliability is easily computed:

$$P(X|\bar{X}_2) = 1 - (1 - R_1R_4)(1 - R_3R_5). \tag{1.15}$$

Combining equations (1.13), (1.14), and (1.15) we have:

$$R = [1 - (1 - R_4)(1 - R_5)]R_2 + [1 - (1 - R_1R_4)(1 - R_3R_5)](1 - R_2)$$
$$= 1 - R_2(1 - R_4)(1 - R_5) - (1 - R_2)(1 - R_1R_4)(1 - R_3R_5). \qquad \#$$

Problems

1. A technique for fault-tolerant software, suggested by Randell [RAND 1978], consists of a primary and an alternate module for each critical task, together with a test for determining whether a module performed its function correctly. Such a construct is called a **recovery block**. Define the events:

   A = "Primary module functions correctly."

   B = "Alternate module functions correctly."

   D = "Detection test following the execution of the primary performs its task correctly."

   Assume that event pairs A and D as well as B and D are independent but events A and B are dependent. Derive an expression for the failure probability of a recovery block [HECH 1976]. [Hint: Use a tree diagram].

2. Consider the non-series-parallel system of four independent components shown in Figure 1.P.3. The system is considered to be functioning properly if all components along at least one path from input to output are functioning properly. Determine an expression for system reliability as a function of component reliabilities.

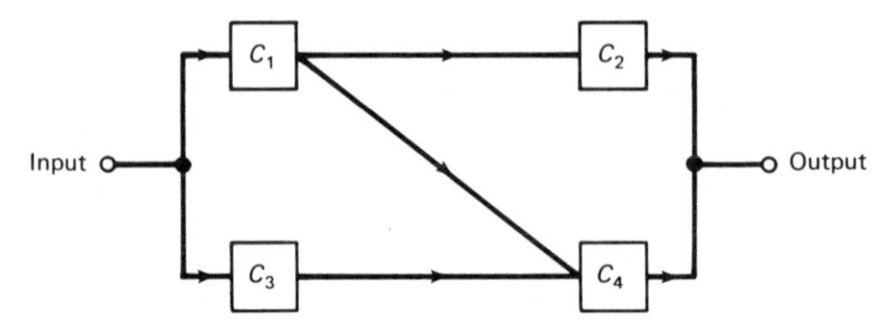

Figure 1.P.3 Another non-series-parallel system

3. A lot of transistors contains 0.6 percent defectives. Each transistor is subjected to a test that correctly identifies a defective but also misidentifies as defective about two in every 100 good transistors. Given that a randomly chosen transistor is declared defective by the tester, compute the probability that it is actually defective.

4. A certain firm has plants A, B, and C producing, respectively, 35 percent, 15 percent, and 50 percent, of the total output. The probabilities of a nondefective product are, respectively, 0.75, 0.95, and 0.85. A customer receives a defective product. What is the probability that it came from plant C?

5. [STAR 1979] Consider a trinary communication channel whose channel diagram is shown in Figure 1.P.4. For $i = 1, 2, 3$, let T_i denote the event "Digit i is transmitted" and let R_i denote the event "Digit i is received." Assume that a 3 is transmitted three times more frequently than a 1, and a 2 is sent twice as often as 1. If a 1 has been received, what is the expression for the probability that a 1 was sent? Derive an expression for the probability of a transmission error.

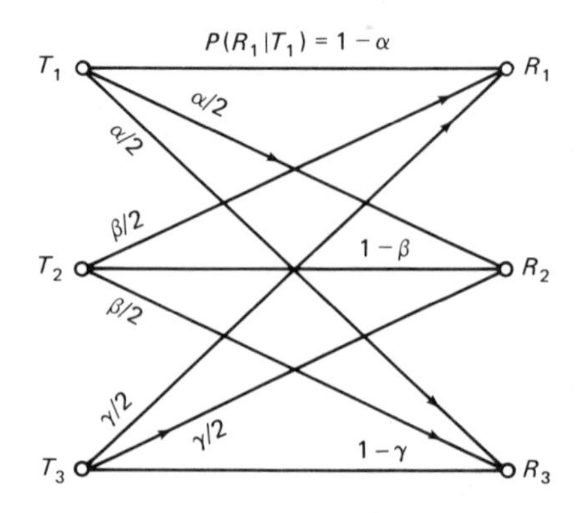

Figure 1.P.4 A trinary communication channel: channel diagram

6. Of all the graduate students in a university, 70 percent are women and 30 percent are men. Suppose that 20 percent and 25 percent of the female and male population, respectively, smoke cigarettes. What is the probability that a randomly selected graduate student is:
   (a) A woman who smokes?
   (b) A man who smokes?
   (c) A smoker?

7. Compute the reliability of the system discussed in Example 1.15 (Figure 1.14), using equation (1.12) and the relation (Rd).

8. Another method of evaluating the reliability of the system discussed in Example 1.15 is to use the methods of switching theory. Noting that X_1, X_2, X_3, X_4, X_5 are Boolean variables and X is a switching function of these variables, we can draw a truth table with $2^5 = 32$ rows. Rows of the truth table represent a collection of mutually independent and collectively exhaustive events. Each row represents an elementary event that is an intersection of independent events, hence its probability can be easily computed. For example, the elementary event $\overline{X}_1 \cap X_2 \cap \overline{X}_3 \cap X_4 \cap X_5$ is assigned the probability $(1 - R_1)\, R_2\, (1 - R_3)\, R_4\, R_5$. Computing $P(X)$ now reduces to adding up probabilities of rows of the truth table with 1's in the function column. Use this method to compute the reliability of the system in Figure 1.14.

1.12 BERNOULLI TRIALS

Consider a random experiment that has two possible outcomes, "success" and "failure" (or "hit" and "miss," or "good" and "defective," or "digit received correctly" and "digit received incorrectly," or the like. Let the probabilities of the two outcomes be p and q, respectively, with $p + q = 1$. Now consider the compound experiment consisting of a sequence of n independent repetitions of this experiment. Such a sequence is known as a **sequence of Bernoulli trials**. This abstract sequence models many physical situations of interest to us:

1. Observe n consecutive executions of an **if** statement, with success = "**then** clause is executed" and failure = "**else** clause is executed."
2. Examine components produced on an assembly line, with success = "acceptable" and failure = "defective."
3. Transmit binary digits through a communication channel, with success = "digit received correctly" and failure = "digit received incorrectly."
4. Consider a time-sharing computer system that allocates a finite quantum (or time slice) to a job scheduled for processor service, in an attempt to give fast service to requests for trivial processing. Observe n time-slice terminations, with success = "job has completed processing" and failure = "job still requires processing and joins the tail end of the ready queue of processes." This situation may be depicted as in Figure 1.15.

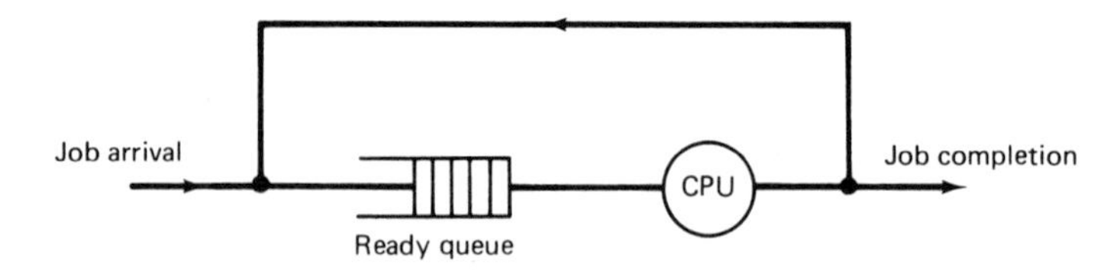

Figure 1.15 A CPU queue with time-slicing

Let 0 denote failure and 1 denote success. Let S_n be the sample space of an experiment involving n Bernoulli trials, defined by:

$$S_1 = \{0, 1\},$$

$$S_2 = \{(0, 0),\ (0, 1),\ (1, 0),\ (1, 1)\}$$

$$S_n = \{2^n\ n\text{-tuples of 0's and 1's}\}.$$

The probability assignment over the sample space S_1 is already specified: $P(0) = q \geqslant 0$, $P(1) = p \geqslant 0$, and $p + q = 1$. We wish to assign probabilities to the points in S_n.

Let A_i = "Success on trial i" and $\overline{A}_i$ = "Failure on trial i," then $P(A_i) = p$ and $P(\overline{A}_i) = q$. Now consider s an element of S_n such that $s = (1, 1, \ldots, 1, 0, 0, \ldots, 0)$ [k 1's and $(n-k)$ 0's]. Then the elementary event $\{s\}$ can be written as:

$$\{s\} = A_1 \cap A_2 \cap \cdots \cap A_k \cap \overline{A}_{k+1} \cap \cdots \cap \overline{A}_n$$

and:

$$\begin{aligned} P(s) &= P(A_1 \cap A_2 \cdots \cap A_k \cap \overline{A}_{k+1} \cap \cdots \cap \overline{A}_n) \\ &= P(A_1)P(A_2) \cdots P(A_k)P(\overline{A}_{k+1}) \cdots P(\overline{A}_n) \end{aligned}$$

by independence. Therefore:

$$P(s) = p^k q^{n-k}. \tag{1.16a}$$

Similarly, any sample point with k 1's and $(n-k)$ 0's is assigned probability $p^k q^{n-k}$. Noting that there are $\binom{n}{k}$ such sample points, the probability of obtaining exactly k successes in n trials is:

$$p(k) = \binom{n}{k} p^k q^{n-k}, \qquad k = 0, 1, \ldots, n. \tag{1.16b}$$

We may verify that (1.16a) is a legitimate probability assignment over the sample space S_n since:

$$\begin{aligned} \sum_{k=0}^{n} \binom{n}{k} p^k q^{n-k} &= (p+q)^n \\ &= 1 \end{aligned}$$

by the binomial theorem.

Consider the set of events $\{B_0, B_1, \ldots, B_n\}$ where $B_k = \{s \in S_n$ such that s has exactly k 1's and $(n-k)$ 0's$\}$. It is clear that this is a mutually exclusive and collectively exhaustive family of events. Furthermore:

$$P(B_k) = \binom{n}{k} p^k q^{n-k} \geqslant 0 \quad \text{and} \quad \sum_{k=0}^{n} P(B_k) = 1.$$

Therefore, this collection of events is an event space with $(n+1)$ events. Compare this with 2^n sample points in S_n. Thus, when in a physical situation, if we are not concerned with the actual sequence of successes and failures but merely with the number of successes and the number of failures, it is profitable to use the event space rather than the original sample space.

Example 1.16

In connection with reliability computation, we have considered series and parallel systems. Now we consider a system with n components that requires m $(\leqslant n)$ or more components to function for the correct operation of the system. Such systems are often called m-out-of-n systems. If we let $m = n$, then we have a series system; if we let $m = 1$, then we have a system with parallel redundancy. Assume that all n components are statistically identical and function independently of each other. If we let R denote the reliability of a component (and $q = 1 - R$ gives its unreliability), then the experiment of observing the statuses of n components can be thought of as a sequence of n Bernoulli trials with the probability of success equal to R. Now the reliability of the system is:

$$\begin{aligned} R_{m|n} &= P(\text{``}m\text{ or more components functioning properly''}) \\ &= P\Big(\bigcup_{i=m}^{n} \{\text{``exactly } i \text{ components functioning properly''}\}\Big) \\ &= \sum_{i=m}^{n} P(\text{``exactly } i \text{ components functioning properly''}) \\ &= \sum_{i=m}^{n} p(i), \end{aligned}$$

$$R_{m|n} = \sum_{i=m}^{n} \binom{n}{i} R^i (1-R)^{n-i}. \tag{1.17}$$

Verify that $R_{1|n} = R(\text{parallel})$:

$$\begin{aligned} R_{1|n} &= \sum_{i=1}^{n} \binom{n}{i} R^i (1-R)^{n-i} \\ &= \sum_{i=0}^{n} \binom{n}{i} R^i (1-R)^{n-i} - \binom{n}{0} R^0 (1-R)^n \\ &= [R + (1-R)]^n - (1-R)^n \\ &= 1 - (1-R)^n. \end{aligned}$$

Verify that $R_{n|n} = R(\text{series})$:

$$R_{n|n} = \sum_{i=n}^{n} \binom{n}{i} R^i (1-R)^{n-i}$$

$$= \binom{n}{n} R^n (1-R)^0$$

$$= R^n. \qquad \#$$

As another special case of formula (1.17) above, consider a system with triple modular redundancy, often known as TMR (see Figure 1.16). In such a system there are three components, two of which are required to be in working order for the system to function properly (i.e., $n = 3$ and $m = 2$). This is achieved by feeding the outputs of the three components into a majority voter. Then:

$$R_{\text{TMR}} = \sum_{i=2}^{3} \binom{3}{i} R^i (1-R)^{3-i}$$

$$= \binom{3}{2} R^2 (1-R) + \binom{3}{3} R^3 (1-R)^0$$

$$= 3R^2(1-R) + R^3,$$

and thus:

$$R_{\text{TMR}} = 3R^2 - 2R^3. \tag{1.18}$$

Note that:

$$R_{\text{TMR}} \begin{cases} > R, & \text{if } R > 1/2, \\ = R, & \text{if } R = 1/2, \\ < R, & \text{if } R < 1/2. \end{cases}$$

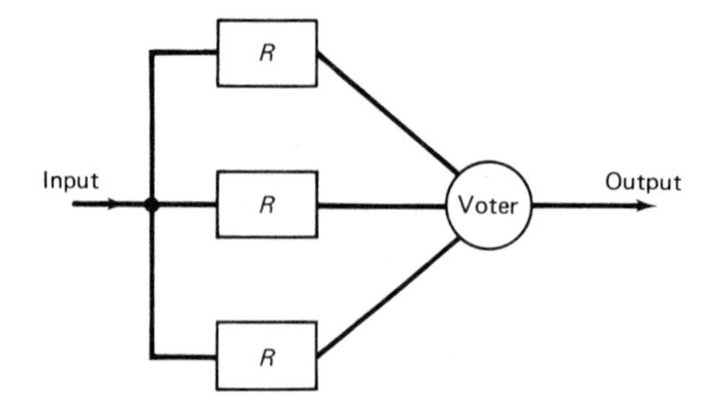

Figure 1.16 A triple modular redundant system

Thus TMR increases reliability over the simplex system only if the simplex reliability is greater than 0.5; otherwise this type of redundancy actually *decreases* reliability.

It should be noted that the voter output simply corresponds to the majority, and therefore it is possible for two or more malfunctioning units to agree, producing an erroneous voter output. Additional detection logic is required to avoid this situation. Also, the unreliability of the voter will further degrade the TMR reliability.

Example 1.17

Consider a binary communication channel transmitting coded words of n bits each. Assume that the probability of successful transmission of a single bit is p (and the probability of an error is $q = 1 - p$), and that the code is capable of correcting up to e (where $e \geqslant 0$) errors. For example, if no coding or parity checking is used, then $e = 0$. If a single error-correcting Hamming code is used then $e = 1$. For more details on this topic see [HAMM 1980]. If we assume that the transmission of successive bits is independent, then the probability of successful word transmission is:

$$P_w = P(\text{``}e \text{ or fewer errors in } n \text{ trials''})$$

$$= \sum_{i=0}^{e} \binom{n}{i} (1-p)^i p^{n-i}.$$

#

Next we consider **generalized Bernoulli trials**. Here we have a sequence of n independent trials, and on each trial the result is exactly one of the k possibilities $b_1, b_2, \ldots, b_k$. On a given trial, let b_i occur with probability p_i, $i = 1, 2, \ldots, k$ such that:

$$p_i \geqslant 0 \quad \text{and} \quad \sum_{i=1}^{k} p_i = 1.$$

The sample space S consists of all k^n n-tuples with components $b_1, b_2, \ldots, b_k$. To a point $s \in S$:

$$s = (\underbrace{b_1, b_1, \ldots, b_1}_{n_1}, \underbrace{b_2, b_2, \ldots, b_2}_{n_2}, \ldots, \underbrace{b_k, \ldots, b_k}_{n_k})$$

we assign the probability of $p_1^{n_1} p_2^{n_2} \cdots p_k^{n_k}$, where $\sum_{i=1}^{k} n_i = n$. This is the probability assigned to any n-tuple having n_i occurrences of b_i, where $i = 1, 2, \ldots, k$. The number of such n-tuples is given by the multinomial coefficient (see [LIU 1968]):

$$\binom{n}{n_1\, n_2 \cdots n_k} = \frac{n!}{n_1! n_2! \cdots n_k!}$$

As before, the probability that b_1 will occur n_1 times, b_2 will occur n_2 times, . . . , and b_k will occur n_k times is given by:

$$p(n_1, n_2, \ldots, n_k) = \frac{n!}{n_1!\, n_2! \cdots n_k!} p_1^{n_1} p_2^{n_2} \cdots p_k^{n_k} \tag{1.19}$$

and:

$$\sum_{n_i \geq 0} p(n_1, n_2, \ldots, n_k) = (p_1 + p_2 + \cdots + p_k)^n$$
$$= 1$$

(where $\sum n_i = n$) by the multinomial theorem.

If we let $k = 2$, then generalized Bernoulli trials reduce to ordinary Bernoulli trials, where $b_1 =$ "success," $b_2 =$ "failure," $p_1 = p$, $p_2 = q = 1 - p$, $n_1 = k$, and $n_2 = n - k$.

Two situations of importance are examples of generalized Bernoulli trials:

1. We are given that at the end of a CPU burst, a program will request service from an I/O device i with probability p_i, where $i = 1, 2, \ldots, k$ and $\sum_i p_i = 1$. If we assume that successive CPU bursts are independent of each other, then the observation of n CPU burst terminations corresponds to a sequence of generalized Bernoulli trials. This situation may be pictorially visualized by the queuing network shown in Figure 1.17.

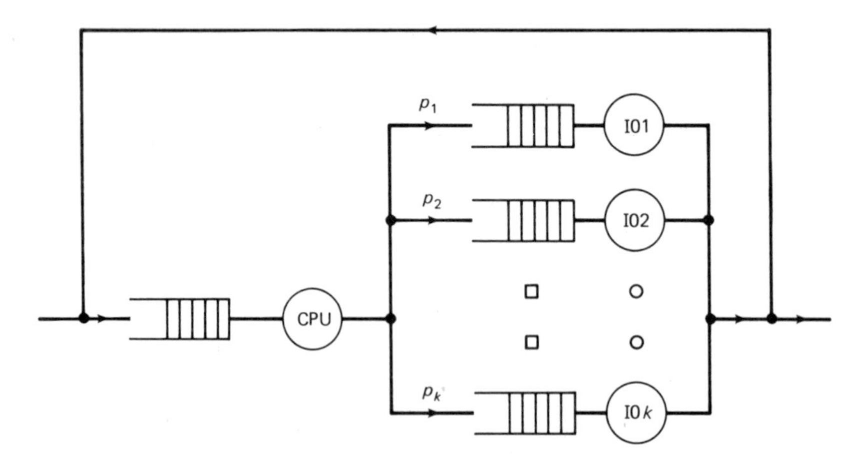

Figure 1.17 A processor to peripheral device queuing scheme

2. If we observe n consecutive independent executions of a **case** statement (see below), then we have a sequence of generalized Bernoulli trials where p_i is the probability of executing the statement group S_i on an individual trial.

```
case I of
  begin
    1:   S1;
    2:   S2;
          .
          .
          .
    k:   Sk
  end
```

Example 1.18

Out of every 100 jobs received at a computing center, 50 are of class 1, 30 of class 2, and 20 of class 3. A sample of 30 jobs is taken with replacement.

1. Find the probability that the sample will contain ten jobs of each class.
2. Find the probability that there will be exactly twelve jobs of class 2.

This is an example of generalized Bernoulli trials with $k=3$, $n=30$, $p_1=0.5$, $p_2=0.3$, and $p_3=0.2$. The answer to part (1) is:

$$p(10, 10, 10) = \frac{30!}{10! \cdot 10! \cdot 10!} \cdot 0.5^{10} \cdot 0.3^{10} \cdot 0.2^{10}$$

$$= 0.003278.$$

The answer to part (2) is obtained more easily if we collapse class 1 and class 3 together and consider this as an example of an ordinary Bernoulli trial with $p = 0.3$ (success corresponds to a class 2 job), $q = 1 - p = 0.7$ (failure corresponds to a class 1 or class 3 job). Then the required answer is:

$$p(12) = \binom{30}{12} \cdot 0.3^{12} \cdot 0.7^{18}$$

$$= \frac{30!}{12! \cdot 18!} \cdot 0.3^{12} \cdot 0.7^{18}$$

$$= 0.07485.$$

#

Example 1.19

So far, we have assumed that a component is either functioning properly or has malfunctioned. Sometimes we will find it is useful to consider more than two states. For example, a diode functions properly with probability p_1, develops a short circuit with probability p_2, and develops an open circuit with probability p_3 such that $p_1 + p_2 + p_3 = 1$. Thus there are two types of malfunctions, an open circuit and a closed circuit. In order to protect against such malfunctions, we investigate three types of redundancy schemes (refer to Figure 1.18): (a) a series connection, (b) a parallel connection, and (c) a series-parallel configuration.

First we analyze the series configuration. Let s_1, s_2, and s_3, respectively, denote the probabilities of correct functioning, a short circuit, and an open circuit for the

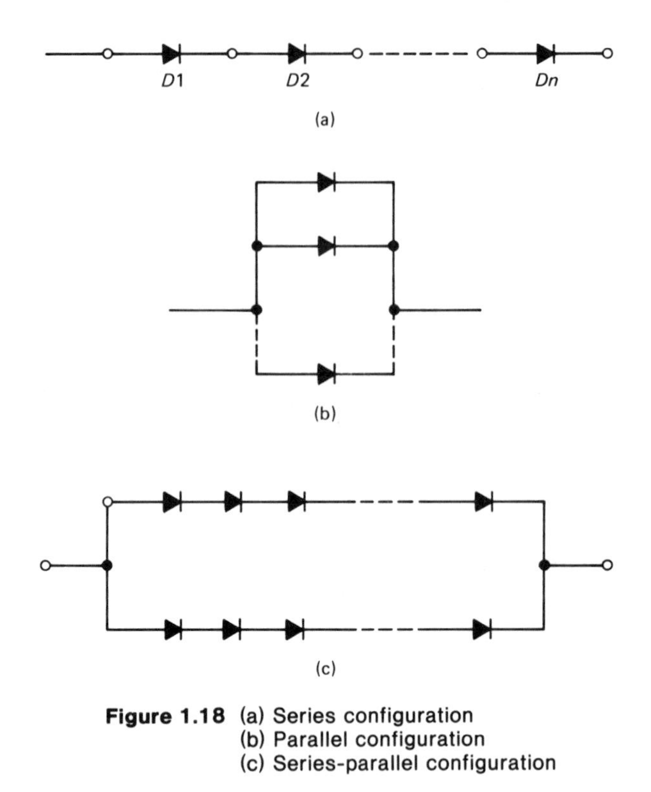

Figure 1.18 (a) Series configuration
(b) Parallel configuration
(c) Series-parallel configuration

series configuration as a whole. The experiment of observing n diodes corresponds to a sequence of n generalized Bernoulli trials. Let n_1 diodes be functioning properly, n_2 diodes be short-circuited, and n_3 diodes be open-circuited. Then the event "The series configuration is functioning properly" is described by "None of the diodes is open-circuited and at least one of the diodes is functioning properly". This event consists of the sample points $\{(n_1, n_2, n_3) \mid n_1 \geqslant 1,\ n_2 \geqslant 0,\ n_3 = 0,\ n_1 + n_2 = n\}$. Therefore:

$$s_1 = \sum_{\substack{n_1 \geqslant 1 \\ n_2 \geqslant 0 \\ n_1+n_2=n}} p(n_1, n_2, 0)$$

$$= \sum \binom{n}{n_1, n_2, 0} p_1^{n_1} p_2^{n_2} p_3^0$$

$$= \sum_{n_1 \geqslant 1} \frac{n!}{n_1!\,(n-n_1)!} p_1^{n_1} p_2^{n-n_1}$$

$$= \sum_{n_1=0}^{n} \binom{n}{n_1} p_1^{n_1} p_2^{n-n_1} - \frac{n!}{0!n!} p_1^0 p_2^n$$

$$= (p_1 + p_2)^n - p_2^n$$
$$= (1 - p_3)^n - p_2^n.$$

Note that $(1 - p_3)^n$ is the probability that none of the diodes is open and p_2^n is the probability that all diodes are shorted. Similarly:

$$\begin{aligned} s_2 &= P(\text{"Series combination is short-circuited"}) \\ &= P(\text{"All diodes are shorted"}) \\ &= P(\{(n_1, n_2, n_3) \mid n_2 = n\}) \\ &= p_2^n. \end{aligned}$$

Also,

$$\begin{aligned} s_3 &= P(\text{"Series combination is open-circuited"}) \\ &= P(\text{"At least one diode is open-circuited"}) \\ &= P(\{(n_1, n_2, n_3) \mid n_3 \geqslant 1, n_1 + n_2 + n_3 = n\}) \\ &= 1 - P(\{(n_1, n_2, n_3) \mid n_3 = 0,\ n_1 + n_2 = n\}) \\ &= 1 - \sum_{n_1+n_2=n} \binom{n}{n_1, n_2} p_1^{n_1} p_2^{n_2} \\ &= 1 - (p_1 + p_2)^n \\ &= 1 - (1 - p_3)^n \\ &= 1 - P(\text{"No diodes are open-circuited"}). \end{aligned}$$

Check that $s_1 + s_2 + s_3 = 1$.

Next consider the parallel configuration, with P_i $(i = 1, 2, 3)$ respectively denoting the probabilities of properly functioning, short-circuit, and open-circuit situations. Then:

$$\begin{aligned} P_1 &= P(\text{"Parallel combination working properly"}) \\ &= P(\text{"At least one diode functioning and none shorted"}) \\ &= P(\{(n_1, n_2, n_3) \mid n_1 \geqslant 1,\ n_2 = 0,\ n_1 + n_3 = n\}) \\ &= (1 - p_2)^n - p_3^n \\ &= P(\text{"No diodes shorted"}) - P(\text{"All diodes are open"}), \\ P_2 &= P(\{(n_1, n_2, n_3) \mid n_2 \geqslant 1,\ n_1 + n_2 + n_3 = n\}) \\ &= 1 - (1 - p_2)^n, \\ P_3 &= P(\{(n_1, n_2, n_3) \mid n_3 = n\}) \\ &= p_3^n. \end{aligned}$$

To analyze the series-parallel configuration, we first reduce each one of the series configurations to an "equivalent" diode with respective probabilities s_1, s_2, and s_3. The total configuration is then a parallel combination of two "equivalent" diodes. Thus the probability that series-parallel diode configuration functions properly is given by:

$$\begin{aligned} R_1 &= (1 - s_2)^2 - s_3^2 \\ &= s_1^2 + 2s_1 s_3 \\ &= s_1\,(s_1 + 2s_3) \\ &= [(1 - p_3)^n - p_2^n]\,[(1 - p_3)^n - p_2^n + 2 - 2\,(1 - p_3)^n] \\ &= [(1 - p_3)^n - p_2^n]\,[2 - (1 - p_3)^n - p_2^n]. \end{aligned}$$

#

Problems

1. Consider the following program segment:

```
if B then
        repeat S1 until B1
else
        repeat S2 until B2
```

Assume that $P(B = \text{true}) = p$, $P(B_1 = \text{true}) = 3/5$, and $P(B_2 = \text{true}) = 2/5$. Exactly one statement is common to statement groups S_1 and S_2: write ('good day'). After many repeated executions of the program segment, it has been estimated that the probability of printing exactly three 'good day' messages is 3/25. Derive the value of p.

2. Given that the probability of error in transmitting a bit over a communication channel is 8×10^{-4}, compute the probability of error in transmitting a block of 1,024 bits. Note that this model assumes that bit errors occur at random, but in practice errors tend to occur in bursts. Actual block error rate will be considerably lower than that estimated here.

3. In order to increase the probability of correct transmission of a message over a noisy channel, a *repetition* code is often used. Assume that the "message" consists of a single bit, and that the probability of a correct transmission on a single trial is p. With a repetition code of rate $1/n$, the message is transmitted a fixed number (n) of times and a majority voter at the receiving end is used for decoding. Assuming $n = 2m - 1$, $m = 1, 2, \ldots$, determine the error probability P_e of a repetition code as a function of m.

4. An application requires that at least two processors in a multiprocessor system be available with more than 95 percent probability. The cost of a processor with 60 percent reliability is $10,000 and each 10 percent increase in reliability will cost $8,000. Determine the number of processors (n) and the reliability (p) of each processor (assume that all processors have the same reliability) that minimizes the total system cost.

***5.** Show that the number of terms in the multinomial expansion:

$$\left[\sum_{i=1}^{k} p_i\right]^n \quad \text{is} \quad \binom{n+k-1}{n}.$$

Note that the required answer is the number of unordered sets of size n chosen from a set of k distinct objects with repetition allowed [LIU 1968].

6. A communication channel receives independent pulses at the rate of 12 pulses per microsecond. The probability of a transmission error is 0.001 for each pulse. Compute the probabilities of:
(a) No errors per microsecond.
(b) One error per microsecond.
(c) At least one error per microsecond.
(d) Exactly two errors per microsecond.

7. Plot the reliabilities of an m-out-of-n system as a function of the simplex reliability R $(0 \leqslant R \leqslant 1)$ using $n = 3$ and $m = 1, 2, 3$ (parallel redundancy, TMR, and a series system, respectively).

8. Determine the conditions under which diode configurations in Figure (1.18a), (b) and (c) will improve reliability over that of a single diode. Use $n = 2$ to simplify the problem.

Review Problems

1. In the computation of TMR reliability, we assumed that when two units have failed they will both produce incorrect results and hence, after voting, the wrong answer will be produced by the TMR configuration. If the two faulty units produce the opposite answers (one correct and the other incorrect), the overall result will be correct. Assuming that the probability of such a compensating error is c, derive the reliability expression for the TMR configuration.

2. [RAMA 1975] In order to use parallel redundancy in digital logic, we have to associate an on-line detector with each unit, giving us **detector-redundant** systems. However, a detector may itself fail. Compare the reliability of a three-unit detector-redundant system with a TMR system (without on-line detectors). Assume that the reliability of a simplex unit is r, the reliability of a detector is d and the reliability of a voter is v. A detector redundant system is said to have failed when all unit-detector pairs have failed; a unit-detector pair is a series combination of the unit and its associated detector.

3. In manufacturing a certain component, two types of defects are likely to occur with respective probabilities 0.05 and 0.1. What is the probability that a randomly chosen component:
(a) Does not have either kinds of defects?
(b) Is defective?
(c) Has only one kind of defect, given that it is found to be defective?

4. Assume that the probability of successful transmission of a single bit over a binary communication channel is p. We desire to transmit a four-bit word over the channel. To increase the probability of successful word transmission, we may use seven-bit Hamming code (four data bits + three check bits). Such a code is known to be able to correct single-bit errors. Derive the probabilities of successful word transmission under the two schemes, and derive the condition under which the use of Hamming code will improve performance.

5. We want to compare two different schemes for increasing the reliability of a system by using redundancy. Suppose the system needs s identical components in series for proper operation. Further suppose that we are given $m \cdot s$ components. Out of the two schemes shown in Figure 1.P.5, which one will provide a higher reliability?

   Given that the reliability of an individual component is r, derive the expressions for the reliabilities of two configurations. For $m = 3$ and $s = 2$, compare the two expressions.

6. In three boxes there are capacitors as shown in the following table:

	Number in box		
Value in μF	1	2	3
1.0	10	90	25
0.1	50	30	80
0.01	70	90	120

   An experiment consists of first randomly selecting a box, (assume each box has the same probability of selection) and then randomly selecting a capacitor from the chosen box.
   (a) What is the probability of selecting a 0.1-μF capacitor, given that box 3 is chosen?
   (b) If a 0.1-μF capacitor is chosen, what is the probability that it came from box 1?
   (c) List all nine conditional probabilities of capacitor selections, given a box selection.

References

[AHO 1974] A. V. AHO, J. E. HOPCROFT, and J. D. ULLMAN, *The Design and Analysis of Computer Algorithms,* Addison-Wesley, Reading, Mass.

[ASH 1970] R. B. ASH, *Basic Probability Theory*, John Wiley & Sons, New York.

[GOOD 1977] S. E. GOODMAN and S. T. HEDETNIEMI, *Introduction to the Design and Analysis of Algorithms,* McGraw-Hill, New York.

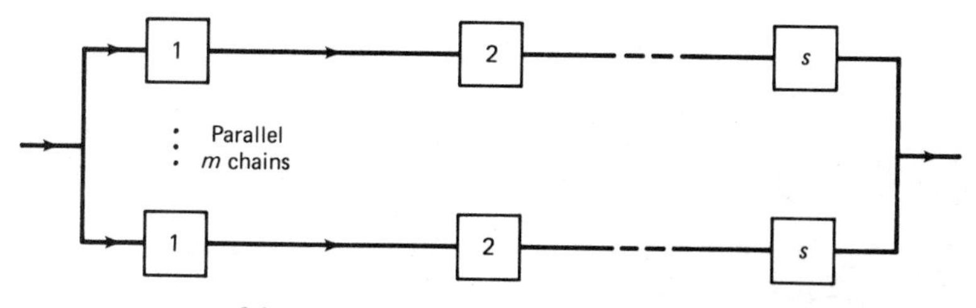

Scheme I : Redundancy at the system level

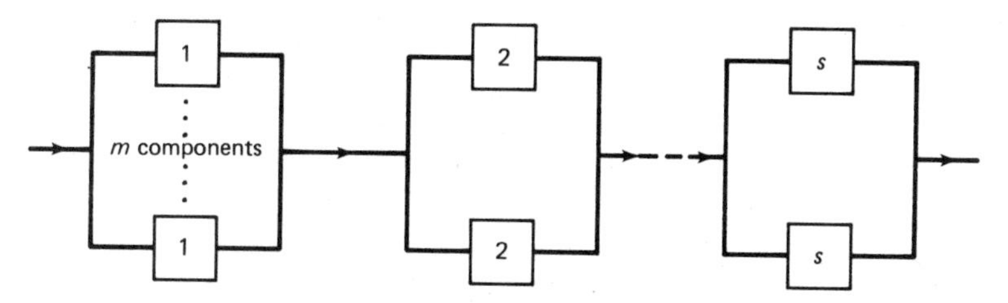

Scheme II : Redundancy at the subsystem level

Figure 1.P.5

[HAMM 1980] R. W. HAMMING, *Coding and Information Theory,* Prentice-Hall, Englewood Cliffs, N.J.

[HECH 1976] H. HECHT "Fault-Tolerant Software for Real-Time Applications," *ACM Computing Surveys*, December 1976, pp. 391–408.

[LIU 1968] C. L. LIU, *Introduction to Combinatorial Mathematics*, McGraw-Hill, New York.

[RAMA 1975] C. V. RAMAMOORTHY and Y. -W. HAN, "Reliability Analysis of Systems with Concurrent Error Detection," *IEEE Trans. on Comput.*, September 1975, pp. 868–78.

[RAND 1978] B. RANDELL and others, "Reliability Issues in Computing System Design" *ACM Computing Surveys*, June 1978, pp. 123–66.

[STAR 1979] H. STARK and F. B. TUTEUR, *Modern Electrical Communications*, Prentice-Hall, Englewood Cliffs, N.J.

Chapter 2

Discrete Random Variables

2.1 INTRODUCTION

Thus far we have treated the sample space as the set of all possible outcomes of a random experiment. Some examples of sample spaces we have considered are:

$$S_1 = \{0, 1\},$$
$$S_2 = \{(0, 0),\ (0, 1),\ (1, 0),\ (1, 1)\},$$
$$S_3 = \{\text{success, failure}\}.$$

Thus some experiments yield sample spaces whose elements are numbers, but some other experiments do not yield numerically valued elements. For mathematical convenience, it is often desirable to associate one or more numbers (in addition to probabilities) with each possible outcome of an experiment. Such numbers might naturally correspond, for instance, to the cost of each experimental outcome, the total number of defective items in a batch, or the time to failure of a component.

Through the notion of *random variables*, this and the following chapter extend our earlier work to develop methods for the study of experiments whose outcomes may be described numerically. Besides this convenience, random variables also provide a more compact description of an experiment than the finest-grain description of the sample space. For example, in the inspection of manufactured products, we may be interested only in the total number of defective items and not in the nature of the defects; in a sequence of Bernoulli trials, we may be interested only in the number of successes and not in the actual sequence of successes and failures. The notion of random variables provides us the power of abstraction and thus allows us to discard unimportant details in the outcome of an experiment. Virtually all serious probabilistic computations are performed in terms of random variables.

2.2 RANDOM VARIABLES AND THEIR EVENT SPACES

A **random variable** is a rule that assigns a numerical value to each possible outcome of an experiment. The term "random variable" is actually a misnomer, since a random variable X is a function whose domain is the sample space S, and whose range is the set of all real numbers, $\mathbb{R}$. The set of all values taken by X, called the **image** of X, will then be a subset of the set of all real numbers.

Definition (Random Variable). A random variable X on a sample space S is a function $X: S \longrightarrow \mathbb{R}$ that assigns a real number $X(s)$ to each sample point $s \in S$.

Example 2.1

As an example, consider a random experiment defined by a sequence of three Bernoulli trials. The sample space S consists of eight triples of 0's and 1's. We may define any number of random variables on this sample space. For our example, define a random variable X to be the total number of successes from the three trials.

The tree diagram of this sequential sample space is shown in Figure 2.1, where S_n and F_n respectively denote a success and a failure on the nth trial, and the probability of success, p, is equal to 0.5. The value of random variable X assigned to each sample point is also included.

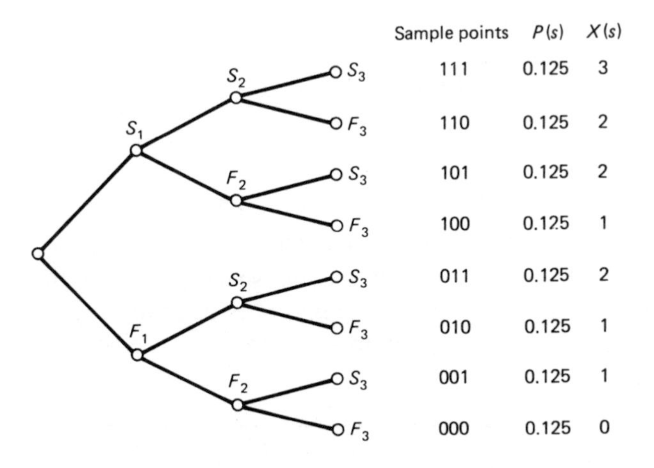

Figure 2.1 Tree diagram of a sequential sample space

If the outcome of one performance of the experiment were $s = (0, 1, 0)$, then the resulting experimental value of the random variable X would be 1—that is, $X(0, 1, 0) = 1$. Note that two or more sample points might give the same value for X (i.e., X may not be a one-to-one function), but that two different numbers in the range cannot be assigned to the same sample point (i.e., X is a well-defined function). For example:

$$X(1, 0, 0) = X(0, 1, 0) = X(0, 0, 1) = 1.$$ #

A random variable partitions its sample space into a mutually exclusive and collectively exhaustive set of events. Thus for a random variable X and a real number x, we define the event A_x (commonly called the **inverse image** of the set $\{x\}$) to be the subset of S consisting of all sample points s to which the random variable X assigns the value x:

$$A_x = \{s \in S \mid X(s) = x\}.$$

It is clear that $A_x \cap A_y = \varnothing$ if $x \neq y$, and that:

$$\bigcup_{x \in \mathbb{R}} A_x = S$$

(see problem 1 at the end of this section). Thus the collection of events A_x for all x defines an **event space**. We may find it more convenient to work in this event space (rather than the original sample space), provided our only interest in performing the experiment has to do with the resulting experimental value of random variable X. The notation $[X = x]$ will be used as an abbreviation for the event A_x. Thus:

$$[X = x] = \{s \in S \mid X(s) = x\}.$$

In Example 2.1 the random variable X defines four events:

$$A_0 = \{s \in S \mid X(s) = 0\} = \{(0, 0, 0)\},$$
$$A_1 = \{(0, 0, 1),\ (0, 1, 0),\ (1, 0, 0)\}$$
$$A_2 = \{(0, 1, 1),\ (1, 0, 1),\ (1, 1, 0)\}$$
$$A_3 = \{(1, 1, 1)\}.$$

For all values of x outside the image of X (i.e., values of x other than 0, 1, 2, 3), A_x is the null set. The resulting event space contains four event points (see Figure 2.2). For a sequence of n Bernoulli trials the event space defined by X will have $(n + 1)$ points, compared with 2^n sample points in the original sample space!

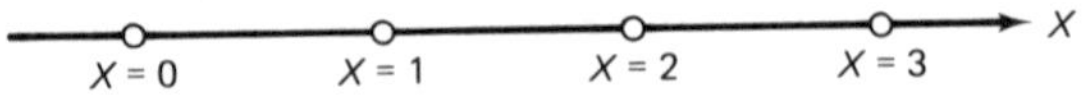

Figure 2.2 Event space for three Bernoulli trials

The random variable discussed in our example could take on values from a discrete set of numbers; hence the image of the random variable is either finite or countable. Such random variables, known as **discrete random variables**, are the subject of this chapter, while continuous random variables are discussed in the next chapter. A random variable defined on a discrete sample space will be discrete, but it is possible to define a discrete random variable on a continuous sample space. For instance, for a continuous sample space S, the random variable defined by $X(s) = 4$ for all $s \in S$ is discrete.

Problems

1. Given a discrete random variable X, define the event A_x by:

$$A_x = \{s \in S \mid X(s) = x\}.$$

Show that the family of events $\{A_x\}$ defines an event space.

2.3 THE PROBABILITY MASS FUNCTION

We have defined the event A_x as the set of all sample points $\{s \mid X(s) = x\}$. Consequently:

$$\begin{aligned} P(A_x) &= P([X = x]) \\ &= P(\{s \mid X(s) = x\}) \\ &= \sum_{X(s)=x} P(s). \end{aligned}$$

This formula provides us with a method of computing $P(X = x)$ for all $x \in \mathbb{R}$. Thus we have defined a function with its domain consisting of the event space of the random variable X, and with its range in the closed interval $[0, 1]$. This function is known as the **probability mass function** (or pmf) or the **discrete density function** of the random variable X and will be denoted by $p_X(x)$. Thus:

$$\begin{aligned} p_X(x) &= P(X = x) \\ &= \sum_{X(s)=x} P(s) \end{aligned}$$

= probability that the value of the random variable X obtained on a performance of the experiment is equal to x.

It should be noted that the argument x of the pmf $p_X(x)$ is a dummy variable, hence it can be changed to any other dummy variable y with no effect on the definition.

The following properties hold for the pmf:

(p1) $0 \leqslant p_X(x) \leqslant 1$ for all $x \in \mathbb{R}$. This must be true, since $p_X(x)$ is a probability.

(p2) Since the random variable assigns some value $x \in \mathbb{R}$ to each sample point $s \in S$, we must have:

$$\sum_{x \in \mathbb{R}} p_X(x) = 1.$$

(p3) For a discrete random variable X, the set $\{x \mid p_X(x) \neq 0\}$ is a finite or countably infinite subset of real numbers (this set is defined to be the image of X). Let this set be denoted by $\{x_1, x_2, \ldots\}$. Then the property (p2) above may be restated as:

$$\sum_{i} p_X(x_i) = 1.$$

A real-valued function $p_X(x)$ defined on $\mathbb{R}$ is the pmf of some random variable X provided that it satisfies properties (p1) to (p3). Continuing with Example 2.1, we can easily obtain $p_X(x)$ for $x = 0, 1, 2, 3$ from the above definitions:

$$p_X(0) = 0.125,$$
$$p_X(1) = 0.375,$$
$$p_X(2) = 0.375,$$
$$p_X(3) = 0.125.$$

Check that all the properties listed above hold. This pmf may be visualized as a bar graph drawn over the event space for the random variable (see Figure 2.3).

Example 2.2

Returning to the example of a computer system with five tape drives from Chapter 1, and defining the random variable X = the number of available tape drives, we have:

$$p_X(0) = 1/32, \qquad p_X(1) = 5/32, \qquad p_X(2) = 10/32$$
$$p_X(3) = 10/32, \qquad p_X(4) = 5/32, \qquad p_X(5) = 1/32.$$

#

2.4 DISTRIBUTION FUNCTIONS

So far we have restricted our attention to computing $P(X = x)$, but often we may be interested in computing the probability of the set $\{s \mid X(s) \in A\}$ for

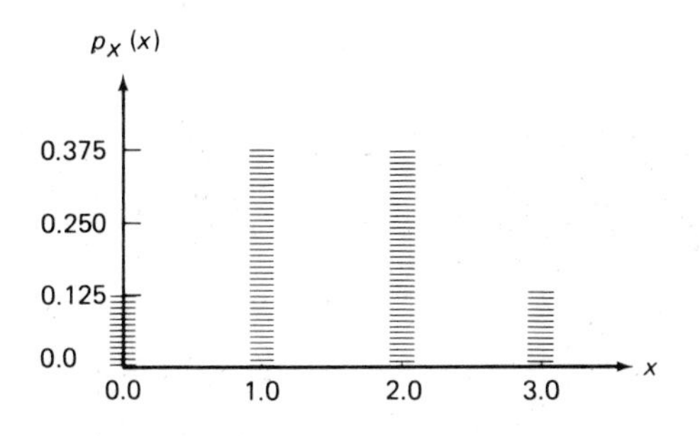

Figure 2.3 Bar graph of pmf for Example 2.1

some subset A of $\mathbb{R}$ other than a one-point set. It is clear that:

$$\{s \mid X(s) \in A\} = \bigcup_{x_i \in A} \{s \mid X(s) = x_i\}. \tag{2.1}$$

Usually this event is denoted as $[X \in A]$ and its probability by $P(X \in A)$. If $-\infty < a < b < \infty$ and A is an interval with end points a and b, say $A = (a, b)$, then we usually write $P(a < X < b)$ instead of $P[X \in (a, b)]$. Similarly, if $A = (a, b]$, then $P(X \in A)$ will be written as $P(a < X \leqslant b)$. The semi-infinite interval $A = (-\infty, x]$ will be of special interest, and in this case we denote the event $[X \in A]$ by $[X \leqslant x]$.

If $p_X(x)$ denotes the pmf of random variable X, then from equation (2.1) above we have:

$$P(X \in A) = \sum_{x_i \in A} p_X(x_i).$$

Thus in Example 2.2, the probability that two or fewer tape drives will be available may now be evaluated quite simply as:

$$\begin{aligned} P(X \leqslant 2) &= P(X = 0) + P(X = 1) + P(X = 2) \\ &= p_X(0) + p_X(1) + p_X(2) \\ &= 1/32 + 5/32 + 10/32 \\ &= 16/32 \\ &= 1/2. \end{aligned}$$

The function $F_X(t)$, $-\infty < t < \infty$, defined by:

$$\begin{aligned} F_X(t) &= P(-\infty < X \leqslant t) \\ &= P(X \leqslant t) \\ &= \sum_{x \leqslant t} p_X(x) \end{aligned}$$

is called the **probability distribution function** or the **cumulative distribution function (CDF)** of the random variable X. We will omit the subscript X whenever no confusion arises. It follows from this definition that:

$$\begin{aligned} P(a < X \leqslant b) &= P(X \leqslant b) - P(X \leqslant a) \\ &= F(b) - F(a). \end{aligned}$$

If X is an integer-valued random variable, then:

$$F(t) = \sum_{-\infty < x \leqslant \lfloor t \rfloor} p_X(x)$$

where $\lfloor t \rfloor$ denotes the greatest integer less than or equal to t (also known as the **floor** of t).

Several properties of $F_X(x)$ follow directly from its definition.

(F1) $0 \leqslant F(x) \leqslant 1$ for $-\infty < x < \infty$. This follows because $F(x)$ is a probability.

(F2) $F(x)$ is a monotone nondecreasing function of x; that is, if $x_1 \leqslant x_2$, then $F(x_1) \leqslant F(x_2)$. This follows by first observing that the interval $(-\infty, x_1]$ is contained in the interval $(-\infty, x_2]$ whenever $x_1 \leqslant x_2$ and hence:
$P(-\infty < X \leqslant x_1) \leqslant P(-\infty < X \leqslant x_2)$. That is, $F(x_1) \leqslant F(x_2)$.

(F3) $\lim_{x \to \infty} F(x) = 1$, and $\lim_{x \to -\infty} F(x) = 0$. If the random variable X has a finite image, then $F(x) = 0$ for all x sufficiently small and $F(x) = 1$ for all x sufficiently large.

(F4) $F(x)$ has a positive jump equal to $p_X(x_i)$ at $i = 1, 2, \ldots,$ and in the interval $[x_i, x_{i+1})$ $F(x)$ has a constant value. Thus:

$$F(x) = F(x_i) \qquad \text{for } x_i \leqslant x < x_{i+1}$$

and

$$F(x_{i+1}) = F(x_i) + p_X(x_{i+1}).$$

It can be shown that any function $F(x)$ satisfying properties (F1) to (F4) is the distribution function of some discrete random variable.

We note that distribution functions of discrete random variables grow only by jumps, whereas the distribution functions of continuous random variables are continuous functions and hence have no jumps. A random variable X is said to be of **mixed type** if its distribution function has both jumps and continuous growth. The domains and ranges of the four functions (the probability function, the random variable X, the pmf, and the CDF) we have studied so far are summarized in Figure 2.4.

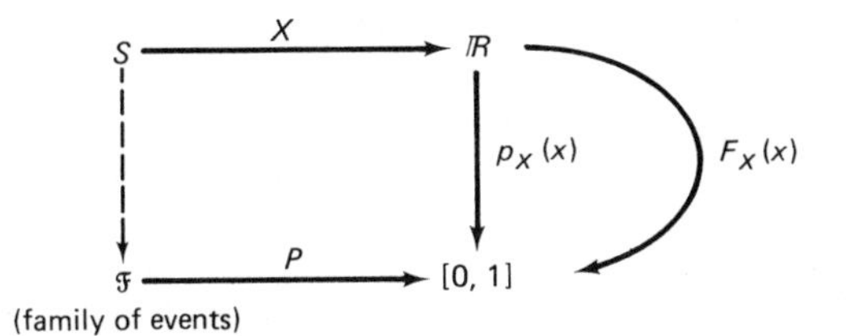

Figure 2.4 Domain and range of P, X, pmf, and CDF

The cumulative distribution function contains most of the interesting information about the underlying probability system, and it will be used extensively. Often the concepts of sample space, event space, and probability measure, which are fundamental in building the theory of probability, will fade into the background, and functions such as the distribution function or the probability mass function become the most important entities. It is important, nevertheless, to keep this background in mind. You will often see the statement "Let X be a discrete random variable with pmf p_X," with no reference to the underlying probability space. We can always construct an appropriate space, as follows. Take $S = \mathbb{R}$; $X(s) = s$, for $s \in S$; and

$$P(A) = \sum_{x \in A} p_X(x)$$

for a subset, A, of $\mathbb{R}$. In this case, the event space of the random variable X is identical to the sample space defined above. Similarly, the statement, "Let X be a discrete random variable with the CDF F," always makes sense.

Example 2.3

The CDF of the running example of the sequence of three Bernoulli trials is shown in Figure 2.5. The properties (F1)–(F4) above are easily seen to hold. #

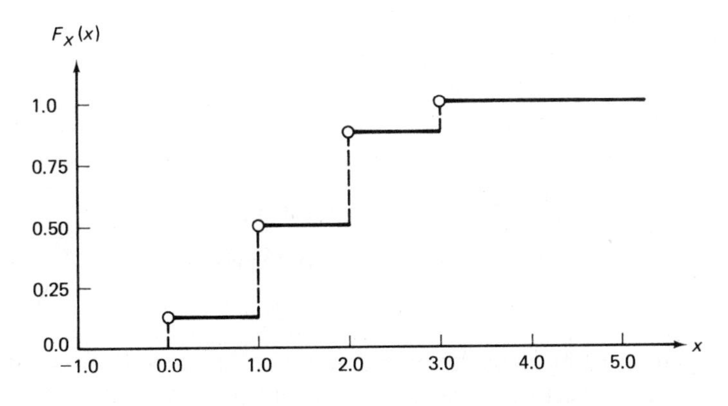

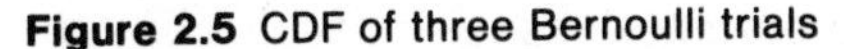
Figure 2.5 CDF of three Bernoulli trials

2.5 SPECIAL DISCRETE DISTRIBUTIONS

In many theoretical and practical problems, several probability mass functions appear frequently enough that they are worth exploring here.

2.5.1 The Bernoulli pmf

The Bernoulli pmf is the density function of a discrete random variable X having 0 and 1 as its only possible values; it originates from the experiment consisting of a single Bernoulli trial. It is given by:

$$p_X(0) = p_0 = P(X = 0) = q,$$
$$p_X(1) = p_1 = P(X = 1) = p,$$

where $p + q = 1$. The corresponding CDF is given by (see Figure 2.6):

$$F(x) = \begin{cases} 0 & \text{for } x < 0, \\ q & \text{for } 0 \leqslant x < 1, \\ 1 & \text{for } x \geqslant 1. \end{cases}$$

2.5.2 The Binomial pmf

To generate the Bernoulli pmf, we considered a single Bernoulli trial. Now we consider a sequence of n independent Bernoulli trials with the probability of success equal to p on each trial. Let Y_n denote the number of successes in n trials. The domain of the random variable Y_n is all the n-tuples of 0's and 1's, and the image is $\{0, 1, \ldots, n\}$. The value assigned to a sample point (an n-tuple) by Y_n simply corresponds to the number of 1's in the n-tuple. As shown in Section 1.12, the pmf of Y_n is:

$$\begin{aligned} p_k &= P(Y_n = k) \\ &= p_{Y_n}(k) \\ &= \begin{cases} \binom{n}{k} p^k (1-p)^{n-k} & \text{for } 0 \leqslant k \leqslant n, \\ 0 & \text{otherwise.} \end{cases} \end{aligned} \tag{2.2}$$

That is, equation (2.2) gives the probability of k "successes" in n independent trials of an experiment that has probability p of success on each trial. One of the more important densities in probability theory, this is called the

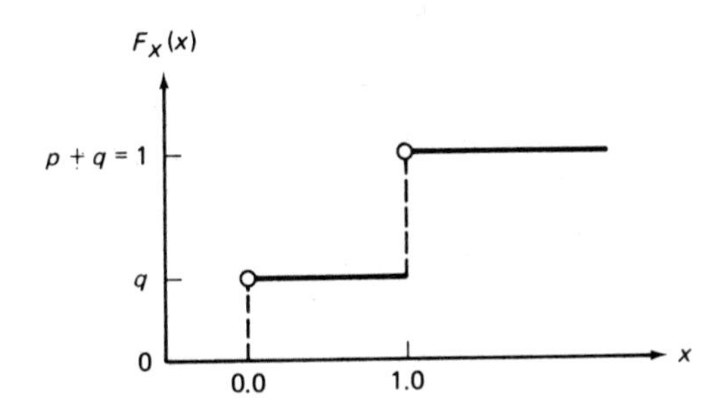

Figure 2.6 CDF of Bernoulli random variable

binomial density with parameters n and p often denoted by $b(k; n, p)$. An example of $b(k; 3, 0.5)$ occurred earlier in this chapter (see Figure 2.3).

It may be easily verified using the binomial theorem that:

$$\begin{aligned}\sum_{i=0}^{n} p_i &= \sum_{i=0}^{n} \binom{n}{i} p^i (1-p)^{n-i} \\ &= [p + (1-p)]^n \\ &= 1.\end{aligned}$$

This is the reason for the name *binomial* pmf. We often refer to a random variable Y_n's having a binomial pmf by saying that Y_n has a **binomial distribution** (with parameters n and p if we want to be more precise). Similar phraseology will be used for other random variables having a named density. The distribution function of a binomial random variable will be denoted by $B(t; n, p)$ and is given by:

$$\begin{aligned}B(t; n, p) &= F_{Y_n}(t) \\ &= \sum_{i=0}^{\lfloor t \rfloor} \binom{n}{i} p^i (1-p)^{n-i}.\end{aligned} \tag{2.3}$$

The binomial distribution is applicable whenever a series of trials is made satisfying the following conditions.

1. Each trial has exactly two mutually exclusive outcomes, usually labeled "success" and "failure."
2. The probability of "success" on each trial is a constant, denoted by p. The probability of "failure" is $q = 1 - p$.
3. The outcomes of successive trials are mutually independent.

A typical situation in which these conditions will apply (at least approximately) occurs when several components are selected at random (with replacement) from a batch of components and examined to see if there are any defective components (that is, failures). The number of defectives in a sample of size n is a random variable, denoted by Y_n, which is binomially distributed.

These assumptions constitute what is called a *binomial model*, which is a typical example of a mathematical model in that it attempts to describe a physical situation in mathematical terms. Models such as these depend on one or more **parameters** that govern their behavior. The binomial model has two parameters, n and p. If the values of model parameters are known, then it is relatively easy to evaluate the probabilities of the events of interest.

We emphasize that the three properties listed above are **assumptions** and need not always hold. We may wish to analyze empirically observed data and may hypothesize that the assumptions of the binomial model (or any other such model) hold. This hypothesis needs to be tested and can be either rejected or accepted on the basis of the test. Hypothesis testing is discussed in Chapter 10.

Example 2.4

As an example of binomial distribution, consider a plant manufacturing IC chips, of which 10 percent are expected to be defective. The quality control procedure consists of counting the number of defective IC's in a sample of size 35. Suppose after 800 applications of this procedure we find that our experience is reflected in the following table. Though we do not expect exactly 10 percent defectives every time, are the observations consistent with our hypothesis that 10 percent are defective?

Number of defects	*Number of samples showing this number of defects*	*Fraction (of 800 samples) showing this number of defects*
0	11	0.01375
1	95	0.11875
2	139	0.17375
3	213	0.26625
4	143	0.17875
5	113	0.14125
6	49	0.06125
7	27	0.03375
8	6	0.00750
9	4	0.00500
10	0	0.00000
	800	1.00000

This situation is typical of those fitting a binomial model. "Success" is finding a defective IC, and we are counting the number of successes in 35 trials. Since the probability of success $p = 0.1$, the observed fraction defective should be close to the binomial pmf:

$$b(k; 35, 0.1) = \binom{35}{k} \cdot 0.1^k \cdot 0.9^{35-k}.$$

The observed data and the binomial pmf are compared in the following table as well as in Figure 2.7. In Chapter 10 we will study statistical tests that will allow us to quantify the goodness of fit of the above data to the binomial model.

k = defects/sample	*Data*	*b(k; 35, 0.1)*
0	0.01375	0.0250
1	0.11875	0.0974
2	0.17375	0.1839
3	0.26625	0.2248
4	0.17875	0.1998
5	0.14125	0.1376
6	0.06125	0.0765
7	0.03375	0.0352
8	0.00750	0.0137
9	0.00500	0.0046
10	0.00000	0.0013
11	0.00000	0.0003
12	0.00000	0.0000

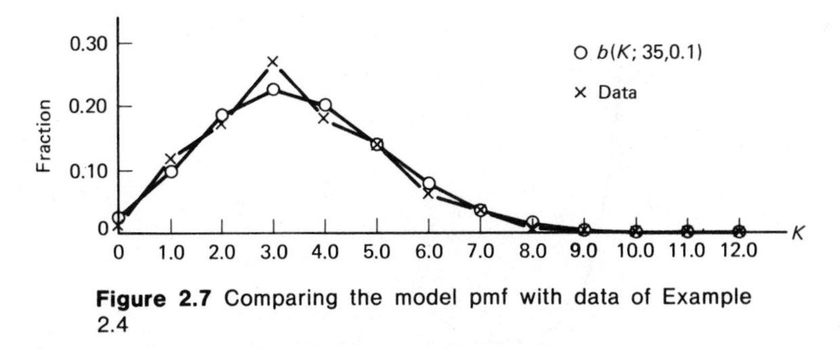

Figure 2.7 Comparing the model pmf with data of Example 2.4

Example 2.5

The number of surviving components, Y_n, out of a given number of n identical and independent components has a binomial distribution $b(m; n, R)$, where R is the reliability of a single component. Thus the reliability of an m out of n system is given by:

$$\begin{aligned} R_{m|n} &= P(\text{``}m \text{ or more components have not failed''}) \\ &= 1 - \sum_{i=0}^{m-1} p_{Y_n}(i) \\ &= 1 - F_{Y_n}(m-1) \\ &= \sum_{i=m}^{n} \binom{n}{i} R^i (1-R)^{n-i}. \end{aligned} \tag{2.4}$$

#

Example 2.6

While transmitting binary digits through a communication channel, the number of digits received correctly, C_n, out of n transmitted digits has a binomial distribution $b(k; n, p)$, where p is the probability of successfully transmitting one digit. The probability of exactly i errors is given by:

$$P_e(i) = p_{C_n}(n-i) = \binom{n}{i} p^{n-i}(1-p)^i,$$

and thus the probability of an error-free transmission is given by:

$$P_e(0) = p^n.$$

#

Example 2.7

Consider taking a random sample of ten IC chips from a very large batch. If no chips in the sample are found to be defective, then we accept the entire batch; otherwise we reject the batch. The number of defective chips in a sample has the pmf $b(k; 10, p)$, where p denotes the probability that a randomly chosen chip is defective. Thus:

$$\begin{aligned} P(\text{``No defectives''}) &= (1-p)^{10} \\ &= \text{probability that a batch is accepted.} \end{aligned}$$

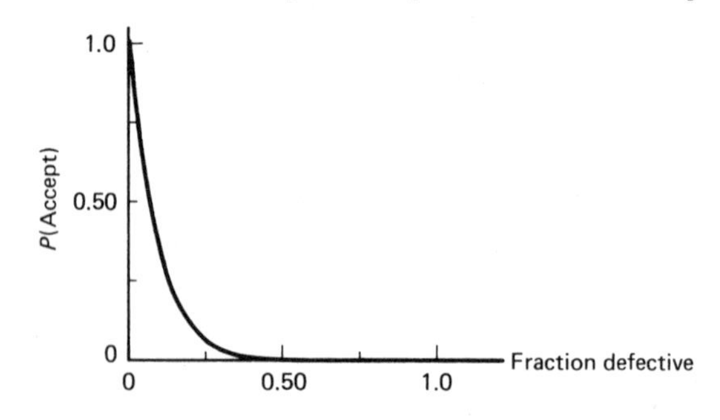

Figure 2.8 Probability of acceptance versus fraction defective

If $p = 0$, the batch is certain to be accepted, and if $p = 1$, the batch will certainly be rejected. The expression for the probability of acceptance is plotted in Figure 2.8. #

The student should not be misled by these examples into thinking that quality control problems can always be solved by simply plugging numbers into the binomial pmf.

Example 2.8 ***(Simpson's Reversal Paradox)***

Consider two shipments (labeled I and II) of IC chips from each of the two manufacturers A and B. Suppose that the proportions of defectives among the four shipments are as follows:

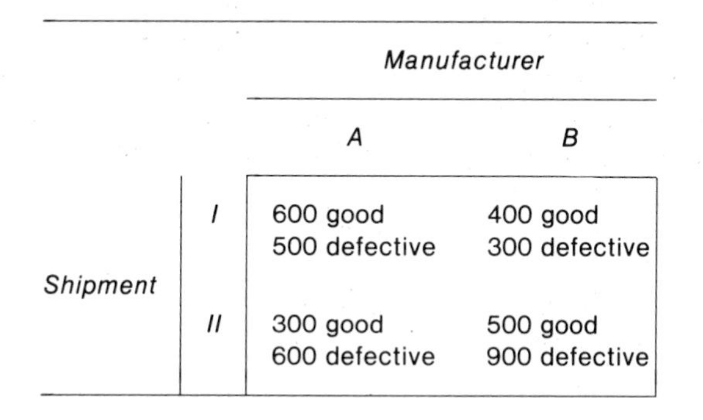

		Manufacturer	
		A	B
Shipment	I	600 good 500 defective	400 good 300 defective
	II	300 good 600 defective	500 good 900 defective

If the quality control engineer inspects shipment I, he will find:

$$P(\text{selecting a defective chip from A}) = 5/11$$
$$> P(\text{selecting a defective chip from B}) = 3/7.$$

If he then considers shipment II, again:

$$P(\text{selecting a defective chip from A}) = 6/9$$
$$> P(\text{selecting a defective chip from B}) = 9/14.$$

The engineer will presumably conclude that the manufacturer B is sending better chips than the manufacturer A. Suppose, however, the engineer mixes the two shipments from A together and similarly for B. A subsequent test leads him to a reverse conclusion, since

$$P(\text{selecting a defective chip from A}) = 11/20$$
$$< P(\text{selecting a defective chip from B}) = 12/21 \ !$$

The problem here is that we are tempted to add the fractions $5/11 + 6/9$ and compare the sum with $3/7 + 9/14$; unfortunately, what is called for is adding numerators and adding denominators, which is *not* the way we add fractions. #

When n becomes very large, computation using the binomial formula becomes unmanageable. In the limit as n approaches infinity, it can be shown that:

$$b(k; n, p) \simeq \frac{1}{\sqrt{2\pi npq}} \cdot e^{-(k-np)^2/(2npq)}. \tag{2.5}$$

This is known as the Laplace (or normal) approximation to the binomial pmf, and the agreement between the two formulas depends on the values of n and p. Take $n = 5$ and $p = 0.5$; then:

k	$b(k;\ 5,\ 0.5)$	Laplace approximation to $b(k;\ 5,\ 0.5)$
0	0.03125	0.02929
1	0.15625	0.14507
2	0.31250	0.32287
3	0.31250	0.32287
4	0.15625	0.14507
5	0.03125	0.02929

As p moves away from 0.5, larger values of n are needed. Larson [LARS 1974] suggests that for $n \geqslant 10$, if:

$$\frac{9}{n+9} \leqslant p \leqslant \frac{n}{n+9},$$

then the Laplace formula provides a good approximation to the binomial pmf. Yet another approximation to the binomial pmf is the Poisson pmf, which we will study later.

Owing to the importance of the binomial distribution, the binomial CDF

$$B(k; n, p) = \sum_{i=0}^{k} b(i; n, p)$$

has been tabulated for $n = 2$ to $n = 49$ by the National Bureau of Standards [NBS 1950] and for $n = 50$ to $n = 100$ by H. G. Romig [ROMI 1953]. [In Appendix C, we have tabulated $B(k; n, p)$ for $n = 2$ to 20]. Three different possible shapes of binomial pmf's are illustrated in Figures 2.9, 2.10, and 2.11. If $p = 0.5$, the bar chart of the binomial pmf is **symmetrical**, as in Figure 2.9. If $p < 0.5$, then a **positively skewed** binomial pmf is obtained (see Figure 2.10), and if $p > 0.5$, a **negatively skewed** binomial pmf (see Figure 2.11). Here, a bar chart is said to be positively skewed if the long "tail" is on the right, and it is said to be negatively skewed if the long "tail" is on the left.

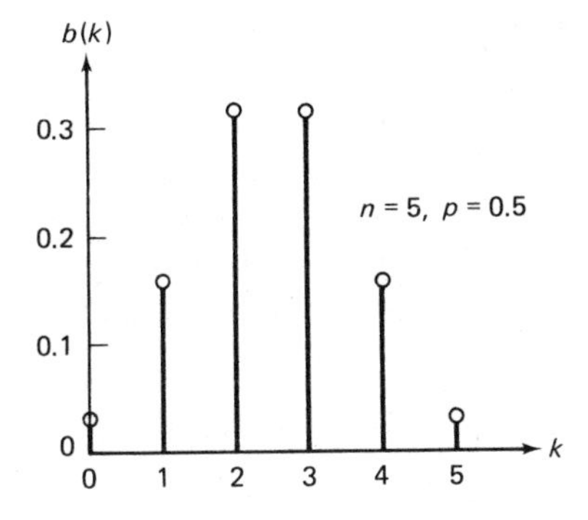

Figure 2.9 Symmetric binomial pmf

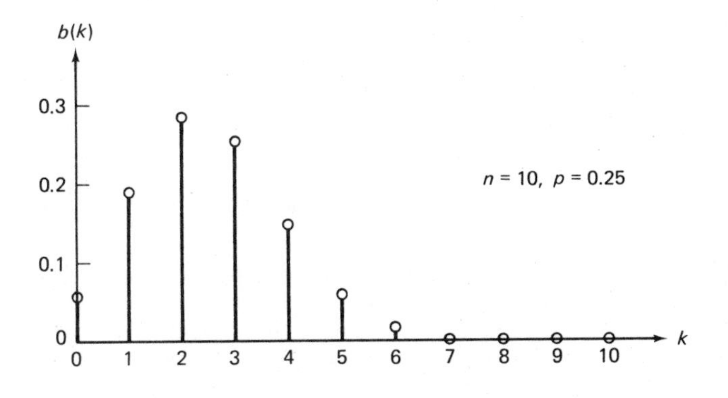

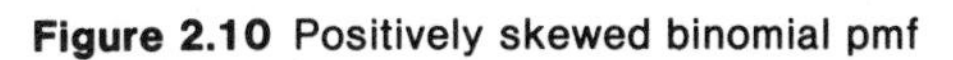

Figure 2.10 Positively skewed binomial pmf

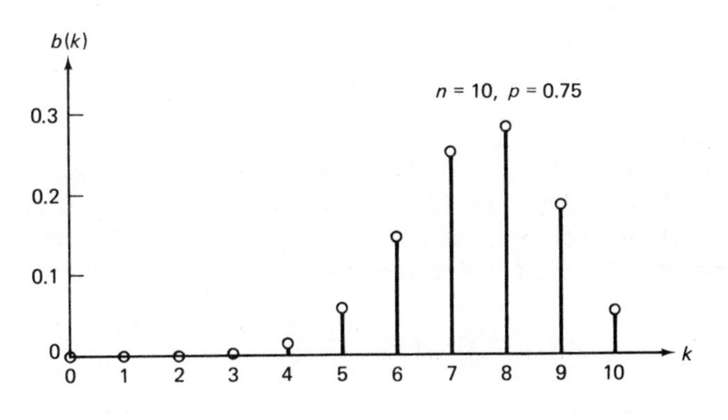

Figure 2.11 Negatively skewed binomial pmf

2.5.3 The Geometric pmf

Once again we consider a sequence of Bernoulli trials, but instead of counting the number of successes in a fixed number n of trials, we count the number of trials until the first "success" occurs. If we let 0 denote a failure and let 1 denote a success, then the sample space of this experiment consists of the set of all binary strings with an arbitrary number of 0's followed by a single 1:

$$S = \{0^{i-1}1 \mid i = 1, 2, 3, \ldots\}.$$

Note that this sample space has a countably infinite number of sample points. Define a random variable Z on this sample space so that the value assigned to the sample point $0^{i-1}1$ is i. Thus Z is the number of trials up to and including the first success. Therefore, Z is a random variable with image $\{1, 2, \ldots\}$, which is a countably infinite set. To find the pmf of Z, we note that the event $[Z = i]$ occurs if and only if we have a sequence of $i - 1$ failures followed by one success. This is a sequence of independent Bernoulli trials with the probability of success equal to p. Hence, we have:

$$\begin{aligned} p_Z(i) &= q^{i-1}p \\ &= p(1-p)^{i-1} \qquad \text{for } i = 1, 2, \ldots, \end{aligned} \tag{2.6}$$

where $q = 1 - p$. By the formula for the sum of a geometric series, we have:

$$\begin{aligned} \sum_{i=1}^{\infty} p_Z(i) &= \sum_{i=1}^{\infty} pq^{i-1} \\ &= \frac{p}{1-q} \\ &= \frac{p}{p} \\ &= 1. \end{aligned}$$

Any random variable Z with the image $\{1, 2, \ldots\}$ and pmf given by a formula of the form of equation (2.6) is said to have a **geometric distribution**, and the function given by (2.6) is termed a **geometric pmf** with parameter p. The distribution function of Z is given by:

$$\begin{aligned} F_Z(t) &= \sum_{i=1}^{\lfloor t \rfloor} p(1-p)^{i-1} \\ &= 1 - (1-p)^{\lfloor t \rfloor} \qquad \text{for } t \geqslant 0. \end{aligned} \tag{2.7}$$

Graphs of the geometric pmf for two different values of parameter p are sketched in Figure 2.12.

The random variable Z counts the total number of trials up to and including the first success. We are often interested in counting the number of failures before the first success. Let this number be called the random variable X with the image $\{0, 1, 2, \ldots\}$. Clearly, $Z = X + 1$. The random variable X is said to have a **modified geometric pmf**, specified by:

$$p_X(i) = p(1-p)^i \qquad \text{for } i = 0,\ 1,\ 2,\ \ldots. \tag{2.8}$$

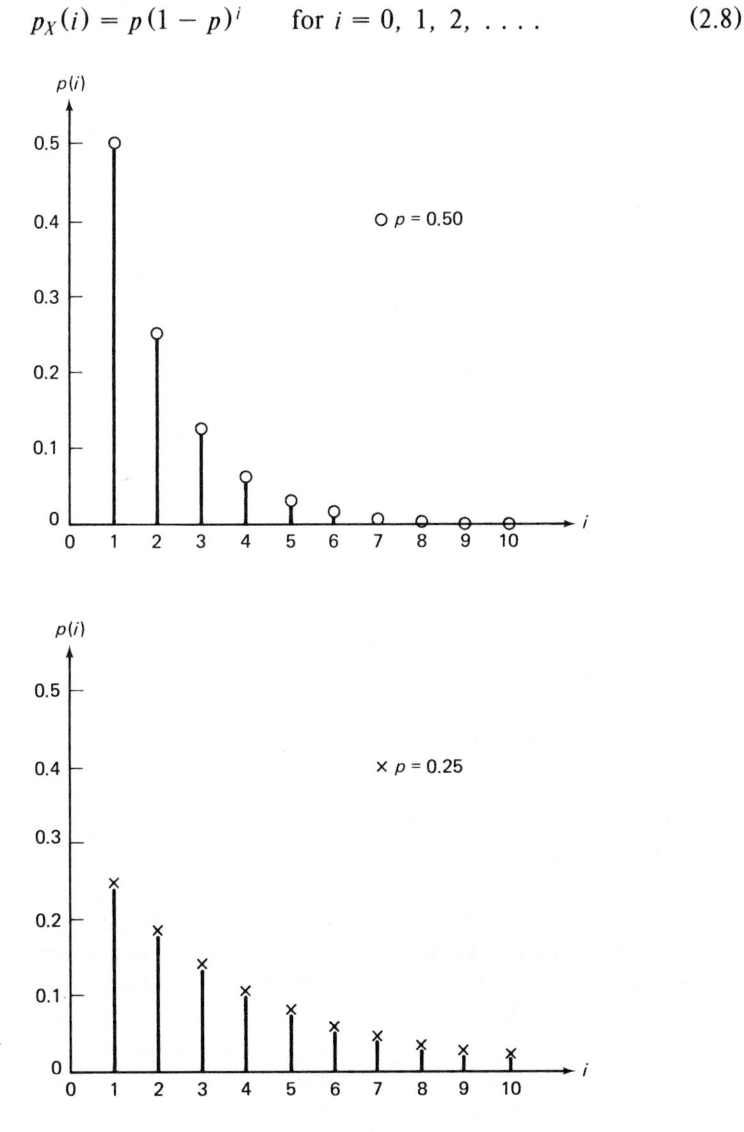

Figure 2.12 Geometric pmf

The distribution function of X is given by:

$$F_X(t) = \sum_{i=0}^{\lfloor t \rfloor} p(1-p)^i$$

$$= 1 - (1-p)^{\lfloor t+1 \rfloor} \qquad \text{for } t \geqslant 0. \tag{2.9}$$

The geometric (and modified geometric) distribution is encountered in some problems in queuing theory. Several examples where this distribution occurs are:

1. A series of components is made by a certain manufacturer. The probability that any given component is defective is a constant p, which does not depend on the quality of the previous components. The probability that the ith item is the first defective one is given by formula (2.6).
2. Consider the operation of a time-sharing computer system with a fixed time slice (see Figure 1.15). At the end of a time slice, the program would have completed execution with a probability p; thus there is a probability $q = 1 - p > 0$ that it needs to perform more computation. The pmf of the random variable denoting the number of time slices needed to complete the execution of a program is given by formula (2.6), if we assume that the operation of the computer satisfies the usual independence assumptions.
3. Consider the following program segment consisting of a **while** loop:

while ¬B **do** S

Assume that the Boolean expression B takes the value **true** with probability p and the value **false** with probability q. If the successive tests on B are independent, then the number of times the body (or the statement group S) of the loop is executed will be a random variable having a modified geometric distribution with parameter p.

4. With the assumptions as in example 3 above, consider a **repeat** loop:

repeat S **until** B

The number of times the body of the **repeat** loop is executed will be a geometrically distributed random variable with parameter p.

The geometric distribution has an important property, known as the **memoryless** or **Markov property**; it is the only discrete distribution with this property. To illustrate this property, consider a sequence of Bernoulli trials and let Z represent the number of trials until the first success. Now assume that we have observed a fixed number n of these trials and found them all to be failures. Let Y denote the number of additional trials that must be performed until the first success. Then $Y = Z - n$, and the conditional probability is:

$$
\begin{aligned}
q_i &= P(Y = i \mid Z > n) \\
&= P(Z - n = i \mid Z > n) \\
&= P(Z = n + i \mid Z > n) \\
&= \frac{P(Z = n + i \text{ and } Z > n)}{P(Z > n)}
\end{aligned}
$$

by using the definition of conditional probability. But for $i = 1, 2, 3, \ldots$, $Z = n + i$ implies that $Z > n$. Thus the event $[Z = n + i \text{ and } Z > n]$ is the same as the event $[Z = n + i]$. Therefore:

$$
\begin{aligned}
q_i &= P(Y = i \mid Z > n) \\
&= \frac{P(Z = n + i)}{P(Z > n)} \\
&= \frac{p_Z(n + i)}{1 - F_Z(n)} \\
&= \frac{pq^{n+i-1}}{1 - (1 - q^n)} \\
&= \frac{pq^{n+i-1}}{q^n} \\
&= pq^{i-1} \\
&= p_Z(i).
\end{aligned}
$$

Thus we see that, conditioned on $Z > n$, the number of trials remaining until the first success, $Y = Z - n$, has the same pmf as Z had originally. If a run of failures is observed in a sequence of Bernoulli trials, we need not "remember" how long the run was to determine the probabilities for the number of additional trials needed until the first success. The proof that any discrete random variable Z with image $\{1, 2, 3, \ldots\}$ and having the Markov property must have the geometric distribution is left as an exercise.

2.5.4 The Negative Binomial pmf

To obtain the geometric pmf, we observed the number of trials until the first success in a sequence of Bernoulli trials. Now let us observe the number of trials until the rth success, and let T_r be the random variable denoting this number. It is clear that the image of T_r is $\{r, r + 1, r + 2, \ldots\}$. To compute $p_{T_r}(n)$, define the events:

A = "$T_r = n$."

B = "Exactly $r - 1$ successes occur in $n - 1$ trials."

C = "The nth trial results in a success."

Then clearly:

$$A = B \cap C$$

and the events B and C are independent. Therefore:

$$P(A) = P(B)P(C).$$

To compute $P(B)$, consider a particular sequence of $n - 1$ trials with $r - 1$ successes and $n - 1 - (r - 1) = n - r$ failures. The probability associated with such a sequence is $p^{r-1}q^{n-r}$ and there are $\binom{n-1}{r-1}$ such sequences. Therefore:

$$P(B) = \binom{n-1}{r-1}p^{r-1}q^{n-r}.$$

Now since $P(C) = p$,

$$\begin{aligned} p_{T_r}(n) &= P(T_r = n) \\ &= P(A) \\ &= \binom{n-1}{r-1}p^r q^{n-r} \\ &= \binom{n-1}{r-1}p^r(1-p)^{n-r}, \qquad n = r, r+1, r+2, \ldots . \end{aligned}$$

Using some combinatorial identities [KNUT 1973; p.57], an alternative form of this pmf can be established:

$$p_{T_r}(n) = p^r\binom{-r}{n-r}(-1)^{n-r}(1-p)^{n-r},$$
$$n = r, r+1, r+2, \ldots . \qquad (2.10a)$$

This pmf is known as the **negative binomial pmf**, and although we derived it assuming an integral value of r, any positive real value of r is allowed (of course the interpretation of r as a number of successes is no longer applicable). Quite clearly, if we let $r = 1$ in the above formula, then we get the geometric pmf.

To verify that $\sum_{n=r}^{\infty} p_{T_r}(n) = 1$, we recall that the Taylor series expansion of $(1-t)^{-r}$ for $-1 < t < 1$ is:

$$(1-t)^{-r} = \sum_{n=r}^{\infty} \binom{-r}{n-r} (-t)^{n-r}.$$

Substituting $t = 1 - p$, we have:

$$p^{-r} = \sum_{n=r}^{\infty} \binom{-r}{n-r} (-1)^{n-r} (1-p)^{n-r},$$

which gives us the required result.

As in the case of geometric distribution, there is a modified version of the negative binomial distribution. Let the random variable Z denote the number of failures before the occurrence of the rth success. Then Z is said to have **modified negative binomial distribution** with the pmf:

$$p_Z(n) = \binom{n+r-1}{r-1} p^r (1-p)^n, \quad n \geqslant 0. \tag{2.10b}$$

The pmf (2.10b) above reduces to the modified geometric pmf when $r = 1$.

2.5.5 The Poisson pmf

Let us consider another problem related to the binomial distribution. Suppose we are observing the arrival of jobs to a large computation center for the time interval $(0, t]$. It is reasonable to assume that for each small interval of time Δt the probability of a new job arrival is $\lambda \cdot \Delta t$, where λ is a constant that depends upon the user population of the computation center. If Δt is sufficiently small, then the probability of two or more jobs arriving in the interval of duration Δt may be neglected. We are interested in calculating the probability of k jobs arriving in the interval of duration t.

Suppose that the interval $(0, t]$ is divided into n subintervals of length t/n, and suppose further that the arrival of a job in any given interval is independent of the arrival of a job in any other interval. Then for a sufficiently large n, we can think of the n intervals as constituting a sequence of Bernoulli trials with the probability of success $p = \lambda t/n$. It follows that the probability of k arrivals in a total of n intervals each with a duration t/n is approximately given by:

$$b\left(k; n, \frac{\lambda t}{n}\right) = \binom{n}{k} \left(\frac{\lambda t}{n}\right)^k \left(1 - \frac{\lambda t}{n}\right)^{n-k}, \qquad k = 0, 1, \ldots, n.$$

Since the assumption that the probability of more than one arrival per interval can be neglected is reasonable only if t/n is very small, we will take the limit of the above pmf as n approaches ∞. Now

$$b(k; n, \frac{\lambda t}{n}) = \frac{n(n-1)(n-2)\ldots(n-k+1)}{k!n^k}(\lambda t)^k \cdot (1 - \frac{\lambda t}{n})^{(n-k)}$$

$$= \frac{n}{n} \cdot \frac{n-1}{n} \cdot \ldots \frac{n-k+1}{n} \cdot \frac{(\lambda t)^k}{k!} \cdot (1 - \frac{\lambda t}{n})^{-k} \cdot (1 - \frac{\lambda t}{n})^n.$$

We are interested in what happens to this expression as n increases, because then the subinterval width aproaches zero, and the approximation involved gets better and better. In the limit as n approaches infinity, the first k factors approach unity, the next factor is fixed, the next approaches unity, and the last factor becomes:

$$\lim_{n \to \infty} \{[1 - \frac{\lambda t}{n}]^{-n/(\lambda t)}\}^{-\lambda t}.$$

Setting $-\lambda t/n = h$, this factor is:

$$[\lim_{h \to 0} (1 + h)^{1/h}]^{-\lambda t} = e^{-\lambda t},$$

since the limit in the bracket is the common definition of e. Thus, the binomial pmf approaches:

$$\frac{e^{-\lambda t}(\lambda t)^k}{k!}, \qquad k = 0, 1, 2, \ldots.$$

Now replacing λt by a single parameter α, we get the well-known Poisson pmf:

$$f(k; \alpha) = e^{-\alpha}\frac{\alpha^k}{k!}, \qquad k = 0, 1, 2, \ldots. \tag{2.11}$$

Thus the Poisson pmf can be used as a convenient approximation to the binomial pmf when n is large and p is small:

$$\binom{n}{k} p^k q^{n-k} \simeq e^{-\alpha}\frac{\alpha^k}{k!}, \qquad \text{where } \alpha = np.$$

An acceptable rule of thumb is to use the Poisson approximation for binomial probabilities if $n \geqslant 20$ and $p \leqslant 0.05$. The table that follows compares $b(k; 5, 0.2)$ and $b(k; 20, 0.05)$ with $f(k; 1)$. Observe that the approximation is better in the case of larger n and smaller p.

k	$b(k; 5, 0.2)$	$b(k; 20, 0.05)$	$f(k; 1)$
0	0.328	0.359	0.368
1	0.410	0.377	0.368
2	0.205	0.189	0.184
3	0.051	0.060	0.061

Example 2.9

A manufacturer produces IC chips, 1 percent of which are defective. Find the probability that in a box containing 100 chips, no defectives are found.

Since $n = 100$ and $p = 0.01$, the required answer is:

$$\begin{aligned} b(0; 100, 0.01) &= \binom{100}{0} \cdot 0.01^0 \cdot 0.99^{100} \\ &= 0.99^{100} \\ &= 0.366. \end{aligned}$$

Using the Poisson approximation, $\alpha = 100 \cdot 0.01 = 1$, and the required answer is:

$$\begin{aligned} f(0; 1) &= e^{-1} \\ &= 0.3679. \end{aligned}$$

#

It is easily verified that the probabilities from equation (2.11) are nonnegative and sum to 1:

$$\begin{aligned} \sum_{k=0}^{\infty} f(k; \alpha) &= \sum_{k=0}^{\infty} \frac{\alpha^k}{k!} e^{-\alpha} \\ &= e^{-\alpha} \sum_{k=0}^{\infty} \frac{\alpha^k}{k!} \\ &= e^{-\alpha} \cdot e^{\alpha} \\ &= 1. \end{aligned}$$

The probabilities $f(k; \alpha)$ are easy to calculate, starting with:

$$f(0; \alpha) = e^{-\alpha}$$

and using the recurrence relation:

$$f(k + 1; \alpha) = \frac{\alpha f(k; \alpha)}{k + 1}. \tag{2.12}$$

The Poisson probabilities are tabulated in [PEAR 1966] for $\alpha = 0.1$ to 15, in the increments of 0.1 [In Appendix C, we have tabulated the Poisson CDF]. In Figure 2.13 we have plotted the Poisson pmf with parameters $\alpha = 1$ and $\alpha = 4$. Note that this pmf is positively skewed; in fact, it can be shown that the Poisson pmf is positively skewed for any $\alpha > 0$.

Apart from its ability to approximate a binomial pmf, the Poisson pmf is found to be useful in many other situations. In reliability theory, it is quite

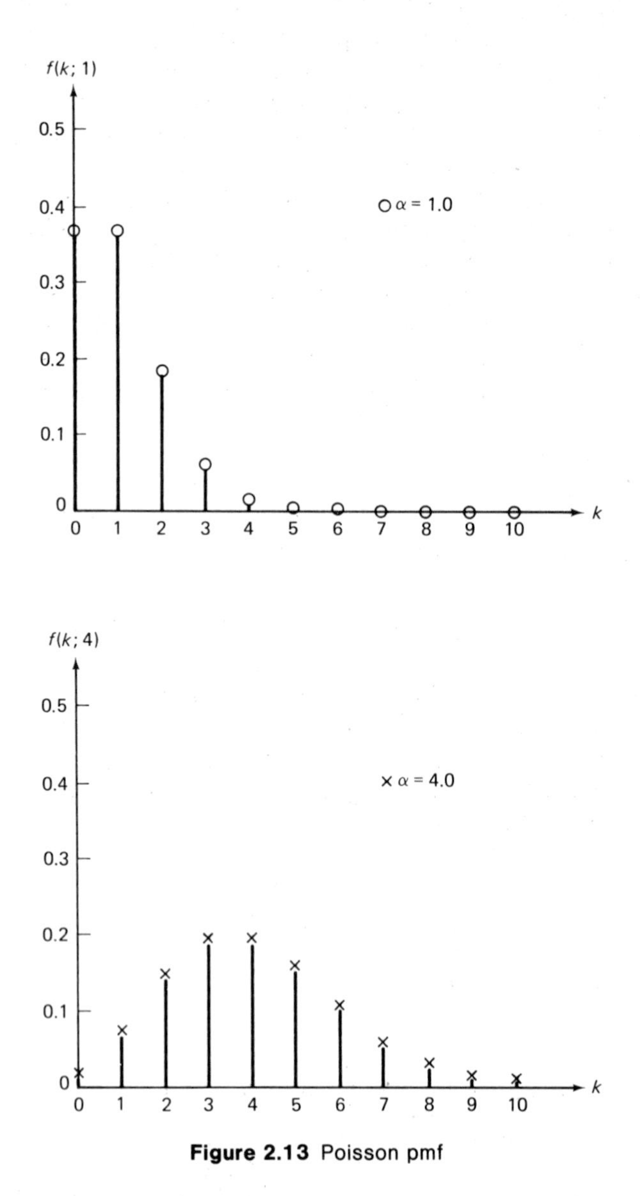

Figure 2.13 Poisson pmf

reasonable to assume that the probability of k components malfunctioning within an interval of time t in a system with a large number of components is given by the Poisson pmf [here λ is known as the component failure rate]:

$$f(k; \lambda t) = e^{-\lambda t}\frac{(\lambda t)^k}{k!}, \qquad k = 0, 1, 2, \ldots \tag{2.13}$$

In studying systems with congestion (or queuing), we find that the number of jobs arriving, the number of jobs completing service, or the number of messages transmitted through a communication channel in a fixed interval of time is approximately Poisson distributed.

2.5.6 The Hypergeometric pmf

We have noted earlier that the binomial pmf is obtained while "sampling with replacement." The hypergeometric pmf is obtained while "sampling without replacement." Suppose we select a random sample of n components from a box containing N components, d of which are known to be defective. For the first component selected, the probability that it is defective is given by d/N, but for the second selection it remains d/N only if the first component selected is replaced. Otherwise, this probability is $(d-1)/(N-1)$ or $d/(N-1)$, depending on whether or not a defective component was selected in the first drawing. Thus the assumption of a constant probability of success, as in a sequence of Bernoulli trials, is not satisfied.

We are interested in computing the **hypergeometric pmf**, $h(k; n, d, N)$, defined to be the probability of choosing k defective components in a random sample of n components, chosen without replacement, from a total of N components, d of which are defective. The sample space of this experiment consists of $\binom{N}{n}$ sample points. The k defectives can be selected from d defectives in $\binom{d}{k}$ ways, and the $n-k$ nondefective components may be selected from $N-d$ nondefectives in $\binom{N-d}{n-k}$ ways. Therefore, the whole sample of n components with k defectives can be selected in $\binom{d}{k}\cdot\binom{N-d}{n-k}$ ways. Assuming an equiprobable sample space, the required probability is:

$$h(k; n, d, N) = \frac{\binom{d}{k}\cdot\binom{N-d}{n-k}}{\binom{N}{n}}, \qquad \max\{0, d+n-N\} \le k \le \min\{d, n\}. \tag{2.14}$$

Example 2.10

Compute the probability of obtaining three defectives in a sample of size ten taken without replacement from a box of twenty components containing four defectives.

We are required to compute:

$$h(3;\ 10,\ 4,\ 20) = \frac{\binom{4}{3} \cdot \binom{16}{7}}{\binom{20}{10}}$$

$$= \frac{4 \cdot 11,440}{184,756}$$

$$= 0.247678.$$

If we were to approximate the above probability using a binomial distribution with $n = 10$ and $p = 4/20 = 0.20$, we will get $b(3;\ 10,\ 0.20) = 0.2013$, a considerable underestimate of the actual probability. #

In situations where the sample size n is small compared to the lot size N, the binomial distribution provides a good approximation to the hypergeometric distribution; that is, $h(k;\ n,\ d,\ N) \simeq b(k;\ n,\ d/N)$ for large N.

2.5.7 The Uniform pmf

Let X be a discrete random variable with a finite image $\{x_1, x_2, \ldots, x_N\}$. One of the simplest pmf's to consider in this case is one in which each value in the image has equal probability. If we require that $p_X(x_i) = p$ for all i, then, since:

$$1 = \sum_{i=1}^{N} p_X(x_i) = \sum_{i=1}^{N} p = Np,$$

it follows that:

$$p_X(x_i) = \begin{cases} \dfrac{1}{N}, & x_i \text{ in the image of } X, \\ 0, & \text{otherwise.} \end{cases}$$

Such a random variable is said to have a **discrete uniform distribution**. This distribution plays an important role in the theory of random numbers and its applications to Monte Carlo simulation. In the average-case analysis of programs, it is often assumed that the input data are uniformly distributed over the input space.

Note that the concept of uniform distribution cannot be extended to a discrete random variable with a countably infinite image, $\{x_1,\ x_2,\ \ldots\}$. The requirements that $\sum_i p_X(x_i) = 1$ and $p_X(x_i) = \text{constant}$ (for $i = 1,\ 2, \ldots$) are incompatible.

If we let X take on the values $\{1, 2, \ldots, N\}$ with $p_X(i) = 1/N,\ 1 \leqslant i \leqslant N$, then its distribution function is given by:

$$F(t) = \sum_{i=1}^{\lfloor t \rfloor} p_X(i)$$

$$= \frac{\lfloor t \rfloor}{N}, \qquad 1 \leqslant t \leqslant N$$

A graph of this distribution is given in Figure 2.14.

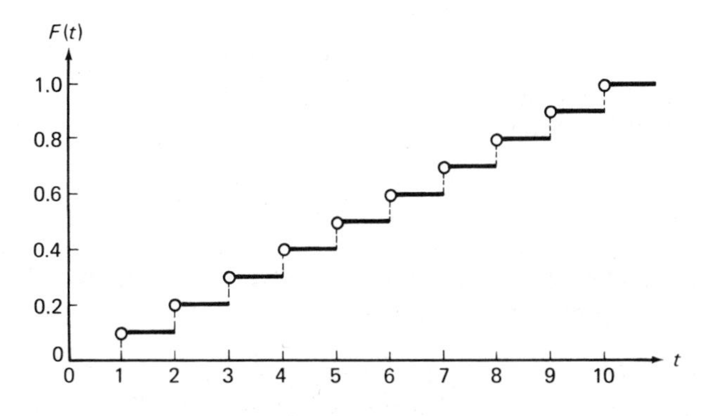

Figure 2.14 Discrete uniform distribution

2.5.8 Constant Random Variable

For a real number c, the function X defined by $X(s) = c$ for all s in S is a discrete random variable. Clearly, $P(X = c) = 1$. Therefore the pmf of this random variable is given by:

$$p_X(x) = \begin{cases} 1, & \text{if } x = c, \\ 0, & \text{otherwise.} \end{cases}$$

Such a random variable is called a **constant random variable**.

The distribution function of X is given by:

$$F_X(t) = \begin{cases} 0 & \text{for } t < c, \\ 1 & \text{for } t \geqslant c, \end{cases}$$

and is shown in Figure 2.15.

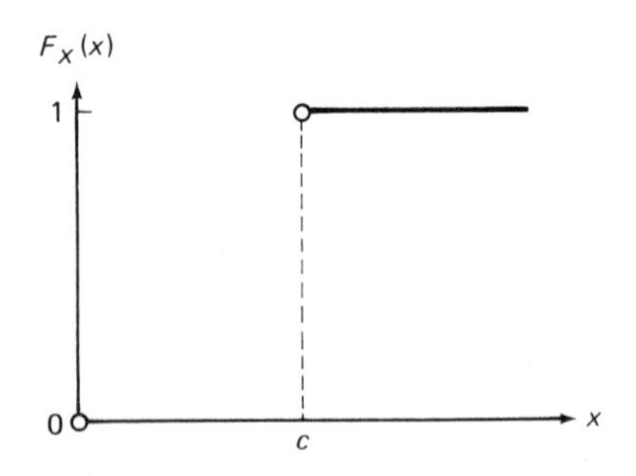

Figure 2.15 CDF of constant random variable

2.5.9 Indicator Random Variable

An event A partitions the sample space S into two mutually exclusive and collectively exhaustive subsets, A and $\bar{A}$. The **indicator** of the event A is a random variable I_A defined by:

$$I_A(s) = \begin{cases} 1, & \text{if } s \in A, \\ 0, & \text{if } s \in \bar{A}. \end{cases}$$

Then the event A occurs if and only if $I_A = 1$. This may be visualized as in Figure 2.16. The pmf of I_A is given by:

$$\begin{aligned} p_{I_A}(0) &= P(\bar{A}) \\ &= 1 - P(A) \end{aligned}$$

and

$$p_{I_A}(1) = P(A).$$

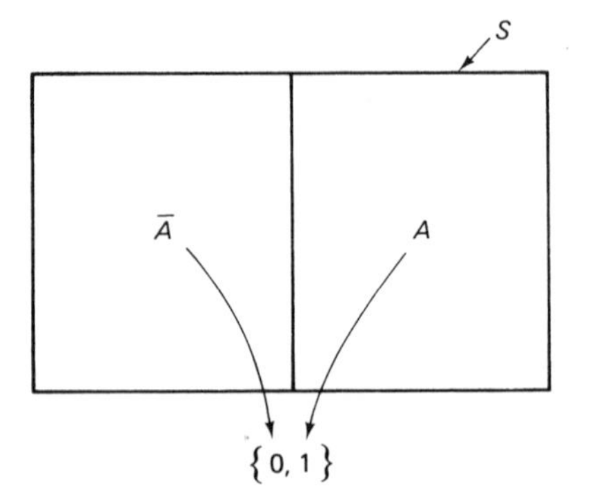

Figure 2.16 Indicator random variable

The concept of the indicator function in certain cases allows us to make efficient computations, without the detailed knowledge of distribution functions. This is quite useful, particularly in cases where the distribution is difficult to calculate. Now if X is a Bernoulli random variable with parameter p and image $\{0, 1\}$, then X is the indicator of the event:

$$A = \{s \mid X(s) = 1\}$$

and

$$\begin{aligned} p_X(0) &= P(\bar{A}) \\ &= 1 - P(A) \\ &= 1 - p \end{aligned}$$

and

$$\begin{aligned} p_X(1) &= P(A) \\ &= p. \end{aligned}$$

Problems

1. Out of a job population of ten jobs with six jobs of class 1 and four of class 2, a random sample of size n is selected. Let X be the number of class 1 jobs in the sample. Calculate the pmf of X if the sampling is (a) without replacement, (b) with replacement.

2. A mischievous student wants to break into a computer file, which is password protected. Assume that there are n equally likely passwords, and that the student chooses passwords independently and at random and tries them. Let N_n be the number of trials required to break into the file. Determine the pmf of N_n (a) if unsuccessful passwords are not eliminated from further selections, and (b) if they are.

3. A telephone call may pass through a series of trunks before reaching its destination. If the destination is within the caller's own local exchange, then no trunks will be used. Assume that the number of trunks used, X, is a modified geometric random variable with parameter p. Define Z to be the number of trunks used for a call directed to a destination outside the caller's local exchange. What is the pmf of Z? Given that a call requires at least three trunks, what is the conditional pmf of the number of trunks required?

4. Assume that the probability of error-free transmission of a message over a communication channel is p. If a message is not received correctly, a retransmission is initiated. This procedure is repeated until a correct transmission occurs. Such a channel is often called a **feedback channel**. Assuming that successive transmissions are independent, what is the probability that no retransmissions are required? What is the probability that exactly two retransmissions are required?

5. One percent of jobs arriving at a computer system need to wait until weekends for scheduling, owing to core-size limitations. Find the probability that among a sample of 200 jobs there are no jobs that have to wait until the weekend for scheduling. [*Hint*: You may use the Poisson approximation to the binomial distribution.]

6. Five percent of the disk controllers produced by a plant are known to be defective. A sample of fifteen controllers is drawn randomly from each month's production and the number of defectives noted. What proportion of these monthly samples would have at least two defective controllers?

7. The probability of error in the transmission of a bit over a communication channel is $p = 10^{-4}$. What is the probability of more than three errors in transmitting a block of 1,000 bits?

8. Assume that the number of messages input to a communication channel in an interval of duration t seconds is Poisson distributed with parameter $0.3t$. Compute the probabilities of the following events:
   (a) Exactly three messages will arrive during a ten-second interval,
   (b) At most twenty messages arrive in a period of twenty seconds,
   (c) The number of message arrivals in an interval of duration five seconds is between three and seven.

9. SSI chips, essential to the running of a computer system, fail in accordance with a Poisson distribution with the rate of one chip in about five weeks. If there are two spare chips on hand, and if a new supply will arrive in eight weeks, what is the probability that during the next eight weeks the system will be down for a week or more, owing to a lack of chips?

2.6 ANALYSIS OF PROGRAM MAX

We will now apply some of the techniques of the preceding sections to the analysis of a typical algorithm. Given an array of n elements, $B[1], B[2], \ldots, B[n]$, we will find m and j such that $m = B[j] = \max\{B[k] \mid 1 \leqslant k \leqslant n\}$, and for which j is as large as possible. In other words, the **Pascal program MAX**, on page 85, finds the largest element in the given array B. Our discussion here closely parallels that in [KNUT 1973; pp. 94–101].

There are at least two aspects of the analysis of an algorithm: the storage space required and the time of execution. Since the storage space required by the program MAX is fixed, we will analyze only the time required for its execution. The time of execution depends, in general, upon the machine on which it is executed, the compiler used to translate the program, and the input data supplied to it. We are interested in studying the effect of the input data on the time of execution. It is convenient to abstract and study the frequency counts for each step. In this way, we need not consider the details of the machine and the compiler used. Counting the number of times each step

```
program MAX(input, output);
    label 1, 2, 3, 4, 5, 6;
    const n = 100;
    var j, k, m: integer;
            B: array [1..n] of integer;
begin
1: j := n;   k := n - 1;   m := B[n];
2: while (k > 0) do
     begin
3:       if B[k] > m
           then
4:           begin
               j := k;
               m := B[k]
             end;
5:       k := k - 1
     end;
6: writeln(j,m)
end.
```

is executed is facilitated by drawing a flowchart as in Figure 2.17. Noting that the amount of flow into each node must equal the amount of flow out of the node, we obtain the following table:

Step number	*Frequency count*	*Number of statements*
M1	1	3
M2	n	1
M3	$n - 1$	1
M4	X	2
M5	$n - 1$	1
M6	1	1

This table gives us the information necessary to determine the execution time of program MAX on a given computer. In this table, everything except the quantity X is known. Here X is the number of times we must change the value of the current maximum. The value of X will depend upon the pattern of numbers constituting the elements of the array B. As these sets of numbers vary over some specified set, the value of X will also change. Each such pattern of numbers may be considered a sample point with a fixed assigned probability. Then X may be thought of as a random variable over the sample space. We are interested in studying the distribution of the random variable X for a given assignment of probabilities over the sample space.

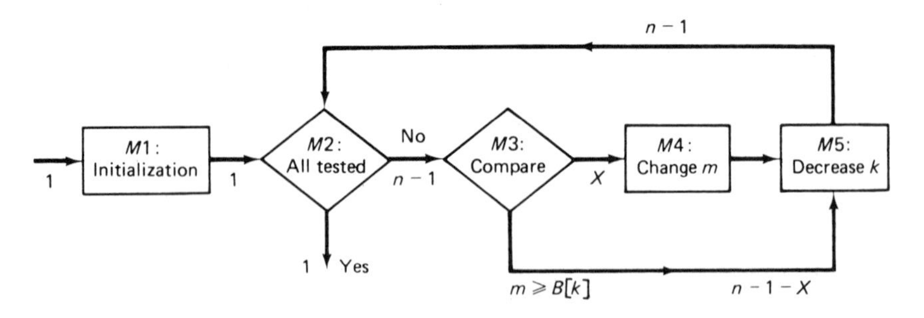

Figure 2.17 Flowchart of MAX

Clearly, the image of the random variable X is $\{0, 1, \ldots, n-1\}$. The minimum value of X occurs when $B[n] = \max\{B[k] \mid 1 \leqslant k \leqslant n\}$, and the maximum value of X occurs when $B[1] > B[2] > \ldots > B[n]$.

To simplify our analysis, we will assume that the $B[k]$ are distinct values. Furthermore, without a loss of generality, assume that the vector of elements $(B[1], B[2], \ldots, B[n])$ is any one of $n!$ permutations of the integers $\{1, 2, \ldots, n\}$. Thus the sample space S_n is the set of all permutations of the n integers $\{1, 2, \ldots, n\}$. Finally, we assume that all $n!$ permutations are equally likely. Therefore, for all s in S_n, $P(s) = 1/n!$. We may define the random variable X_n as a function with domain S_n and the image $\{0, 1, \ldots, n-1\}$. As n changes, we have a sequence of random variables $X_1, X_2, \ldots$, where X_i is defined on the sample space S_i.

The probability mass function of X_n, $p_{X_n}(k)$, will be denoted by p_{nk}. Then:

$$\begin{aligned} p_{nk} &= P(X_n = k) \\ &= \frac{\text{number of permutations of } n \text{ objects for which } X_n = k}{n!}. \end{aligned}$$

We will establish a recurrence relation for p_{nk}.

Consider a sample point $s = (b_1, b_2, \ldots, b_n)$, a permutation on $\{1, 2, \ldots, n\}$, and consider two mutually exclusive and collectively exhaustive events:

$$A = \text{“}b_1 = n\text{”}$$

and

$$\bar{A} = \text{“}b_1 \neq n.\text{”}$$

If event A occurs, then a comparison with b_1 (in program MAX) will force a change in the value of m. Therefore the value obtained for X_n will be one higher than a similar value obtained while examining the previous $n-1$ elements $(b_2, \ldots, b_n)$. Note that $(b_2, \ldots, b_n)$ is a permutation on $\{1, 2, \ldots, n-1\}$. Therefore, the number of times the value of m gets changed while examining $(b_2, \ldots, b_n)$ is X_{n-1}. From these observations, we conclude that:

$$P(X_n = k \mid A) = P(X_{n-1} = k - 1)$$
$$= p_{n-1,k-1}.$$

On the other hand, the occurrence of event $\bar{A}$ implies that the count of exchanges does not change when we examine b_1, that is:

$$P(X_n = k \mid \bar{A}) = P(X_{n-1} = k)$$
$$= p_{n-1,k}.$$

Now by the assumption of equiprobable sample space, we have $P(A) = 1/n$ and $P(\bar{A}) = (n-1)/n$. Then by the theorem of total probability we conclude that:

$$\begin{aligned} p_{nk} &= P(X_n = k) \\ &= P(X_n = k \mid A)P(A) + P(X_n = k \mid \bar{A})P(\bar{A}) \\ &= \frac{1}{n}p_{n-1,k-1} + \frac{n-1}{n}p_{n-1,k}. \end{aligned} \tag{2.15}$$

This equation will allow us to recursively compute p_{nk} if we provide the initial conditions. Since the image of X_n is $\{0, 1, \ldots, n-1\}$, we know that $p_{nk} = 0$ if $k < 0$. Next consider the random variable X_1. With $n = 1$, we observe that the **while** loop in program MAX will never be executed. Therefore $X_1 = 0$; that is, X_1 is a constant random variable, with $P(X_1 = 0) = p_{1,0} = 1$ and $P(X_1 = 1) = p_{1,1} = 0$. Thus the complete specification to evaluate p_{nk} is:

$$p_{1,0} = 1,$$
$$p_{1,1} = 0, \tag{2.16}$$

$$p_{nk} = \begin{cases} \frac{1}{n}p_{n-1,k-1} + \frac{n-1}{n}p_{n-1,k}, & 0 \leqslant k \leqslant n-1,\ n \geqslant 2, \\ 0, & \text{otherwise.} \end{cases} \tag{2.17}$$

Generating functions are a convenient tool for evaluation of quantities defined by such recurrence relations. In the context of discrete random variables, these functions will be referred to as **probability generating functions (PGF)**. They will be discussed in the next section. For the moment, let us study the pmf of the random variable X_2. The image of X_2 is $\{0, 1\}$, and from the above recurrence relation:

$$p_{2,0} = ½ \quad \text{and} \quad p_{2,1} = ½.$$

Thus X_2 is a Bernoulli random variable with the parameter $p = ½$.

Another quantity of interest is the probability $p_{n0} = P(X_n = 0)$. From the above recurrence relation:

$$
\begin{aligned}
p_{n0} &= \frac{1}{n} p_{n-1,-1} + \frac{n-1}{n} p_{n-1,0} \\
&= \frac{n-1}{n} p_{n-1,0} \\
&= \frac{n-1}{n} \cdot \frac{n-2}{n-1} \cdots \frac{1}{2} p_{1,0} \\
&= \frac{(n-1)!}{n!} \\
&= \frac{1}{n}.
\end{aligned}
$$

Alternatively, this result could be obtained by observing that no changes to the value of m will be required if $b_n = B[n] = n$, since m will be set equal to the largest value n before entering the **while** loop. Now out of $n!$ permutations, $(n-1)!$ of them have $b_n = n$; therefore we get the required result.

Problems

1. Explicitly determine the pmf of random variable X_3, the number of exchanges in program MAX with array size $n = 3$.

2.7 THE PROBABILITY GENERATING FUNCTION

The notion of probability generating functions (PGF) is a convenient tool that simplifies computations involving integer-valued, discrete random variables. Given a nonnegative integer-valued discrete random variable X with $P(X = k) = p_k$, define the PGF of X by:

$$
\begin{aligned}
G_X(z) &= \sum_{i=o}^{\infty} p_i z^i \\
&= p_0 + p_1 z + p_2 z^2 + \cdots + p_k z^k + \cdots .
\end{aligned}
$$

$G_X(z)$, also known as the z-transform of X, converges for any complex number z such that $|z| < 1$. It may be easily verified that

$$G_X(1) = 1 = \sum_{i=0}^{\infty} p_i .$$

In many problems we will know the PGF $G_X(z)$, but we will not have explicit knowledge for the pmf of X. Later we will see that we can determine interesting quantities such as the mean and variance of X from the PGF itself. One reason for the usefulness of PGF is found in the following theorem, which we quote without proof.

THEOREM 2.1. If two discrete random variables X and Y have the same PGF's, then they must have the same distributions and pmf's.

If we can show that a random variable that is under investigation has the same PGF as that of another random variable with a known pmf, then this theorem assures us that the pmf of the original random variable must be the same.

Continuing with our analysis of program MAX, define the PGF of X_n as:

$$G_{X_n}(z) = \sum_{k \geqslant 0} p_{nk} \cdot z^k.$$

$G_{X_n}(z)$ is actually a polynomial, even though an infinite sum is specified for convenience. From equation (2.16) we have:

$$\begin{aligned} G_{X_1}(z) &= p_{1,0} + p_{1,1} \cdot z \\ &= 1. \end{aligned}$$

Multiplying the recurrence relation (2.15) by z^k and summing for $k = 1$ to infinity:

$$\sum_{k \geqslant 1} p_{nk} \cdot z^k = \frac{z}{n} \sum_{k \geqslant 1} p_{n-1,k-1} \cdot z^{k-1} + \frac{n-1}{n} \sum_{k \geqslant 1} p_{n-1,k} \cdot z^k.$$

Thus:

$$G_{X_n}(z) - p_{n0} = \frac{z}{n} G_{X_{n-1}}(z) + \frac{n-1}{n} [G_{X_{n-1}}(z) - p_{n-1,0}].$$

Noting that $p_{n0} = 1/n$ and simplifying, we get:

$$\begin{aligned} G_{X_n}(z) &= \frac{(z+n-1)}{n} G_{X_{n-1}}(z) \\ &= \frac{(z+n-1)}{n} \cdot \frac{(z+n-2)}{n-1} \cdots \frac{(z+1)}{2} G_{X_1}(z) \\ &= \frac{(z+n-1)(z+n-2) \cdots (z+1)}{n!}. \end{aligned} \qquad (2.18)$$

To obtain an explicit expression for p_{nk}, we must expand $G_{X_n}(z)$ into a power series of z. Stirling numbers of the first kind, denoted by $\left[{n \atop k}\right]$, can be used for this purpose. Stirling numbers are defined by [KNUT 1973, p. 65]:

$$x(x-1)\cdots(x-n+1) = \left[{n \atop n}\right]x^n - \left[{n \atop n-1}\right]x^{n-1} + \cdots + (-1)^n\left[{n \atop 0}\right]$$

$$= \sum_{k=0}^{n}(-1)^{n-k}\left[{n \atop k}\right]x^k.$$

Substituting $x = -z$ in the above formula, we get:

$$z(z+1)\cdots(z+n-1) = \sum_{k=0}^{n}\left[{n \atop k}\right]z^k.$$

Then, using (2.18), we have:

$$G_{X_n}(z) = \frac{(z+1)(z+2)\cdots(z+n-1)}{n!}$$

$$= \frac{1}{n!}\sum_{k=0}^{n}\left[{n \atop k}\right]z^{k-1}.$$

Therefore:

$$p_{nk} = \frac{\left[{n \atop k+1}\right]}{n!}. \tag{2.19}$$

Thus the pmf of the random variable X_n is described by the Stirling numbers of the first kind.

At this point it is useful to derive the PGF's of some of the distributions studied in Section 2.5. When the random variable X is understood, we will use the abbreviated notation $G(z)$ for its PGF.

1. *The Bernoulli random variable*

$$G(z) = qz^0 + pz^1$$

$$= q + pz$$

$$= 1 - p + pz \tag{2.20}$$

2. *The binomial random variable*

$$G(z) = \sum_{k=0}^{n} \binom{n}{k} p^k (1-p)^{n-k} z^k$$

$$= (pz + 1 - p)^n. \tag{2.21}$$

3. *The modified geometric random variable*

$$G(z) = \sum_{k=0}^{\infty} p(1-p)^k z^k$$

$$= \frac{p}{1 - z(1-p)}. \tag{2.22}$$

4. *The Poisson random variable*

$$G(z) = \sum_{k=0}^{\infty} \frac{\alpha^k}{k!} e^{-\alpha} z^k$$

$$= e^{-\alpha} e^{\alpha z}$$

$$= e^{\alpha(z-1)}$$

$$= e^{-\alpha(1-z)}. \tag{2.23}$$

5. *The uniform random variable*

$$G(z) = \sum_{k=1}^{N} \frac{1}{N} z^k$$

$$= \frac{1}{N} \sum_{k=1}^{N} z^k. \tag{2.24}$$

6. *The constant random variable.* Let $X = i$ for some $0 \leqslant i < \infty$; then:

$$G(z) = z^i. \tag{2.25}$$

7. *The indicator random variable.* Let $P(I_A = 0) = P(\bar{A}) = 1 - p$ and $P(I_A = 1) = P(A) = p$; then:

$$G(z) = (1-p)z^0 + pz$$

$$= 1 - p + pz$$

$$= P(\bar{A}) + P(A)z. \tag{2.26}$$

Problems

1. Let X denote the execution time of a job rounded to the nearest second. The charges are based on a linear function $Y = mX + n$ of the execution time for suitably chosen nonnegative integers m and n. Given the PGF and pmf of X, find the PGF and pmf of Y.

2. Show that the PGF of a geometric random variable with parameter p is given by $pz/(1 - qz)$ where $q = 1 - p$.

3. Let X be a negative binomial random variable with parameters n, p, and r. Show that its PGF is given by

$$\left[\frac{pz}{1 - z(1 - p)}\right]^r.$$

2.8 DISCRETE RANDOM VECTORS

It often happens that we are interested in studying relationships between two or more random variables defined on a given sample space. For example, consider a program consisting of two modules with execution times X and Y, respectively. Since the execution times will depend upon input data values and since the execution times will be discrete, we may assume that X and Y are discrete random variables. If the program is organized such that two modules are executed serially, one after the other, then the random variable $Z_1 = X + Y$ gives the total execution time of the program. Alternatively, if the program's two modules execute independently and concurrently, then the total program execution time is given by the random variable $Z_2 = \max\{X, Y\}$, and the time until the completion of the faster module is given by $Z_3 = \min\{X, Y\}$.

Let $X_1, X_2, \ldots, X_r$ be r discrete random variables defined on a sample space S. Then for each sample point s in S, each of the random variables $X_1, X_2, \ldots, X_r$ takes on one of its possible values, as:

$$X_1(s) = x_1, \quad X_2(s) = x_2, \quad \ldots, \quad X_r(s) = x_r.$$

The random vector $\mathbf{X} = (X_1, X_2, \ldots, X_r)$ is an r-dimensional vector-valued function $\mathbf{X}: S \longrightarrow \mathbb{R}^r$ with $\mathbf{X}(s) = \mathbf{x} = (x_1, x_2, \ldots, x_r)$. Thus, a discrete r-dimensional random vector $\mathbf{X}$ is a function from S to $\mathbb{R}^r$ taking on a finite or countably infinite set of vector values $\mathbf{x}_1, \mathbf{x}_2, \ldots$.

Definition (Joint pmf). The **compound** (or **joint**) **pmf** for a random vector $\mathbf{X}$ is defined to be:

$$\begin{aligned} p_{\mathbf{X}}(\mathbf{x}) &= P(\mathbf{X} = \mathbf{x}) \\ &= P(X_1 = x_1, X_2 = x_2, \ldots, X_r = x_r). \end{aligned}$$

As in the one-dimensional case, the compound pmf has the following four properties:

(j1) $p_{\mathbf{X}}(\mathbf{x}) \geqslant 0, \ \mathbf{x} \in \mathbb{R}^r$.

(j2) $\{\mathbf{x} \mid p_{\mathbf{X}}(\mathbf{x}) \neq 0\}$ is a finite or countably infinite subset of $\mathbb{R}^r$, which will be denoted by $\{\mathbf{x}_1, \ \mathbf{x}_2, \ldots\}$.

(j3) $P(\mathbf{X} \in A) = \sum_{\mathbf{x} \in A} p_{\mathbf{X}}(\mathbf{x})$.

(j4) $\sum_i p_{\mathbf{X}}(\mathbf{x}_i) = 1$.

It can be shown that any real-valued function defined on $\mathbb{R}^r$ having these four properties is the compound pmf of some discrete r-dimensional random vector.

Suppose we consider a program with two modules, and module execution times are X and Y, respectively. The images of the discrete random variables X and Y are given by $\{1, 2\}$ and $\{1, 2, 3, 4\}$. The compound pmf is described by the following table:

	y = 1	y = 2	y = 3	y = 4
x = 1	1/4	1/16	1/16	1/8
x = 2	1/16	1/8	1/4	1/16

Each possible event $[X = x, \ Y = y]$ can be pictured as an event point on an (x, y) coordinate system with the value $p_{X,Y}(x, y)$ indicated as a bar perpendicular to the (x, y) plane above the event point (x, y). This can be visualized as in Figure 2.18, where we indicate the value $p_{X,Y}(x, y)$ associated with each event by writing it beside the event point (x, y).

In situations where we are concerned with more than one random variable, the pmf of a single variable, such as $p_X(x)$, is referred to as a **marginal pmf**. Since the eight events shown in Figure 2.18 are collectively exhaustive and mutually exclusive, the marginal pmf is:

$$\begin{aligned} p_X(x) &= P(X = x) \\ &= P(\bigcup_j \{X = x, \ Y = y_j\}) \\ &= \sum_j P(X = x, \ Y = y_j) \\ &= \sum_j p_{X,Y}(x, y_j). \end{aligned}$$

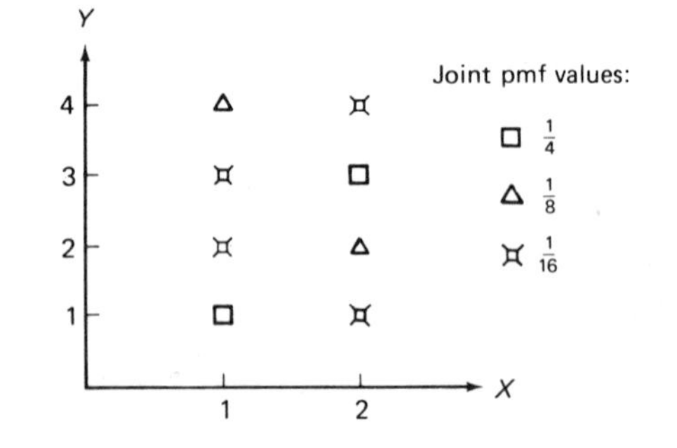

Figure 2.18 Joint pmf for two-module execution problem

In other words, to obtain the marginal pmf $p_X(x)$, we erect a vertical column at $X = x$ and sum the probabilities of all event points touched by the column. Similarly:

$$p_Y(y) = \sum_i p_{X,Y}(x_i, y).$$

In the example above, we get:

$$p_X(1) = 1/2, \qquad p_X(2) = 1/2, \qquad p_Y(1) = 5/16,$$
$$p_Y(2) = 3/16, \qquad p_Y(3) = 5/16, \qquad p_Y(4) = 3/16.$$

Check that

$$\sum_{i=1}^{2} p_X(i) = 1 \quad \text{and} \quad \sum_{i=1}^{4} p_Y(i) = 1.$$

The above formula for computing the marginal pmf's from the compound pmf can be easily generalized to the r-dimensional case.

We have seen that the task of obtaining the marginal pmf's from the compound pmf is relatively straightforward. Note that given the marginal pmf's there is no way to go back, in general, to determine the compound pmf. However, an exception occurs when the random variables are independent. The notion of independence of random variables will be developed in the next section.

Example 2.11

Let X and Y be two random variables, each with image $\{1, 2\}$ and with the compound pmf:

$$p_{X,Y}(1, 1) = p_{X,Y}(2, 2) = a,$$
$$p_{X,Y}(1, 2) = p_{X,Y}(2, 1) = 1/2 - a \qquad \text{for } 0 \leqslant a \leqslant 1/2.$$

It is easy to see that $p_X(1) = p_X(2) = p_Y(1) = p_Y(2) = 1/2$, independent of the value of a. Thus, we have uncountably many distinct compound pmf's associated with the same marginal pmf's. #

An interesting example of a compound pmf is the **multinomial pmf**, which is a generalization of the binomial pmf. Consider a sequence of n generalized Bernoulli trials with a finite number r of distinct outcomes having probabilities $p_1, p_2, \ldots, p_r$ where $\sum_{i=1}^{r} p_i = 1$. Define the random vector $\mathbf{X} = (X_1, X_2, \ldots, X_r)$ such that X_i is the number of trials that resulted in the ith outcome. Then the compound pmf of $\mathbf{X}$ is given by:

$$p_{\mathbf{X}}(\mathbf{n}) = P(X_1 = n_1, X_2 = n_2, \ldots, X_r = n_r)$$
$$= \binom{n}{n_1\, n_2 \ldots n_r} p_1^{n_1} p_2^{n_2} \ldots p_r^{n_r}, \tag{2.27}$$

where $\mathbf{n} = (n_1, n_2, \ldots, n_r)$ and $\sum_{i=1}^{r} n_i = n$. The marginal pmf of X_i may be computed by:

$$p_{X_i}(n_i) = \sum_{\mathbf{n}: \sum_{j \neq i} n_j = n - n_i} \binom{n}{n_1\, n_2 \cdots n_r} p_1^{n_1} p_2^{n_2} \ldots p_r^{n_r}$$
$$= \frac{n!\, p_i^{n_i}}{(n-n_i)!\,(n_i)!} \sum_{\sum_{j \neq i} n_j = n - n_i} \frac{(n-n_i)!\, p_1^{n_1} \cdots p_{i-1}^{n_{i-1}}\, p_{i+1}^{n_{i+1}} \cdots p_r^{n_r}}{n_1!\, n_2! \cdots n_{i-1}!\, n_{i+1}! \cdots n_r!}$$
$$= \binom{n}{n_i} p_i^{n_i} (p_1 + \cdots + p_{i-1} + p_{i+1} + \cdots + p_r)^{n-n_i}$$
$$= \binom{n}{n_i} p_i^{n_i} (1 - p_i)^{n-n_i}.$$

Thus the marginal pmf of X_i is binomial with parameters n and p_i.

Many practical situations give rise to a multinomial distribution. For example, a program requires I/O service from device i with probability p_i at the end of a CPU burst, with $\sum_{i=1}^{r} p_i = 1$. This situation is depicted in Figure 2.19. If we observe n CPU burst terminations, then the probability that n_i of these will be directed to I/O device i (for $i = 1, 2, \ldots, r$) is given by the mul-

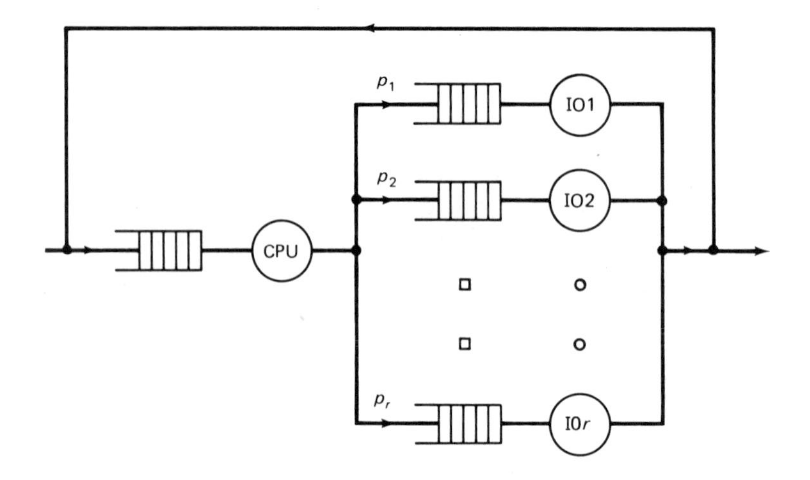

Figure 2.19 I/O queuing at the end of a CPU burst

tinomial pmf. Now if we are just interested in the number of I/O requests (out of n) directed to a specific device j, then it has a binomial distribution with parameters n and p_j. We may also replace the phrase "I/O device" by "file" in this example. Another example of a multinomial distribution occurs when we consider a paging system and we model a program using the so-called **independent reference model** [COFF 1973]. In this model we assume that successive page references are independent, and the probability of referencing page i is fixed at p_i.

Example 2.12

An inspection plan calls for inspecting five chips and for either accepting each chip, rejecting each chip, or submitting it for reinspection, with probabilities of $p_1 = 0.70$, $p_2 = 0.20$, $p_3 = 0.10$ respectively. What is the probability that all five chips must be reinspected? What is the probability that none of the chips must be reinspected? What is the probability that at least one of the chips must be reinspected?

Let X = number of chips accepted, Y = number of chips rejected; then the remaining $Z = 5 - X - Y$ are sent for reinspection. The compound pmf is:

$$p_{X,Y,Z}(i,j,k) = \frac{5!}{i!\,j!\,k!} \cdot 0.7^i \cdot 0.2^j \cdot 0.1^k.$$

The answer to the first question is $p_{X,Y,Z}(0, 0, 5) = 10^{-5}$. The second question pertains to the event:

$$\begin{aligned}[X + Y = 5] &= \{s \mid X(s) + Y(s) = 5\} \\ &= \bigcup_{i+j=5} \{s \mid X(s) = i \text{ and } Y(s) = j \text{ and } Z(s) = 0\}\end{aligned}$$

and therefore:

$$
\begin{aligned}
P(X + Y = 5) &= \sum_{i+j=5} p_{X,Y,Z}(i, j, k) \\
&= \sum_{i+j=5} \frac{5!}{i!\,j!} p_1^i p_2^j p_3^0 \\
&= \sum_{i=0}^{5} \binom{5}{i} p_1^i p_2^{5-i} \\
&= (p_1 + p_2)^5 \\
&= (0.7 + 0.2)^5 \\
&= 0.59.
\end{aligned}
$$

To answer the third question, note that the event {"at least one chip reinspected"} = S − {"none reinspected"}, but P ("none reinspected") = $P(X + Y = 5) = 0.59$. Therefore, the required answer is:

$$1 - 0.59 = 0.41.$$

#

Problems

1. Two discrete random variables X and Y have joint pmf given by the following table:

		Y		
		1	2	3
X	1	1/12	1/6	1/12
	2	1/6	1/4	1/12
	3	1/12	1/12	0

Compute the probability of each of the following events.
(a) $X \leqslant 1\frac{1}{2}$.
(b) X is odd.
(c) XY is even.
(d) Y is odd given that X is odd.

2. Each telephone call passes through a number of switching offices before reaching its destination. The number and types of switching offices in the United States in 1971 were estimated to be as follows:

	$i = 1$	2	3	4
Office type	Step-by-step	Panel	Crossbar	No. 1 ESS
Number	8,600	500	5,700	286

Note that the type of switching office in the local exchange of the originating call is fixed. Let n be the number of switching offices encountered (other than the local exchange) by a telephone call. Assuming that each switching office is randomly and independently chosen from the population, determine the probability that the call passes through exactly n_i ($i = 1, 2, 3, 4$) switching offices of type i, where $n_1 + n_2 + n_3 + n_4 = n$. Determine the marginal pmf for each type of office.

2.9 INDEPENDENT RANDOM VARIABLES

We have noted that the problem of determining the compound pmf given the marginal pmf does not have a unique solution, unless the random variables are independent.

Definition (Independent Random Variables). Two discrete random variables X and Y are defined to be independent provided their joint pmf is the product of their marginal pmf's:

$$p_{X,Y}(x,y) = p_X(x)p_Y(y) \qquad \text{for all } x \text{ and } y. \tag{2.28}$$

If X and Y are two independent random variables, then for any two subsets A and B of $\mathbb{R}$, the events "X is an element of A" and "Y is an element of B" are independent:

$$P(X \in A \cap Y \in B) = P(X \in A)P(Y \in B).$$

To see this, note that:

$$\begin{aligned} P(X \in A \cap Y \in B) &= \sum_{x \in A} \sum_{y \in B} p_{X,Y}(x, y) \\ &= \sum_{x \in A} \sum_{y \in B} p_X(x)p_Y(y) \\ &= \sum_{x \in A} p_X(x) \sum_{y \in B} p_Y(y) \\ &= P(X \in A)P(Y \in B). \end{aligned}$$

To further clarify the notion of independent random variables, assume that on a particular performance of the experiment, the event $[Y = y]$ has been observed, and we want to know the probability that a certain value of X will occur. We write:

$$
\begin{aligned}
P(X = x \mid Y = y) &= \frac{P(X = x \cap Y = y)}{P(Y = y)} \\
&= \frac{p_{X,Y}(x, y)}{p_Y(y)} \\
&= \frac{p_X(x)p_Y(y)}{p_Y(y)} \qquad \text{by independence} \\
&= p_X(x).
\end{aligned}
$$

Thus if X, Y are independent, then the knowledge that a particular value of Y has been observed does not affect the probability of observing a particular value of X. The notion of independence of twc random variables can be easily generalized to r random variables.

Definition. Let $X_1, X_2, \ldots, X_r$ be r discrete random variables with pmf's $p_{X_1}, p_{X_2}, \ldots, p_{X_r}$, respectively. These random variables are said to be **mutually independent** if their compound pmf p is given by:

$$
p_{X_1, X_2, \ldots, X_r}(x_1, x_2, \ldots, x_r) = p_{X_1}(x_1)p_{X_2}(x_2)\ldots p_{X_r}(x_r).
$$

In situations involving many random variables, the assumption of mutual independence usually leads to considerable simplification. We note that it is possible for every pair of random variables in the set $\{X_1, X_2, \ldots, X_r\}$ to be independent (**pairwise independent**) without the entire set being mutually independent.

Example 2.13

Consider a sequence of two Bernoulli trials and define X_1 and X_2 as the number of successes on the first and second trials respectively. Let X_3 define the number of matches on the two trials. Then it can be shown that the pairs (X_1, X_2), (X_1, X_3), and (X_2, X_3) are each independent, but that the set $\{X_1, X_2, X_3\}$ is not mutually independent. #

Returning to our earlier example of a program with two modules, let us determine the pmf's of the random variables Z_1, Z_2, and Z_3, given that X and Y are independent. Consider the event $[Z_1 = X + Y = t]$. On a two-dimensional (x, y) event space, this event is represented by all the event points on the line $X + Y = t$ (see Figure 2.20). The probability of this event may be computed by adding the probabilities of all the event points on this line. Therefore:

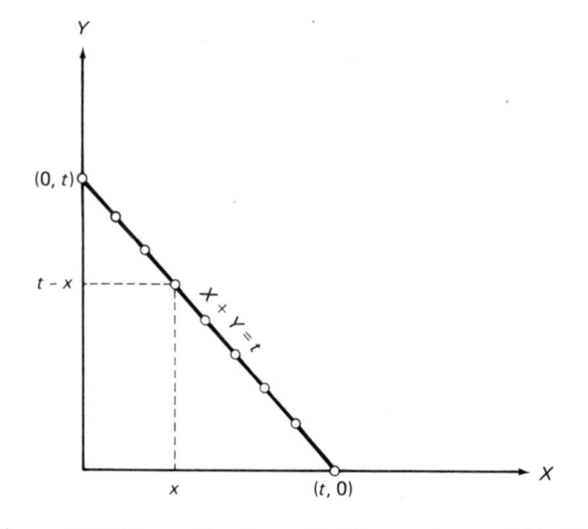

Figure 2.20 Computing the pmf of the random variable $Z_1 = X + Y$

$$
\begin{aligned}
P(Z_1 = t) &= \sum_{x=0}^{t} P(X = x,\ X + Y = t) \\
&= \sum_{x=0}^{t} P(X = x,\ Y = t - x) \\
&= \sum_{x=0}^{t} P(X = x)P(Y = t - x)
\end{aligned}
$$

by independence. Thus:

$$
\begin{aligned}
p_{Z_1}(t) &= p_{X+Y}(t) \\
&= \sum_{x=0}^{t} p_X(x)p_Y(t - x).
\end{aligned}
$$

This summation is said to represent the **discrete convolution**, and it gives the formula for the pmf of the sum of two nonnegative independent discrete random variables. In case X and Y are allowed to take negative values as well, the lower index of summation is changed from 0 to $-\infty$.

Restricting attention to nonnegative integer-valued random variables and recalling the definition of the probability generating function, the PGF of the sum of two independent random variables is the product of their PGF's:

$$
\begin{aligned}
G_{Z_1}(z) &= G_{X+Y}(z) \\
&= G_X(z)G_Y(z).
\end{aligned}
$$

To see this, note that:

$$\begin{aligned} G_{Z_1}(z) &= \sum_{t=0}^{\infty} p_{Z_1}(t) z^t \\ &= \sum_{t=0}^{\infty} z^t \sum_{x=0}^{t} p_X(x) p_Y(t-x) \\ &= \sum_{x=0}^{\infty} p_X(x) z^x \sum_{t=x}^{\infty} p_Y(t-x) z^{t-x} \\ &= \sum_{x=0}^{\infty} p_X(x) z^x \sum_{y=0}^{\infty} p_Y(y) z^y \\ &= G_X(z) G_Y(z), \end{aligned}$$

which is the desired result.

It follows by induction that if $X_1, X_2, \ldots, X_r$ are mutually independent nonnegative integer-valued random variables, then:

$$G_{X_1+X_2+\cdots+X_r}(z) = G_{X_1}(z) G_{X_2}(z) \ldots G_{X_r}(z). \tag{2.29}$$

This result is useful in proving the following theorem.

THEOREM 2.2. Let $X_1, X_2, \ldots, X_r$ be mutually independent.

(a) If X_i has the binomial distribution with parameters n_i and p, then $\sum_{i=1}^{r} X_i$ has the binomial distribution with parameters $n_1 + n_2 + \ldots + n_r$ and p.

(b) If X_i has a (modified) negative binomial distribution with parameters α_i and p, then $\sum_{i=1}^{r} X_i$ has the (modified) negative binomial distribution with parameters $\alpha_1 + \alpha_2 + \ldots + \alpha_r$ and p.

(c) If X_i has Poisson distribution with parameter α_i, then $\sum_{i=1}^{r} X_i$ has a Poisson distribution with parameter $\sum_{i=1}^{r} \alpha_i$.

This theorem can be visualized with Figure 2.21.

Proof: First note that:

$$G_{\sum_{i=1}^{r} X_i}(z) = G_{X_1}(z) G_{X_2}(z) \ldots G_{X_r}(z).$$

(a) If X_i is $b(k; n_i, p)$, then:

$$G_{X_i}(z) = (pz + 1 - p)^{n_i}.$$

Therefore:

$$G_{\Sigma X_i}(z) = (pz + 1 - p)^{\sum_{i=1}^{r} n_i}.$$

But this implies that $\sum_{i=1}^{r} X_i$ is $b(k; \sum_{i=1}^{r} n_i, p)$, as was to be shown. The proofs of parts (b) and (c) are left as an exercise.

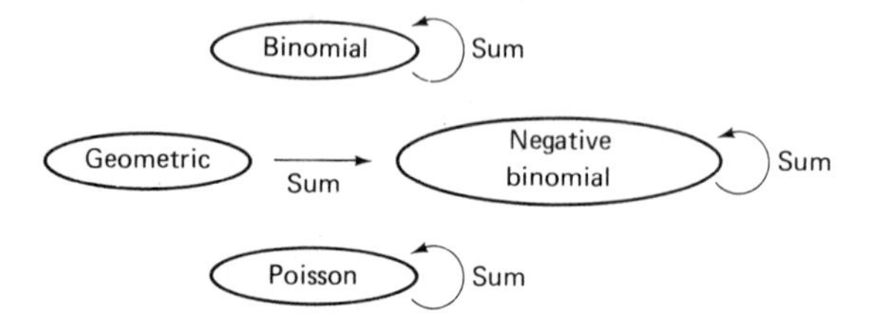

Figure 2.21 Theorem 2.2

Returning now to our example of a program with two modules, if we assume that X and Y are both geometrically distributed with parameter p, then we know that the total serial execution time $Z_1 = X + Y$ is negative binomially distributed with parameters 2 and p (note that the geometric distribution is a negative binomial distribution with parameters 1 and p).

Let us proceed to compute the distribution of $Z_2 = \max\{X, Y\}$ for the above example. Note that the event

$$\begin{aligned}[Z_2 \leqslant t] &= \{s \mid \max\{X(s), Y(s)\} \leqslant t\} \\ &= \{s \mid X(s) \leqslant t \text{ and } Y(s) \leqslant t\} = [X \leqslant t \text{ and } Y \leqslant t].\end{aligned}$$

Therefore:

$$\begin{aligned}F_{Z_2}(t) &= P(Z_2 = \max\{X, Y\} \leqslant t) \\ &= P(X \leqslant t \text{ and } Y \leqslant t) \\ &= P(X \leqslant t)P(Y \leqslant t) \qquad \text{by independence} \\ &= F_X(t)F_Y(t). \end{aligned} \tag{2.30}$$

Thus the CDF of $\max\{X, Y\}$ of two independent random variables X, Y is the product of their CDF's.

Next consider the random variable $Z_3 = \min\{X, Y\}$. First compute:

$$\begin{aligned}P(Z_3 = \min\{X, Y\} > t) &= P(X > t \text{ and } Y > t) \qquad (2.31) \\ &= P(X > t)P(Y > t) \qquad \text{by independence.}\end{aligned}$$

But $P(X > t) = 1 - F_X(t)$. Therefore:

$$1 - F_{Z_3}(t) = [1 - F_X(t)][1 - F_Y(t)]$$

or

$$F_{Z_3}(t) = F_X(t) + F_Y(t) - F_X(t)F_Y(t). \tag{2.32}$$

This last expression can be alternatively derived by first defining the events $A = [X \leqslant t]$, $B = [Y \leqslant t]$, and $C = [Z_3 \leqslant t]$ and noting that

$$P(C) = P(A \cup B) = P(A) + P(B) - P(A \cap B).$$

If we assume that X and Y are (modified) geometrically distributed with parameter p, then:

$$p_X(i) = p(1-p)^i$$

and

$$p_Y(j) = p(1-p)^j.$$

Also:

$$\begin{aligned} F_X(k) &= \sum_{i=0}^{k} p(1-p)^i \\ &= \frac{p[1-(1-p)^{k+1}]}{[1-(1-p)]} \\ &= 1-(1-p)^{k+1} \end{aligned}$$

and

$$F_Y(k) = 1 - (1-p)^{k+1}. \tag{2.33}$$

Then by (2.32) we have:

$$\begin{aligned} F_{Z_3}(k) &= 2[1-(1-p)^{k+1}] - [1 - 2(1-p)^{k+1} + (1-p)^{2(k+1)}] \\ &= 1 - (1-p)^{2(k+1)} \qquad (2.34) \\ &= 1 - [(1-p)^2]^{k+1}. \end{aligned}$$

From this, we conclude that Z_3 is also (modified) geometrically distributed with parameter $1 - (1-p)^2 = 2p - p^2$. In general, min $\{X_1, X_2, \ldots, X_r\}$ is geometrically distributed if each X_i $(1 \leqslant i \leqslant r)$ is geometrically distributed, given that $X_1, X_2, \ldots, X_r$ are mutually independent.

Let us consider the event that the module 2 takes longer to finish than module 1; that is, $[Y \geqslant X]$. Then from Figure 2.22:

$$P(Y \geqslant X) = \sum_{x=0}^{\infty} P(X = x,\ Y \geqslant x)$$

$$= \sum_{x=0}^{\infty} P(X = x)P(Y \geqslant x) \qquad \text{by independence.}$$

Now since Y has a modified geometric distribution, we have:

$$P(Y \geqslant x) = 1 - F_Y(x - 1) = (1 - p)^x.$$

Therefore:

$$P(Y \geqslant X) = \sum_{x=0}^{\infty} p(1 - p)^x(1 - p)^x \tag{2.35}$$

$$= p\sum_{x=0}^{\infty} [(1 - p)^2]^x$$

$$= \frac{p}{1 - (1 - p)^2}$$

$$= \frac{p}{2p - p^2}$$

$$= \frac{1}{2 - p}.$$

We may also compute $P(Y = X)$:

$$P(Y = X) = \sum_{x=0}^{\infty} P(X = x,\ Y = x)$$

$$= \sum_{x=0}^{\infty} P(X = x)P(Y = x)$$

since X and Y are independent,

$$= \sum_{x=0}^{\infty} p(1 - p)^x p(1 - p)^x$$

$$= \frac{p^2}{2p - p^2}$$

$$= \frac{p}{2 - p}.$$

Thus if $p = ½$, then there is a 33 percent chance that both modules will take exactly the same time to finish. Similar events will be seen to occur with probability zero when X and Y are continuous random variables.

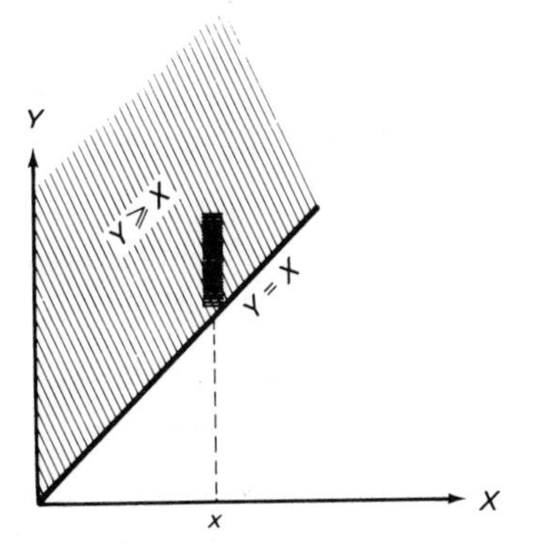

Figure 2.22 Graph for two-module execution problem

Finally, consider the conditional probability of the event $[Y = y]$ given that $[X + Y = t]$:

$$P(Y = y \mid X + Y = t) = \frac{P(Y = y \text{ and } X + Y = t)}{P(X + Y = t)}$$

$$= \frac{P(X = t - y,\ Y = y)}{P(X + Y = t)}$$

$$= \frac{P(X = t - y)P(Y = y)}{P(X + Y = t)} \qquad \text{by independence.}$$

Recall that $X + Y$ has a modified negative binomial pmf with parameters 2 and p [see formula (2.10b)] while X and Y have the pmf given by (2.8). Therefore:

$$P(Y = y \mid X + Y = t) = \frac{p(1-p)^{t-y}p(1-p)^y}{p^2(1+t)(1-p)^t}$$

$$= \frac{1}{t+1}.$$

Thus given that the total serial execution time was t units, the execution time of the second module is distributed uniformly over $\{0, 1, \ldots, t\}$.

Problems

1. Consider two program segments:

```
S1: while B1 do
    begin
          writeln('hey you!');
          writeln('finished')
    end
```

and

```
S2: if B2 then writeln('hey you!')
         else writeln('finished')
```

Assuming that B_1 is true with probability p_1 and B_2 is true with probability p_2, compute the pmf of the number of times "hey you!" is printed and compute the pmf of the number of times "finished" is printed by the following program:

```
begin S1; S2 end
```

2. Complete the proofs of parts (b) and (c) of Theorem 2.2.

3. Prove Theorem 2.2 for $r = 2$ without using generating functions—that is, directly using the convolution formula for the pmf of the sum of two independent random variables.

4. Reconsider the example of a program with two modules and assume that respective module execution times X and Y are independent random variables uniformly distributed over $\{1, 2, \ldots, n\}$. Find:
   (a) $P(X \geqslant Y)$.
   (b) $P(X = Y)$.
   (c) The pmf and the PGF of $Z_1 = X + Y$.
   (d) The pmf of $Z_2 = \max\{X, Y\}$.
   (e) The pmf of $Z_3 = \min\{X, Y\}$.

5. Compute the pmf and the CDF of $\max\{X, Y\}$ where X and Y are independent random variables such that X and Y are both Poisson distributed with parameter α.

*6. Consider a program that needs two stacks. We want to compare two different ways to allocate storage for the two stacks. The first is to separately allocate n locations to each stack. The second is to let the two stacks grow toward each other in a common area of memory consisting of N locations. If the required value of N is smaller than $2n$, then the latter solution is preferable to the former. Determine the required values of n and N so as to keep the probability of overflow below 5 percent, assuming:
   (a) The size of each stack is geometrically distributed with parameter p (use $p = \frac{1}{4}, \frac{1}{2}$, and $\frac{3}{4}$).
   (b) The size of each stack is Poisson distributed with parameter $\alpha = 0.5$.
   (c) The size of each stack is uniformly distributed over $\{1, 2, \ldots, 20\}$.

Review Problems

1. Consider the combinational switching circuit shown in Figure 2.P.1, with four inputs and one output. The switching function realized by the circuit is easily shown to be:

$$y = (x_1 \text{ and } x_2) \text{ or } \overline{(x_3 \text{ and } x_4)}.$$

Associate the random variable X_i with the switching variable x_i. Assuming that X_i $(i = 1, 2, 3, 4)$ is a Bernoulli random variable with parameter p_i, compute the pmf of the output random variable Y.

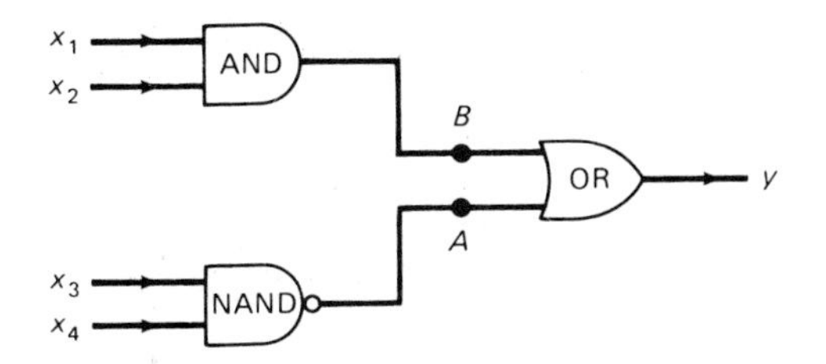

Figure 2.P.1 A combinational circuit

If a fault develops then the pmf of Y will change. Assume that only single faults of stuck-at-1 or stuck-at-0 type occur at any one of the input points, at the output point, or at internal points A and B. Compute the pmf of Y for each of these fourteen faulty conditions, assuming that $p_i = b$ for each $i = 1, 2, 3, 4$.

2. [WETH 1980] Consider a context-free language $L = \{a^n b^n \mid n \geqslant 0\}$ as the sample space of an experiment and define the random variable X that maps the string (sample point) $a^n b^n$ into the integer n. Determine the value of the constant k such that the pmf of X is $p_X(n) = k/n!$.

*3. [BURK 1963] In designing a parallel binary adder, we are interested in analyzing the length of the longest carry sequence. Assume that the two n-bit integer operands X_n and Y_n are independent random variables, uniformly distributed over $\{0, 1, \ldots, 2^n - 1\}$. Let the random variable V_n denote the length of the longest carry sequence while adding X_n and Y_n. Let the pmf of V_n be denoted by $p_n(v)$ and let $R_n(v) = \sum_{j=v}^{n} p_n(j)$. Define $R_n(v) = 0$ if $v > n$. Show that:

$$R_n(v) = R_{n-1}(v) + \frac{1-R_{n-v}(v)}{2^{v+1}}, \qquad v \leqslant n.$$

Further show that $R_n(v) \leqslant \min\{1, (n - v + 1)/2^{v+1}\}$.

References

[BURK 1963] A. W. BURKS, H. H. GOLDSTINE, and J. VON NEUMANN, "Preliminary Discussion of the Logical Design of an Electronic Computing Instrument," in A. H. Taub, ed., *Collected Works of John von Neumann*, Vol. 5, The Macmillan Company, New York.

[COFF 1973] E. G. COFFMAN, Jr., and P. J. DENNING, *Operating System Theory*, Prentice-Hall, Englewood Cliffs, N.J.

[KNUT 1973] D.E. KNUTH, *The Art of Computer Programming*, Vol. I, *Fundamental Algorithms*, Addison-Wesley, Reading, Mass.

[LARS 1974] H. J. LARSON, *Introduction to Probability Theory and Statistical Inference*, John Wiley & Sons, New York.

[NBS 1950] NATIONAL BUREAU OF STANDARDS, *Tables of the Binomial Distribution*, U.S. Government Printing Office, Washington, D.C.

[PEAR 1966] E. S. PEARSON and H. O. HARTLEY, *Biometrika Tables for Statisticians*, Cambridge University Press, Cambridge.

[ROMI 1953] H. G. ROMIG, *50-100 Binomial Tables*, John Wiley & Sons, New York.

[WETH 1980] C. J. WETHERELL, "Probabilistic Languages: A Review and Some Open Questions," *ACM Computing Surveys*, 12:4 (December 1980), 361–80.

Chapter 3

Continuous Random Variables

3.1 INTRODUCTION

So far, we have considered discrete random variables and their distributions. In applications, such random variables denote the number of objects of certain type, such as the number of job arrivals to a computing center in one hour or the number of calls into a telephone exchange in one minute.

Many situations, both applied and theoretical, require the use of random variables that are "continuous" rather than discrete. As described in the last chapter, a random variable is a real-valued function on the sample space S. When the sample space S is nondenumerable (as mentioned in Section 1.7), not every subset of the sample space is an event that can be assigned a probability. As before, let $\mathcal{F}$ denote the class of measurable subsets of S. Now, if X is to be a random variable, it is natural to require that $P(X \leqslant x)$ be well defined for every real number x. In other words, if X is to be a random variable defined on a probability space $(S, \mathcal{F}, P)$, we require that $\{s \mid X(s) \leqslant x\}$ be an event (that is, a member of $\mathcal{F}$). We are, therefore, led to the following extension of our earlier definition:

Definition (Random Variable). A random variable X on a probability space $(S, \mathcal{F}, P)$ is a function $X: S \longrightarrow \mathbb{R}$ that assigns a real number $X(s)$ to each sample point $s \in S$, such that for every real number x, the set $\{s \mid X(s) \leqslant x\}$ is an event; that is, a member of $\mathcal{F}$.

Definition (Distribution Function). The (cumulative) distribution function F_X of a random variable X is defined to be the function

$$F_X(x) = P(X \leqslant x), \qquad -\infty < x < \infty.$$

The subscript X is used here to indicate the random variable under consideration. When there is no ambiguity the subscript will be dropped, and $F_X(x)$ will be denoted by $F(x)$.

As we saw in Chapter 2, the distribution function of a discrete random variable grows only by jumps. By contrast, the distribution function of a continuous random variable has no jumps but grows continuously. Thus, a **continuous random variable** X is characterized by a distribution function $F_X(x)$ that is a continuous function of x for all $-\infty < x < \infty$. Most continuous random variables that we encounter will have an absolutely continuous distribution function, $F(x)$, that is, one for which the derivative, $dF(x)/dx$, exists everywhere (except perhaps at a finite number of points). Such a random variable is called **absolutely continuous**. Thus, for instance, the continuous uniform distribution, given by

$$F(x) = \begin{cases} 0, & x < 0, \\ x, & 0 \leqslant x < 1, \\ 1, & x \geqslant 1. \end{cases}$$

possesses a derivative at all points except at $x = 0$ and $x = 1$. Therefore, it is an absolutely continuous distribution. All continuous random variables that we will study are absolutely continuous and hence the adjective will be dropped.

Definition (Probability Density Function). For a continuous random variable, X, $f(x) = dF(x)/dx$ is called the probability density function (pdf or density function) of X.

The pdf enables us to obtain the CDF by integrating under the pdf:

$$F_X(x) = P(X \leqslant x) = \int_{-\infty}^{x} f_X(t)\, dt, \qquad -\infty < x < \infty.$$

We can also obtain other probabilities of interest such as,

$$\begin{aligned} P(X \in (a, b]) &= P(a < X \leqslant b) \\ &= P(x \leqslant b) - P(X \leqslant a) \\ &= \int_{-\infty}^{b} f_X(t)\, dt - \int_{-\infty}^{a} f_X(t)\, dt \\ &= \int_{a}^{b} f_X(t)\, dt. \end{aligned}$$

The pdf, $f(x)$, satisfies the following properties:

(f1) $f(x) \geqslant 0$ for all x.

(f2) $\int_{-\infty}^{\infty} f(x)\, dx = 1$.

It should be noted that, unlike the pmf, the values of the pdf are not probabilities and thus it is perfectly acceptable if $f(x) > 1$ at a point x.

As is the case for the CDF of a discrete random variable, the CDF of a continuous random variable, $F(x)$, satisfies the following properties:

(F1) $0 \leqslant F(x) \leqslant 1, \ -\infty < x < \infty.$

(F2) $F(x)$ is a monotone nondecreasing function of x.

(F3) $\lim_{x \to -\infty} F(x) = 0$ and $\lim_{x \to +\infty} F(x) = 1.$

Unlike the CDF of a discrete random variable, the CDF of a continuous random variable does not have any jumps. Therefore, the probability associated with the event $[X = c] = \{s \mid X(s) = c\}$ is zero:

$$\text{(F4}'\text{)} \quad P(X = c) = P(c \leqslant X \leqslant c) = \int_c^c f_X(y)\, dy = 0.$$

This does not imply that the set $\{s \mid X(s) = c\}$ is empty, but that the probability assigned to this set is zero. As a consequence of the fact that $P(X = c) = 0$, we have:

$$\begin{aligned} P(a \leqslant X \leqslant b) &= P(a < X \leqslant b) = P(a \leqslant X < b) \\ &= P(a < X < b) \\ &= \int_a^b f_X(x)\, dx \\ &= F_X(b) - F_X(a). \end{aligned} \tag{3.1}$$

The relation between the functions f and F is illustrated in Figure 3.1. Probabilities are represented by areas under the pdf curve. The total area under the curve is unity.

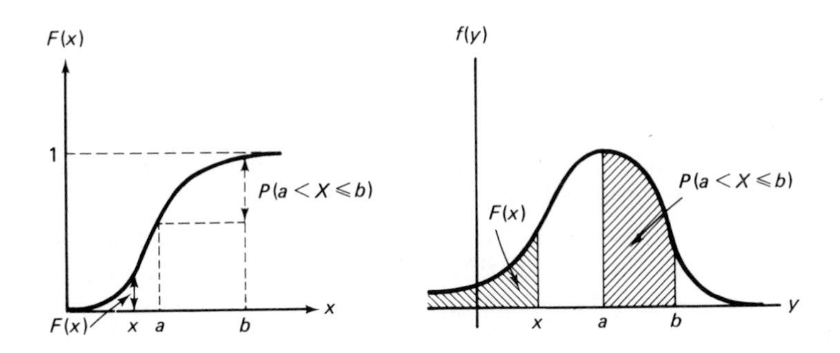

Figure 3.1 Relation between CDF and pdf

Example 3.1

The time (measured in years), X, required to complete a software project has a pdf of the form:

$$f_X(x) = \begin{cases} kx(1-x), & 0 \leqslant x \leqslant 1, \\ 0, & \text{otherwise.} \end{cases}$$

Since f_X satisfies property (f1), we know $k \geqslant 0$. In order for f_X to be a pdf, it must also satisfy property (f2); hence:

$$\int_0^1 kx(1-x)\,dx = 1$$

$$k\left(\frac{x^2}{2} - \frac{x^3}{3}\right)\Big|_0^1 = 1.$$

Therefore:

$$k = 6.$$

Now the probability that the project will be completed in less than four months is given by:

$$P(X < 4/12) = F_X(1/3) = \int_0^{1/3} f_X(x)\,dx = 7/27$$

or about a 26 percent chance. #

Most random variables we consider will either be discrete (as in Chapter 2) or continuous, but *mixed* random variables do occur sometimes. For example, there may be a nonzero probability, say, p_0, of initial failure of a component at time 0 due to manufacturing defects. In this case, the time to failure, X, of the component is neither a discrete nor a continuous random variable. A possible CDF of X (shown in Figure 3.2) is then:

$$F_X(x) = \begin{cases} 0, & x < 0, \\ p_0, & x = 0, \\ p_0 + (1-p_0)(1-e^{-\lambda x}), & x > 0. \end{cases} \tag{3.2}$$

The CDF of a mixed random variable satisfies properties (F1)–(F3) but it does not satisfy property (F4) of Chapter 2 or the property (F4′) above.

The distribution function of a mixed random variable can be written as a linear combination of two distribution functions, denoted by $F^{(d)}(.)$ and $F^{(c)}(.)$, which are discrete and continuous, respectively, so that for every real number x,

$$F_X(x) = \alpha_d F^{(d)}(x) + \alpha_c F^{(c)}(x)$$

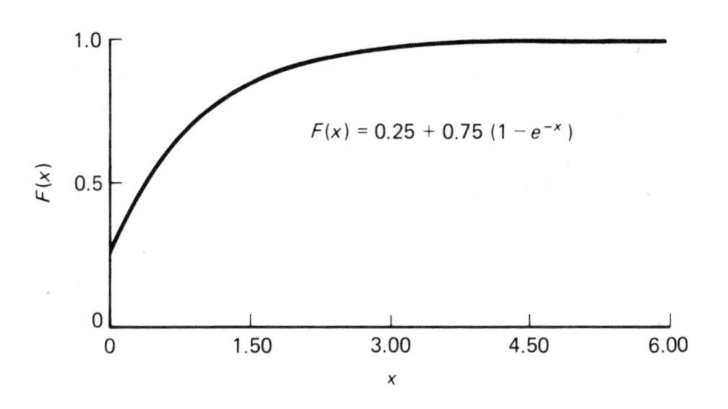

Figure 3.2 CDF of a mixed random variable

where $0 \leqslant \alpha_d$, $\alpha_c \leqslant 1$ and $\alpha_d + \alpha_c = 1$. Thus the mixed distribution (3.2) can be represented in this way if we let

$$F^{(d)}(x) = \begin{cases} 0, & x < 0, \\ 1, & x \geqslant 0, \end{cases}$$

$$F^{(c)}(x) = 1 - e^{-\lambda x}, \qquad x \geqslant 0,$$

$$\alpha_d = p_0\,, \text{ and } \alpha_c = 1 - p_0.$$

(A unified treatment of discrete, continuous, and mixed random variables can also be given through the use of the Riemann-Stieltjes Integral [BREI 1968], [RUDI 1964].)

Problems

1. Find the value of the constant k so that:

$$f(x) = \begin{cases} kx^2(1 - x^3), & 0 < x < 1, \\ 0, & \text{otherwise,} \end{cases}$$

is a proper density function of a continuous random variable.

2. Let X be the random variable denoting the time to failure of a component. Suppose the distribution function of X is $F(x)$. Use this distribution function to express the probability of the following events:

   a. $9 < X < 90$.
   b. $X < 90$.
   c. $X > 90$, given that $X > 9$.

3. Consider a random variable X defined by the CDF:

$$F_X(x) = \begin{cases} 0, & x < 0, \\ \tfrac{1}{2}\sqrt{x} + \tfrac{1}{2}(1 - e^{-\sqrt{x}}), & 0 \leqslant x \leqslant 1, \\ \tfrac{1}{2} + \tfrac{1}{2}(1 - e^{-\sqrt{x}}), & x > 1. \end{cases}$$

Show that this function satisfies properties (F1)–(F3) and (F4′). Note that F_X is a continuous function but it does not have a derivative at $x = 1$. (That is, the pdf of X has a discontinuity at $x = 1$.) Plot the CDF and the pdf of X.

4. [HAMM 1973] Consider a normalized floating-point number in base (or radix) β so that the mantissa, X, satisfies the condition $1/\beta \leqslant X < 1$. Experience shows that X has the **reciprocal density**:

$$f_X(x) = \frac{k}{x}, \qquad k > 0.$$

Determine:

a. The value of k.
b. The distribution function of X.
c. The probability that the leading digit of X is i for $1 \leqslant i < \beta$.

3.2 THE EXPONENTIAL DISTRIBUTION

This distribution, sometimes called the negative exponential distribution, occurs in applications such as reliability theory and queuing theory. Reasons for its use include its memoryless (Markov) property (and resulting analytical tractability) and its relation to the (discrete) Poisson distribution. Thus the following random variables will often be modeled as exponential:

1. Time between two successive job arrivals to a computing center (often called interarrival time).
2. Service time at a server in a queuing network; the server could be a resource such as the CPU, I/O device, or a communication channel.
3. Time to failure (lifetime) of a component.
4. Time required to repair a component that has malfunctioned.

Note that the assertion "Above distributions are exponential" is not a given fact but an assumption. Experimental verification of this assumption must be sought before relying on the results of the analysis (see Chapter 10 for further elaboration on this topic).

The **exponential distribution function**, shown in Figure 3.3, is given by:

$$F(x) = \begin{cases} 1 - e^{-\lambda x}, & \text{if } 0 \leqslant x < \infty, \\ 0, & \text{otherwise.} \end{cases} \tag{3.3}$$

If a random variable X possesses CDF given by equation (3.3), we use the notation $X \sim \text{EXP}(\lambda)$, for brevity. The pdf of X has the shape shown in Figure 3.4 and is given by:

$$f(x) = \begin{cases} \lambda e^{-\lambda x}, & \text{if } x > 0 \\ 0, & \text{otherwise.} \end{cases} \tag{3.4}$$

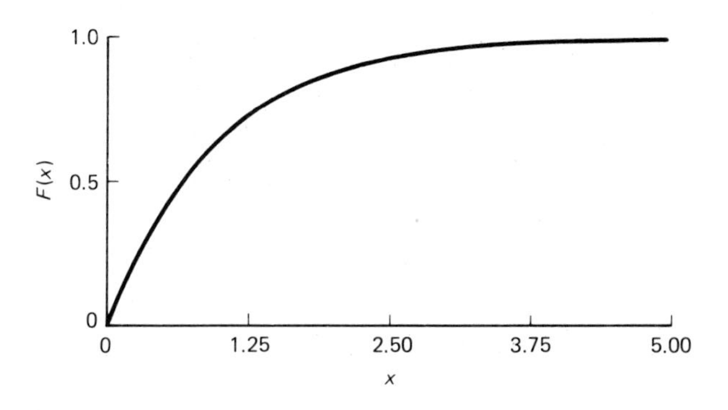

Figure 3.3 The CDF of an exponentially distributed random variable with parameter $\lambda = 1$

While specifying a pdf, usually we state only the nonzero part, and it is understood that the pdf is zero over any unspecified region. Since $\lim_{x \to \infty} F(x) = 1$, it follows that the total area under the exponential pdf is unity. Also:

$$P(X \geqslant t) = \int_t^\infty f(x)\,dx \qquad (3.5)$$

$$= e^{-\lambda t}$$

and

$$P(a \leqslant X \leqslant b) = F(b) - F(a)$$

$$= e^{-\lambda a} - e^{-\lambda b}.$$

Now let us investigate the so-called **memoryless** or **Markov property** of the exponential distribution. Suppose we know that X exceeds some given value t; that is, $X > t$. For example, let X be the lifetime of a component,

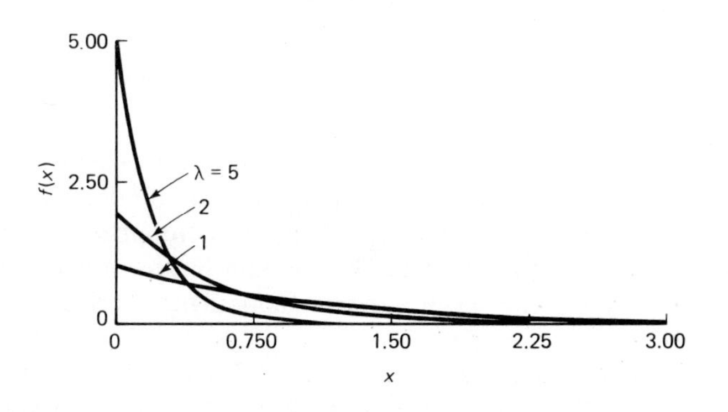

Figure 3.4 Exponential pdf

and suppose we have observed that this component has already been operating for t hours. We may then be interested in the distribution of $Y = X - t$, the remaining (residual) lifetime. Let the conditional probability of $Y \leqslant y$, given that $X > t$, be denoted by $G_t(y)$. Thus for $y \geqslant 0$ we have:

$$\begin{aligned} G_t(y) &= P(Y \leqslant y | X > t) \\ &= P(X - t \leqslant y | X > t) \\ &= P(X \leqslant y + t | X > t) \\ &= \frac{P(X \leqslant y + t \text{ and } X > t)}{P(X > t)} \end{aligned}$$

by the definition of conditional probability,

$$= \frac{P(t < X \leqslant y + t)}{P(X > t)}.$$

Thus (see Figure 3.5):

$$\begin{aligned} G_t(y) &= \frac{\int_t^{y+t} f(x)\, dx}{\int_t^{\infty} f(x)\, dx} \\ &= \frac{\int_t^{y+t} \lambda e^{-\lambda x}\, dx}{\int_t^{\infty} \lambda e^{-\lambda x}\, dx} \\ &= \frac{e^{-\lambda t}(1 - e^{-\lambda y})}{e^{-\lambda t}} \\ &= 1 - e^{-\lambda y}. \end{aligned}$$

Thus $G_t(y)$ is independent of t and is identical to the original exponential distribution of X. The distribution of the remaining life does not depend on how long the component has been operating. The component does not "age" (it is as good as new) or it "forgets" how long it has been operating, and its eventual breakdown is the result of some suddenly appearing failure, not of gradual deterioration.

If the interarrival times are exponentially distributed, then the memoryless property implies that the time we must wait for a new arrival is statistically independent of how long we have already spent waiting for it.

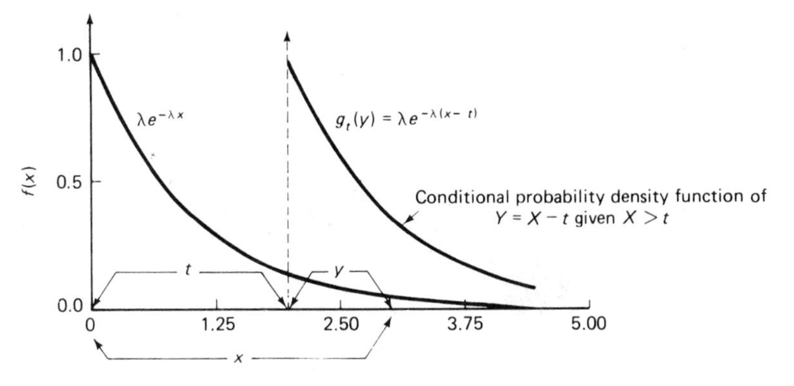

Figure 3.5 Memoryless property of the exponential distribution ($\lambda = 1$)

If X is a nonnegative continuous random variable with the Markov property, then we can show that the distribution of X must be exponential:

$$\frac{P(t < X \leqslant y + t)}{P(X > t)} = P(X \leqslant y) = P(0 < X \leqslant y)$$

or

$$F_X(y + t) - F_X(t) = [1 - F_X(t)]\,[F_X(y) - F_X(0)].$$

Since $F_X(0) = 0$, we rearrange the above equation to get:

$$\frac{F_X(y + t) - F_X(y)}{t} = \frac{F_X(t)\,[1 - F_X(y)]}{t}.$$

Taking the limit as t approaches zero, we get:

$$F_X'(y) = F_X'(0)\,[1 - F_X(y)],$$

where F_X' denotes the derivative of F_X. Let $R_X(y) = 1 - F_X(y)$; then the above equation reduces to:

$$R_X'(y) = R_X'(0)\,R_X(y).$$

The solution to this differential equation is given by:

$$R_X(y) = Ke^{R_X'(0)y},$$

where K is a constant of integration and $-R_X'(0) = F_X'(0) = f_X(0)$, the pdf evaluated at 0. Noting that the $R_X(0) = 1$, and denoting $f_X(0)$ by the constant λ, we get:

$$R_X(y) = e^{-\lambda y}$$

and hence

$$F_X(y) = 1 - e^{-\lambda y}, \qquad y > 0.$$

Therefore X must have the exponential distribution.

The exponential distribution can be obtained from the Poisson distribution by considering the interarrival times rather than the number of arrivals.

Example 3.2

Let the discrete random variable N_t denote the number of jobs arriving to a computer system in the interval $(0, t]$. Let X be the time of the next arrival. Further assume that N_t is Poisson distributed with parameter λt, so that λ is the arrival rate. Then:

$$\begin{aligned} P(X > t) &= P(N_t = 0) \\ &= \frac{e^{-\lambda t}(\lambda t)^0}{0!} \\ &= e^{-\lambda t} \end{aligned}$$

and

$$F_X(t) = 1 - e^{-\lambda t}.$$

Therefore, the time to the next arrival is exponentially distributed. More generally, it can be shown that the interarrival times of Poisson events are exponentially distributed [BHAT 1972, p. 126]. #

Example 3.3

Consider a university computer center with an average rate of job submission $\lambda = 0.1$ jobs per second. Assuming that the number of arrivals per unit time is Poisson distributed, the interarrival time, X, is exponentially distributed with parameter λ. The probability that an interval of 10 seconds elapses without job submission is then given by

$$\begin{aligned} P(X \geqslant 10) &= \int_{10}^{\infty} 0.1 e^{-0.1t}\,dt = \lim_{t \to \infty} [-e^{-0.1t} - e^{-1}] \\ &= e^{-1} = 0.368. \end{aligned}$$

#

Problems

1. Jobs arriving at a computer system have been found to require CPU time that can be modelled by an exponential distribution with parameter 1/140 per millisecond. The CPU scheduling discipline is quantum-oriented so that a job not completing within a quantum of 100 milliseconds will be routed back to the tail of the queue of waiting jobs. Find the probability that an arriving job will be forced to wait for a second quantum. Of the 800 jobs coming in during a day, how many are expected to finish within the first quantum ?

3.3 THE RELIABILITY, FAILURE DENSITY, AND HAZARD FUNCTION

Let the random variable X be the lifetime or the time to failure of a component. The probability that the component survives until some time t is called the **reliability** $R(t)$ of the component. Thus, $R(t) = P(X > t) = 1 - F(t)$, where F is the distribution function of the component lifetime, X. The component is assumed to be working properly at time $t = 0$ [i.e., $R(0) = 1$] and no component can work forever without failure [i.e., $\lim_{t \to +\infty} R(t) = 0$]. Also, $R(t)$ is a monotone nonincreasing function of t. For t less than zero, reliability has no meaning, but we let $R(t) = 1$ for $t < 0$. $F(t)$ will often be called the **unreliability**.

Consider a fixed number of identical components, N_0, under test. After time t, $N_f(t)$ components have failed and $N_s(t)$ components have survived with $N_f(t) + N_s(t) = N_0$. The estimated probability of survival may be written (using the frequency interpretation of probability) as:

$$\hat{P}(\text{survival}) = \frac{N_s(t)}{N_0}.$$

In the limit as $N_0 \to \infty$, we expect $\hat{P}$(survival) to approach $R(t)$. As the test progresses, $N_s(t)$ gets smaller and $R(t)$ decreases:

$$\begin{aligned} R(t) &\simeq \frac{N_s(t)}{N_0} \\ &= \frac{N_0 - N_f(t)}{N_0} \\ &= 1 - \frac{N_f(t)}{N_0}. \end{aligned}$$

The total number of components N_0 is constant, while the number of failed components N_f increases with time. Taking derivatives on both sides of the above equation, we get:

$$R'(t) \simeq -\frac{1}{N_0} N_f'(t). \tag{3.6}$$

In equation (3.6), $N_f'(t)$ is the rate at which components fail. Therefore, as $N_0 \to \infty$, the right hand side may be interpreted as the negative of the failure density function, $f_X(t)$:

$$R'(t) = -f_X(t). \tag{3.7}$$

Note that $f(t)\Delta t$ is the (unconditional) probability that a component will fail in the interval $(t, t + \Delta t]$. However, if we have observed the component functioning up to some time t, we expect the (conditional) probability of its failure in the interval to be different from $f(t)\Delta t$. This leads us to the notion of instantaneous failure rate as follows.

Notice that the conditional probability that the component does not survive for an (additional) interval of duration x given that it has survived until time t can be written as:

$$G_t(x) = \frac{P(t < X < t + x)}{P(X > t)} = \frac{F(t + x) - F(t)}{R(t)}. \tag{3.9}$$

Definition (Instantaneous Failure Rate). The instantaneous failure rate $h(t)$ at time t is defined to be:

$$h(t) = \lim_{x \to 0} \frac{1}{x} \frac{F(t + x) - F(t)}{R(t)} = \lim_{x \to 0} \frac{R(t) - R(t + x)}{xR(t)}$$

so that

$$h(t) = \frac{f(t)}{R(t)}. \tag{3.10}$$

Thus, $h(t)\Delta t$ represents the conditional probability that a component surviving to age t will fail in the interval $(t, \; t + \Delta t]$. Alternate names for $h(t)$ are hazard rate, force of mortality, intensity rate, conditional failure rate, or simply failure rate.

It should be noted that the exponential distribution is characterized by a constant failure rate, since:

$$h(t) = \frac{f(t)}{R(t)} = \frac{\lambda e^{-\lambda t}}{e^{-\lambda t}} = \lambda.$$

By integrating both sides of equation (3.10), we get:

$$\begin{aligned}\int_0^t h(x)\, dx &= \int_0^t \frac{f(x)}{R(x)}\, dx \\ &= \int_0^t -\frac{R'(x)}{R(x)}\, dx \qquad \text{using equation (3.7)} \\ &= -\int_{R(0)}^{R(t)} \frac{dR}{R}\end{aligned}$$

or

$$-\ln R(t) = \int_0^t h(x)\, dx$$

using the boundary condition, $R(0) = 1$. Therefore:

$$R(t) = \exp\left[-\int_0^t h(x)\, dx\right]. \tag{3.11}$$

This formula holds even when the distribution of the time to failure is not exponential.

The cumulative failure rate, $H(t) = \int_0^t h(x)\,dx$, is referred to as the **cumulative hazard**. Equation (3.11) gives a useful theoretical representation of reliability as a function of the failure rate. An alternate representation gives the reliability in terms of cumulative hazard:

$$R(t) = e^{-H(t)}. \tag{3.12}$$

Note that if the lifetime is exponentially distributed, then $H(t) = \lambda t$ and we obtain the exponential reliability function.

We should note the difference between $f(t)$ and $h(t)$. The quantity $f(t)\Delta t$ is the unconditional probability that the component will fail in the interval $(t, t + \Delta t]$, whereas $h(t)\,\Delta t$ is the conditional probability that the component will fail in the same time interval, given that it has survived until time t. Also, $h(t) = f(t)/R(t)$ is always greater than or equal to $f(t)$, because $R(t) \leqslant 1$. $f(t)$ is a probability density whereas $h(t)$ is not. By analogy, the probability that a newborn child will die at an age between 99 and 100 years [corresponding to $f(t)\,\Delta t$] is quite small because few of them will survive that long. But the probability of dying in that same period, provided that the child has survived until age 99 [corresponding to $h(t)\,\Delta t$] is much greater.

To further see the difference between the failure rate $[h(t)]$ and failure density $[f(t)]$, we need the notion of conditional probability density. Let $V_t(x)$ denote the conditional distribution of the lifetime X given that the component has survived past fixed time t. Then:

$$V_t(x) = \frac{\int_t^x f(y)\,dy}{P(X > t)}$$

$$= \begin{cases} \dfrac{F(x) - F(t)}{1 - F(t)}, & x \geqslant t, \\ 0, & x < t\,. \end{cases}$$

[Note that $V_t(x) = G_t(x - t)$.]

Then the conditional failure density is:

$$v_t(x) = \begin{cases} \dfrac{f(x)}{1 - F(t)}, & x \geqslant t, \\ 0 & x < t. \end{cases}$$

The conditional density $v_t(x)$ satisfies properties (f1) and (f2) and hence is a probability density while the failure rate $h(t)$ does not satisfy property (f2) since:

$$0 = \lim_{t \to \infty} R(t) = \exp\left[-\int_0^\infty h(t)\,dt\right].$$

[Note that $h(t) = v_t(t)$.]

Define the **conditional reliability** $R_t(y)$ to be the probability that the component survives an (additional) interval of duration y given that it has survived until time t. Thus:

$$R_t(y) = \frac{R(t+y)}{R(t)}. \qquad (3.13)$$

[Note that $R_t(y) = 1 - G_t(y)$.]

Now consider a component that does not age stochastically. In other words, its survival probability over an additional period of length y is the same regardless of its present age:

$$R_t(y) = R(y) \qquad \text{for all } y,t \geqslant 0.$$

Then, using formula (3.13), we get:

$$R(y+t) = R(t)\,R(y), \qquad (3.14)$$

and rearranging, we get:

$$\frac{R(y+t) - R(y)}{t} = \frac{[R(t) - 1]\,R(y)}{t}.$$

Taking the limit as t approaches zero and noting that $R(0) = 1$, we obtain:

$$R'(y) = R'(0)\,R(y).$$

So $R(y) = e^{yR'(0)}$. Letting $R'(0) = -\lambda$, we get:

$$R(y) = e^{-\lambda y}, \qquad y > 0,$$

which implies that the lifetime $X \sim \text{EXP}(\lambda)$. In this case the failure rate $h(t)$ is equal to λ, which is a constant, independent of component age t. Conversely, the exponential lifetime distribution is the only distribution with a **constant failure rate** [BARL 1975]. If a component has exponential lifetime distribution, then it follows that:

1. Since a used component is (stochastically) as good as new, a policy of a scheduled replacement of used components (known to be still functioning) does not accrue any benefit.
2. In estimating mean life, reliability, and other such quantities, data may be collected consisting only of the number of hours of observed life and of

the number of observed failures; the ages of components under observation are of no concern.

Now consider a component that ages adversely in the sense that the conditional survival probability decreases with the age t; that is, $R_t(y)$ is decreasing in $0 < t < \infty$ for all $y \geqslant 0$. As a result:

$$h(t) = \lim_{y \to 0} \frac{R(t) - R(t+y)}{y\ R(t)}$$

is an increasing function of t for $t \geqslant 0$. The corresponding distribution function $F(t)$ is known as an **IFR (increasing failure rate) distribution**.

Alternately, if aging is beneficial in the sense that the conditional survival probability increases with age, then the failure rate will be a decreasing function of age, and the corresponding distribution is known as a **DFR (decreasing failure rate) distribution**.

The behavior of the failure rate as a function of age is known as the **mortality curve, hazard function, life characteristic,** or **lambda characteristic**. The mortality curve is empirically observed to have the so-called bathtub shape shown in Figure 3.6. During the early life period (infant mortality phase, burn-in period, debugging period, or break-in period), failures are of the **endogenous** type and arise from inherent defects in the system attributed to faulty design, manufacturing, or assembly. During this period, the failure rate is expected to drop with age.

When the system has been debugged, it is prone to chance or random failure (also called **exogenous** failure). Such failures are usually associated with environmental conditions under which the component is operating. They are the results of severe, unpredictable stresses arising from sudden environmental shocks, the failure rate being determined by the severity of the

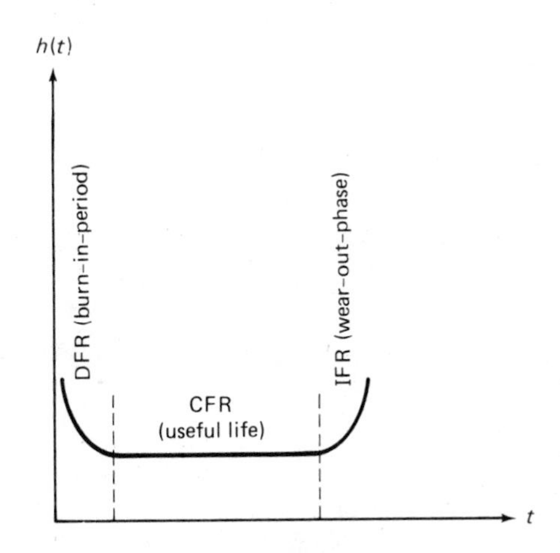

Figure 3.6 Failure rate as a function of time

environment. During this useful-life phase, failure rate is approximately constant and the exponential model is usually acceptable.

The rationale for the choice of exponential failure law is provided by assuming that the component is operating in an environment that subjects it to a stress varying in time. A failure occurs when the applied stress exceeds the maximum allowable stress, S_{max} (see Figure 3.7). Such "peak" stresses may be assumed to follow a Poisson distribution with parameter λt, where λ is a constant rate of occurrence of peak loads. Denoting the number of peak stresses in the interval $(0, t]$ by N_t, we get:

$$P(N_t = r) = \frac{e^{-\lambda t}(\lambda t)^r}{r!}, \qquad \lambda > 0, \; r = 0, 1, 2, \ldots.$$

Now the event $[X > t]$, where X is the component lifetime, corresponds to the event $[N_t = 0]$, and thus:

$$\begin{aligned} R(t) &= P(X > t) \\ &= P(N_t = 0) \\ &= e^{-\lambda t}, \end{aligned}$$

the exponential reliability function.

When components begin to reach their "rated life," the system failure rate begins to increase and it is said to have entered the wear-out phase. The wear-out failure is the outcome of accumulated wear and tear, a result of a depletion process due to abrasion, fatigue, creep, and the like.

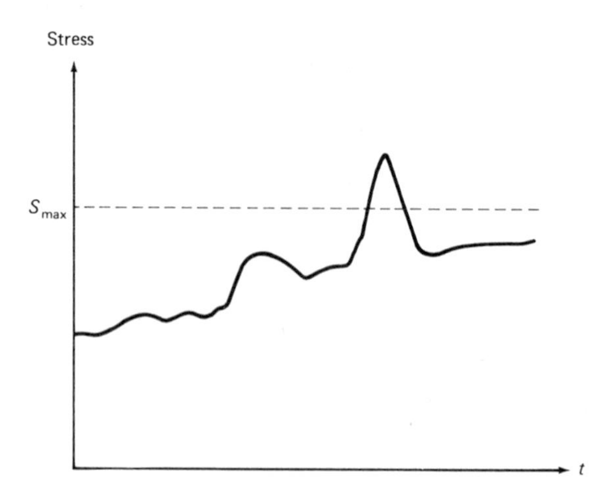

Figure 3.7 Stress as a function of time

Problems

1. The failure rate of a certain component is $h(t) = \lambda_0 t$, where $\lambda_0 > 0$ is a given constant. Determine the reliability, $R(t)$, of the component. Repeat for $h(t) = \lambda_0 t^{1/2}$.

2. The failure rate of a computer system for on-board control of a space vehicle is estimated to be the following function of time:

$$h(t) = \alpha\mu t^{\alpha-1} + \beta\gamma t^{\beta-1}.$$

Derive an expression for the reliability $R(t)$ of the system. Plot $h(t)$ and $R(t)$ as functions of time with parameter values $\alpha = 1/4$, $\beta = 1/7$, $\mu = 0.0004$ failures/hr, and $\gamma = 0.0007$ failures/hr.

3.4 SOME IMPORTANT DISTRIBUTIONS

3.4.1 Hypoexponential Distribution

Many processes in nature can be divided into sequential phases. If the time the process spends in each phase is independent and exponentially distributed, then it can be shown that the overall time is hypoexponentially distributed. It has been empirically observed that the service times for input-output operations in a computer system often possess this distribution. The distribution has r parameters, one for each of its distinct phases. A two-stage hypoexponential random variable, X, with parameters λ_1 and λ_2 ($\lambda_1 \neq \lambda_2$), will be denoted by $X \sim$ HYPO (λ_1, λ_2) and its pdf is given by (see Figure 3.8):

$$f(t) = \frac{\lambda_1\lambda_2}{\lambda_2 - \lambda_1}(e^{-\lambda_1 t} - e^{-\lambda_2 t}), \qquad t > 0. \tag{3.15}$$

The corresponding distribution function is (see Figure 3.9):

$$F(t) = 1 - \frac{\lambda_2}{\lambda_2 - \lambda_1} e^{-\lambda_1 t} + \frac{\lambda_1}{\lambda_2 - \lambda_1} e^{-\lambda_2 t}, \qquad t \geqslant 0. \tag{3.16}$$

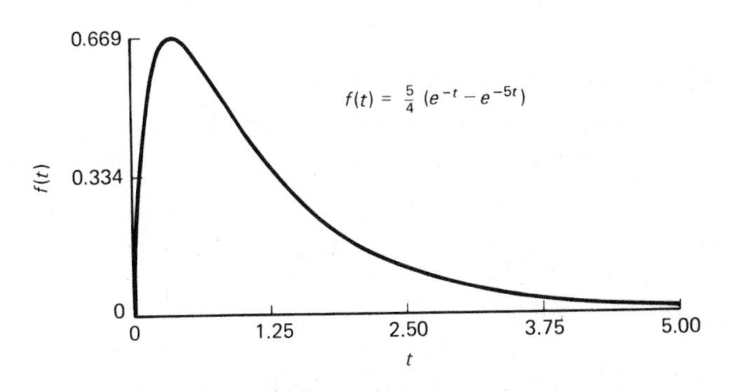

Figure 3.8 The pdf of the hypoexponential distribution

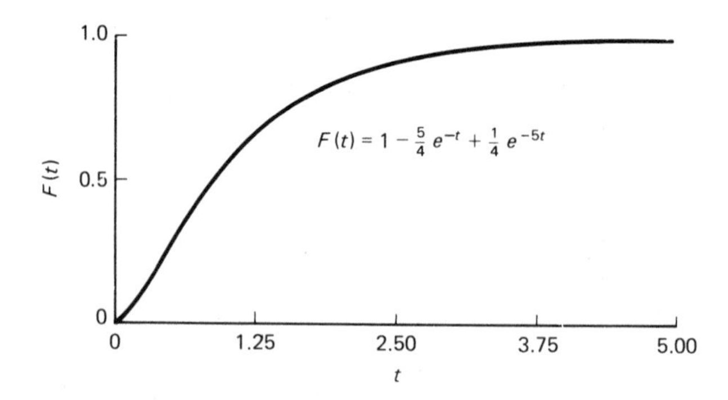

Figure 3.9 The CDF of the hypoexponential distribution

The hazard rate of this distribution is given by:

$$h(t) = \frac{\lambda_1 \lambda_2 (e^{-\lambda_1 t} - e^{-\lambda_2 t})}{\lambda_2 e^{-\lambda_1 t} - \lambda_1 e^{-\lambda_2 t}}. \tag{3.17}$$

It is not difficult to see that this is an IFR distribution with the failure rate increasing from 0 up to min $\{\lambda_1, \lambda_2\}$ (see Figure 3.10).

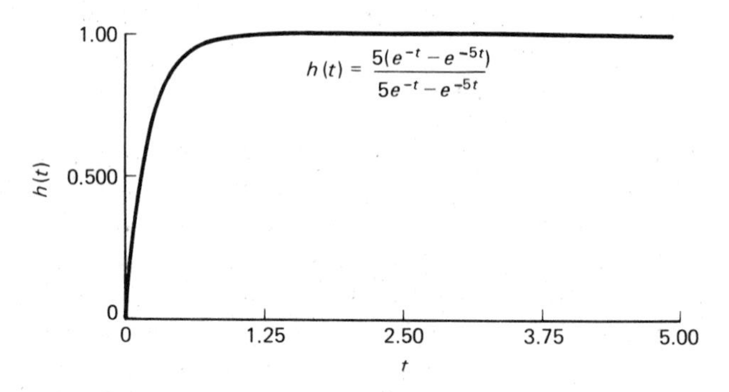

Figure 3.10 The failure rate of the hypoexponential distribution

3.4.2 Erlang and Gamma Distribution

When r sequential phases have independent identical exponential distributions, then the resulting density is known as r-stage (or r-phase) Erlang and is given by:

$$f(t) = \frac{\lambda^r t^{r-1} e^{-\lambda t}}{(r-1)!}, \qquad t > 0,\ \lambda > 0,\ r = 1, 2, \ldots. \tag{3.18}$$

The distribution function is:

$$F(t) = 1 - \sum_{k=0}^{r-1} \frac{(\lambda t)^k}{k!} e^{-\lambda t}, \qquad t \geqslant 0,\ \lambda > 0,\ r = 1, 2, \ldots. \tag{3.19}$$

Also:

$$h(t) = \frac{\lambda^r t^{r-1}}{(r-1)! \sum_{k=0}^{r-1} \frac{(\lambda t)^k}{k!}}, \qquad t > 0,\ \lambda > 0,\ r = 1, 2, \ldots. \tag{3.20}$$

The exponential distribution is a special case of the Erlang distribution with $r = 1$. The physical interpretation of this distribution is a natural extension of that for the exponential. Consider a component subjected to an environment so that N_t, the number of peak stresses in the interval $(0, t]$, is Poisson distributed with parameter λt. Suppose further that the component can withstand $(r-1)$ peak stresses and the rth occurrence of a peak stress causes a failure. Then the component lifetime X is related to N_t so that the following two events are equivalent:

$$[X > t] = [N_t < r]$$

thus

$$\begin{aligned} R(t) &= P(X > t) \\ &= P(N_t < r) \\ &= \sum_{k=0}^{r-1} P(N_t = k) \\ &= e^{-\lambda t} \sum_{k=0}^{r-1} \frac{(\lambda t)^k}{k!}. \end{aligned}$$

Then $F(t) = 1 - R(t)$ yields the formula (3.19) above. We conclude that the component lifetime has an r-stage Erlang distribution.

If we let r (call it α) take nonintegral values, then we get the gamma density:

$$f(t) = \frac{\lambda^\alpha t^{\alpha-1} e^{-\lambda t}}{\Gamma\alpha}, \qquad \alpha > 0,\ t > 0. \tag{3.21}$$

where the **gamma function** is defined by the integral:

$$\Gamma\alpha = \int_0^\infty x^{\alpha-1} e^{-x}\, dx, \qquad \alpha > 0. \tag{3.22}$$

The following properties of the gamma function will be useful in the sequel. Integration by parts shows that for $\alpha > 1$:

$$\Gamma\alpha = (\alpha - 1)\,\Gamma(\alpha-1). \tag{3.23}$$

In particular, if α is a positive integer, denoted by n, then:

$$\Gamma n = (n-1)! \tag{3.24}$$

Other useful formulas related to the gamma function are:

$$\Gamma(\tfrac{1}{2}) = \sqrt{\pi} \tag{3.25}$$

and

$$\int_0^\infty x^{\alpha-1} e^{-\lambda x}\, dx = \frac{\Gamma\alpha}{\lambda^\alpha}. \tag{3.26}$$

A random variable X with pdf (3.21) will be denoted by $X \sim \text{GAM}(\lambda, \alpha)$. This distribution has two parameters. The parameter α is called a **shape parameter** since, as α increases, the density becomes more peaked. The parameter λ is a **scale parameter**; that is, the distribution depends on λ only through the product λt. The gamma distribution is DFR for $0 < \alpha < 1$ and IFR for $\alpha > 1$ (see Figure 3.11). For $\alpha = 1$, the distribution degenerates to the exponential distribution; that is, $\text{EXP}(\lambda) = \text{GAM}(\lambda, 1)$.

The gamma distribution is the continuous counterpart of the (discrete) negative binomial distribution. The chi-square distribution, useful in mathematical statistics, is a special case of the gamma distribution with $\alpha = n/2$ (n is a positive integer) and $\lambda = \tfrac{1}{2}$. Thus, if $X \sim \text{GAM}(\tfrac{1}{2}, n/2)$, then it is said to have a chi-square distribution with n degrees of freedom; that is, $X \sim X_n^2$. Figure 3.12 illustrates possible shapes of gamma density with $\alpha =$ 0.5, 1, and 10.

In Chapter 4 we will show that if a sequence of k random variables $X_1, X_2, \ldots, X_k$ are mutually independent and identically distributed as GAM (λ, α) then their sum $\sum_{i=1}^{k} X_i$ is distributed as $\text{GAM}(\lambda, k\alpha)$.

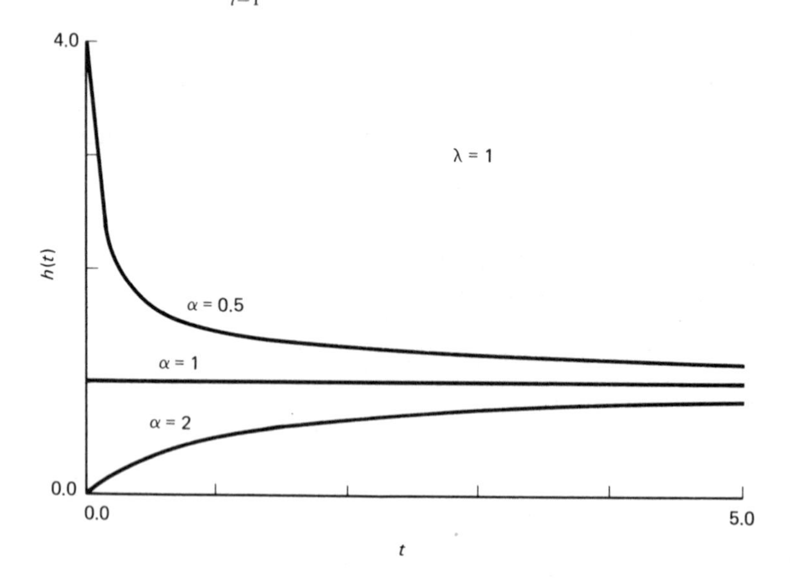

Figure 3.11 The failure rate of the gamma distribution

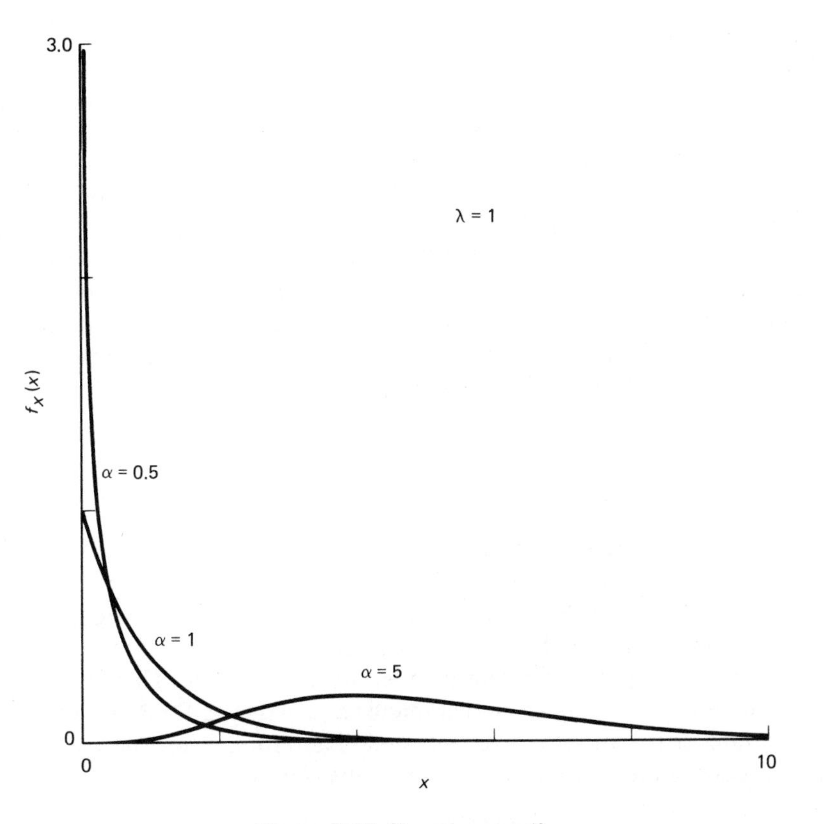

Figure 3.12 The gamma pdf

3.4.3 Hyperexponential Distribution

A process with sequential phases gives rise to a hypoexponential or an Erlang distribution, depending upon whether or not the phases have identical distributions. Instead, if a process consists of alternate phases—that is, during any single experiment, the process experiences one and only one of the many alternate phases—and these phases have independent exponential distributions, then the overall distribution is hyperexponential. The density function of a k-phase hyperexponential random variable is:

$$f(t) = \sum_{i=1}^{k} \alpha_i \lambda_i e^{-\lambda_i t}, \qquad t > 0,\ \lambda_i > 0,\ \alpha_i > 0,\ \sum_{i=1}^{k} \alpha_i = 1 \quad (3.27)$$

and the distribution function is:

$$F(t) = \sum_i \alpha_i (1 - e^{-\lambda_i t}), \qquad t \geqslant 0. \quad (3.28)$$

The failure rate is:

$$h(t) = \frac{\sum \alpha_i \lambda_i e^{-\lambda_i t}}{\sum \alpha_i e^{-\lambda_i t}}, \qquad t > 0, \tag{3.29}$$

which is a decreasing failure rate from $\sum \alpha_i \lambda_i$ down to min $\{\lambda_1, \lambda_2, \ldots\}$.

The hyperexponential distribution exhibits more variability than the exponential. CPU service-time distribution in a computer system has often been observed to possess such a distribution (see Figure 3.13). Similarly, if a product is manufactured in several parallel assembly lines and the outputs are merged, then the failure density of the overall product is likely to be hyperexponential. The hyperexponential is a special case of mixture distributions that often arise in practice—that is, of the form:

$$F(x) = \sum_i \alpha_i F_i(x), \qquad \sum \alpha_i = 1, \ \alpha_i \geqslant 0. \tag{3.30}$$

3.4.4 Weibull Distribution

The Weibull distribution has been used to describe fatigue failure, vacuum-tube failure, and ball-bearing failure. At present, it is perhaps the most widely used parametric family of failure distributions. The reason is that by a proper choice of its shape parameter α, an IFR, a DFR, or a constant failure rate distribution can be obtained. Therefore, it can be used for all three phases of the mortality curve. The density is given by:

$$f(t) = \lambda \alpha t^{\alpha-1} e^{-\lambda t^{\alpha}}, \tag{3.31}$$

the distribution function by:

$$F(t) = 1 - e^{-\lambda t^{\alpha}}, \tag{3.32}$$

the hazard rate by:

$$h(t) = \lambda \alpha t^{\alpha-1}, \tag{3.33}$$

and the cumulative hazard is a power function, $H(t) = \lambda t^{\alpha}$. For all the above formulas, $t \geqslant 0$, $\lambda > 0$, $\alpha > 0$. Figure 3.14 shows $h(t)$ plotted as a function of t, for various values of α.

Example 3.4

The lifetime X in hours of a component is modeled by a Weibull distribution with $\alpha = 2$. Starting with a large number of components, it is observed that 15 percent of the components that have lasted 90 hours fail before 100 hours. Determine the parameter λ.

Note that:

$$F_X(x) = 1 - e^{-\lambda x^2}$$

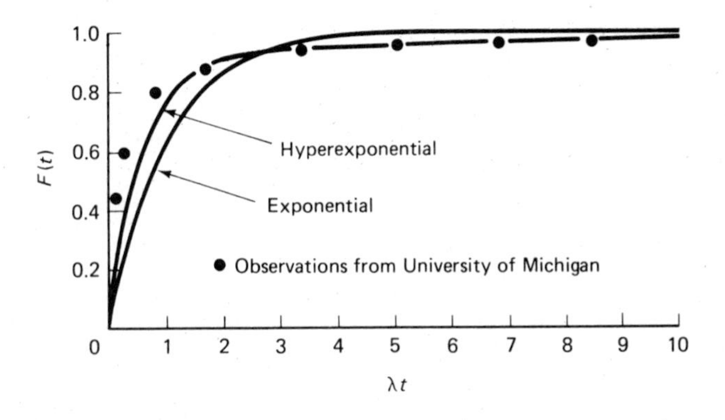

Figure 3.13 The CPU service-time distribution compared with the hyperexponential distribution. (Reproduced from R. F. Rosin, "Determining a Computing Center Environment," *CACM*, 1965; with permission of the Association of Computing Machinery.)

and we are given that

$$P(X < 100|X > 90) = 0.15.$$

Also:

$$P(X < 100|X > 90) = \frac{P(90 < X < 100)}{P(X > 90)}$$

$$= \frac{F_X(100) - F_X(90)}{1 - F_X(90)}$$

$$= \frac{e^{-\lambda(90)^2} - e^{-\lambda(100)^2}}{e^{-\lambda(90)^2}}.$$

Equating the two expressions and solving for λ, we get:

$$\lambda = -\ln(0.85)/1900 = 0.1625/1900 = 0.00008554.$$

#

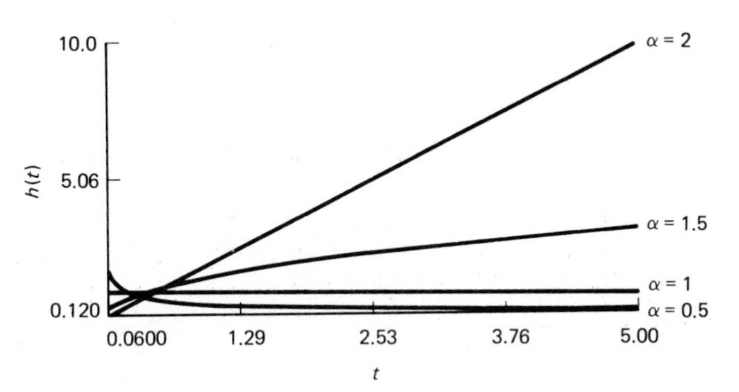

Figure 3.14 Failure rate of the Weibull distribution with various values of α and $\lambda = 1$

3.4.5 Normal or Gaussian Distribution

This distribution is extremely important in statistical applications because of the central limit theorem, which states that, under very general assumptions, the mean of a sample of n mutually independent random variables (having distributions with finite mean and variance) is normally distributed in the limit $n \to \infty$. It has been observed that errors of measurement often possess this distribution. Experience also shows that during the wear-out phase, component lifetime follows a normal distribution.

The normal density has the well-known bell-shaped curve (see Figure 3.15) and is given by:

$$f(x) = \frac{1}{\sigma\sqrt{2\pi}} \exp\left[-\frac{1}{2}\left(\frac{x-\mu}{\sigma}\right)^2\right], \qquad -\infty < x < \infty, \tag{3.34}$$

where $-\infty < \mu < \infty$ and $\sigma > 0$ are two parameters of the distribution. (We will see in Chapter 4 that these parameters are, respectively, the mean and the standard deviation of the distribution.) If a random variable X has the pdf (3.34), then we write $X \sim N(\mu, \sigma^2)$.

Since the distribution function $F(x)$ has no closed form, between every pair of limits a and b, probabilities relating to normal distributions are usually obtained numerically and recorded in special tables (see Appendix C). Such tables pertain to the **standard normal distribution** $[Z \sim N(0, 1)]$—a normal distribution with parameters $\mu = 0$ and $\sigma = 1$—and their entries are the values of:

$$F_Z(z) = \frac{1}{\sqrt{2\pi}} \int_{-\infty}^{z} e^{-t^2/2}\, dt. \tag{3.35}$$

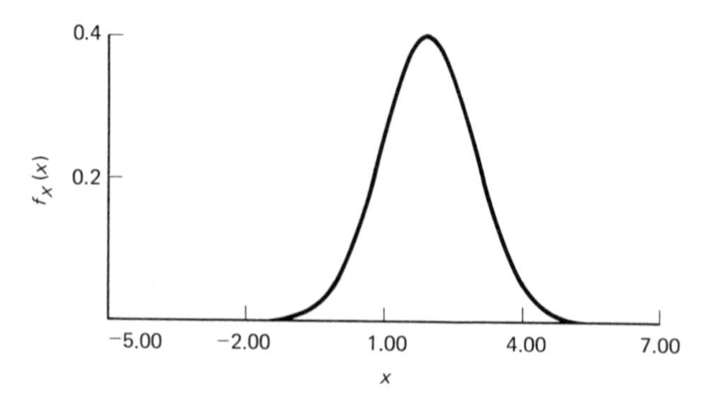

Figure 3.15 Normal density with parameters $\mu = 2$ and $\sigma = 1$

Since the standard normal density is clearly symmetric, it follows that for $z > 0$ we have:

$$
\begin{aligned}
F_Z(-z) &= \int_{-\infty}^{-z} f_Z(t)\, dt \\
&= \int_{z}^{\infty} f_Z(-t)\, dt \\
&= \int_{z}^{\infty} f_Z(t)\, dt \\
&= \int_{-\infty}^{\infty} f_Z(t)\, dt - \int_{-\infty}^{z} f_Z(t)\, dt \\
&= 1 - F_Z(z). \qquad (3.36)
\end{aligned}
$$

Therefore, the tabulations of the normal distribution are made only for $z \geqslant 0$. To find $P(a \leqslant Z \leqslant b)$, we use $F(b) - F(a)$.

For a particular value, x, of a normal random variable X, the corresponding value of the standardized variable Z is given by $z = (x - \mu)/\sigma$. The distribution function of X can now be found by using the relation:

$$
\begin{aligned}
F_Z(z) &= P(Z \leqslant z) \\
&= P(\frac{X - \mu}{\sigma} \leqslant z) \\
&= P(X \leqslant \mu + z\sigma) \\
&= F_X(\mu + z\sigma).
\end{aligned}
$$

Alternatively:

$$F_X(x) = F_Z(\frac{x - \mu}{\sigma}). \qquad (3.37)$$

Example 3.5

An analog signal received at a detector (measured in microvolts) may be modeled as a Gaussian random variable $N(200, 256)$ at a fixed point in time. What is the probability that the signal will exceed 240 microvolts? What is the probability that the signal is larger than 240 microvolts, given that it is larger than 210 microvolts?

$$
\begin{aligned}
P(X > 240) &= 1 - P(X \leqslant 240) \\
&= 1 - F_Z\left(\frac{240-200}{16}\right), \qquad \text{using equation (3.37)} \\
&= 1 - F_Z(2.5) \\
&\simeq 0.00621.
\end{aligned}
$$

Next:

$$P(X \geqslant 240 | X \geqslant 210) = \frac{P(X \geqslant 240)}{P(X \geqslant 210)}$$

$$= \frac{1 - F_Z\left(\dfrac{240 - 200}{16}\right)}{1 - F_Z\left(\dfrac{210 - 200}{16}\right)}$$

$$= \frac{0.00621}{0.26599}$$

$$\simeq 0.02335. \qquad \#$$

If X denotes the measured quantity in a certain experiment, then the probability of an event such as $\mu - k\sigma \leqslant X \leqslant \mu + k\sigma$ is an indicator of the measurement error:

$$P(\mu - k\sigma \leqslant X \leqslant \mu + k\sigma) = F_X(\mu + k\sigma) - F_X(\mu - k\sigma)$$

$$= F_Z(k) - F_Z(-k)$$

$$= 2F_Z(k) - 1$$

$$= \frac{2}{\sqrt{2\pi}} \int_{-\infty}^{k} e^{-t^2/2}\, dt - \frac{1}{\sqrt{2\pi}} \int_{-\infty}^{\infty} e^{-t^2/2}\, dt$$

$$= \frac{2}{\sqrt{2\pi}} \int_{-\infty}^{k} e^{-t^2/2}\, dt - \frac{2}{\sqrt{2\pi}} \int_{-\infty}^{0} e^{-t^2/2}\, dt$$

$$= \frac{2}{\sqrt{2\pi}} \int_{0}^{k} e^{-t^2/2}\, dt.$$

By the variable transformation, $t = \sqrt{2}y$, we get

$$P(\mu - k\sigma \leqslant X \leqslant \mu + k\sigma) = \frac{2}{\sqrt{\pi}} \int_{0}^{k/\sqrt{2}} e^{-y^2}\, dy.$$

The **error function** (or error integral) is defined by:

$$\operatorname{erf}(u) = \frac{2}{\sqrt{\pi}} \int_{0}^{u} e^{-y^2}\, dy. \tag{3.38}$$

Thus:

$$P(\mu - k\sigma \leqslant X \leqslant \mu + k\sigma) = \operatorname{erf}\left(\frac{k}{\sqrt{2}}\right). \tag{3.39}$$

For example, for $k = 3$, we obtain:

$$P(\mu - 3\sigma \leqslant X \leqslant \mu + 3\sigma) = 0.997.$$

Thus, a Gaussian random variable deviates from its mean by more than ± 3 standard deviations in only 0.3 percent of the trials, on the average. We often find tables of the error function rather than that of the CDF of the standard normal random variable.

Many physical experiments result in a nonnegative random variable, whereas the normal random variable takes negative values, as well. Therefore, it is of interest to define a **truncated normal density**:

$$f(x) = \begin{cases} 0, & x < 0, \\ \dfrac{1}{\alpha\sigma\sqrt{2\pi}} \exp\left[\dfrac{-(x-\mu)^2}{2\sigma^2}\right], & x \geqslant 0, \end{cases} \tag{3.40}$$

where:

$$\alpha = \int_0^{\infty} \frac{1}{\sigma\sqrt{2\pi}} \exp\left[\frac{-(t-\mu)^2}{2\sigma^2}\right] dt.$$

The introduction of α insures that $\int_{-\infty}^{\infty} f(t)\,dt = 1$, so that f is the density of a nonnegative random variable. For $\mu > 3\sigma$, the value of α is close to 1, and for most practical purposes it may be omitted, so that the truncated normal density reduces to the usual normal density.

The normal distribution is IFR (see Figure 3.16), which implies that it can be used to model the behavior of components during the wear-out phase.

Example 3.6

Assuming that the life of a given subsystem, in the wear-out phase, is normally distributed with $\mu = 10{,}000$ hours and $\sigma = 1{,}000$ hours, determine the reliability for an operating time of 500 hours given that (a) the age of the component is 9,000 hours, (b) the age of the component is 11,000 hours.

The required quantity under (a) is $R_{9,000}(500)$ and under (b) is $R_{11,000}(500)$. Note that with the usual exponential assumption these two quantities will be identical. But in the present case:

$$R_{9,000}(500) = \frac{R(9,500)}{R(9,000)} = \frac{\int_{9,500}^{\infty} f(t)\,dt}{\int_{9,000}^{\infty} f(t)\,dt}$$

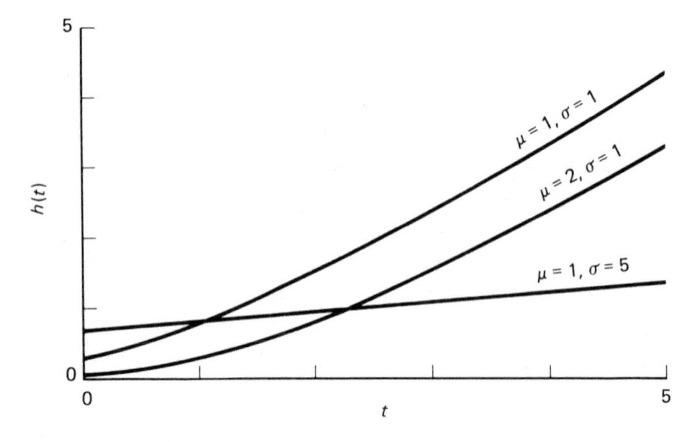

Figure 3.16 Failure rate of the normal distribution

Noting that $\mu - 0.5\sigma = 9{,}500$ and $\mu - \sigma = 9{,}000$, we have:

$$R_{9,000}(500) = \frac{\int_{\mu-0.5\sigma}^{\infty} f(t)\,dt}{\int_{\mu-\sigma}^{\infty} f(t)\,dt}$$

$$= \frac{1 - F_X(\mu - 0.5\sigma)}{1 - F_X(\mu - \sigma)}$$

$$= \frac{1 - F_Z(-0.5)}{1 - F_Z(-1)}$$

$$= \frac{F_Z(0.5)}{F_Z(1)}$$

$$= \frac{0.6915}{0.8413} \qquad \text{(from tables)}$$

$$= 0.8219.$$

Similarly, since $\mu + 1.5\sigma = 11{,}500$ and $\mu + \sigma = 11{,}000$, we have:

$$R_{11,000}(500) = \frac{1 - F_X(\mu + 1.5\sigma)}{1 - F_X(\mu+\sigma)}$$

$$= \frac{0.0668}{0.1587} \qquad \text{(from tables)}$$

$$= 0.4209.$$

Thus, unlike the exponential assumption, $R_{11,000}(500) < R_{9,000}(500)$; that is, the subsystem has aged. #

It can be shown that the normal distribution is a good approximation to the (discrete) binomial distribution for large n, provided p is not close to 0 or 1. The corresponding parameters are $\mu = np$ and $\sigma^2 = np(1 - p)$.

Example 3.7

Twenty percent of IC chips made in a certain plant are defective. Assuming that a binomial model is acceptable, the probability of at most thirteen rejects in a lot of 100 chosen for inspection may be computed by:

$$\sum_{x=0}^{13} b(x; 100, 0.20) = B(13; 100, 0.20).$$

Let us approximate this probability by using the normal distribution with $\mu = np = 20$ and $\sigma^2 = np(1 - p) = 16$. From Figure 3.17, observe that we are actually approximating the sum of the areas of the first fourteen rectangles of the histogram of the binomial distribution by means of the shaded area under the continuous curve. Thus, it is preferable to compute the area under the curve between -0.5 and 13.5 rather than between 0 and 13. Making this so-called **continuity correction**, we get:

$$\begin{aligned} B(13; 100, 0.2) &\simeq F_X(13.5) - F_X(-0.5) \\ &= F_Z\left(\frac{13.5 - 20}{4}\right) - F_Z\left(\frac{-20.5}{4}\right) \\ &= F_Z(-1.625) - F_Z(-5.125) \\ &= 0.0521 - 0, \end{aligned}$$

which compares favorably to the exact value of 0.046912. #

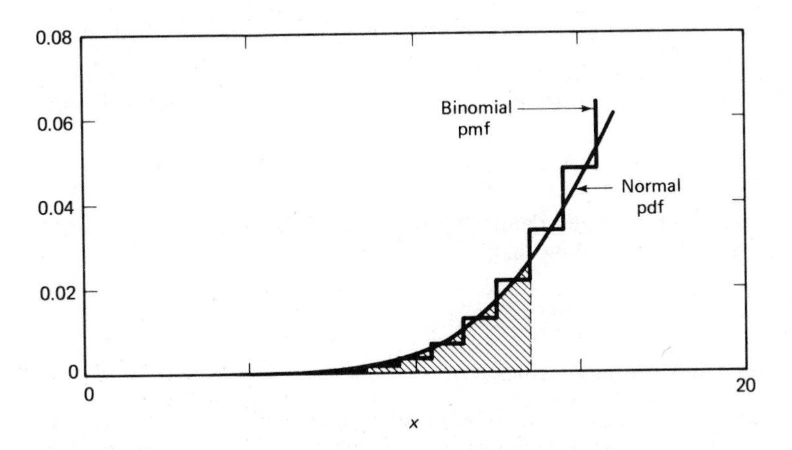

Figure 3.17 Normal approximation to the binomial pmf

3.4.6 The Uniform or Rectangular Distribution

A continuous random variable X is said to have a uniform distribution over the interval (a, b) if its density is given by:

$$f(x) = \begin{cases} \dfrac{1}{b-a}, & a < x < b, \\ 0, & \text{otherwise,} \end{cases} \tag{3.41}$$

and the distribution function is given by:

$$F(x) = \begin{cases} 0, & x < a, \\ \dfrac{x-a}{b-a}, & a \leqslant x < b, \\ 1, & x \geqslant b. \end{cases} \tag{3.42}$$

Problems

1. Lifetimes of IC chips manufactured by a semiconductor manufacturer are approximately normally distributed with $\mu = 5 \cdot 10^6$ hours and $\sigma = 5 \cdot 10^5$ hours. A mainframe manufacturer requires that at least 95 percent of a batch should have a lifetime greater than $4 \cdot 10^6$ hours. Will the deal be made?

2. Errors occur in data transmission over a binary communication channel due to white Gaussian noise. The probability of an error P_e, can be shown to be:

$$P_e = \frac{1}{2} - \frac{1}{\sqrt{\pi}} \int_0^u e^{-y^2}\, dy = \frac{1}{2}\,[1 - \text{erf}(u)]$$

where $z = u^2$ is a measure of the ratio of the signal power to noise power. The variable u is usually specified in terms of $10 \log_{10} z$ in units called dB. Plot P_e as a function of $10 \log_{10} z$.

3. Show that the failure rate $h(t)$ of the hypoexponential distribution has the property:

$$\lim_{t \to \infty} h(t) = \min\{\lambda_1, \lambda_2\}.$$

4. Show that a two-stage Erlang pdf is the limiting case of two-stage hypoexponential pdf. In other words show that:

$$\lim_{\lambda_1 \to \lambda_2} \frac{\lambda_1 \lambda_2}{\lambda_2 - \lambda_1} (e^{-\lambda_1 t} - e^{-\lambda_2 t}) = \lambda_2^2\, t e^{-\lambda_2 t}.$$

[Hint: Use L'Hospital's rule.]

5. The CPU time requirement of a typical program measured in minutes is found to follow a three-stage Erlang distribution with $\lambda = \frac{1}{2}$. What is the probability that the CPU demand of a program will exceed 1 minute?

3.5 FUNCTIONS OF A RANDOM VARIABLE

Situations often arise in systems analysis where knowledge of some characteristic of the system, together with the knowledge of the input, will allow some estimate of the behavior at the output. For example, the input random variable X and its density $f(x)$ are known and the input-output behavior is characterized by:

$$Y = \Phi(X).$$

We are interested in computing the density of the random variable Y. Note that for a given random variable X and a function Φ, Y may not satisfy the definition of a random variable. But if we assume that Φ is continuous or piecewise continuous, then $Y = \Phi(X)$ will be a random variable [ASH 1970].

Example 3.8

Let $Y = \Phi(X) = X^2$. As an example, X could denote the measurement error in a certain physical experiment, and Y would then be the square of the error (recall the method of least squares).

Note that $F_Y(y) = 0$ for $y \leqslant 0$. For $y > 0$:

$$\begin{aligned} F_Y(y) &= P(Y \leqslant y) \\ &= P(X^2 \leqslant y) \\ &= P(-\sqrt{y} \leqslant X \leqslant \sqrt{y}) \\ &= F_X(\sqrt{y}) - F_X(-\sqrt{y}), \end{aligned}$$

and by differentiation the density of Y is:

$$f_Y(y) = \begin{cases} \dfrac{1}{2\sqrt{y}} [f_X(\sqrt{y}) + f_X(-\sqrt{y})], & y > 0, \\ 0, & \text{otherwise}. \end{cases} \tag{3.43}$$

#

Example 3.9

As a special case of the above example, let X have the standard normal distribution [N(0, 1)] so that:

$$f_X(x) = \frac{1}{\sqrt{2\pi}} e^{-\frac{x^2}{2}}, \qquad -\infty < x < \infty.$$

Then:

$$f_Y(y) = \begin{cases} \dfrac{1}{2\sqrt{y}} \left[\dfrac{1}{\sqrt{2\pi}} e^{-y/2} + \dfrac{1}{\sqrt{2\pi}} e^{-y/2} \right], & y > 0, \\ 0, & y \leqslant 0, \end{cases}$$

or

$$f_Y(y) = \begin{cases} \dfrac{1}{\sqrt{2\pi y}} e^{-\frac{y}{2}}, & y > 0, \\ 0, & y \leqslant 0. \end{cases}$$

By comparing this formula with formula (3.21) and remembering (3.25), we conclude that Y has a gamma distribution with $\alpha = 1/2$ and $\lambda = 1/2$. Now, since $\mathrm{GAM}(1/2, n/2) = X_n^2$, it follows that if X is standard normal then $Y = X^2$ is chi-square distributed with one degree of freedom. #

Example 3.10

Let X be uniformly distributed on (0, 1). We show that $Y = -\lambda^{-1} \ln(1 - X)$ has an exponential distribution with parameter $\lambda > 0$.

Observe that Y is a nonnegative random variable implying $F_Y(y) = 0$ for $y \leqslant 0$. For $y > 0$, we have:

$$\begin{aligned} F_Y(y) &= P(Y \leqslant y) = P[-\lambda^{-1} \ln(1 - X) \leqslant y] \\ &= P[\ln(1 - X) \geqslant -\lambda y] \\ &= P[(1 - X) \geqslant e^{-\lambda y}] \end{aligned}$$

since e^x is an increasing function of x,

$$\begin{aligned} &= P(X \leqslant 1 - e^{-\lambda y}) \\ &= F_X(1 - e^{-\lambda y}). \end{aligned}$$

But since X is uniform over (0, 1), $F_X(x) = x,\ 0 \leqslant x \leqslant 1$. Thus:

$$F_Y(y) = 1 - e^{-\lambda y}.$$

Therefore Y is exponentially distributed with parameter λ. #

The above fact can be used in Monte Carlo simulation. In such simulation programs it is important to be able to generate values of variables with known distribution functions. Such values are known as **random deviates**. Most computer systems provide built-in functions to generate random deviates from the uniform distribution over (0, 1), say u. Such random deviates are called **random numbers**. Now if we are interested in generating a random deviate, y, of an exponentially distributed random variable Y with the parameter λ, then from the above example it follows that $y = -\lambda^{-1} \ln(1 - u)$. For a discussion of random number generation, the reader is referred to [KNUT 1969; chap. 3].

Examples 3.9 and 3.10 are special cases of problems that can be solved using the following theorem.

> **THEOREM 3.1.** Let X be a continuous random variable with density f_X that is nonzero on a subset I of real numbers (that is, $f_X(x) > 0,\ x \in I$ and $f_X(x) = 0,\ x \notin I$). Let Φ be a differentiable monotone function whose domain is I and whose range is

the set of reals. Then $Y = \Phi(X)$ is a continuous random variable with the density, f_Y, given by:

$$f_Y(y) = \begin{cases} f_X[\Phi^{-1}(y)]\,[|(\Phi^{-1})'(y)|], & y \in \Phi(I), \\ 0, & \text{otherwise}, \end{cases} \tag{3.44}$$

where Φ^{-1} is the uniquely defined inverse of Φ and $(\Phi^{-1})'$ is the derivative of the inverse function.

Proof: We prove the theorem assuming that $\Phi(x)$ is an increasing function of x. The proof for the other case follows in a similar way.

$$\begin{aligned} F_Y(y) &= P(Y \leqslant y) = P[\Phi(X) \leqslant y] \\ &= P[X \leqslant \Phi^{-1}(y)], \qquad \text{since } \Phi \text{ is monotone increasing} \\ &= F_X[\Phi^{-1}(y)]. \end{aligned}$$

Taking derivatives and using the chain rule, we get the required result.

Example 3.11

Now let Φ be the distribution function, F, of a random variable X, with density f. Applying the above theorem, $Y = F(X)$ and $F_Y(y) = F_X(F_X^{-1}(y)) = y$. Therefore, the random variable $Y = F(X)$ has the density given by:

$$f_Y(y) = \begin{cases} 1, & 0 < y < 1, \\ 0, & \text{otherwise.} \end{cases}$$

In other words, if X is a continuous random variable with CDF F, then the new random variable $Y = F(X)$ is uniformly distributed over the interval $(0, 1)$.

This idea can be used to generate a random deviate x of X by first generating a random number u from a uniform distribution over $(0, 1)$ and then using the relation $x = F^{-1}(u)$ as illustrated in Figure 3.18. The generation of the $(0, 1)$ random number can be accomplished rather easily. The real question, however, is whether $F^{-1}(u)$ can be expressed in a closed mathematical form. This is possible for distributions such as the exponential, while for distributions such as the normal, other techniques must be used. #

Another interesting special case of Theorem 3.1 occurs when Φ is linear; that is, $Y = aX + b$. In other words, Y differs from X only in origin and scale of measurement. In fact, we have already made use of such a transformation when relating $N(\mu, \sigma^2)$ to the standard normal distribution, $N(0, 1)$. The use of the above theorem yields:

$$f_Y(y) = \begin{cases} \dfrac{1}{|a|} f_X\left(\dfrac{y-b}{a}\right), & y \in aI + b, \\ 0, & \text{otherwise}, \end{cases} \tag{3.45}$$

where I is the interval over which $f(x) \neq 0$.

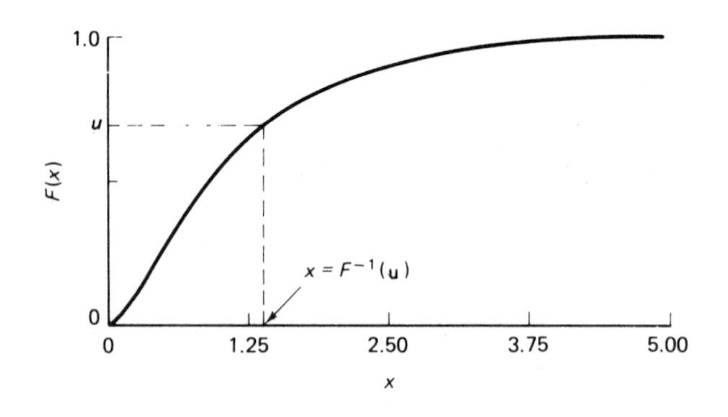

Figure 3.18 Generating a random deviate

Example 3.12

Let X be exponentially distributed; that is, $X \sim \text{EXP}(\lambda)$. Consider the random variable $Y = rX$, where r is a positive real number. Note that the interval I over which $f_X \neq 0$ is $(0, \infty)$. Also:

$$f_X(x) = \lambda e^{-\lambda x}.$$

Using formula (3.45), we have:

$$f_Y(y) = \frac{1}{r} \lambda e^{\frac{-\lambda y}{r}}, \qquad y > 0.$$

It follows that Y is exponentially distributed with parameter λ/r. This result will be used later in this book. #

Example 3.13

Let X be normally distributed and consider $Y = e^X$. Since:

$$f_X(x) = \frac{1}{\sigma\sqrt{2\pi}} \exp\left[-\frac{1}{2}\left(\frac{x-\mu}{\sigma}\right)^2\right],$$

and

$$\Phi^{-1}(y) = \ln(y) \quad \text{implies that} \quad [\Phi^{-1}]'(y) = \frac{1}{y};$$

then, using Theorem 3.1, the density of Y is:

$$\begin{aligned} f_Y(y) &= \frac{f(\ln y)}{y} \\ &= \frac{1}{\sigma y \sqrt{2\pi}} \exp\left[-\frac{(\ln y - \mu)^2}{2\sigma^2}\right], \qquad y > 0. \end{aligned}$$

The random variable Y is said to have log-normal distribution. The importance of this distribution stems from another form of the central limit theorem, which states that the product of n mutually independent random variables has a log-normal distribution in the limit $n \to \infty$. #

Problems

1. Show that if X has the k-stage Erlang distribution with parameter λ, then:

$$Y = 2\lambda X$$

has the chi-square distribution with $2k$ degrees of freedom.

2. Consider a nonlinear amplifier whose input X and the output Y are related by its transfer characteristic:

$$Y = \begin{cases} X^{1/2}, & X \geqslant 0, \\ -|X|^{1/2}, & X < 0. \end{cases}$$

Assuming that $X \sim N(0,1)$ compute the pdf of Y and plot your result.

3. The phase X of a sine wave is uniformly distributed over $(-\pi/2, \pi/2)$; that is:

$$f_X(x) = \begin{cases} \dfrac{1}{\pi}, & -\dfrac{\pi}{2} < x < \dfrac{\pi}{2}, \\ 0, & \text{otherwise.} \end{cases}$$

Let $Y = \sin X$ and show that:

$$f_Y(y) = \frac{1}{\pi} \frac{1}{\sqrt{1-y^2}}, \qquad -1 < y < 1.$$

4. Let X be a chi-square random variable with n ($\geqslant 1$) degrees of freedom with the pdf:

$$f(x) = \begin{cases} \dfrac{x^{n/2-1}}{2^{n/2}\Gamma(n/2)} e^{-x/2}, & x > 0, \\ 0, & \text{elsewhere.} \end{cases}$$

Find the pdf of the random variable $Y = \sqrt{X/n}$.

5. Consider an IBM 2314 type disk file. Assume that the number of tracks N to be traversed between two disk requests is a normally distributed random variable with mean $\mu = n/3$ and $\sigma^2 = n^2/18$ where n is the total number of tracks. The seek time T is a random variable related to the seek distance N by:

$$T = a + bN$$

In the particular case of the IBM 2314, experimental data show $a = 45$, $b = 0.43$, and $n = 200$. Determine the pdf of T. Recalling that T is a nonnegative random variable whereas the normal model allows negative values, make appropriate corrections.

6. Consider a normalized floating point number in base (or radix) β so that the mantissa X satisfies the condition $1/\beta \leqslant X < 1$. Assume that the density of the continuous random variable X is $1/(x \ln \beta)$.
   (a) Show that a random deviate of X is given by the formula, β^{u-1} where u is a random number.
   (b) Determine the pdf of the normalized reciprocal $Y = 1/(\beta X)$.

3.6 JOINTLY DISTRIBUTED RANDOM VARIABLES

So far, we have been concerned with the properties of a single random variable. In many practical problems, however, it is important to consider two or more random variables defined on the same probability space.

Let X and Y be two random variables defined on the same probability space $(S, \mathcal{F}, P)$. The event $[X \leqslant x, \; Y \leqslant y] = [X \leqslant x] \cap [Y \leqslant y]$ consists of all sample points $s \in S$ such that $X(s) \leqslant x$ and $Y(s) \leqslant y$.

Definition (Joint Distribution Function). The joint (or compound) distribution function of random variables X and Y is defined by:

$$F_{X,Y}(x, y) = P(X \leqslant x, \; Y \leqslant y), \qquad -\infty < x < \infty, \; -\infty < y < \infty.$$

The subscripts will be dropped whenever the two random variables under consideration are clear from the context; that is, $F_{X,Y}(x, y)$ will be written as $F(x, y)$. Such a function satisfies the following properties:

(J1) $0 \leqslant F(x, y) \leqslant 1, \; -\infty < x < \infty, \; -\infty < y < \infty$. This is evident since $F(x, y)$ is a probability.

(J2) $F(x, y)$ is monotone nondecreasing in both the variables; that is, if $x_1 \leqslant x_2$ and $y_1 \leqslant y_2$, then $F(x_1, y_1) \leqslant F(x_2, y_2)$. This follows since the event $[X \leqslant x_1$ and $Y \leqslant y_1]$ is contained in the event $[X \leqslant x_2$ and $Y \leqslant y_2]$.

(J3) If either x or y approaches $-\infty$, then $F(x, y)$ approaches 0, and if both x and y approach $+\infty$, then $F(x, y)$ approaches 1.

(J4) $F(x, y)$ is right continuous in general, and if X and Y are continuous random variables then $F(x, y)$ is continuous.

(J5) $P(a < X \leqslant b$ and $c < Y \leqslant d) = F(b, d) - F(a, d) - F(b, c) + F(a, c)$. This relation follows from Figure 3.19.

Note that in the limit $y \to \infty$, the event $[X \leqslant x, \; Y \leqslant y]$ approaches the event $[X \leqslant x, \; Y < \infty] = [X \leqslant x]$. Therefore, $\lim_{y \to \infty} F_{X,Y}(x, y) = F_X(x)$. Also $\lim_{x \to \infty} F_{X,Y}(x, y) = F_Y(y)$. These two formulas show how to compute the **individual** or **marginal distribution functions** of X and Y given their joint distribution function.

If both X and Y are continuous random variables, then we can often find a function $f(x, y)$ such that:

$$F(x, y) = \int_{-\infty}^{y} \int_{-\infty}^{x} f(u, v) \, du \, dv. \tag{3.46}$$

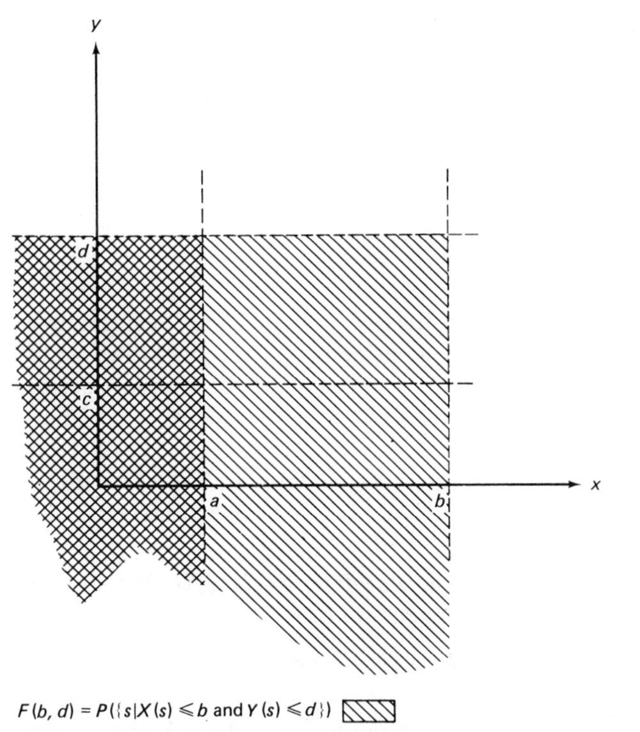

Figure 3.19 Properties of joint CDF

Such a function is known as the **joint** or the **compound probability density function** of X and Y. It follows that:

$$\int_{-\infty}^{\infty} \int_{-\infty}^{\infty} f(u, v)\, du\, dv = 1 \tag{3.47}$$

and

$$P(a < X \leqslant b,\ c < Y \leqslant d) = \int_{a}^{b} \int_{c}^{d} f(x, y)\, dy\, dx. \tag{3.48}$$

Also:

$$f(x, y) = \frac{\partial^2 F(x, y)}{\partial x\, \partial y}, \tag{3.49}$$

assuming that the partial derivative exists. Now since:

$$F_X(x) = \int_{-\infty}^{x} \int_{-\infty}^{\infty} f(u, y)\, dy\, du,$$

we obtain the marginal density f_X as:

$$f_X(x) = \int_{-\infty}^{\infty} f(x, y)\, dy. \tag{3.50}$$

Similarly:

$$f_Y(y) = \int_{-\infty}^{\infty} f(x, y)\, dx. \tag{3.51}$$

Thus the marginal densities, $f_X(x)$ and $f_Y(y)$, can easily be determined from the knowledge of the joint density, $f(x, y)$. However, the knowledge of the marginal densities does not, in general, uniquely determine the joint density. The exception occurs when the two random variables are independent.

Intuitively, if X and Y are independent random variables, then we expect that events such as $[X \leqslant x]$ and $[Y \leqslant y]$ will be independent events.

Definition (Independent Random Variables). We define two random variables X and Y to be **independent** if:

$$F(x, y) = F_X(x)\, F_Y(y), \qquad -\infty < x < \infty,\ -\infty < y < \infty.$$

Thus the independence of random variables X and Y implies that their joint CDF factors into the product of the marginal CDF's. This definition applies whatever the types of the random variables involved. In the case that X and Y are discrete, the above definition of independence is equivalent to the definition given in Chapter 2:

$$p(x, y) = p_X(x)\, p_Y(y).$$

In the case that X and Y are continuous random variables, having a joint pdf, the above definition of independence is equivalent to the condition:

$$f(x, y) = f_X(x) f_Y(y), \qquad -\infty < x < \infty,\ -\infty < y < \infty.$$

We will also have occasion to consider the joint distribution of X and Y when one of them is a discrete random variable while the other is a continuous random variable. In case X is discrete and Y is continuous, the condition for their independence becomes:

$$P(X = x,\ Y \leqslant y) = p_X(x) F_Y(y), \qquad \text{all } x \text{ and } y.$$

The definition of joint distribution, joint density, and independence of two random variables can be easily generalized to a set of n random variables $X_1, X_2, \ldots, X_n$.

Example 3.14

Assume that the lifetime X and the brightness Y of a light bulb are being modeled as continuous random variables. Let the joint pdf be given by

$$f(x, y) = \lambda_1\lambda_2 e^{-(\lambda_1 x+\lambda_2 y)}, \qquad 0 < x < \infty,\ 0 < y < \infty$$

(this is known as the bivariate exponential density). The marginal density of X is

$$\begin{aligned} f_X(x) &= \int_{-\infty}^{\infty} f(x,y)\,dy \\ &= \int_0^{\infty} \lambda_1\lambda_2 e^{-(\lambda_1 x+\lambda_2 y)}\,dy \\ &= \lambda_1 e^{-\lambda_1 x}, \qquad 0 < x < \infty. \end{aligned}$$

Similarly:

$$f_Y(y) = \lambda_2 e^{-\lambda_2 y}, \qquad 0 < y < \infty.$$

It follows that X and Y are independent random variables. The joint distribution function can be computed to be:

$$\begin{aligned} F(x, y) &= \int_{-\infty}^{x}\int_{-\infty}^{y} f(u,v)\,du\,dv \\ &= \int_0^{x}\int_0^{y} \lambda_1\lambda_2 e^{-(\lambda_1 u+\lambda_2 v)}\,du\,dv \\ &= (1 - e^{-\lambda_1 x})(1 - e^{-\lambda_2 y}), \qquad 0 < x < \infty,\ 0<y<\infty. \end{aligned}$$

#

Problems

1. 1K RAM IC chips are purchased from two different semiconductor houses. Let X and Y denote the times to failure of the chips purchased from the two suppliers. The joint probability density of X and Y is estimated by:

$$f_{X,Y}(x, y) = \begin{cases} \lambda\mu e^{-(\lambda x + \mu y)}, & x > 0,\ y > 0 \\ 0, & \text{otherwise.} \end{cases}$$

Assume $\lambda = 10^{-5}$ per hour and $\mu = 10^{-6}$ per hour.

Determine the probability that the time to failure is greater for chips characterized by X than it is for chips characterized by Y.

2. Let X and Y have joint pdf:

$$f(x, y) = \begin{cases} \dfrac{1}{\pi}, & x^2 + y^2 \leqslant 1 \\ 0, & \text{otherwise.} \end{cases}$$

Determine the marginal pdf's of X and Y. Are X and Y independent?

3. Consider a series connection of two components, with respective lifetimes X and Y. The joint pdf of the lifetimes is given by:

$$f(x, y) = \begin{cases} 1/200, & (x, y) \in A, \\ 0, & \text{elsewhere,} \end{cases}$$

where A is the triangular region in the (x, y) plane with the vertices (100, 100), (100, 120), and (120, 120). Find the reliability expression for the entire system.

4. If the random variables B and C are independent and uniformly distributed over (0, 1), compute the probability that the roots of the equation:

$$x^2 + 2\,B\,x + C = 0$$

are real.

5. Let the joint pdf of X and Y be given by:

$$f_{X,Y}(x, y) = \frac{1}{2\pi\sqrt{1-\rho^2}} \exp\left[-\frac{x^2 - 2\rho\, xy + y^2}{2(1-\rho^2)}\right],$$

where $|\rho| < 1$. Show that the marginal pdf's are:

$$f_X(x) = \frac{1}{\sqrt{2\pi}}\, e^{-x^2/2},$$

$$f_Y(y) = \frac{1}{\sqrt{2\pi}}\, e^{-y^2/2}.$$

The random variables X and Y are said to have a two-dimensional (or bivariate) normal pdf. Also note that X and Y are *not* independent unless $\rho = 0$.

3.7 ORDER STATISTICS

Let $X_1, X_2, \ldots, X_n$ be mutually independent, identically distributed continuous random variables, each having the distribution function F and density f. Let $Y_1, Y_2, \ldots, Y_n$ be random variables obtained by permuting the set $X_1, X_2, \ldots, X_n$ so as to be in increasing order. To be specific:

$$Y_1 = \min\{X_1, X_2, \ldots, X_n\}$$

and

$$Y_n = \max\{X_1, X_2, \ldots, X_n\}.$$

The random variable Y_k is called the **kth-order statistic.** Since $X_1, X_2, \ldots, X_n$ are continuous random variables, it follows that $Y_1 < Y_2 < \cdots < Y_n$ (as opposed to $Y_1 \leqslant Y_2 \leqslant \cdots \leqslant Y_n$) with a probability of one.

As examples of use of order statistics, let X_i be the lifetime of the ith component in a system of n independent components. If the system is a series system, then Y_1 will be the overall system lifetime. Similarly, Y_n will denote the lifetime of a parallel system and Y_{n-m+1} will be the lifetime of an m-out-of-n system (the so-called N-tuply Modular Redundant or NMR system).

To derive the distribution function of Y_k, we note that the probability that exactly j of the X_i's lie in $(-\infty, y]$ and $(n - j)$ lie in (y, ∞) is:

$$\binom{n}{j} F^j(y)[1 - F(y)]^{n-j},$$

since the binomial distribution with parameters n and $p = F(y)$ is applicable. Then:

$$\begin{aligned} F_{Y_k}(y) &= P(Y_k \leqslant y) \\ &= P(\text{"at least } k \text{ of the } X_i\text{'s lie in the interval } (-\infty, y]\text{"}) \\ &= \sum_{j=k}^{n} \binom{n}{j} F^j(y)[1 - F(y)]^{n-j}, \qquad -\infty < y < \infty. \end{aligned} \tag{3.52}$$

In particular, the distribution functions of Y_n and Y_1 can be obtained from (3.52) as:

$$F_{Y_n}(y) = [F(y)]^n, \qquad -\infty < y < \infty,$$

and

$$F_{Y_1}(y) = 1 - [1 - F(y)]^n, \qquad -\infty < y < \infty.$$

From this we obtain:

$$\begin{aligned} R_{\text{series}}(t) &= R_{Y_1}(t) \\ &= 1 - F_{Y_1}(t) \\ &= [1 - F(t)]^n \\ &= [R(t)]^n, \end{aligned}$$

and

$$\begin{aligned} R_{\text{parallel}}(t) &= R_{Y_n}(t) \\ &= 1 - F_{Y_n}(t) \\ &= 1 - [F(t)]^n \\ &= 1 - [1 - R(t)]^n. \end{aligned}$$

Both these formulas easily generalize to the case when the lifetime distributions of individual components are distinct:

$$R_{\text{series}}(t) = \prod_{i=1}^{n} R_i(t) \tag{3.53}$$

and

$$R_{\text{parallel}}(t) = 1 - \prod_{i=1}^{n} [1 - R_i(t)]. \tag{3.54}$$

Example 3.15

Let the lifetime distribution of the ith component be exponential with parameter λ_i. Then equation (3.53) reduces to:

$$R_{\text{series}}(t) = \exp\left[-(\sum_{i=1}^{n} \lambda_i)t\right],$$

so that the lifetime distribution of a series system whose components have independent exponentially distributed lifetimes is itself exponentially distributed with parameter $\sum_{i=1}^{n} \lambda_i$.

This fact is responsible for the "parts-count method" of system reliability analysis often used in practice. Under this method, the analyst counts the number n_i of parts of type i each with a failure rate λ_i. Now if there are k such part types, then the system failure rate λ is computed by:

$$\lambda = \sum_{i=1}^{k} \lambda_i n_i. \tag{3.55}$$

#

Clary and others [CLAR 1978] present the following example of the parts-count method of reliability analysis.

Example 3.16

One implementation of the CPU-cache-main memory subsystem of PDP 11/70 consists of the following chip types:

Chip type	*Number of chips,* n_i	*Failure rate per chip (number of failures/10^6 hours),* λ_i
SSI	1,202	0.1218
MSI	668	0.242
ROM	58	0.156
RAM	414	0.691
MOS	256	1.0602
BIP	2,086	0.1588

It is assumed that times to failure of all chip types are exponentially distributed with the failure rate shown above. All the chips must be fault free in order for the system as a whole to be fault free (that is, a series system). The system time to failure is then exponentially distributed with parameter:

$$\lambda = \sum_{\text{all chip types}} n_i \lambda_i$$

$$= 146.40 + 161.66 + 9.05 + 286.07 + 261.41 + 331.27$$

$$= 1205.85 \text{ failures per } 10^6 \text{ hours.}$$ #

Example 3.17

We have seen that the lifetime of a series system is exponentially distributed, provided that the component lifetimes are independent and exponentially distributed. Thus a series system whose components have constant failure rate itself has a constant failure rate. *This does not apply to a parallel system.* The failure rate of a parallel system is a variable function of its age, even though the failure rates of individual components are constant. It can be shown that the corresponding distribution is IFR. In particular, the reliability of a parallel system of n independent components, each with an exponential failure law (parameter λ), is given by:

$$\begin{aligned} R_p(t) &= 1 - (1 - e^{-\lambda t})^n \\ &= \binom{n}{1} e^{-\lambda t} - \binom{n}{2} e^{-2\lambda t} + \cdots + (-1)^{n-1} e^{-n\lambda t}. \end{aligned} \tag{3.56}$$

Figure 3.20 shows the reliability improvement obtained by parallel redundancy. #

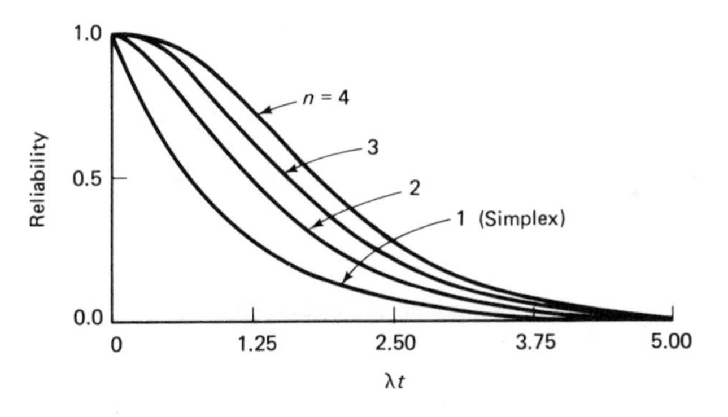

Figure 3.20 Reliability of a parallel-redundant system

Example 3.18

Another interesting case of order statistics occurs when we consider the triple modular redundant (TMR) system ($n = 3$ and $m = 2$).. Y_2 then denotes the time until the second failure. Using equation (3.52), we get:

$$R_{\text{TMR}}(t) = 3R^2(t) - 2R^3(t). \tag{3.57}$$

Assuming that the reliability of a single component is given by $R(t) = e^{-\lambda t}$, we get:

$$R_{\text{TMR}}(t) = 3e^{-2\lambda t} - 2e^{-3\lambda t}. \tag{3.58}$$

In Figure 3.21, we have plotted $R_{\text{TMR}}(t)$ against t as well as $R(t)$ against t. Note that:

$$R_{\text{TMR}}(t) \geqslant R(t), \qquad 0 \leqslant t \leqslant t_0,$$

and

$$R_{\text{TMR}}(t) \leqslant R(t), \qquad t_0 \leqslant t < \infty,$$

where t_0 is the solution to the equation:

$$3e^{-2\lambda t_0} - 2e^{-3\lambda t_0} = e^{-\lambda t_0},$$

which is:

$$t_0 = \frac{\ln 2}{\lambda} \simeq \frac{0.7}{\lambda}.$$

Thus if we define a "short" mission by the mission time $t \leqslant t_0$, then it is clear that TMR type of redundancy improves reliability only for short missions. For long missions, this type of redundancy actually degrades reliability. The same type of behavior is exhibited by any NMR system—that is, m out of n system with $n = 2m - 1$ (see Figure 3.22). #

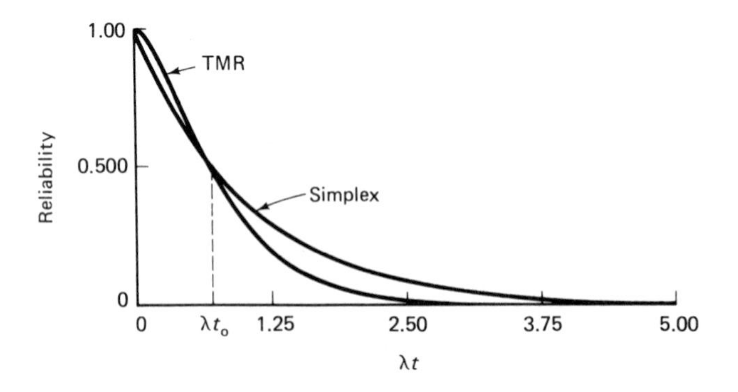

Figure 3.21 Comparison of TMR and simplex reliabilities

Example 3.19

Consider a computer system with jobs arriving from several independent sources. Let $N_i(t)$ $(1 \leqslant i \leqslant n)$ denote the number of jobs arriving in the interval $(0, t]$ from source i. Assume that $N_i(t)$ is Poisson distributed with parameter $\lambda_i t$. Then the time between two arrivals from source i, denoted by X_i, is exponentially distributed with parameter λ_i (refer to Example 3.2). Also from Theorem 2.2(c), the total number of jobs arriving from all sources in the interval $(0, t]$, $N(t) = \sum_{i=1}^{n} N_i(t)$, is Poisson distributed with parameter $\sum_{i=1}^{n} \lambda_i t$. Then, again recalling Example 3.2, the interarrival time, Y_1, for jobs from all sources will be exponentially distributed with parameter $\sum_{i=1}^{n} \lambda_i$. But this could also be derived using order statistics. Note that:

$$Y_1 = \min\{X_1, X_2, \ldots, X_n\},$$

since Y_1 is the time until the next arrival from any one of the sources. Now since:

$$F_{X_i}(t) = 1 - e^{-\lambda_i t},$$

it follows that:

$$\begin{aligned} F_{Y_1}(t) &= 1 - \prod_{i=1}^{n} [1 - F_{X_i}(t)] \\ &= 1 - \prod_{i=1}^{n} e^{-\lambda_i t} \\ &= 1 - \exp[-\sum_{i=1}^{n} \lambda_i t]. \end{aligned}$$

Thus Y_1 is exponentially distributed with parameter $\sum_{i=1}^{n} \lambda_i$. #

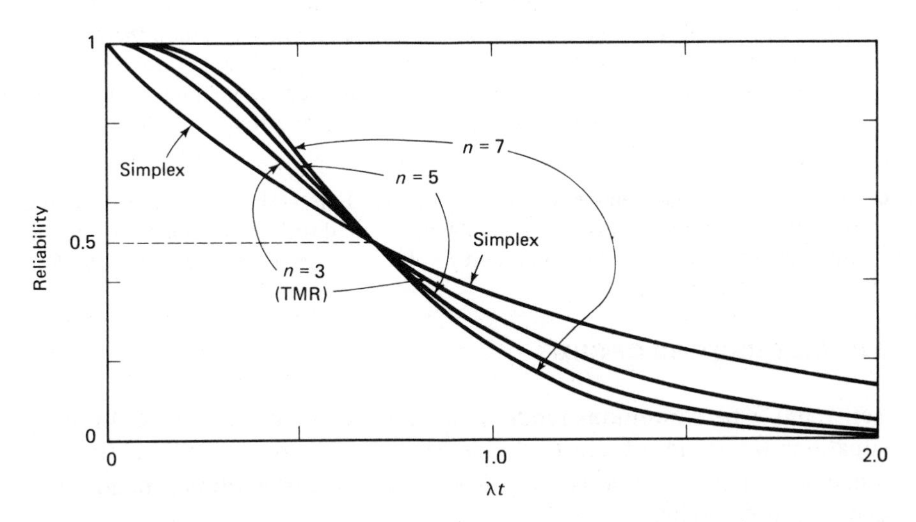

Figure 3.22 The reliability of NMR system ($n = 2m - 1$)

Problems

1. A system with three independent components works correctly if at least one component is functioning properly. Failure rates of the individual components are $\lambda_1 = 0.01$, $\lambda_2 = 0.02$, and $\lambda_3 = 0.04$ (assume exponential lifetime distributions).
   (a) Determine the probability that the system will work for 100 hours.
   (b) Determine the density function of the lifetime X of the system.

2. A multiprocessor system has n processors. Service time of a process executing on the ith processor is exponentially distributed with parameter μ_i $(i = 1, 2, \ldots, n)$. Given that all n processors are active and that they are executing mutually independent processes, what is the distribution of time until a processor becomes idle.

3. Consider a series system consisting of n independent components. Assuming that the lifetime of the ith component is Weibull distributed with parameters λ_i and α, show that the system lifetime also has a Weibull distribution.

4. Consider a random access memory card consisting of d IC chips each containing w bits. Each chip supplies one bit position in a d bit word for a total of w, d bit words. Assuming a failure rate of λ per chip (and an exponential lifetime distribution), derive the reliability expression $R_0(t)$ for the memory card. Suppose that we now introduce single error correction, so that up to one chip may fail without a system failure. Note that this will require c extra chips where c must satisfy the relation $c \geqslant \log_2 (c + d + 1)$.(See [KUCK 1978].) Derive the new reliability expression $R_1(t)$. Plot $R_0(t)$ and $R_1(t)$ as functions of t on the same plot. Derive expressions for the failure rates $h_0(t)$ and $h_1(t)$ and plot these as functions of time t. For these plots assume $d = 16$ (hence $c = 5$) and $\lambda = 10$ per million hours.

5. The memory requirement distribution for jobs in a computer system is exponential with parameter λ. The memory scheduler scans all eligible jobs on a job queue and loads the job with the smallest memory requirement first, then the job with the next smallest memory requirement, and so on. Given that there are n jobs on the job queue, write down the distribution function for the memory requirement of the job with the smallest memory requirement, and that of the job with the largest memory requirement.

6. A series system has n independent components. For $i = 1, 2, \ldots, n$, the lifetime X_i of the ith component is exponentially distributed with parameter λ_i. Compute the probability that a given component $j = (1, 2, \ldots, n)$ is the cause of the system failure.

3.8 DISTRIBUTION OF SUMS

Let X and Y be continuous random variables with joint density f. In many situations we are interested in the density of a random variable Z that is a function of X and Y—that is, $Z = \Phi\,(X, Y)$. The distribution function of Z may be computed by:

$$F_Z(z) = P(Z \leqslant z) = \iint_{A_z} f(x, y)\, dx\, dy, \tag{3.59}$$

where A_z is a subset of $\mathbb{R}^2$ given by:

$$A_z = \{(x, y) \mid \Phi(x, y) \leqslant z\}. = \Phi^{-1}((-\infty, z]).$$

One function of special interest is $Z = X + Y$ with:

$$A_z = \{(x, y) \mid x + y \leqslant z\},$$

which is the half-plane to the lower left of the line $x + y = z$ (see Figure 3.23). Then:

$$F_Z(z) = \iint_{A_z} f(x, y)\, dx\, dy$$

$$= \int_{-\infty}^{\infty} \int_{-\infty}^{z-x} f(x, y)\, dy\, dx.$$

Making a change of variable $y = t - x$, we get:

$$F_Z(z) = \int_{-\infty}^{\infty} \int_{-\infty}^{z} f(x, t - x)\, dt\, dx$$

$$= \int_{-\infty}^{z} \int_{-\infty}^{\infty} f(x, t - x)\, dx\, dt$$

$$= \int_{-\infty}^{z} f_Z(t)\, dt$$

by the definition of density. Thus the density of $Z = X + Y$ is given by:

$$f_Z(z) = \int_{-\infty}^{\infty} f(x, z - x)\, dx, \qquad -\infty < z < \infty. \tag{3.60}$$

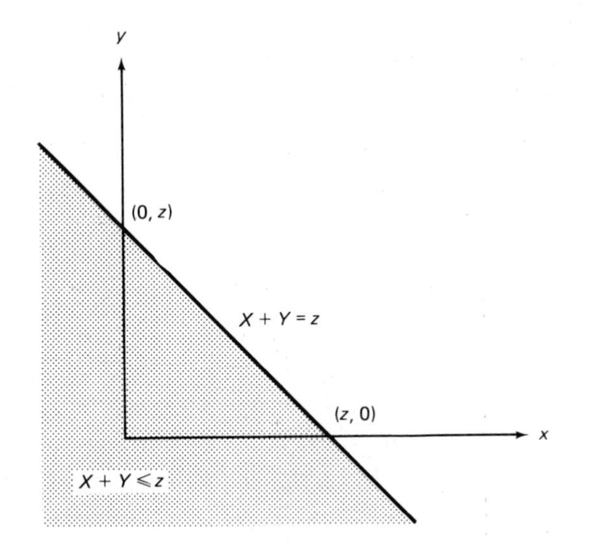

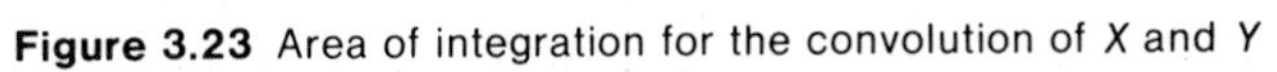
Figure 3.23 Area of integration for the convolution of X and Y

Now if X and Y are assumed to be independent, then $f(x, y) = f_X(x)f_Y(y)$, and the above formula reduces to:

$$f_Z(z) = \int_{-\infty}^{\infty} f_X(x)\, f_Y(z-x)\, dx, \qquad -\infty < z < \infty. \tag{3.61}$$

Furthermore, if X and Y are nonnegative random variables, then:

$$f_Z(z) = \int_{0}^{z} f_X(x)\, f_Y(z-x)\, dx, \qquad 0 < z < \infty. \tag{3.62}$$

The above integral is often called the **convolution** of f_X and f_Y. Thus the density of the sum of two nonnegative independent random variables is the convolution of the individual densities.

Example 3.20

Consider a job consisting of three tasks. Tasks 1 and 2 are noninterfering and hence can be executed in parallel. Task 3 cannot be started until both task 1 and task 2 have completed. If T_1, T_2, and T_3, respectively, denote the times of execution of three tasks, then the time of execution of the entire job is given by:

$$\begin{aligned} T &= \max\{T_1, T_2\} + T_3 \\ &= M + T_3. \end{aligned}$$

Assume that T_1 and T_2 are continuous random variables with uniform distribution over $[t_1 - t_0,\ t_1 + t_0]$ and T_3 is a continuous random variable uniformly distributed over $[t_3 - t_0,\ t_3 + t_0]$. Also assume that T_1, T_2, and T_3 are mutually independent. We are asked to compute the probability that $T > t_1 + t_3$.

Note that:

$$\begin{aligned} f_{T_1}(t) &= f_{T_2}(t) \\ &= \begin{cases} \dfrac{1}{2t_0}, & t_1 - t_0 < t < t_1 + t_0, \\ 0, & \text{otherwise}, \end{cases} \end{aligned}$$

and

$$f_{T_3}(t) = \begin{cases} \dfrac{1}{2t_0}, & t_3 - t_0 < t < t_3 + t_0, \\ 0, & \text{otherwise}. \end{cases}$$

First compute the distribution of the random variable M:

$$\begin{aligned} F_M(m) &= P(M \leqslant m) \\ &= P(\max\{T_1, T_2\} \leqslant m) \\ &= P(T_1 \leqslant m \text{ and } T_2 \leqslant m) \\ &= P(T_1 \leqslant m)\, P(T_2 \leqslant m) \qquad \text{by independence} \\ &= F_{T_1}(m) F_{T_2}(m) \end{aligned}$$

Now observe that:

$$F_{T_1}(t) = \begin{cases} 0, & t < t_1 - t_0, \\ \dfrac{t - t_1 + t_0}{2t_0}, & t_1 - t_0 \leqslant t < t_1 + t_0, \\ 1, & \text{otherwise.} \end{cases}$$

Thus:

$$F_M(m) = \begin{cases} 0, & m < t_1 - t_0, \\ \dfrac{(m - t_1 + t_0)^2}{4t_0^2}, & t_1 - t_0 \leqslant m < t_1 + t_0, \\ 1, & m \geqslant t_1 + t_0. \end{cases}$$

Also:

$$f_M(m) = \begin{cases} \dfrac{m - t_1 + t_0}{2t_0^2}, & t_1 - t_0 < m < t_1 + t_0 \\ 0, & \text{otherwise.} \end{cases}$$

Now consider the (M, T_3) plane as shown in Figure 3.24 and let A denote the shaded region. Then:

$$\begin{aligned} P(T > t_1 + t_3) &= \int_A\int f_{M,T_3}(m, t)\, dm\, dt \\ &= \int_A\int f_M(m) f_{T_3}(t)\, dm\, dt \end{aligned}$$

since M and T_3 are independent,

$$= \int_{t_1-t_0}^{t_1+t_0} \left[\int_{t_1+t_3-m}^{t_3+t_0} \frac{m - t_1 + t_0}{4t_0^3}\, dt \right] dm$$

$$= \frac{2}{3}. \qquad \#$$

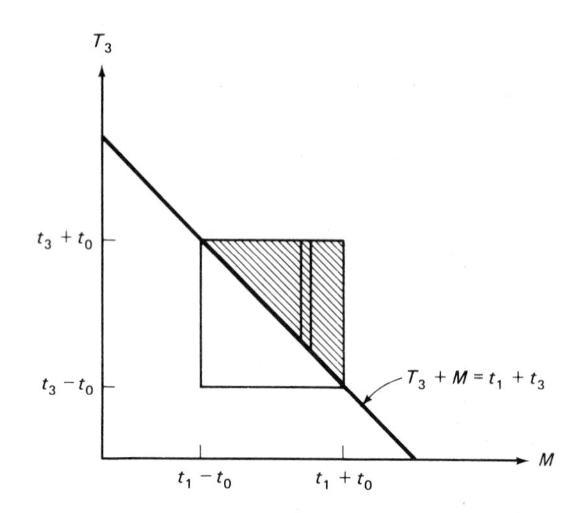

Figure 3.24 Area of integration for Example 3.20

In the rest of this section we will concentrate on sums of independent exponentially distributed random variables.

Example 3.21

Consider a system with two statistically identical components, each with an exponential failure law with parameter λ. Only one component is required to be operative for the system to function properly. One method of utilizing the second component is to use parallel redundancy, in which both components are initially operative simultaneously. Alternately, we could initially keep the spare component in a power-off state (deenergized) and later, upon the failure of the operative component, replace it by the spare. Assuming that a deenergized component does not fail and that the failure detection and switching equipment is perfect, we can characterize the lifetime Z of such a system in terms of the lifetimes X and Y of individual components by $Z = X + Y$. Such a system is known to possess **standby redundancy** in contrast to a system with parallel redundancy. Now if the random variables X and Y are assumed to be independent, then the density of Z is the convolution of the densities of X and Y.

Let X and Y have exponential distributions with parameter λ; that is, $f_X(x) = \lambda e^{-\lambda x}, x > 0$, and $f_Y(y) = \lambda e^{-\lambda y}, y > 0$. Then, using the convolution formula (3.62), we have:

$$f_Z(z) = \int_0^z \lambda e^{-\lambda x} \lambda e^{-\lambda (z-x)}\, dx$$

$$= \lambda^2 e^{-\lambda z} \int_0^z dx$$

$$= \lambda^2 z e^{-\lambda z}, \qquad z > 0.$$

Thus Z has a gamma density with parameters λ and $\alpha = 2$ (or equivalently, Z has a two-stage Erlang distribution). An expression for the reliability of a two-component standby redundant system is obtained by using equation (3.19):

$$R(t) = 1 - F(t) = \sum_{k=0}^{1} \frac{(\lambda t)^k}{k!} e^{-\lambda t}$$

$$= (1 + \lambda t)e^{-\lambda t}, \qquad t \geqslant 0. \tag{3.63}$$

Figure 3.25 compares the simplex reliability with a two-component standby system reliability. #

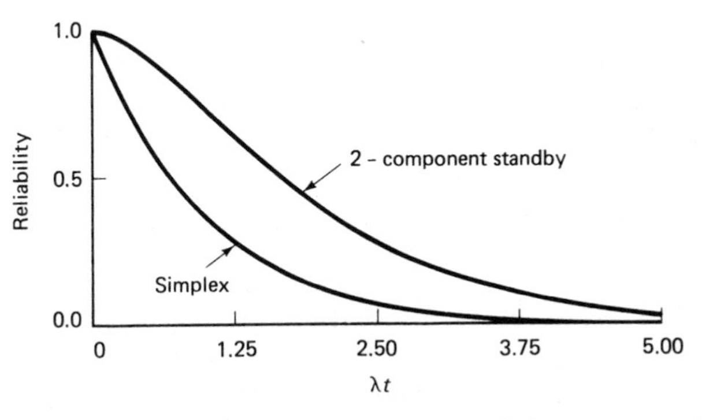

Figure 3.25 Reliabilities of simplex and two-component standby systems

The above example is a special case of the following theorem, which will be proved in Chapter 4.

THEOREM 3.2. If $X_1, X_2, \ldots, X_r$ are mutually independent, identically distributed random variables so that $X_i \sim \text{EXP}(\lambda)$ for each i, then the random variable $X_1 + X_2 + \cdots + X_r$ has an r-stage Erlang distribution with parameter λ.

As a consequence of this theorem, the reliability expression for a standby redundant system with a total of n independent components, each of which has an exponentially distributed lifetime with parameter λ, is given by:

$$R_{\text{standby}}(t) = \sum_{k=0}^{n-1} \frac{(\lambda t)^k}{k!} e^{-\lambda t}, \qquad t \geqslant 0. \tag{3.64}$$

Example 3.22

Consider a system consisting of n processors. One way to operate the system is to use the concept of standby sparing, and the system reliability will be given by the expression (3.64) above. Note that in that case only one processor is active at a time, while the others are idle. Now let us consider another way of utilizing the system. In the beginning we let all n processors be active, performing different computations, so that the total computing capacity is n (where a unit of computing capacity corresponds to that of one active processor).

Let $X_1, X_2, \ldots, X_n$ be the times to failure of the n processors. Then after a period of time $Y_1 = \min\{X_1, X_2, \ldots, X_n\}$ only $n - 1$ processors will be active and the computing capacity of the system will have dropped to $n - 1$. The cumulative computing capacity that the system supplies until all processors have failed is then given by the random variable:

$$C_n = nY_1 + (n-1)(Y_2 - Y_1) + \cdots + (n-j)(Y_{j+1} - Y_j) + \cdots + (Y_n - Y_{n-1}).$$

From Figure 3.26, we note that C_n is the area under the curve. Beaudry [BEAU 1978] has coined the phrase "computation before failure" for C_n, while Meyer [MEYE 1980] prefers the term "performability."

In order to obtain the distribution of C_n, we first obtain the distribution of $Y_{j+1} - Y_j$. If we assume processor lifetimes are mutually independent EXP (λ) random variables, then we claim that the distribution of $Y_{j+1} - Y_j$ is EXP $[(n-j)\lambda]$. Define $Y_0 = 0$. Then, since we know that

$$Y_1 = \min\{X_1, X_2, \ldots, X_n\} \sim \text{EXP}(n\lambda),$$

our claim holds for $j = 0$. After j processors have failed, the residual lifetimes of the remaining $(n - j)$ processors, denoted by $W_1, W_2, \ldots, W_{n-j}$, are each exponentially distributed with parameter λ due to the memoryless property of the exponential distribution. Note that $Y_{j+1} - Y_j$ is simply the time between the $(j + 1)$st and the jth failure; that is,

$$Y_{j+1} - Y_j = \min\{W_1, W_2, \ldots, W_{n-j}\}.$$

It follows that $Y_{j+1} - Y_j \sim \text{EXP}[(n-j)\lambda]$. Hence, using Example 3.12 we get:

$$(n-j)(Y_{j+1} - Y_j) \sim \text{EXP}(\lambda).$$

Therefore, C_n is the sum of n independent identically distributed exponential random variables. It follows from Theorem 3.2 that C_n is n-stage Erlang distributed with parameter λ. Now, since the standby redundant system of expression (3.64) has a unit processing capacity while functioning and the total duration of its lifetime is n-stage Erlang with parameter λ, we conclude that the computation capacity before failure has the same distribution in both modes of operation. (Remember the assumptions behind our model, however.) #

Example 3.23

Consider a computer system with job interarrival times that are independent and exponentially distributed with parameter λ. Let X_i be the random variable denoting the time between the $(i - 1)$st and ith arrival. Then $Z_r = X_1 + X_2 + \ldots + X_r$ is the time until the rth arrival and has an r-stage Erlang distribution. Another way to obtain this result is to consider N_t, the number of arrivals in the interval $(0, t]$. As pointed out earlier, N_t has a Poisson distribution with parameter λt. Now the events $[Z_r > t]$ and $[N_t < r]$ are equivalent. Therefore,

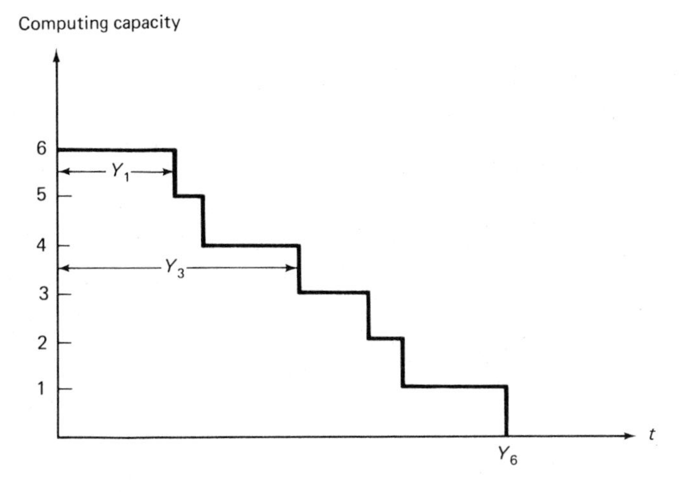

Figure 3.26 Computing capacity as a function of time

$$P(Z_r > t) = P(N_t < r) = \sum_{j=0}^{r-1} P(N_t = j) = \sum_{j=0}^{r-1} e^{-\lambda t}\left[\frac{(\lambda t)^j}{j!}\right],$$

which implies that:

$$F_{Z_r}(t) = P(Z_r \leqslant t) = 1 - \sum_{j=0}^{r-1} \frac{(\lambda t)^j}{j!} e^{-\lambda t},$$

which is the r-stage Erlang distribution function. #

In Example 3.21 of standby redundancy, we assumed that the failure rates of the two components were the same. Now let the failure rates be distinct; that is, let X and Y be exponentially distributed with parameters λ_1 and λ_2, respectively.

THEOREM 3.3. If $X \sim$ EXP (λ_1), $Y \sim$ EXP (λ_2), X and Y are independent, and $\lambda_1 \neq \lambda_2$, then $Z = X + Y$ has a two-stage hypoexponential distribution with parameters λ_1 and λ_2; that is, $Z \sim$ HYPO (λ_1, λ_2).

Proof:

$$f_Z(z) = \int_0^z f_X(x) f_Y(z-x)\, dx, \qquad z > 0 \qquad \text{[by equation (3.62)]}$$

$$= \int_0^z \lambda_1 e^{-\lambda_1 x} \lambda_2 e^{-\lambda_2 (z-x)}\, dx$$

$$= \lambda_1 \lambda_2\, e^{-\lambda_2 z} \int_0^z e^{(\lambda_2 - \lambda_1) x}\, dx$$

$$= \lambda_1 \lambda_2 e^{-\lambda_2 z} \left[\frac{e^{-(\lambda_1 - \lambda_2) x}}{-(\lambda_1 - \lambda_2)} \right]_{x=0}^{x=z}$$

$$= \frac{\lambda_1 \lambda_2}{\lambda_1 - \lambda_2} e^{-\lambda_2 z} + \frac{\lambda_1 \lambda_2}{\lambda_2 - \lambda_1} e^{-\lambda_1 z}.$$

Comparing the above density with equation (3.15), we conclude $X + Y$ has a two-stage hypoexponential distribution with parameters λ_1 and λ_2.

A more general version of the above theorem is stated without proof.

THEOREM 3.4 Let $Z = \sum_{i=1}^{r} X_i$, where $X_1, X_2, \ldots, X_r$ are mutually independent and X_i is exponentially distributed with parameter λ_i ($\lambda_i \neq \lambda_j$ for $i \neq j$). Then the density of Z, which is an r-stage hypoexponentially distributed random variable, is given by:

$$f_Z(z) = \sum_{i=1}^{r} a_i \lambda_i e^{-\lambda_i z}, \qquad z > 0, \tag{3.66}$$

where:

$$a_i = \prod_{\substack{j=1 \\ j \neq i}}^{r} \frac{\lambda_j}{\lambda_j - \lambda_i}, \qquad 1 \leqslant i \leqslant r. \tag{3.67}$$

Such a stage type distribution is often visualized as in Figure 3.27.

Another related result is stated by the following:

COROLLARY 3.4. If $X_1 \sim$ HYPO $(\lambda_1, \lambda_2, \ldots, \lambda_k)$, X_2 HYPO $(\lambda_{k+1}, \ldots, \lambda_r)$, and X_1 and X_2 are independent, then $X_1 + X_2 \sim$ HYPO $(\lambda_1, \lambda_2, \ldots, \lambda_k, \lambda_{k+1}, \ldots, \lambda_r)$.

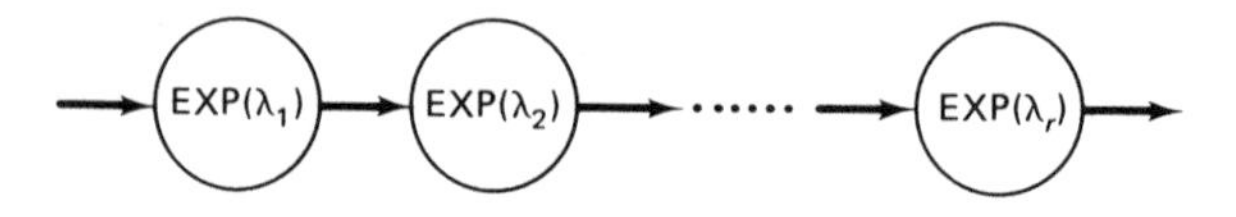

Figure 3.27 Hypoexponential as a series of exponential stages

Example 3.24

We have noted that a TMR system has higher reliability than simplex for short missions only. To improve upon the performance of TMR, we observe that after one of the three units have failed, both of the two remaining units must function properly for the classical TMR configuration to function properly. Thus after one failure, the system reduces to a series system of two components, from the reliability point of view. An improvement over this simple scheme, known as TMR/simplex, detects a single component failure, discards the failed component, and reverts to one of the nonfailing simplex components. In other words, not only the failed component but also one of the good components is discarded.

Let X, Y, Z denote the times to failure of the three components. Also let W denote the residual time to failure of the selected surviving component. Let X, Y, Z be mutually independent and exponentially distributed with parameter λ. If L denotes the time to failure of TMR/simplex, then it is clear that

$$L = \min\{X, Y, Z\} + W.$$

Now since the exponential distribution is memoryless, it follows that the lifetime W of the surviving component is exponentially distributed with parameter λ. Also, from our discussion of order statistics, it follows that $\min\{X, Y, Z\}$ is exponentially distributed with parameter 3λ. Then L has a two-stage hypoexponential distribution with parameters 3λ and λ (using Theorem 3.3). Therefore, using equation (3.16), we have:

$$F_L(t) = 1 - \frac{3\lambda}{2\lambda} e^{-\lambda t} + \frac{\lambda}{2\lambda} e^{-3\lambda t}, \qquad t \geqslant 0$$

$$= 1 - \frac{3e^{-\lambda t}}{2} + \frac{e^{-3\lambda t}}{2}$$

Thus the reliability expression of TMR/simplex is given by:

$$R(t) = \frac{3e^{-\lambda t}}{2} - \frac{e^{-3\lambda t}}{2}. \tag{3.68}$$

It is not difficult to see that TMR/simplex has a higher reliability than either a simplex or an ordinary TMR system for all $t \geqslant 0$. Figure 3.28 compares the simplex reliability with that of TMR and that of TMR/simplex. #

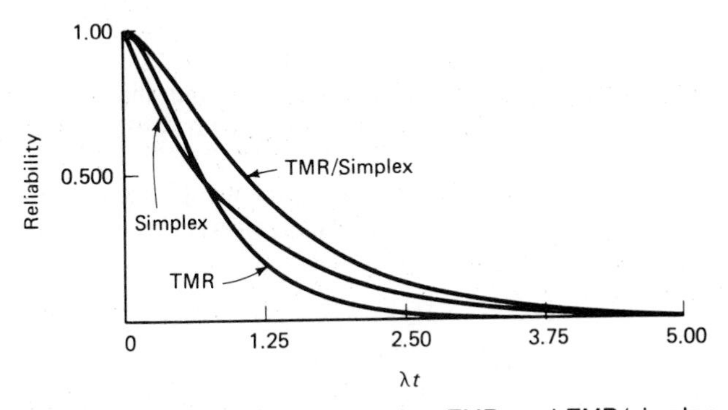

Figure 3.28 Comparison of simplex, TMR, and TMR/simplex reliabilities

Example 3.25

Consider a module shown in Figure 3.29, consisting of a functional unit (e.g., an adder) together with an on-line fault detector (e.g., a modulo-3 checker). Let T and C, respectively, denote the times to failure of the unit and the detector. After the unit fails, a finite time D (called the detection latency) is required to detect the failure. Failure of the detector, however, is detected instantaneously. Let X denote the time to failure indication and Y denote the time to failure occurrence (of either the detector or the unit). Clearly, $X = \min\{T + D, C\}$ and $Y = \min\{T, C\}$. If the detector fails before the unit, then a false alarm is said to have occurred. If the unit fails before the detector, then the unit keeps producing erroneous output during the detection phase and thus propagates the effect of the failure. The purpose of the detector is to reduce the detection time D.

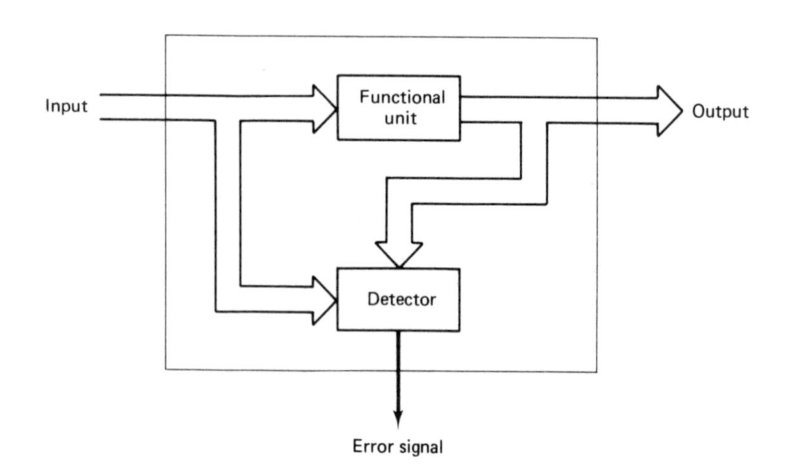

Figure 3.29 A module with on-line fault detector

We define the real reliability $R_r(t) = P(Y \geqslant t)$ and the apparent reliability $R_a(t) = P(X \geqslant t)$. A powerful detector will tend to narrow the gap between $R_r(t)$ and $R_a(t)$.

Assume that T, D and C are mutually independent and exponentially distributed with parameters λ, δ, and α, respectively. Then, clearly, Y is exponentially distributed with parameter $\lambda + \alpha$ and:

$$R_r(t) = e^{-(\lambda+\alpha)t}.$$

Also, $T + D$ is hypoexponentially distributed, so that:

$$F_{T+D}(t) = 1 - \frac{\delta}{\delta - \lambda} e^{-\lambda t} + \frac{\lambda}{\delta - \lambda} e^{-\delta t}.$$

Next the apparent reliability:

$$\begin{aligned}
R_a(t) &= P(X \geqslant t) \\
&= P(\min\{T + D,\ C\} \geqslant t) \\
&= P(T + D \geqslant t \text{ and } C \geqslant t) \\
&= P(T + D \geqslant t)\ P(C \geqslant t) \qquad \text{by independence} \\
&= [1 - F_{T+D}(t)]\ e^{-\alpha t} \\
&= \frac{\delta}{\delta - \lambda}\ e^{-(\lambda+\alpha)t} - \frac{\lambda}{\delta - \lambda}\ e^{-(\delta+\alpha)t}.
\end{aligned}$$

#

Many of the examples in the previous sections can be interpreted as hypoexponential random variables.

Example 3.26

Consider the TMR system and let $X,\ Y,\ Z$ denote the lifetimes of the three components. Assume that these random variables are mutually independent and exponentially distributed with parameter λ. Let L denote the lifetime of the TMR system. Then:

$$L = \min\{X,\ Y,\ Z\} + \min\{U,\ V\}.$$

Here U and V denote the residual lifetimes of the two surviving components after the first failure. By the memoryless property of the exponential distribution, we conclude that U and V are exponentially distributed with parameter λ. Therefore min $\{X,\ Y,\ Z\}$ has exponential distribution with parameter 3λ and min $\{U,\ V\}$ has exponential distribution with parameter 2λ. Therefore, L has hypoexponential distribution with parameters 3λ and 2λ. Then the density of L is:

$$\begin{aligned}
f_L(t) &= \frac{6\lambda^2}{3\lambda - 2\lambda}\ e^{-2\lambda t} + \frac{6\lambda^2}{2\lambda - 3\lambda}\ e^{-3\lambda t} \\
&= 6\lambda e^{-2\lambda t} - 6\lambda e^{-3\lambda t}.
\end{aligned}$$

The distribution function of L is:

$$\begin{aligned}
F_L(t) &= \frac{6}{2}\ (1 - e^{-2\lambda t}) - \frac{6}{3}\ (1 - e^{-3\lambda t}) \\
&= 1 - 3e^{-2\lambda t} + 2e^{-3\lambda t}.
\end{aligned}$$

Finally, the reliability of TMR is:

$$\begin{aligned}
R_{\text{TMR}}(t) &= 1 - F_L(t) \\
&= 3e^{-2\lambda t} - 2e^{-3\lambda t}.
\end{aligned}$$

This agrees with expression (3.58) derived earlier.

#

THEOREM 3.5. The order statistic Y_{n-k+1} (of $X_1, X_2, \ldots, X_n$) is hypoexponentially distributed with $(n - k + 1)$ phases, that is:

$$Y_{n-k+1} \sim \text{HYPO}\,[n\lambda,\ (n-1)\lambda,\ \ldots,\ k\lambda]$$

if $X_i \sim \text{EXP}(\lambda)$ for each i, and if $X_1, X_2, \ldots, X_n$ are mutually independent random variables.

Proof: We prove this theorem by induction. Let $n - k + 1 = 1$, then:

$$Y_1 = \min\{X_1,\ X_2,\ \ldots,\ X_n\}$$

and clearly $Y_1 \sim \text{EXP}(n\lambda)$, which can be interpreted as a one-stage hypoexponential, HYPO $(n\lambda)$. Next assume that Y_{n-j+1} is hypoexponentially distributed with parameters $n\lambda,\ (n-1)\lambda,\ \ldots,\ j\lambda$. It is clear that:

$$Y_{n-j+2} = Y_{n-j+1} + \min\{W_{n-j+2},\ \ldots,\ W_n\},$$

where the W_i $(n - j + 2 \leqslant i \leqslant n)$ denote the residual lifetimes of the surviving components. By the memoryless property of the exponential distribution, W_i has exponential distribution with parameter λ. Therefore:

$$\min\{W_{n-j+2},\ \ldots,\ W_n\}$$

has an exponential distribution with parameter:

$$[n - (n - j + 2) + 1]\lambda = (j - 1)\lambda.$$

The proof of the theorem then follows using Corollary 3.4. The result of the theorem may be visualized as in Figure 3.30.

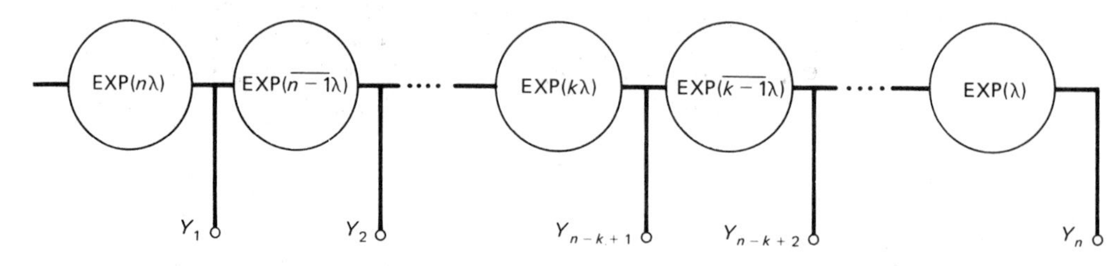

Figure 3.30 The order statistics of the exponential distribution

Example 3.27

Consider an m out of n system, each of whose components follow an exponential failure law with parameter λ. Then, with the usual assumption of independence, the lifetime of the system is given by $L(m|n) = Y_{n-m+1}$ and is hypoexponentially distributed with parameters $n\lambda,\ (n-1)\lambda,\ \ldots,\ m\lambda$. The density of $L(m|n)$ is given by:

$$f(t) = \sum_{i=m}^{n} a_i\,\lambda_i\,e^{-\lambda_i t}, \text{ where } \lambda_i = i\lambda, \text{ and}$$

$$
\begin{aligned}
a_i &= \prod_{\substack{j=m \\ j \neq i}}^{n} \frac{\lambda_j}{\lambda_j - \lambda_i} \\
&= \prod_{\substack{j=m \\ j \neq i}}^{n} \frac{j}{j - i} \\
&= \frac{m(m+1)\cdots(i-1)(i+1)\cdots(n-1)n}{(m-i)\cdots(-1)(1)\cdots(n-i)} \\
&= (-1)^{i-m} \frac{(i-1)!n!}{(m-1)!(i-m)!i!(n-i)!} \\
&= (-1)^{i-m} \binom{n}{i} \binom{i-1}{m-1}.
\end{aligned}
$$

Then:

$$
\begin{aligned}
R(t) &= \sum_{i=m}^{n} a_i e^{-\lambda_i t} \\
&= \sum_{i=m}^{n} \binom{n}{i} \binom{i-1}{m-1} (-1)^{i-m} e^{-i\lambda t} \qquad (3.69)
\end{aligned}
$$

[Note that by substituting $m = 1$ in (3.69), we get (3.56).] It can be verified [see problem 4 at the end of this section], using a set of combinatorial identities, that the above expression is equivalent to:

$$
\sum_{i=m}^{n} \binom{n}{i} e^{-i\lambda t} (1 - e^{-\lambda t})^{n-i}
$$

as derived earlier. #

Example 3.28

Consider a hybrid NMR system with $N + S$ independent components, N of which are initially put into operation with the remaining S components in a deenergized standby status. An active component has an exponential failure law with parameter λ. Unlike our earlier examples, we assume that a component can fail in a deenergized state with a constant failure rate μ (presumably $0 \leqslant \mu \leqslant \lambda$). Let X_i $(1 \leqslant i \leqslant N)$ denote the lifetime of an energized component and let Y_j $(1 \leqslant j \leqslant S)$ denote the lifetime of a deenergized component. Then the system lifetime $L(m|N, S)$ is given by:

$$
\begin{aligned}
L(m|N, S) &= \min(X_1, X_2, \ldots, X_N; Y_1, Y_2, \ldots, Y_S) + L(m|N, S-1) \\
&= W(N, S) + L(m|N, S-1).
\end{aligned}
$$

This follows since $W(N, S)$ is the time to first failure among the $N + S$ components and after the removal of the failed component, the system has N energized and $S - 1$

deenergized components. Note that these $N + S - 1$ components have not aged by the exponential assumption. Therefore:

$$L(m|N, S) = L(m|N, 0) + \sum_{i=1}^{S} W(N, i). \tag{3.70}$$

Here $L(m|N, 0) = L(m|N)$ is simply the lifetime of an m-out-of-N system and is therefore the kth order statistic with $k = N - m + 1$ as shown in Example 3.31. The distribution of $L(m|N, 0)$ is therefore an $(N - m + 1)$-phase hypoexponential with parameters $N\lambda$, $(N-1)\lambda, \ldots, m\lambda$. Also, $W(N, i)$ has an exponential distribution with parameter $N\lambda + i\mu$. Then, using corollary 3.4, we conclude that $L(m|N, S)$ has an $(N + S - m + 1)$ stage hypoexponential distribution with parameters $N\lambda + S\mu$, $N\lambda + (S-1)\mu, \ldots, N\lambda + \mu$, $N\lambda$, $(N-1)\lambda, \ldots, m\lambda$. This can be visualized as in Figure 3.31.

Let $R_{[m|N,S]}(t)$ denote the reliability of such a system; then:

$$R_{[m|N,S]}(t) = \sum_{i=1}^{S} a_i\, e^{-(N\lambda + i\mu)t} + \sum_{i=m}^{N} b_i\, e^{-i\lambda t}, \tag{3.71}$$

where:

$$a_i = \prod_{\substack{j=1\\ j\neq i}}^{S} \frac{N\lambda + j\mu}{j\mu - i\mu} \prod_{j=m}^{N} \frac{j\lambda}{j\lambda - N\lambda - i\mu}, \tag{3.72}$$

and

$$b_i = \prod_{j=1}^{S} \frac{N\lambda + j\mu}{(N-i)\lambda + j\mu} \prod_{\substack{j=m\\ j\neq i}}^{N} \frac{j\lambda}{j\lambda - i\lambda}. \tag{3.73}$$

Letting $K = \lambda/\mu$:

$$\begin{aligned}
a_i &= \frac{(NK+S)\cdots(NK+1)}{(NK+i)(S-i)\cdots(1)(-1)\cdots(1-i)} \\
&\quad \cdot \frac{(-1)^{N-m+1}N(N-1)\cdots m}{\left(\frac{i}{K} + N - m\right)\cdots\left(\frac{i}{K}+1\right)\left(\frac{i}{K}\right)} \\
&= (-1)^{i-1}\frac{(NK+S)!S!i}{(NK+i)(NK)!S!(S-i)!i!} \\
&\quad \cdot (-1)^{N-m+1}\frac{N(N-1)!\left(\frac{i}{K}\right)!(N-m)!}{\frac{i}{K}(m-1)!\left[\left(\frac{i}{K}\right)N-m\right]!(N-m)!} \\
&= (-1)^{N-m+i}\frac{\binom{NK+S}{S}\binom{S}{i}\binom{N-1}{m-1}}{\left(1+\frac{i}{NK}\right)\binom{\frac{i}{K}+N-m}{N-m}}.
\end{aligned} \tag{3.74}$$

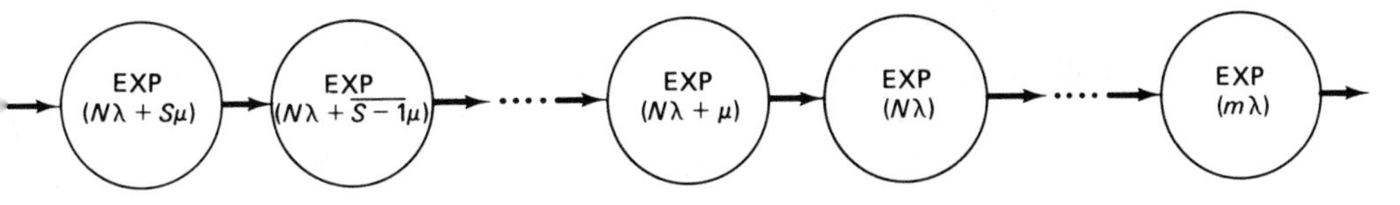

Figure 3.31 Lifetime distribution of a hybrid NMR system

[Note that if K is not an integer, we use the generalized definition of factorial above. Thus, for a real number α, $\alpha! = \Gamma(\alpha + 1)$.]
Similarly:

$$\begin{aligned} b_i &= \frac{(NK + S) \cdots (NK + 1)}{[(N - i)K + S] \cdots [(N - i)K + 1]} \\ &\quad \cdot \frac{N \cdots m}{i[(N - i) \cdots (1)(-1) \cdots (m - i)]} \\ &= \frac{(NK + S)![(N - i)K]!N!(-1)^{i-m}}{(NK)![(N - i)K + S]!i(m - 1)!(N - i)!(i - m)!} \\ &= (-1)^{i-m} \frac{(NK + S)!S!((N - i)K)!N!i!m}{S!(NK)![(N - i)K + S]!(N - i)!i!(i - m)!m!i} \\ &= (-1)^{i-m} \frac{\binom{NK + S}{S}\binom{N}{i}\binom{i}{m}}{\frac{i}{m}\binom{(N - i)K + S}{S}}. \end{aligned} \tag{3.75}$$

Once again, using combinatorial identities, we can verify that our expression for hybrid NMR reliability matches with that given by Mathur and Avizienis [MATH 1970]. Also see Ng's thesis [NG 1976] in this connection. #

Problems

1. Given n random numbers $u_1, u_2, \ldots, u_n$, derive an expression for a random deviate of an n-stage hypoexponential distribution with parameters $\lambda_1, \lambda_2, \ldots, \lambda_n$.

2. Compare the TMR/Simplex reliability with two-component and three-component redundant systems having standby redundancy. Graph the expressions on the same plot.

3. Repeat problem 2 for two- and three-component parallel-redundant systems.

4. Show that the reliability expression (3.69) for NMR reliability reduces to the expression:

$$\sum_{i=m}^{n} \binom{n}{i} e^{-i\lambda t}(1 - e^{-\lambda t})^{n-i}.$$

5. Using (3.71) obtain an explicit expression for the reliability of an hybrid-TMR system with one spare. Compare the reliability of this system with those of a TMR system and a simplex system by plotting. Use $\lambda = 1/1000$ per hour and $\mu = 1/10000$ per hour.

6. Compare (by plotting) reliability expressions for the simplex system, two component parallel redundant system, and two component standby redundant system. Assume that the failure rate of an active component is constant at $1/1000$ per hour, the failure rate of a spare is zero, and that the switching mechanism is fault-free.

3.9 FUNCTIONS OF NORMAL RANDOM VARIABLES

The normal distribution has great importance in mathematical statistics because of the central limit theorem alluded to earlier. This distribution also plays an important role in communication and information theory. We will now study distributions derivable from the normal distribution. The use of most of these distributions will be deferred until Chapters 10 and 11.

THEOREM 3.6. Let $X_1, X_2, \ldots, X_n$ be mutually independent random variables such that $X_i \sim N(\mu_i, \sigma_i^2)$, $i = 1, 2, \ldots, n$. Then $S_n = \sum_{i=1}^{n} X_i$ is normally distributed, that is, $S_n \sim N(\mu, \sigma^2)$, where:

$$\mu = \sum_{i=1}^{n} \mu_i \quad \text{and} \quad \sigma^2 = \sum_{i=1}^{n} \sigma_i^2.$$

Owing to this theorem, we say that the normal distribution has the **reproductive** property. A proof of this theorem will be given in Chapter 4. The theorem can be further generalized as in problem 1 at the end of Section 4.4, so that if $X_1, X_2, \ldots, X_n$ are mutually independent random variables with $X_i \sim N(\mu_i, \sigma_i^2)$, $i = 1, 2, \ldots, n$ and $a_1, a_2, \ldots, a_n$ are real constants, then $Y_n = \sum_{i=1}^{n} a_i X_i$ is normally distributed; that is, $Y_n \sim N(\mu, \sigma^2)$, where:

$$\mu = \sum_{i=1}^{n} a_i\mu_i \quad \text{and} \quad \sigma^2 = \sum_{i=1}^{n} a_i^2\sigma_i^2.$$

In particular, if we let $n = 2$, $a_1 = +1$, and $a_2 = -1$, then we conclude that the difference $Y = X_1 - X_2$ of two independent normal random variables $X_1 \sim N(\mu_1, \sigma_1^2)$ and $X_2 \sim N(\mu_2, \sigma_2^2)$ is normally distributed, that is, $Y \sim N(\mu_1 - \mu_2, \sigma_1^2 + \sigma_2^2)$.

Example 3.29

It has been empirically determined that the memory requirement of a program (called the working-set size of the program) is approximately normal. In a multiprogramming system, the number of programs sharing the main memory simultaneously (called the degree of multiprogramming) is found to be n. Now if X_i denotes the working-set size of the ith program with $X_i \sim N(\mu_i, \sigma_i^2)$, then it follows that the sum total memory demand, S_n, of the n programs is normally distributed with parameters $\mu = \sum_{i=1}^{n} \mu_i$ and $\sigma^2 = \sum_{i=1}^{n} \sigma_i^2$. #

Example 3.30

A sequence of independent, identically distributed random variables, X_1, X_2, ..., X_n is known in mathematical statistics, as a *random sample* of size n. In many problems of statistical sampling theory, it is reasonable to assume that the underlying distribution is the normal distribution. Thus let $X_i \sim N(\mu, \sigma^2)$, $i = 1, 2, \ldots, n$. Then from Theorem 3.6, $S_n = \sum_{i=1}^{n} X_i \sim N(n\mu, n\sigma^2)$.

One important function known as the sample mean is quite useful in problems of statistical inference. *Sample mean* $\overline{X}$ is given by:

$$\overline{X} = \frac{S_n}{n} = \sum_{i=1}^{n} \frac{X_i}{n}. \tag{3.76}$$

To obtain the pdf of the sample mean $\overline{X}$, we use equation (3.45) to obtain:

$$f_{\overline{X}} = n f_{S_n}(nx).$$

But since $S_n \sim N(n\mu, n\sigma^2)$, we have:

$$f_{\overline{X}}(x) = n \frac{1}{\sqrt{2\pi}(\sqrt{n}\ \sigma)} e^{-\frac{(nx-n\mu)^2}{2n\sigma^2}}, \qquad -\infty < x < \infty,$$

$$= \frac{1}{\sqrt{2\pi}\,[\sigma (n)^{-1/2}]} e^{-\frac{(x-\mu)^2}{2(\sigma^2/n)}}, \qquad -\infty < x < \infty.$$

It follows that $\overline{X} \sim N(\mu, \sigma^2/n)$. Similarly, it can be shown that the random variable $(\overline{X} - \mu)\sqrt{n}/\sigma$ has the standard normal distribution, $N(0, 1)$. #

If X is $N(0, 1)$, we know from Example 3.9 that $Y = X^2$ is gamma distributed with $Y \sim GAM(½, ½)$, which is the chi-square distribution with one degree of freedom. Now consider X_1, X_2 that are independent standard normal random variables and $Y = X_1^2 + X_2^2$.

Example 3.31

If $X_1 \sim N(0, 1)$, $X_2 \sim N(0, 1)$ and X_1 and X_2 are independent, then $Y = X_1^2 + X_2^2$ is exponentially distributed so that $Y \sim EXP(½)$.

To see this, we obtain the distribution function of Y:

$$F_Y(y) = P(X_1^2 + X_2^2 \leqslant y)$$

$$= \iint\limits_{x_1^2 + x_2^2 \leqslant y} f_{X_1X_2}(x_1, x_2)\, dx_1\, dx_2.$$

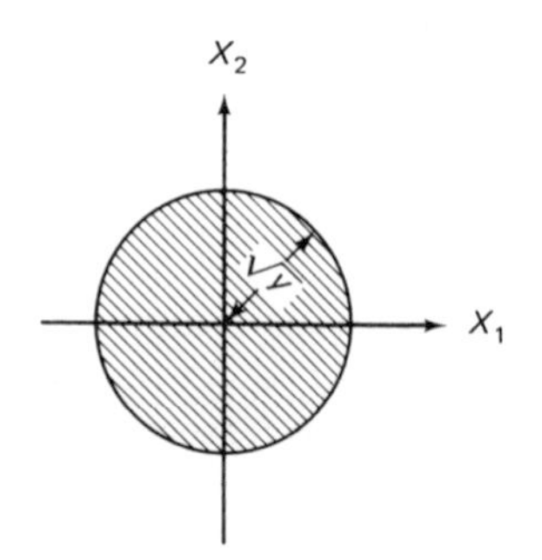

Figure 3.32 The area of integration for Example 3.31

Note that the surface of integration is a circular area about the origin with the radius $\sqrt{y}$ (see Figure 3.32). Using the fact that X_1 and X_2 are independent, and standard normal, we have:

$$F_Y(y) = \iint_{x_1^2 + x_2^2 \leqslant y} \frac{1}{2\pi} e^{-\frac{x_1^2 + x_2^2}{2}} dx_1\, dx_2.$$

Making a change of variables (to polar coordinates), $x_1 = r \cos \theta$, $x_2 = r \sin \theta$, so that $r^2 = x_1^2 + x_2^2$ and $\theta = \tan^{-1}(x_2/x_1)$, we have:

$$F_Y(y) = \int_{\theta=0}^{2\pi} \int_{r=0}^{\sqrt{y}} \frac{r}{2\pi} e^{-r^2/2}\, dr\, d\theta$$

$$= \begin{cases} 1 - e^{-y/2}, & y > 0, \\ 0, & \text{otherwise.} \end{cases}$$

Therefore, Y is exponentially distributed with parameter ½. #

The above example is a special case of the following theorem:

THEOREM 3.7. If $X_1, X_2, \ldots, X_n$ is a sequence of mutually independent, standard normal random variables, then:

$$Y = \sum_{i=1}^{n} X_i^2$$

has the gamma distribution, GAM (½, n/2), or the chi-square distribution with n degrees of freedom, X_n^2.

The above theorem follows from the reproductive property of the gamma distribution:

THEOREM 3.8. Let $X_1, X_2, \ldots, X_n$ be a sequence of mutually independent gamma random variables such that $X_i \sim$ GAM (λ, α_i) for $i = 1, 2, \ldots, n$. Then $S_n = \sum_{i=1}^{n} X_i$ has the gamma distribution GAM (λ, α), where $\alpha = \alpha_1 + \alpha_2 + \cdots + \alpha_n$. This theorem will be proved in Chapter 4.

Since $\chi_n^2 \sim$ GAM $(\frac{1}{2}, n/2)$, we have the following corollary:

COROLLARY 3.8. Let $Y_1, Y_2, \ldots, Y_n$ be mutually independent chi-square random variables such that $Y_i \sim \chi_{k_i}^2$. Then $Y_1 + Y_2 + \cdots + Y_n$ has the χ_k^2 distribution, where:

$$k = \sum_{i=1}^{n} k_i.$$

Example 3.32

Assume that $X_1, X_2, \ldots, X_n$ are mutually independent, identically distributed normal random variables such that $X_i \sim N(\mu, \sigma^2)$, $i = 1, 2, \ldots, n$. It follows that $Z_i = (X_i - \mu)/\sigma$ is standard normal. Thus $Z_1, Z_2, \ldots, Z_n$ are independent standard normal random variables. Hence, using Theorem 3.7, we have:

$$Y = \sum_{i=1}^{n} Z_i^2 = \sum_{i=1}^{n} \frac{(X_i - \mu)^2}{\sigma^2} \tag{3.77}$$

has the chi-square distribution with n degrees of freedom. Note that the random variable $\sum_{i=1}^{n} (X_i - \mu)^2/n$ may be used as an estimator of the parameter σ^2. #

Example 3.33

In the last example, we suggested that $\sum_{i=1}^{n} (X_i - \mu)^2/n$ may be used as an estimator of the parameter σ^2 assuming that $X_1, X_2, \ldots, X_n$ are independent observations from a normal distribution $N(\mu, \sigma^2)$. However, this expression assumes that the parameter μ of the distribution is already known. This is rarely the case in practice, and the sample mean $\bar{X} = \sum_{i=1}^{n} X_i/n$ is usually substituted in its place. Thus, the random variable:

$$U = \frac{\sum_{i=1}^{n} (X_i - \bar{X})^2}{n - 1} \tag{3.78}$$

is usually used as an estimator of the parameter σ^2 and is often denoted by S^2. (The reason for the value $n - 1$ rather than n in the denominator will be seen in Chapter 10.)

Rewriting:

$$S^2 = U = \frac{\sigma^2}{n - 1} \sum_{i=1}^{n} \left(\frac{X_i - \bar{X}}{\sigma}\right)^2, \tag{3.79}$$

we note that n random variables $\{(X_i - \overline{X})/\sigma \mid 1 \leqslant i \leqslant n\}$ in the above expression satisfy the relation:

$$\sum_{i=1}^{n} \frac{X_i - \overline{X}}{\sigma} = 0 \tag{3.80}$$

(from the definition of the sample mean, $\overline{X}$). Thus they are linearly dependent. It can be shown that the random variable:

$$W = \sum_{i=1}^{n} \left(\frac{X_i - \overline{X}}{\sigma}\right)^2 \tag{3.81}$$

can be transformed to a sum of $(n-1)$ independent chi-square random variables, and hence $W = (n-1)U/\sigma^2 = (n-1)S^2/\sigma^2$ has a chi-square distribution with $n-1$ degrees of freedom (rather than n degrees of freedom). #

Just as the sums of chi-square random variables are of interest, so is the ratio of two chi-square random variables. First assume that X and Y are independent, positive-valued random variables and let Z be their quotient; that is:

$$Z = \frac{Y}{X}. \tag{3.82}$$

Then the distribution function of Z is obtained using the formula:

$$F_Z(z) = \iint_{A_z} f(x, y)\, dx\, dy,$$

where the set:

$$A_z = \{(x, y) \mid y/x \leqslant z\}$$

is shown in Figure 3.33. Therefore:

$$\begin{aligned} F_Z(z) &= \int_0^{\infty} \left[\int_0^{xz} f(x, y)\, dy\right] dx \\ &= \int_0^{\infty} \left[\int_0^{z} x\, f(x, xv)\, dv\right] dx, \end{aligned} \tag{3.83}$$

after a change of variables $y = xv$.

It follows that the pdf of Z is given by:

$$\begin{aligned} f_Z(z) &= \int_0^{\infty} x\, f(x, xz)\, dx \\ &= \int_0^{\infty} x f_X(x) f_Y(xz)\, dx, \qquad 0 < z < \infty \end{aligned} \tag{3.84}$$

(by independence of X and Y).

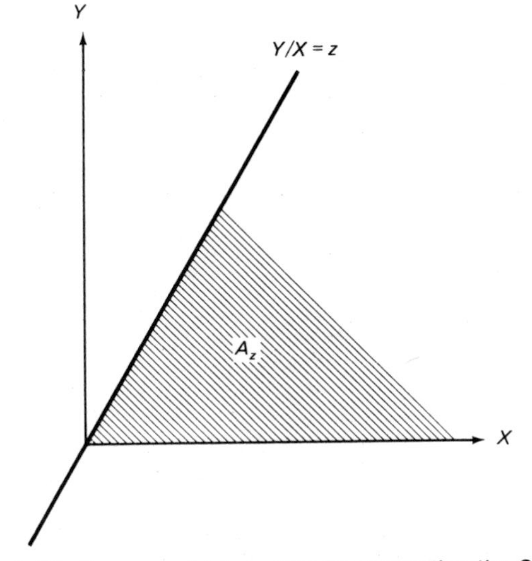

Figure 3.33 The area of integration for computing the CDF of $Y/X = z$

THEOREM 3.9. Let Y_1 and Y_2 be independent random variables with $X^2_{n_1}$ and $X^2_{n_2}$ distributions, respectively. Then:

$$Z = \frac{Y_1/n_1}{Y_2/n_2}$$

has the F distribution, which is characterized by two parameters, $(n_1,\ n_2)$,—that is, $Z \sim F_{n_1,n_2}$. The pdf of Z is given by:

$$f_Z(z) = \begin{cases} \dfrac{(n_1/n_2)\Gamma[(n_1+n_2)/2](n_1 z/n_2)^{(n_1/2)-1}}{\Gamma(n_1/2)\Gamma(n_2/2)[1+(n_1 z/n_2)]^{(n_1+n_2)/2}}, & z > 0, \\ 0, & \text{otherwise.} \end{cases} \tag{3.85}$$

Proof: Recall that:

$$f_{Y_1}(y_1) = \frac{y_1^{(n_1/2)-1} e^{-y_1/2}}{2^{n_1/2}\,\Gamma(n_1/2)}$$

and

$$f_{Y_2}(y_2) = \frac{y_2^{(n_2/2)-1} e^{-y_2/2}}{2^{n_2/2}\,\Gamma(n_2/2)}.$$

Let $Y = Y_1/n_1$ and $X = Y_2/n_2$. Using formula (3.45), it follows that:

$$f_Y(y) = \frac{n_1 (y n_1)^{n_1/2-1} e^{-(n_1 y)/2}}{2^{n_1/2}\Gamma(n_1/2)}$$

and

$$f_X(x) = \frac{n_2(xn_2)^{n_2/2-1} e^{-(n_2x)/2}}{2^{n_2/2}\Gamma(n_2/2)}.$$

Now applying equation (3.84), we get:

$$f_Z(z) = \int_0^\infty x \frac{n_1 n_2}{2^{(n_1+n_2)/2}\,\Gamma\left(\frac{n_1}{2}\right)\Gamma\left(\frac{n_2}{2}\right)} \cdot (xzn_1)^{n_1/2-1}(xn_2)^{n_2/2-1} e^{-\frac{n_1xz+n_2x}{2}}\,dx$$

$$= \frac{n_1n_2(n_1)^{n_1/2-1}(n_2)^{n_2/2-1}z^{n_1/2-1}}{2^{(n_1+n_2)/2}\Gamma\left(\frac{n_1}{2}\right)\Gamma\left(\frac{n_2}{2}\right)} \cdot \int_0^\infty x^{n_1/2+n_2/2-1} e^{-x(n_1z+n_2)/2}\,dx. \tag{3.86}$$

Using equation (3.26), we evaluate the last integral as:

$$\frac{\Gamma\left(\dfrac{n_1+n_2}{2}\right)}{\left(\dfrac{n_1z+n_2}{2}\right)^{\frac{n_1+n_2}{2}}}.$$

Subsituting this in (3.86), we get the required result as in (3.85).

Example 3.34

Suppose that $X_1, X_2, \ldots, X_m, X_{m+1}, \ldots, X_n$ are mutually independent normal random variables with the common distribution, $N(0, \sigma^2)$. Then by Theorem 3.7:

$$Y = \frac{\sum_{i=1}^{m} X_i^2}{\sigma^2} \quad \text{and} \quad X = \frac{\sum_{i=m+1}^{n} X_i^2}{\sigma^2}$$

are chi-square distributed with m and $(n - m)$ degrees of freedom, respectively. Furthermore, X and Y are independent. It follows by Theorem 3.9 that:

$$Z = \frac{\dfrac{\sum_{i=1}^{m} X_i^2}{m}}{\dfrac{\sum_{i=m+1}^{n} X_i^2}{(n-m)}} \tag{3.87}$$

has the $F_{m,n-m}$ distribution. #

The last distribution we introduce here is the Student's t distribution.

THEOREM 3.10. If V and W are independent random variables such that $V \sim N(0, 1)$ and $W \sim X_n^2$, then the random variable:

$$T = \frac{V}{\sqrt{W/n}} \tag{3.88}$$

has the t distribution with n degrees of freedom. The pdf of this random variable is given by:

$$f_T(t) = \frac{\Gamma\left(\frac{n+1}{2}\right)}{\sqrt{n\pi}\,\Gamma\left(\frac{n}{2}\right)}\left(1 + \frac{t^2}{n}\right)^{-(n+1)/2}, \quad -\infty < t < \infty. \tag{3.89}$$

For $n = 1$, the above pdf reduces to:

$$f_T(t) = \frac{1}{\pi(1 + t^2)}, \tag{3.90}$$

which is known as the Cauchy pdf.

The pdf (3.89) is plotted for various degrees of freedom in Figure 3.34. It may be noted that as n approaches infinity, the t distribution approaches the standard normal distribution.

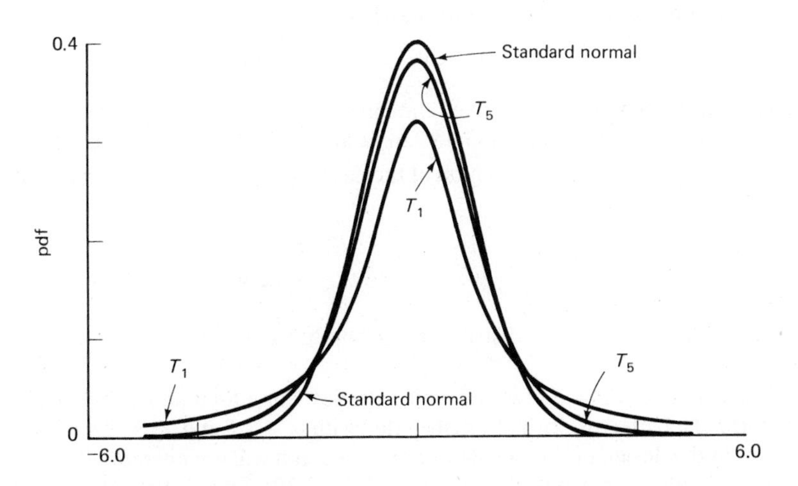

Figure 3.34 Students t pdf and its comparison with the standard normal pdf

Example 3.35

Assume that $X_1, X_2, \ldots, X_n$ are mutually independent identically distributed normal random variables such that $X_i \sim N(\mu, \sigma^2)$. Then from Example 3.30, it follows that:

$$V = \frac{(\bar{X} - \mu)\sqrt{n}}{\sigma} \tag{3.91}$$

has the standard normal distribution. Also, from Example 3.33:

$$\frac{(n-1)S^2}{\sigma^2} = W = \sum_{i=1}^{n} \left(\frac{X_i - \bar{X}}{\sigma}\right)^2 \tag{3.92}$$

has the X_{n-1}^2 distribution. It follows that:

$$T = \frac{V}{\sqrt{\dfrac{W}{(n-1)}}} = \frac{(\bar{X} - \mu)\sqrt{n}/\sigma}{\left(S\dfrac{\sqrt{n-1}}{\sigma}\right)} \cdot \sqrt{n-1}$$

$$= \frac{\bar{X} - \mu}{S/\sqrt{n}} \tag{3.93}$$

has the t distribution with $(n - 1)$ degrees of freedom. #

Problems

1. In communication theory, waveforms of the form:

   $$A(t) = x(t)\cos(\omega t) - y(t)\sin(\omega t)$$

   appear quite frequently. At a fixed time instant, $t = t_1$, $X = x(t_1)$, and $Y = y(t_1)$ are known to be independent Gaussian random variables, specifically, $N(0, \sigma^2)$. Show that the distribution function of the **envelope** $Z = \sqrt{X^2 + Y^2}$ is given by:

   $$F_Z(z) = \begin{cases} 1 - e^{-z^2/2\sigma^2}, & z > 0, \\ 0, & \text{otherwise.} \end{cases}$$

   This is called the Rayleigh distribution. Compute and plot its pdf.

2. A calculator operates on two 1.5-volt batteries (for a total of 3 volts). The actual voltage of a battery is normally distributed with $\mu = 1.5$ and $\sigma^2 = 0.45$. The tolerances in the design of the calculator are such that it will not operate satisfactorily if the total voltage falls outside the range (2.70, 3.30) volts. What is the probability that the calculator will function correctly?

Review Problems

1. Show that the pdf of the product $Z = XY$ of two independent random variables with respective densities f_X and f_Y is given by:

$$f_Z(z) = \int_{-\infty}^{\infty} \frac{1}{|x|} f_X(x) f_Y\left(\frac{z}{x}\right) dx.$$

*2. [HAMM 1973] Consider the problem of multiplying the mantissas X and Y of two floating point numbers in base β. Assume X and Y are independent random variables with densities f_X and f_Y respectively and $1/\beta \leqslant X,\ Y < 1$. Note that the product XY satisfies $1/\beta^2 \leqslant XY < 1$. If $1/\beta^2 \leqslant XY < 1/\beta$ then a left shift is required in order to normalize the result. Let Z_N denote the normalized product (that is, $Z_N = XY$ if $XY \geqslant 1/\beta$ and $Z_N = \beta\ XY$ otherwise). Show that the pdf of Z_N is given by:

$$f_{Z_N}(z) = \frac{1}{\beta} \int_{1/\beta}^{z} \frac{f_X(x)}{x} f_Y\left(\frac{z}{\beta x}\right) dx + \int_{z}^{1} \frac{f_X(x)}{x} f_Y\left(\frac{z}{x}\right) dx, \qquad \frac{1}{\beta} \leqslant z < 1.$$

Assuming that $f_Y(y) = 1/(y \ln \beta)$, show that Z_N also has the same reciprocal density. Thus, in a long sequence of multiplications, if at least one factor has the reciprocal density, then the normalized product has the reciprocal density. Assuming that both X and Y have the reciprocal density compute the probability that a left shift is required for normalization.

*3. [HAMM 1973] Consider the quotient Y/X of two independent normalized floating point mantissas in base β. Since $1/\beta \leqslant Y/X < \beta$, a one-digit right shift may be required to obtain the normalized quotient Q_N. Show that the pdf of Q_N is given by:

$$f_{Q_N}(z) = \frac{1}{z^2} \int_{1/\beta}^{z} x f_X(x) f_Y\left(\frac{x}{z}\right) dx + \frac{1}{\beta z^2} \int_{z}^{1} x f_X(x) f_Y\left(\frac{x}{\beta z}\right) dx, \qquad \frac{1}{\beta} \leqslant z < 1.$$

Show that if the dividend Y has the reciprocal density, then the normalized quotient also has the same density. Also compute the probability that a shift is required, assuming that both X and Y have the reciprocal density.

References

[ASH 1970] R. B. ASH, *Basic Probability Theory*, John Wiley & Sons, New York.

[BARL 1975] R. E. BARLOW and F. PROSCHAN, *Statistical Theory of Reliability and Life Testing: Probability Models,* Holt, Rinehart and Winston, New York.

[BEAU 1978] M. D. BEAUDRY, "Performance Related Reliability for Computing Systems," *IEEE Transactions on Computers,* June 1978, pp. 540–47.

[BHAT 1972] U. N. BHAT, *Elements of Applied Stochastic Processes,* John Wiley & Sons, New York.

[BREI 1968] L. BREIMAN, *Probability*, Addison-Wesley, Reading, Mass.

[CLAR 1978] J. CLARY, A. JAI, S. WEIKEL, R. SAEKS, and D. SIEWIOREK, "A Preliminary Study of Built-In-Test for the Military Computer Family," Technical Report, Research Triangle Institute, Research Triangle Park, N.C.

[HAMM 1973] R. W. HAMMING, *Numerical Methods for Scientists and Engineers*, McGraw-Hill, New York.

[KNUT 1969] D. E. KNUTH, *The Art of Computer Programming*, Vol.II, *Seminumerical Algorithms*, Addison-Wesley, Reading, Mass.

[KUCK 1978] D. J. KUCK, *The Structure of Computers and Computations*, Vol.I, John Wiley & Sons, New York.

[MATH 1970] F. P. MATHUR and A. AVIZIENIS, "Reliability Analysis and Architecture of a Hybrid Redundant Digital System: Generalized Triple Modular Redundancy with Self-Repair," *AFI PS Conference Proc., Spring Joint Comp. Conf.*, Vol. 36, pp. 375–83.

[MEYE 1980] J. F. MEYER, D. G. FURCHTGOTT, and L. T. WU, "Performability Evaluation of the SIFT Computer," *IEEE Transactions on Computers,* June 1980, pp. 501–509.

[NG 1976] Y-W. NG, "Reliability Modeling and Analysis for Fault-Tolerant Computers," Ph.D. Dissertation, Computer Science Department, University of California at Los Angeles.

[RUDI 1964] W. RUDIN, *Principles of Mathematical Analysis*, McGraw-Hill, New York.

Chapter 4

Expectation

4.1 INTRODUCTION

The distribution function $F(x)$ or the density $f(x)$ [pmf $p(x_i)$ for a discrete random variable] completely characterizes the behavior of a random variable X. Frequently, however, we need a more concise description such as a single number or a few numbers, rather than an entire function. One such number is the **expectation** or the **mean**, denoted by $E[X]$. Similarly, the **median**, which is defined as any number x such that $P(X < x) \leqslant 1/2$ and $P(X > x) \leqslant 1/2$, and the **mode**, defined as the number x, for which $f(x)$ or $p(x_i)$ attains its maximum, are two other quantities sometimes used to describe a random variable X. The mean, median, and mode are often called **measures of central tendency** of a random variable X.

Definition (Expectation). The expectation, $E[X]$, of a random variable X is defined by:

$$E[X] = \begin{cases} \sum_i x_i\, p(x_i), & \text{if } X \text{ is discrete,} \\ \int_{-\infty}^{\infty} x f(x)\, dx, & \text{if } X \text{ is continuous,} \end{cases} \tag{4.1}$$

provided that the relevant sum or integral is absolutely convergent; that is, $\sum_i |x_i|\, p(x_i) < \infty$ and $\int_{-\infty}^{\infty} |x|\, f(x)\, dx < \infty$. If the right hand side in (4.1) is not absolutely convergent, then $E[X]$ does not exist. Most common random variables have finite expectation; however, problem 1 at the end of this section provides an example of a random variable whose expectation does not exist. Definition (4.1) can be extended to the case of mixed random variables through the use of Riemann-Stieltjes integral. Alternatively, the formula given in problem 2 at the end of this section can be used in the general case.

Example 4.1

Consider the problem of searching for a specific name in a table of names. A simple method is to scan the table sequentially, starting from one end, until we either find the name or reach the other end, indicating that the required name is missing from the table. The following is a Pascal program fragment for sequential search:

```
var T: array[0..n] of NAME;
Z: NAME;
I: 0..n;
begin {Z has been initialized elsewhere}
  T[0]:=Z; {T[0] is used as a sentinel or marker}
  I:=n;

  while Z ≠ T[I] do
    I:=I − 1;

  if I > 0 then {found; I points to Z}
   else {not found}.
end
```

In order to analyze the time required for sequential search, let X be the discrete random variable denoting the number of comparisons "$Z \neq T[I]$" made. Clearly, the set of all possible values of X is $\{1, 2, ..., n+1\}$, and $X = n+1$ for unsuccessful searches. Since the value of X is fixed for unsuccessful searches, it is more interesting to consider a random variable Y that denotes the number of comparisons on a successful search. The set of all possible values of Y is $\{1, 2, ..., n\}$. To compute the average search time for a successful search, we must specify the pmf of Y. In the absence of any specific information, it is natural to assume that Y is uniform over its range; that is:

$$p_Y(i) = \frac{1}{n}, \qquad 1 \leqslant i \leqslant n.$$

Then

$$E[Y] = \sum_{i=1}^{n} i p_Y(i) = \frac{1}{n} \frac{n(n+1)}{2} = \frac{n+1}{2}.$$

Thus, on the average, approximately half the table needs to be searched. #

Example 4.2

The assumption of uniform distribution, used in Example 4.1, rarely holds in practice. It is possible to collect statistics on access patterns and use empirical distributions to reorganize the table so as to reduce the average search time. Unlike Example 4.1, we now assume for convenience that table search starts from the front. If α_i denotes the access probability for name $T[i]$, then the average successful search time $E[Y] = \sum i\alpha_i$. Then $E[Y]$ is minimized when names in the table are in the order of nonincreasing access probabilities; that is, $\alpha_1 \geqslant \alpha_2 \geqslant \cdots \geqslant \alpha_n$. As an example, many tables in practice follow Zipf's law [ZIPF 1949]:

$$\alpha_i = \frac{c}{i}, \qquad 1 \leqslant i \leqslant n,$$

where the constant c is determined from the normalization requirement, $\sum_{i=1}^{n} \alpha_i = 1$. Thus:

$$c = \frac{1}{\sum_{i=1}^{n} \frac{1}{i}} = \frac{1}{H_n} \simeq \frac{1}{\ln(n)},$$

where H_n is the partial sum of a harmonic series; that is: $H_n = \sum_{i=1}^{n} \frac{1}{i}$.

Now, if the names in the table are ordered as above, then the average search time is

$$E[Y] = \sum_{i=1}^{n} i\alpha_i = \frac{1}{H_n} \sum_{i=1}^{n} 1 = \frac{n}{H_n} \simeq \frac{n}{\ln(n)},$$

which is considerably less than the previous value $(n+1)/2$, for large n. #

Example 4.3

Recall the example of a computer system with five tape drives (Examples 1.1 and 2.2) and let X be the number of available tape drives. Then:

$$\begin{aligned} E[X] &= \sum_{i=0}^{5} i p_X(i) \\ &= 0 \cdot 1/32 + 1 \cdot 5/32 + 2 \cdot 10/32 + 3 \cdot 10/32 + 4 \cdot 5/32 + 5 \cdot 1/32 \\ &= 2.5. \end{aligned}$$ #

The example above illustrates that $E[X]$ need not correspond to a possible value of the random variable X. The expected value denotes the "center" of a probability distribution in the sense of a weighted average, or better, in the sense of a center of gravity.

Example 4.4

Let X be a continuous random variable with an exponential density given by:

$$f(x) = \lambda e^{-\lambda x}, \qquad x > 0.$$

Then

$$E[X] = \int_{-\infty}^{\infty} x f(x)\, dx = \int_{0}^{\infty} \lambda x e^{-\lambda x}\, dx.$$

Let $u = \lambda x$, then $du = \lambda\, dx$, and:

$$E[X] = \frac{1}{\lambda}\int_0^\infty ue^{-u}du = \frac{1}{\lambda}\Gamma 2 = \frac{1}{\lambda}, \qquad \text{using formula (3.26).}$$

Thus, if a component obeys an exponential failure law with parameter λ (known as the **failure rate**), then its expected life, or its mean time to failure (MTTF), is $1/\lambda$. Similarly, if the interarrival times of jobs to a computer center are exponentially distributed with parameter λ (known as the **arrival rate**), then the mean (average) interarrival time is $1/\lambda$. Finally, if the service-time requirement of a job is an exponentially distributed random variable with parameter μ (known as the **service rate**), then the mean (average) service time is $1/\mu$. #

Problems

1. Consider a discrete random variable X with the pmf:

$$p_X(x) = \begin{cases} \dfrac{1}{x(x+1)}, & x = 1, 2, \ldots, \\ 0, & \text{otherwise.} \end{cases}$$

Show that the function defined satisfies the properties of a pmf. Show that the formula (4.1) of expectation does not converge in this case and hence $E[X]$ is undefined. [*Hint:* Rewrite

$$\frac{1}{x(x+1)} \quad \text{as} \quad \frac{1}{x} - \frac{1}{x+1}.]$$

***2.** Using integration by parts, show (assuming that $E[X]$, $\int_0^\infty [1-F(x)]\, dx$, and $\int_{-\infty}^0 F(x)\, dx$ are all finite) that for a continuous random variable X:

$$E[X] = \int_0^\infty [1 - F(x)]\, dx - \int_{-\infty}^0 F(x)\, dx.$$

This result states that the expectation of a random variable X equals the difference of the areas of the right-hand and left-hand shaded regions in Figure 4.P.1. (This formula applies to the case of discrete and mixed random variables as well.)

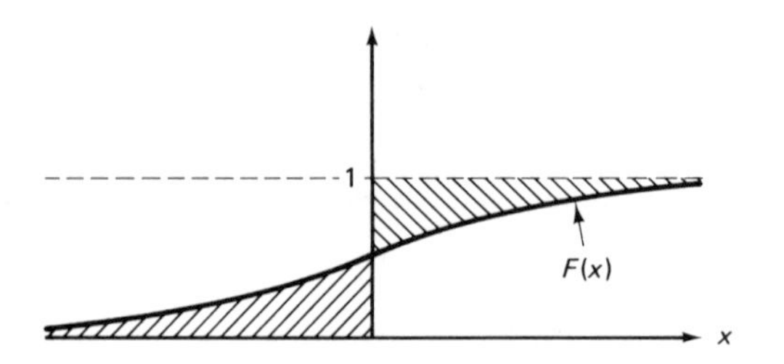

Figure 4.P.1 An alternative method of computing $E[X]$

3. For a given event A show that the expectation of its indicator random variable I_A (refer to Section 2.5.9) is given by:

$$E[I_A] = P(A).$$

4.2 MOMENTS

Let X be a random variable, and define another random variable Y as a function of X so that $Y = \phi(X)$. Suppose we wish to compute $E[Y]$. In order to apply Definition (4.1), we must first compute the pmf (or pdf in the continuous case) of Y by methods of Chapter 2 (or Chapter 3 in the continuous case). An easier method of computing $E[Y]$ is to use the following result:

$$E[Y] = E[\phi(X)] = \begin{cases} \sum_i \phi(x_i) p_X(x_i), & \text{if } X \text{ is discrete,} \\ \int_{-\infty}^{\infty} \phi(x) f_X(x)\, dx, & \text{if } X \text{ is continuous,} \end{cases} \tag{4.2}$$

(provided the sum or the integral on the right-hand side is absolutely convergent).

A special case of interest is the power function $\phi(X) = X^k$. For $k = 1, 2, 3, \ldots$, $E[X^k]$ is known as the kth moment of the random variable X. Note that the first moment, $E[X]$, is the ordinary expectation or the mean of X.

It is possible to show that if X and Y are random variables that have matching corresponding moments of *all* orders; that is, $E[X^k] = E[Y^k]$ for $k = 1, 2, \ldots$, then X and Y have the same distribution.

To center the origin of measurement, it is convenient to work with powers of $X - E[X]$. We define the kth central moment, μ_k, of the random variable X by $\mu_k = E[(X - E[X])^k]$. Of special interest is the quantity:

$$\mu_2 = E[(X - E[X])^2], \tag{4.3}$$

known as the variance of X, Var $[X]$, often denoted by σ^2.

Definition (Variance). The variance of a random variable X is

$$\text{Var}[X] = \mu_2 = \sigma^2 = \begin{cases} \int_{-\infty}^{\infty} (x - E[X])^2 f(x)\, dx & \text{if } X \text{ is continuous,} \\ \sum_i (x_i - E[X])^2 p(x_i) & \text{if } X \text{ is discrete.} \end{cases} \tag{4.4}$$

It is clear that Var $[X]$ is always a positive number. The square root, σ, of the variance is known as the **standard deviation.** The variance and the stan-

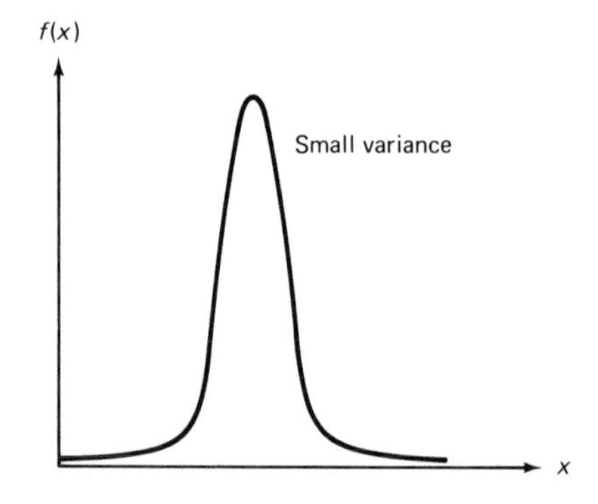

Figure 4.1 The pdf of a "concentrated" distribution

dard deviation are measures of the "spread" or "dispersion" of a distribution. If X has a "concentrated" distribution so that X takes values near to $E[X]$ with a large probability, then the variance is small (see Figure 4.1). Figure 4.2 shows a diffuse distribution—one with a large value of σ^2. Note that variance need not always exist (see problem 3 at the end of Section 4.7).

Example 4.5

Let X be an exponentially distributed random variable with parameter λ. Then, since $E[X] = 1/\lambda$ and $f(x) = \lambda e^{-\lambda x}$:

$$\sigma^2 = \int_0^\infty (x - \frac{1}{\lambda})^2 \lambda e^{-\lambda x} dx$$

$$= \int_0^\infty \lambda x^2 e^{-\lambda x}\, dx - 2\int_0^\infty x e^{-\lambda x} dx + \frac{1}{\lambda}\int_0^\infty e^{-\lambda x} dx$$

$$= \frac{1}{\lambda^2}\Gamma 3 - \frac{2}{\lambda^2}\Gamma 2 + \frac{1}{\lambda^2}\Gamma 1 = \frac{1}{\lambda^2}, \qquad \text{using formula (3.26).} \qquad \#$$

The standard deviation is expressed in the same units as the individual value of the random variable. If we divide it by the mean, then we obtain a

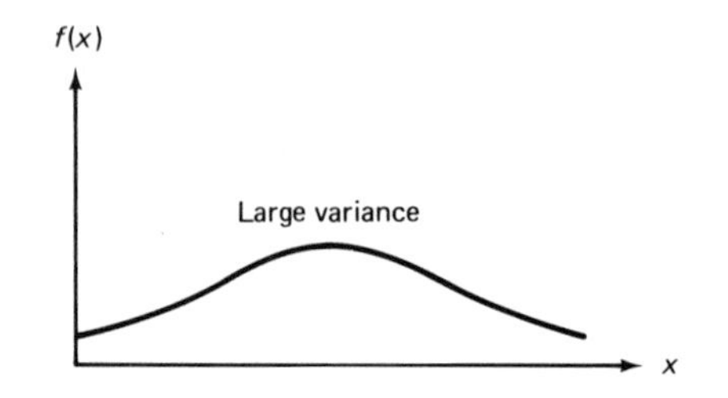

Figure 4.2 The pdf of a diffuse distribution

relative measure of the spread of the distribution of X. The coefficient of variation of a random variable X is denoted by C_X and defined by:

$$C_X = \frac{\sigma_X}{E[X]}. \tag{4.5}$$

Note that the coefficient of variation of an exponential random variable is 1, so C_X is a measure of deviation from the exponential distribution.

Yet another function of X that is often of interest is $Y = aX + b$, where a and b are constants. It is not difficult to show that:

$$E[Y] = E[aX + b] = aE[X] + b. \tag{4.6}$$

In particular, if a is zero, then $E[b] = b$; that is, the expectation of a constant random variable is that constant. If we take $a = 1$ and $b = -E[X]$, then we conclude that the first central moment, $\mu_1 = E[X - E[X]] = E[X] - E[X] = 0$.

Problems

*1. [WOLM 1965]. The problem of dynamic storage allocation in the main memory of a computer system can be simplified by choosing a fixed node size, k, for allocation. Out of the k units (bytes, say) of storage allocated to a node, only $k - b$ bytes are available to the user, since b bytes are required for control information. Let the random variable L denote the length in bytes of a user request. Thus $\lceil L/(k-b) \rceil$ nodes must be allocated to satisfy the user request. Thus the total number of bytes allocated is $X = k \lceil L/(k-b) \rceil$. Find $E[X]$ as a function of k and $E[L]$. Then, by differentiating $E[X]$ with respect to k, show that the optimal value of k is approximately $b + \sqrt{2bE[L]}$.

2. Recall the problem of the mischievous student trying to open a password-protected file, and determine the expected number of trials $E[N_n]$ and the variance Var $[N_n]$ for both techniques described in problem 2, Section 2.5.

3. The number of hardware failures of a computer system in a week of operation has the following pmf:

No. of Failures	0	1	2	3	4	5	6
Probability	.18	.28	.25	.18	.06	.04	.01

(a) Find the expected number of failures in a week.
(b) Find the variance of the number of failures in a week.

4. In a Bell System study made in 1961 regarding the dialing of calls between White Plains, N.Y., and Sacramento, Calif., the pmf of the number of trunks, X, required for a connection was found to be:

i	1	2	3	4	5
$p_X(i)$	.50	.30	.12	.07	.01

Determine the distribution function of X. Compute $E[X]$, Var $[X]$ and mode $[X]$. Let Y denote the number of telephone switching exchanges the above call has to pass through. Then $Y = X + 1$. Determine the pmf, the distribution function, the mean, and the variance of Y.

5. Let X, Y, and Z respectively denote EXP(1), 2-stage hyperexponential with $\alpha_1 = .5 = \alpha_2$, $\lambda_1 = 2$, and $\lambda_2 = 2/3$, and 2-stage Erlang with parameter 2 random variables. Note that $E[X] = E[Y] = E[Z]$. Find the mode, the median, the variance, and the coefficient of variation of each of the random variables. Compare the densities of X, Y, and Z by plotting on the same graph. Similarly compare the three distribution functions.

6. Given a random variable X and two functions $h(x)$ and $g(x)$ satisfying the condition $h(x) \leqslant g(x)$ for all x, show that:

$$E[h(X)] \leqslant E[g(X)]$$

whenever both expectations exist.

4.3 EXPECTATION OF FUNCTIONS OF MORE THAN ONE RANDOM VARIABLE

Let $X_1, X_2, \ldots, X_n$ be n random variables defined on the same probability space and let $Y = \phi(X_1, X_2, \ldots, X_n)$. Then:

$$E[Y] = E[\phi(X_1, X_2, \ldots, X_n)]$$

$$= \begin{cases} \displaystyle\int_{-\infty}^{\infty}\int_{-\infty}^{\infty} \cdots \int_{-\infty}^{\infty} \phi(x_1,x_2,\ldots,x_n) f(x_1,x_2,\ldots,x_n)\,dx_1 dx_2 \cdots dx_n & \text{(continuous case)}, \\ \displaystyle\sum_{x_1}\sum_{x_2} \cdots \sum_{x_n} \phi(x_1,x_2,\ldots,x_n) p(x_1,x_2,\ldots,x_n) & \text{(discrete case)}. \end{cases} \quad (4.7)$$

Example 4.6

Consider a moving head disk with the innermost cylinder of radius a and the outermost cylinder of radius b. We assume that the number of cylinders is very large and the cylinders are very close to each other, so that we may assume a continuum of cylinders. Let the random variables X and Y respectively, denote the current and the desired position of the head. Further assume that X and Y are independent and uniformly distributed over the interval (a, b). Therefore:

$$f_X(x) = f_Y(y) = \frac{1}{b-a}, \qquad a < x,\ y < b,$$

and

$$f(x,y) = \frac{1}{(b-a)^2}, \qquad a < x,\ y < b.$$

Head movement for a seek operation traverses a distance that is a random variable given by $|X - Y|$. The expected seek distance is then given by (see Figure 4.3):

$$E[|X-Y|] = \int_a^b \int_a^b |x-y| f(x,y)\, dx\, dy$$

$$= \int_a^b \int_a^b |x-y| \frac{1}{(b-a)^2}\, dx\, dy$$

$$= \iint_{a \leqslant y < x \leqslant b} \frac{(x-y)}{(b-a)^2}\, dy\, dx + \iint_{a \leqslant x \leqslant y \leqslant b} \frac{(y-x)}{(b-a)^2}\, dy\, dx$$

$$= \frac{2}{(b-a)^2} \int_a^b \int_a^x (x-y)\, dy\, dx, \qquad \text{by symmetry}$$

$$= \frac{2}{(b-a)^2} \int_a^b \left(xy - \frac{y^2}{2}\right) \Big|_a^x dx$$

$$= \frac{2}{(b-a)^2} \int_a^b \left(x^2 - ax - \frac{x^2}{2} + \frac{a^2}{2}\right) dx$$

$$= \frac{2}{(b-a)^2} \left[\frac{b^3 - a^3}{6} - \frac{a}{2}(b^2 - a^2) + \frac{a^2(b-a)}{2}\right]$$

$$= \frac{b-a}{3}.$$

Thus, the expected seek distance is one third the maximum seek distance. Intuition may have led us to the incorrect conclusion that the expected seek distance is

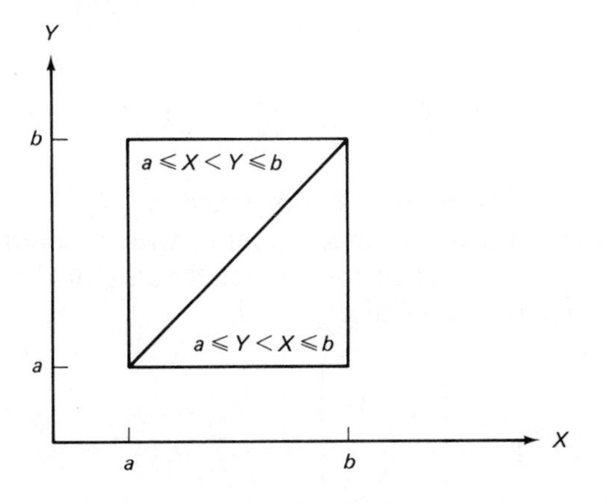

Figure 4.3 Two areas of integration for Example 4.6

half of the maximum. (In practice, the expected seek distance is even smaller due to correlations between successive requests [HUNT 1980].) #

Certain functions of random variables (e.g., sums), are of special interest and are of considerable use.

THEOREM 4.1 (THE LINEARITY PROPERTY OF EXPECTATION). Let X and Y be two random variables. Then the expectation of their sum is the sum of their expectations; that is, if $Z = X + Y$, then $E[Z] = E[X + Y] = E[X] + E[Y]$.

Proof:
We will prove the theorem assuming that X, Y, and hence Z are continuous random variables. Proof for the discrete case is very similar.

$$E[X + Y] = \int_{-\infty}^{\infty}\int_{-\infty}^{\infty} (x+y)\, f(x, y)\, dx\, dy$$

$$= \int_{-\infty}^{\infty} x \int_{-\infty}^{\infty} f(x, y)\, dy\, dx + \int_{-\infty}^{\infty} y \int_{-\infty}^{\infty} f(x, y)\, dx\, dy$$

$$= \int_{-\infty}^{\infty} x f_X(x)\, dx + \int_{-\infty}^{\infty} y f_Y(y)\, dy$$

(by definition of the marginal densities)

$$= E[X] + E[Y].$$

Note that the above theorem *does not* require that X and Y be independent. It can be generalized to the case of n variables—that is:

$$E[\sum_{i=1}^{n} X_i] = \sum_{i=1}^{n} E[X_i]$$

and to

$$E[\sum_{i=1}^{n} a_i X_i] = \sum_{i=1}^{n} a_i E[X_i], \tag{4.8}$$

where $a_1, \ldots, a_n$ are constants. For instance, let $X_1, X_2, \ldots, X_n$ be random variables (not necessarily independent) with a common mean $\mu = E[X_i]$ $(i = 1, 2, \ldots, n)$. Then the expected value of their sample mean (defined in Section 3.9) is equal to μ:

$$E[\bar{X}] = E\left[\frac{\sum_{i=1}^{n} X_i}{n}\right] = \frac{1}{n} \sum_{i=1}^{n} E[X_i] = \mu. \tag{4.9}$$

Example 4.7

We have noted that the variance:

$$\begin{aligned}\sigma^2 &= E[(X - E[X])^2]\\ &= E[X^2 - 2XE[X] + (E[X])^2] \qquad \text{by (4.8)}\\ &= E[X^2] - E[2XE[X]] + (E[X])^2,\\ &= E[X^2] - 2E[X]E[X] + (E[X])^2\end{aligned}$$

(noting that $E[X]$ is a constant).
Thus,

$$\sigma^2 = E[X^2] - (E[X])^2. \tag{4.10}$$

This formula for Var $[X]$ is usually preferred over the original Definition (4.4). #

Unlike the case of expectation of a sum, the expectation of a product of two random variables does not have a simple form, unless the two random variables are independent.

THEOREM 4.2 $E[XY] = E[X]E[Y]$, if X and Y are independent random variables.

Proof:
We give a proof of the theorem under the assumption that X and Y are discrete random variables. The proof for the continuous case is similar.

$$\begin{aligned}E[XY] &= \sum_i \sum_j x_i y_j p(x_i, y_j)\\ &= \sum_i \sum_j x_i y_j p_X(x_i) p_Y(y_j) \qquad \text{by independence}\\ &= \sum_i x_i p_X(x_i) \sum_j y_j p_Y(y_j) = E[X]E[Y].\end{aligned}$$

Note that converse of Theorem 4.2 does not hold; that is, random variables X and Y may satisfy the relation $E[XY] = E[X]\,E[Y]$ without being independent.

The theorem above can be easily generalized to a mutually independent set of n random variables $X_1, X_2, \ldots, X_n$:

$$E[\prod_{i=1}^{n} X_i] = \prod_{i=1}^{n} E[X_i] \tag{4.11}$$

and further to

$$E[\prod_{i=1}^{n} \phi_i(X_i)] = \prod_{i=1}^{n} E[\phi_i(X_i)].$$

Again with the assumption of independence, the variance of a sum takes a simpler form also:

THEOREM 4.3. Var $[X + Y] =$ Var $[X]$ + Var $[Y]$, if X and Y are independent random variables.

Proof:
From the definition of variance:

$$\begin{aligned}
\text{Var } [X+Y] &= E[((X+Y) - E[X+Y])^2] \\
&= E[((X+Y) - E[X] - E[Y])^2] \\
&= E[\,(X-E[X])^2 + (Y-E[Y])^2 + 2(X-E[X])(Y-E[Y])\,] \\
&= E[(X-E[X])^2] + E[(Y-E[Y])^2] + 2E[(X-E[X])(Y-E[Y])] \\
&= \text{Var } [X] + \text{Var } [Y] + 2E[(X-E[X])(Y-E[Y])],
\end{aligned}$$

by the linearity property of expectation.

The quantity $E[(X-E[X])(Y-E[Y])]$ is defined to be the **covariance** of X and Y and is denoted by Cov (X, Y). It is easy to see that Cov (X, Y) is zero when X and Y are independent:

$$\begin{aligned}
\text{Cov } (X,Y) &= E[(X-E[X])(Y-E[Y])] \\
&= E[\,XY - YE[X] - XE[Y] + E[X]E[Y]] \\
&= E[XY] - E[Y]E[X] - E[X]E[Y] + E[X]E[Y]
\end{aligned}$$

by the linearity of expectation,

$$\begin{aligned}
&= E[XY] - E[X]E[Y] \\
&= 0,
\end{aligned}$$

by Theorem 4.2, since X and Y are independent.

Therefore, Var $[X + Y] =$ Var $[X]$ + Var $[Y]$ if X and Y are independent random variables.

In case X and Y are not independent we obtain the formula:

$$\text{Var } [X + Y] = \text{Var } [X] + \text{Var } [Y] + 2 \text{ Cov } (X, Y). \tag{4.12}$$

The theorem above can be generalized for a set of n mutually independent random variables $X_1, X_2, \ldots, X_n$; and constants $a_1, a_2, \ldots, a_n$:

$$\text{Var } [\sum_{i=1}^{n} a_i X_i] = \sum_{i=1}^{n} a_i^2 \text{ Var } [X_i]. \tag{4.13}$$

Thus if $X_1, X_2, \ldots, X_n$ are mutually independent random variables with a common variance $\sigma^2 =$ Var $[X_i]$ $(i = 1, 2, \ldots, n)$ then the variance of their sum is given by:

$$\text{Var}\,[\sum_{i=1}^{n} X_i] = n\ \text{Var}\,[X_i] = n\sigma^2 \tag{4.14}$$

and the variance of their sample mean is:

$$\text{Var}\,[\bar{X}] = \text{Var}\left[\frac{\sum_{i=1}^{n} X_i}{n}\right] = \frac{1}{n^2}\ \text{Var}\,[\sum_{i=1}^{n} X_i] \tag{4.15}$$

$$= \frac{\sigma^2}{n}.$$

We have noted that Cov $(X, Y) = 0$, if X and Y are independent random variables. However, it is possible for two random variables to satisfy the condition Cov $(X, Y) = 0$ without being independent.

Definition (Uncorrelated Random Variables). Random variables X and Y are said to be uncorrelated provided Cov $(X, Y) = 0$.

Since Cov $(X, Y) = E[XY] - E[X]E[Y]$, an equivalent definition of uncorrelated random variables is the condition $E[XY] = E[X]\,E[Y]$. It follows that independent random variables are uncorrelated, but the converse need not hold.

Example 4.8

Let X be uniformly distributed over the interval $(-1, 1)$ and let $Y = X^2$, so Y is completely dependent on X. Noting that for all odd values of $k > 0$, the kth moment $E[X^k] = 0$, we have:

$$E[XY] = E[X^3] = 0 \quad \text{and} \quad E[X]E[Y] = 0 \cdot E[Y] = 0.$$

Therefore X and Y are uncorrelated! #

We have declared that Cov $(X, Y) = 0$ means X and Y are uncorrelated. On the other hand, if X and Y are linearly related—that is, $X = aY$ for some constant $a \neq 0$—then, since $E[X] = aE[Y]$, we have:

$$\text{Cov}\,(X, Y) = a\ \text{Var}\,[Y] = \frac{1}{a}\ \text{Var}\,[X]$$

or

$$\text{Cov}^2\,(X, Y) = \text{Var}\,[X]\ \text{Var}\,[Y].$$

In the general case, it can be shown that:

$$0 \leqslant \text{Cov}^2\,(X, Y) \leqslant \text{Var}\,[X]\ \text{Var}\,[Y] \tag{4.16}$$

using the following Cauchy-Schwarz inequality:

$$(E[XY])^2 \leqslant E[X^2]E[Y^2]. \tag{4.17}$$

Cov (X, Y) measures the degree of linear dependence (or the degree of correlation) between the two random variables. Recalling Example 4.8, we note that the notion of covariance completely misses the quadratic dependence. It is often useful to define a measure of this dependence in a scale-independent fashion. The correlation coefficient $\rho(X, Y)$ is defined by:

$$\rho(X, Y) = \frac{\text{Cov}(X, Y)}{\sqrt{\text{Var}[X]\,\text{Var}[Y]}} = \frac{\text{Cov}(X, Y)}{\sigma_X \sigma_Y} \tag{4.18}$$

whenever σ_X and σ_Y are defined.

Using the relation (4.16), we conclude that:

$$-1 \leqslant \rho(X, Y) \leqslant 1. \tag{4.19}$$

Also:

$$\rho(X, Y) = \begin{cases} -1, & \text{if } X = -aY \ (a>0), \\ 0, & \text{if } X \text{ and } Y \text{ are uncorrelated}, \\ +1, & \text{if } X = aY \ \ (a>0)\ . \end{cases} \tag{4.20}$$

Problems

1. [BLAK 1979] Consider discrete random variables X and Y with the joint pmf as shown below:

X	Y: −1	0	1
−2	1/16	1/16	1/16
−1	1/8	1/16	1/8
1	1/8	1/16	1/8
2	1/16	1/16	1/16

Are X and Y independent? Are they uncorrelated?

2. Consider the discrete version of Example 4.6 and assume that the records of a file are evenly scattered over n tracks of a moving-head disk. Compute the expected number of gaps X between tracks that the head will pass over between two reads. Also compute the variance of X. [Hint:

$$\sum_{i=1}^{N} i^3 = (\sum_{i=1}^{N} i)^2$$

is a useful identity.] Next assume that the seek time T is a function $\theta(X)$ of the number of gaps passed over. Compute $E[T]$ and Var $[T]$, assuming $T = 30.0 + 0.5X$.

3. Consider a directed graph G with n nodes. Let X_{ij} be a variable defined so that:

$$X_{ij} = \begin{cases} 0 & \text{if there is no edge between node } i \text{ and node } j, \\ 1 & \text{otherwise.} \end{cases}$$

Assume that the X_{ij}'s are mutually independent Bernoulli random variables with parameter p. The corresponding graph is called a **p-random-graph.** Find the pmf, the expected value, and the variance of the total number of edges X in the graph.

4. Let $X_1, X_2, \ldots, X_n$ be mutually independent and identically distributed random variables with mean μ and variance σ^2. Let $\bar{X} = (\sum_{i=1}^{n} X_i)/n$. Show that:

$$\sum_{k=1}^{n} (X_k - \bar{X})^2 = \sum_{k=1}^{n} (X_k - \mu)^2 - n(\bar{X} - \mu)^2$$

and hence:

$$E[\sum_{k=1}^{n} (X_k - \bar{X})^2] = (n-1)\sigma^2.$$

5. A certain telephone company charges for calls in the following way: \$0.20 for the first three minutes or less; \$0.08 per minute for any additional time. Thus if X is the duration of a call, the cost Y is given by:

$$Y = \begin{cases} 0.20, & 0 \leqslant X \leqslant 3, \\ 0.20 + 0.08(X - 3), & X \geqslant 3. \end{cases}$$

Find the expected value of the cost of a call ($E[Y]$), assuming that the duration of a call is exponentially distributed with a mean of $1/\lambda$ minutes. Use $1/\lambda = 1, 2, 3, 4$, and 5 minutes.

6. Show that $\text{Cov}^2 (X, Y) \leqslant \text{Var}[X] \text{Var}[Y]$.

7. Random variables X and Y are said to be **orthogonal** if and only if $E[XY] = 0$.
   (a) If X and Y are orthogonal determine the conditions under which they are uncorrelated.
   (b) If X and Y are uncorrelated, determine the conditions under which they are orthogonal.

8. Consider random variables X and Y with the joint pdf (bivariate Gaussian):

$$f(x,y) = \frac{1}{2\pi\sigma_X\sigma_Y\sqrt{1-\rho^2}} \times \exp\left\{-\frac{1}{2(1-\rho^2)}\left[\left(\frac{x-\mu_X}{\sigma_X}\right)^2 - \frac{2\rho(x-\mu_X)(y-\mu_Y)}{\sigma_X\sigma_Y} + \left(\frac{y-\mu_Y}{\sigma_Y}\right)^2\right]\right\}$$

where $\rho \neq \pm 1$. Show that $\text{Cov}(X,Y) = \rho\sigma_X\sigma_Y$. Hence show that if X, Y are jointly Gaussian and uncorrelated (i.e., $\rho = 0$), then they are also independent. Note that this is *not* true in general.

4.4 TRANSFORM METHODS

In many probability problems, the form of the density function (or the pmf in the discrete case) may be so complex so as to make computations difficult, if not impossible. As an example, recall the analysis of the program MAX. A transform can provide a compact description of a distribution, and it is relatively easy to compute the mean, the variance, and other moments directly from a transform rather than resorting to a tedious sum (discrete case) or an equally tedious integral (continuous case). The transform methods are particularly useful in problems involving sums of independent random variables and in solving difference equations (discrete case) and differential equations (continuous case) related to a stochastic process. We will introduce the z-transform (also called the probability generating function), the Laplace transform, and the characteristic function (also called the Fourier transform). We will first define the moment generating function and derive the above three transforms as special cases.

For a random variable X, $e^{X\theta}$ is another random variable. The expectation $E[e^{X\theta}]$ will be a function of θ. Define the moment generating function (MGF) $M_X(\theta)$, abbreviated $M(\theta)$, of the random variable X by:

$$M(\theta) = E[e^{X\theta}] \tag{4.21}$$

so that:

$$M(\theta) = \begin{cases} \sum_j e^{x_j\theta} p(x_j), & \text{if } X \text{ is discrete,} \\ \int_{-\infty}^{\infty} e^{x\theta} f(x)\, dx, & \text{if } X \text{ is continuous.} \end{cases} \tag{4.22}$$

The expectation defining $M(\theta)$ may not exist for all real numbers θ, but for most problems that we encounter, there will be an interval of θ values within which $M(\theta)$ does exist. Note that the Definition (4.21) allows us to define the moment-generating function for a mixed random variable as well.

The closely related characteristic function of a random variable X is given by

$$N_X(\tau) = N(\tau) = M_X(i\tau). \tag{4.23}$$

Here i denotes $\sqrt{-1}$. $N(\tau)$ is known among electrical engineers as the Fourier transform. The advantage here is that for any X, its characteristic function, $N_X(\tau)$, is always defined for all τ. If X is a nonnegative continuous random variable, then we define the (one-sided) Laplace transform:

$$L_X(s) = L(s) = M_X(-s) = \int_0^{\infty} e^{-sx} f(x)\, dx. \tag{4.24}$$

Finally, if X is a nonnegative integer-valued discrete random variable, then we define its z-transform:

$$G_X(z) = G(z) = E[z^X] = M_X(\ln z) = \sum_{i=0}^{\infty} p_X(i) z^i. \tag{4.25}$$

The reasons for the usefulness of transform methods will be summarized by the following properties of the transforms. We will give the properties of the moment generating function, but by an appropriate substitution for θ, a similar property can be stated for all the other transforms as well.

THEOREM 4.4 (LINEAR TRANSLATION) Let $Y = aX + b$; then:

$$M_Y(\theta) = e^{b\theta} M_X(a\theta).$$

Proof:

$$\begin{aligned} E[e^{Y\theta}] &= E[e^{(aX+b)\theta}] \\ &= E[e^{b\theta} e^{aX\theta}] \\ &= e^{b\theta} E[e^{aX\theta}] \end{aligned}$$

by the linearity property of expectation.

THEOREM 4.5 (THE CONVOLUTION THEOREM) Let $X_1, X_2, \ldots, X_n$ be mutually independent random variables on a given probability space, and let $Y = \sum_{i=1}^{n} X_i$. If $M_{X_i}(\theta)$ exists for all i, then $M_Y(\theta)$ exists, and:

$$M_Y(\theta) = M_{X_1}(\theta) M_{X_2}(\theta) \cdots M_{X_n}(\theta).$$

Thus the moment generating function of a sum of independent random variables is the product of the moment generating functions.

Proof:

$$\begin{aligned} M_Y(\theta) = E[e^{Y\theta}] &= E\left[e^{(X_1 + X_2 + \cdots + X_n)\theta}\right] \\ &= E[\prod_{i=1}^{n} e^{X_i\theta}] \\ &= \prod_{i=1}^{n} E[e^{X_i\theta}], \qquad \text{by independence,} \\ &= \prod_{i=1}^{n} M_{X_i}(\theta). \end{aligned}$$

Thus we may find the transform of a sum of independent random variables without any n-dimensional integration. But the technique will be of little value unless we can recover the distribution function from the transform. The following theorem, which we state without proof, is useful in this regard.

THEOREM 4.6 (CORRESPONDENCE THEOREM OR UNIQUENESS THEOREM) If $M_{X_1}(\theta) = M_{X_2}(\theta)$ for all θ, then $F_{X_1}(x) = F_{X_2}(x)$ for all x.

In other words, if two random variables have the same transform, then they have the same distribution function.

Next we study the **moment generating property** of the MGF. $e^{X\theta}$ can be expanded into a power series:

$$e^{X\theta} = 1 + X\theta + \frac{X^2\theta^2}{2!} + \cdots + \frac{X^k\theta^k}{k!} + \cdots.$$

Taking expectations on both sides (assuming all the expectations exist):

$$\begin{aligned} M(\theta) &= E[e^{X\theta}] \\ &= 1 + E[X]\theta + \cdots + \frac{E[X^k]\theta^k}{k!} + \cdots \\ &= \sum_{k=0}^{\infty} \frac{E[X^k]\theta^k}{k!}. \end{aligned}$$

Therefore, the coefficient of $\theta^k/k!$ in the power-series expansion of its MGF yields the kth moment $E[X^k]$ of the random variable X. Alternatively:

$$E[X^k] = \frac{d^kM}{d\theta^k}\Big|_{\theta=0}, \qquad k = 1, 2, \ldots. \tag{4.26}$$

Note that $E[X^0] = M(0) = 1$.

The corresponding property for the Laplace transform is:

$$E[X^k] = (-1)^k \frac{d^kL(s)}{ds^k}\Big|_{s=0}, \qquad k = 1, 2, \ldots. \tag{4.27}$$

For the z-transform:

$$E[X] = \frac{dG}{dz}\Big|_{z=1}, \tag{4.28}$$

$$E[X^2] = \frac{d^2G}{dz^2}\Big|_{z=1} + \frac{dG}{dz}\Big|_{z=1}, \tag{4.29}$$

and for the characteristic function:

$$E[X^k] = (-i)^k \frac{d^kN}{d\tau^k}\Big|_{\tau=0}, \qquad k = 1, 2, \ldots. \tag{4.30}$$

Example 4.9

Let X be exponentially distributed with parameter λ. Then:

$$f_X(x) = \lambda e^{-\lambda x}, \qquad x > 0$$

and

$$L_X(s) = \int_0^\infty e^{-sx}\lambda e^{-\lambda x}\, dx$$

$$= \frac{\lambda}{\lambda + s}\int_0^\infty (\lambda + s)e^{-(s+\lambda)x}\, dx$$

$$= \frac{\lambda}{\lambda + s}. \tag{4.31}$$

Now, using (4.27), we have:

$$E[X] = (-1)\frac{dL_X}{ds}\Big|_{s=0} = (-1)\frac{-\lambda}{(\lambda + s)^2}\Big|_{s=0} = \frac{1}{\lambda},$$

as derived earlier in Example 4.4. Also:

$$E[X^2] = \frac{d^2L_X}{ds^2}\Big|_{s=0} = \frac{2\lambda}{(\lambda + s)^3}\Big|_{s=0} = \frac{2}{\lambda^2}$$

and

$$\text{Var}\,[X] = \frac{2}{\lambda^2} - \frac{1}{\lambda^2} = \frac{1}{\lambda^2}.$$

#

Example 4.10

We are now in a position to complete the analysis of program MAX (Section 2.6). Recall that the PGF of the random variable X_n was shown to have the recurrence:

$$G_{X_n}(z) = \frac{(z + n - 1)}{n}\, G_{X_{n-1}}(z)$$

with

$$G_{X_1}(z) = 1.$$

Here X_n denotes the number of executions of the **then** clause in program MAX. Then the expected number of executions of the **then** clause is derived using the property (4.28):

$$E[X_n] = \frac{dG_{X_n}}{dz}\Big|_{z=1}$$

$$= \frac{1}{n}\, G_{X_{n-1}}\,(1) + \frac{z + n - 1}{n}\Big|_{z=1} \cdot \frac{dG_{X_{n-1}}(z)}{dz}\Big|_{z=1}$$

$$= \frac{1}{n} + E[X_{n-1}]$$

(since $G_X(1) = 1$ for any PGF). With $E[X_1] = 0$, we have:

$$E[X_n] = \sum_{i=2}^{n} \frac{1}{i} = H_n - 1 \simeq \ln n.$$

To compute the variance of X_n, first observe that if Y_k is a Bernoulli random variable with parameter $p = 1/k$, then

$$G_{Y_k}(z) = (1 - \frac{1}{k}) + \frac{z}{k} = \frac{z + k - 1}{k}.$$

If $Y_2, Y_3, \ldots, Y_n$ are mutually independent, then, using the convolution theorem, $W = \sum_{k=2}^{n} Y_k$ has the PGF:

$$\begin{aligned} G_W(z) &= \prod_{k=2}^{n} G_{Y_k}(z) \\ &= \prod_{k=2}^{n} \frac{z + k - 1}{k} \\ &= G_{X_n}(z). \end{aligned}$$

So we conclude, by the correspondence theorem, that X_n has the same distribution as W; that is:

$$X_n = \sum_{k=2}^{n} Y_k.$$

(Note that although X_n is the sum of $n - 1$ mutually independent Bernoulli random variables, it is not a binomial random variable; why?) Now since the Y_k's are mutually independent, we use formula (4.13) to obtain:

$$\begin{aligned} \text{Var}\,[X_n] &= \sum_{k=2}^{n} \text{Var}\,[Y_k] \\ &= \sum_{k=2}^{n} \frac{1}{k}\left(1 - \frac{1}{k}\right) \\ &= H_n - H_n^{(2)} \end{aligned}$$

where $H_n^{(2)}$ is defined to be $\sum_{k=1}^{n} \frac{1}{k^2}$.

The power of the notion of transforms should now be clear, since we could compute the mean and the variance without the explicit knowledge of pmf, which in this case is quite complex (it involves Stirling numbers!). #

Example 4.11 (Analysis of Straight Selection Sort)

We are given an array:

var a: **array** [1..n] **of** item, where **type** item = **record** key: integer; info: τ **end.**

We are required to sort the array so that keys are in nondecreasing order. We can use the following procedure [WIRT 1976]:

```
for i:= n downto 2 do
begin
1: "assign the index of the item with the largest
    key among the items a[1], a[2], ..., a[i] to k";
2: "exchange a[i] and a[k]"
end.
```

Assume that each element of the array is a large record and, therefore, exchanging (or moving) items is expensive. The total number of moves due to the second statement is easily computed and seen to be a fixed number. But the number of moves in the first statement is variable. Assume that the program MAX is used to perform this operation. Then the number of moves for a fixed value of i will be given by X_i, which was studied in Chapter 2 and in the previous example. Now the total number of moves, W_n, contributed by the first statement is given by:

$$W_n = \sum_{i=2}^{n} X_i.$$

Then:

$$E[W_n] = \sum_{i=2}^{n} E[X_i] = \sum_{i=2}^{n} (H_i - 1)$$

$$\simeq n(\ln n)$$

Example 4.12

Let X_i $(i = 1,2)$ be independent exponentially distributed random variables with parameters λ_i. If $\lambda_1 = \lambda_2 = \lambda$, then $X = X_1 + X_2$ will be a two-stage Erlang random variable. Assume $\lambda_1 \neq \lambda_2$, implying that X is a hypoexponentially distributed random variable. Using formula (4.31), we have:

$$L_{X_1}(s) = \frac{\lambda_1}{\lambda_1 + s} \quad \text{and} \quad L_{X_2}(s) = \frac{\lambda_2}{\lambda_2 + s}.$$

By the convolution theorem:

$$L_X(s) = \frac{\lambda_1\lambda_2}{(\lambda_1 + s)(\lambda_2 + s)}. \tag{4.32}$$

We expand this expression into a partial fraction:

$$L_X(s) = \frac{a_1\lambda_1}{\lambda_1 + s} + \frac{a_2\lambda_2}{\lambda_2 + s},$$

where:

$$a_1 = \frac{\lambda_2}{\lambda_2 - \lambda_1} \quad \text{and} \quad a_2 = \frac{\lambda_1}{\lambda_1 - \lambda_2}.$$

Recalling that if Y is EXP (λ), then $L_Y(s) = \lambda/(\lambda + s)$, we conclude (using the uniqueness theorem of Laplace transforms) that:

$$\begin{aligned} f_X(x) &= a_1 \lambda_1 e^{-\lambda_1 x} + a_2 \lambda_2 e^{-\lambda_2 x} \\ &= \frac{\lambda_1 \lambda_2}{\lambda_2 - \lambda_1} e^{-\lambda_1 x} + \frac{\lambda_1 \lambda_2}{\lambda_1 - \lambda_2} e^{-\lambda_2 x}, \end{aligned}$$

which is the hypoexponential density. #

More generally, if the X_i's ($i = 1, 2, \ldots, n$) are mutually independent and exponentially distributed with parameters λ_i ($\lambda_i \neq \lambda_j$, $i \neq j$), then $X = \sum_{i=1}^{n} X_i$ is an n-stage hypoexponential random variable and

$$L_X(s) = \prod_{i=1}^{n} \frac{\lambda_i}{\lambda_i + s}.$$

Using the technique of partial fraction expansion [KOBA 1978], the Laplace transform of X can be rewritten as:

$$L_X(s) = \sum_{i=1}^{n} \frac{a_i \lambda_i}{\lambda_i + s}, \tag{4.33}$$

where:

$$a_i = \prod_{\substack{j=1 \\ j \neq i}}^{n} \frac{\lambda_j}{\lambda_j - \lambda_i}. \tag{4.34}$$

Again, from the uniqueness theorem of Laplace transforms, it follows that

$$f_X(x) = \sum_{i=1}^{n} a_i \lambda_i e^{-\lambda_i x} \tag{4.35}$$

(Although, this form of f_X appears like a hyperexponential density function, it is quite a different hypoexponential density; why?)

Example 4.13

Let X be normally distributed with parameters μ and σ^2. Then:

$$f_X(x) = \frac{1}{\sigma\sqrt{2\pi}} \exp\left[-\frac{1}{2} \left(\frac{x - \mu}{\sigma}\right)^2\right], \qquad -\infty < x < \infty.$$

The characteristic function of X is given by:

$$N_X(\tau) = \int_{-\infty}^{\infty} e^{i\tau x} f_X(x)\, dx.$$

Making the change of variables $y = (x - \mu)/\sigma$, we obtain:

$$N_X(\tau) = \int_{-\infty}^{\infty} e^{i(\sigma y+\mu)\tau} \frac{e^{-(1/2)y^2} dy}{\sqrt{2\pi}}$$

$$= e^{i\tau\mu} \int_{-\infty}^{\infty} \frac{e^{-\frac{y^2}{2}}}{\sqrt{2\pi}} e^{i\tau\sigma y}\, dy$$

$$= e^{i\tau\mu + \frac{(i\tau\sigma)^2}{2}} \int_{-\infty}^{\infty} e^{-1/2(y-i\tau\sigma)^2} \frac{dy}{\sqrt{2\pi}},$$

(noting that $i^2 = -1$). Thus, the characteristic function of a normal random variable is given by:

$$N_X(\tau) = e^{i\tau\mu - \tau^2\sigma^2/2}, \tag{4.36}$$

since it can be shown that:

$$\int_{-\infty}^{\infty} e^{-1/2(y-i\tau\sigma)^2} \frac{dy}{\sqrt{2\pi}} = 1.$$

[It is the area under the normal density $N(i\tau\sigma, 1)$]. Check that

$$N_X(0) = e^0 = 1.$$

To compute the expected value, we use equation (4.30):

$$E[X] = \frac{1}{i} \frac{dN_X}{d\tau} \Big|_{\tau=0}$$

$$= \frac{1}{i} [(i\mu - \tau\sigma^2)\, e^{i\tau\mu - \frac{\tau^2\sigma^2}{2}}] \Big|_{\tau=0}$$

$$= \frac{1}{i} [i\mu e^0] = \mu.$$

Similarly, it can be shown that:

$$E[X^2] = \frac{1}{i^2} \frac{d^2 N_X}{d\tau^2} \Big|_{\tau=0}$$

$$= \sigma^2 + \mu^2$$

(after the computations are worked out). #

Thus the normal distribution $N(\mu, \sigma^2)$ has mean μ and variance σ^2. This distribution is completely specified by the two parameters.

Example 4.14 (Proof of Theorem 3.6)

Let $X_1, X_2, \ldots, X_n$ be mutually independent Gaussian random variables so that X_j is $N(\mu_j, \sigma_j^2)$, $j = 1, 2, \ldots, n$. Then from formula (4.36) we have:

$$N_{X_j}(\tau) = e^{i\tau\mu_j - \tau^2\sigma_j^2/2}, \qquad j = 1, 2, \ldots, n.$$

Let $Y = \sum_{i=1}^{n} X_i$; then, using the convolution theorem, we have:

$$N_Y(\tau) = \prod_{j=1}^{n} N_{X_j}(\tau)$$

$$= e^{i\tau\mu - \frac{\tau^2\sigma^2}{2}},$$

where

$$\mu = \sum_{j=1}^{n} \mu_j \quad \text{and} \quad \sigma^2 = \sum_{j=1}^{n} \sigma_j^2.$$

Comparing the characteristic function above with that in (4.36), we conclude that Y is $N(\mu, \sigma^2)$. #

Characteristic functions are somewhat more complex than the MGF, but they have two advantages. First, $N_X(\tau)$ is finite for all random variables X and for all real numbers τ. Second, the characteristic function possesses the inversion property, so that the density $f_X(s)$ may be derived from $N_X(\tau)$ by the inversion formula:

$$f_X(x) = \frac{1}{2\pi} \int_{-\infty}^{\infty} e^{-ix\tau} N_X(\tau) \, d\tau. \tag{4.37}$$

Inversion of a Laplace transform is usually performed using a table lookup. It is helpful first to perform a partial fraction expansion of the transform. See Appendix D for further details.

Problems

1. Show that if $X_1 \sim N(\mu_1, \sigma_1^2)$ and $X_2 \sim N(\mu_2, \sigma_2^2)$ are independent random variables, then the random variable:

$$Y = X_1 - X_2$$

is also normally distributed with:

$$E[Y] = \mu_1 - \mu_2 \quad \text{and} \quad \text{Var}[Y] = \sigma_1^2 + \sigma_2^2.$$

Generalize to the case of n mutually independent random variables with:

$$X_i \sim N(\mu_i, \sigma_i^2) \quad \text{and} \quad Y = \sum_{i=1}^{n} a_i X_i.$$

2. Take the program to find the maximal element of a given (one dimensional) array B of size n (discussed in Chapter 2 and also on p. 95 of Knuth, Vol. I). Call this

subroutine MAX. Write a driver for this subroutine that generates all $n!$ permutations of the set $\{1, 2, \ldots, n\}$ and, for each such permutation, loads it into the array B and calls subroutine MAX. Count the number of exchanges made in subroutine MAX. Add the number of exchanges over all permutations and divide the sum by $(n!)$. Check whether the result equals $H_n - 1$. Similarly compute the variance and check it against the expression $H_n - H_n^{(2)}$. Use $n = 1, 3, 5$, and 10.

To generate $n!$ permutations systematically, you may refer to the following article in the computing surveys: R. Sedgewick, "Permutation Generation Methods," *Computing Surveys,* June 1977, pp. 137-66.

4.5 MOMENTS AND TRANSFORMS OF SOME IMPORTANT DISTRIBUTIONS

4.5.1 Discrete Uniform Distribution

The pmf is given by:

$$p_X(i) = \frac{1}{n}, \qquad 1 \leqslant i \leqslant n .$$

Therefore:

$$E[X^k] = \sum_{i=1}^{n} \frac{i^k}{n}.$$

Then, the mean:

$$E[X] = \frac{n+1}{2},$$

and the variance:

$$\begin{aligned}\text{Var}\,[X] &= E[X^2] - (E[X])^2 \\ &= \frac{(n+1)(2n+1)}{6} - \frac{(n+1)^2}{4} \\ &= \frac{n+1}{12}\,[2(2n+1) - 3(n+1)] \\ &= \frac{(n+1)(n-1)}{12} = \frac{n^2-1}{12}.\end{aligned}$$

The coefficient of variation:

$$C_X = \sqrt{\frac{n^2-1}{3(n+1)^2}} = \sqrt{\frac{1}{3}(1 - \frac{2}{n+1})},$$

so

$$0 \leqslant C_X < \frac{1}{\sqrt{3}}.$$

The generating function in this case is:

$$G_X(z) = \sum_{i=1}^{n} \frac{1}{n} z^i = \frac{1}{n} \sum_{i=1}^{n} z^i.$$

4.5.2 Bernoulli pmf

$$p_X(0) = q, \qquad p_X(1) = p, \qquad p + q = 1.$$

$$E[X^k] = 0^k \cdot q + 1^k \cdot p = p, \qquad k = 1, 2, \ldots.$$

Therefore, the mean:

$$E[X] = p,$$

and the variance:

$$\text{Var}\ [X] = E[X^2] - (E[X])^2 = p - p^2 = p(1-p) = pq.$$

The coefficient of variation is:

$$C_X = \sqrt{\frac{q}{p}},$$

and the generating function:

$$G_X(z) = (1-p) + pz = q + pz.$$

4.5.3 Binomial Distribution

Note that a binomial random variable X is the sum of n mutually independent Bernoulli random variables $X_1, X_2, \ldots, X_n$. Thus:

$$X = \sum_{i=1}^{n} X_i,$$

and the linearity property of the expectation yields the result:

$$E[X] = \sum_{i=1}^{n} E[X_i] = np.$$

Similarly, using formula (4.13), we get the variance:

$$\text{Var}\ [X] = \sum_{i=1}^{n} \text{Var}\ [X_i] = npq.$$

The coefficient of variation:

$$C_X = \sqrt{\frac{npq}{n^2p^2}} = \sqrt{\frac{q}{np}}.$$

Thus, the expected number of successes in a sequence of n Bernoulli trials is np. Also note that the coefficient of variation reduces as n increases, and it approaches zero in the limit as $n \to \infty$. This observation is related to the weak law of large numbers, as will be seen later. We can easily obtain the generating function, using the convolution theorem:

$$G_X(z) = \prod_{i=1}^{n} G_{X_i}(z) = (q + pz)^n.$$

4.5.4 Geometric Distribution

The pmf is given by:

$$p_X(i) = pq^{i-1}, \qquad i = 1, 2, \ldots .$$

The mean is computed by:

$$\begin{aligned}
E[X] &= \sum_{i=1}^{\infty} ipq^{i-1} \\
&= p\sum_{i=1}^{\infty} iq^{i-1} \\
&= p \sum_{i=0}^{\infty} \frac{d}{dq} (q^i) \\
&= p \frac{d}{dq} (\sum_{i=0}^{\infty} q^i) \\
&= p \frac{d}{dq} \left(\frac{1}{1-q}\right) \\
&= \frac{p}{(1-q)^2} \\
&= \frac{1}{p}.
\end{aligned}$$

Therefore, if we assume that at the end of a CPU burst, a program requests an I/O operation with probability q and it finishes execution with probability p, then the average number of CPU bursts per program is given by $1/p$. Similarly, if a communication channel transmits a message correctly, on each trial, with probability p, then the average number of trials required for a successful transmission is $1/p$.

The generating function of X is given by:

$$\begin{aligned} G_X(z) &= \sum_{i=1}^{\infty} pq^{i-1}z^i \\ &= pz \sum_{i=1}^{\infty} (qz)^{i-1} \\ &= pz \sum_{j=0}^{\infty} (qz)^j \\ &= \frac{pz}{1-qz}. \end{aligned}$$

From this, $E[X]$ can be derived in an easier fashion:

$$\begin{aligned} E[X] &= \frac{dG}{dz}\Big|_{z=1} \\ &= \frac{p(1-qz) - pz(-q)}{(1-qz)^2}\Big|_{z=1} \\ &= \frac{p(1-q)+pq}{(1-q)^2} \\ &= \frac{p}{p^2} \\ &= \frac{1}{p}. \end{aligned}$$

The variance is computed in a fashion similar to that used for the mean; we get:

$$\text{Var}[X] = \frac{q}{p^2} \quad \text{and} \quad C_X = \sqrt{\frac{qp^2}{p^2}} = \sqrt{q} = \sqrt{1-p}.$$

For the modified geometric distribution, with the pmf $p_Y(i) = pq^i$, $i = 0, 1, 2, \ldots$:

$$E[Y] = \frac{q}{p}, \qquad \text{Var}[Y] = \frac{q}{p^2}, \qquad C_Y = \sqrt{\frac{qp^2}{p^2q^2}} = \frac{1}{\sqrt{q}},$$

and the generating function is:

$$G_Y(z) = \frac{p}{1-qz}.$$

4.5.5 Poisson pmf

$$p_X(i) = \frac{\alpha^i e^{-\alpha}}{i!}, \qquad 0 \leqslant i < \infty.$$

Then:

$$E[X] = \sum_{i=0}^{\infty} \frac{i\alpha^i}{i!} e^{-\alpha}$$

$$= \alpha e^{-\alpha} \sum_{i=1}^{\infty} \frac{\alpha^{i-1}}{(i-1)!}$$

$$= \alpha e^{-\alpha} e^{\alpha} = \alpha.$$

If the number of job arrivals to a computer center in interval $(0, t]$ is Poisson distributed with parameter $\alpha = \lambda t$, then the average number of arrivals in that interval is λt. Thus, the average arrival rate of jobs is λ.

The Var $[X]$ is easily computed to be α. Therefore:

$$C_X = \frac{1}{\sqrt{\alpha}}.$$

The generating function is given by:

$$G_X(z) = \sum_{k=0}^{\infty} e^{-\alpha} \frac{\alpha^k}{k!} z^k = e^{-\alpha} \sum_{k=0}^{\infty} \frac{(\alpha z)^k}{k!} = e^{-\alpha} e^{\alpha z} = e^{-\alpha(1-z)}.$$

4.5.6 Continuous Uniform Distribution

The density function is given by:

$$f_X(x) = \frac{1}{b-a}, \qquad a < x < b.$$

Then:

$$E[X] = \int_a^b \frac{x}{b-a}\, dx = \frac{b^2 - a^2}{2(b-a)} = \frac{b+a}{2},$$

the midpoint of the interval (a, b). The kth moment is computed as:

$$E[X^k] = \frac{1}{b-a} \int_a^b x^k\, dx = \frac{b^{k+1} - a^{k+1}}{(k+1)(b-a)}.$$

Therefore:

$$\text{Var}\ [X] = E[X^2] - (E[X])^2$$

$$= \frac{b^3 - a^3}{3(b-a)} - \frac{(b+a)^2}{4}$$

$$= \frac{(b-a)^2}{12}$$

and

$$C_X = \frac{b-a}{b+a} \sqrt{\frac{1}{3}}.$$

Assuming $0 \leqslant a < b$, the Laplace transform of X is:

$$L_X(s) = \int_a^b e^{-sx} \frac{1}{b-a} dx$$

$$= \frac{e^{-as} - e^{-bs}}{s(b-a)}.$$

4.5.7 Exponential Distribution

We have already determined that if the density is given by:

$$f_X(x) = \lambda e^{-\lambda x}, \qquad x > 0,$$

then the mean:

$$E[X] = \frac{1}{\lambda},$$

the variance:

$$\text{Var }[X] = \frac{1}{\lambda^2},$$

the coefficient of variation:

$$C_X = 1,$$

and the Laplace transform:

$$L_X(s) = \frac{\lambda}{\lambda + s}.$$

4.5.8 Gamma Distribution

The density function of the random variable X is given by:

$$f_X(x) = \frac{\lambda^\alpha x^{\alpha-1} e^{-\lambda x}}{\Gamma(\alpha)}, \qquad x > 0.$$

Then, making the substitution $u = \lambda x$, we compute the mean:

$$E[X] = \int_0^\infty \frac{x^\alpha \lambda^\alpha e^{-\lambda x}}{\Gamma(\alpha)} = \frac{1}{\lambda \Gamma(\alpha)} \int_0^\infty u^\alpha e^{-u} du,$$

and hence, using formula (3.26):

$$E[X] = \frac{\Gamma(\alpha+1)}{\lambda \Gamma(\alpha)} = \frac{\alpha}{\lambda}.$$

Similarly, the variance is computed to be:

$$\text{Var }[X] = \frac{\alpha}{\lambda^2}$$

and thus:

$$C_X = \frac{1}{\sqrt{\alpha}}.$$

Note that if α is an integer, then the above results could be shown by the properties of sums, since X will be the sum of α exponential random variables. Note also that the coefficient of variation of a gamma random variable is less than 1 if $\alpha > 1$; it is equal to 1 if $\alpha = 1$; and otherwise the coefficient of variation is greater than 1. Thus the gamma family is capable of modeling a very powerful class of random variables exhibiting from almost none to a very high degree of variability.

The Laplace transform is given by:

$$\begin{aligned} L_X(s) &= \int_0^\infty e^{-sx} \frac{\lambda^\alpha x^{\alpha-1} e^{-\lambda x}}{\Gamma(\alpha)} dx \\ &= \frac{\lambda^\alpha}{(\lambda + s)^\alpha} \int_0^\infty \frac{(\lambda+s)^\alpha x^{\alpha-1} e^{-(\lambda+s)x}}{\Gamma(\alpha)} dx \\ &= \frac{\lambda^\alpha}{(\lambda + s)^\alpha} \end{aligned}$$

since the last integral is the area under a gamma density with parameter $\lambda + s$ and α—that is, 1. If α were an integer, this result could be derived using the convolution property of the Laplace transforms.

4.5.9 Hypoexponential Distribution

We have seen that if $X_1, X_2, \ldots, X_n$ are mutually independent exponentially distributed random variables with parameters $\lambda_1, \lambda_2, \ldots, \lambda_n$ ($\lambda_i \neq \lambda_j$, $i \neq j$), respectively, then:

$$X = \sum_{i=1}^{n} X_i$$

is hypoexponentially distributed with parameters $\lambda_1, \lambda_2, \ldots, \lambda_n$; that is, X is HYPO $(\lambda_1, \lambda_2, \ldots, \lambda_n)$. The mean of X can then be obtained using the linearity property of expectation, so that:

$$E[X] = \sum_{i=1}^{n} E[X_i] = \sum_{i=1}^{n} \frac{1}{\lambda_i}.$$

Also, because of independence of the X_i's, we get the variance of X as:

$$\text{Var}\ [X] = \sum_{i=1}^{n} \text{Var}\ [X_i] = \sum_{i=1}^{n} \frac{1}{\lambda_i^2},$$

$$C_X = \sqrt{\frac{\sum_{i=1}^{n} \frac{1}{\lambda_i^2}}{(\sum_{i=1}^{n} \frac{1}{\lambda_i})^2}} \quad \text{and} \quad L_X(s) = \prod_{i=1}^{n} \frac{\lambda_i}{\lambda_i + s}.$$

Note that $C_X \leqslant 1$, and thus this distribution can model random variables with variability less than or equal to that of the exponential distribution.

It has been observed that service times at I/O devices are generally hypoexponentially distributed. Also, programs are often organized into a set of sequential phases (or job steps). If the execution time of the ith step is exponentially distributed with parameter λ_i, then the total program execution time is hypoexponentially distributed and its parameters are specified by the formulas above.

4.5.10 Hyperexponential Distribution

The density in this case is given by:

$$f(x) = \sum_{i=1}^{n} \alpha_i \lambda_i e^{-\lambda_i x}, \qquad \sum_{i=1}^{n} \alpha_i = 1, \qquad x > 0.$$

Then the mean:

$$\begin{aligned} E[X] &= \int_0^\infty (\sum_{i=1}^{n} x\alpha_i \lambda_i e^{-\lambda_i x})\, dx \\ &= \sum_{i=1}^{n} \alpha_i \int_0^\infty x\lambda_i e^{-\lambda_i x}\, dx \\ &= \sum_{i=1}^{n} \frac{\alpha_i}{\lambda_i} \end{aligned}$$

(since the last integral represents the expected value of an exponential random variable with parameter λ_i). Similarly:

$$E[X^2] = \sum_{i=1}^{n} \frac{2\alpha_i}{\lambda_i^2}$$

and

$$\begin{aligned} \text{Var}\,[X] &= E[X^2] - (E[X])^2 \\ &= 2\sum_{i=1}^{n} \frac{\alpha_i}{\lambda_i^2} - \left[\sum_{i=1}^{n} \frac{\alpha_i}{\lambda_i}\right]^2. \end{aligned}$$

Finally:

$$C_X^2 = \frac{\text{Var}[X]}{(E[X])^2} = \frac{2\sum_{i=1}^{n} \frac{\alpha_i}{\lambda_i^2} - \left[\sum_{i=1}^{n} \frac{\alpha_i}{\lambda_i}\right]^2}{\left[\sum_{i=1}^{n} \frac{\alpha_i}{\lambda_i}\right]^2}$$

$$= 2\frac{\sum_{i=1}^{n} \frac{\alpha_i}{\lambda_i^2}}{(\sum_{i=1}^{n} \frac{\alpha_i}{\lambda_i})^2} - 1.$$

Using the well-known Cauchy-Schwartz inequality [an alternative form of (4.17)], we can show that $C_X > 1$ for $n > 1$. The inequality states that:

$$(\sum_{i=1}^{n} a_i b_i)^2 \leqslant (\sum_{i=1}^{n} a_i^2)\ (\sum_{i=1}^{n} b_i^2). \tag{4.38}$$

Substitute $a_i = \sqrt{\alpha_i}$, and $b_i = (\sqrt{\alpha_i})/\lambda_i$; then:

$$(\sum_{i=1}^{n} \frac{\alpha_i}{\lambda_i})^2 \leqslant (\sum_{i=1}^{n} \alpha_i)\ (\sum_{i=1}^{n} \frac{\alpha_i}{\lambda_i^2})$$

$$= \sum_{i=1}^{n} \frac{\alpha_i}{\lambda_i^2},$$

which implies that:

$$C_X^2 = 2\frac{\sum_{i=1}^{n} \frac{\alpha_i}{\lambda_i^2}}{(\sum_{i=1}^{n} \frac{\alpha_i}{\lambda_i})^2} - 1 \geqslant 1.$$

Thus the hyperexponential distribution models random variables with more variability than the exponential distribution. As has been pointed out, CPU service times usually follow this distribution.

The Laplace transform:

$$L_X(s) = \int_0^{\infty} e^{-sx} \sum_{i=1}^{n} \alpha_i \lambda_i e^{-\lambda_i x}\, dx$$

$$= \sum_{i=1}^{n} \alpha_i \int_0^{\infty} \lambda_i e^{-\lambda_i x} e^{-sx}\, dx$$

$$= \sum_{i=1}^{n} \frac{\alpha_i \lambda_i}{\lambda_i + s}.$$

4.5.11 Weibull Distribution

Recall that this is the random variable with the pdf $f(x) = H'(x)\, e^{-H(x)}$ where $H(x) = \lambda x^\alpha$; that is:

$$f(x) = \lambda \alpha x^{\alpha-1} e^{-\lambda x^\alpha}, \qquad x > 0.$$

The mean:

$$E[X] = \int_0^\infty \lambda x \alpha x^{\alpha-1} e^{-\lambda x^\alpha}\, dx.$$

Now, making the substitution $u = \lambda x^\alpha$, we obtain:

$$\begin{aligned} E[X] &= \int_0^\infty \left(\frac{u}{\lambda}\right)^{1/\alpha} e^{-u}\, du \\ &= \left(\frac{1}{\lambda}\right)^{1/\alpha} \int_0^\infty u^{1/\alpha}\, e^{-u}\, du = \left(\frac{1}{\lambda}\right)^{1/\alpha} \Gamma\left(1 + \frac{1}{\alpha}\right) \end{aligned}$$

This reduces to the value $1/\lambda$ when $\alpha = 1$, since then the Weibull distribution becomes the exponential distribution. Similarly:

$$\begin{aligned} E[X^2] &= \int_0^\infty \lambda x^2 \alpha x^{\alpha-1} e^{-\lambda x^\alpha}\, dx \\ &= \left(\frac{1}{\lambda}\right)^{2/\alpha} \int_0^\infty u^{2/\alpha}\, e^{-u}\, du \\ &= \left(\frac{1}{\lambda}\right)^{2/\alpha} \Gamma\left(1 + \frac{2}{\alpha}\right), \end{aligned}$$

so:

$$\begin{aligned} \text{Var}\,[X] &= \left(\frac{1}{\lambda}\right)^{2/\alpha} \Gamma\left(1 + \frac{2}{\alpha}\right) - \left(\frac{1}{\lambda}\right)^{2/\alpha} \left[\Gamma\left(1 + \frac{1}{\alpha}\right)\right]^2 \\ &= \left(\frac{1}{\lambda}\right)^{2/\alpha} \left[\Gamma\left(1 + \frac{2}{\alpha}\right) - \left[\Gamma\left(1 + \frac{1}{\alpha}\right)\right]^2\right] \end{aligned}$$

and

$$C_X = \sqrt{\frac{\Gamma\left(1 + \frac{2}{\alpha}\right) - \left[\Gamma\left(1 + \frac{1}{\alpha}\right)\right]^2}{\left[\Gamma\left(1 + \frac{1}{\alpha}\right)\right]^2}} = \sqrt{\frac{\Gamma\left(1 + \frac{2}{\alpha}\right)}{\left[\Gamma\left(1 + \frac{1}{\alpha}\right)\right]^2} - 1}.$$

4.5.12 The Normal Distribution

The density:

$$f(x) = \frac{1}{\sigma\sqrt{2\pi}} e^{-\frac{(x-\mu)^2}{2\sigma^2}}, \qquad -\infty < x < \infty.$$

The mean, the variance, and the characteristic function have been derived earlier:

$$E[X] = \mu, \qquad \text{Var}[X] = \sigma^2, \qquad N_X(\tau) = e^{i\tau\mu - \frac{\tau^2\sigma^2}{2}}.$$

Problems

1. Consider an interactive system designed to handle a maximum of fifteen transactions per second. During the peak hour of its activity, transactions arrive at the average rate of ten per second. Assuming that the number of transactions arriving per second follows a Poisson distribution, compute the probability that the system will be overloaded during a peak hour.

2. The CPU time requirement X of a typical job can be modeled by the following hyperexponential distribution:

$$P(X \leqslant t) = \alpha(1 - e^{-\lambda_1 t}) + (1 - \alpha)(1 - e^{-\lambda_2 t}),$$

   where $\alpha = 0.6$, $\lambda_1 = 10$, and $\lambda_2 = 1$. Compute (a) the probability density function of X, (b) the mean service time $E[X]$, (c) the variance of service time Var $[X]$, and (d) the coefficient of variation. Plot the distribution and the density function of X.

3. The CPU time requirement, T, for jobs has a gamma distribution with mean of 40 seconds and variance of 400 seconds^2.
   (a) Find the shape parameter α and the scale parameter λ.
   (b) A short job ($T < 20$ seconds) gets priority. Compute the probability that a randomly chosen job is a short job.

4. A telephone exchange can handle at most twenty simultaneous conversations. It has been observed that an incoming call finds an "all busy" signal 1 percent of the time. Assuming that the number of incoming calls, X, per unit time has a Poisson distribution, find the parameter α (or the average call arrival rate) of the distribution.

4.6 COMPUTATION OF MEAN TIME TO FAILURE

Let X denote the lifetime of a component so that its reliability $R(t) = P(X > t)$ and $R'(t) = -f(t)$. Then the **expected life** or the **mean time to failure** (MTTF) of the component is given by:

$$E[X] = \int_0^\infty tf(t)\, dt = -\int_0^\infty tR'(t)\, dt.$$

Integrating by parts we obtain:

$$E[X] = -tR(t)\,|_0^\infty + \int_0^\infty R(t)\, dt.$$

Now, since $R(t)$ approaches zero faster than t approaches ∞, we have:

$$E[X] = \int_0^\infty R(t)\, dt. \tag{4.39}$$

(The above formula is a special case of problem 2 at the end of Section 4.1.) This latter expression for MTTF is in more common use in reliability theory. More generally:

$$\begin{aligned} E[X^k] &= \int_0^\infty t^k f(t)\, dt \\ &= -\int_0^\infty t^k R'(t)\, dt \\ &= -t^k R(t)\,|_0^\infty + \int_0^\infty k t^{k-1} R(t)\, dt. \end{aligned}$$

Thus:

$$E[X^k] = \int_0^\infty k t^{k-1}\, R(t)\, dt. \tag{4.40}$$

In particular:

$$\text{Var}\ [X] = \int_0^\infty 2tR(t)\, dt - [\int_0^\infty R(t)\, dt]^2. \tag{4.41}$$

If the component lifetime is exponentially distributed, then $R(t) = e^{-\lambda t}$ and:

$$\begin{aligned} E[X] &= \int_0^\infty e^{-\lambda t}\, dt = \frac{1}{\lambda}, \\ \text{Var}\ [X] &= \int_0^\infty 2te^{-\lambda t}\, dt - \frac{1}{\lambda^2} \\ &= \frac{2}{\lambda^2} - \frac{1}{\lambda^2} = \frac{1}{\lambda^2}, \end{aligned}$$

as derived earlier. Next we consider a system consisting of n components connected in several different ways (we continue to make the usual assumptions of independence).

4.6.1 Series System

Assume that the lifetime of the ith component is exponentially distributed with parameter λ_i. Then system reliability [using equation (3.53)] is given by:

$$R(t) = \prod_{i=1}^{n} R_i(t) = \prod_{i=1}^{n} e^{-\lambda_i t} = \exp\left[-(\sum_{i=1}^{n} \lambda_i)t\right].$$

Thus, the lifetime of the system is also exponentially distributed with parameter $\lambda = \sum_{i=1}^{n} \lambda_i$. Therefore the series system MTTF is:

$$\frac{1}{\sum_{i=1}^{n} \lambda_i}. \tag{4.42}$$

The MTTF of a series system is much smaller than the MTTF of its components.

If X_i denotes the lifetime of component i (not necessarily exponentially distributed), and X denotes the series system lifetime, then we can show that

$$0 \leqslant E[X] \leqslant \min_i\{E[X_i]\} \tag{4.43}$$

which gives rise to the common remark that a system is weaker than its weakest link.

To prove inequality (4.43), note that:

$$R_X(t) = \prod_{i=1}^{n} R_{X_i}(t) \leqslant \min_i \{R_{X_i}(t)\},$$

since $0 \leqslant R_{X_i}(t) \leqslant 1$. Then:

$$E[X] = \int_0^\infty R_X(t)\, dt \leqslant \min_i \{\int_0^\infty R_{X_i}(t)\, dt\}$$

$$= \min_i\{E[X_i]\}.$$

4.6.2 Parallel System

Consider a parallel system of n independent components, with X_i denoting the lifetime of component i and X denoting the lifetime of the system. Then:

$$X = \max\{X_1, X_2, \ldots, X_n\}$$

and, using formula (3.54):

$$R_X(t) = 1 - \prod_{i=1}^{n} [1 - R_{X_i}(t)] \geqslant 1 - [1 - R_{X_i}(t)], \qquad \text{for all } i, \tag{4.44}$$

which implies that the reliability of a parallel redundant system is larger than that of any of its components. Therefore:

$$E[X] = \int_0^\infty R_X(t)\, dt \geqslant \max_i \{\int_0^\infty R_{X_i}(t)\, dt\}$$

$$= \max_i \{E[X_i]\}. \tag{4.45}$$

Now assume that X_i is exponentially distributed with parameter λ (all components have the same parameter). Then:

$$R_X(t) = 1 - (1 - e^{-\lambda t})^n$$

and

$$E[X] = \int_0^\infty [1 - (1 - e^{-\lambda t})^n]\, dt.$$

Let $u = 1 - e^{-\lambda t}$; then $dt = 1/\lambda(1-u)\ du$. Thus:

$$E[X] = \frac{1}{\lambda} \int_0^1 \frac{1-u^n}{1-u}\, du.$$

Now since the integrand above is the sum of a finite geometric series:

$$E[X] = \frac{1}{\lambda} \int_0^1 (\sum_{i=1}^{n} u^{i-1})\, du$$

$$= \frac{1}{\lambda} \sum_{i=0}^{n-1} \int_0^1 u^{i-1}\, du.$$

Note that:

$$\int_0^1 u^{i-1}\, du = \frac{u^i}{i}\Big|_0^1 = \frac{1}{i}.$$

Thus, with the usual exponential assumptions, the MTTF of a parallel redundant system is given by:

$$E[X] = \frac{1}{\lambda} \sum_{i=1}^{n} \frac{1}{i} = \frac{H_n}{\lambda} \simeq \frac{\ln(n)}{\lambda}. \tag{4.46}$$

Figure 4.4 shows the expected life of a parallel system as a function of n. It should be noted that beyond $n = 2$ or 3, the gain in expected life (due to adding one additional component), is not very significant. Note that the rate of increase in the MTTF is $1/(n\lambda)$.

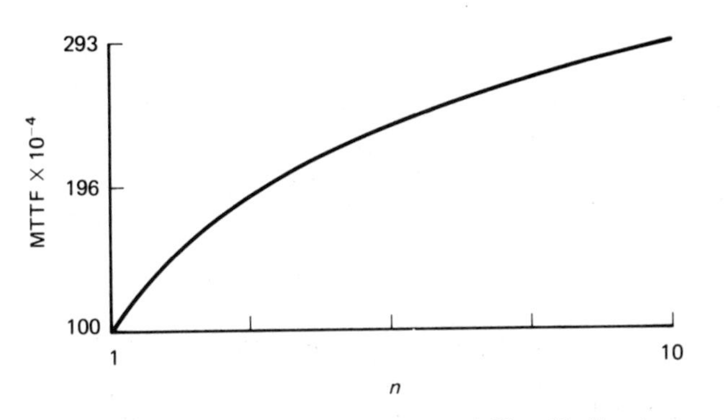

Figure 4.4 The variation in the expected life with the degree of (parallel) redundancy (simplex failure rate $\lambda = 10^{-6}$)

Alternatively, the formula above for $E[X]$ can be derived by noting that X is hypoexponentially distributed with parameters $n\lambda, (n-1)\lambda, \ldots, \lambda$. (See Theorem 3.5.) In other words, $X = \sum_{i=1}^{n} Y_i$, where Y_i is exponentially distributed with parameter $i\lambda$. Then, using the linearity property of expectation:

$$E[X] = \sum_{i=1}^{n} E[Y_i] = \sum_{i=1}^{n} \frac{1}{i\lambda} = \frac{H_n}{\lambda}.$$

Also, since the Y_i's are mutually independent:

$$\text{Var}[X] = \sum_{i=1}^{n} \text{Var}[Y_i] = \sum_{i=1}^{n} \frac{1}{i^2\lambda^2} = \frac{1}{\lambda^2} H_n^{(2)}. \tag{4.47}$$

Note that $C_X < 1$; hence, not only does the parallel configuration increase the MTTF but it also reduces the variability of the system lifetime.

4.6.3 Standby Redundancy

Assume that the system has one component operating and $(n-1)$ cold (unpowered) spares. The failure rate of an operating component is λ, and a cold spare does not fail. Furthermore, the switching equipment is failure free. Let X_i be the lifetime of the ith component from the point it is put into operation until its failure. Then the system lifetime X is given by:

$$X = \sum_{i=1}^{n} X_i.$$

Thus X has an n-stage Erlang distribution, and therefore:

$$E[X] = \frac{n}{\lambda} \quad \text{and} \quad \text{Var}[X] = \frac{n}{\lambda^2}. \tag{4.48}$$

Note that the gain in expected life is linear as a function of the number of components, unlike the case of parallel redundancy. Of course, the price paid is the added complexity of the detection and switching mechanism. Furthermore, if we allow the detection and switching equipment to fail, the gain will be much less. Note that the computation of $E[X]$ does not require the independence assumption (but the computation of variance does).

4.6.4 TMR and TMR/Simplex Systems

We have noted that the reliability, $R(t)$, of a TMR system consisting of components with independent exponentially distributed lifetimes (with parameter λ) is given by [formula (3.58)]:

$$R(t) = 3e^{-2\lambda t} - 2e^{-3\lambda t}.$$

Then the expected life is given by:

$$E[X] = \int_0^\infty 3e^{-2\lambda t}\, dt - \int_0^\infty 2e^{-3\lambda t}\, dt.$$

Thus, the TMR MTTF is:

$$E[X] = \frac{3}{2\lambda} - \frac{2}{3\lambda} = \frac{5}{6\lambda}. \tag{4.49}$$

Compare this with the expected life of a single component ($= 1/\lambda$). Thus, TMR actually *reduces* (by about 16 percent) the MTTF over the simplex system. This fact points out that in certain cases MTTF can be a misleading measure of performance. Although TMR has a lower MTTF than simplex, we know that TMR has higher reliability than simplex for "short" missions, defined by mission time $t < (\ln 2)/\lambda$.

Next consider the case when the voter used in TMR is not perfect and it has reliability $r \leqslant 1$. Then:

$$R(t) = r(3e^{-2\lambda t} - 2e^{-3\lambda t})$$

and the MTTF of a TMR system with imperfect voter is:

$$E[X] = \frac{5r}{6\lambda}. \tag{4.50}$$

Thus TMR MTTF is degraded even further.

Next consider the improvement of TMR known as TMR/simplex. In Example 3.24 the lifetime X of this system was shown to be the sum of two exponential random variables, one with parameter 3λ and the other with parameter λ. Then the MTTF of TMR/simplex is given by:

$$E[X] = \frac{1}{3\lambda} + \frac{1}{\lambda} = \frac{4}{3\lambda}. \tag{4.51}$$

Thus the TMR/simplex has 33 percent longer expected life than the simplex.

4.6.5 The NMR System

We showed in Example 3.27 that the lifetime $L(m|n)$ of an m-out-of-n system with components having independent exponential lifetimes (with parameter λ) is the sum of $(n-m+1)$ exponential random variables with parameters $m\lambda$, $(m+1)\lambda$, $\ldots$, $n\lambda$. Therefore, the MTTF of an m-out-of-n system is given by:

$$E[L(m|n)] = \sum_{i=m}^{n} \frac{1}{i\lambda} = \frac{H_n - H_{m-1}}{\lambda}. \tag{4.52}$$

Also, the variance of the lifetime of an NMR system is:

$$\text{Var}\ [L(m|n)] = \sum_{i=m}^{n} \frac{1}{i^2\lambda^2} = \frac{H_n^{(2)} - H_{m-1}^{(2)}}{\lambda^2}. \tag{4.53}$$

It may be verified that TMR is a special case of the above with $n = 3$ and $m = 2$.

4.6.6 The Hybrid NMR System

Consider a system of N operating components and S cold spares. Of the N active components M are required for the system to function correctly. An active component has an exponential lifetime distribution with parameter λ, and now we will let the cold spare also fail with failure rate $\mu < \lambda$. The lifetime $L(M|N,S)$ was shown in Example 3.28 to be the sum of $N - M + 1 + S$ exponential random variables with parameters $N\lambda + S\mu$, $N\lambda + (S-1)\mu$, $\ldots$, $N\lambda + \mu$, $N\lambda$, $\ldots$, $M\lambda$. Then the MTTF of a hybrid NMR system is given by:

$$E[L(M|N,S)] = \sum_{i=1}^{S} \frac{1}{N\lambda + i\mu} + \sum_{i=M}^{N} \frac{1}{i\lambda}. \tag{4.54}$$

Also, the variance of the lifetime of such a system is given by:

$$\text{Var}\ [L(M|N,S)] = \sum_{i=1}^{S} \frac{1}{(N\lambda + i\mu)^2} + \sum_{i=M}^{N} \frac{1}{i^2\lambda^2}. \tag{4.55}$$

All the previous cases we have considered are special cases of hybrid NMR. For example, the series system corresponds to $S = 0$, $M = N$. The parallel system corresponds to $S = 0$, $M = 1$. The standby system corresponds to $S = n-1$, $N = 1$, $M = 1$, $\mu = 0$. The NMR system corresponds to $S = 0$.

Problems

1. The time to failure T of a device is known to follow a normal distribution with mean μ and $\sigma = 10{,}000$ hours. Determine the value of μ if the device is to have a reliability equal to 0.90 for a mission time of 50,000 hours.

*2. Consider a series system of two independent components. The lifetime of the first component is exponentially distributed with parameter λ, and the lifetime of the second component is normally distributed with parameters μ and σ^2. Determine the reliability $R(t)$ of the system and show that the expected life of the system is:

$$\frac{1}{\lambda}\left[1 - \exp\left(-\lambda\mu + \frac{\lambda^2\sigma^2}{2}\right)\right].$$

3. The failure rate for a certain type of component is $\lambda(t) = at$ $(t \geqslant 0)$ where $a > 0$ and is constant. Find the component's reliability, and it's expected life (or MTTF).

4. Two alternative microcomputer systems are being considered for acquisition. System A consists of ten chips, each with a constant failure rate of 10^{-5} per hour and all ten chips must function properly for the system to function. System B consists of five chips, each having a time-dependent failure rate given by at per hour, for some constant a and all five chips must function properly for the system to function. If both systems have the same mean time to failure, which one should be recommended? Assume that the reliability for a mission time of 1,000 hours is the criterion for selection.

5. The data obtained from testing a device indicate that the expected life is five hours and the variance is approximately 1 hour^2. Compare the reliability functions obtained by assuming (a) a Weibull failure law with $\alpha = 5$, (b) a normal failure pdf with $\mu = 5$ and $\sigma^2 = 1$, and (c) a gamma pdf with appropriate values of α and λ. Plot your results.

6. The failure rate of a device is given by:

$$h(t) = \begin{cases} at, & 0 < t < 1{,}000 \text{ hours}, \\ b, & t \geqslant 1{,}000 \text{ hours}. \end{cases}$$

Choose b so that $h(t)$ is continuous, and find an expression for device reliability.

7. Consider the system shown in Figure 4.P.2. Each component has an exponential failure law with parameter λ. All components behave independently, except that whenever C_4 fails it triggers an immediate failure of C_5 and vice versa. Find the reliability and the expected life of the system.

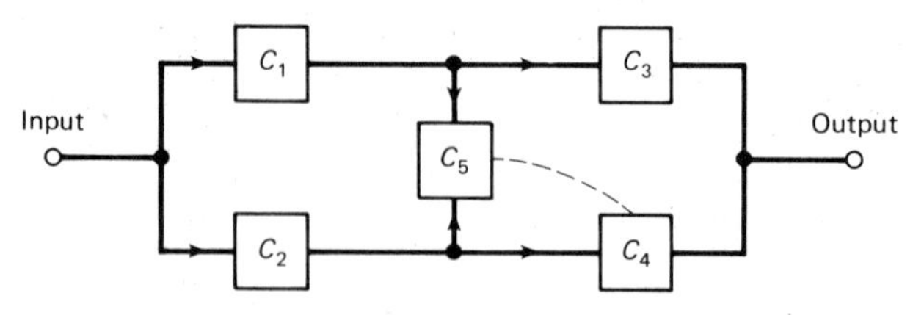

Figure 4.P.2 A system with dependent failures

8. Carnegie-Mellon multiprocessor C.mmp consists of processors, switches, and memory units. Consider a configuration with sixteen processors, sixteen 64K memories, and one switch. At least four processors and four memories are required for a given task. Assume that we have a (constant) failure rate for each

processor of 68.9 failures/10^6 hr, 224 failures/10^6 hr for each memory, and a failure rate for the switch of 202 failures/10^6hr. Compute the reliability function for the system. Also compute the MTTF of the system.

9. Consider a parallel redundant system of two independent components with the lifetime of ith component $X_i \sim \text{EXP}(\lambda_i)$. Show that system MTTF is given by:

$$\text{MTTF} = \frac{1}{\lambda_1} + \frac{1}{\lambda_2} - \frac{1}{\lambda_1+\lambda_2}.$$

Generalize to the case of n components. Next consider a standby redundant system made of these two components. Assuming that the component in the spare status does not fail, obtain the reliability and the MTTF of the system.

4.7 INEQUALITIES AND LIMIT THEOREMS

We have mentioned that the distribution function (equivalently the density function or the pmf) provides a complete characterization of a random variable, and that the probability of any event concerning the random variable can be determined from it. Numbers such as the mean or the variance provide a limited amount of information about the random variable. We have discussed methods to compute various moments (including mean and variance), given the distribution function. Conversely, if all the moments, $E[X^k]$, $k = 1, 2, \ldots$, are given, then we can reconstruct the distribution function via the transform. In case all moments are not available, we are not able to recover the distribution function in general. However, it may still be possible to obtain bounds on the probabilities of various events based on the limited information.

First assume that we are given the mean $E[X] = \mu$ of a nonnegative random variable X, where μ is assumed to be finite. Then the Markov inequality states that for $t > 0$:

$$P(X \geqslant t) \leqslant \frac{\mu}{t}. \tag{4.56}$$

To prove this inequality, fix $t > 0$ and define the random variable Y by

$$Y = \begin{cases} 0, & \text{if } X < t, \\ t, & X \geqslant t. \end{cases}$$

Then Y is a discrete random variable with the pmf:

$$p_Y(0) = P(X < t)$$

and

$$p_Y(t) = P(X \geqslant t)$$

Thus:

$$E[Y] = 0 \cdot p_Y(0) + t \cdot p_Y(t) = tP(X \geqslant t).$$

Now, since $X \geqslant Y$, we have $E[X] \geqslant E[Y]$ and hence:

$$E[X] \geqslant E[Y] = tP(X \geqslant t),$$

which gives the desired inequality.

Example 4.15

Consider a system with MTTF = 100 hours. We can get a bound on the reliability of the system for a mission time t using the Markov inequality:

$$R(t) = P(X \geqslant t) \leqslant \frac{100}{t}.$$

Thus if the mission time exceeds $100/0.9 \simeq 111$ hours, we know that the reliability will be less than or equal to 0.9. This suggests that if the required level of reliability is 0.9, then mission time can be no more than 111 hours (it may have to be restricted further). #

It is not difficult to see that the inequality (4.56) is quite crude, since only the mean is assumed to be known. For example, let the lifetime, X, of a system be exponentially distributed with mean $1/\lambda$. Then the reliability $R(t) = e^{-\lambda t}$, and the Markov inequality asserts that:

$$R(t) \leqslant \frac{1}{\lambda t} \quad \text{or} \quad \frac{1}{R(t)} \geqslant \lambda t$$

—that is:

$$e^{\lambda t} \geqslant \lambda t.$$

which is quite poor in this case. On the other hand, using our knowledge of the distribution of X, let us reconsider Example 4.15. Now the mission time at which the required level of reliability is certainly lost is computed from $e^{-t/100} \leqslant 0.9$, or:

$$t \geqslant 10.5 \text{ hours},$$

which allows for much shorter missions than that suggested by the Markov inequality.

Next assume that both the mean μ and variance σ^2 are given. We can now get a better estimate of the probability of events of interest by using the Chebychev inequality:

$$P(|X - \mu| \geqslant t) \leqslant \frac{\sigma^2}{t^2}, \qquad t > 0. \tag{4.57}$$

This inequality formalizes the intuitive meaning of the variance given earlier: if σ is small, there is a high probability for getting a value close to the mean, and if σ is large there is a high probability for getting values farther away from the mean.

To prove the Chebychev inequality (4.57), we apply the Markov inequality (4.56) to the nonnegative random variable $(X - \mu)^2$ and number t^2 to obtain:

$$P[(X - \mu)^2 \geqslant t^2] \leqslant \frac{E[(X - \mu)^2]}{t^2} = \frac{\sigma^2}{t^2}, \tag{4.58}$$

noting that the event $[(X - \mu)^2 \geqslant t^2] = [|X - \mu| \geqslant t]$ yields the Chebychev inequality (4.57).

The importance of Chebychev's inequality lies in its generality. No assumption on the nature of random variable X is made other than that it has a finite variance. For most distributions, there are sharper bounds for $P(|X - \mu| \geqslant t)$ than that given by Chebychev's inequality; however, examples show that in general the bound given by this inequality cannot be improved (see problem 4 at the end of this section).

Example 4.16

Let X be the execution time of a job in a computer center, and assume that X is exponentially distributed with mean $1/\lambda$ and variance $1/\lambda^2$. Then, using the Chebychev inequality, we have:

$$P\left(\left|X - \frac{1}{\lambda}\right| \geqslant t\right) \leqslant \frac{1}{\lambda^2 t^2}.$$

In particular, if we let $t = 1/\lambda$, this inequality does not give us any information, since it yields:

$$P\left(\left|X - \frac{1}{\lambda}\right| \geqslant \frac{1}{\lambda}\right) \leqslant 1.$$

But if we compute this probability from the distribution function $F_X(x) = 1 - e^{-\lambda x}$, we get:

$$\begin{aligned} P\left(\left|X - \frac{1}{\lambda}\right| \geqslant t\right) &= P\left(0 \leqslant X \leqslant \frac{1}{\lambda} - t \quad \text{or} \quad \frac{1}{\lambda} + t \leqslant X < \infty\right) \\ &= F\left(\frac{1}{\lambda} - t\right) + 1 - F\left(\frac{1}{\lambda} + t\right) \\ &= 1 - e^{\lambda t - 1} + e^{-\lambda t - 1} \end{aligned}$$

and thus:

$$P\left(\left|X - \frac{1}{\lambda}\right| \geqslant \frac{1}{\lambda}\right) = e^{-2} < 1.$$

#

Two alternate forms of Chebychev's inequality are easily derived from (4.57):

$$P(|X - \mu| \geqslant k\sigma) \leqslant \frac{1}{k^2} \tag{4.59}$$

and

$$P(|X - \mu| \leqslant k\sigma) \geqslant 1 - \frac{1}{k^2}. \tag{4.60}$$

Another important result can be obtained by applying Chebychev's inequality to the binomial distribution. Substituting $\mu = np$ and $\sigma = \sqrt{np(1-p)}$ into (4.60) above, we get:

$$P(|X - np| < k\sqrt{np(1-p)}) \geqslant 1 - \frac{1}{k^2}$$

and

$$P\left(\left|\frac{X}{n} - p\right| < k\sqrt{\frac{p(1-p)}{n}}\right) \geqslant 1 - \frac{1}{k^2}.$$

Substitute ϵ for $k\sqrt{\frac{p(1-p)}{n}}$ to obtain:

$$P\left(\left|\frac{X}{n} - p\right| < \epsilon\right) \geqslant 1 - \frac{p(1-p)}{n\epsilon^2} \tag{4.61}$$

which implies that:

$$\lim_{n\to\infty} P\left(\left|\frac{X}{n} - p\right| < \epsilon\right) = 1 \tag{4.62}$$

for any given value of $\epsilon > 0$. Recalling that X denotes the observed number of successes in a sequence of Bernoulli trials, we conclude that as the number of trials, n, increases, the probability that the observed proportion of successes differs from p by less than any positive number ϵ (however small) approaches unity. Formula (4.62) is known as Bernoulli's theorem and is a special case of the weak law of large numbers, which is discussed next.

Let $X_1, X_2, \ldots, X_n$ be n mutually independent identically distributed random variables. An n-tuple of values $(x_1, x_2, \ldots, x_n)$, where x_i is a specific value of X_i, may be thought of as n independent measurements of some quantity that is distributed according to their common distribution. In this sense, we sometimes speak of the n-tuple $(x_1, x_2, \ldots, x_n)$ as a random sample of size n from this distribution.

Assume that the common distribution of these random variables has a finite mean μ. Then for a sufficiently large value of n, we would expect that their arithmetic mean:

$$\bar{x} = \frac{x_1 + x_2 + \cdots + x_n}{n}$$

will be close to μ. Let $S_n = \sum_{i=1}^{n} X_i$ and the sample mean $\bar{X} = S_n/n$. If X_i has a finite variance σ^2, then:

$$\text{Var}\,[\bar{X}] = \text{Var}\left[\frac{S_n}{n}\right] = \frac{n\sigma^2}{n^2} = \frac{\sigma^2}{n}. \tag{4.63}$$

Thus Var $[\bar{X}]$ approaches 0 as n approaches infinity, implying that the distribution of $\bar{X}$ becomes more concentrated about its mean μ. In fact, by applying Chebychev's inequality to $\bar{X}$ we obtain:

$$P(|\bar{X} - \mu| \geqslant \delta) \leqslant \frac{\text{Var}\,[\bar{X}]}{\delta^2} = \frac{\sigma^2}{n\delta^2} \tag{4.64}$$

from which we get:

$$\lim_{n\to\infty} P(|\bar{X} - \mu| \geqslant \delta) = 0. \tag{4.65}$$

Here number δ can be thought of as the desired accuracy in the approximation of μ by $\bar{X}$. Equation (4.65) assures us that no matter how small δ is, the probability that $\bar{X}$ approximates μ to within δ converges to 1 .

Equation (4.65) is known as the **weak law of large numbers.** Although our derivation required that the X_i's have finite variance, the law holds just under the assumption that the X_i's have finite mean.

For the final limit theorem, recall from Chapter 3 that sums of independent normal random variables are themselves normally distributed. The following **central limit theorem** tells us that sums of independent random variables tend to be normally distributed even though the summands are not.

THEOREM 4.7 (THE CENTRAL LIMIT THEOREM). Let $X_1, X_2, \ldots, X_n$ be independent random variables with a finite mean $E[X_i] = \mu_i$ and a finite variance Var $[X_i] = \sigma_i^2$ $(i = 1, 2, \ldots, n)$. We form the normalized random variable:

$$Z_n = \frac{\sum_{i=1}^{n} X_i - \sum_{i=1}^{n} \mu_i}{\sqrt{\sum_{i=1}^{n} \sigma_i^2}} \tag{4.66}$$

so that $E[Z_n] = 0$ and Var $[Z_n] = 1$. Then, under certain regularity conditions, the limiting distribution of Z_n is standard normal, denoted $Z_n \rightarrow N(0, 1)$. That is:

$$\lim_{n\to\infty} F_{Z_n}(t) = P(Z_n \leqslant t) = \int_{-\infty}^{t} \frac{1}{\sqrt{2\pi}}\, e^{-y^2/2}\, dy. \tag{4.67}$$

Example 4.17

As a special case of the central limit theorem, assume that $X_1, X_2, \ldots, X_n$ are mutually independent and identically distributed with the common mean $\mu = E[X_i]$ and common variance $\sigma^2 = \text{Var}[X_i]$. Then equation (4.66) reduces to:

$$Z_n = \frac{(\bar{X} - \mu)\sqrt{n}}{\sigma}, \tag{4.68}$$

where $\bar{X}$ is the sample mean. Therefore, the sample mean from random samples (after standardization) tends toward normality as the sample size n increases. #

The central limit theorem should not be used indiscriminately, since there are distributions that do not obey it. For example, the Cauchy random variable X with pdf:

$$f(x) = \frac{1}{\pi(1+x^2)} \tag{4.69}$$

does not have a finite variance; hence the standard form of Z_n [of equation (4.66)] cannot be written.

It is difficult to give the value of n (the sample size) beyond which the normal approximation is accurate, since it depends upon the form of the underlying distributions (F_{X_i}). Moderate sample sizes, such as ten, commonly are considered adequate.

Problems

1. The average CPU time per session is known to be 4.39 sec for a time-sharing system. We classify a session as a trivial session if it takes less than 1 sec of CPU time, an editing session if it takes between 1 and 5 sec of CPU time, and a number-crunching session otherwise.

   (a) Obtain a bound on the probability that a given session is a number-crunching session.

   (b) Obtain a bound on the probability that a given session is not a trivial session.

   Now, assume that the CPU time per session is exponentially distributed with mean 4.39 sec. Recompute the two bounds.

2. Using the normal tables, plot $P(|X| \geqslant \delta)$ for $0 < \delta < 3$, where $X \sim N(0, 1)$. On the same graph, plot the upper bound on the above probability given by Chebychev's inequality, and compare the two plots.

3. Consider a random variable X with the Cauchy pdf:

   $$f(x) = \frac{1}{\pi(1+x^2)}, \qquad -\infty < x < \infty.$$

   (a) Show that neither $E[X]$ nor $\text{Var}[X]$ exists in this case.

   (b) Show that the characteristic function is given by $N_X(\tau) = e^{-|\tau|}$.

(c) Now consider $Z = \sum_{i=1}^{n} X_i$ where the X_i's are Cauchy and mutually independent. Thus, $N_Z(\tau) = e^{-n|\tau|}$; hence show that Z/n has the Cauchy distribution.

Comment: Z/n is not Gaussian in the limit, since Var $[X_i]$ is not finite, and hence the central limit theorem does not apply.

4. Construct an example of a discrete random variable X that takes on each of the values $-b$, 0, b with nonzero probability, so that the Chebychev inequality becomes an equality when applied to the expression:

$$P(|X - E[X]| \geqslant b).$$

In particular determine $p_X(-b)$, $p_X(0)$, and $p_X(b)$.

***5.** In order to represent a nonnegative real number X in a computer with finite precision, the number is either rounded to obtain X_r or chopped to obtain X_c [STER 1974]. The representation errors in the two cases are bounded by:

$$-½ \leqslant Y_r = X - X_r \leqslant ½$$

and

$$0 \leqslant Y_c = X - X_c < 1$$

(measured in the units of the last digit). It is common to assume that Y_r and Y_c are uniformly distributed over their respective ranges. Now assume two independent numbers X_1 and X_2 are being added. Compute the pdf, the mean, and the variance of the cumulative error in the sum $X_1 + X_2$ both in the case of rounding and in the case of chopping.

Next assume that n mutually independent real numbers are to be added, each subject to round-off or chopping. What are the mean and the variance of the cumulative error in the two cases? Compare the mean with the worst-case errors. For $n = 4, 9, 16, 25, 36, 49, 100$, estimate the probability that the computed sum will differ from the sum of the original numbers by more than 0.5. [*Hint:* Use the central limit theorem.]

Review Problems

1. A program has potentially N distinct input data sets indexed from 1 to N. Suppose the program is run on n randomly chosen data sets with repetition allowed. Let X be the largest index out of the n data sets used. Derive the pmf and the expected value of X. Assume that each of the N data sets is equally likely.

2. Given a **for** statement:

for i:=1 **to** n **do** S

Derive expressions for the distribution, the expected value, and the variance of the execution time T of the **for** loop, assuming that the distribution of the execution time of a single execution of the statement group S is known and that successive executions of S are independent.

3. [WEID 1978] Let the execution time X of a fixed instance of a problem using some randomized algorithm have the distribution function:

$$F(x) = x^{\delta}, \qquad 0 \leqslant x \leqslant 1 \text{ for some } \delta > 0.$$

If we ran the algorithm on that problem instance on a multiprocessor with two processors and ran the same algorithm on each one, the expected solution time would be equal to the expected value of the minimum of two independent random variables (denoted by Y), each having the distribution function F. Determine the conditions under which the speedup (defined by the ratio $E[X]/E[Y]$) exceeds the number of required processors; that is, under what conditions is $2E[Y] < E[X]$?

*4. Returning to the problem of adder design (Chapter 2, review problem 3), show that the expected length of the longest carry sequence is given by:

$$E[V_n] = \sum_{\nu=0}^{n} \nu[R_n(\nu) - R_n(\nu+1)]$$
$$\leqslant \log_2 n.$$

Thus, although in the worst case the length of a carry sequence can be as large as n, it is much smaller on the average. This fact can be used in speeding up the average addition time.

*5. [TSAO 1974] Consider the representation error in storing a real number in a machine with m-digit base β normalized floating-point arithmetic. Let the original mantissa X have the reciprocal pdf: $f_X(x) = 1/(x \ln \beta)$, $1/\beta \leqslant x < 1$. Let X_c and X_r denote the machine representations of X assuming chopping and rounding, respectively. Then the respective (relative) representation errors Δ_c and Δ_r are given by:

$$\Delta_c = \frac{X - X_c}{X} \quad \text{and} \quad \Delta_r = \frac{X - X_r}{X}.$$

Assuming that the absolute error $Y_c = X - X_c$ is a continuous random variable, uniformly distributed over the interval $(0, \beta^{-m})$, and that Y_c is independent of X (a questionable assumption), show that the pdf of the relative representation error Δ_c is given by:

$$f_{\Delta_c}(\delta) = \begin{cases} \dfrac{\beta^{m-1}(\beta - 1)}{\ln \beta}, & \beta^{-m} > \delta \geqslant 0, \\ \dfrac{1/\delta - \beta^{m-1}}{\ln \beta}, & \beta^{1-m} > \delta \geqslant \beta^{-m}. \end{cases}$$

Similarly, assuming that $Y_r = X - X_r$ is uniformly distributed over $(-\beta^{-m}/2, \beta^{-m}/2)$ and that Y_r and X are independent, show that:

$$f_{\Delta_r}(\delta) = \begin{cases} \dfrac{\beta^{m-1}(\beta - 1)}{\ln \beta}, & |\delta| \leqslant \beta^{-m}/2, \\ \dfrac{\dfrac{1}{2|\delta|} - \beta^{m-1}}{\ln \beta}, & \beta^{-m}/2 < |\delta| < \beta^{1-m}/2. \end{cases}$$

Plot the two densities and compute the average representation errors $E[\Delta_c]$ and $E[\Delta_r]$. Compare these with the respective maximum representation errors β^{-m+1} and $\frac{1}{2}\beta^{-m+1}$. Also compute the variances of Δ_c and Δ_r.

References

[BLAK 1979] I. F. BLAKE, *An Introduction to Applied Probability*, John Wiley and Sons, New York.

[HUNT 1980] D. HUNTER, "Modeling Real DASD Configurations," IBM T. J. Watson Center Report, RC-8606, Yorktown Heights, N. Y.

[KOBA 1978] H. KOBAYASHI, *Modeling and Analysis*, Addison-Wesley, Reading, Mass.

[STER 1974] P. H. STERBENZ, *Floating-Point Computation*, Prentice-Hall, Englewood Cliffs, N.J.

[TSAO 1974] N. -K. TSAO, "On the Distributions of Significant Digits and Roundoff Errors," *CACM*, 17:5 (May 1974), 269–71.

[WEID 1978] B. WEIDE, "Statistical Methods in Algorithm Design and Analysis," Ph.D. thesis, Department of Computer Science, Carnegie-Mellon University, Pittsburgh, Pa.

[WIRT 1976] N. WIRTH, *Algorithms + Data Structures = Programs*, Prentice-Hall, Englewood Cliffs, N.J.

[WOLM 1965] E. WOLMAN, "A Fixed Optimum Cell-Size for Records of Variable Length," *JACM*, 12, pp. 53–70.

[ZIPF 1949] G. K. ZIPF, *Human Behavior and the Principle of Least Effort, An Introduction to Human Ecology*, Addison-Wesley, Reading, Mass.

Chapter 5

Conditional Distribution and Conditional Expectation

5.1 INTRODUCTION

We have seen that if two random variables are independent, then their joint distribution can be determined from their marginal distribution functions. In the case of dependent random variables, however, the joint distribution can not be determined in this simple fashion. This leads us to the notions of conditional pmf, conditional pdf, and conditional distribution.

Recalling the definition of conditional probability, $P(A|B)$, for two events A and B, we can define the **conditional probability** $P(A|X = x)$ of event A, given that the event $[X = x]$ has occurred, as:

$$P(A|X = x) = \frac{P(A \text{ occurs and } X = x)}{P(X = x)}, \tag{5.1}$$

whenever $P(X = x) \neq 0$. In Chapter 3 we noted that if X is a continuous random variable, then $P(X = x) = 0$ for all x. In this case, Definition (5.1) is not satisfactory. On the other hand, if X is a discrete random variable, then Definition (5.1) is adequate:

Definition (Conditional pmf). Let X and Y be discrete random variables having a joint pmf $p(x, y)$. The conditional pmf of Y given X is defined by:

$$\begin{aligned} p_{Y|X}(y|x) &= P(Y = y|X = x) \\ &= \frac{P(Y=y,\ X=x)}{P(X=x)} \\ &= \frac{p(x, y)}{p_X(x)}, \end{aligned} \tag{5.2}$$

if $p_X(x) \neq 0$.

Note that the conditional pmf, as defined above, satisfies properties (p1)–(p3) of a pmf, discussed in Chapter 2. Rewriting the above definition another way, we have:

$$p(x, y) = p_X(x)p_{Y|X}(y|x) = p_Y(y)p_{X|Y}(x|y). \tag{5.3}$$

This is simply another form of the multiplication rule (of Chapter 1), and it gives us a way to compute the joint pmf whether or not X and Y are independent. *If X and Y are independent,* then from (5.3) and the definition of independence (in Chapter 2) we conclude that:

$$p_{Y|X}(y|x) = p_Y(y). \tag{5.4}$$

From (5.3) we also have the marginal probability:

$$p_Y(y) = \sum_{\text{all } x} p(x, y) = \sum_{\text{all } x} p_{Y|X}(y|x)p_X(x). \tag{5.5}$$

This is another form of the theorem of total probability, discussed in Chapter 1.

We can also define the conditional distribution function $F_{Y|X}(y|x)$ of a random variable Y, given a discrete random variable X by:

$$F_{Y|X}(y|x) = P(Y \leqslant y|X = x) = \frac{P(Y \leqslant y \text{ and } X = x)}{P(X = x)} \tag{5.6}$$

for all values of y and for all values of x such that $P(X = x) > 0$.

Definition (5.6) applies even for the case when Y is not discrete.

Note that the conditional distribution function can be obtained from the conditional pmf (in case Y is discrete):

$$F_{Y|X}(y|x) = \frac{\sum_{t \leqslant y} p(x, t)}{p_X(x)} = \sum_{t \leqslant y} p_{Y|X}(t|x). \tag{5.7}$$

Example 5.1

A computer center has two computer systems labeled A and B. Incoming jobs are independently routed to system A with probability p and to system B with probability $(1 - p)$. The number of jobs, X, arriving per unit time is Poisson distributed with parameter λ. Determine the distribution function of the number of jobs, Y, received by system A, per unit time.

Let us determine the conditional probability of the event $[Y = k]$ given that event $[X = n]$ has occurred. Note that routing of the n jobs can be thought of as a sequence of n independent Bernoulli trials. Hence, the conditional probability that $[Y = k]$ given $[X = n]$ is binomial with parameters n and p:

$$p_{Y|X}(k|n) = \begin{cases} P(Y = k \mid X = n) = \binom{n}{k} p^k (1 - p)^{n-k}, & 0 \leqslant k \leqslant n \\ 0, & \text{otherwise.} \end{cases}$$

Recalling that $P(X = n) = e^{-\lambda}\lambda^n/n!$ and using formula (5.5), we get:

$$p_Y(k) = \sum_{n=k}^{\infty} \binom{n}{k} p^k(1-p)^{n-k} \frac{\lambda^n e^{-\lambda}}{n!}$$

$$= \frac{(\lambda p)^k e^{-\lambda}}{k!} \sum_{n=k}^{\infty} \frac{(\lambda(1-p))^{n-k}}{(n-k)!}$$

$$= \frac{(\lambda p)^k e^{-\lambda}}{k!} e^{\lambda(1-p)}$$

(since the last sum is the Taylor series expansion of $e^{\lambda(1-p)}$)

$$= \frac{(\lambda p)^k e^{-\lambda p}}{k!}.$$

Thus, Y is Poisson distributed with parameter λp. For this reason we often say that the Poisson distribution is preserved under random selection. #

If X and Y are jointly continuous, then we define the conditional pdf of Y given X in a way analogous to the definition of the conditional pmf.

Definition (Conditional pdf). Let X and Y be continuous random variables with joint pdf $f(x, y)$. The conditional density $f_{Y|X}$ is defined by:

$$f_{Y|X}(y|x) = \frac{f(x, y)}{f_X(x)}, \qquad \text{if } 0 < f_X(x) < \infty. \tag{5.8}$$

It can be easily verified that the function defined in (5.8) satisfies properties (f1) and (f2) of a pdf.

It follows from the definition of conditional density that:

$$f(x, y) = f_X(x) f_{Y|X}(y|x) = f_Y(y)\, f_{X|Y}(x|y). \tag{5.9}$$

This is the continuous analog of the multiplication rule, MR, of Chapter 1. If X and Y are independent, then:

$$f(x, y) = f_X(x)\, f_Y(y),$$

which implies that:

$$f_{Y|X}(y|x) = f_Y(y). \tag{5.10}$$

Conversely, if equation (5.10) holds, then it follows that X and Y are independent random variables. Thus (5.10) is a necessary and sufficient condition for two random variables X and Y having a joint density to be independent.

From the expression of joint density (5.9), we can obtain an expression for the marginal density of Y in terms of conditional density by integration:

$$f_Y(y) = \int_{-\infty}^{\infty} f(x, y)\, dx$$

$$= \int_{-\infty}^{\infty} f_X(x) f_{Y|X}(y|x)\, dx. \tag{5.11}$$

This is the continuous analog of the theorem of total probability.

Further, in the definition of conditional density, we can reverse the role of X and Y to define (whenever $f_Y(y) > 0$):

$$f_{X|Y}(x|y) = \frac{f(x,y)}{f_Y(y)}.$$

Using the expression (5.11) for $f_Y(y)$ and noting that $f(x, y) = f_X(x)f_{Y|X}(y|x)$, we obtain:

$$f_{X|Y}(x|y) = \frac{f_X(x)\, f_{Y|X}(y|x)}{\int_{-\infty}^{\infty} f_X(x)\, f_{Y|X}(y|x)\, dx}. \tag{5.12}$$

This is the continuous analog of Bayes' rule discussed in Chapter 1.

The conditional pdf can be used to obtain the conditional probability:

$$P(a \leqslant Y \leqslant b \mid X = x) = \int_a^b f_{Y|X}(y|x)\, dy, \qquad a \leqslant b. \tag{5.13}$$

In particular, the conditional distribution function $F_{Y|X}(y|x)$ is *defined*, analogously to (5.6), as:

$$F_{Y|X}(y|x) = P(Y \leqslant y \mid X = x) = \frac{\int_{-\infty}^{y} f(x, t)\, dt}{f_X(x)}$$

$$= \int_{-\infty}^{y} f_{Y|X}(t|x)\, dt. \tag{5.14}$$

As motivation for Definition (5.14) we observe that:

$$F_{Y|X}(y|x) = \lim_{h \to 0} P(Y \leqslant y \mid x \leqslant X \leqslant x + h)$$

$$= \lim_{h \to 0} \frac{P(x \leqslant X \leqslant x + h \text{ and } Y \leqslant y)}{P(x \leqslant X \leqslant x + h)}$$

$$= \lim_{h \to 0} \frac{\int_x^{x+h} \int_{-\infty}^{y} f(s, t)\, dt\, ds}{\int_x^{x+h} f_X(s)\, ds}$$

$$= \lim_{h \to 0} \frac{h \int_{-\infty}^{y} f(x_1^*, t)\, dt}{h f_X(x_2^*)}$$

for some x_1^*, x_2^* with $x \leqslant x_1^*, x_2^* \leqslant x + h$

(by the mean value theorem of integrals)

$$= \lim_{h \to 0} \frac{\int_{-\infty}^{y} f(x_1^*, t)\, dt}{f_X(x_2^*)}$$

$$= \int_{-\infty}^{y} \frac{f(x, t)}{f_X(x)}\, dt$$

(since both x_1^* and x_2^* approach x as h approaches 0.)

$$= \int_{-\infty}^{y} f_{Y|X}(t|x)\, dt.$$

Example 5.2

Consider a series system of two independent components with the respective lifetime distributions $X \sim \text{EXP}(\lambda_1)$ and $Y \sim \text{EXP}(\lambda_2)$. We wish to determine the probability that component 2 is the cause of system failure. Let A denote the event that component 2 is the cause of system failure; then:

$$P(A) = P(X \geqslant Y).$$

To compute this probability, first consider the conditional distribution function:

$$F_{X|Y}(t|t) = P(X \leqslant t | Y = t) = F_X(t)$$

(by the independence of X and Y). Now by the continuous version of the theorem of total probability:

$$P(A) = \int_0^{\infty} P(X \geqslant t | Y = t) f_Y(t)\, dt$$

$$= \int_0^{\infty} [1 - F_X(t)]\, f_Y(t)\, dt$$

$$= \int_0^{\infty} e^{-\lambda_1 t} \lambda_2 e^{-\lambda_2 t}\, dt$$

$$= \frac{\lambda_2}{\lambda_1 + \lambda_2}.$$

This result generalizes to a series system of n independent components, each with a respective constant failure rate λ_j ($j = 1, 2, \ldots, n$). The probability that the jth component is the cause of system failure is given by:

$$\frac{\lambda_j}{\sum_{i=1}^{n} \lambda_i}. \tag{5.15}$$

#

Example 5.3 [BARL 1975]

Thus far in our reliability computations, we have considered failure mechanisms of components to be independent. We have derived the exponential lifetime distribution from a Poisson shock model. We now model the behavior of a system of two nonindependent components using a bivariate exponential distribution. Assume three independent Poisson shock sources. A shock from source 1 destroys component 1, and the time to the occurrence U_1 of such a shock is exponentially distributed with parameter λ_1, so that $P(U_1 > t) = e^{-\lambda_1 t}$. A shock from source 2 destroys component 2, and $P(U_2 > t) = e^{-\lambda_2 t}$. Finally, a shock from source 3 destroys both components and it occurs at random time U_{12}, so that $P(U_{12} > t) = e^{-\lambda_{12} t}$. Thus the lifetime X of component 1 satisfies:

$$X = \min\{U_1, U_{12}\}$$

and is exponentially distributed with parameter $\lambda_1 + \lambda_{12}$. The lifetime Y of component 2 is given by:

$$Y = \min\{U_2, U_{12}\}$$

and is exponentially distributed with parameter $\lambda_2 + \lambda_{12}$. Therefore:

$$f_X(x) = (\lambda_1 + \lambda_{12})e^{-(\lambda_1+\lambda_{12})x}, \qquad x > 0,$$

and

$$f_Y(y) = (\lambda_2 + \lambda_{12})e^{-(\lambda_2+\lambda_{12})y}, \qquad y > 0.$$

To compute the joint distribution function $F(x, y) = P(X \leqslant x,\ Y \leqslant y)$, we first compute:

$$\begin{aligned} R(x, y) &= P(X > x,\ Y > y) \\ &= P(\min\{U_1, U_{12}\} > x,\ \min\{U_2, U_{12}\} > y) \\ &= P(U_1 > x,\ U_{12} > \max\{x, y\},\ U_2 > y) \\ &= P(U_1 > x)P(U_{12} > \max\{x, y\})P(U_2 > y) \\ &= e^{-\lambda_1 x - \lambda_2 y - \lambda_{12}\max\{x,y\}}, \qquad x \geqslant 0,\ y \geqslant 0. \end{aligned}$$

This is true since U_1, U_2, and U_{12} are mutually independent. It is interesting to note that $R(x, y) \geqslant R_X(x)\, R_Y(y)$. Now $F(x, y)$ can be obtained using the relation (see Figure 5.1):

$$\begin{aligned} F(x, y) &= R(x, y) + F_X(x) + F_Y(y) - 1 \\ &= 1 + e^{-\lambda_1 x - \lambda_2 y - \lambda_{12}\max\{x,y\}} - e^{-(\lambda_1+\lambda_{12})x} - e^{-(\lambda_2+\lambda_{12})y}. \end{aligned}$$

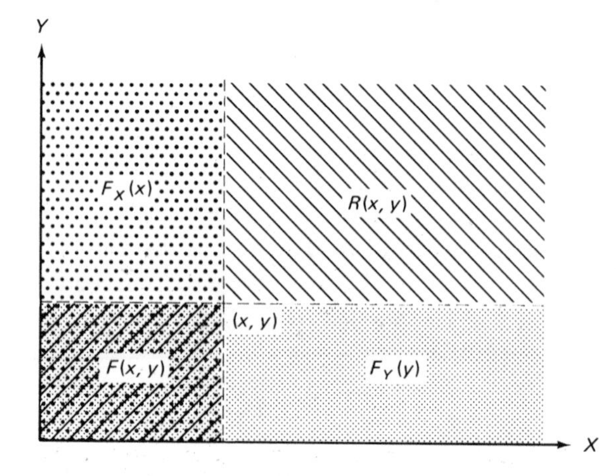

Figure 5.1 Illustration for $R(x, y) + F_x(x) + F_y(y) = 1 + F(x, y)$

In particular:

$$F(x, y) \neq F_X(x)F_Y(y)$$

since:

$$F_X(x)F_Y(y) = 1 - e^{-(\lambda_1+\lambda_{12})x} - e^{-(\lambda_2+\lambda_{12})y} + e^{-(\lambda_1+\lambda_{12})x-(\lambda_2+\lambda_{12})y}$$

Thus X and Y are indeed dependent random variables.

The joint density $f(x, y)$ may be obtained by taking partial derivatives:

$$f(x, y) = \frac{\partial^2 F(x, y)}{\partial x \partial y}$$

$$= \begin{cases} \lambda_1(\lambda_2 + \lambda_{12})e^{-\lambda_1 x-\lambda_2 y-\lambda_{12}y}, & x \leqslant y, \\ \lambda_2(\lambda_1 + \lambda_{12})e^{-\lambda_1 x-\lambda_2 y-\lambda_{12}x}, & x > y, \end{cases}$$

and the conditional density:

$$f_{Y|X}(y|x) = \begin{cases} \dfrac{\lambda_1\,(\lambda_2 + \lambda_{12})}{\lambda_1 + \lambda_{12}}\, e^{-(\lambda_2+\lambda_{12})y+\lambda_{12}x}, & x \leqslant y, \\ \lambda_2\, e^{-\lambda_2 y}, & x > y. \end{cases}$$

Once again, this confirms that X and Y are not independent. #

Problems

1. Consider again the problem of 1K RAM chips supplied by two semiconductor houses (problem 1, Section 3.6). Determine the conditional probability density of the lifetime X, given that the lifetime Y does not exceed 10^6 hours.

2. [MEND 1979] Consider the operation of an on-line file updating system. Let p_i be the probability that a transaction inserts a record into file i ($i = 1, 2, \ldots, n$), so that $\sum_{i=1}^{n} p_i = 1$. The record size (in bytes) of file i is a random variable denoted by Y_i. Determine:
   (a) The average number of bytes added to file i per transaction.
   (b) The variance of the number of bytes added to file i per transaction.

   [*Hint:* You may define the Bernoulli random variable:

$$A_i = \begin{cases} 1, & \text{transaction updates file } i, \\ 0, & \text{otherwise,} \end{cases}$$

   and let the random variable $V_i = A_i\, Y_i$ be the number of bytes added to file i in a transaction.]

3. X_1 and X_2 are independent random variables with Poisson distributions, having respective parameters α_1 and α_2. Show that the conditional pmf of X_1, given $X_1 + X_2$, $p_{X_1|X_1+X_2}(X_1 = x_1 | X_1 + X_2 = y)$, is binomial. Determine its parameters.

4. Let the execution times X and Y of two independent parallel processes be uniformly distributed over $(0, t_X)$ and $(0, t_Y)$, respectively, with $t_X \leqslant t_Y$. Find the probability that the former process finishes execution before the latter.

5.2 MIXTURE DISTRIBUTIONS

The definition of conditional density (and conditional pmf) can be naturally extended to the case where X is a discrete random variable and Y is a continuous random variable (or vice versa).

Example 5.4

Consider a computer system whose workload may be divided into r distinct classes. For job class i ($1 \leqslant i \leqslant r$), the CPU service time is exponentially distributed with parameter λ_i. Let Y denote the service time of a job and let X be the job class. Then:

$$f_{Y|X}(y|i) = \lambda_i e^{-\lambda_i y}, \qquad y > 0.$$

Now let α_i ($\geqslant 0$) be the probability that a randomly chosen job belongs to class i; that is:

$$p_X(i) = \alpha_i, \qquad \sum_{i=1}^{r} \alpha_i = 1.$$

Then the joint density is:

$$\begin{aligned} f(i, y) &= f_{Y|X}(y|i) p_X(i) \\ &= \alpha_i \lambda_i e^{-\lambda_i y}, \qquad y > 0, \end{aligned}$$

and the marginal density is:

$$
\begin{aligned}
f_Y(y) &= \sum_{i=1}^{r} f(i, y) \\
&= \sum_{i=1}^{r} \alpha_i\, f_{Y|X}(y|i) \\
&= \sum_{i=1}^{r} \alpha_i\, \lambda_i\, e^{-\lambda_i y}, \qquad y > 0.
\end{aligned}
$$

Thus Y has an r-stage hyperexponential distribution, denoted by a set of parallel exponential servers as in Figure 5.2. #

Of course, the conditional distribution of Y does not have to be exponential. In general, if we let:

$$f_{Y|X}(y|i) = f_i(y) = f_{Y_i}(y)$$

and

$$F_{Y|X}(y|i) = F_i(y),$$

then we have the unconditional pdf of Y:

$$f_Y(y) = \sum_{i=1}^{r} \alpha_i f_i(y), \tag{5.16}$$

and the unconditional CDF of Y:

$$F_Y(y) = \sum_{i=1}^{r} \alpha_i F_i(y). \tag{5.17}$$

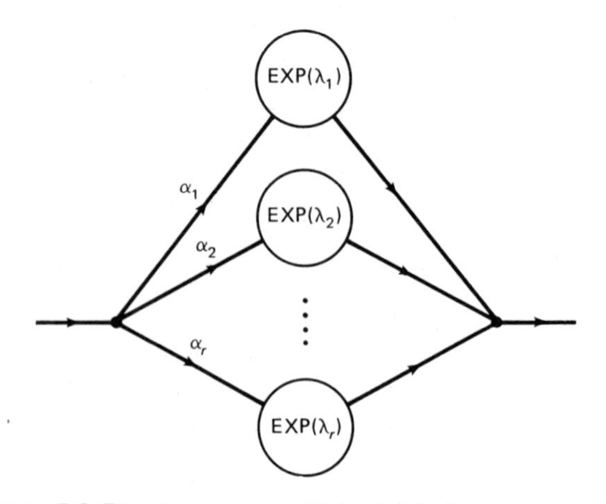

Figure 5.2 The hyperexponential distribution as a set of parallel exponential stages

Taking Laplace transforms on both sides of (5.16), we also have:

$$L_Y(s) = \sum_{i=1}^{r} \alpha_i L_{Y_i}(s). \tag{5.18}$$

Finally applying the definitions of the mean and higher moments to (5.16), we have:

$$E[Y] = \sum_{i=1}^{r} \alpha_i E[Y_i], \tag{5.19}$$

$$E[Y^k] = \sum_{i=1}^{r} \alpha_i E[Y_i^k]. \tag{5.20}$$

Such mixture distributions often arise in a number of reliability situations. For example, suppose a manufacturer produces α_i fraction of a certain product in assembly line i, and the life length of a unit produced in assembly line i has a distribution F_i. Now if the outputs of the assembly lines are merged, then a randomly chosen unit from the merged stream will possess the life-length distribution given by equation (5.17) above.

Example 5.5

Assume that in a mixture of two groups, one group consists of components in the chance-failure period (with constant hazard rate λ_1) and the other of aging items (modeled by an r-stage Erlang lifetime distribution with parameter λ_2). If α is the fraction of group-one components, then the distribution of the lifetime Y of a component from the merged stream is given by:

$$F_Y(y) = \alpha(1 - e^{-\lambda_1 y}) + (1-\alpha)(1 - \sum_{k=0}^{r-1} \frac{(\lambda_2 y)^k}{k!} e^{-\lambda_2 y})$$

and

$$f_Y(y) = \alpha\lambda_1 e^{-\lambda_1 y} + (1-\alpha) \frac{\lambda_2^r y^{r-1}}{(r-1)!} e^{-\lambda_2 y}.$$

This density and the corresponding hazard rate are shown in Figures 5.3 and 5.4. Note that this distribution has a nonmonotonic hazard function. #

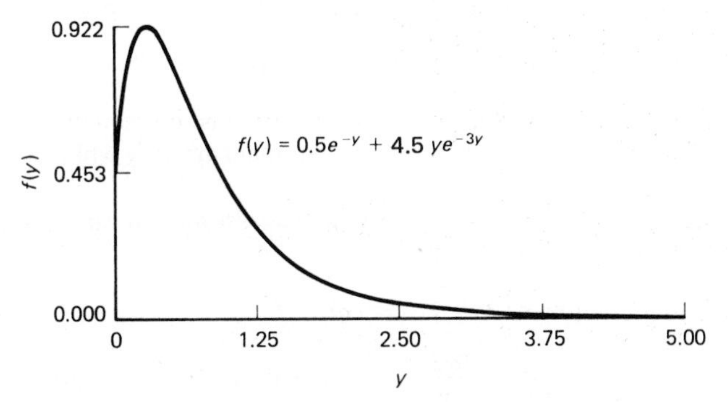

Figure 5.3 The pdf of a mixture of exponential and Erlang

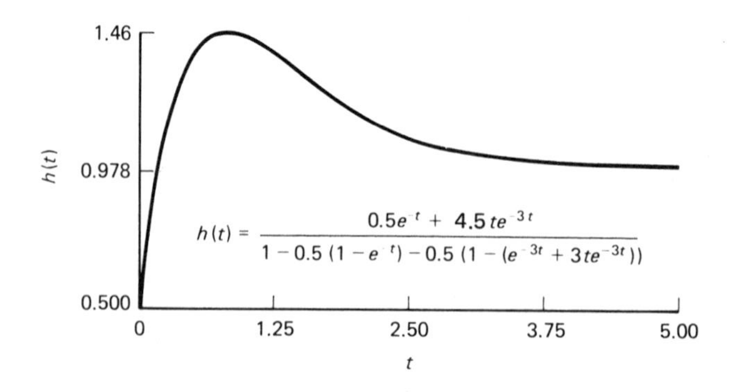

Figure 5.4 Hazard rate of a mixture of exponental and Erlang

More generally, the distributions being mixed may be uncountably infinite in number; that is, X may be a continuous random variable. For instance, the lifetime of a product may depend upon the amount X of impurity present in the raw material. Let the conditional distribution of the lifetime Y be given by:

$$F_{Y|X}(y|x) = G_x(y) = \int_{-\infty}^{y} \frac{f(x, t)\,dt}{f_X(x)},$$

where the impurity X has a density function $f_X(x)$. Then the resultant lifetime distribution F_Y is given by:

$$F_Y(y) = \int_{-\infty}^{\infty}\int_{-\infty}^{y} f(x, t)\ dt\ dx = \int_{-\infty}^{\infty} f_X(x)\ G_x(y)\ dx.$$

In the next example we let Y be discrete and X continuous.

Example 5.6 [CLAR 1970]

Let X be the service time of a customer in a computing center and let it be exponentially distributed with parameter μ, so that:

$$f_X(x) = \mu e^{-\mu x}, \qquad x > 0.$$

Let the number of customers arriving in the interval $(0, t]$ be Poisson distributed with parameter λt. Finally, let Y be the number of customers arriving while one is being served.

If we fix the value of X to be x, the Poisson arrival assumption can be used to obtain the conditional pmf of Y given $[X = x]$:

$$\begin{aligned} p_{Y|X}(y|x) &= P(Y = y|X = x) \\ &= e^{-\lambda x}\frac{(\lambda x)^y}{y!}, \qquad y = 0, 1, 2, \ldots. \end{aligned}$$

The joint probability density function of X and Y is then given by:

$$f(x, y) = f_X(x) p_{Y|X}(y|x)$$
$$= \frac{\mu e^{-(\lambda+\mu)x}(\lambda x)^y}{y!}, \qquad y = 0, 1, 2, \ldots; \qquad x > 0.$$

The unconditional (or marginal) pmf of Y can now be obtained by integration:

$$p_Y(y) = P(Y = y)$$
$$= \int_0^\infty f(x, y)\, dx$$
$$= \frac{\mu}{y!} \int_0^\infty e^{-(\lambda+\mu)x}(\lambda x)^y\, dx.$$

Substituting $(\lambda + \mu)x = w$, we get:

$$p_Y(y) = \frac{\mu\lambda^y}{y!(\lambda + \mu)^{y+1}} \int_0^\infty e^{-w} w^y\, dw$$
$$= \frac{\mu\lambda^y y!}{y!(\lambda + \mu)^{y+1}}$$

[since the last integral is equal to $\Gamma(y + 1) = y!$ by formulas (3.26) and (3.24)]. Thus:

$$p_Y(y) = \frac{\rho^y}{(1 + \rho)^{y+1}}, \qquad \text{where } \rho = \frac{\lambda}{\mu},$$
$$= \left(\frac{\rho}{1 + \rho}\right)^y \frac{1}{1 + \rho}, \qquad y = 0, 1, 2, \ldots.$$

Thus Y has a modified geometric distribution with parameter $\frac{1}{1 + \rho}$; hence the expected value is:

$$E[Y] = \frac{\frac{\rho}{1 + \rho}}{\frac{1}{1 + \rho}} = \rho = \frac{\lambda}{\mu}.$$

This is an example of the so-called $M/M/1$ queuing system to be discussed in a later chapter. We may argue that an undesirable backlog of customers will not occur provided the average number of customers arriving in the interval representing the service time of a typical customer is less than 1. In other words, the queuing system will remain stable provided:

$$E[Y] = \rho < 1 \quad \text{or} \quad \lambda < \mu.$$

This last condition says that the rate at which customers arrive is less than the rate at which work can be completed. #

Example 5.7 [GAVE 1973]

Consider a series system with n components, each with a lifetime distribution function $G(t)$ and density $g(t)$. Because of options offered, the number of components, Y, in a specific system is a random variable. Let X denote the lifetime of the series system. Then clearly:

$$F_{X|Y}(t|n) = 1 - [1 - G(t)]^n, \quad n = 0, 1, 2, \ldots, \quad t > 0,$$

$$f_{X|Y}(t|n) = n[1 - G(t)]^{n-1} g(t), \quad n = 0, 1, 2, \ldots, \quad t > 0.$$

Assume that the number of components, Y, has a Poisson distribution with parameter α. Then:

$$p_Y(n) = e^{-\alpha} \frac{\alpha^n}{n!}, \qquad \alpha > 0, \quad n = 0, 1, 2, \ldots,$$

and the joint density is:

$$\begin{aligned} f(t, n) &= f_{X|Y}(t \mid n)\, p_Y(n) \\ &= \begin{cases} e^{-\alpha} \dfrac{\alpha^n}{(n-1)!} [1 - G(t)]^{n-1} g(t), & t > 0, \quad n = 1, 2, \ldots, \\ 0, & \text{otherwise.} \end{cases} \end{aligned}$$

We can now determine the marginal density:

$$f_X(t) = \sum_{n=1}^{\infty} [1 - G(t)]^{n-1} g(t)\, e^{-\alpha} \frac{\alpha^n}{(n-1)!}.$$

The system reliability is given by:

$$\begin{aligned} R_X(t) &= P(X > t) \\ &= \sum_{n=0}^{\infty} [1 - F_{X|Y}(t|n)]\, p_Y(n) \end{aligned}$$

(by the theorem of total probability)

$$\begin{aligned} &= \sum_{n=0}^{\infty} [1 - G(t)]^n e^{-\alpha} \frac{\alpha^n}{n!} \\ &= e^{-\alpha} \sum_{n=0}^{\infty} \frac{\{\alpha[1 - G(t)]\}^n}{n!} \\ &= e^{-\alpha} e^{\alpha[1 - G(t)]} \\ &= e^{-\alpha G(t)}. \end{aligned}$$

Now suppose that the system has survived until time t. We are interested in computing the conditional pmf of the number of components Y it has:

$$P(Y = n | X > t) = \frac{P(X > t,\ Y = n)}{P(X > t)}$$

$$= \frac{[1 - F_{X|Y}(t|n)]\, p_Y(n)}{R_X(t)}$$

$$= e^{-\alpha[1-G(t)]} \frac{[\alpha(1 - G(t))]^n}{n!}.$$

Thus the conditional pmf of Y, given that no failure has occurred until time t, is Poisson with parameter $\alpha[1 - G(t)]$. Since $G(t)$ is a monotonically increasing function of t, the parameter of the Poisson distribution decreases with t. In other words, the longer the system survives, the greater is the evidence that it has a small number of components. #

Yet another case of a mixture distribution occurs when we mix two distributions, one discrete and the other continuous. The mixture distribution then represents a mixed random variable (see the distribution (3.2) in Chapter 3).

Problems

1. Consider the **if** statement:

   if B **then** S_1 **else** S_2.

   Let the random variables X_1 and X_2 respectively, denote the execution times of the statement groups S_1 and S_2. Assuming the probability that the Boolean expression $B=$ true is p, derive an expression for the distribution of the total execution time X of the **if** statement. Compute $E[X]$ and Var $[X]$ as functions of the means and variances of X_1 and X_2. Generalize your results to a **case** statement with k clauses.

2. Describe a method of generating a random deviate of a two-stage hyperexponential distribution.

3. One of the inputs to a certain program is a random variable whose value is a nonnegative real number; call it Λ. The probability density function of Λ is given by:

   $$f_\Lambda(\lambda) = \lambda\, e^{-\lambda}, \qquad \lambda > 0.$$

   Conditioned on $\Lambda = \lambda$, the execution time of the program is an exponentially distributed random variable with parameter λ. Compute the distribution function of the program execution time X.

5.3. CONDITIONAL EXPECTATION

If X and Y are continuous random variables, then the conditional density $f_{Y|X}$ is given by formula (5.8). Since $f_{Y|X}$ is a density of a continuous random variable, we can talk about its various moments. Its mean (if it exists) is called the **conditional expectation** of Y given $[X = x]$ and will be denoted by $E[Y|X = x]$ or $E[Y|x]$. Thus:

$$E[Y|x] = \int_{-\infty}^{\infty} y\, f(y|x)\, dy$$

$$= \frac{\int_{-\infty}^{\infty} y\, f(x, y)\, dy}{f_X(x)}, \qquad 0 < f_X(x) < \infty. \tag{5.21}$$

We will define $E[Y|x] = 0$ elsewhere. The quantity $m(x) = E[Y|x]$, considered as a function of x, is known as the **regression function** of Y on X.

In case the random variables X and Y are discrete, the conditional expectation $E[Y|x]$ is defined as:

$$E[Y|X = x] = \sum_y yP(Y = y|X = x)$$

$$= \sum_y y p_{Y|X}(y|x). \tag{5.22}$$

Similar definitions can be given in mixed situations. These definitions can be easily generalized to define the conditional expectation of a function $\phi(Y)$:

$$E[\phi(Y)|X = x] = \begin{cases} \int_{-\infty}^{\infty} \phi(y) f_{Y|X}(y|x)\, dy, & \text{if } Y \text{ is continuous,} \\ \sum_i \phi(y_i) p_{Y|X}(y_i|x), & \text{if } Y \text{ is discrete.} \end{cases} \tag{5.23}$$

As a special case of definition (5.23), we have the conditional kth moment of Y, $E[Y^k|X = x]$, and the conditional moment generating function of Y, $M_{Y|X}(\theta|x) = E[e^{\theta Y}|X = x]$. From the conditional moment generating function we also obtain the definition of the conditional Laplace transform, $L_{Y|X}(s|x) = E[e^{-sY}|X = x]$, and the conditional PGF, $G_{Y|X}(z|x) = E[z^Y|X = x]$.

We may take the expectation of the regression function $m(X)$ to obtain the unconditional expectation of Y:

$$E[m(X)] = E[E[Y|X]] = E[Y]$$

that is to say:

$$E[Y] = \begin{cases} \sum_x E[Y|X = x] p_X(x), & \text{if } X \text{ is discrete,} \\ \int_{-\infty}^{\infty} E[Y|X = x] f_X(x)\, dx, & \text{if } X \text{ is continuous.} \end{cases} \tag{5.24}$$

This last formula, known as the **theorem of total expectation**, is found to be quite useful in practice. A similar result called the theorem of total moments is given by:

$$E[Y^k] = \begin{cases} \sum_x E[Y^k|X = x]p_X(x), & \text{if } X \text{ is discrete,} \\ \int_{-\infty}^{\infty} E[Y^k|X = x]f_X(x)\, dx, & \text{if } X \text{ is continuous.} \end{cases} \tag{5.25}$$

Similarly, we have theorems of total transforms. For example, the theorem of total Laplace transform is (assuming Y is a nonnegative continuous random variable):

$$L_Y(s) = \begin{cases} \sum_x L_{Y|X}(s|x)p_X(x), & \text{if } X \text{ is discrete,} \\ \int_{-\infty}^{\infty} L_{Y|X}(s|x)f_X(x)\, dx, & \text{if } X \text{ is continuous.} \end{cases} \tag{5.26}$$

Example 5.8

Consider the Example 5.4 of r job classes in a computer system. Since:

$$f_{Y|X}(y|i) = \lambda_i e^{-\lambda_i y},$$

then:

$$E[Y|X = i] = \frac{1}{\lambda_i}$$

and

$$E[Y^2|X = i] = \frac{2}{\lambda_i^2}.$$

Then by the theorem of total expectation:

$$E[Y] = \sum_{i=1}^{r} \frac{\alpha_i}{\lambda_i}$$

and

$$E[Y^2] = \sum_{i=1}^{r} \frac{2\alpha_i}{\lambda_i^2}.$$

Then:

$$\text{Var}\,[Y] = \sum_{i=1}^{r} \frac{2\alpha_i}{\lambda_i^2} - (\sum_{i=1}^{r} \frac{\alpha_i}{\lambda_i})^2.$$

#

Example 5.9

Refer to Example 5.7 of a series system with a random number of components, where:

$$f_{X|Y}(t|n) = n\,[1 - G(t)]^{n-1}\, g(t), \qquad t > 0.$$

Let:

$$G(t) = 1 - e^{-\lambda t}, \qquad \lambda > 0,\ t \geqslant 0.$$

Then:

$$\begin{aligned} f_{X|Y}(t|n) &= n e^{-\lambda(n-1)t} \lambda e^{-\lambda t} \\ &= n\lambda e^{-n\lambda t}, \end{aligned}$$

which is the exponential pdf with parameter $n\lambda$. It follows that:

$$E[X|Y = n] = \frac{1}{n\lambda}$$

and

$$E[X] = \sum_{n=1}^{\infty} \frac{1}{n\lambda} e^{-\alpha} \frac{\alpha^n}{n!}. \qquad \#$$

Example 5.10

(Analysis of uniform hashing) [KNUT 1973b]. A popular method of storing tables for fast searching is known as **hashing.** The table has M entries indexed from 0 to $M-1$. Given a search key k, an application of the hash function h produces an index, $h(k)$, into the table, where we generally expect to find the required entry. Since there are distinct keys $k_i \neq k_j$ that hash to the same value $h(k_i) = h(k_j)$, a situation known as collision, we have to derive some method for producing secondary indices for search.

Assume that k entries out of M in the table are currently occupied. As a consequence of the assumption that h distributes values uniformly over the table, all $\binom{M}{k}$ possible configurations are equally likely. Let the random variable X denote the number of probes necessary to insert the next item in the table, and let Y denote the number of occupied entries in the table. For a given number of occupied entries $Y = k$, if the number of probes is equal to r, then $(r-1)$ given cells are known to be occupied and the last inspected cell is known to be unoccupied. Out of the remaining $M - r$ cells, $(k - r + 1)$ can be occupied in $\binom{M-r}{k-r+1}$ ways. Therefore:

$$\begin{aligned} P(X = r | Y = k) &= p_{X|Y}(r|k) \\ &= \frac{\binom{M-r}{k-r+1}}{\binom{M}{k}}, \qquad 1 \leqslant r \leqslant M. \end{aligned} \tag{5.27}$$

This implies that:

$$\begin{aligned} E[X|Y = k] &= \sum_{r=1}^{M} r p_{X|Y}(r|k) \\ &= \sum_{r=1}^{M} (M+1) p_{X|Y}(r|k) - \sum_{r=1}^{M} (M + 1 - r) p_{X|Y}(r|k). \end{aligned}$$

Now since $p_{X|Y}$ is a pmf, the first sum on the right-hand side equals $M+1$. We substitute expression (5.27) in the second sum to obtain:

$$E[X|Y=k] = (M+1) - \sum_{r=1}^{M} (M+1-r) \frac{\binom{M-r}{k-r+1}}{\binom{M}{k}}$$

$$= (M+1) - \sum_{r=1}^{M} \frac{(M+1-r)\,(M-r)!}{(k-r+1)!(M-k-1)!\cdot\binom{M}{k}}$$

$$= (M+1) - \sum_{r=1}^{M} \frac{(M-r+1)!(M-k)}{(k-r+1)!(M-k)!\cdot\binom{M}{k}}$$

$$= (M+1) - \sum_{r=1}^{M} \frac{(M-k)\binom{M-r+1}{M-k}}{\binom{M}{k}}$$

Now the sum:

$$\sum_{r=1}^{M} \binom{M-r+1}{M-k} = \sum_{i=1}^{M} \binom{i}{M-k} = \sum_{i=0}^{M} \binom{i}{M-k} = \binom{M+1}{M-k+1}$$

(using formula (11) from [KNUT 1973a, p. 54]).

After substitution and simplification, we have:

$$E[X|Y=k] = \frac{M+1}{M-k+1}, \qquad 0 \leqslant k \leqslant M-1.$$

Now, assuming that Y is uniformly distributed over $0 \leqslant k < N \leqslant M$, we get:

$$p_Y(k) = \frac{1}{N},$$

$$E[X] = \sum_{k=0}^{N-1} \frac{1}{N} E[X|Y=k]$$

$$= \frac{M+1}{N}\left[\frac{1}{M+1} + \frac{1}{M} + \cdots + \frac{1}{M-N+2}\right]$$

$$= \frac{M+1}{N}(H_{M+1} - H_{M-N+1})$$

$$\simeq \frac{1}{\alpha} \ln\frac{1}{1-\alpha},$$

where $\alpha = N/(M+1)$, the table occupancy factor. This is the expected number of probes necessary to locate an entry in the table, provided the search is successful.

Note that if the table occupancy factor is low (below 80 percent), the average number of probes is nearly equal to 1. In other words, where applicable, this is an efficient method of search. #

Problems

*1. The notion of a recovery block was introduced by Randell [RAND 1975] to facilitate software fault tolerance in presence of software design errors. This construct provides a "normal" algorithm to perform the required function together with an acceptance test of its results. If the test results are unsatisfactory then an alternative algorithm is executed. Assume that X is the execution time of the normal algorithm and Y is the execution time of the alternative algorithm. Assume p is the probability that the results of the normal execution satisfy the acceptance test. Determine the distribution function of the total execution time T of the recovery block assuming that X and Y are uniformly distributed over (a, b). Repeat, assuming that X and Y are exponentially distributed with parameters λ_1 and λ_2, respectively. In each case determine $E[T]$, Var $[T]$, and in the latter case $L_T(s)$.

2. Consider the flowchart model of fault recovery in a computer system (such as Bell System's Electronic Switching system) as shown in Figure 5.P.1. Assuming that the random variables D, L, R, M_D, and M_L are exponentially distributed with

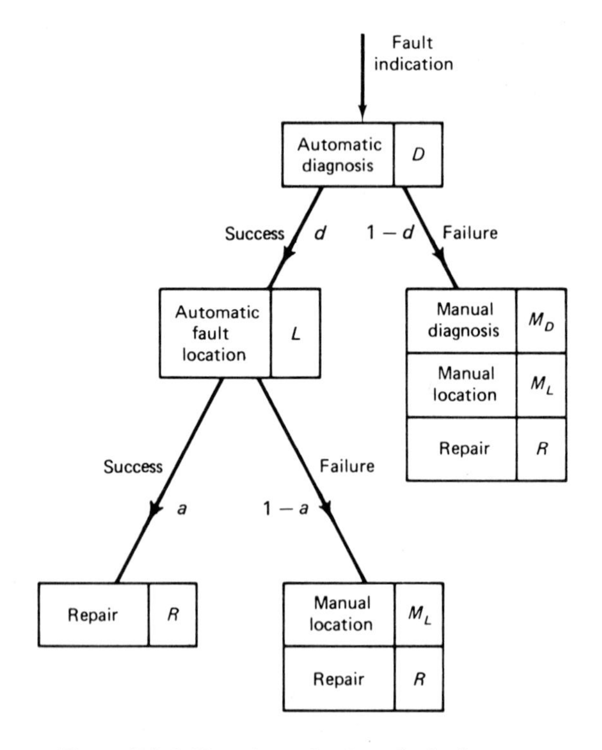

Figure 5.P.1 Flowchart of automatic fault recovery

parameters δ, λ, ρ, μ_1 and μ_2, determine the distribution function of the random variable X, denoting the total recovery time. Also compute $E[X]$ and Var $[X]$.

3. (Linear searching problem) We are given an unordered list with n distinct keys. We are searching linearly for a specific key that has a probability p of being present in the list (and probability q of being absent). Given that the key is in the list, the probability of its being in position i is $1/n$, $i = 1, 2, \ldots, n$. Compute the expected number of comparisons for:
   (a) A successful search.
   (b) An unsuccessful search.
   (c) A search (unconditionally).

4. [MEND 1979] Let V_1 be the random variable denoting the length (in bytes) of a source program. Let p be the probability of successful compilation of the program. Let V_2 be the length of the compiled code (load module). Clearly, V_2 and V_1 will not be independent. Assume $V_2 = BV_1$ where B is a random variable, and B and V_1 are independent. After the compilation, the load module will be entered into a library. Let X be the length of a request for space allocation to the library due to the above source program. Determine $E[X]$ and Var $[X]$ in terms of $E[B]$, $E[V_1]$, Var $[B]$, and Var $[V_1]$.

5.4 IMPERFECT FAULT COVERAGE AND RELIABILITY

Reliability models of systems with dynamic redundancy (e.g., standby redundancy, hybrid NMR) developed earlier are not very realistic. It has been demonstrated that the reliability of such systems depends strongly on the effectiveness of recovery mechanisms. In particular, it may be impossible to switch in an existing spare module and thus recover from a failure. Faults such as these are said to be "uncovered", and the probability that a given fault belongs to this class is denoted by $1 - c$, where c denotes the probability of occurrence of "covered" faults, and is known as the **coverage parameter** [BOUR 1969].

Example 5.11

Let X denote the lifetime of a system with two units, one active and the other a cold standby spare. The failure rate of an active unit is λ, and a cold spare does not fail. Let Y be the indicator random variable of the fault class; that is:

$Y = 0$ if the fault is uncovered,
$Y = 1$ if the fault is covered.

Then:

$$p_Y(0) = 1 - c \quad \text{and} \quad p_Y(1) = c.$$

To compute the MTTF of this system, we first obtain the conditional expectation of X given Y by noting that if an uncovered fault occurs, the mean life of the system equals the mean life of the initially active unit. That is:

$$E[X|Y=0] = \frac{1}{\lambda}.$$

On the other hand, if a covered fault occurs, then the mean life of the system is the sum of the mean lives of the two units:

$$E[X|Y=1] = \frac{2}{\lambda}.$$

Now, using the theorem of total expectation, we obtain the system MTTF as:

$$E[X] = \frac{1-c}{\lambda} + \frac{2c}{\lambda} = \frac{1+c}{\lambda}. \tag{5.28}$$

Thus when $c = 0$, the standby module does not contribute anything to system reliability, and when $c = 1$, the full potential of this module is realized. For $c < 0.5$, MTTF of a *parallel redundant configuration* with two units (static redundancy) is *higher* than that of a two-unit *standby redundant system.*

Given that the fault was covered ($Y = 1$), the system lifetime, X, is the sum of two independent exponentially distributed variables, each with parameter λ. Thus the conditional pdf of X given $Y = 1$ is the two-stage Erlang density:

$$f_{X|Y}(t|1) = \lambda^2 t e^{-\lambda t}.$$

On the other hand, given that an uncovered fault occurred, the system lifetime X is simply the lifetime of the initially active component. Hence:

$$f_{X|Y}(t|0) = \lambda e^{-\lambda t}.$$

Then the joint density is computed by $f(t, y) = f_{X|Y}(t|y)p_Y(y)$ as:

$$f(t, y) = \begin{cases} \lambda(1-c)e^{-\lambda t}, & t > 0,\ y = 0, \\ \lambda^2 c t e^{-\lambda t}, & t > 0,\ y = 1, \end{cases}$$

and the marginal density of X is computed by summing over the joint density:

$$f_X(t) = \lambda^2 c t e^{-\lambda t} + \lambda(1-c)e^{-\lambda t}.$$

Therefore, the system reliability is given by:

$$\begin{aligned} R_X(t) &= (1-c)e^{-\lambda t} + ce^{-\lambda t}(1+\lambda t) \\ &= e^{-\lambda t} + c\lambda t e^{-\lambda t} \\ &= (1 + c\lambda t)e^{-\lambda t}. \end{aligned} \tag{5.29}$$

Figure 5.5 shows $R_X(t)$ as a function of t for various values of the coverage parameter.

The conditional Laplace transform of the lifetime X is given by:

$$L_{X|Y}(s|0) = \frac{\lambda}{s+\lambda} \quad \text{and} \quad L_{X|Y}(s|1) = \left(\frac{\lambda}{s+\lambda}\right)^2.$$

Then the unconditional transform is computed using the theorem of total transform:

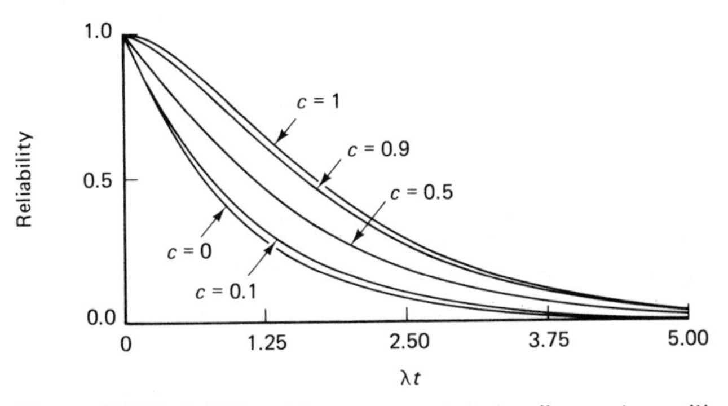

Figure 5.5 Reliability of two-component standby system with imperfect coverage

$$L_X(s) = c\frac{\lambda^2}{(s+\lambda)^2} + (1-c)\frac{\lambda}{s+\lambda}$$

$$= \frac{\lambda}{s+\lambda}\left[c\,\frac{\lambda}{s+\lambda} + (1-c)\right]. \tag{5.30}$$

Let us rewrite this as:

$$L_X(s) = L_{Y_1}(s)L_{Y_2}(s), \tag{5.31}$$

where:

$$L_{Y_1}(s) = \frac{\lambda}{s+\lambda} \tag{5.32}$$

and

$$L_{Y_2}(s) = \frac{c\lambda}{s+\lambda} + (1-c). \tag{5.33}$$

Using the convolution theorem, we conclude that we can regard system lifetime X as the sum of two independent random variables Y_1 and Y_2. From the Laplace transform of Y_1 we see that $Y_1 \sim \text{EXP}(\lambda)$. Noting that:

$$\lim_{\mu\to\infty} \frac{\mu}{s+\mu} = 1, \tag{5.34}$$

we can rewrite:

$$L_{Y_2}(s) = \lim_{\mu\to\infty}\left[c\,\frac{\lambda}{s+\lambda} + (1-c)\,\frac{\mu}{s+\mu}\right]. \tag{5.35}$$

Thus, Y_2 may be thought of as the limit of a two-stage hyperexponentially distributed random variable with parameters λ and μ. Further thought reveals that in the limit $\mu \longrightarrow \infty$, the distribution function of $Z \sim \text{EXP}(\mu)$ becomes:

$$\lim_{\mu\to\infty} F_Z(z) = \lim_{\mu\to\infty} 1 - e^{-\mu z}$$

$$= \begin{cases} 1, & z > 0, \\ 0, & \text{otherwise.} \end{cases} \tag{5.36}$$

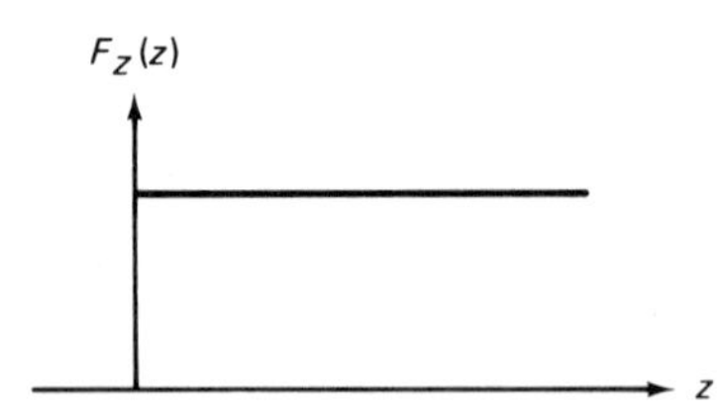

Figure 5.6 The unit-step function: CDF of a constant random variable, $Z = 0$

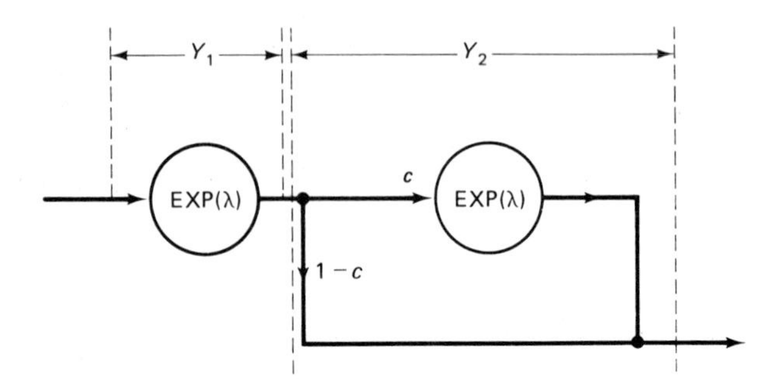

Figure 5.7 The stage-type distribution of system lifetime for the system in Example 5.11

This function is the unit-step function shown in Figure 5.6, and the corresponding random variable is the constant random variable $Z = 0$.

Based on the discussion above we can visualize the system lifetime X as composed of exponential stages as shown in Figure 5.7. #

Example 5.12

In the last example we assumed that a unit in a cold standby status does not fail. Let us now assume that such a unit can fail with a constant failure rate μ (presumably, $0 \leqslant \mu \leqslant \lambda$). Let c_1 be the probability of successful recovery upon the failure of an active unit and let c_2 be the probability of successful recovery following the failure of a spare unit. Note that Bouricius and others [BOUR 1969] assumed that $c_1 = c_2 = c$. Keeping the same notations and assumptions as before, we can compute the Laplace transform of the system lifetime X as follows:

Let X_1 and X_2 denote the time to failure of the powered and unpowered units, respectively. Also let W be the residual lifetime of the unit in operation after a covered fault has occurred. We observe that $X_1 \sim$ EXP (λ), $X_2 \sim$ EXP (μ), and, because of the memoryless property of the exponential distribution, $W \sim$ EXP (λ). Define the random variable Y so that:

$$Y = \begin{cases} 0, & \text{uncovered failure in the active unit,} \\ 1, & \text{covered failure in the active unit,} \\ 2, & \text{uncovered failure in the standby unit,} \\ 3, & \text{covered failure in the standby unit.} \end{cases}$$

First, we compute the pmf of Y by noting that the probability of the active unit failing first is $\lambda/(\lambda+\mu)$ while the probability of the spare unit failing first is $\mu/(\lambda+\mu)$ (refer to Example 5.2). Then (see the tree diagram of Figure 5.8):

$$p_Y(1) = \frac{\lambda\, c_1}{\lambda+\mu}, \quad p_Y(0) = \frac{\lambda(1-c_1)}{\lambda+\mu},$$

$$p_Y(3) = \frac{\mu\, c_2}{\lambda+\mu}, \quad \text{and } p_Y(2) = \frac{\mu(1-c_2)}{\lambda+\mu}.$$

Now, if an uncovered fault has occurred (that is, $Y=0$ or $Y=2$) the lifetime of the system is simply $\min\{X_1, X_2\}$ while a covered fault (that is, $Y=1$ or $Y=3$) implies a system lifetime of:

$$\min\{X_1, X_2\} + W.$$

Note that since X_1 and X_2 are exponentially distributed, $\min\{X_1, X_2\}$ is exponentially distributed with parameter $\lambda+\mu$. The conditional Laplace transform of X for each type of uncovered fault is therefore:

$$L_{X|Y}(s|Y{=}0) = \frac{\lambda+\mu}{s+(\lambda+\mu)} = L_{X|Y}(s|Y=2),$$

and for a covered fault, X is the sum of two independent exponentially distributed random variables and hence:

$$L_{X|Y}(s|Y{=}1) = \frac{\lambda+\mu}{s+\lambda+\mu}\cdot\frac{\lambda}{s+\lambda} = L_{X|Y}(s|Y=3).$$

The unconditional Laplace transform of X is then computed using the theorem of total transform:

$$L_X(s) = \frac{\lambda+\mu}{s+\lambda+\mu}\cdot\frac{\lambda(1-c_1)+\mu(1-c_2)}{\lambda+\mu} + \frac{(\lambda+\mu)\lambda}{(s+\lambda+\mu)(s+\lambda)}\cdot\frac{\lambda c_1+\mu c_2}{\lambda+\mu}. \tag{5.37}$$

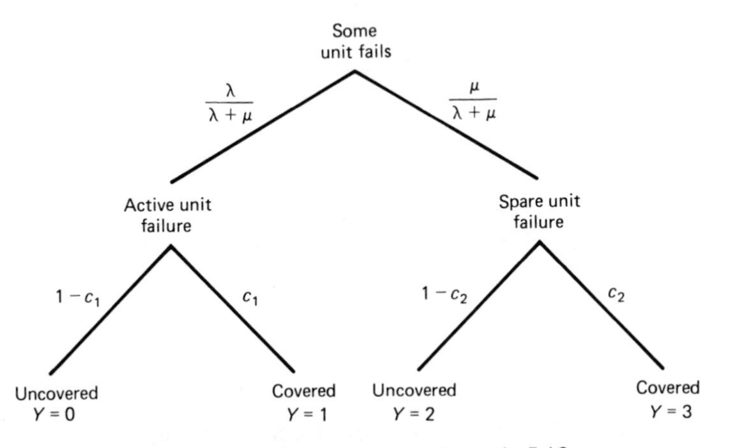

Figure 5.8 Tree diagram for Example 5.12

Thus,

$$L_X(s) = \frac{\lambda + \mu}{s + \lambda + \mu} \left[\frac{\lambda c_1 + \mu c_2}{\lambda + \mu} \cdot \frac{\lambda}{s + \lambda} + \frac{\lambda(1 - c_1) + \mu(1 - c_2)}{\lambda + \mu} \right] \tag{5.38}$$

$$= L_{Y_1}(s)\, L_{Y_2}(s),$$

where:

$$L_{Y_1}(s) = \frac{\lambda + \mu}{s + \lambda + \mu}$$

and

$$L_{Y_2}(s) = \frac{\lambda c_1 + \mu c_2}{\lambda + \mu} \cdot \frac{\lambda}{s + \lambda} + \frac{\lambda(1 - c_1) + \mu(1 - c_2)}{\lambda + \mu}$$

$$= c\,\frac{\lambda}{s + \lambda} + (1 - c),$$

where the "equivalent" coverage c is given by

$$c = \frac{\lambda c_1 + \mu c_2}{\lambda + \mu}. \tag{5.39}$$

We conclude that the system lifetime $X = Y_1 + Y_2$, and it can be thought of as a stage-type random variable as shown in Figure 5.9. Now, since:

$$E[Y_1] = \frac{1}{\lambda + \mu}$$

and

$$E[Y_2] = -\frac{dL_{Y_2}}{ds}\Big|_{s=0} = \frac{c}{\lambda},$$

we conclude that the MTTF of the system is given by:

$$E[X] = \frac{1}{\lambda + \mu} + \frac{c}{\lambda}. \tag{5.40}$$

Comparing the form (5.37) of the Laplace transform of X with the Laplace transform (5.18) of a mixture distribution, we conclude that X is a mixture of

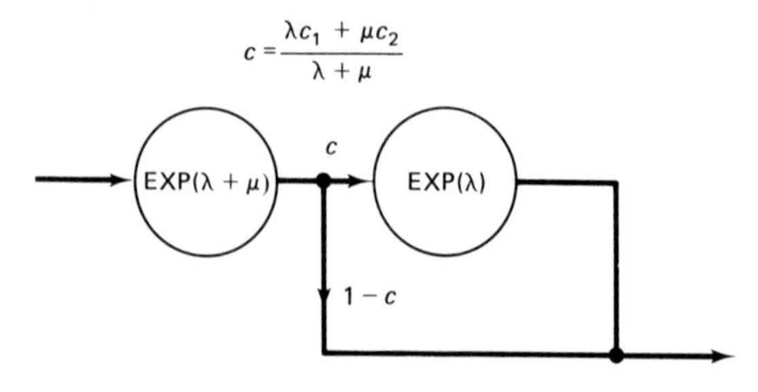

Figure 5.9 Lifetime distribution of a two-component standby redundant system with imperfect coverage

$100(1 - c)$ percent of an exponential, EXP $(\lambda + \mu)$, with $100c$ percent of a hypoexponential, HYPO $(\lambda + \mu, \lambda)$. Therefore:

$$f_X(t) = (1 - c)(\lambda + \mu)e^{-(\lambda+\mu)t} + c\frac{\lambda(\lambda + \mu)}{\mu}(e^{-\lambda t} - e^{-(\lambda+\mu)t}), \qquad t > 0, \quad (5.41)$$

and the system reliability is given by:

$$R_X(t) = (1 - c)e^{-(\lambda+\mu)t} + c\frac{\lambda(\lambda + \mu)}{\mu}[\frac{1}{\lambda}e^{-\lambda t} - \frac{1}{\lambda + \mu}e^{-(\lambda+\mu)t}]$$

$$= (1 - c)e^{-(\lambda+\mu)t} + \frac{c}{\mu}[(\lambda + \mu)e^{-\lambda t} - \lambda e^{-(\lambda+\mu)t}], \qquad t \geqslant 0, \quad (5.42)$$

where c is given by (5.39). #

Cox [COX 1955] has analyzed the more general stage-type distribution shown in Figure 5.10. The Laplace transform of such a stage-type random variable X is given by:

$$L_X(s) = \gamma_1 + \sum_{i=1}^{r} \beta_1\beta_2 \cdots \beta_i\gamma_{i+1} \prod_{j=1}^{i} \frac{\mu_j}{s + \mu_j}, \quad (5.43)$$

where $\gamma_i + \beta_i = 1$ for $1 \leqslant i \leqslant r$ and $\gamma_{r+1} = 1$.

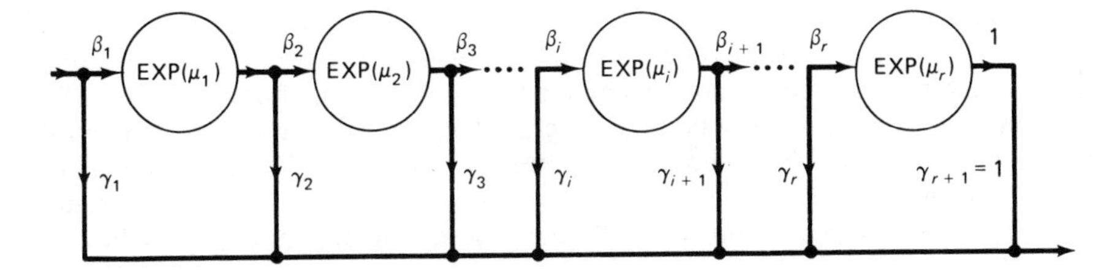

Figure 5.10 Coxian stage-type distribution

Example 5.13

Consider a standby redundant system with n units where one unit is initially active and $(n - 1)$ units are in a standby status. Assume that the failure rate of an active unit is λ while that of a standby unit is μ. For simplicity, we assume that the coverage factor is the same for active and spare failures and is denoted by c. By analogy with Example 5.12, we can say that the lifetime distribution of this system will be stage-type, as shown in Figure 5.11. Using the notation of Figure 5.10, the distribution of Figure 5.11 corresponds to the parameters:

$$\beta_1 = 1, \qquad \gamma_1 = 0, \qquad \gamma_{n+1} = 1,$$

$$\beta_i = c, \qquad \gamma_i = 1 - c, \qquad 2 \leqslant i \leqslant n,$$

$$\mu_i = \lambda + (n - i)\mu, \qquad 1 \leqslant i \leqslant n.$$

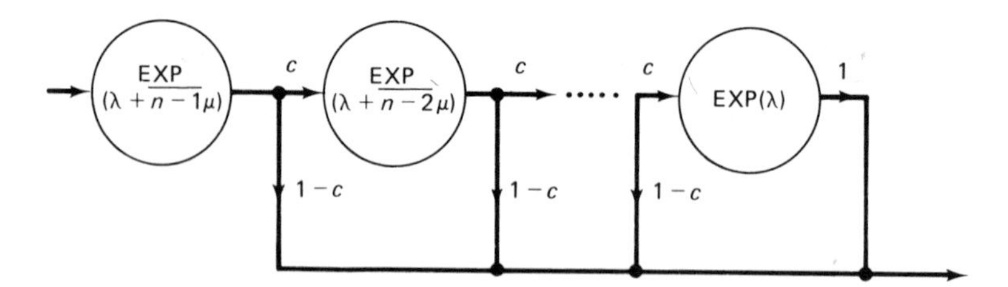

Figure 5.11 Lifetime distribution of an n-component standby redundant system with imperfect coverage

Using equation (5.43), we obtain the Laplace transform of the system lifetime as:

$$L_X(s) = \sum_{i=1}^{n-1} c^{i-1}(1-c) \prod_{j=1}^{i} \frac{\lambda + (n-j)\mu}{s + \lambda + (n-j)\mu} + c^{n-1} \prod_{j=1}^{n} \frac{\lambda + (n-j)\mu}{s + \lambda + (n-j)\mu}. \tag{5.44}$$

The system MTTF can be easily computed from expression (5.44) as:

$$E[X] = \sum_{i=1}^{n-1} c^{i-1}(1-c) \sum_{j=1}^{i} \frac{1}{\lambda + (n-j)\mu} + c^{n-1} \sum_{j=1}^{n} \frac{1}{\lambda + (n-j)\mu}. \tag{5.45}$$

#

Example 5.14

Now we can consider a hybrid NMR system with imperfect coverage. Assume the failure rate of an active unit is λ and the failure rate of a standby spare is μ. We continue with the assumption that the coverage factor is the same for an active-unit failure as that for a standby-unit failure. Initially, there are N active units and S spares. The lifetime of such a system has the stage-type distribution shown in Figure 5.12. Therefore, in the notation of the Coxian distribution of Figure 5.10, we have:

$$\beta_1 = 1, \quad \gamma_1 = 0,$$

$$\beta_i = c, \quad \gamma_i = 1 - c, \quad 2 \leqslant i \leqslant S + 1,$$

$$\beta_i = 1, \quad \gamma_i = 0, \quad S + 2 \leqslant i \leqslant (N - M) + S + 1,$$

$$\gamma_{N-M+S+2} = 1,$$

$$\mu_j = N\lambda + (S - j + 1)\mu, \quad 1 \leqslant j \leqslant S,$$

$$\mu_j = N\lambda + (S - j + 1)\lambda, \quad S + 1 \leqslant j \leqslant N - M + S + 1.$$

Then, using formula (5.43), we have:

$$L_X(s) = \sum_{i=1}^{S} c^{i-1}(1-c) \prod_{j=1} \frac{N\lambda + (S-j+1)\mu}{s + N\lambda + (S-j+1)\mu} + c^S \prod_{j=1}^{S} \frac{N\lambda + j\mu}{s + (N\lambda + j\mu)} \prod_{j=1}^{N-M+1} \frac{(N-j+1)\lambda}{s + (N-j+1)\lambda}. \tag{5.46}$$

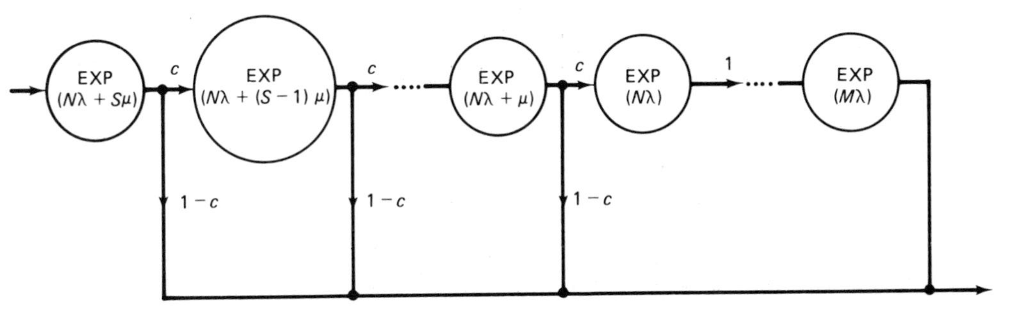

Figure 5.12 Lifetime distribution of hybrid NMR system with imperfect coverage

After a partial fraction expansion and inversion of the transform above, we can get the required expression for reliability of hybrid NMR with imperfect coverage. We omit the details here; the interested reader is referred to [NG 1976]. We can easily compute the mean life from the expression of the Laplace transform:

$$E[X] = \sum_{i=1}^{S} c^{i-1}(1-c) \sum_{j=1}^{i} \frac{1}{N\lambda + (S-j+1)\mu}$$

$$+ c^S \left[\sum_{j=1}^{S} \frac{1}{N\lambda + j\mu} + \sum_{j=1}^{N-M+1} \frac{1}{(N-j+1)\lambda} \right]$$

$$= \sum_{i=1}^{S} c^{i-1}(1-c) \sum_{j=S-i+1}^{S} \frac{1}{N\lambda + j\mu}$$

$$+ c^S \left[\sum_{j=1}^{S} \frac{1}{N\lambda + j\mu} + \sum_{j=M}^{N} \frac{1}{j\lambda} \right]. \tag{5.47}$$

#

The method of stages works quite well for the reliability analysis of non-repairable systems. When we deal with repairable systems, the number of stages will become infinite, and this method becomes cumbersome to use. We will analyze such systems in Chapter 8 using the theory of Markov chains.

Problems

1. We are given a system with three components. Whenever a component is energized, it has an exponential failure law with parameter λ. When a component is deenergized, it has an exponential failure law with parameter μ ($< \lambda$). The system will function correctly if two components are in proper working order. Consider two ways of using the spare unit: one way is to keep all three units energized, another way is to keep the third unit as a deenergized spare, and, when one of the operating units fails, switch the spare in. But the switching equipment need not be perfect; suppose it has reliability 0.9. Find the reliability expressions for the two schemes and find conditions under which one scheme will be better than the other.

2. Starting from formula (5.44), use the moment generating property of the Laplace transform to show (5.45). Also obtain a formula for the variance of system lifetime for the system of Example 5.13.

5.5 RANDOM SUMS

We have considered sums of N mutually independent random variables when N is a fixed constant. Here we are interested in the case where N itself is a random variable. Given a list $X_1, X_2, \ldots,$ of mutually independent identically distributed random variables with distribution function $F(x)$, mean $E[X]$, and variance Var $[X]$, consider the random sum:

$$T = X_1 + X_2 + \cdots + X_N. \tag{5.48}$$

Here the pmf of the discrete random variable $p_N(n)$ is assumed to be given.

For a fixed value $N = n$, the conditional expectation of T is easily obtained:

$$\begin{aligned} E[T|N = n] &= \sum_{i=1}^{n} E[X_i] \\ &= nE[X]. \end{aligned} \tag{5.49}$$

Then, using the theorem of total expectation, we get:

$$\begin{aligned} E[T] &= \sum_{n} nE[X]p_N(n) \\ &= E[X] \sum_{n} np_N(n) \\ &= E[X]E[N]. \end{aligned} \tag{5.50}$$

In order to obtain the Var $[T]$, we first compute $E[T^2]$. Note that:

$$E[T^2|N = n] = \text{Var}\ [T|N = n] + (E[T|N = n])^2 \tag{5.51}$$

but:

$$\begin{aligned} \text{Var}\ [T|N = n] &= \sum_{i=1}^{n} \text{Var}\ [X_i] \\ &= n\ \text{Var}\ [X] \qquad \text{(by independence of the } X_i\text{'s).} \end{aligned} \tag{5.52}$$

Substituting (5.49) and (5.52) in (5.51):

$$E[T^2|N = n] = n\ \text{Var}\ [X] + n^2(E[X])^2.$$

Now, using the theorem of total moments, we have:

$$E[T^2] = \sum_n [n \text{ Var } [X] + n^2(E[X])^2] \, p_N(n)$$
$$= \text{Var } [X]E[N] + E[N^2](E[X])^2.$$

Finally:

$$\text{Var } [T] = E[T^2] - (E[T])^2$$
$$= \text{Var } [X]E[N] + E[N^2](E[X])^2 - (E[X])^2(E[N])^2$$
$$= \text{Var } [X]E[N] + (E[X])^2 \text{ Var } [N]. \tag{5.53}$$

Assuming that the X_i's are continuous random variables with Laplace transform $L_X(s)$, we obtain the conditional Laplace transform of T as,

$$L_{T|N}(s|n) = [L_X(s)]^n.$$

Then, using the theorem of total Laplace transform, we have:

$$L_T(s) = \sum_n L_{T|N}(s|n)p_N(n)$$
$$= \sum_n [L_X(s)]^n \, p_N(n) \tag{5.54}$$
$$= G_N(L_X(s)).$$

As a special case, assume that N has a geometric distribution with parameter p, so that:

$$p_N(n) = (1 - p)^{n-1} \, p$$

and

$$L_T(s) = \sum_{n=1}^{\infty} [L_X(s)]^n \, (1 - p)^{n-1} \, p$$
$$= \frac{pL_X(s)}{1 - (1 - p) \, L_X(s)}. \tag{5.55}$$

Next assume that the X_i's are discrete with the common generating function $G_X(z)$. Then, the conditional PGF of T is:

$$G_{T|N}(z|n) = [G_X(z)]^n,$$

and using the theorem of total generating functions, we have the unconditional PGF of T:

$$G_T(z) = \sum_n [G_X(z)]^n \, p_N(n) = G_N[G_X(z)]. \tag{5.56}$$

Now if N is geometrically distributed with parameter p, then the above formula reduces to:

$$G_T(z) = \frac{pG_X(z)}{1-(1-p)G_X(z)}. \tag{5.57}$$

Example 5.15

Consider a memory module with a time to failure X. Since the reliability $R(t)$ of the memory was found to be inadequate, addition of hardware to incorporate error-correcting capability was desired. Assume that the probability of an undetected and/or uncorrected error (uncovered, for short) is p (and hence the conditional probability that a given error is covered by the coding scheme is $1-p$). Now let N be the number of errors that are corrected before an uncovered error occurs, then N has a geometric distribution with parameter p and $E[N] = 1/p$. Let T denote the time to occurrence of an uncovered error. Denoting X_i to be the time between the occurrence of the $(i-1)$ st and ith error, $T = X_1 + X_2 + \ldots + X_N$. Then, using formula (5.50) above, we get:

$$\begin{aligned}\text{MTTF}_{\text{with code}} &= \frac{\text{MTTF}_{\text{without code}}}{p}\\ &= \frac{\text{MTTF}_{\text{without code}}}{\text{probability of an uncovered error}}.\end{aligned}$$

If we assume that X is exponentially distributed with parameter λ, then

$$L_X(s) = \frac{\lambda}{s+\lambda},$$

and, assuming that $X_1, X_2, \ldots,$ are mutually independent, and using formula (5.55), we have:

$$\begin{aligned}L_T(s) &= \frac{p\dfrac{\lambda}{s+\lambda}}{1-(1-p)\dfrac{\lambda}{s+\lambda}}\\ &= \frac{p\lambda}{s+p\lambda}.\end{aligned} \tag{5.58}$$

This implies that T is exponentially distributed with parameter $p\lambda$. #

Example 5.16

Consider the following program segment consisting of a **repeat** loop:

repeat S **until** B

Let X_i denote the execution time for the ith iteration of statement group S. Assume that the sequence of tests of the Boolean expression B defines a sequence of Bernoulli trials with parameter p. Clearly, the number N of iterations of the loop is a geometric random variable with parameter p so that $E[N] = 1/p$. Letting T denote the total execution time of the loop, and using equation (5.50), the average execution time is easily determined to be:

$$E[T] = \frac{E[X]}{p}. \tag{5.59}$$

The variance of the execution time T is determined using (5.53), noting that Var $[N] = q/p^2$:

$$\text{Var}\ [T] = \frac{\text{Var}\ [X]}{p} + (E[X])^2 \frac{q}{p^2}. \tag{5.60}$$

Next assume that the X_i's are exponentially distributed with parameter λ, so that:

$$L_X(s) = \frac{\lambda}{s + \lambda},$$

and, using formula (5.58), we get:

$$L_T(s) = \frac{p\lambda}{s + p\lambda}. \tag{5.61}$$

Thus the total execution time of the **repeat** loop is also exponentially distributed with parameter $p\lambda$. In this case, $E[T] = \frac{1}{p\lambda}$ which agrees with (5.59), and from (5.60):

$$\text{Var}\ [T] = \frac{1}{\lambda^2 p} + \frac{1}{\lambda^2} \frac{q}{p^2} = \frac{1}{p^2 \lambda^2}$$

as expected. #

Example 5.17

In measuring the execution time of the **repeat** loop above, we use a real-time clock with a resolution of 1 microsecond. In this case, the execution times will be discrete random variables. Assume that the X_i's are geometrically distributed with parameter p_1. Then:

$$G_X(z) = \frac{zp_1}{1 - z(1 - p_1)},$$

and, using formula (5.57) above, we get:

$$G_T(z) = \frac{pzp_1}{1 - z(1 - p_1) - (1 - p)zp_1}$$

$$= \frac{zpp_1}{1 - z(1 - pp_1)}.$$

Thus T is a geometrically distributed random variable with parameter pp_1. #

We are now in a position to obtain the distribution of the execution time of a **repeat** loop and, in a similar fashion, that of a **while** loop (see problem 1 at the end of this section). Our earlier methods allow us to compute the distribution of the execution time of a compound statement, that of a **for** loop (see the discussion of sums of independent random variables in Chapter 3), and that of an **if** and a **case** statement (see the discussion of mixture distributions in this chapter). Thus, we are now in a position to analyze a *structured* *program*—a program that uses only combinations of the above-listed control structures. (Linger and others [LING 1979] use the phrase *proper program* for such programs.) We have summarized these results in Appendix E. Beizer [BEIZ 1978] presents many examples of program analysis along these lines. We can also deal with the concurrent control statement **cobegin**. Programs that use unrestricted **goto**'s can be analyzed by the methods of Chapter 7.

Example 5.18

Consider the following program:

```
begin
  COMP;
  while B do
  begin
    case j of
      1: I/O1;
      2: I/O2;
      .
      .
      .
      m: I/Om;
    end;
    COMP
  end
end.
```

Let the random variable C denote the time to execute the statement group COMP and let I_j $(1 \leqslant j \leqslant m)$ denote the time to execute the statement group I/Oj. Assume that the condition test on B is a sequence of independent Bernoulli trials with the probability of failure p_0, and let p'_j be the probability of executing the jth case, given that the **case** statement is executed. Note that:

$$\sum_{j=1}^{m} p'_j = 1.$$

Let the random variable I denote the execution time of the **case** statement. Note that I is a mixture of random variables $I_j (1 \leqslant j \leqslant m)$. Given the Laplace transforms of C and I_j, we proceed to compute the Laplace transform of the overall execution time T of the above program.

Using the table in Appendix E, we have

$$L_I(s) = \sum_{j=1}^{m} p'_j L_{I_j}(s),$$

$$L_{\text{whilebody}}(s) = L_I(s) \cdot L_C(s),$$

$$L_{\text{whileloop}}(s) = \sum_{n=0}^{\infty} (1 - p_0)^n \, p_0 \, [L_{\text{whilebody}}(s)]^n$$

$$= \frac{p_0}{1-(1-p_0)L_{\text{whilebody}}(s)},$$

$$L_T(s) = \frac{p_0 L_C(s)}{1-(1-p_0)L_{\text{whilebody}}(s)}$$

$$= \frac{p_0 L_C(s)}{1-(1-p_0)L_C(s)L_I(s)} = \frac{U(s)}{V(s)}. \tag{5.62}$$

For a continuous random variable X it is known that $L_X(0) = 1$, $-L'_X(0) = E[X]$, and $L''_X(0) = E[X^2]$; hence:

$$V(0) = 1 - (1-p_0)L_C(0)L_I(0) = p_0 \ ,$$

$$U(0) = p_0 L_C(0) = p_0 \ ,$$

$$V'(0) = -(1-p_0)[L_C(0)L'_I(0) + L'_C(0)L_I(0)] = (1-p_0)[E[C] + E[I]] \ ,$$

$$U'(0) = p_0 L'_C(0) = -p_0 E[C],$$

$$V''(0) = -(1-p_0)[L''_C(0)L_I(0) + 2L'_C(0)L'_I(0) + L_C(0)L''_I(0)]$$

$$= -(1-p_0)(E[C^2] + 2E[C]E[I] + E[I^2]),$$

$$U''(0) = p_0 L''_C(0) = p_0 E[C^2].$$

Now we can compute the first two moments of T:

$$E[T] = -L'_T(0)$$

$$= \frac{-V(0)U'(0) + U(0)V'(0)}{V^2(0)}$$

$$= \frac{p_0(p_0 E[C]) + (1-p_0)p_0(E[C] + E[I])}{p_0^2}$$

$$= \frac{E[C]}{p_0} + \frac{(1-p_0)E[I]}{p_0}. \tag{5.63}$$

The terms on the right-hand side of (5.63) are easily interpreted: the first term is the expected value of the total time to execute the statement COMP, since $1/p_0$ is the average number of times COMP is executed, while $E[C]$ is the average time per execution. The second term is the expected total time of all I/O statements, since this term can be written as:

$$\sum_{j=1}^{m} \frac{p_j}{p_0} E[I_j],$$

where p_j is defined to be $(1-p_0)p'_j$. Note that p_j/p_0 is the average number of executions of statement I/Oj, and $E[I_j]$ is the average time per execution.

Next we proceed to compute $E[T^2]$:

$$
\begin{aligned}
E[T^2] &= L''_T(0) \\
&= \frac{V^2(0)\,U''(0) - U(0)\,V(0)\,V''(0) - 2V(0)\,V'(0)\,U'(0) + 2U(0)[V'(0)]^2}{V^3(0)} \\
&= \frac{E[C^2]}{p_0} + \frac{2(1-p_0)}{p_0^2}(E[C])^2 + \frac{4(1-p_0)}{p_0^2}E[C]E[I] \qquad (5.64) \\
&\quad + \frac{1-p_0}{p_0}E[I^2] + \frac{2(1-p_0)^2}{p_0^2}(E[I])^2.
\end{aligned}
$$

Now from (5.63) and (5.64) we compute the variance:

$$
\text{Var}\,[T] = \frac{\text{Var}\,[C]}{p_0} + \frac{1-p_0}{p_0}\,\text{Var}\,[I] + \frac{1-p_0}{p_0^2}\,(E[C] + E[I])^2. \qquad (5.65)
$$

We will use these formulas in Chapter 9 in analyzing a queuing network in which individual programs will behave as discussed in this example. #

We should caution the reader that several unrealistic assumptions have been made here. The assumption of independence, for example, is questionable. More importantly, we have associated a fixed probability with each conditional branch, independent of the current state of the program. For more involved analyses that attempt to remove such assumptions, see [RAMS 1979]. Our treatment of program analysis in this section was control-structure based. Alternatively, we could perform a data-structure-oriented analysis as in the analysis of program MAX (Chapter 2). Further examples of this technique will be presented in Chapter 7. In practice, these two techniques need to be used in conjunction with each other.

Problems

1. Carry out an analysis of the execution time of the **while** loop:

   while B **do** S,

   following the analysis of the **repeat** loop given in Example 5.16.

2. A CPU burst of a task is exponentially distributed with mean $1/\mu$. At the end of a burst, the task requires another burst with probability p and finishes execution with probability $1 - p$. Thus the number of CPU bursts required for task completion is a random variable with the image $\{1, 2, \ldots\}$. Find the distribution function of the total service time of a task. Also compute its mean and variance.

3. The number of messages, N, arriving to a communications channel per unit time is Poisson distributed with parameter λ. The number of characters, X_i, in the ith message is geometrically distributed with parameter θ. Determine the distribution of the total number of characters, Y, that arrive per unit time. [*Hint:* Y is a random sum.] Determine $G_Y(z)$, $E[Y]$, and $\text{Var}\,[Y]$.

References

[BARL 1975] R. E. BARLOW and F. PROSCHAN, *Statistical Theory of Reliability and Life Testing*, Holt, Rinehart and Winston, New York.

[BEIZ 1978] B. BEIZER, *Micro-Analysis of Computer System Performance*, Van Nostrand-Reinhold, New York.

[BOUR 1969] W. G. BOURICIUS, W. C. CARTER, and P. R. SCHNEIDER, "Reliability Modeling Techniques for Self-Repairing Computer Systems," *Proceedings of 24th National Conference of the ACM*, August 1969, pp 295–309.

[CLAR 1970] A. B. CLARKE and R. L. DISNEY, *Probability and Random Processes for Engineers and Scientists*, John Wiley & Sons, New York.

[COX 1955] D. R. COX, "A Use of Complex Probabilities in Theory of Stochastic Processes," *Proc. Cambridge Phil. Soc.*, 51, 313–319.

[GAVE 1973] D. P. GAVER and G. L. THOMPSON, *Programming and Probability Models in Operations Research*, Brooks/Cole Publishing Company, Monterey, Calif.

[KNUT 1973a] D. E. KNUTH, *The Art of Computer Programming, Vol. I: Fundamental Algorithms*, Addison-Wesley, Reading, Mass..

[KNUT 1973b] D. E. KNUTH, *The Art of Computer Programming, Vol. III: Sorting and Searching*, Addison-Wesley, Reading, Mass..

[MEND 1979] H. MENDELSON, J. S. PLISKIN, and U. YECHIALI, "Optimal Storage Allocation for Serial Files," *CACM*, February 1979, pp. 124–130.

[LING 1979] R. C. LINGER, H. D. MILLS, and B. I. WITT, *Structured Programming: Theory and Practice*, Addison-Wesley, Reading, Mass.

[NG 1976] Y -W. NG, "Reliability Modeling and Analysis for Fault-Tolerant Computers," Ph.D. Dissertation, Computer Science Department, University of California at Los Angeles.

[RAMS 1979] L. H. RAMSHAW, "Formalizing the Analysis of Algorithms," Stanford University Computer Science Department technical report STAN-CS-79-741.

[RAND 1975] B. RANDELL, "System Structure for Software Fault Tolerance," *IEEE Transactions of Soft. Engg.*, SE-1, pp. 202-232.

Chapter 6

Stochastic Processes

6.1 INTRODUCTION

In the previous chapters we have seen the need to consider a collection or a family of random variables instead of a single random variable. A family of random variables that is indexed by a parameter such as time is known as a **stochastic process** (or **chance** or **random process**).

Definition (Stochastic Process). A stochastic process is a family of random variables $\{X(t) \mid t \in T\}$, defined on a given probability space, indexed by the parameter t, where t varies over an index set T.

The values assumed by the random variable $X(t)$ are called **states**, and the set of all possible values forms the **state space** of the process. The state space will be denoted by I.

Recall that a random variable is a function defined on the sample space S of the underlying experiment. Thus the above family of random variables is a family of functions $\{X(t, s) \mid s \in S, t \in T\}$. For a fixed $t = t_1$, $X_{t_1}(s) = X(t_1, s)$ is a random variable [denoted by $X(t_1)$] as s varies over the sample space S. At some other fixed instant of time t_2, we have another random variable $X_{t_2}(s) = X(t_2, s)$. For a fixed sample point $s_1 \in S$, $X_{s_1}(t) = X(t, s_1)$ is a single function of time t, called a **sample function** or a **realization** of the process. When both s and t are varied, we have the family of random variables constituting a stochastic process.

Example 6.1 [STAR 1979]

Consider the experiment of randomly choosing a resistor s from a set S of thermally agitated resistors and measuring the noise voltage $X(t, s)$ across the resistor at time t. Sample functions for two different resistors are shown in Figure 6.1.

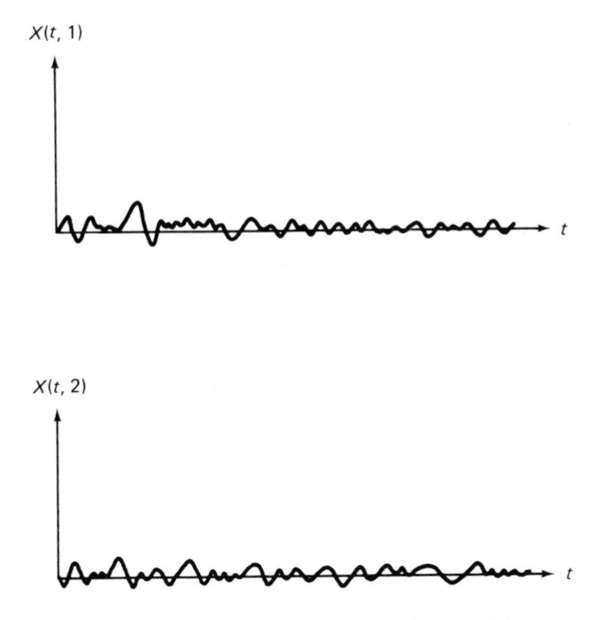

Figure 6.1 Noise voltages across two resistors

At a fixed time $t = t_1$, suppose we measure the voltages across all the resistors in the set S, count the number of resistors with a voltage level less than or equal to x_1 and divide this count by the total number of resistors in S. Using the frequency interpretation of probability, this will give the distribution function, $F_{X(t_1)}(x_1) = P[X(t_1) \leqslant x_1]$, of the random variable $X(t_1)$. This calculation can be repeated at other instants of time $t_2, t_3, \ldots,$ to obtain the distribution functions of $X(t_2), X(t_3), \ldots$. The joint distribution function of $X(t_1)$ and $X(t_2)$ can similarly be obtained by computing the relative frequency of the event $[X(t_1) \leqslant x_1$ and $X(t_2) \leqslant x_2]$. Continuing in this fashion, we can compute the joint distribution function of $X(t_1), X(t_2), \ldots, X(t_n)$. #

If the state space of a stochastic process is discrete, then it is called a **discrete-state process**, often referred to as a **chain**. In this case, the state space is often assumed to be $\{0, 1, 2, \ldots\}$. Alternatively, if the state space is continuous, then we have a **continuous-state process**. Similarly, if the index set T is discrete, then we have a **discrete (time)-parameter process**; otherwise we have a **continuous parameter process**. A discrete-parameter process is also called a stochastic sequence and is denoted by $\{X_n \mid n \in T\}$. This gives us four different types of stochastic processes, as shown in Table 6.1.

The theory of queues (or waiting lines) provides many examples of stochastic processes. Before introducing these processes, we present a notation to describe the queues. A queue may be generated when customers (jobs) arrive at a station (computing center) to receive service (see Figure 6.2). Assume that successive interarrival times $Y_1, Y_2, \ldots,$ between jobs are independent identically distributed random variables having a distribution

Table 6.1 A Classification of Stochastic Processes

		Index Set T	
		Discrete	*Continuous*
State Space I	Discrete	Discrete-parameter stochastic chain	Continuous-parameter stochastic chain
	Continuous	Discrete-parameter continuous-state process	Continuous-parameter continuous-state process

F_Y. Similarly, the service times $S_1, S_2, \ldots$, are assumed to be independent identically distributed random variables having a distribution F_S. Let m denote the number of servers (computer systems) in the station (computing center). We use the notation $F_Y/F_S/m$ to describe the queuing system. To denote the specific types of interarrival-time and service-time distributions, we use the following symbols:

M (for memoryless) for the exponential distribution

D for a deterministic or constant interarrival or service time

E_k for a k-stage Erlang distribution

H_k for a k-stage hyperexponential distribution

G for a general distribution

GI for general independent interarrival times

Thus $M/G/1$ denotes a single-server queue with exponential interarrival times and an arbitrary service-time distribution. The most frequent example of a queue that we will use is $M/M/1$. Besides the nature of the interarrival-time and service-time distributions, we also need to specify a scheduling discipline that decides how the server is to be allocated to the jobs waiting for service. Unless otherwise specified, we will assume that jobs are selected for

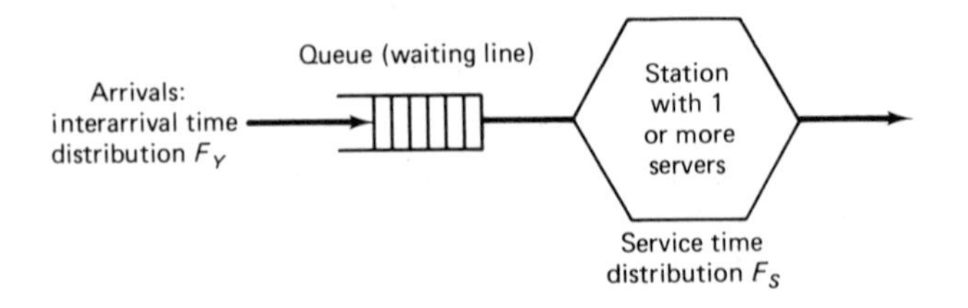

Figure 6.2 A queuing system

service in the order of their arrivals; that is, we will assume FCFS (first-come, first-served) scheduling discipline. Now we will describe various stochastic processes associated with a queue.

Example 6.2

Consider a computer system with jobs arriving at random points in time, queuing for service, and departing from the system after service completion.

Let N_k be the number of jobs in the system at the time of the departure of the kth customer (after service completion). The stochastic process $\{N_k \mid k = 1, 2, \ldots\}$ is a discrete-parameter, discrete-state process with the state space $I = \{0, 1, 2, \ldots\}$ and the index set $T = \{1, 2, 3, \ldots\}$. A realization of this process is shown in Figure 6.3.

Next let $X(t)$ be the number of jobs in the system at time t. Then $\{X(t) \mid t \in T\}$ is a continuous-parameter, discrete-state process with $I = \{0, 1, 2, \ldots\}$ and $T = \{t \mid 0 \leqslant t < \infty\}$. A realization of this process is shown in Figure 6.4.

Let W_k be the time that the kth customer has to wait in the system before receiving service. Then $\{W_k \mid k \in T\}$, with $I = \{x \mid 0 \leqslant x < \infty\}$ and $T = \{1, 2, 3, \ldots\}$, is a discrete-parameter, continuous-state process. A realization of this process is shown in Figure 6.5.

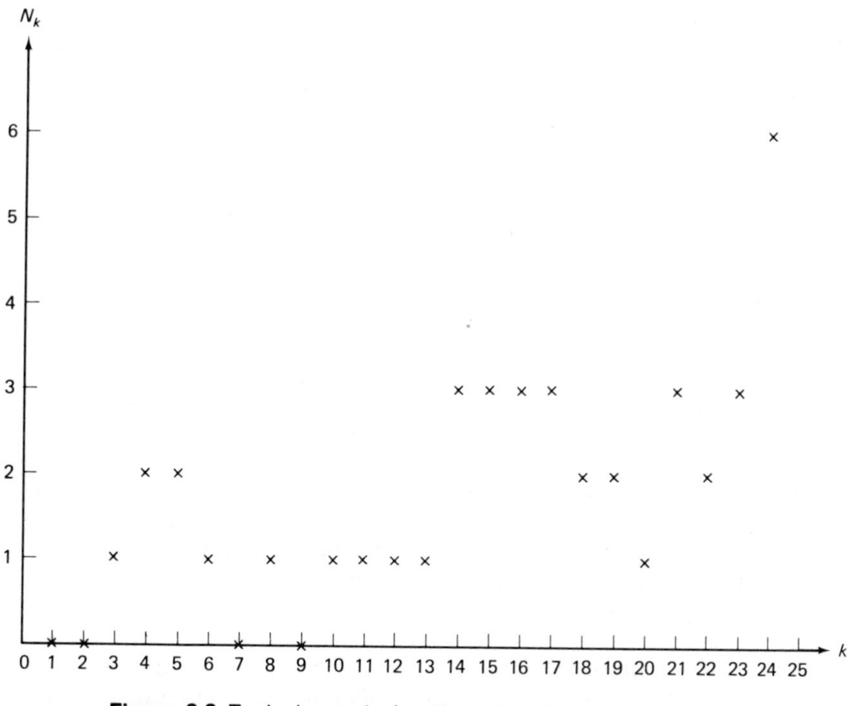

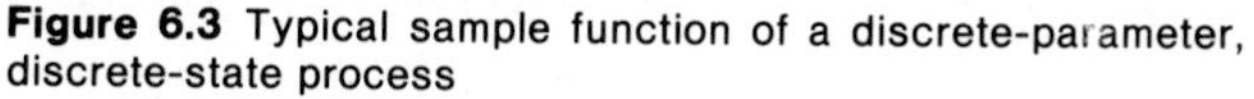

Figure 6.3 Typical sample function of a discrete-parameter, discrete-state process

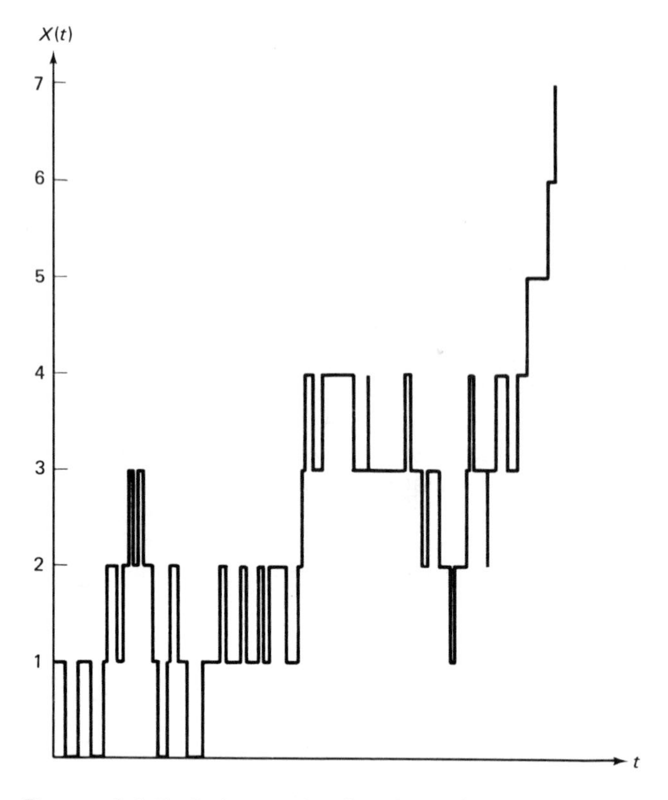

Figure 6.4 Typical sample function of a continuous-parameter, discrete-state process

Finally, let $Y(t)$ denote the cumulative service requirement of all jobs in the system at time t. Then $\{Y(t) \mid 0 \leqslant t < \infty\}$ is a continuous-parameter, continuous-state process with $I = [0, \infty)$. A realization of this process is shown in Figure 6.6. #

Problems

1. Write and run a program to simulate an $M/E_2/1$ queue and obtain realizations of the four stochastic processes defined in Example 6.2. Plot these realizations. You may use a simulation language such as SIMULA or GPSS or you may use one of the standard high-level languages. You will have to generate random deviates of the interarrival-time distribution (assume arrival rate $\lambda = 1$) and the service-time distribution (assume mean service time 0.8) using methods of Chapter 3.

   Study the process $\{N_k \mid k = 1, 2, \ldots\}$ in detail as follows:

   By varying the seeds for generating random numbers you get different realizations. For a fixed k, different observed values of N_k for these distinct realizations can be used to estimate the mean and variance of N_k. Using a sample size of 30, estimate $E[N_k]$, Var $[N_k]$ for $k = 1, 5, 10, 100, 200, 1000$. What can you conclude from this experiment?

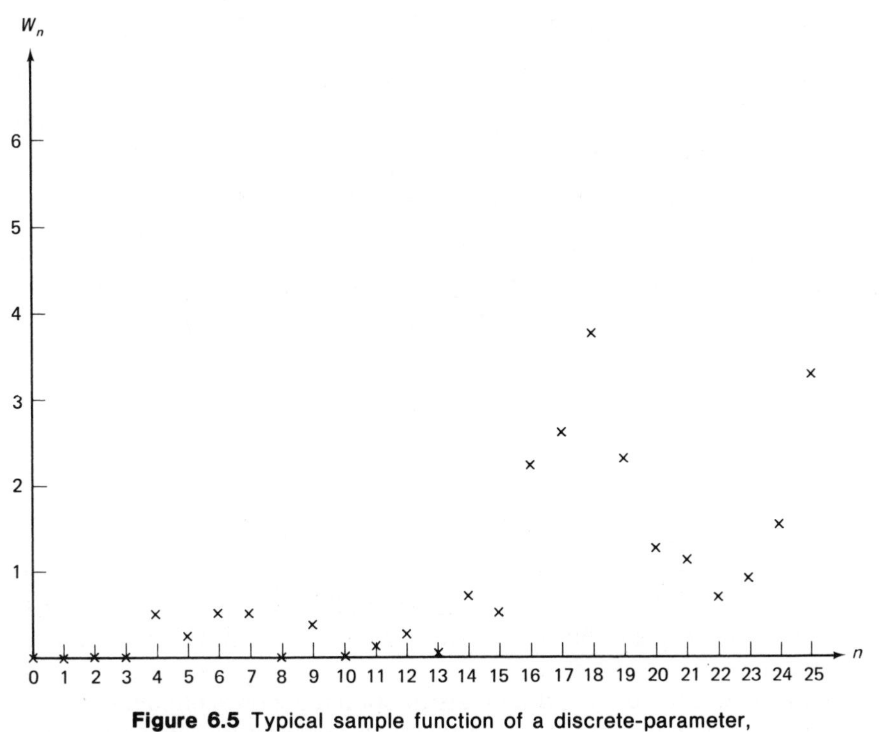

Figure 6.5 Typical sample function of a discrete-parameter, continuous-state process

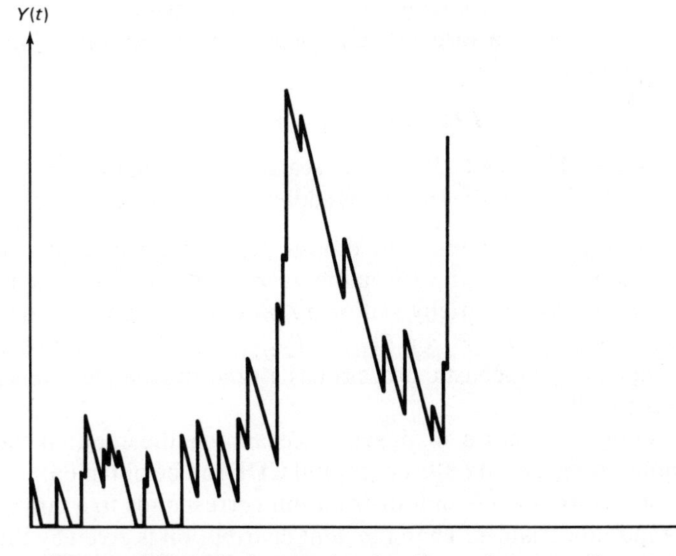

Figure 6.6 Typical sample function of a continuous-parameter, continuous-state process

6.2 CLASSIFICATION OF STOCHASTIC PROCESSES

For a fixed time $t = t_1$, $X(t_1)$ is a simple random variable that describes the state of the process at time t_1. For a fixed number x_1, the probability of the event $[X(t_1) \leqslant x_1]$ gives the CDF of the random variable $X(t_1)$, denoted by:

$$F(x_1; t_1) = F_{X(t_1)}(x_1) = P[X(t_1) \leqslant x_1].$$

$F(x_1; t_1)$ is known as the first-order distribution of the process $\{X(t)\}$. Given two time instants t_1 and t_2, $X(t_1)$ and $X(t_2)$ are two random variables on the same probability space. Their joint distribution is known as the second-order distribution of the process and is given by:

$$F(x_1, x_2; t_1, t_2) = P(X(t_1) \leqslant x_1 , X(t_2) \leqslant x_2).$$

In general, we define the nth order joint distribution of the stochastic process $\{X(t),\ t \in T\}$ by:

$$F(\mathbf{x}; \mathbf{t}) = P[X(t_1) \leqslant x_1, \ldots, X(t_n) \leqslant x_n] \tag{6.1}$$

for all $\mathbf{x} = (x_1, \ldots, x_n) \in \mathbb{R}^n$ and $\mathbf{t} = (t_1, t_2, \ldots, t_n) \in T^n$ such that $t_1 < t_2 < \cdots < t_n$. Such a complete description of a process is no small task. Many processes of practical interest, however, permit a much simpler description.

For instance, the nth order joint distribution function is often found to be invariant under shifts of the time origin. Such a process is said to be a strict-sense stationary stochastic process.

Definition (Strictly Stationary Process). A stochastic process $\{X(t)\}$ is said to be stationary in the strict sense if for $n \geqslant 1$, its nth-order joint CDF satisfies the condition:

$$F(\mathbf{x}; \mathbf{t}) = F(\mathbf{x}; \mathbf{t} + \tau)$$

for all vectors $\mathbf{x} \in \mathbb{R}^n$ and $\mathbf{t} \in T^n$, and all scalars τ such that $t_i + \tau \in T$. The notation $\mathbf{t} + \tau$ implies that the scalar τ is added to all components of vector $\mathbf{t}$.

We let $\mu(t) = E[X(t)]$ denote the time-dependent mean of the stochastic process. $\mu(t)$ is often called the **ensemble average** of the stochastic process. Applying the definition of strictly stationary process to the first-order CDF, we get $F(x; t) = F(x; t + \tau)$ or $F_{X(t)} = F_{X(t+\tau)}$ for all τ. It follows that a strict-sense stationary stochastic process has a time-independent mean; that is, $\mu(t) = \mu$ for all $t \in T$.

By restricting the nature of dependence among the random variables $\{X(t)\}$, a simpler form of the nth-order joint CDF can be obtained.

The simplest form of the joint distribution corresponds to a family of independent random variables. Then the joint distribution is given by the product of individual distributions.

Definition (Independent Process). A stochastic process $\{X(t) \mid t \in T\}$ is said to be an **independent process** provided its nth-order joint distribution satisfies the condition:

$$F(\mathbf{x}; \mathbf{t}) = \prod_{i=1}^{n} F(x_i; t_i)$$

$$= \prod_{i=1}^{n} P[X(t_i) \leqslant x_i]. \tag{6.2}$$

As a special case we have:

Definition (Renewal Process). A **renewal process** is defined to be a discrete-parameter independent process $\{X_n \mid n = 1, 2, \ldots\}$ where $X_1, X_2, \ldots$, are independent, identically distributed, nonnegative random variables.

As an example of such a process, consider a system in which the repair (or replacement) after a failure is performed, requiring negligible time. Now the times between successive failures might well be independent, identically distributed random variables $\{X_n \mid n = 1, 2, \ldots\}$ of a renewal process.

Though the assumption of an independent process considerably simplifies analysis, such an assumption is often unwarranted, and we are forced to consider some sort of dependence among these random variables. The simplest and the most important type of dependence is the first-order dependence or **Markov dependence**.

Definition (Markov Process). A stochastic process $\{X(t) \mid t \in T\}$ is called a Markov process if for any $t_0 < t_1 < t_2 < \cdots < t_n < t$, the conditional distribution of $X(t)$ for given values of $X(t_0), X(t_1), \ldots, X(t_n)$ depends only on $X(t_n)$; that is:

$$\begin{aligned} P[X(t) \leqslant x \mid X(t_n) = x_n, X(t_{n-1}) = x_{n-1}, \ldots, X(t_0) = x_0] \\ = P[X(t) \leqslant x \mid X(t_n) = x_n]. \end{aligned} \tag{6.3}$$

Although this definition applies to Markov processes with continuous state space, we will be mostly concerned with discrete-state Markov processes—that is, Markov chains. We will study both discrete-parameter and continuous-parameter Markov chains.

In many problems of interest, the conditional distribution function (6.3) has the property of invariance with respect to the time origin t_n; that is:

$$P[X(t) \leqslant x \mid X(t_n) = x_n] = P[X(t - t_n) \leqslant x \mid X(0) = x_n].$$

In this case the Markov chain is said to be **(time)-homogeneous.** Note that the stationarity of the conditional distribution function (6.3) does not imply the stationarity of the joint distribution function (6.1). Thus, a homogeneous Markov process need not be a stationary stochastic process.

For a homogeneous Markov chain, the past history of the process is completely summarized in the current state; therefore, the distribution for the time Y the process spends in a given state must be memoryless. That is:

$$P(Y \leqslant r + t \mid Y \geqslant t) = P(Y \leqslant r). \tag{6.4}$$

But this implies that the time that a homogeneous, continuous-parameter Markov chain spends in a given state has an exponential distribution:

From (6.4) we have:

$$P(Y \leqslant r) = \frac{P(t \leqslant Y \leqslant t + r)}{P(Y \geqslant t)};$$

that is:

$$F_Y(r) = \frac{F_Y(t + r) - F_Y(t)}{1 - F_Y(t)}.$$

If we divide by r and take the limit as r approaches zero, we get:

$$F'_Y(0) = \frac{F'_Y(t)}{1 - F_Y(t)},$$

a differential equation with unique solution:

$$F_Y(t) = 1 - e^{-F'_Y(0)t}.$$

Similarly, the time that a homogeneous, discrete-parameter Markov chain spends in a given state has a geometric distribution.

In modeling practical situations, the restriction on times between state transitions may not hold. A semi-Markov process is a generalization of a Markov process where the distribution of time the process spends in a given state is allowed to be general.

As a generalization in another direction, consider the number of renewals (repairs or replacements) $N(t)$ required in the interval $(0, t]$, always a quantity of prime interest in renewal processes. The continuous-parameter process $\{N(t) \mid t \geqslant 0\}$ is called a **renewal counting process**. Note that, if we restrict the times between renewals to have an exponential distribution, then the corresponding renewal counting process is a special case of a continuous-parameter Markov chain, known as the **Poisson process**.

A measure of dependence among the random variables of a stochastic process is provided by its **autocorrelation function** R, defined by:

$$R(t_1, t_2) = E[X(t_1) \cdot X(t_2)]$$

Note that:

$$R(t_1, t_1) = E[X^2(t_1)]$$

and

$$\text{Cov}\,[X(t_1), X(t_2)] = R(t_1, t_2) - \mu(t_1)\,\mu(t_2).$$

The autocorrelation function $R(t_1, t_2)$ of a stationary process depends only on the time difference. (See problem 4 at the end of this section.) Thus, $R(t_1, t_2)$ is a one-dimensional function in this case and is written as $R(\tau)$.

Definition (Wide-Sense Stationary Process). A stochastic process is said to be wide-sense stationary if:

1. $\mu(t) = E[X(t)]$ is independent of t,
2. $R(t_1, t_2) = R(0, t_2 - t_1) = R(\tau), \;\; t_2 \geqslant t_1 \geqslant 0,$
3. $R(0) = E[X^2(t)] < \infty$ (finite second moment).

Note that a strict-sense stationary process is also wide-sense stationary, but the converse does not hold.

Example 6.3

[STAR 1979]. Consider the so-called random-telegraph process. This is a discrete-state, continuous-parameter process $\{X(t) \mid -\infty < t < \infty\}$ with the state space $\{-1, 1\}$. Assume that these two values are equally likely; that is:

$$P[X(t) = -1] = \tfrac{1}{2} = P[X(t) = 1], \qquad -\infty < t < \infty. \tag{6.5}$$

[Equation (6.5) implies that the first-order distribution function is stationary in time, but since higher-order distributions may be nonstationary, the stochastic process $X(t)$ need not be stationary in the strict sense.] A typical sample function of the process is shown in Figure 6.7.

Assume that the number of flips, $N(\tau)$, from one value to another occurring in an interval of duration τ is Poisson distributed with parameter $\lambda\tau$. Thus:

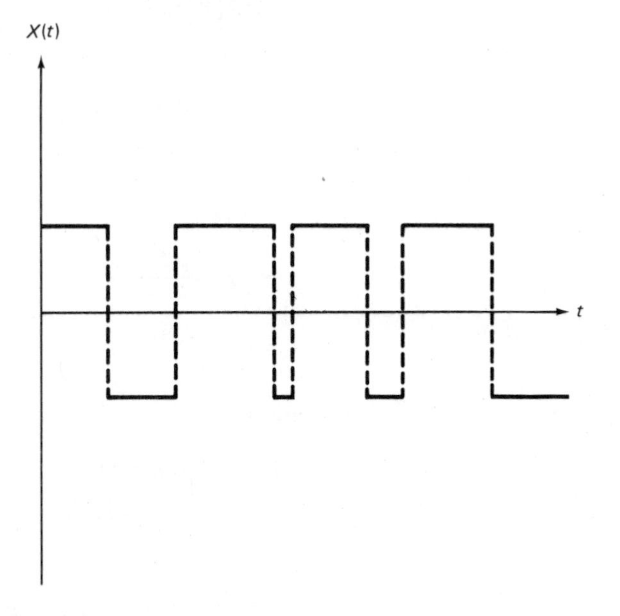

Figure 6.7 Typical sample function of the telegraph process

$$P[N(\tau) = k] = \frac{(\lambda\tau)^k e^{-\lambda\tau}}{k!}, \qquad k = 0, 1, 2, \ldots,$$

where λ is the average number of flips per unit time. Finally assume that the number of flips in a given time interval is statistically independent of the value assumed by the stochastic process $X(t)$ at the beginning of the interval.

For the telegraph process:

$$\mu(t) = E[X(t)] = -1 \cdot \tfrac{1}{2} + 1 \cdot \tfrac{1}{2} = 0, \qquad \text{for all } t.$$

$$\begin{aligned} R(t_1, t_2) &= E[X(t_1)\, X(t_2)] \\ &= P[X(t_1) = 1,\ X(t_2) = 1] - P[X(t_1) = 1,\ X(t_2) = -1] \\ &\quad - P[X(t_1) = -1,\ X(t_2) = 1] + P[X(t_1) = -1,\ X(t_2) = -1]. \end{aligned}$$

Since the marginal distribution functions of $X(t_1)$ and $X(t_2)$ are specified by equation (6.5), it can be shown that the events:

$$[X(t_1) = 1,\ X(t_2) = 1] \quad \text{and} \quad [X(t_1) = -1,\ X(t_2) = -1]$$

are equally likely. Similarly, the events:

$$[X(t_1) = 1,\ X(t_2) = -1] \quad \text{and} \quad [X(t_1) = -1, X(t_2) = 1]$$

are equally likely. It follows that the autocorrelation function:

$$\begin{aligned} R(t_1, t_2) &= 2\{P[X(t_1) = 1,\ X(t_2) = 1] - P[X(t_1) = 1,\ X(t_2) = -1]\} \\ &= 2\{P[X(t_2) = 1 | X(t_1) = 1] P[X(t_1) = 1] \\ &\quad - P[X(t_2) = -1 | X(t_1) = 1] P[X(t_1) = 1]\} \\ &= P[X(t_2) = 1 | X(t_1) = 1] - P[X(t_2) = -1 | X(t_1) = 1] \end{aligned}$$

To evaluate the conditional probability $P[X(t_2) = 1 | X(t_1) = 1]$ we observe that the corresponding event is equivalent to the event, "An even number of flips in the interval $(t_1, t_2]$." Let $\tau = t_2 - t_1$. Then:

$$\begin{aligned} P[X(t_2) = 1 | X(t_1) = 1] &= P[N(\tau) = \text{even}] \\ &= \sum_{k \text{ even}} \frac{e^{-\lambda\tau}(\lambda\tau)^k}{k!} \\ &= \frac{1 + e^{-2\lambda\tau}}{2} \end{aligned}$$

(by problem 2 at the end of this section). Similarly:

$$\begin{aligned} P[X(t_2) = -1 | X(t_1) = 1] &= P[N(\tau) = \text{odd}] \\ &= \sum_{k \text{ odd}} \frac{e^{-\lambda\tau}(\lambda\tau)^k}{k!} \\ &= \frac{1 - e^{-2\lambda\tau}}{2}. \end{aligned}$$

Substituting, we get:

$$R(t_1, t_2) = e^{-2\lambda\tau}, \qquad \tau > 0.$$

Furthermore, since $R(0) = E[X^2(t)] = 1 \cdot ½ + 1 \cdot ½ = 1$ is finite, we conclude that the random-telegraph process is stationary in the wide sense. In Figure 6.8 we have plotted $R(\tau)$ as a function of τ. #

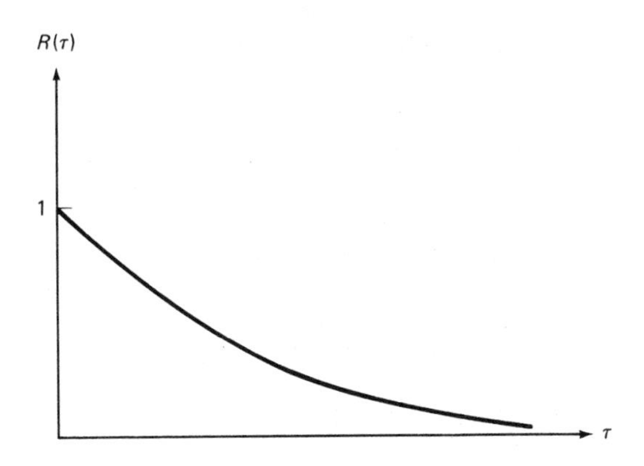

Figure 6.8 Autocorrelation function of the telegraph process

Problems

1. Show that the time that a discrete-parameter homogeneous Markov chain spends in a given state has a geometric distribution.

2. Assuming that the number of arrivals in the interval $(0, t]$ is Poisson distributed with parameter λt, compute the probability of an even number of arrivals. Also compute the probability of an odd number of arrivals.

3. Consider a stochastic process defined on a finite sample space with three sample points. Its description is provided by the specifications of the three sample functions:

$$X(t, s_1) = 3, \qquad X(t, s_2) = 3\cos(t), \qquad X(t, s_3) = 4\sin(t).$$

Also given is the probability assignment:

$$P(s_1) = P(s_2) = P(s_3) = ⅓.$$

Compute $\mu(t) = E[X(t)]$ and the autocorrelation function $R(t_1, t_2)$. Now answer the following questions: Is the process strict-sense stationary? Is it wide-sense stationary?

*4. Show that the autocorrelation function $R(t_1, t_2)$ of a strict-sense stationary stochastic process depends only on the time difference $(t_2 - t_1)$, if it exists.

6.3 THE BERNOULLI PROCESS

Consider a sequence of independent Bernoulli trials and let the discrete random variable Y_i denote the result of the ith trial, so that the event $[Y_i = 1]$ denotes a success on the ith trial and the event $[Y_i = 0]$ denotes a failure on the ith trial. Further assume that the probability of success on the ith trial, $P(Y_i = 1)$, is p, which is independent of the index i. Then $\{Y_i \mid i = 1, 2, ...\}$ is a discrete-state, discrete-parameter, stochastic process, which is stationary in the strict sense. Since the Y_i's are mutually independent, the above process is an independent process known as the **Bernoulli process**. We saw many examples of the Bernoulli process in Chapter 1. Since Y_i is a Bernoulli random variable, we recall that:

$$E[Y_i] = p,$$

$$E[Y_i^2] = p,$$

$$\text{Var}\,[Y_i] = p\,(1 - p),$$

and

$$G_{Y_i}(z) = (1 - p) + pz.$$

Based on the Bernoulli process, we may form another stochastic process by considering the sequence of **partial sums** $\{S_n \mid n = 1, 2, \ldots\}$, where $S_n = Y_1 + Y_2 + \cdots + Y_n$. By rewriting $S_n = S_{n-1} + Y_n$, it is not difficult to see that $\{S_n\}$ is a discrete-state, discrete-parameter Markov process, since:

$$\begin{aligned} P(S_n = k \mid S_{n-1} = k) &= P(Y_n = 0) \\ &= 1 - p, \end{aligned}$$

and

$$\begin{aligned} P(S_n = k \mid S_{n-1} = k - 1) &= P(Y_n = 1) \\ &= p. \end{aligned}$$

We showed in Chapter 2 that S_n is a binomial random variable, so that $\{S_n \mid n = 1, 2, \ldots\}$ is often called a **binomial process**. Clearly:

$$P(S_n = k) = \binom{n}{k} p^k (1 - p)^{n-k},$$

$$E[S_n] = np,$$

$$\text{Var}\,[S_n] = np\,(1 - p),$$

and

$$G_{S_n}(z) = (1 - p + pz)^n.$$

If we refer to successes in a Bernoulli process as arrivals, then we are led to the study of the number of trials between successes or **interarrival times**.

Define the discrete random variable T_1, called the **first-order interarrival time**, to be the number of trials up to and including the first success. Clearly, T_1 is geometrically distributed, so that:

$$P(T_1 = i) = p\,(1-p)^{i-1}, \qquad i = 1, 2, \ldots,$$

$$E[T_1] = \frac{1}{p},$$

$$\text{Var}\,[T_1] = \frac{1-p}{p^2},$$

and

$$G_{T_1}(z) = \frac{zp}{1 - z(1-p)}.$$

Now the total number of trials from the beginning of the process until and including the first success is a geometric random variable, and, owing to the mutual independence of successive trials, the number of trials after the $(i-1)$st success up to and including the ith success has the same distribution as T_1.

Recall that the geometric distribution possesses the memoryless property, so that the conditional pmf for the remaining number of trials up to and including the next success, given that there were no successes in the first m trials, is still geometric with parameter p. Since an arrival, as defined here, signals a change in state of the sum process $\{S_n\}$, we have that the occupancy time in state S_n is memoryless.

The notion of the first-order interarrival time can be generalized to higher-order interarrival times. Define the **rth-order interarrival time**, T_r, as the number of trials up to and including the rth success. Clearly, T_r is the r-fold convolution of T_1 with itself, and therefore T_r has the negative binomial distribution [using Theorem 2.1(b)]. Then:

$$P(T_r = i) = \binom{i-1}{r-1} p^r (1-p)^{i-r}, \qquad i = r, r+1, \ldots, \; r = 1, 2, \ldots,$$

$$E[T_r] = \frac{r}{p},$$

$$\text{Var}\,[T_r] = \frac{r(1-p)}{p^2},$$

and

$$G_{T_r}(z) = [\frac{zp}{1 - z(1-p)}]^r.$$

Example 6.4

Consider a file management system that employs blocking of several logical records on a single physical block [MADN 1974]. Now if records are accessed sequentially, a logical I/O will require a physical I/O operation only occasionally. As-

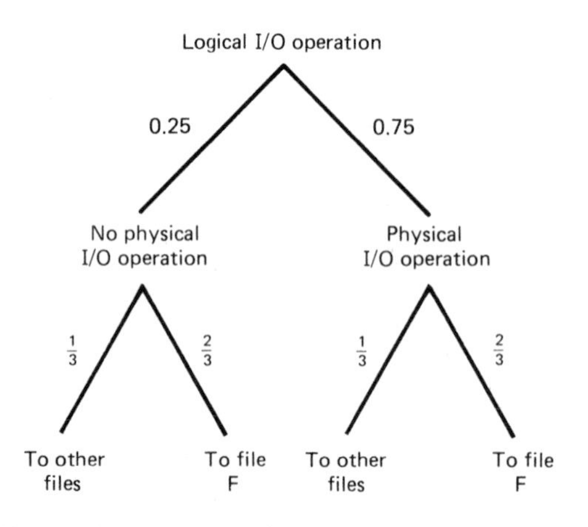

Figure 6.9 The tree diagram for accessing a file

sume that the probability that a logical I/O will generate a physical I/O is known to be 0.75. We are interested in studying the usage profile of a specific file F. Assume that the probability that a given I/O operation refers to a record of the file F is ⅔. Further assume that the two events "file F requested" and "logical I/O gives rise to physical I/O" are independent, and that successive I/O operations are independent.

Using the tree diagram of Figure 6.9, we may consider the sequence of logical I/O operations as a Bernoulli process with the probability of success (a logical I/O causes a physical I/O to file F) equal to $\frac{3}{4} \cdot \frac{2}{3} = 0.5$. The number of logical I/O operations between two physical I/O operations directed to file F will then be geometrically distributed with parameter $p = 0.5$. #

Several generalizations of the Bernoulli process are possible. One possibility is to allow each Bernoulli variable Y_i a distinct parameter p_i. We retain the assumption that $Y_1, Y_2, \ldots,$ are independent random variables. Then the process $\{Y_i \mid i = 1, 2, \ldots\}$ is an independent process called the **nonhomogeneous Bernoulli process.**

As an example, let us return to the analysis of program MAX (Chapter 2). Recall that the number of executions, X_n, of the **then** clause for a given array size n is a discrete random variable with the generating function:

$$G_{X_n}(z) = \prod_{i=2}^{n} \left(\frac{z + i - 1}{i}\right).$$

By the convolution property of transforms, we can write:

$$X_n = \sum_{i=2}^{n} Y_i,$$

where $Y_2, Y_3, \ldots,$ are independent random variables such that:

$$G_{Y_i}(z) = \frac{z + i - 1}{i}.$$

But this implies that Y_i is a Bernoulli random variable with parameter $1/i$. Thus, $\{Y_i \mid i = 2, 3, \ldots, n\}$ is a nonhomogeneous Bernoulli process, and $\{X_i \mid i = 2, 3, \ldots, n\}$ is the corresponding sum process.

Another generalization of the Bernoulli process is to assume that each trial has more than two possible outcomes. Let $\{Y_i \mid i = 1, 2, 3, \ldots, n\}$ be a sequence of independent discrete random variables, and define the partial sum $S_n = \sum_{i=1}^{n} Y_i$. Then the sum process $\{S_n \mid n = 1, 2, \ldots\}$ is a Markov chain known as a **random walk.**

Yet another generalization of the Bernoulli process is to study the limiting behavior of the discrete-parameter process into a continuous-parameter process. Recall, from Chapter 2, that the Poisson distribution can be derived as a limiting case of the binomial distribution. Now, since the sum process corresponding to the Bernoulli process is the binomial process, the pmf of S_n is $b(k; n, p)$. If n is large and p is small, $b(k; n, p)$ approaches the Poisson pmf $f(k; np)$. Thus, the number of successes $N(t)$ is approximately Poisson distributed with parameter $\lambda t = np$.

Problems

1. Show that the first-order interarrival time of the nonhomogeneous Bernoulli process is *not* memoryless.

6.4 THE POISSON PROCESS

The Poisson process is a continuous-parameter, discrete-state process that is a good model for many practical situations. Here, the interest is in counting the number of events $N(t)$ occurring in the time interval $(0, t]$. The event of interest may, for example, correspond to:

1. The number of incoming telephone calls to a switchboard or a trunk.
2. The number of job arrivals to a computer system.
3. The number of failed components in a large group of initially fault-free components.

We define the Poisson process as follows. Suppose that the events occur successively in time, so that the intervals between successive events are independent and identically distributed according to an exponential distribution $F(x) = 1 - e^{-\lambda x}$. Let the number of events in the interval $(0, t]$ be denoted by $N(t)$. Then the stochastic process $\{N(t) \mid t \geq 0\}$ is a **Poisson process** with mean rate λ. In the first two situations listed above, λ is called the average arrival rate, while in the third situation λ is called the failure rate. From this definition, it is clear that a Poisson process is a renewal counting process for which the underlying distribution is exponential.

An alternative (and equivalent) definition of the Poisson process is as follows: As before, let $N(t)$ be the number of events that have occurred in the interval $(0, t]$. Let the event A denote the occurrence of exactly one event in the interval $(t, t+h]$. Similarly let B and C, respectively, denote the occurrences of none and more than one events in the same interval. Let $P(A) = p(h)$, $P(B) = q(h)$, and $P(C) = \epsilon(h)$. $N(t)$ forms a Poisson process, provided the following four conditions are met:

1. $N(0) = 0$.
2. Events occurring in nonoverlapping intervals of time are mutually independent.
3. Probabilities $p(h)$, $q(h)$, and $\epsilon(h)$ depend only on the length h of the interval and not on the time origin t.
4. For sufficiently small values of h, we can write (for some positive constant λ):

$$p(h) = P[\text{one event in the interval } (t, t+h]] = \lambda h + o(h),$$

$$q(h) = P[\text{no events in the interval } (t, t+h]] = 1 - \lambda h + o(h),$$

and

$$\epsilon(h) = P[\text{more than one event in the interval } (t, t+h]] = o(h)$$

where $o(h)$ denotes any quantity having an order of magnitude smaller than h; that is:

$$\lim_{h \to 0} \frac{o(h)}{h} = 0.$$

Let $p_n(t) = P[N(t) = n]$ be the pmf of $N(t)$. Because of condition 1 above:

$$p_0(0) = 1 \quad \text{and} \quad p_n(0) = 0 \qquad \text{for } n > 0. \tag{6.6}$$

Now consider two successive nonoverlapping intervals $(0, t]$ and $(t, t+\tau]$. To compute $p_n(t+\tau)$, the probability that n events occur in the interval $(0, t+\tau]$, we note that:

$$P(n \text{ events in } (0, t+\tau])$$

$$= \sum_{k=0}^{n} P(k \text{ events in } (0,t] \text{ and } n-k \text{ events in } (t, t+\tau])$$

$$= \sum_{k=0}^{n} P[k \text{ events in } (0, t]]\, P[n-k \text{ events in } (t, t+\tau]],$$

by condition 2 above. Then by condition 3 we have:

$$p_n(t+\tau) = \sum_{k=0}^{n} p_k(t)\, p_{n-k}(\tau). \tag{6.7}$$

State transitions of the Poisson random process may be visualized as in Figure 6.10. Using equation (6.7) for $n > 0$ and $\tau = h$,

$$\begin{aligned}
p_n(t+h) &= P[N(t+h) = n] \\
&= P[N(t) = n]\, P[\text{no events in } (t,\ t+h]] \\
&\quad + P[N(t) = n-1]\, P[\text{one event in } (t,\ t+h]] \\
&\quad + \sum_{i=0}^{n-2} P[N(t) = i]\, P[n-i \text{ events in } (t,\ t+h]] \\
&= p_n(t)\,[1 - \lambda h + o(h)] + p_{n-1}(t)\,[\lambda h + o(h)] \\
&\quad + \sum_{i=0}^{n-2} p_i(t)\, o(h) \\
&= (1 - \lambda h)\, p_n(t) + \lambda h\, p_{n-1}(t) + o(h), \qquad n > 0.
\end{aligned}$$

Similarly:

$$p_0(t+h) = (1 - \lambda h)\, p_0(t) + o(h).$$

After some algebra, we get:

$$\lim_{h \to 0} \frac{p_0(t+h) - p_0(t)}{h} = -\lambda\, p_0(t)$$

and

$$\lim_{h \to 0} \frac{p_n(t+h) - p_n(t)}{h} = -\lambda\, p_n(t) + \lambda\, p_{n-1}(t).$$

This gives rise to the following differential equations:

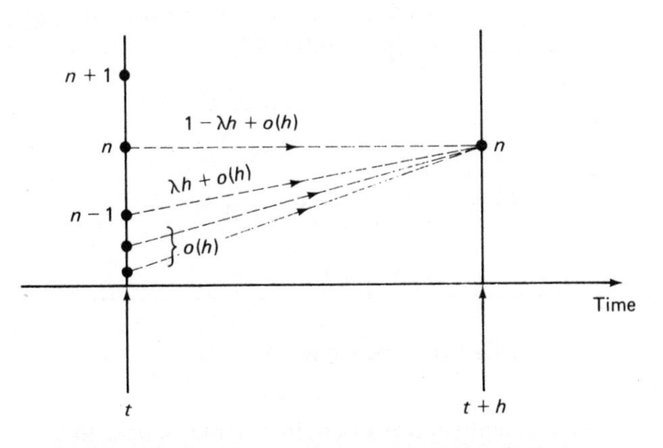

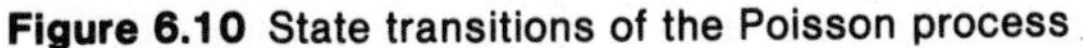
Figure 6.10 State transitions of the Poisson process

$$\frac{dp_0(t)}{dt} = -\lambda\, p_0(t)$$

and
$$\frac{dp_n(t)}{dt} = -\lambda\, p_n(t) + \lambda p_{n-1}(t), \qquad n > 0. \tag{6.8}$$

It is not difficult to show by induction on n that the solution to equations (6.8) with initial condition (6.6) is:

$$p_n(t) = e^{-\lambda t} \frac{(\lambda t)^n}{n!}.$$

Therefore, the number of events $N(t)$ in the interval $(0, t]$ has a Poisson distribution with parameter λt. (Note that this implies that the Poisson process is not a stationary stochastic process.) From Chapter 4 we know that the mean and variance of this distribution are both equal to λt. Therefore, as t approaches infinity, $E[N(t)/t]$ approaches λ and Var $[N(t)/t]$ approaches zero. In other words, $N(t)/t$ converges to λ as t approaches infinity. Because of this, the parameter λ is called the arrival rate of the Poisson process. The Poisson process plays an important role in queuing theory and reliability theory. One reason for its importance is its analytical tractability, and another reason is a result due to Palm [PALM 1943] and Khinchin [KHIN 1960], which states that under very general assumptions the sum of a large number of independent renewal processes behaves like a Poisson process.

Often we are interested in the **superposition of independent Poisson processes**. For example, suppose there are two independent Poisson job-arrival streams into a computing center, with respective arrival rates λ_1 and λ_2. We are interested in the pooled job-arrival stream. Figure 6.11 shows the arrival times of job stream 1, of job stream 2, and of the pooled stream.

Recall from Chapter 2 that the sum of n independent Poisson random variables is itself Poisson. Based on this result, is can be shown that the superposition of n independent Poisson processes with respective average rates $\lambda_1, \lambda_2, \ldots, \lambda_n$ is also a Poisson process with the average rate $\lambda = \lambda_1 + \lambda_2 + \cdots + \lambda_n$ [BARL 1975]. The notion of superposition of Poisson processes is illustrated in Figure 6.12.

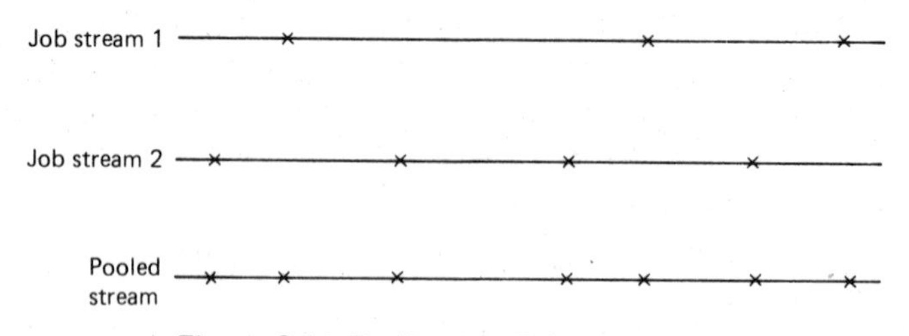

Figure 6.11 Pooling two Poisson streams

Example 6.5

There are n independent sources of environmental shocks to a component. The number of shocks from the ith source in the interval $(0, t]$, denoted $N_i(t)$, is

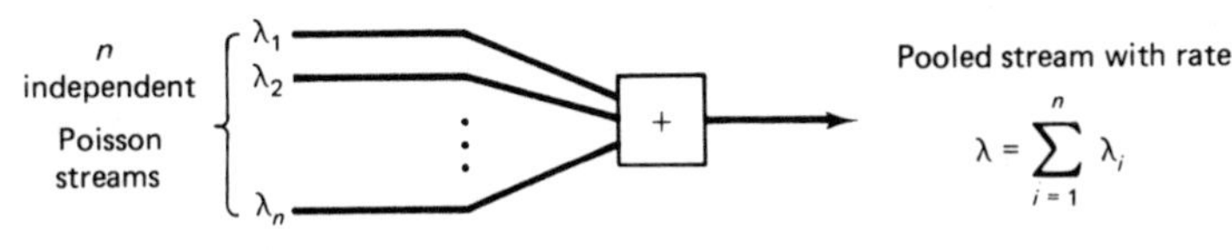

Figure 6.12 Superposition of independent Poisson processes

governed by a Poisson process with rate λ_i. Then the total number of shocks of all kinds in the interval $(0, t]$ forms a Poisson process with the rate $\lambda = \lambda_1 + \lambda_2 + \cdots + \lambda_n$. #

Example 6.6

Consider a series system of n independent components. The lifetime X_i of the ith component is exponentially distributed with parameter λ_i $(i = 1, 2, \ldots, n)$. Assume that upon failure of the ith component, it is instantaneously replaced by a spare. Now, recalling the relation between the Poisson process and the exponential inter-event times, we can conclude that for $i = 1, 2, \ldots, n$, the number of failures, $N_i(t)$, of the ith component in the interval $(0,t]$ form a Poisson process with rate λ_i. Then the total number of system failures $N(t)$ in the interval $(0, t]$ is a superposition of n independent Poisson processes, and hence it is a Poisson process with rate

$$\lambda = \sum_{i=1}^{n} \lambda_i.$$

This implies that the times between system failures are exponentially distributed with parameter λ. But this result can be independently verified from our discussion in Chapter 3, where we demonstrated that the lifetime $X = \min\{X_1, X_2, \ldots, X_n\}$ of a series system of n independent components with exponential lifetime distribution is itself exponentially distributed with parameter $\lambda = \sum_{i=1}^{n} \lambda_i$. #

A similar result holds with respect to the **decomposition of a Poisson process**. Assume that a Poisson process with mean arrival rate λ branches out into n output paths as shown in Figure 6.13. We assume that the successive selections of an output stream form a sequence of generalized Bernoulli trials with $p_i (1 \leqslant i \leqslant n)$ denoting the probability of the selection of output stream i. Let $\{N(t) \mid t \geqslant 0\}$ be the input Poisson process and let $\{N_k(t) \mid t \geqslant 0\}$ for $1 \leqslant k \leqslant n$ denote the output processes. The conditional distribution of $N_k(t)$, $1 \leqslant k \leqslant n$, given that $N(t) = m$, is the multinomial distribution (see Chapter 2):

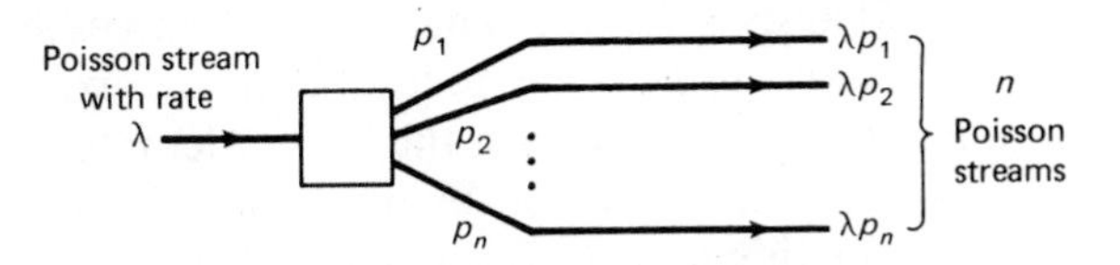

Figure 6.13 Decomposition of a Poisson process

$$P(N_1(t) = m_1, N_2(t) = m_2, \ldots, N_n(t) = m_n \mid N(t) = m)$$
$$= \frac{m!}{m_1! m_2! \cdots m_n!} p_1^{m_1} p_2^{m_2} \cdots p_n^{m_n},$$

where:

$$\sum_{i=1}^{n} p_i = 1 \quad \text{and} \quad \sum_{i=1}^{n} m_i = m.$$

Now, since:

$$P[N(t) = m] = e^{-\lambda t} \frac{(\lambda t)^m}{m!},$$

we get the unconditional pmf:

$$P[N_1(t) = m_1, N_2(t) = m_2, \ldots, N_n(t) = m_n]$$
$$= \frac{m!}{m_1! m_2! \ldots m_n!} p_1^{m_1} p_2^{m_2} \cdots p_n^{m_n} e^{-\lambda t} \frac{(\lambda t)^m}{m!}$$
$$= \prod_{i=1}^{n} e^{-p_i \lambda t} \frac{(p_i \lambda t)^{m_i}}{m_i!}.$$

But this implies that the random variables $N_1(t), N_2(t), \ldots, N_n(t)$ are mutually independent (for all $t \geqslant 0$) and have Poisson distributions with respective parameters $p_1\lambda, p_2\lambda, \ldots, p_n\lambda$. This in turn verifies that the n output processes are all Poisson with the parameters listed above.

Example 6.7

The number of environmental shocks $N(t)$ experienced by a component in the interval $(0, t]$ is governed by a Poisson process with a rate λ. With probability p_1 the component will continue to function in spite of the shock, and with probability p_2 $(= 1 - p_1)$ the shock is fatal. Then the random process corresponding to the arrival sequence of fatal shocks is a Poisson process with rate $p_2\lambda$. #

Example 6.8

The number of transactions arriving into a database system forms a Poisson process with rate λ. The database consists of n distinct files. An arriving transaction requests the ith file with probability p_i. With the usual independence assumption, the number of requests directed to the ith file $(1 \leqslant i \leqslant n)$ forms a Poisson process of rate $p_i\lambda$. #

We have noted in Chapter 3 that the times between successive events of a Poisson process with rate λ are mutually independent and exponentially distributed with parameter λ. In other words, the first-order interarrival time T_1 of a Poisson process is exponentially distributed. It follows that the rth order interarrival time, T_r, for this process is an r-stage Erlang random variable with parameter λ.

Thus, if T_k denotes the time of the kth event (from the beginning), then T_k is a k-stage Erlang random variable. Now suppose we know that n arrivals have occurred in the interval $(0, t]$. We wish to compute the conditional distribution of T_k $(1 \leqslant k \leqslant n)$.

Suppose that exactly one arrival has occurred in the interval $(0, t]$. Then, because of the properties of the Poisson process, we can show that the conditional distribution of arrival time T_1 is uniform over $(0, t)$:

$$P[T_1 \leqslant x | N(t) = 1] = \frac{P[N(x) = 1 \text{ and } N(t) - N(x) = 0]}{P[N(t) = 1]} = \frac{x}{t}.$$

This result is generalized in the following theorem:

THEOREM 6.1. Given that $n \geqslant 1$ arrivals have occurred in the interval $(0, t]$, the conditional joint pdf of the arrival times $T_1, T_2, \ldots, T_n$ is given by:

$$f[t_1, t_2, \ldots, t_n \,|\, N(t) = n] = \frac{n!}{t^n}, \qquad 0 \leqslant t_1 \leqslant \cdots \leqslant t_n \leqslant t.$$

Proof: Let T_{n+1} be the time of the $n + 1$st arrival (which occurs after time t). Define the random variables:

$$Y_i = T_i - T_{i-1}, \qquad i = 1, \ldots, n + 1,$$

where T_0 is defined to be equal to zero. It is clear that $Y_1, Y_2, \ldots, Y_{n+1}$ are independent identically distributed random variables such that $Y_i \sim \text{EXP}(\lambda)$ for $i = 1, 2, \ldots, n + 1$.

Define events A and B such that:

$$A = \text{“}t_i < T_i \leqslant t_i + h_i \text{ for } i = 1, 2, \ldots, n\text{”},$$

and

$$B = [N(t) = n] = [T_n \leqslant t < T_{n+1}].$$

Now, using the definition of Y_i, the event A can be rewritten as:

$$A = \text{“}t_i - t_{i-1} - h_{i-1} \leqslant Y_i \leqslant t_i - t_{i-1} + h_i, \text{ for } i = 1, 2, \ldots, n\text{”}$$

where t_0 and h_0 are defined to be equal to zero. Now:

$$\begin{aligned} A \cap B &= \text{“}t_i - t_{i-1} - h_{i-1} \leqslant Y_i \leqslant t_i - t_{i-1} + h_i, \text{ for } i = 1, 2, \ldots, n\text{”} \\ &\quad \text{and “}T_n \leqslant t < T_{n+1}\text{”} \\ &= \text{“}t_i - t_{i-1} - h_{i-1} \leqslant Y_i \leqslant t_i - t_{i-1} + h_i, \text{ for } i = 1, 2, \ldots, n\text{”} \\ &\quad \text{and “}Y_{n+1} \geqslant t - t_n - h_n\text{”}. \end{aligned}$$

Since the Y_i's are independent, we have:

$$P(A|B) = \frac{P(A \cap B)}{P(B)}$$

$$= \frac{[\prod_{i=1}^{n} P(t_i - t_{i-1} - h_{i-1} \leqslant Y_i \leqslant t_i - t_{i-1} + h_i)] P(Y_{n+1} \geqslant t - t_n - h_n)}{P[N(t) = n]}.$$

Dividing both sides by $h_1 h_2 \cdots h_n$, taking the limit as $h_i \longrightarrow 0$ $(i = 1, 2, \ldots n)$, and recalling that the Y_i's are exponentially distributed with parameter λ, we have:

$$f[t_1, t_2, \ldots, t_n \mid N(t) = n] = \frac{\left[\prod_{i=1}^{n} \lambda e^{-\lambda(t_i - t_{i-1})}\right] e^{-\lambda(t - t_n)}}{(\lambda t)^n e^{-\lambda t}/n!}$$

$$= \frac{n!}{t^n}.$$

A more general version of the above theorem also holds [BARL 1975]:

THEOREM 6.2. Given that n events have occurred in the interval $(0, t]$, the times of occurrence $S_1, S_2, \ldots, S_n$, when *unordered*, are independent, uniformly distributed over the interval $(0, t]$. In fact, the random variables $T_1, T_2, \ldots, T_n$ of Theorem 6.1 are the order statistics of the random variables $S_1, S_2, \ldots, S_n$.

Example 6.9 (The *M/G/∞* queue)

Suppose that jobs arrive at a computer center in accordance with a Poisson process of rate λ. The computer center has an abundance of microcomputers, so that a job is serviced immediately upon its arrival (that is, no queuing takes place). For analysis, we may assume that the number of servers is infinitely large. Job service times are assumed to be independent general random variables with a common distribution function G.

Let $X(t)$ denote the number of jobs in the system at time t and let $N(t)$ denote the total number of job arrivals in the interval $(0, t]$. The number of departures $D(t) = N(t) - X(t)$. First we determine the conditional pmf of $X(t)$ given $N(t) = n$.

Consider a job that arrived at time $0 \leqslant y \leqslant t$. By Theorem 6.2, the time of arrival Y of the job is uniformly distributed over $(0, t)$; that is, $f_Y(y) = 1/t$, $0 < y < t$. The probability that this job is still undergoing service at time t given that it arrived at time y is $1 - G(t - y)$. Then the unconditional probability that the job is undergoing service at time t is (by the continuous version of the theorem of total probability):

$$p = \int_0^t [1 - G(t - y)] f_Y(y)\, dy$$

$$= \int_0^t \frac{1 - G(t - y)}{t}\, dy$$

$$= \int_0^t \frac{1 - G(x)}{t}\, dx.$$

Since n jobs have arrived and each has the probability p of independently not completing by time t, we have a sequence of n Bernoulli trials. Thus:

$$P[X(t) = j \mid N(t) = n] = \begin{cases} \binom{n}{j} p^j (1-p)^{n-j}, & j = 0, 1, \ldots, n, \\ 0, & \text{otherwise.} \end{cases}$$

Then by the theorem of total probability we have:

$$\begin{aligned} P[X(t) = j] &= \sum_{n=j}^{\infty} \binom{n}{j} p^j (1-p)^{n-j} e^{-\lambda t} \frac{(\lambda t)^n}{n!} \\ &= e^{-\lambda t} \frac{(\lambda tp)^j}{j!} \sum_{n=j}^{\infty} \frac{[\lambda t(1-p)]^{n-j}}{(n-j)!} \\ &= e^{-\lambda tp} \frac{(\lambda tp)^j}{j!}. \end{aligned}$$

Thus, the number of jobs in the system at time t has the Poisson distribution with parameter:

$$\lambda' = \lambda tp = \lambda \int_0^t [1 - G(x)]\, dx.$$

In particular, if the service times are exponentially distributed with parameter μ, then:

$$G(x) = 1 - e^{-\mu x},$$

$$\int_0^t [1 - G(x)]\, dx = \frac{1}{\mu} - \frac{e^{-\mu t}}{\mu},$$

and hence, in the limit as t approaches infinity, $\lambda' = \lambda/\mu$. This implies that after a sufficiently long time, the number of jobs in an $M/M/\infty$ queue is Poisson distributed with parameter λ/μ. #

Problems

1. Consider a computer system with Poisson job-arrival stream at an average rate of 60 per hour. Determine the probability that the time interval between successive job arrivals is:
   (a) Longer than four minutes.
   (b) Shorter than eight minutes.
   (c) Between two and six minutes.

2. Spare Parts Problem [BARL 1975]. A system requires k components of a certain type to function properly. All components are assumed to have a constant failure rate λ, and their lifetimes are statistically independent. During a mission, component j is put into operation for t_j time units, and a component can fail only while

in operation. Determine the number of spares needed (in a common supply of spares) in order to achieve a probability greater than α for the mission to succeed. As an example, let $k = 3$, $t_1 = 1{,}300$, $t_2 = 1{,}500$, $t_3 = 1{,}200$, $\lambda = 0.002$, and $\alpha = 0.90$, and determine the number of needed spares n. Now consider the alternative strategy of keeping a separate supply of spares for each of the three component types. Find the number of spares n_1, n_2, and n_3 required to provide an assurance of more than 90 percent that no shortage for any component type will occur. Show that the former strategy is the better one.

3. When we considered the decomposition of a Poisson process in the text, we assumed that a generalized Bernoulli trial was performed to select the output stream an arriving job should be directed to. Let us now consider a cyclic method of decomposition in which each output stream receives the nth arrival so that the first , $(n + 1)$ th, $(2n + 1)$ th , . . . , arrivals are directed to output stream 1, the second, $(n + 2)$ th, $(2n + 2)$ th , . . . , arrivals are directed to stream 2, and so on. Show that the interarrival times of any output substream comprise an n-stage Erlang variable. Note that none of the output streams is Poisson!

4. We are given two independent Poisson arrival streams $\{X_t \mid 0 \leqslant t < \infty\}$ and $\{Y_t \mid 0 \leqslant t < \infty\}$ with respective arrival rates λ_x and λ_y. Show that the number of arrivals of the Y_t process occurring between two successive arrivals of X_t process has a modified geometric distribution with parameter $\lambda_x/(\lambda_x + \lambda_y)$.

5. Consider the generalization of the ordinary Poisson process, called the **compound Poisson process.** In an ordinary Poisson process, we assumed that the probability of occurrence of multiple events in a small interval is negligible with respect to the length of the interval. If the arrival of a message in a computer communications system is being modeled, the counting process may represent the number of bytes (or packets) in a message. In this case suppose that the pmf of the number of bytes in a message is specified:

$$P(\text{number of bytes in a message} = k) = a_k, \qquad k \geqslant 1.$$

Further assume that the number of message-arrivals form an ordinary Poisson process with rate λ. Then the process $\{X(t) \mid t \geqslant 0\}$, where $X(t)$ = number of bytes arriving in the interval $(0, t]$, is a compound Poisson process. Show that the generating function of $X(t)$ is given by:

$$G_{X(t)}(z) = e^{\lambda t[G_A(z)-1]},$$

where:

$$G_A(z) = \sum_{k \geqslant 1} a_k z^k.$$

*6 Prove Theorem 6.1 starting with Theorem 6.2. [*Hint:* Refer to the section on Order Statistics in Chapter 3.]

6.5 RENEWAL PROCESSES

We have noted that successive interevent times of a Poisson process are independent exponentially distributed random variables. A natural generalization of the Poisson process is obtained by removing the restriction of the exponential interevent times. Let X_i be the time between the $(i-1)$st and the ith events. Let $\{X_i \mid i=1, 2, \ldots\}$ be a sequence of independent nonnegative identically distributed random variables. This general independent process, $\{X_i \mid i=1, 2, \ldots\}$, is a **renewal process** or a **recurrent process** as defined in Section 6.2. The random variable X_i is interpreted as the time between the occurrence of the ith event and the $(i-1)$ st event. Note that the restriction of exponential distribution is removed.

The recurrent events (also called renewals) may correspond to a job arrival in a computer system or a telephone-call arrival to a switchboard. The event may also correspond to the failure of a component in an environment with inexhaustible spares and an instant replacement of a faulty component with a spare one. Similarly, an event may correspond to a reference to a specific page in a paging system, where $\{X_i\}$ will then represent successive intervals between references to this specific page [COFF 1973]. In this case, there will be a distinct renewal process corresponding to each page in the address space of the program being modeled. Here it may be more appropriate to think of X_i as a discrete random variable counting the number of page references between two references to the specific page. In such a case, we have a discrete-state, discrete-time renewal process.

Our development here will assume that X_i is a continuous random variable with the distribution function $F(x)$, called the **underlying distribution** of the renewal process.

Let S_k denote the time from the beginning until the occurrence of the kth event; that is:

$$S_k = X_1 + X_2 + \cdots + X_k,$$

and let $F^{(k)}(t)$ denote the distribution function of S_k. Clearly, $F^{(k)}$ is the k-fold convolution of F with itself. For notational convenience, we define:

$$F^{(0)}(t) = \begin{cases} 1, & t \geqslant 0, \\ 0, & t < 0. \end{cases}$$

Our primary interest here is in the number of renewals $N(t)$ in the interval $(0, t]$. The discrete-state, continuous-parameter process $\{N(t) \mid t \geqslant 0\}$ is called a **renewal counting process**. This is the generalization of the Poisson process that we alluded to at the beginning of the section. $N(t)$ is called the **renewal random variable** and is easily related to the random variables S_k and S_{k+1} by observing that $N(t) = n$ if and only if $S_n \leqslant t < S_{n+1}$. But then:

$$\begin{aligned} P[N(t) = n] &= P(S_n \leqslant t < S_{n+1}) \\ &= P(S_n \leqslant t) - P(S_{n+1} \leqslant t) \\ &= F^{(n)}(t) - F^{(n+1)}(t). \end{aligned} \tag{6.9}$$

Define the **renewal function** $M(t)$ as the average number of renewals in the interval $(0, t]$:

$$M(t) = E[N(t)].$$

Thus, for example, if $N(t)$ is a Poisson process with rate λ, then its renewal function $M(t) = \lambda t$. Using relation (6.9) above, we have:

$$\begin{aligned}
M(t) &= \sum_{n=0}^{\infty} n\, P[N(t) = n] \\
&= \sum_{n=0}^{\infty} n\, F^{(n)}(t) - \sum_{n=0}^{\infty} n\, F^{(n+1)}(t) \\
&= \sum_{n=0}^{\infty} n\, F^{(n)}(t) - \sum_{n=1}^{\infty} (n-1)\, F^{(n)}(t) \qquad (6.10) \\
&= \sum_{n=1}^{\infty} F^{(n)}(t) \\
&= F(t) + \sum_{n=1}^{\infty} F^{(n+1)}(t).
\end{aligned}$$

Noting that $F^{(n+1)}$ is the convolution of $F^{(n)}$ and F, and letting f be the density function of F, we can write:

$$F^{(n+1)}(t) = \int_0^t F^{(n)}(t-x)\, f(x)\, dx,$$

and therefore:

$$\begin{aligned}
M(t) &= F(t) + \sum_{n=1}^{\infty} \int_0^t F^{(n)}(t-x)\, f(x)\, dx \\
&= F(t) + \int_0^t \left[\sum_{n=1}^{\infty} F^{(n)}(t-x)\right] f(x)\, dx \\
&= F(t) + \int_0^t M(t-x)\, f(x)\, dx. \qquad (6.11)
\end{aligned}$$

This last equation is known as the **fundamental renewal equation.** [In the above and subsequent derivations we assume that $\sum F^{(n)}$ satisfies appropriate convergence conditions.]

Define the **renewal density** $m(t)$ to be the derivative of the renewal function $M(t)$; that is:

$$m(t) = \frac{dM(t)}{dt}.$$

For small h, $m(t)h$ is interpreted as the probability of occurrence of a renewal in the interval $(t, t+h]$. Thus in the case of a Poisson process, renewal density $m(t)$ equals the Poisson rate λ. From equation (6.10) we have:

$$m(t) = \sum_{n=1}^{\infty} f^{(n)}(t),$$

and, using equation (6.11), we have:

$$m(t) = f(t) + \int_0^t m(t-x) f(x)\, dx. \tag{6.12}$$

This is known as the **renewal equation**. Under appropriate conditions, the asymptotic renewal rate can be shown to be equal to $1/E[X]$:

$$\lim_{t \to \infty} m(t) = \frac{1}{E[X]}.$$

This is a form of the **key renewal theorem** [ROSS 1970].

To solve the renewal equation (6.12), we will use Laplace transforms. Our notation here will be somewhat different from that in Chapter 4, since we now associate a Laplace transform with a function (distribution, density, renewal) rather than a random variable. Thus:

$$L_f(s) = \int_0^{\infty} e^{-sx} f(x)\, dx$$

and

$$L_m(s) = \int_0^{\infty} e^{-sx} m(x)\, dx.$$

Now, using the convolution property of transforms, we get from equation (6.12):

$$L_m(s) = L_f(s) + L_m(s)\, L_f(s),$$

so that

$$L_m(s) = \frac{L_f(s)}{1 - L_f(s)} \tag{6.13}$$

and

$$L_f(s) = \frac{L_m(s)}{1 + L_m(s)}. \tag{6.14}$$

Thus $m(t)$ may be determined from $f(t)$, and conversely, $f(t)$ can be determined from $m(t)$.

Example 6.10

The solution to equation (6.13) may be obtained in closed form for special cases. Assume that the interevent times are exponentially distributed so that:

$$f(x) = \lambda e^{-\lambda x}.$$

Then:

$$L_f(s) = \frac{\lambda}{s + \lambda},$$

and, using (6.13):

$$L_m(s) = \frac{\lambda}{s + \lambda - \lambda}$$
$$= \frac{\lambda}{s}.$$

Since the Laplace transform of a constant k is k/s, we conclude that:

$$m(t) = \lambda, \qquad t \geqslant 0,$$
$$M(t) = \lambda t, \qquad t \geqslant 0.$$

These results should be expected, since the renewal counting process in this case is a Poisson process where the average number of events in the interval $(0, t]$ is λt.

In this case, $F^{(n)}(t)$ is a convolution of n identical exponential distributions and hence is an n-stage Erlang distribution:

$$F^{(n)}(t) = 1 - \left[\sum_{k=0}^{n-1} \frac{(\lambda t)^k}{k!}\right] e^{-\lambda t}.$$

Then:

$$P[N(t) = n] = F^{(n)}(t) - F^{(n+1)}(t)$$
$$= \frac{(\lambda t)^n}{n!} e^{-\lambda t}.$$

Thus $N(t)$ has a Poisson distribution with parameter λt, as expected. #

Problems

***1.** [SEVC 1977]. In Section 6.4 we showed that the decomposition of a Poisson process using Bernoulli selection produces Poisson processes (see Figure 6.13). Now consider a general renewal counting process $N(t)$ with an underlying distribution $F(x)$. Suppose we send an arrival into one of two streams using a Bernoulli filter with respective probabilities p and $1 - p$. Show that the Laplace transform of the interarrival times for the first stream is given by:

$$L_{X_1}(s) = \frac{pL_X(s)}{1 - (1 - p)\, L_X(s)}.$$

[*Hint:* The section on random sums in Chapter 5 will be useful.]

Now show that the coefficient of variation of the interarrival time X_1 of the first output stream is given by:

$$C_{X_1}^2 = 1 + p(C_X^2 - 1)$$

and the mean by

$$E[X_1] = \frac{1}{p} E[X].$$

*2. [BHAT 1972] Consider a computer system during peak load where the CPU is saturated. Assume that the processing requirement of a job is exponentially distributed with parameter μ. Further assume that a fixed time t_{sys} per job is spent performing overhead functions in the operating system. Let $N(t)$ be the number of jobs completed in the interval $(0, t]$. Show that:

$$P[N(t) < n] = \sum_{k=0}^{n-1} e^{-\mu(t - nt_{sys})} \frac{[\mu(t-nt_{sys})]^k}{k!}, \qquad \text{if } t \geqslant nt_{sys}.$$

6.6 AVAILABILITY ANALYSIS

Assume that upon the failure of a component, it is repaired and restored to be "as good as new." Let T_i be the duration of the ith functioning period and let D_i be the system downtime for the ith repair or replacement. We assume that the sequence of random variables $\{X_i = T_i + D_i\}$ $(i = 1, 2, \ldots)$ is mutually independent. Further assume that the T_i's are identically distributed with the common CDF $W(t)$ and common pdf $w(t)$, and that the D_i's are identically distributed with CDF $G(t)$ and pdf $g(t)$. Then X_i's are also identically distributed, hence $\{X_i \mid i = 1, 2, \ldots\}$ is a renewal process. A renewal point of this process corresponds to the event of the completion of a repair. The underlying density $f(t)$ of the renewal process is the convolution of w and g (assuming T_i and D_i are independent). Thus:

$$L_f(s) = L_w(s)\, L_g(s) \tag{6.15}$$

and, using equation (6.13), we have:

$$L_m(s) = \frac{L_w(s)\, L_g(s)}{1 - L_w(s)\, L_g(s)}. \tag{6.16}$$

The average number of repairs or replacements $M(t)$ in the interval $(0, t]$ has the Laplace transform:

$$L_M(s) = \frac{L_w(s)\, L_g(s)}{s[1 - L_w(s)\, L_g(s)]}. \tag{6.17}$$

Now we define the instantaneous availability $A(t)$ of a component (or a system) as the probability that the component is properly functioning at time t. Note that in the absence of a repair or a replacement, availability $A(t)$ is simply equal to the reliability $R(t) = 1 - W(t)$ of the component. The component may be functioning at time t by reason of two mutually exclusive cases: either the component has not failed from the beginning (no renewals in the period $(0, t]$) with the associated probability $R(t)$, or the last renewal (repair) occurred at time x, $0 < x < t$, and the component has continued to function since that time. The probability associated with the second case is:

$$\int_0^t R(t - x)\, m(x)\, dx.$$

Thus:

$$A(t) = R(t) + \int_0^t R(t-x)\, m(x)\, dx. \tag{6.18}$$

Note that the instantaneous availability is always greater than or equal to the reliability. Taking Laplace transforms on both sides of the above equation, we get:

$$\begin{aligned} L_A(s) &= L_R(s) + L_R(s) L_m(s) \\ &= L_R(s)[1 + L_m(s)] \\ &= L_R(s)[1 + \frac{L_w(s) L_g(s)}{1 - L_w(s) L_g(s)}] \\ &= \frac{L_R(s)}{1 - L_w(s) L_g(s)}, \end{aligned}$$

using equation (6.17). Now, since $R(t) = 1 - W(t)$:

$$\begin{aligned} L_R(s) &= \frac{1}{s} - L_W(s) \\ &= \frac{1}{s} - \frac{L_w(s)}{s} \\ &= \frac{1 - L_w(s)}{s}. \end{aligned}$$

Substituting, we get:

$$L_A(s) = \frac{1 - L_w(s)}{s[1 - L_w(s)\, L_g(s)]}. \tag{6.19}$$

If we are given the failure-time and repair-time distributions, the above equation enables us to compute the instantaneous availability $A(t)$ as a function of time.

Often we are interested in the state of the system after a sufficiently long time has elapsed. For this purpose, we define the limiting availability (or simply availability) A as the limiting value of $A(t)$ as t approaches infinity. Here we point out another distinction between the notions of reliability and availability. The "limiting reliability" is given by:

$$\lim_{t \to \infty} R(t) = \lim_{t \to \infty} 1 - W(t) = 0,$$

whereas the limiting availability $\lim_{t \to \infty} A(t)$ is usually nonzero.

In order to derive an expression for the limiting availability, we make use of the following result, known as the final-value theorem of Laplace transforms:

Let $H(t) = \int_0^t h(x)dx + H(0^-)$. Then, using a table of Laplace transforms (see Appendix D), we get:

$$sL_H(s) - H(0^-) = L_h(s) = \int_0^\infty e^{-st}h(t)\, dt$$

and hence:

$$\lim_{s \to 0} sL_H(s) = \int_0^\infty h(t)\, dt + H(0^-)$$

$$= \lim_{t \to \infty} [\int_0^t h(x)\, dx] + H(0^-) = \lim_{t \to \infty} H(t).$$

It follows that the limiting availability A is given by:

$$A = \lim_{t \to \infty} A(t)$$

$$= \lim_{s \to 0} sL_A(s).$$

Now for small values of s, the following approximations can be used [APOS 1974]:

$$e^{-st} \simeq 1 - st,$$

so:

$$L_w(s) = \int_0^\infty e^{-st}w(t)dt$$

$$\simeq \int_0^\infty w(t)\, dt - s\int_0^\infty tw(t)\, dt$$

$$\simeq 1 - \frac{s}{\lambda},$$

where $1/\lambda$ is the mean time to failure (MTTF). Also:

$$L_g(s) \simeq 1 - \frac{s}{\mu},$$

where $1/\mu$ is the mean time to repair (MTTR). Then the limiting availability is:

$$A = \lim_{s \to 0} \left[\frac{1 - \left(1 - \frac{s}{\lambda}\right)}{1 - \left(1 - \frac{s}{\lambda}\right)\left(1 - \frac{s}{\mu}\right)} \right]$$

$$= \frac{\dfrac{1}{\lambda}}{\dfrac{1}{\lambda} + \dfrac{1}{\mu}}$$

$$= \frac{\text{MTTF}}{\text{MTTF} + \text{MTTR}}.$$

This shows that the limiting availability depends only on the mean time to failure and mean time to repair, and not on the nature of the distributions of failure times and repair times.

Example 6.11

Assume exponential failure and repair distributions. Then:

$$w(t) = \lambda e^{-\lambda t},$$

$$g(t) = \mu e^{-\mu t},$$

$$L_w(s) = \frac{\lambda}{s + \lambda},$$

$$L_g(s) = \frac{\mu}{s + \mu},$$

and from (6.16) we have:

$$L_m(s) = \frac{L_w(s)\, L_g(s)}{1 - L_w(s)\, L_g(s)}$$

$$= \frac{\lambda \mu}{s\,[s + (\lambda + \mu)]}.$$

This can be rewritten as:

$$L_m(s) = \frac{\lambda\mu}{(\lambda + \mu)s} - \frac{\lambda\mu}{(\lambda + \mu)^2}\,\frac{\lambda + \mu}{s + \lambda + \mu}.$$

Inverting:

$$m(t) = \frac{\lambda\mu}{\lambda + \mu} - \frac{\lambda\mu}{\lambda + \mu}\, e^{-(\lambda + \mu)t}, \qquad t \geqslant 0.$$

Thus the limiting rate of repairs is given by:

$$\lim_{t \to \infty} m(t) = \frac{\lambda\mu}{\lambda + \mu}$$

$$= \frac{1}{\text{MTTF} + \text{MTTR}}.$$

Now from equation (6.19) we get:

$$L_A(s) = \frac{1 - \dfrac{\lambda}{(s+\lambda)}}{s\left[1 - \dfrac{\lambda\mu}{(s+\lambda)(s+\mu)}\right]}$$

$$= \frac{s+\mu}{s[s+(\lambda+\mu)]}.$$

The last expression can be rewritten as:

$$L_A(s) = \frac{\dfrac{\mu}{(\lambda+\mu)}}{s} + \frac{\dfrac{\lambda}{(\lambda+\mu)}}{s+(\lambda+\mu)}.$$

Inverting, we get:

$$A(t) = \frac{\mu}{\lambda+\mu} + \frac{\lambda}{\lambda+\mu} e^{-(\lambda+\mu)t},$$

and the limiting availability:

$$\lim_{t\to\infty} A(t) = \frac{\mu}{\lambda+\mu},$$

exactly the value obtained by the earlier analysis based on a small value of s. $A(t)$ and $R(t) = e^{-\lambda t}$ are plotted as functions of time in Figure 6.14. Note that the availability $A(t)$ approaches the reliability $R(t)$ as μ approaches zero; thus in a nonmaintained system the notions of reliability and instantaneous availability are synonymous. #

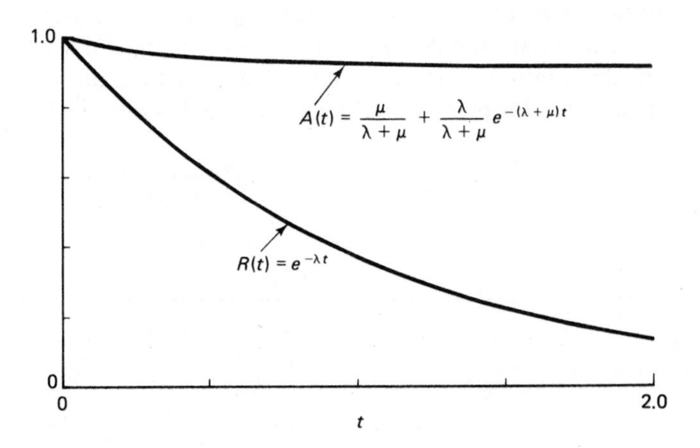

Figure 6.14 Reliability and instantaneous availability as functions of time ($\lambda = 0.1$ and $\mu = 1.0$)

Problems

1. You are given a system with s components. The mean time between failures for each component is 100 hours and the mean time to repair is five hours, and each component has its own repair facility. Find the limiting availability of the system when:
   (a) All s components are required for the system to function.
   (b) At least one of the s components should function for the system to function correctly.

*2. Following the development of Section 6.6, derive an expression for $A(t)$, assuming:
   (a) $T_i \sim$ EXP (λ) but the repair times D_i's are constant at $1/\mu$.
   (b) T_i's are two-stage Erlang and D_i EXP (μ).

6.7 RANDOM INCIDENCE

We have noted that the first-order interevent times of a renewal counting process are independent identically distributed random variables with the density $f_X(x)$. Now consider the experiment of **random incidence** where we pick a random time instant and wait until the occurrence of the next event. Let the random variable Y denote the waiting time until the next event following random entry (see Figure 6.15). Y is often called the **residual lifetime** or the **forward recurrence time.** Let T be the time of the random entry measured from the last event. The random variable T is known as the **backward recurrence time.** For the special case of a Poisson process, X is exponentially distributed and the memoryless property implies that Y is also exponentially distributed. We are interested in deriving the density function $f_Y(y)$ of the residual lifetime for the general renewal counting process.

We proceed to obtain the density $f_Y(y)$ in two steps. First we compute the density of the random variable W, denoting the length of the interevent time into which we enter by random incidence. Having obtained $f_W(w)$, we

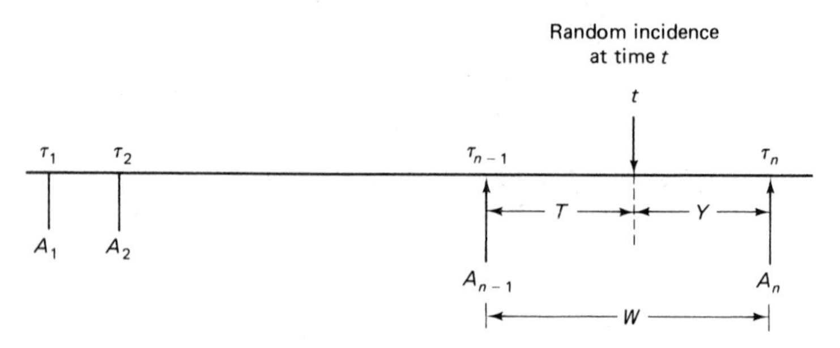

Figure 6.15 Random incidence

next compute the conditional density $f_{Y|W}$. From these two densities we can obtain the joint density $f_{W,Y}$ and subsequently the marginal (or the unconditional) density $f_Y(y)$.

Since both W and X are first-order interevent times, we may be tempted to conclude that they have the same distributions. In fact, W and X do not have identical distributions, since the experiments on which they are defined are different. The interevent interval W in which our random entry occurs is not a typical interval. A long interval is more likely to be "intercepted" than a short one. We assume that the probability that our random entry falls in an interevent gap of length w is proportional to the gap length w, and to the relative occurrence of such intervals [which is given by $f_X(w)\,dw$]. Then:

$$f_W(w)\,dw = \frac{w\,f_X(w)\,dw}{E[X]},$$

where the denominator provides the normalization factor. Thus the density of W is given by:

$$f_W(w) = \frac{w\,f_X(w)}{E[X]} = \frac{w\,f_X(w)}{\int_0^\infty x\,f_X(x)\,dx}. \tag{6.20}$$

Example 6.12

Assume X has an exponential density:

$$f_X(x) = \lambda e^{-\lambda x}, \qquad x > 0.$$

Then, using equation (6.20),

$$f_W(w) = \frac{w\lambda\, e^{-\lambda w}}{\frac{1}{\lambda}} = \lambda^2 w\, e^{-\lambda w}, \qquad w > 0.$$

Thus W has the two-stage Erlang distribution. The expected value of W is $E[W] = 2/\lambda$, which is twice as large as $E[X]$. This confirms the assumption that the larger values of w are more likely to be "intercepted" by the random entry. We could have obtained this result from the memoryless property of the exponential distribution by observing that both the forward and the backward recurrence times have exponential distributions with parameter λ, and W is the sum of these two independent random variables. #

Next we proceed to determine the conditional density of the residual lifetime. Assume that an interevent gap of duration w is intercepted by the random entry. Since a randomly chosen point is selected in this interval, it must be uniformly distributed. Thus:

$$f_{Y|W}(y|w) = \begin{cases} \dfrac{1}{w}, & 0 < y \leqslant w, \\ 0, & \text{otherwise.} \end{cases}$$

Now using formula (6.20), we get the joint density:

$$\begin{aligned} f_{W,Y}(w,y) &= f_W(w)\, f_{Y|W}(y|w) \\ &= \frac{w\, f_X(w)}{w\, E[X]} \\ &= \frac{f_X(w)}{E[X]}, \qquad 0 < y \leqslant w < \infty. \end{aligned}$$

Remembering that $y \leqslant w$, and integrating with respect to w, we get:

$$\begin{aligned} f_Y(y) &= \int_{w=y}^{\infty} \frac{f_X(w)\, dw}{E[X]} \\ &= \frac{1 - F_X(y)}{E[X]} \\ &= \frac{R_X(y)}{E[X]}. \end{aligned} \tag{6.21}$$

For the special case of the Poisson process:

$$f_X(y) = \lambda e^{-\lambda y},$$

and from equation (6.21):

$$\begin{aligned} f_Y(y) &= \frac{e^{-\lambda y}}{\dfrac{1}{\lambda}} \\ &= \lambda e^{-\lambda y} \\ &= f_X(y), \end{aligned}$$

confirming our earlier derivation of the memoryless property!

Using a very similar procedure, we can show that the distribution of the backward recurrence time T is identical to the distribution of the forward recurrence time Y.

Our discussion has been directed to continuous-parameter renewal processes. Corresponding results for discrete-parameter renewal processes can be easily derived. For example, the discrete analog of equation (6.20) is the pmf:

$$p_W(w) = \frac{w\, p_X(w)}{E[X]}. \tag{6.22}$$

The discrete analog of equation (6.21) is:

$$p_Y(y) = \frac{1 - F_X(y)}{E[X]}. \tag{6.23}$$

Problems

*1. [SEVC 1977] We have shown that if two independent Poisson streams are merged, we still get a Poisson stream. Now consider two independent renewal counting processes $N_1(t)$ and $N_2(t)$ being merged into the process $N(t)$. Let the underlying distribution functions be F_{X_1}, F_{X_2} and F_X. First show that, conditioned on the fact that the last event in the output stream was contributed by the first stream, the time to next event in the output stream is the minimum of X_1 and $Y_2 = Y(X_2)$ where $Y(X_2)$ is the residual time to next event in input stream 2. Next show that

$$F_X(t) = \frac{E[X_2]\, F_{Z_1}(t) + E[X_1]\, F_{Z_2}(t)}{E[X_1] + E[X_2]}$$

where:

$$F_{Z_1}(t) = F_{X_1}(t) + \frac{1 - F_{X_1}(t)}{E[X_2]} \int_0^t [1 - F_{X_2}(x)]\, dx$$

and

$$F_{Z_2}(t) = F_{X_2}(t) + \frac{1 - F_{X_2}(t)}{E[X_1]} \int_0^t [1 - F_{X_1}(x)]\, dx.$$

2. Using the results of problem 1 above, verify that the result of merging two independent Poisson processes produces a Poisson process.

3. Derive equations (6.22) and (6.23).

6.8 RENEWAL MODEL OF PROGRAM BEHAVIOR

We are interested in modeling the memory-referencing behavior of a program executing in a paged virtual memory system. Let $N = \{1, 2, \ldots, n\}$ denote the logical address space of an n-page program. For the present model, the dynamic behavior of the program is captured by the reference string w, which is a sequence:

$$w = r_1\, r_2 \cdots r_t \cdots,$$

where each r_t is in N. If $r_t = i$ then a reference is made to the page indexed by i at the tth reference. Clearly, the reference string w is a discrete-parameter, discrete-state stochastic process.

It is convenient to decompose the stochastic process w into n distinct stochastic processes. For the ith stochastic process ($i = 1, 2, \ldots, n$), the event of interest is a reference to the ith page. Assume that the time intervals between references to page i are independent identically distributed random variables with distribution function F_i. Let X_{ij} denote the time between the jth and the $(j-1)$st reference to page i. Then for each $i = 1, 2, \ldots, n$, $\{X_{ij} \mid j = 1, 2, \ldots\}$ is a discrete-parameter renewal process with underlying distribution F_i. We assume that the n renewal processes are independent.

The first-order interevent times of the ith renewal process are interpreted as the interreference intervals for the ith page. $F_i(t)$ and $p_i(t)$ are the corresponding interreference distribution and interreference pmf. The mean interreference interval for page i is given by:

$$E[X_i] = \sum_x x\, p_i(x).$$

Let $m_i(t)$ denote the renewal pmf of the ith process; that is, $m_i(t)$ is the probability of a reference to page i at time t. By using the discrete analogue of the key renewal theorem, we obtain the asymptotic value of the renewal rate:

$$m_i = \frac{1}{E[X_i]}. \tag{6.24}$$

Here m_i may be interpreted as the long-term average number of references to page i per unit time. We impose the normalizing condition:

$$\sum_{i=1}^{n} m_i(t) = 1, \qquad t \geqslant 0 \tag{6.25}$$

which assures that one reference (to some page) occurs at every time instant t.

Virtual memory systems usually retain only a portion of a program's logical address space in main memory. For each instant of time, the subset of the address space to retain in main memory is determined by the **paging algorithm**. A popular paging algorithm is the working-set (WS) algorithm, which we shall analyze [COFF 1973].

A program's working set $W(t, \tau)$ at time t is defined to be the set of distinct pages referenced in the time interval $(t - \tau, t]$. For $t < \tau$, $W(t, \tau)$ contains the distinct pages among $r_1, r_2, \ldots, r_t$. The parameter τ is known as the **window size**. The working-set (WS) paging algorithm assures us that a program's working set at time t is in main memory at time t. Now if the next page to be referenced is not in the working set—that is, $r_{t+1} \notin W(t, \tau)$—then a **page fault** is said to have occurred. Following a page fault, the required page will be loaded by the operating system. Let $w(t, \tau)$ be the size of the working set $W(t, \tau)$. Important measures of the WS algorithm are the asymptotic average working-set size $s(\tau)$:

$$s(\tau) = \lim_{t \to \infty} E[w(t, \tau)]$$

and the asymptotic average page-fault rate $q(\tau)$.

To compute the average working-set size $s(\tau)$, consider a random incidence in an interreference interval of page i. Let T_i be the backward recurrence time. Then, using (the backward analogue of) equation (6.23), we obtain the pmf of T_i as:

$$p_{T_i}(j) = \frac{1 - F_i(j)}{E[X_i]}. \tag{6.26}$$

Given a window size τ, the probability that page i is in memory (that is, in the working set) at the time of random entry is given by:

$$\begin{aligned} P(T_i < \tau) &= \sum_{j=0}^{\tau-1} p_{T_i}(j) \\ &= \sum_{j=0}^{\tau-1} \frac{1 - F_i(j)}{E[X_i]}. \end{aligned} \tag{6.27}$$

Define the Bernoulli random variable Y_{it} = "page i is in memory at time t". Then the expected value of Y_{it} is given by $E[Y_{it}] = P(Y_{it} = 1)$ which, in the limit, equals $P(T_i < \tau)$. Now the average working-set size is:

$$\begin{aligned} s(\tau) &= \sum_{i=1}^{n} P(T_i < \tau) \\ &= \sum_{i=1}^{n} \sum_{j=0}^{\tau-1} \frac{1 - F_i(j)}{E[X_i]}. \end{aligned} \tag{6.28}$$

To compute the average page-fault rate $q(\tau)$, we first compute the conditional probability of a page fault given that the ith page is referenced at time t. The required probability is $1 - F_i(\tau)$, since this is the probability that the interreference interval of the page exceeds the window size; that is, the page has not been referenced during the last τ references. Thus:

$$P(\text{"page fault at time } t\text{"} \mid r_t = i) = 1 - F_i(\tau).$$

Using the theorem of total probability:

$$\begin{aligned} P(\text{"page fault at time } t\text{"}) &= \sum_{i=1}^{n} [1 - F_i(\tau)]\, P(r_t = i) \\ &= \sum_{i=1}^{n} [1 - F_i(\tau)]\, m_i(t). \end{aligned}$$

Taking the limit as t approaches infinity, the left-hand side is the asymptotic average page-fault rate $q(\tau)$, and, using equation (6.24), we get:

$$q(\tau) = \sum_{i=1}^{n} m_i[1 - F_i(\tau)] = \sum_{i=1}^{n} \frac{1 - F_i(\tau)}{E[X_i]}. \tag{6.29}$$

References

[APOS 1974]. G. APOSTOLAKIS, "Mathematical Methods of Probabilistic Safety Analysis," Technical Report, School of Engineering and Applied Science, University of California at Los Angeles.

[BARL 1975] R. E. BARLOW and F. PROSCHAN, *Statistical Theory of Reliability and Life Testing: Probability Models,* Holt, Rinehart and Winston, New York.

[BHAT 1972] U. N. BHAT, *Elements of Applied Stochastic Processes,* John Wiley & Sons, New York.

[COFF 1973] E. G. COFFMAN, Jr., and P. J. DENNING, *Operating System Theory,* Prentice-Hall, Englewood Cliffs, N.J.

[KHIN 1960] A. Y. KHINTCHINE, *Mathematical Methods in Queuing Theory,* Griffen, London.

[MADN 1974] S. E. MADNICK and J. J. DONOVAN, *Operating Systems,* McGraw-Hill, New York.

[PALM 1943] C. PALM, "Intensitätsschwankungen im Fernsprechverkehr," *Ericsson Technics,* 44:3, 189.

[ROSS 1970] S. ROSS, *Applied Probability Models with Optimization Applications,* Holden-Day, San Francisco.

[SEVC 1977] K. C. SEVCIK, A. LEVY, S. TRIPATHI, and J. ZAHORJAN, "Improving Approximations of Aggregated Queuing Network Subsystems," in Reiser and Chandy, eds., *Computer Performance,* North-Holland, Amsterdam.

[STAR 1974] H. STARK and F. B. TUTEUR, *Modern Electrical Communications,* Prentice-Hall, Englewood Cliffs, N. J.

Chapter 7

Discrete-Parameter Markov Chains

7.1 INTRODUCTION

A **Markov process** is a stochastic process whose dynamic behavior is such that probability distributions for its future development depend only on the present state and not on how the process arrived in that state. If we assume that the state space, I, is discrete (finite or countably infinite), then the Markov process is known as a **Markov chain.** If we further assume that the parameter space, T, is also discrete, then we have a **discrete-parameter Markov chain**. Such processes are the subject of this chapter. Since the parameter space is discrete, we will let $T = \{0, 1, 2, \ldots\}$ without loss of generality.

We choose to observe the state of a system at a discrete set of times. The successive observations define the random variables $X_0, X_1, X_2, \ldots, X_n, \ldots$, at time steps $0, 1, 2, \ldots$, respectively. If $X_n = j$, then the state of the system at time step n is j. X_0 is the initial state of the system. The Markov property can then be succinctly stated as:

$$P(X_n = i_n | X_0 = i_0, X_1 = i_1, \ldots, X_{n-1} = i_{n-1}) = P(X_n = i_n | X_{n-1} = i_{n-1}). \tag{7.1}$$

Intuitively, equation (7.1) implies that given the "present" state of the system, the "future" is independent of its "past."

We let $p_j(n)$ denote the pmf of the random variable X_n; that is:

$$p_j(n) = P(X_n = j), \tag{7.2}$$

and let the conditional pmf:

$$p_{jk}(m, n) = P(X_n = k | X_m = j), \qquad 0 \leqslant m \leqslant n \tag{7.3}$$

denote the probability that the process makes a transition from state j at step m to state k at step n. Thus, $p_{jk}(m, n)$ is known as the **transition probability function** of the Markov chain. We will only be concerned with **homogeneous Markov chains**—those in which $p_{jk}(m, n)$ depends only on the difference $n - m$ (in this case, the Markov chain is said to have stationary transition probabilities). For such chains, we use the notation

$$p_{jk}(n) = P(X_{m+n} = k | X_m = j) \tag{7.4}$$

to denote the ***n*-step transition probabilities.** In words, $p_{jk}(n)$ is the probability that a homogeneous Markov chain will move from state j to state k in exactly n steps. The one-step transition probabilities $p_{jk}(1)$ are simply written as p_{jk}, thus:

$$p_{jk} = P(X_n = k | X_{n-1} = j), \qquad n \geqslant 1. \tag{7.5}$$

It is convenient to define 0-step transition probabilities by:

$$p_{jk}(0) = \begin{cases} 1, & \text{if } j = k, \\ 0, & \text{otherwise.} \end{cases}$$

Since equation (7.1) holds for all values of n, we can use the generalized multiplication rule (GMR of Chapter 1) to obtain the joint probability:

$$\begin{aligned} P(X_0 = i_0,\ X_1 = i_1,\ \ldots,\ X_n = i_n) & \\ &= P(X_0 = i_0,\ X_1 = i_1,\ \ldots,\ X_{n-1} = i_{n-1}) \\ &\quad \cdot P(X_n = i_n | X_0 = i_0,\ \ldots,\ X_{n-1} = i_{n-1}) \\ &= P(X_0 = i_0,\ X_1 = i_1,\ \ldots,\ X_{n-1} = i_{n-1}) \\ &\quad \cdot P(X_n = i_n | X_{n-1} = i_{n-1}) \\ &= P(X_0 = i_0,\ X_1 = i_1,\ \ldots,\ X_{n-1} = i_{n-1})\, p_{i_{n-1}, i_n} \\ &\ \ \vdots \\ &= p_{i_0}(0) p_{i_0, i_1} \cdots p_{i_{n-1}, i_n}. \end{aligned} \tag{7.6}$$

This implies that all joint probabilities of interest are determined from the initial pmf $p_{i_0}(0) = P(X_0 = i_0)$, and the one-step transition probabilities p_{ij}.

The pmf of the random variable X_0, often called the **initial distribution,** is specified by the **probability vector:**

$$\mathbf{p}(0) = [p_0(0),\ p_1(0),\ \ldots,\ \ldots].$$

The one-step transition probabilities are compactly specified in the form of a **transition probability matrix**:

$$P = [p_{ij}] = \begin{bmatrix} p_{00} & p_{01} & p_{02} & \cdot & \cdot \\ p_{10} & p_{11} & p_{12} & \cdot & \cdot \\ \cdot & \cdot & \cdot & \cdot & \cdot \\ \cdot & \cdot & \cdot & \cdot & \cdot \\ \cdot & \cdot & \cdot & \cdot & \cdot \end{bmatrix}.$$

The entries of the matrix P satisfy the following two properties:

$$0 \leqslant p_{ij} \leqslant 1 \quad \text{and} \quad \sum_j p_{ij} = 1.$$

Any such square matrix that has nonnegative entries with row sums all equal to unity is called a **stochastic matrix.**

An equivalent description of the one-step transition probabilities can be given by a directed graph called the **state-transition diagram** (state diagram, for short) of the Markov chain. A node labeled i of the state diagram represents state i of the Markov chain and a branch labeled p_{ij} from node i to j implies that the conditional probability (or the one-step transition probability) is:

$$P(X_n = j \mid X_{n-1} = i) = p_{ij}.$$

Example 7.1

We observe the state of a system (or a component) at discrete points in time. We say that the system is in state 0 if it is operating properly. If the system is undergoing repair (following a breakdown), then the system state is denoted by state 1. If we assume that the system possesses the Markov property, then we have a two-state discrete-parameter Markov chain. Further assuming that the Markov chain is homogeneous, we could specify its transition probability matrix by

$$P = \begin{bmatrix} 1-a & a \\ b & 1-b \end{bmatrix}, \qquad 0 \leqslant a,\ b \leqslant 1.$$

The actual values of the entries will have to be estimated from the measurements made on the system using statistical techniques (see Chapter 10). #

Example 7.2

Another example of a two-state Markov chain is provided by a communication net consisting of a sequence (or a cascade) of stages of binary communication channels. Here X_n denotes the digit leaving the nth stage of the system and X_0 denotes the digit entering the first stage. Assume that the binary communication channels are stochastically identical. The transition probability matrix of the corresponding Markov chain of the communication net can be read off from the channel diagram (see Figure 7.1). #

7.2 COMPUTATION OF *n*-STEP TRANSITION PROBABILITIES

We are interested in obtaining an expression for evaluating the n-step transition probability $p_{ij}(n)$ from the one-step transition probabilities $p_{ij}(1) = p_{ij}$. Recall that:

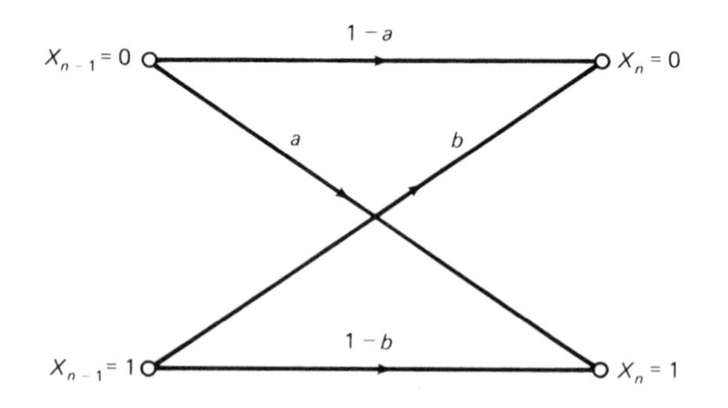

Figure 7.1 A channel diagram

$$p_{ij}(n) = P(X_{m+n} = j \mid X_m = i).$$

Now the probability that the process goes to state k at the (m^{th}) step, given that $X_0 = i$, is $p_{ik}(m)$; and the probability that the process reaches state j at step $(m + n)$, given that $X_m = k$, is given by $p_{kj}(n)$. The Markov property implies that these two events are independent. Now, using the theorem of total probability [or the fact that the chain must be in some state at step m], we get:

$$p_{ij}(m + n) = \sum_k p_{ik}(m)\, p_{kj}(n). \tag{7.7}$$

The above equation is one form of the well-known **Chapman-Kolmogorov equation,** which provides an efficient means of calculating the n-step transition probabilities. This equation need *not* apply to the more general stochastic processes discussed in Chapter 6.

If we let $P(n)$ be the matrix whose (i, j) entry is $p_{ij}(n)$—that is, $P(n)$ is the matrix of n-step transition probabilities—then equation (7.7) can be written in matrix form (with $m = 1$ and n replaced by $n - 1$):

$$P(n) = P \cdot P(n - 1) = P^n. \tag{7.8}$$

Thus the matrix of n-step transition probabilities is obtained by multiplying the matrix of one-step transition probabilities by itself $n - 1$ times. In other words, the problem of finding the n-step transition probabilities is reduced to one of forming powers of a given matrix. It should be clear that the matrix $P(n)$ consists of probabilities and its row sums are equal to unity, so it is a stochastic matrix (see problem 4 at the end of this section).

We can obtain the (marginal) pmf of the random variable X_n from the n-step transition probabilities and the initial distribution as follows:

$$p_j(n) = P(X_n = j) = \sum_i P(X_0 = i)P(X_n = j \mid X_0 = i) \tag{7.9}$$

$$= \sum_i p_i(0)\, p_{ij}(n).$$

If the pmf of X_n (the state of the system at time n) is expressed as the row vector:

$$\mathbf{p}(n) = [p_0(n),\ p_1(n),\ \ldots,\ p_j(n),\ \ldots],$$

then from (7.9) we get:

$$\mathbf{p}(n) = \mathbf{p}(0)P(n),$$

and from (7.8) we have:

$$\mathbf{p}(n) = \mathbf{p}(0)P^n. \tag{7.10}$$

This implies that the probability distributions of a homogeneous Markov chain are completely determined from the one-step transition probability matrix P and the initial probability vector $\mathbf{p}(0)$.

If the state space I of a Markov chain $\{X_n\}$ is finite, then computing P^n is relatively straightforward, and an expression for the pmf of X_n (for $n \geqslant 0$) can be obtained using equation (7.10). For Markov chains with a countably infinite state space, computation of P^n poses problems. Therefore, alternative methods for determining the asymptotic behavior (i.e., as n approaches infinity) of P^n and $\mathbf{p}(n)$ have been developed (see Sections 7.3, 7.6, and 7.7).

To illustrate, we will compute P^n for the two-state Markov chain of Examples 7.1 and 7.2, with the transition probability matrix P given by:

$$P = \begin{bmatrix} 1-a & a \\ b & 1-b \end{bmatrix}, \qquad 0 \leqslant a,\ b \leqslant 1.$$

A graphical description of the Markov chain is provided by its state diagram shown in Figure 7.2.

The following theorem gives an explicit expression for P^n and hence for $\mathbf{p}(n)$. We will impose the condition $|1 - a - b| < 1$ on the one-step transition probabilities. Since a and b are probabilities, this last condition can be violated only if $a = b = 0$ or $a = b = 1$. These two cases will be treated separately.

THEOREM 7.1

Given a two state Markov chain with the transition probability matrix

$$P = \begin{bmatrix} 1-a & a \\ b & 1-b \end{bmatrix}, \quad 0 \leqslant a,\ b \leqslant 1,\ |1 - a - b| < 1, \tag{7.11}$$

the n-step transition probability matrix $P(n) = P^n$ is given by:

$$P(n) = \begin{bmatrix} \dfrac{b + a(1-a-b)^n}{a+b} & \dfrac{a - a(1-a-b)^n}{a+b} \\[2ex] \dfrac{b - b(1-a-b)^n}{a+b} & \dfrac{a + b(1-a-b)^n}{a+b} \end{bmatrix}.$$

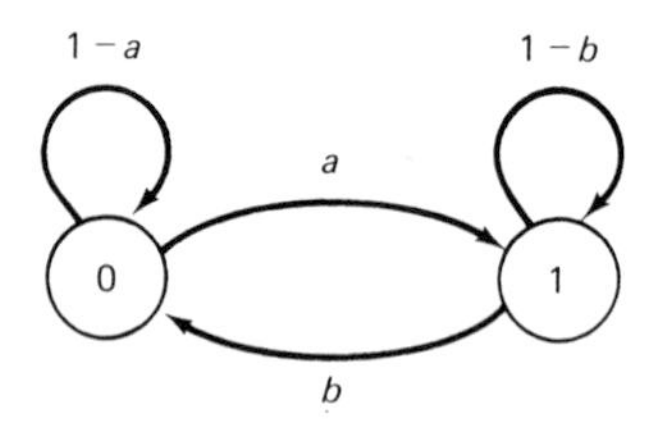

Figure 7.2 The state diagram of a two-state Markov chain

Proof:

It is common to give a proof by induction, but we prefer a constructive proof here [BHAT 1972]. Note that:

$$p_{00}(1) = p_{00} = 1 - a, \qquad p_{01}(1) = p_{01} = a,$$

$$p_{10}(1) = p_{10} = b, \qquad p_{11}(1) = p_{11} = 1 - b.$$

Using equation (7.7), we get:

$$\begin{aligned} p_{00}(1) &= 1 - a, \\ p_{00}(n) &= (1 - a)p_{00}(n - 1) + bp_{01}(n - 1), \qquad n > 1. \end{aligned} \tag{7.12}$$

Now since the row sums of P^{n-1} are unity, we have:

$$p_{01}(n - 1) = 1 - p_{00}(n - 1),$$

hence (7.12) reduces to:

$$\begin{aligned} p_{00}(1) &= 1 - a, \\ p_{00}(n) &= b + (1 - a - b)p_{00}(n - 1), \qquad n > 1. \end{aligned} \tag{7.13}$$

This implies that:

$$\begin{aligned} p_{00}(n) &= b + b(1 - a - b) + b(1 - a - b)^2 + \cdots \\ &\quad + b(1 - a - b)^{n-2} + (1 - a)(1 - a - b)^{n-1} \\ &= b[\sum_{k=0}^{n-2} (1 - a - b)^k] + (1 - a)(1 - a - b)^{n-1}. \end{aligned}$$

By the formula for the sum of a finite geometric series:

$$\sum_{k=0}^{n-2} (1 - a - b)^k = \frac{1 - (1 - a - b)^{n-1}}{1 - (1 - a - b)} = \frac{1 - (1 - a - b)^{n-1}}{a + b}.$$

Thus we get:

$$p_{00}(n) = \frac{b}{a + b} + \frac{a(1 - a - b)^n}{a + b}.$$

Now $p_{01}(n)$ can be obtained by subtracting $p_{00}(n)$ from unity. Expressions for the two remaining entries can be derived in a similar way. Students familiar with determinants, however, can use the following simpler derivation.

Since $\det(P) = 1 - a - b$, $\det(P^n) = (1 - a - b)^n$ but since P^n is a stochastic matrix:

$$\det(P^n) = p_{00}(n) - p_{10}(n),$$

so:

$$p_{10}(n) = p_{00}(n) - (1 - a - b)^n.$$

Finally:

$$p_{11}(n) = 1 - p_{10}(n).$$

Example 7.3

Consider a cascade of binary communication channels as in Example 7.2. Assume that $a = 1/4$ and $b = 1/2$. Then, since $|1 - a - b| = 1/4 < 1$, Theorem 7.1 applies, and:

$$P(n) = P^n = \begin{bmatrix} 2/3 + 1/3(1/4)^n & 1/3 - 1/3(1/4)^n \\ 2/3 - 2/3(1/4)^n & 1/3 + 2/3(1/4)^n \end{bmatrix}, \qquad n \geqslant 0.$$

Since:

$$P(X_2 = 1 | X_0 = 1) = p_{11}(2) = 3/8$$

and

$$P(X_3 = 1 | X_0 = 1) = p_{11}(3) = 11/32,$$

a digit entering the system as a 1 ($X_0 = 1$) has probability $3/8$ of being correctly transmitted over two stages and probability $11/32$ of being correctly transmitted over three stages.

Assuming the initial distribution, $P(X_0 = 0) = 1/3$ and $P(X_0 = 1) = 2/3$ [i.e., $\mathbf{p}(0) = (1/3, 2/3)$] we get:

$$\mathbf{p}(n) = \mathbf{p}(0)P^n = [2/3 - 1/3(1/4)^n,\ 1/3 + 1/3(1/4)^n].$$

It is interesting to observe that the two rows of P^n match in their corresponding elements in the limit $n \to \infty$, and that $\mathbf{p}(n)$ approaches $(2/3, 1/3)$ as n approaches infinity. In other words, the pmf of X_n becomes independent of n for large values of n. Furthermore, we can verify that with any other initial distribution, the same limiting distribution of $\mathbf{p}(n)$ is obtained. This important property of *some* Markov chains will be studied in the next section. #

Example 7.4

Now we consider a cascade of error-free binary communication channels; that is, $a = b = 0$. Clearly, $|1 - a - b| = 1$, and therefore Theorem 7.1 does not apply. The transition probability matrix P is the identity matrix:

$$P = \begin{bmatrix} 1 & 0 \\ 0 & 1 \end{bmatrix}.$$

The state diagram is shown in Figure 7.3. The two states do not communicate with each other. P^n is easily seen to be the identity matrix. In other words, the chain never changes state, and a transmitted digit is correctly received after an arbitrary number (n) of stages. #

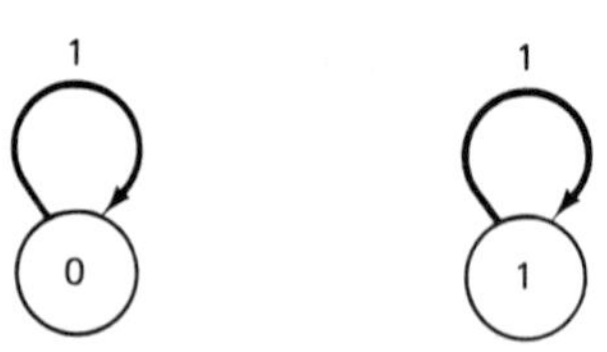

Figure 7.3 The state diagram of the two-state Markov chain of Example 7.4

Example 7.5

Now we consider a cascade of binary channels that are so noisy that the digit transmitted is always complemented. In other words, $a = b = 1$. Once again, Theorem 7.1 does not apply. The matrix P is given by:

$$P = \begin{bmatrix} 0 & 1 \\ 1 & 0 \end{bmatrix},$$

and the state diagram is given in Figure 7.4. It can be verified by induction that:

$$P^n = \begin{cases} \begin{bmatrix} 1 & 0 \\ 0 & 1 \end{bmatrix}, & \text{if } n \text{ is even,} \\ \begin{bmatrix} 0 & 1 \\ 1 & 0 \end{bmatrix}, & \text{if } n \text{ is odd.} \end{cases}$$

This Markov chain has an interesting behavior. Starting in state 0 (or 1), we return to state 0 (state 1) after an even number of steps. Therefore, the time between visits to a given state exhibits a periodic behavior. Such a chain is called a **periodic Markov chain** (with period 2). (Formal definitions are given in the next section.) #

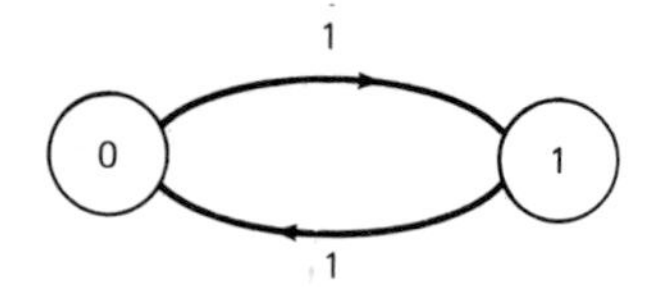

Figure 7.4 The state diagram of a two-state periodic Markov chain

Problems

1. For a cascade of binary communication channels, let $P(X_0 = 1) = \alpha$ and $P(X_0 = 0) = 1 - \alpha$, $\alpha \geqslant 0$, and assume that $a = b$. Compute the probability that a one was transmitted, given that a one was received after the nth stage; that is, compute:

$$P(X_0 = 1 \mid X_n = 1).$$

2. [CLAR 1970] Modify the system of Example 7.1 so that the operating state 0 is split into two states: (a) running and (b) idle. We observe the system only when it changes state. Define X_n as the state of the system after the nth state change, so that:

$$X_n = \begin{cases} 0, & \text{if system is running,} \\ 1, & \text{if system is under repair,} \\ 2, & \text{if system is idle.} \end{cases}$$

Assume that the matrix P is:

$$P = \begin{bmatrix} 0 & \frac{1}{2} & \frac{1}{2} \\ 1 & 0 & 0 \\ 1 & 0 & 0 \end{bmatrix}.$$

Draw the state diagram and compute the matrix P^n.

*3. Define the **vector** z-transform (generating function):

$$G_{\mathbf{p}}(z) = \sum_{n=0}^{\infty} \mathbf{p}(n) z^n.$$

Show that:

(a) $G_{\mathbf{p}}(z) = \mathbf{p}(0)[I - zP]^{-1}$, where I is the identity matrix. Thus P^n is the coefficient of z^n in the matrix power-series expansion of $(I - zP)^{-1}$.

(b) Using the above result, give an alternative proof of Theorem 7.1.

4. Using equation (7.7) and the principle of mathematical induction, show that:

$$\sum_j p_{ij}(n) = 1 \qquad \text{for all } i.$$

7.3 STATE CLASSIFICATION AND LIMITING DISTRIBUTIONS

We observed an interesting property of the two-state Markov chain of Example 7.3, in the last section. As $n \to \infty$, the n-step transition probabilities $p_{ij}(n)$ become independent of both n and i. In other words, all rows of matrix P^n converge toward a common limit (as vectors; that is, matching in corresponding elements). Now, using the definition of $p_{ij}(n)$ and the theorem of total probability, we have:

$$p_j(n) = P(X_n = j) = \sum_i p_i(0) p_{ij}(n),$$

and, since $p_{ij}(n)$ depends neither on n nor on i in the limit, we conclude that $p_j(n)$ approaches a constant as $n \to \infty$. This constant is independent of the initial distribution. We denote the **limiting state probabilities** by:

$$v_j = \lim_{n \to \infty} p_j(n), \qquad j = 0, 1, \ldots.$$

Many (but not all) Markov chains exhibit such a behavior. In order to pursue this topic, we need to classify the states of a Markov chain into those that the system visits infinitely often and those that it visits only a finite number of times. To study the long-run behavior, we need only concentrate on the former type.

Definition (Transient State). A state i is said to be **transient** (or **nonrecurrent**) if and only if there is a positive probability that the process will not return to this state.

For example, if we model a program as a Markov chain, then all but the final state will be transient states. Otherwise, the program has an infinite loop. In general, for a finite Markov chain, we expect that after a sufficient number of steps the probability that the chain is in any transient state approaches zero independent of the initial state.

Let X_{ji} be the number of visits to the state i, starting at j. Then it can be shown [ASH 1970] that:

$$E[X_{ji}] = \sum_{n=0}^{\infty} p_{ji}(n).$$

It follows that if the state i is a transient state, then $\sum_{n=0}^{\infty} p_{ji}(n)$ is finite for all j, hence $p_{ji}(n)$ approaches 0 as n approaches infinity.

Definition (Recurrent State). A state i is said to be **recurrent** if and only if, starting from state i, the process eventually returns to state i with probability one.

An alternative characterization of a recurrent state is that $E[X_{ii}] = \sum_{n=0}^{\infty} p_{ii}(n)$ is infinite. It can be verified from the form of P^n that both states of the chains in Examples 7.3, 7.4, and 7.5 are recurrent.

For recurrent states, the time to reentry is important. Let $f_{ij}(n)$ be the conditional probability that the first visit to state j from state i occurs in exactly n steps. If $i = j$, then we refer to $f_{ij}(n)$ as the probability that the first return to state i occurs in exactly n steps. These probabilities are related to the transition probabilities by [PARZ 1962]:

$$p_{ij}(n) = \sum_{k=1}^{n} f_{ij}(k) p_{jj}(n-k), \qquad n \geqslant 1.$$

Let f_{ij} denote the probability of ever visiting state j, starting from state i. Then:

$$f_{ij} = \sum_{n=1}^{\infty} f_{ij}(n).$$

It follows that state i is recurrent if $f_{ii} = 1$ and transient if $f_{ii} < 1$. If $f_{ii} = 1$, define the mean recurrence time of state i by:

$$\mu_i = \sum_{n=1}^{\infty} n f_{ii}(n).$$

A recurrent state i is said to be **recurrent non-null** (or **positive recurrent**) if its mean recurrence time μ_i is finite and is said to be **recurrent null** if its mean recurrence time is infinite.

Definition. For a recurrent state i, $p_{ii}(n) > 0$ for some $n \geqslant 1$. Define the **period** of state i, denoted by d_i, as the greatest common divisor of the set of positive integers n such that $p_{ii}(n) > 0$.

Definition. A recurrent state i is said to be **aperiodic** if its period $d_i = 1$, and **periodic** if $d_i > 1$.

In Example 7.5, both states 0 and 1 are periodic with period 2. States of Examples 7.3 and 7.4 are all aperiodic.

Definition. A state i is said to be an **absorbing** state if and only if $p_{ii} = 1$.

Both states of the chain in Example 7.4 are absorbing. Once a Markov chain enters such a state, it is destined to remain there forever.

Having defined the properties of individual states, we now define an important property of a Markov chain (as a whole).

Definition (Irreducible Markov Chain). A Markov chain is said to be **irreducible** if every state can be reached from every other state in a finite number of steps. In other words, for all $i, j \in I$, there is an integer $n \geqslant 1$ such that $p_{ij}(n) > 0$.

Markov chains of Examples 7.3 and 7.5 are both irreducible. Feller [FELL 1968] has shown that all states of an irreducible Markov chain are of the same type. Thus if one state of an irreducible chain is aperiodic then so are all the states and such a Markov chain is called **aperiodic**. The Markov chain of Example 7.3 is both irreducible and aperiodic. Similarly, if one state of an irreducible chain is periodic, then all states are periodic and have the same period; if one state is transient, then so are all states; and if one state is recurrent, then so are all states.

The n-step transition probabilities $p_{ij}(n)$ of finite, irreducible, aperiodic Markov chains become independent of i and n as $n \to \infty$. Let $q_j = \lim_{n\to\infty} p_{ij}(n)$. The limiting state probability:

$$\begin{aligned} v_j = \lim_{n\to\infty} p_j(n) &= \lim_{n\to\infty} \sum_i p_i(0) p_{ij}(n) \\ &= \sum_i p_i(0) [\lim_{n\to\infty} p_{ij}(n)] \\ &= \sum_i p_i(0) q_j = q_j \sum_i p_i(0) \\ &= q_j = \lim_{n\to\infty} p_{ij}(n). \end{aligned}$$

But this implies that P^n converges to a matrix V [with identical rows $\mathbf{v} = (v_0, v_1, \ldots)$] as $n \to \infty$.

Assume that for a given Markov chain the limiting probabilities v_j exist for all states $j \in I$ (where v_j does not depend on the initial state i). Then it can be shown [ASH 1976] that $\sum_{j \in I} v_j \leqslant 1$. Furthermore, either all $v_j = 0$ (this can happen only for a chain with an infinite number of states) or $\sum_{j \in I} \nu_j = 1$. In the latter case, the numbers ν_j, $j \in I$, are said to form a **steady-state distribution.** Thus we require that the limiting probabilities exist, that they are independent of the initial state, and that they form a probability distribution. Over a long period the influence of the initial state (or the effect of "start-up" transients) has died down and the Markov chain has reached a **steady state.** The probability ν_j is sometimes interpreted as the **long-run proportion** of time the Markov chain spends in state j.

Now from the theorem of total probability, we have:

$$p_j(n) = \sum_i p_i(n-1) p_{ij}.$$

Then if we have:

$$\lim_{n \to \infty} p_j(n) = \nu_j = \lim_{n \to \infty} p_j(n-1),$$

we get:

$$\nu_j = \sum_i \nu_i p_{ij}, \qquad j = 0, 1, 2, \ldots, \tag{7.14}$$

or in matrix notation:

$$\mathbf{v} = \mathbf{v}P. \tag{7.15}$$

(In other words $\mathbf{v}$ is a left eigenvector of P associated with the eigenvalue $\lambda = 1$.) This gives us a system of linear equations in the unknowns $(\nu_0, \nu_1, \ldots)$. Since $\mathbf{v}$ is a probability vector, we also expect that:

$$\nu_j \geqslant 0, \qquad \sum_j \nu_j = 1. \tag{7.16}$$

Any vector $\mathbf{x}$ that satisfies the properties (7.15) and (7.16) is known as a **stationary probability** vector of the Markov chain.

We state the following important theorems without proof [PARZ 1962]:

THEOREM 7.2 For an aperiodic Markov chain, the limits $\nu_j = \lim_{n \to \infty} p_j(n)$ exist.

THEOREM 7.3 For any irreducible, aperiodic Markov chain, the limiting state probabilities $\nu_j = \lim_{n \to \infty} p_j(n) = \lim_{n \to \infty} p_{ij}(n)$ exist and are independent of the initial distribution $\mathbf{p}(0)$.

THEOREM 7.4 For an irreducible, aperiodic Markov chain, with all states recurrent non-null, the limiting probability vector $\mathbf{v} = (\nu_0, \nu_1, \ldots)$ is the unique stationary probability vector [satisfying equations (7.15) and (7.16) above], hence $\mathbf{v}$ is also the steady-state probability vector.

It can be shown that all states of a finite, irreducible Markov chain are recurrent non-null. Then for a finite, aperiodic, irreducible Markov chain, we can obtain the steady-state probabilities rather easily by solving a system of linear equations, since Theorem 7.4 applies. For chains with an infinite number of states, we can often solve the equations by using the method of generating functions (recall problem 3 at the end of the previous section) or by exploiting the special structure of the matrix P (see, for example, the section on birth-death processes).

Example 7.6

Returning to the periodic Markov chain of Example 7.5, we see that Theorem 7.2 does not apply. In fact, if we let the initial distribution $\mathbf{p}(0) = (p, 1 - p)$, then

$$\mathbf{p}(n) = \begin{cases} (p,\ 1 - p), & \text{if } n \text{ is even,} \\ (1 - p,\ p), & \text{if } n \text{ is odd.} \end{cases}$$

Thus $\mathbf{p}(n)$ does not have a limit. It is interesting to note that, although limiting probabilities do not exist, stationary probabilities are unique and with the use of (7.15) and (7.16) are easily computed to be $v_0 = v_1 = ½$. #

Example 7.7

Returning to the Markov chain of Example 7.4, we see that it is not irreducible (since we cannot go from one state to another) and that Theorem 7.3 does not apply. Although $\mathbf{p}(n)$ has a limit [in fact $\mathbf{p}(n) = \mathbf{p}(0)$], the limit is *dependent* upon the initial distribution $\mathbf{p}(0)$. #

Example 7.8

We consider the two-state Markov chain of the last section (Examples 7.1 and 7.2 and Theorem 7.1) with the condition $0 < a, b < 1$. This implies that $|1 - a - b| < 1$ and Theorem 7.1 applies. From this we conclude that:

$$\lim_{n \to \infty} P^n = \begin{bmatrix} \dfrac{b}{a + b} & \dfrac{a}{a + b} \\ \dfrac{b}{a + b} & \dfrac{a}{a + b} \end{bmatrix} = \begin{bmatrix} v_0 & v_1 \\ v_0 & v_1 \end{bmatrix}.$$

Thus the steady-state probability vector is:

$$\mathbf{v} = (v_0,\ v_1) = \left(\frac{b}{a + b}, \frac{a}{a + b}\right).$$

This result can also be derived using Theorem 7.4, since the chain is irreducible, finite, and aperiodic. Then, using equation (7.15), we have:

$$(v_0,\ v_1) = (v_0,\ v_1) \begin{bmatrix} 1 - a & a \\ b & 1 - b \end{bmatrix}$$

or

$$v_0 = (1 - a)v_0 + bv_1$$

and

$$v_1 = av_0 + (1 - b)v_1.$$

Rearranging:

$$av_0 - bv_1 = 0,$$

$$-av_0 + bv_1 = 0$$

Note that these two equations are linearly dependent, and thus we need one more equation [supplied by condition (7.16)]:

$$v_0 + v_1 = 1.$$

Solving, we get the stationary probability vector:

$$(v_0, \ v_1) = (\frac{b}{a + b}, \frac{a}{a + b})$$

as derived earlier. #

Example 7.9

Consider a two-state Markov chain with $a = 0$ and $b = 1$, so that:

$$P = \begin{bmatrix} 1 & 0 \\ 1 & 0 \end{bmatrix}$$

with the state diagram shown in Figure 7.5.

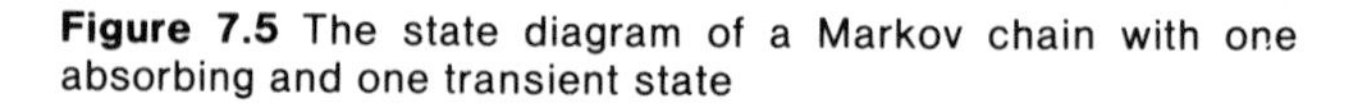

Figure 7.5 The state diagram of a Markov chain with one absorbing and one transient state

In this case the state 1 is transient and state 0 is absorbing. The chain is not irreducible, but the limiting state probabilities exist (since Theorem 7.2 applies) and are given by $v_0 = 1$ and $v_1 = 0$. This says that eventually the chain will remain in state 0 (after at most one transition). #

Example 7.10

Consider a model of a uniprogrammed computer system with m I/O devices and a CPU. For the program currently under execution, the system will be in one of the $m + 1$ states denoted by $0, 1, \ldots, m$, so that in state 0 the program is executing on the CPU, and in state i $(1 \leqslant i \leqslant m)$ the program is performing an I/O operation on device i. Assume that the request for device i occurs at the end of a CPU burst with probability q_i, independent of the past history of the program. The program will finish execution at the end of a CPU burst with probability q_0 so that $\sum_{i=0}^{m} q_i = 1$. We assume that the system is saturated so that upon completion of one program, another statistically identical program will enter the system instantaneously. With these assumptions, the system can be modeled as a discrete-parameter Markov chain with the state diagram shown in Figure 7.6. The transition probability matrix P of the Markov chain is given by:

$$P = \begin{bmatrix} q_0 & q_1 & \cdot & \cdot & \cdot & q_m \\ 1 & 0 & \cdot & \cdot & \cdot & 0 \\ \cdot & \cdot & & & & \cdot \\ \cdot & \cdot & & & & \cdot \\ \cdot & \cdot & & & & \cdot \\ 1 & 0 & 0 & 0 & 0 & 0 \end{bmatrix}.$$

If we assume that $0 < q_i < 1$ $(i = 0, 1, \ldots, m)$, then it is easy to verify that this finite Markov chain is both irreducible and aperiodic. Therefore, Theorem 7.4 applies. The unique steady-state probability vector, $\mathbf{v}$, is obtained by solving the system of linear equations:

$$\mathbf{v} = \mathbf{v}P$$

or

$$v_0 = v_0 q_0 + \sum_{j=1}^{m} v_j, \qquad j = 0,$$

$$v_j = v_0 q_j, \qquad j = 1, 2, \ldots, m.$$

Using the normalization condition:

$$\sum_{j=0}^{m} v_j = 1,$$

we have:

$$v_0 + v_0 \sum_{j=1}^{m} q_j = 1.$$

Noting that $\sum_{j=1}^{m} q_j = 1 - q_0$, we get:

$$v_0(1 + 1 - q_0) = 1$$

or

$$v_0 = \frac{1}{2 - q_0}, \qquad j = 0,$$

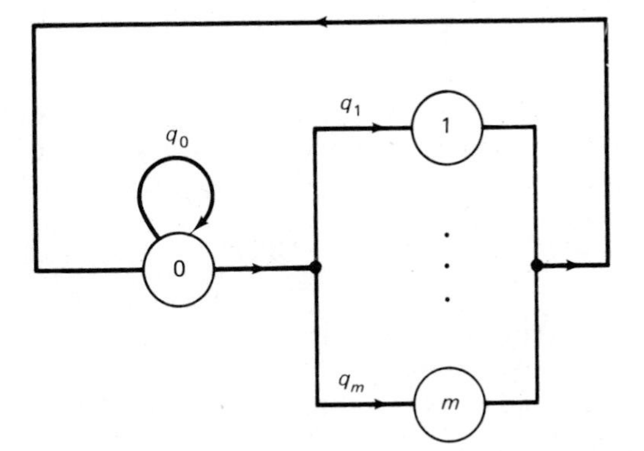

Figure 7.6 A discrete-parameter model of a uniprogrammed computer system

and

$$v_j = \frac{q_j}{2 - q_0}, \qquad j = 1, 2, \ldots, m.$$

The interpretation is that in a real-time interval T, the number of visits to device j will be $v_j T$, on the average, in the long run. #

Problems

1. [ASH 1970] Consider a system with two components. We observe the state of the system every hour. A given component operating at time n has probability p of failing before the next observation at time $n + 1$. A component that was in a failed condition at time n has a probability r of being repaired by time $n + 1$, independent of how long the component has been in a failed state. The component failures and repairs are mutually independent events. Let X_n be the number of components in operation at time n. $\{X_n, n = 0, 1, \ldots\}$ is a discrete-parameter homogeneous Markov chain with the state space $I = \{0, 1, 2\}$. Determine its transition probability matrix P, and draw the state diagram. Obtain the steady-state probability vector, if it exists.

2. Assume that a computer system is in one of three states: busy, idle, or undergoing repair, respectively denoted by states 0, 1, and 2. Observing its state at 2 P.M. each day, we believe that the system approximately behaves like a homogeneous Markov chain with the transition probability matrix:

$$P = \begin{bmatrix} 0.6 & 0.2 & 0.2 \\ 0.1 & 0.8 & 0.1 \\ 0.6 & 0.0 & 0.4 \end{bmatrix}.$$

   Prove that the chain is irreducible, and determine the steady-state probabilities.

3. Any transition probability matrix P is a stochastic matrix; that is, $p_{ij} \geqslant 0$ for all i and all j, and $\sum_j p_{ij} = 1$, for all i. If, in addition, the column sums are also unity—that is:

$$\sum_i p_{ij} = 1, \qquad \text{for all } j,$$

   then matrix P is called **doubly stochastic**. If a Markov chain with doubly stochastic P is irreducible, aperiodic, and finite with n states, show that the steady-state probability is given by:

$$v_j = \frac{1}{n}, \qquad \text{for all } j.$$

4. Show that the Markov chain of Example 7.10 is irreducible and aperiodic if $0 < q_i < 1$ for all i. Also show that if for some j, $1 \leqslant j \leqslant m$, $q_j = 0$, then the chain is not irreducible. Finally show that if $q_j = 1$ for some j, $0 \leqslant j \leqslant m$, then the chain is periodic.

7.4 DISTRIBUTION OF TIMES BETWEEN STATE CHANGES

We have noted that the entire past history of the Markov chain is summarized in its current state. Assume that the state at the nth step is $X_n = i$. But then the probability that the next state is j—that is, $X_{n+1} = j$—should depend only upon the current state i and not upon the time the chain has spent in the current state. Let the random variable T_i denote the time the Markov chain spends in state i during a single visit to state i. (In other words, T_i is one plus the number of transitions $i \rightarrow i$ made before leaving state i.) It follows that the distribution of T_i should be memoryless for $\{X_n, n = 0, 1, \ldots\}$ to form a (homogeneous) Markov chain.

Given that the chain has just entered state i at the nth step, it will remain in this state at the next step with probability p_{ii} and it will leave the state at the next step with probability $\sum_{j \neq i} p_{ij} = 1 - p_{ii}$. Now if the next state is also i—that is, $X_{n+1} = i$—then the same two choices are available at the next step. Furthermore, the probabilities of events at the $(n+1)$st step are independent of the events at the nth step, because $\{X_n\}$ is a Markov chain.

Thus, we have a sequence of Bernoulli trials with the probability of success $1 - p_{ii}$, where success is defined to be the event that the chain leaves state i. The event $T_i = n$ corresponds to n trials up to and including the first success. Thus, T_i has the geometric distribution, so that:

$$P(T_i = n) = (1 - p_{ii})p_{ii}^{n-1}, \qquad i \in I. \tag{7.17}$$

Using the properties of the geometric distribution, the expected number of steps the chain spends in state i, per visit to state i, is given by:

$$E[T_i] = \frac{1}{1 - p_{ii}}, \qquad i \in I, \tag{7.18}$$

and the corresponding variance is:

$$\text{Var}\,[T_i] = \frac{p_{ii}}{(1 - p_{ii})^2}, \qquad i \in I. \tag{7.19}$$

If we define an "event" to be a change of state, then the successive interevent times of a discrete-parameter Markov chain are independent, geometrically distributed random variables. Unlike the special case of the Bernoulli process, however, the successive interevent times do not, in general, have identical distributions.

Example 7.11

We return to our example of a communication net consisting of a cascade of binary communication channels with the matrix P given by:

$$P = \begin{bmatrix} 1 - a & a \\ b & 1 - b \end{bmatrix}.$$

Assuming that a 0 was transmitted—that is, $X_0 = 0$ — $S_0 = T_0 - 1$ is the number of stages before the first error. The average number of stages over which a 0 can be transmitted without an error is given by:

$$E[S_0] = E[T_0] - 1 = \frac{1-a}{a},$$

and the average number of stages over which a 1 can be transmitted without an error is given by:

$$\frac{1-b}{b},$$

Note that T_0 has a geometric distribution with parameter a, while T_1 has a geometric distribution with parameter b. These two interevent times, although possessing the memoryless distribution, have different parameters associated with them. #

7.5 IRREDUCIBLE FINITE CHAINS WITH APERIODIC STATES

In this section we consider some examples of finite Markov chains that satisfy the conditions of Theorem 7.4 so that the unique steady-state probabilities can be obtained by solving the system of linear equations (7.15)–(7.16).

7.5.1 Memory Interference in Multiprocessor Systems

Consider the shared memory multiprocessor system shown in Figure 7.7. The processors' ability to share the entire memory space provides a convenient means of sharing information and provides flexibility in memory allocation. The price of sharing is the contention for the shared resource. To reduce contention, the memory is usually split up into modules, which can be accessed independently and concurrently with other modules. When more than one processor attempts to access the same module, only one processor can be granted access, while other processors must await their turn in a queue. The effect of such contention, or interference, is to increase the average memory access time.

Assume that the time to complete a memory access is a constant and that all modules are synchronized. Processors are assumed to be fast enough to generate a new request as soon as their current request is satisfied. A processor cannot generate a new request when it is waiting for the current request to be completed. The operation of the system can be visualized as a discrete-parameter queuing network as shown in Figure 7.8.

The memory modules are the servers, and the fixed number, n, of processors constitute the "jobs" or "customers" circulating in this closed queuing network. The symbol q_i denotes the probability that a processor-generated request is directed at memory module i, $i = 1, 2, \ldots, m$. Thus: $\sum_{i=1}^{m} q_i = 1$.

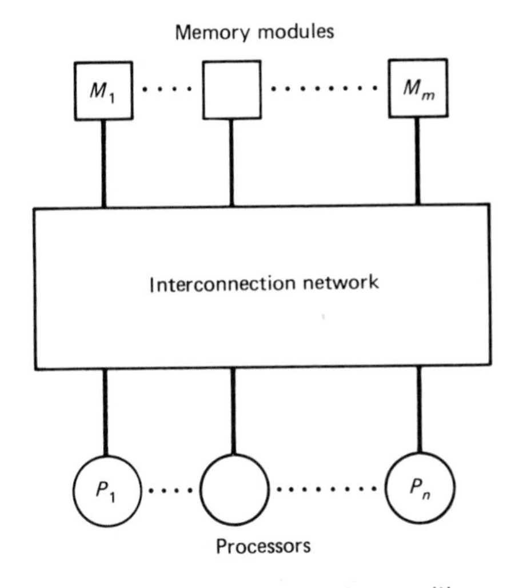

Figure 7.7 A multiprocessor system with multimodule memory

As an example, consider a system with two memory modules and two processors. Let the number of processors waiting or being served at module i ($i = 1, 2$), be denoted by N_i. Clearly, $N_i \geqslant 0$ and $N_1 + N_2 = 2$. The pair (N_1, N_2) denotes the state of the system, and the state space $I = \{(1, 1), (0, 2), (2, 0)\}$. The operation of the system is described by a discrete-parameter Markov chain whose state diagram is shown in Figure 7.9.

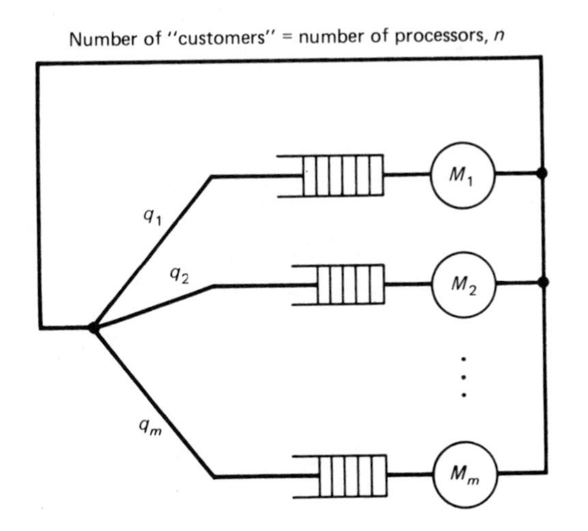

Figure 7.8 A discrete-parameter queuing network representation of multiprocessor memory interference

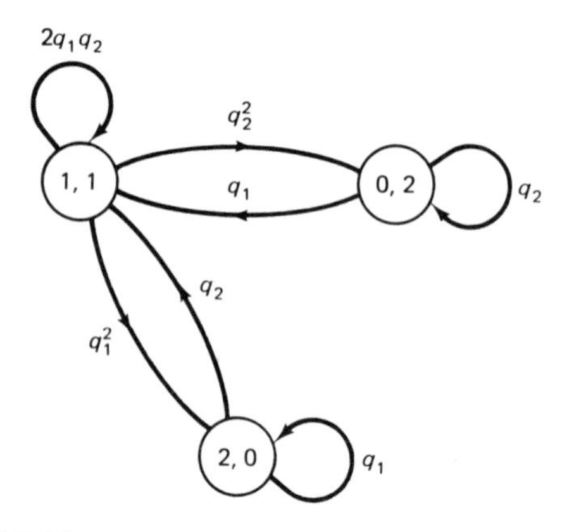

Figure 7.9 The state diagram for an example of the memory interference problem

The transition probability matrix of this chain is given by:

$$P = \begin{matrix} (1,1) \\ (0,2) \\ (2,0) \end{matrix} \overset{\begin{matrix} (1,1) & (0,2) & (2,0) \end{matrix}}{\begin{bmatrix} 2q_1q_2 & q_2^2 & q_1^2 \\ q_1 & q_2 & 0 \\ q_2 & 0 & q_1 \end{bmatrix}}.$$

We will explain the elements in the top row of the above matrix; the remaining entries can be explained in a similar way. Assume that the system is in state (1, 1) at time k, hence both modules and both processors are busy. At the end of this period both processors will be independently generating their new requests. Generation of a new request by a processor may be thought of as a Bernoulli trial with probability q_i of accessing module i. Thus we have a sequence of two Bernoulli trials. The probability that both processors will simultaneously request memory module i is q_i^2. If $i = 1$, the state of the system at time $k + 1$ will be (2, 0) while if $i = 2$, the new state will be (0, 2). If the two processors request access to distinct memory modules, the next state will be (1, 1). The probability of this last event is easily seen to be $2q_1q_2$.

To obtain the steady-state probability vector $\mathbf{v} = (v_{(1,1)}, v_{(0,2)}, v_{(2,0)})$ we use:

$$\mathbf{v} = \mathbf{v}P \quad \text{and} \quad \sum_{(i,j)\in I} v_{(i,j)} = 1$$

or

$$\begin{aligned} v_{(1,1)} &= 2q_1q_2v_{(1,1)} + q_1v_{(0,2)} + q_2v_{(2,0)}, \\ v_{(0,2)} &= q_2^2v_{(1,1)} + q_2v_{(0,2)}, \\ v_{(2,0)} &= q_1^2v_{(1,1)} + q_1v_{(2,0)}, \end{aligned}$$

$$v_{(1,1)} + v_{(0,2)} + v_{(2,0)} = 1.$$

Thus:

$$v_{(2,0)} = \frac{q_1^2}{1-q_1} v_{(1,1)}, \qquad v_{(0,2)} = \frac{q_2^2}{1-q_2} v_{(1,1)},$$

which implies that:

$$v_{(1,1)} = \frac{1}{1 + \dfrac{q_1^2}{1-q_1} + \dfrac{q_2^2}{1-q_2}} = \frac{q_1 q_2}{1 - 2q_1 q_2}.$$

Let the random variable B denote the number of memory requests completed per memory cycle in the steady-state. We are interested in computing the average number, $E[B]$, of memory requests completed per memory cycle. Note that in state (1, 1) two requests are completed, while in states (2, 0) or (0, 2) only one request each is completed. Therefore, the conditional expectations of B are given by:

$$E[B|\text{system in state } (1,\ 1)] = 2,$$

$$E[B|\text{system in state } (2,\ 0)] = 1,$$

$$E[B|\text{system in state } (0,\ 2)] = 1.$$

Now using the theorem of total expectation, we compute:

$$\begin{aligned} E[B] &= 2v_{(1,1)} + v_{(0,2)} + v_{(2,0)} \\ &= \left(2 + \frac{q_1^2}{1-q_1} + \frac{q_2^2}{1-q_2}\right) v_{(1,1)} \\ &= \frac{1 - q_1 q_2}{1 - 2q_1 q_2}. \end{aligned}$$

The quantity $E[B]$ achieves its maximum value, $3/2$, when $q_1 = q_2 = 1/2$. This is considerably smaller than the capacity of the memory system which is two requests per cycle. For a deeper study of this problem, see [BASK 1976, CHAN 1977, FULL 1975].

Problems

1. For the example of multiprocessor memory interference with two processors and two memory modules, explicitly solve the following optimization problem:

$$\begin{aligned} &\text{max: } E[B] \\ &\text{s.t.: } q_1 + q_2 = 1, \\ &\qquad q_1,\ q_2 \geqslant 0. \end{aligned}$$

2. Modify the multiprocessor memory interference example so that processor 1 has associated probabilities r_1 and r_2 respectively, for accessing module 1 and 2, and

processor 2 has distinct probabilities q_1 and q_2 associated with it. Construct the Markov chain state diagram, solve for the steady state probabilities, and compute $E[B]$. For those with extra energy, solve an optimization problem analogous to problem 1 above.

3. Consider another modification to the memory interference example where the processor requires nonzero amount of time to generate a memory request. Simplify the problem by assuming that the processor cycle time is identical to the memory cycle time. Once again go through all the steps as in problem 2 above and compute $E[B]$.

7.5.2 Models of Program Paging Behavior

In Chapter 6 we considered the renewal model of page referencing behavior of programs. In the renewal model, the successive intervals between references to a given page were assumed to be independent identically distributed random variables. The first model we consider in this section is a special case of the renewal model, where the above intervals are geometrically distributed. This is known as the **independent reference model (IRM)** of program behavior. Although simple to analyze, such a model is not very realistic. The LRU stack model, which is a better approximation to the behavior of real programs, is considered next. In-depth treatment of such models are available in [COFF 1973, COUR 1977, SPIR 1977].

7.5.2.1 The Independent Reference Model. In this model we assume that the reference string $w = r_1, r_2, \ldots, r_t, \ldots$ is a sequence of independent, identically distributed random variables with the pmf:

$$P(r_t = i) = \beta_i, \qquad 1 \leqslant i \leqslant n; \qquad \sum_{i=1}^{n} \beta_i = 1,$$

where the address space of the program consists of pages indexed $1, 2, \ldots, n$. Thus w is a discrete-parameter independent process. It is clear then that the interval between two successive references to page i is geometrically distributed with parameter β_i. Using the theory of finite, irreducible, and aperiodic Markov chains developed earlier, we can analyze the performance of several paging algorithms, assuming the independent reference model of program behavior.

We assume that a fixed number, m $(1 \leqslant m \leqslant n)$ of page frames have been allocated to the program. The internal state of the paging algorithm at time t, denoted by $\mathbf{q}(t)$, is an ordered list of the m pages currently in main memory. If the next page referenced (r_{t+1}) is not in main memory, then a **page fault** is said to have occurred, and the required page will be brought from secondary storage into main memory. This will, in general, require the replacement of an existing page from main memory. We will assume that the rightmost page in the ordered list $\mathbf{q}(t)$ will be replaced. On the other hand, if the next page referenced (r_{t+1}) is in main memory, no page fault (and replacement) occurs, but the list $\mathbf{q}(t)$ is updated to $\mathbf{q}(t + 1)$, reflecting the new replacement priorities. It is clear that the sequence of states $\mathbf{q}(0), \mathbf{q}(n), \ldots, \mathbf{q}(t), \ldots$ forms a discrete-parameter homogeneous Markov chain with the state space consisting of $n!/(n - m)!$ permutations over $\{1, 2, \ldots, n\}$. It is assumed that the main memory is preloaded initially with m pages. Since we will be studying the steady-state behavior of the Markov chain, the initial state has no effect on our results.

As an example, consider the LRU (least recently used) paging algorithm with $n = 3$ and $m = 2$. It is natural to let $\mathbf{q}(t)$ be ordered by the recency of usage, so that $\mathbf{q}(t) = (i, j)$ implies that the page indexed i was more recently used than page j, and, therefore, page j will be the candidate for replacement. The state space I is given by:

$$I = \{(1, 2), (2, 1), (1, 3), (3, 1), (2, 3), (3, 2)\}.$$

Let the current state $\mathbf{q}(t) = (i, j)$. Then the next state $\mathbf{q}(t + 1)$ takes one of the values:

$$\mathbf{q}(t+1) = \begin{cases} (i, j), & \text{if } r_{t+1} = i, \text{ with associated probability } \beta_i, \\ (j, i), & \text{if } r_{t+1} = j, \text{ with associated probability } \beta_j, \\ (k, i), & \text{if } r_{t+1} = k,\ k \neq i,\ k \neq j, \text{ with associated probability } \beta_k. \end{cases}$$

Then the transition probability matrix P is given by:

$$P = \begin{array}{c} \\ (1,2) \\ (2,1) \\ (1,3) \\ (3,1) \\ (2,3) \\ (3,2) \end{array} \begin{array}{c} \begin{array}{cccccc} (1,2) & (2,1) & (1,3) & (3,1) & (2,3) & (3,2) \end{array} \\ \left[\begin{array}{cccccc} \beta_1 & \beta_2 & 0 & \beta_3 & 0 & 0 \\ \beta_1 & \beta_2 & 0 & 0 & 0 & \beta_3 \\ 0 & \beta_2 & \beta_1 & \beta_3 & 0 & 0 \\ 0 & 0 & \beta_1 & \beta_3 & \beta_2 & 0 \\ \beta_1 & 0 & 0 & 0 & \beta_2 & \beta_3 \\ 0 & 0 & \beta_1 & 0 & \beta_2 & \beta_3 \end{array}\right] \end{array}$$

It can be verified that the above Markov chain is irreducible and aperiodic, hence a unique steady-state probability vector $\mathbf{v}$ exists. This vector is obtained by solving the system of equations:

$$\mathbf{v} = \mathbf{v}P$$

and

$$\sum_{(i,j)} v_{(i,j)} = 1.$$

Solving this system of equations, we get (the student is urged to verify this):

$$v_{(i,j)} = \frac{\beta_i \beta_j}{1 - \beta_i}.$$

Note that a page fault occurs in state (i, j), provided that a page other than i or j is referenced. The associated conditional probability of this event is $1 - \beta_i - \beta_j$; the steady-state page fault probability is then given by:

$$F(\text{LRU}) = \sum_{(i,j) \in I} (1 - \beta_i - \beta_j) \frac{\beta_i \beta_j}{1 - \beta_i}.$$

More generally, for arbitrary values of $n \geqslant 1$ and $1 \leqslant m \leqslant n$, it can be shown (see problem 3 at the end of this section) that:

$$F(\text{LRU}) = \sum_{\substack{\text{over the} \\ \text{state space}}} D_1^2(\mathbf{q}) \prod_{i=1}^{m} \frac{\beta_{j_i}}{D_i(\mathbf{q})},$$

where $\mathbf{q} = (j_1, j_2, \ldots, j_m)$ and:

$$D_i(\mathbf{q}) = 1 - \sum_{k=1}^{m-i+1} \beta_{j_k}.$$

Similar results can be derived for several other paging algorithms (see problems 1 and 2 at the end of this section).

7.5.2.2 The LRU-Stack Model [SPIR 1977]. Intuitively, we expect the probability of referencing a given page i at time t to depend upon the pages referenced in the immediate past. Thus the independent reference model may be expected to be a poor model of practical reference strings. It has been observed that references to pages tend to cluster together, so that the probability of referencing a page is high for a more recently used page. The LRU-stack model is able to reflect such a behavior of reference strings. Validation experiments have confirmed that this model fits real reference-string behavior much better than IRM.

In the LRU-stack model, we associate a sequence of LRU stacks $\mathbf{s}_0\, \mathbf{s}_1 \cdots \mathbf{s}_t$ with a reference string $w = r_1 r_2 \cdots r_t \cdots$. The stack $\mathbf{s}_t$ is the n-tuple $(j_1, \ldots, j_n)$ in which j_i is the ith most recently referenced page at time t. Let D_t be the position of the page r_t in the stack $\mathbf{s}_{t-1}$. Then, associated with the reference string, we have the distance string $D_1\, D_2 \cdots D_t \cdots$.

The LRU-stack model assumes that the distance string is a sequence of independent identically distributed random variables with the pmf:

$$P(D_t = i) = a_i, \qquad i = 1, 2, \ldots, n, \; t \geqslant 1, \quad \text{and} \quad \sum_{j=1}^{n} a_j = 1.$$

The distribution function:

$$P(D_t \leqslant i) = A_i = \sum_{j=1}^{i} a_j, \qquad i = 1, 2, \ldots, n, \; t \geqslant 1.$$

Without loss of generality, we assume the initial stack

$$\mathbf{s}_0 = (1, 2, \ldots, n).$$

Note that IRM assumes that the reference string is a discrete independent process whereas the LRU-stack model assumes that the distance string is a discrete independent process, the corresponding reference string being a nonindependent stochastic process.

With this model, evaluation of the page-fault rate of the LRU paging algorithm is quite simple. Assume that the program has been allocated m page frames of main memory. Then a page fault will occur at time t provided $D_t > m$. Thus, the page fault probability is given by:

$$F(\text{LRU}) = P(D_t > m) = 1 - P(D_t \leqslant m) = 1 - A_m.$$

Let us study the movement of a tagged page (say x) through the LRU stack as the time progresses. Define the random sequence $E_0\ E_1\ E_2 \cdots E_t \cdots$ such that $E_t = i$ if page x occupies the ith position in stack $\mathbf{s}_t$. Clearly $1 \leqslant E_t \leqslant n$ for all $t \geqslant 1$. Thus the sequence above is a discrete parameter, discrete state stochastic process. By the stack updating procedure shown in Figure 7.10, the position of the page x in stack $\mathbf{s}_{t+1}$ is determined by the next reference r_{t+1} and the position of page x in stack $\mathbf{s}_t$, but not its position in previous stacks. Thus the sequence above is a discrete-parameter Markov chain. Furthermore, the chain is homogeneous.

We obtain the transition probabilities of the chain by observing the stack updating procedure shown in Figure 7.10. Then:

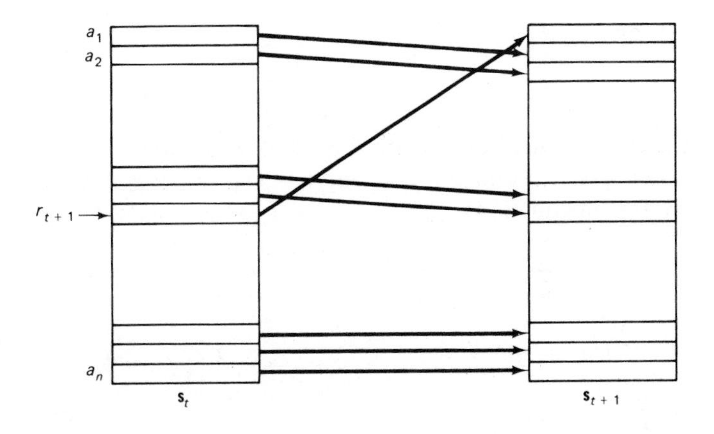

Figure 7.10 LRU-stack updating procedure

$$
\begin{aligned}
p_{i1} &= P(E_{t+1} = 1 | E_t = i) \\
&= P(r_{t+1} = x) = P(D_{t+1} = i) = a_i, && 1 \leqslant i \leqslant n, \\
p_{ii} &= P(E_{t+1} = i | E_t = i) \\
&= P(D_{t+1} < i) = A_{i-1}, && 2 \leqslant i \leqslant n, \\
p_{i,i+1} &= P(E_{t+1} = i + 1 | E_t = i) \\
&= P(D_{t+1} > i) = 1 - A_i, && 1 \leqslant i \leqslant n - 1,
\end{aligned}
$$

and

$$
p_{i,j} = 0, \qquad \text{otherwise.}
$$

The state diagram is given in Figure 7.11. The transition probability matrix is given by:

$$
P = \begin{array}{c}
 \\ 1 \\ 2 \\ \cdot \\ \cdot \\ \cdot \\ i-1 \\ i \\ \cdot \\ \cdot \\ \cdot \\ n-1 \\ n
\end{array}
\begin{array}{c}
\begin{array}{ccccccc}
1 & 2 & \cdots & i & i+1 & \cdots & n
\end{array} \\
\left[\begin{array}{ccccccc}
a_1 & 1 - A_1 & 0 & 0 & \cdot & \cdots & 0 \\
a_2 & A_1 & 1 - A_2 & \cdot & \cdot & & \cdot \\
\cdot & 0 & \cdot & \cdot & \cdot & & \cdot \\
\cdot & 0 & \cdot & \cdot & & & \\
\cdot & 0 & \cdot & \cdot & & & \\
\cdot & 0 & \cdot & 1 - A_{i-1} & & & \\
a_i & \cdot & \cdot & A_{i-1} & 1 - A_i & \cdots & \\
\cdot & \cdot & & \cdot & \cdot & & \cdot \\
\cdot & \cdot & & \cdot & \cdot & & \cdot \\
\cdot & \cdot & & \cdot & \cdot & & \cdot \\
a_{n-1} & 0 & \cdot & 0 & \cdot & & 1 - A_{n-1} \\
a_n & 0 & \cdot & 0 & \cdot & & A_{n-1}
\end{array}\right]
\end{array}
$$

Clearly, the chain is aperiodic and irreducible if we assume that $a_i > 0$ for all i. Then the steady-state probability vector $\mathbf{v} = (v_1, v_2, \ldots, v_n)$ is obtained from the system of equations:

$$
v_1 = \sum_{i=1}^{n} v_i a_i, \tag{7.20}
$$

$$
v_i = v_{i-1}(1 - A_{i-1}) + v_i A_{i-1}, \qquad 2 \leqslant i \leqslant n, \tag{7.21}
$$

$$
\sum_{i=1}^{n} v_i = 1. \tag{7.22}
$$

From equation (7.21) we have, $v_i = v_{i-1} = v_2$, $2 \leqslant i \leqslant n$ and from equation (7.20) we have:

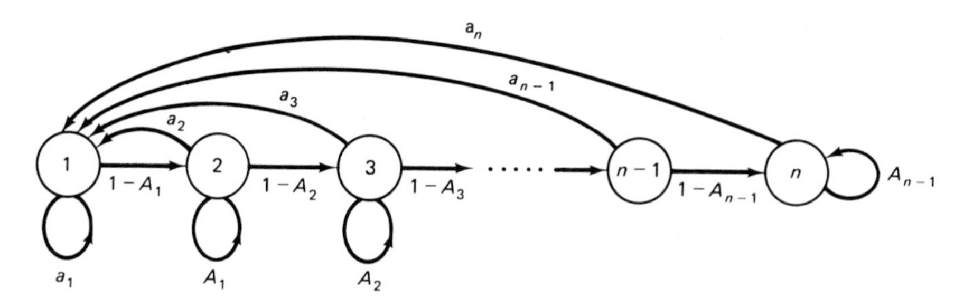

Figure 7.11 The state diagram for the LRU-stack example

$$v_1 = v_1 a_1 + v_2 \sum_{i=2}^{n} a_i = v_1 a_1 + v_2(1 - a_1)$$

and

$$v_1 = v_2.$$

Then from equation (7.22) we conclude:

$$v_i = \frac{1}{n}, \qquad i = 1, 2, \ldots, n.$$

Thus, the position of the tagged page, in the steady-state, is independent of its initial position and it is equally likely to be in any stack position. This implies that each page is equally likely to be referenced in the long run. Therefore, the LRU stack model is not able to cater to the nonuniform page-referencing behavior of real programs although it does reflect the clustering effect. This may be due to the assumption that the distance string is a sequence of independent identically distributed random variables. A natural generalization is to let the distance string be a Markov dependent sequence, as discussed by Shedler and Tung [SHED 1972].

Problems

***1.** Using the independent reference model of program behavior, show that the steady-state page-fault rate of the FIFO (first in first out) paging algorithm is given by:

$$F(\text{FIFO}) = G^{-1} \sum_{\mathbf{q}} D_1(\mathbf{q}) \prod_{i=1}^{m} \beta_{j_i},$$

where $\mathbf{q} = (j_1, j_2, \ldots, j_m)$ and

$$G = \sum_{\mathbf{q}} \prod_{i=1}^{m} \beta_{j_i}.$$

2. [A. Chang] Consider a Markov dependent reference string so that:

$$P(r_t = i \mid r_{t-1} = j) = q_{ij}, \qquad 1 \leqslant i, \ j \leqslant n, \ t > 1,$$

$$P(r_1 = i) = \beta_i, \qquad 1 \leqslant i \leqslant n.$$

Study the steady-state behavior of the page replacement algorithm that selects the page in memory with the smallest probability of being referenced at time $t+1$ conditioned on r_t. As a special case consider:

$$n=3, \quad m=2, \quad q_{11}=0, \quad q_{12}=\epsilon, \quad q_{13}=1-\epsilon,$$
$$q_{21}=\tfrac{1}{2}-\delta, \quad q_{22}=0, \quad q_{23}=\tfrac{1}{2}+\delta,$$
$$q_{31}=0, \quad q_{32}=1, \quad q_{33}=0.$$

Describe the states and state transitions of the paging algorithm, compute steady-state probabilities and steady-state average page-fault rate.

*3. Generalize the result derived for the steady-state page-fault probability of the LRU paging algorithm to the case $n \geqslant 1$ and $1 \leqslant m \leqslant n$.

*7.6 THE *M/G/1* QUEUING SYSTEM

We consider a single-server queuing system whose arrival process is Poisson with the average arrival rate λ. The job service times are independent and identically distributed with the distribution function F_B and pdf f_B. Jobs are scheduled for service in their order of arrival; that is, the scheduling discipline is FCFS. As a special case of the $M/G/1$ system, if we let F_B be the exponential distribution with parameter μ, then we obtain the $M/M/1$ queuing system. If the service times are assumed to be a constant, then we get the $M/D/1$ queuing system.

Let $N(t)$ denote the number of jobs in the system (those in the queue plus any in service) at time t. If $N(t) \geqslant 1$, then a job is in service, and since the general service time distribution need not be memoryless, besides $N(t)$, we also require knowledge of time spent by the job in service in order to predict the future behavior of the system. It follows that the stochastic process $\{N(t),\ t \geqslant 0\}$ is *not* a Markov chain.

To simplify the state description, we take a snapshot of the system at times of departure of jobs. These epochs of departure, called **regeneration points**, are used to specify the index set of a new stochastic process. Let t_n ($n = 1, 2, \ldots$) be the time of departure (immediately following service) of the nth job and let X_n be the number of jobs in the system at time t_n, so that:

$$X_n = N(t_n), \qquad n = 1, 2, \ldots. \tag{7.23}$$

The stochastic process $\{X_n,\ n = 1, 2, \ldots\}$ will be shown to be a discrete-parameter Markov chain, known as the **imbedded Markov chain** of the continuous-parameter stochastic process $\{N(t),\ t \geqslant 0\}$.

The method of the imbedded Markov chain allows us to simplify analysis, since it converts a non-Markovian problem into a Markovian one. We can then use the limiting distribution of the imbedded Markov chain as a measure of the original process $N(t)$, for it can be shown [KLEI 1975] that the limiting distribution of the number of jobs $N(t)$ observed at an arbitrary point in time is identical to the distribution of the number of jobs observed at the departure epochs, that is:

$$\lim_{t \to \infty} P[N(t) = k] = \lim_{n \to \infty} P(X_n = k). \tag{7.24}$$

For $n = 1, 2, \ldots$, let Y_n be the number of jobs arriving during the service time of the nth job. Now the number of jobs immediately following the departure instant of $(n+1)$st job can be written as:

$$X_{n+1} = \begin{cases} X_n - 1 + Y_{n+1}, & \text{if } X_n > 0, \\ Y_{n+1}, & \text{if } X_n = 0. \end{cases} \tag{7.25}$$

In other words, the number of jobs immediately following the departure of the $(n+1)$st job depends on whether the $(n+1)$st job was in the queue when the nth job departed. If $X_n = 0$, the next job to arrive is the $(n+1)$st; during its service time Y_{n+1} jobs arrive, then the $(n+1)$st job departs at time t_{n+1}, leaving Y_{n+1} jobs behind. If $X_n > 0$, then the number of jobs left behind by the $(n+1)$st job equals $X_n - 1 + Y_{n+1}$. Since Y_{n+1} is independent of $X_1, X_2, \ldots, X_n$, it follows that given the value of X_n, we need not know the values of $X_1, X_2, \ldots, X_{n-1}$ in order to determine the probabilistic behavior of X_{n+1}. Thus, $\{X_n, n = 1, 2, \ldots\}$ is a Markov chain.

The transition probabilities of the Markov chain are obtained using equation (7.25):

$$\begin{aligned} p_{ij} &= P(X_{n+1} = j | X_n = i) \\ &= \begin{cases} P(Y_{n+1} = j - i + 1), & \text{if } i \neq 0,\ j \geqslant i - 1, \\ P(Y_{n+1} = j), & \text{if } i = 0,\ j \geqslant 0, \\ 0, & \text{otherwise.} \end{cases} \end{aligned} \tag{7.26}$$

Since all jobs are statistically identical we expect that the Y_n's are identically distributed with the pmf $P(Y_{n+1} = j) = a_j$ so that:

$$\sum_{j=1}^{\infty} a_j = 1.$$

Then the (infinite-dimensional) transition probability matrix of $\{X_n\}$ is given by:

$$P = \begin{bmatrix} a_0 & a_1 & a_2 & a_3 & \cdots \\ a_0 & a_1 & a_2 & a_3 & \cdots \\ 0 & a_0 & a_1 & a_2 & \cdots \\ 0 & 0 & a_0 & a_1 & \cdots \\ 0 & 0 & 0 & a_0 & \cdots \\ \cdot & \cdot & \cdot & \cdot & \cdots \end{bmatrix} \tag{7.27}$$

Let the limiting probability of being in state j be denoted by v_j, so that:

$$v_j = \lim_{n \to \infty} P(X_n = j). \tag{7.28}$$

Using equation (7.15), we get:

$$v_j = v_0\, a_j + \sum_{i=1}^{j+1} v_i\, a_{j-i+1}. \tag{7.29}$$

If we define the generating function $G(z) = \sum_{j=0}^{\infty} v_j z^j$, then since:

$$\sum_{j=0}^{\infty} v_j z^j = \sum_{j=0}^{\infty} v_0 a_j\, z^j + \sum_{j=0}^{\infty} \sum_{i=1}^{j+1} v_i\, a_{j-i+1} z^j,$$

$$G(z) = v_0 \sum_{j=0}^{\infty} a_j z^j + \sum_{i=1}^{\infty} \sum_{j=i-1}^{\infty} v_i a_{j-i+1} z^j$$

(interchanging the order of summation)

$$= v_0 \sum_{j=0}^{\infty} a_j z^j + \sum_{i=1}^{\infty} \sum_{k=0}^{\infty} v_i\, a_k z^{k+i-1}$$

$$= v_0 \sum_{j=0}^{\infty} a_j z^j + \frac{1}{z}\, [\sum_{i=1}^{\infty} v_i z^i \sum_{k=0}^{\infty} a_k z^k].$$

Defining $G_A(z) = \sum_{j=0}^{\infty} a_j z^j$, we have:

$$G(z) = v_0\, G_A(z) + \frac{1}{z}\, [G(z) - v_0] G_A(z)$$

or

$$G(z) = \frac{(z-1)\, v_0\, G_A(z)}{z - G_A(z)}.$$

Since $G(1) = 1 = G_A(1)$, we can use L'Hopital's rule to obtain:

$$G(1) = 1 = \lim_{z \to 1} v_0 \frac{(z-1) G'_A(z) + G_A(z)}{1 - G'_A(z)}$$

$$= \frac{v_0}{1 - G_A'(1)},$$

provided $G'_A(1)$ is finite and less than unity. [Note that $G'_A(1) = E[Y]$.] If we let $\rho = G'_A(1)$, it follows that:

$$v_0 = 1 - \rho, \tag{7.30}$$

and, since v_0 is the probability that the server is idle, ρ is the server utilization in the limit. Also, we then have:

$$G(z) = \frac{(1-\rho)(z-1)\,G_A(z)}{z - G_A(z)}. \tag{7.31}$$

Thus, if we knew the generating function $G_A(z)$, we could compute $G(z)$ from which we could compute the steady-state average number of jobs in the system by using:

$$E[N] = \lim_{n\to\infty} E[X_n] = G'(1). \tag{7.32}$$

In order to evaluate $G_A(z)$, we first compute $a_j = P(Y_{n+1} = j)$. This is the probability that exactly j jobs arrive during the service time of the $(n+1)$st job. Let the random variable B denote job service times. Now we obtain the conditional pmf of Y_{n+1}:

$$P(Y_{n+1} = j \mid B = t) = e^{-\lambda t}\frac{(\lambda t)^j}{j!}$$

by the Poisson assumption. Using the (continuous version) theorem of total probability, we get:

$$\begin{aligned} a_j &= \int_0^\infty P(Y_{n+1} = j \mid t) f_B(t)\,dt \\ &= \int_0^\infty e^{-\lambda t}\frac{(\lambda t)^j}{j!} f_B(t)\,dt. \end{aligned}$$

Therefore:

$$\begin{aligned} G_A(z) &= \sum_{j=0}^{\infty} a_j z^j \\ &= \sum_{j=0}^{\infty}\int_0^\infty e^{-\lambda t}\frac{(\lambda tz)^j}{j!} f_B(t)\,dt \\ &= \int_0^\infty e^{-\lambda t}\left[\sum_{j=0}^{\infty}\frac{(\lambda tz)^j}{j!}\right] f_B(t)\,dt \\ &= \int_0^\infty e^{-\lambda t}\cdot e^{\lambda tz} f_B(t)\,dt \\ &= \int_0^\infty e^{-\lambda t(1-z)} f_B(t)\,dt \\ &= L_B[\lambda(1-z)] \end{aligned} \tag{7.33}$$

where $L_B[\lambda(1-z)]$ is the Laplace transform of the service-time distribution evaluated at $s = \lambda(1-z)$. Note that:

$$\rho = G'_A(1) = \frac{dL_B[\lambda(1-z)]}{dz}\Big|_{z=1}$$
$$= \frac{dL_B}{ds}\Big|_{s=0} \cdot (-\lambda)$$

by the chain rule, so:

$$\rho = \lambda E[B] = \frac{\lambda}{\mu} \tag{7.34}$$

by the moment generating property of the Laplace transform. Here the reciprocal of the service rate μ of the server equals the average service time $E[B]$.

Substituting (7.33) in (7.31), we get the well-known **Pollaczek-Khinchin (P-K) transform** equation:

$$G(z) = \frac{(1-\rho)(z-1)L_B[\lambda(1-z)]}{z - L_B[\lambda(1-z)]}. \tag{7.35}$$

The average number of jobs in the system, in the steady-state, is determined by taking the derivative with respect to z and then taking the limit $z \to 1$:

$$E[N] = \lim_{n\to\infty} E[X_n] = \sum_{j=1}^{\infty} jv_j = \lim_{z\to 1} G'(z). \tag{7.36}$$

As an example consider the $M/M/1$ queue with $f_B(x) = \mu e^{-\mu x}$, $x > 0$, and hence $L_B(s) = \mu/(s+\mu)$. It follows that:

$$G(z) = \frac{\mu - \lambda}{\mu - z\lambda}$$
$$= \frac{1-\rho}{1-\rho z}$$
$$= (1-\rho)\sum_{j=0}^{\infty}(\rho z)^j.$$

The coefficient of z^j in $G(z)$ gives the value of v_j:

$$v_j = (1-\rho)\rho^j, \qquad j = 0, 1, 2, \ldots.$$

It follows that the number of jobs in the system has a modified geometric distribution with parameter $(1-\rho)$. Therefore, the expected number of jobs in the system is given by:

$$E[N] = \frac{\rho}{1-\rho}. \tag{7.37}$$

[This expression for $E[N]$ can also be obtained by taking the derivative of the generating function:

$$E[N] = G'(1) = \frac{\lambda}{\mu - \lambda} = \frac{\rho}{1 - \rho} .]$$

More generally, it can be shown [KLEI 1975] that:

$$E[N] = \rho + \frac{\lambda^2 E[B^2]}{2(1 - \rho)} = \rho + \frac{\rho^2(1 + C_B^2)}{2(1 - \rho)}. \tag{7.38}$$

This is known as the **Pollaczek-Khinchin (P-K) mean-value formula.** Note that the average number of jobs in the system depends only on the first two moments of the service-time distribution. In fact, $E[N]$ grows linearly with the squared coefficient of variation of the service-time distribution. In particular, if we consider the $M/D/1$ system then $C_B^2 = 0$ and:

$$E[N] = \rho + \frac{\rho^2}{2(1 - \rho)} \quad (M/D/1). \tag{7.39}$$

Although we have assumed that the scheduling discipline is FCFS, all the results in this section hold under rather general scheduling disciplines provided we assume that [COFF 1973]:

1. The server is not idle whenever a job is waiting for service.
2. The scheduling discipline does not base job sequencing on any a priori information on job execution times.
3. The scheduling is nonpreemptive; that is, once a job is scheduled for service, it is allowed to complete without interruption.

Problems

1. Jobs submitted to a university computer center can be divided into three classes:

Type	*Relative frequency*	*Mean execution time*
Student jobs	0.8	1 sec
Faculty jobs	0.1	20 sec
Administrative jobs	0.1	5 sec

Assuming that, within a class, execution times are one-stage, three-stage, and two-stage Erlang, respectively, compute the average number of jobs in the center assuming a Poisson overall arrival stream of jobs with average rate of 0.1 jobs per second. Assume that all classes are treated equally by the scheduler.

2. For the $M/G/1$ queue plot the average queue length $E[N]$ as a function of server utilization ρ for several different service-time distributions:
   (a) Deterministic.
   (b) Exponential.
   (c) k-stage Erlang, $k = 2, 5$.
   (d) k-stage hyperexponential, $k = 2; \alpha_1 = 0.5, \alpha_2 = 0.5; \mu_1 = 1, \mu_2 = 10$.

3. Consider a computer system with a CPU and one disk drive. After a burst at the CPU the job completes execution with probability 0.1 and requests a disk I/O with probability 0.9. The time of a single CPU burst is exponentially distributed with mean 0.01 second. The disk service time is broken up into three phases: exponentially distributed seek time with mean 0.03 second, uniformly distributed latency time with mean 0.01 second, and a constant transfer time equal to 0.01 second. After a service completion at the disk the job always requires a CPU burst. The average arrival rate of jobs is 0.8 job per second and the system does not have enough main memory to support multiprogramming. Solve for the average response time using the $M/G/1$ model. In order to compute the mean and the variance of the service-time distribution, you may need the results of the section on random sums in Chapter 5.

4. Starting with the Pollaczek-Khinchin transform equation (7.35), derive expressions for the average number in the system $E[N]$ for an $M/G/1$ queue, assuming:
   (a) Deterministic service times $(M/D/1)$.
   (b) Two-stage Erlang service time distribution $(M/E_2/1)$.
   (c) Two-stage hyperexponential service-time distribution $(M/H_2/1)$.

*5. Consider a modification of the $M/G/1$ queue with FCFS scheduling so that after the completion of a service burst, the job returns to the queue with probability q and completes execution with probability p (see Figure 7.P.1). We wish to obtain the queue-length distribution in the steady state as seen by a completer and as seen by a departer. First consider the departer's distribution. Using the notion of random sums, first derive the Laplace transform for the total service time T of a job as:

$$L_T(s) = \frac{pL_B(s)}{1 - qL_B(s)},$$

where B is the random variable denoting the length of a single service burst. Now show that the generating function of N_d, the number of jobs in the system (in the steady state) as seen by a departer, defined by:

$$G_{N_d}(z) = \sum_{k=0}^{\infty} p_{N_d}(k) z^k,$$

is given by

$$G_{N_d}(z) = \left(1 - \frac{\lambda}{\mu\, p}\right) \frac{p(1-z)L_B[\lambda(1-z)]}{(p+qz)L_B\,[\lambda(1-z)] - z}.$$

Find the average number of jobs $E[N_d]$ as seen by the departer. Specializing to the case in which the service-time distribution is exponential, obtain the distribution of N_d.

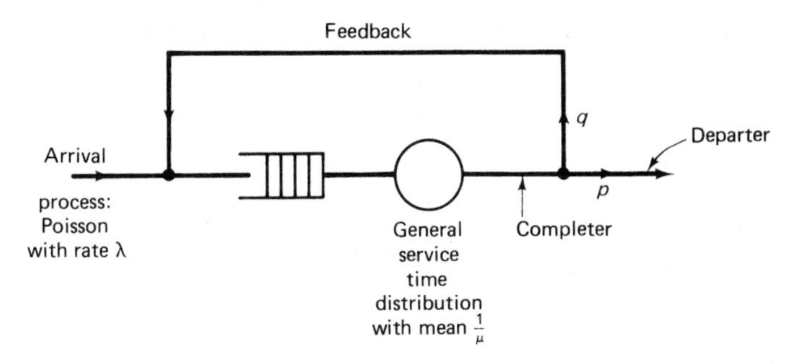

Figure 7.P.1 The $M/G/1$ queue with Bernoulli feedback

Next consider the imbedded Markov chain, where the completion of a service burst is defined to be an epoch. Show that the transition probability matrix of this Markov chain is:

$$\begin{bmatrix} pa_0 & pa_1 + qa_0 & pa_2 + qa_1 & \cdots \\ pa_0 & pa_1 + qa_0 & pa_2 + qa_1 & \cdots \\ 0 & pa_0 & pa_1 + qa_0 & \cdots \\ 0 & 0 & pa_0 & \\ & & 0 & \\ \cdot & \cdot & \cdot & \\ \cdot & \cdot & \cdot & \\ \cdot & \cdot & \cdot & \\ 0 & 0 & 0 & \end{bmatrix}$$

where:

$$a_i = P(Y_{n+1} = i)$$
$$= P(\text{"}i \text{ jobs arrive during the } (n+1)\text{st service burst"}).$$

Now show that the generating function of the steady-state number of jobs in the system as seen by a completer, defined by:

$$G_{N_c}(z) = \sum_{k=0}^{\infty} p_{N_c}(k)\, z^k$$

is given by:

$$G_{N_c}(z) = p\left(1 - \frac{\lambda}{\mu p}\right) \frac{(p + qz)(1 - z)L_B[\lambda(1 - z)]}{(p + qz)L_B[\lambda(1 - z)] - z}.$$

Find the average number in the system $E[N_c]$ as seen by a completer.

7.7 DISCRETE-PARAMETER BIRTH-DEATH PROCESSES

We consider a special type of discrete-parameter Markov chain with all one-step transitions to nearest neighbors only. The transition probability matrix P is a banded matrix with:

$$p_{ij} = 0 \qquad \text{for } |i - j| > 1.$$

To simplify notation, we let:

$$\begin{aligned} b_i &= p_{i,i+1}, & i &\geqslant 0 \quad \{\text{the probability of a birth in state } i\}, \\ d_i &= p_{i,i-1}, & i &\geqslant 1 \quad \{\text{the probability of a death in state } i\}, \\ a_i &= p_{i,i}, & i &\geqslant 0. \end{aligned}$$

Note that $(a_i + b_i + d_i) = 1$ for all i. Thus the (infinite-dimensional) matrix P is given by:

$$P = \begin{bmatrix} a_0 & b_0 & 0 & \cdot & \cdots & & & & \cdots \\ d_1 & a_1 & b_1 & \cdot & \cdots & & & & \cdots \\ 0 & d_2 & a_2 & b_2 & \cdots & & & & \\ \cdot & 0 & \cdot & \cdot & & & & & \\ \cdot & \cdot & \cdot & \cdot & & & & & \\ \cdot & \cdot & \cdot & \cdot & & a_{i-1} & b_{i-1} & 0 & \\ \cdot & \cdot & \cdot & \cdot & & d_i & a_i & b_i & \\ \cdot & \cdot & \cdot & \cdot & & 0 & d_{i+1} & a_{i+1} & b_{i+1} \end{bmatrix}$$

and the state diagram is shown in Figure 7.12. If we assume that $b_i > 0$ and $d_i = 0$ for all i, then all the states of the Markov chain are transient. Similarly, if we let $b_i = 0$ and $d_i > 0$ for all i, then all the states are transient except the state labeled 0, which will be an absorbing state. We will assume that $0 < b_i$, $d_i < 1$ for all $i \geqslant 1$ and $b_0 > 0$, hence the Markov chain is irreducible and aperiodic, which implies by Theorem 7.3 that the limiting probabilities exist and are independent of the initial distribution. To compute the steady-state probability vector $\mathbf{v}$ (if it exists), we use:

$$\mathbf{v} = \mathbf{v}P$$

and get:

$$v_0 = a_0 v_0 + d_1 v_1, \tag{7.40a}$$

$$v_i = b_{i-1} v_{i-1} + a_i v_i + d_{i+1} v_{i+1}, \qquad i \geqslant 1. \tag{7.40b}$$

Since:

$$1 - a_i = b_i + d_i,$$

from (7.40b) we get:

$$b_i v_i - d_{i+1} v_{i+1} = b_{i-1} v_{i-1} - d_i v_i,$$

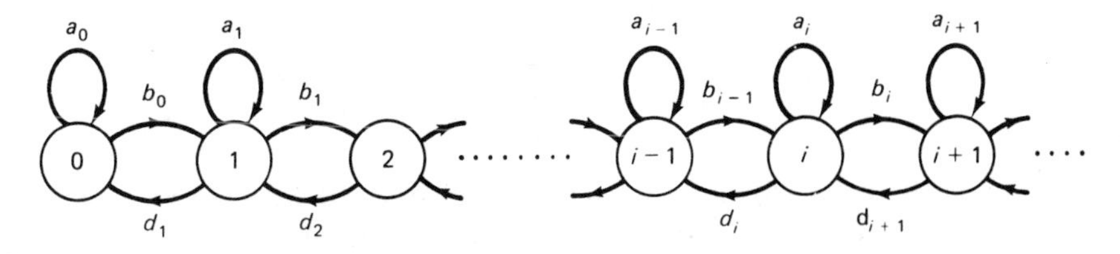

Figure 7.12 The state diagram of the discrete-parameter birth-death process

so, $b_i v_i - d_{i+1} v_{i+1}$ is independent of i, but $b_0 v_0 - d_1 v_1 = (1 - a_0) v_0 - d_1 v_1 =$ 0 from (7.40a). Therefore the solution to the above system of equations is given by:

$$v_i = \frac{b_{i-1}}{d_i} v_{i-1} = \prod_{j=1}^{i} \frac{b_{j-1}}{d_j} v_0. \tag{7.40c}$$

Now, using the condition $\sum_{i \geqslant 0} v_i = 1$, we get:

$$v_0 = \frac{1}{\sum_{i \geqslant 0} \prod_{j=1}^{i} \frac{b_{j-1}}{d_j}}, \tag{7.41}$$

provided the series converges. If the series in the denominator diverges, then we can conclude that all states of the Markov chain are recurrent null. We will assume that the series is convergent; that is, all states are recurrent non-null, hence (7.40c) and (7.41) give the unique steady-state probabilities.

Example 7.12 (Analysis of a Data Structure)

Consider a data structure (such as a linear list) being manipulated in a program. Suppose we are interested only in the amount of memory consumed by the data structure. If the current amount of memory in use is i nodes, then we say that the state of the structure is s_i. Let probabilities associated with the next operation on the data structure be given by:

$$b_i = P(\text{"next operation is an insert"} \mid \text{"current state is } s_i\text{"}),$$

$$d_i = P(\text{"next operation is a delete"} \mid \text{"current state is } s_i\text{"}),$$

$$a_i = P(\text{"next operation is an access"} \mid \text{"current state is } s_i\text{"}).$$

Then the steady state pmf of the number of nodes in use is given by equations (7.40c) and (7.41) above.

As a special case, we let $b_i = b\ (i \geqslant 0)$ and $d_i = d\ (i \geqslant 1)$ for all i. Then, assuming $b < d$ (for the chain to have recurrent non-null states), we have:

$$v_i = \frac{b}{d} v_{i-1} \quad \text{and} \quad v_0 = \frac{1}{\sum_{i \geqslant 0} \left(\frac{b}{d}\right)^i} = 1 - \frac{b}{d}$$

or

$$v_i = \left(1 - \frac{b}{d}\right)\left(\frac{b}{d}\right)^i.$$

Thus the steady-state distribution is modified geometric with parameter $(1 - b/d)$. The expected number of nodes in use in the steady-state is given by $(b/d) / [1 - (b/d)] = \frac{b}{d-b}$. These formulas are valid under the assumption that:

$$\sum_{i \geqslant 0} \left(\frac{b}{d}\right)^i \text{ converges.}$$

This assumption is satisfied provided $b/d < 1$, or the probability of insertion is strictly less than the probability of deletion. If this condition is not satisfied, then the data structure will tend to grow continually, resulting in a memory overflow. #

Example 7.13

In the example above, we assumed that a potentially infinite number of nodes are available for allocation. Next assume that a limited number $m \geqslant 1$ of nodes are available for allocation. Then if m nodes are in use, an insertion operation will give rise to an overflow. We assume that such an operation is simply ignored, leaving the system in state s_m. The state diagram is given in Figure 7.13. The steady-state solution to this system is given by:

$$v_i = \left(\frac{b}{d}\right)^i v_0, \qquad i = 0, 1, \ldots, m$$

and

$$v_0 = \frac{1}{\sum_{i=0}^{m}\left(\frac{b}{d}\right)^i} = \frac{1 - \left(\frac{b}{d}\right)}{1 - \left(\frac{b}{d}\right)^{m+1}}.$$

Now the probability of an overflow is computed by:

$$P_{ov} = bv_m = b\left(\frac{b}{d}\right)^m \frac{1 - \left(\frac{b}{d}\right)}{1 - \left(\frac{b}{d}\right)^{m+1}}$$

$$= \frac{b^{m+1}(d-b)}{d^{m+1} - b^{m+1}}. \tag{7.42}$$

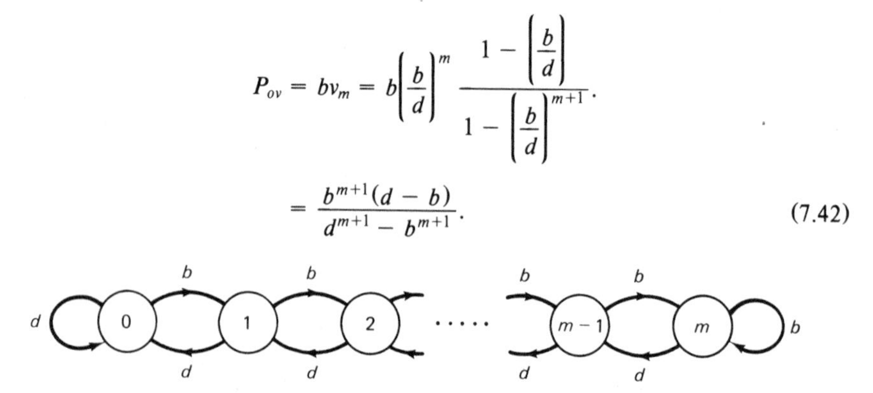

Figure 7.13 The state diagram for Example 7.13

Similarly, the probability of underflow is given by:

$$P_{uf} = dv_0 = d \cdot \frac{1 - \dfrac{b}{d}}{1 - \left(\dfrac{b}{d}\right)^{m+1}} \tag{7.43}$$

$$= \frac{d^{m+1}(d - b)}{d^{m+1} - b^{m+1}}.$$

#

The notion of the birth-death process can be generalized to multidimensional birth-death processes. We will introduce such processes through examples.

Example 7.14

Consider a program that uses two stacks, sharing an area of memory containing m locations. The stacks grow toward each other from the two opposite ends (see Figure 7.14). Clearly, an overflow will occur upon an insertion into either stack when $i + j = m$. Let the state of the system be denoted by the pair (i, j). Then the state space is $I = \{(i, j) | i, j \geqslant 0,\ i + j \leqslant m\}$. We assume that an overflow is simply ignored, leaving the state of the system unchanged. Underflow also does not change the system state as before.

At each instant of time, an operation on one of the stacks takes place with respective probabilities as shown in the probability tree of Figure 7.15. Thus for $i = 1,\ 2,\ p_i$ is the probability that a given operation is directed to stack i, and b_i, d_i and a_i respectively denote the probabilities that this operation is an insertion, deletion, or access.

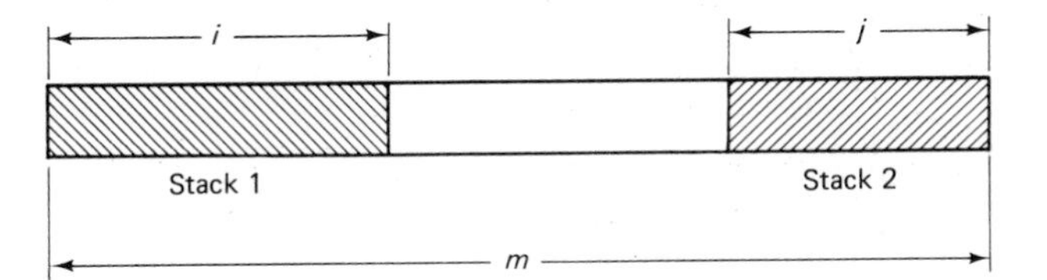

Figure 7.14 Two stacks sharing an area of memory

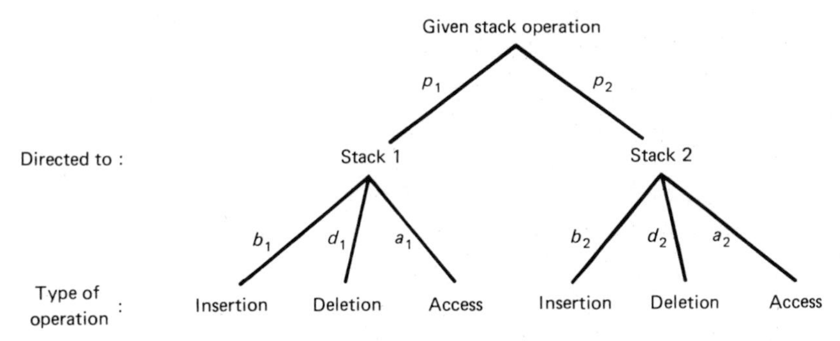

Figure 7.15 Tree diagram for Example 7.14

$a = p_1 a_1 + p_2 a_2$

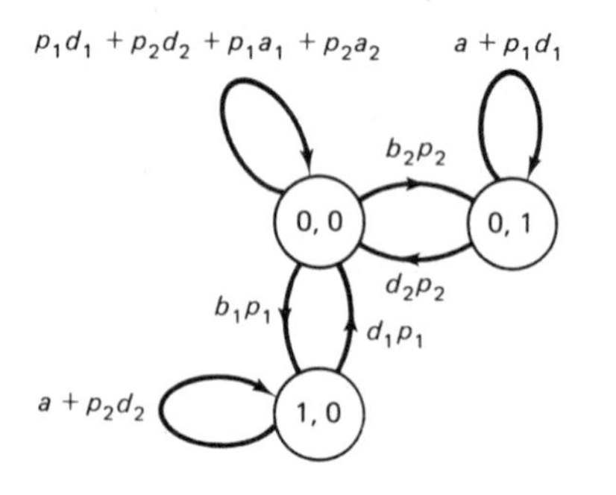

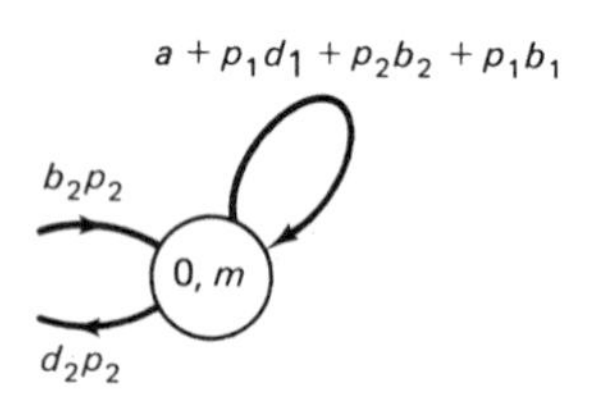

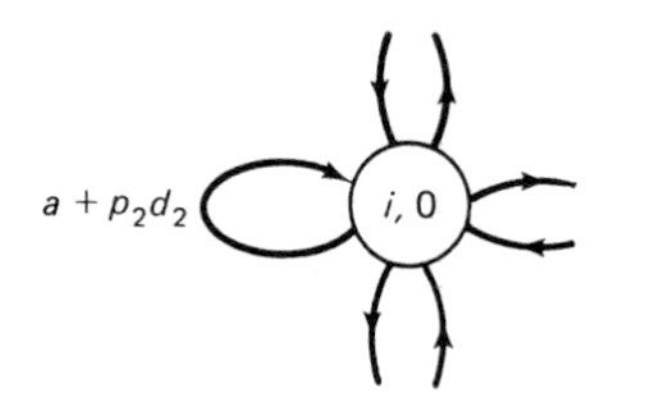

. . . . . .

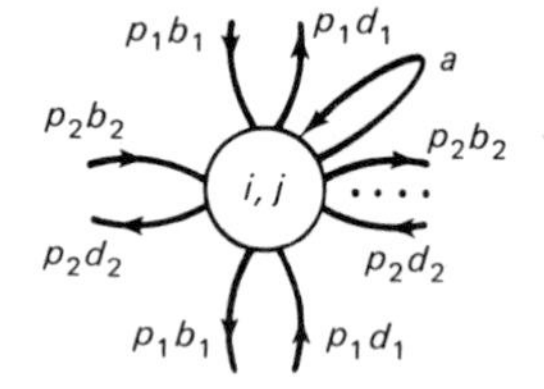

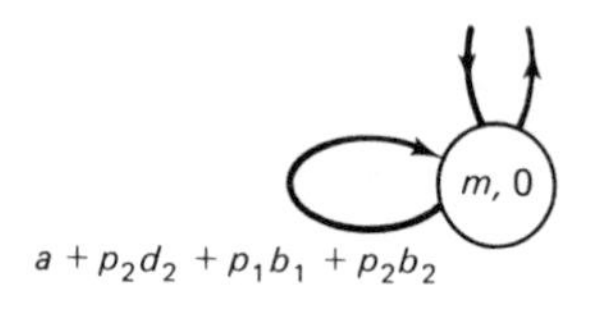

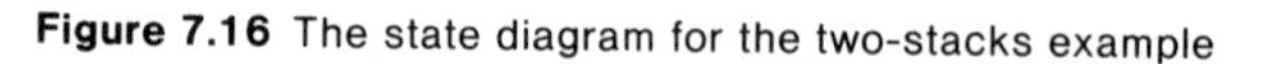

Figure 7.16 The state diagram for the two-stacks example

The system behavior corresponds to a two dimensional birth-death process with the state diagram shown in Figure 7.16. The steady-state probability vector

$$\mathbf{v} = (v_{(0,0)},\ v_{(0,1)},\ \ldots,\ v_{(0,m)},\ \ldots,\ v_{(i,j)},\ \ldots,\ v_{(m,0)})$$

may be obtained by solving:

$$\mathbf{v} = \mathbf{v}P$$

(and using the identities $a_1 + b_1 + d_1 = 1$, $a_2 + b_2 + d_2 = 1$, and $p_1 + p_2 = 1$). Thus:

$$v_{(i,j)} = b_2p_2v_{(i,j-1)} + d_2p_2v_{(i,j+1)} + (a_1p_1+a_2p_2)v_{(i,j)}$$
$$+ b_1p_1v_{(i-1,j)} + d_1p_1p_{(i+1,j)}, \qquad 1 \leqslant i,j;\ i+j < m,$$
$$v_{(i,0)} = (a_1p_1 + a_2p_2 + d_2p_2)v_{(i,0)} + d_2p_2v_{(i,1)}$$
$$+ b_1p_1v_{(i-1,0)} + d_1p_1v_{(i+1,0)}, \qquad 1 \leqslant i \leqslant m-1,$$
$$v_{(0,j)} = (a_1p_1 + a_2p_2 + d_1p_1)v_{(0,j)} + d_1p_1v_{(1,j)}$$
$$+ b_2p_2v_{(0,j-1)} + d_2p_2p_{(0,j+1)}, \qquad 1 \leqslant j \leqslant m-1,$$
$$v_{(0,0)} = (d_1p_1 + d_2p_2 + a_1p_1 + a_2p_2)v_{(0,0)} + d_1p_1v_{(1,0)}$$
$$+ d_2p_2v_{(0,1)},$$
$$v_{(0,m)} = (a_1p_1 + a_2p_2 + d_1p_1 + b_1p_1 + b_2p_2)v_{(0,m)} + b_2p_2v_{(0,m-1)},$$
$$v_{(m,0)} = (a_1p_1 + a_2p_2 + d_2p_2 + b_1p_1 + b_2p_2)v_{(m,0)} + b_1p_1v_{(m-1,0)}.$$

It may be verified by direct substitution that:

$$v_{(i,j)} = v_{(0,0)}\left(\frac{b_1}{d_1}\right)^i\left(\frac{b_2}{d_2}\right)^j, \qquad i,\ j \geqslant 0,\ i+j \leqslant m.$$

We will use the abbreviation $r_1 = b_1/d_1$, $r_2 = b_2/d_2$. Then:

$$v_{(i,j)} = v_{(0,0)}r_1^i r_2^j.$$

The normalization requirement yields:

$$1 = \sum_{i=0}^{m}\sum_{j=0}^{m-i} v_{(i,j)}$$
$$= v_{(0,0)}\sum_{i=0}^{m}\left[\sum_{j=0}^{m-i}(r_2^j)\right] r_1^i$$
$$= v_{(0,0)}\sum_{i=0}^{m}\frac{1-r_2^{m-i+1}}{1-r_2}\, r_1^i$$
$$= \frac{v_{(0,0)}}{1-r_2}\sum_{i=0}^{m}\left[r_1^i - r_2^{m+1}\left(\frac{r_1}{r_2}\right)^i\right]$$
$$= \begin{cases} \dfrac{v_{(0,0)}}{1-r_2}\left[\dfrac{1-r_1^{m+1}}{1-r_1} - r_2^{m+1}\dfrac{1-(r_1/r_2)^{m+1}}{1-(r_1/r_2)}\right], & \text{if } r_1 \neq r_2;\ r_1 \neq 1,\ r_2 \neq 1 \\[2ex] \dfrac{v_{(0,0)}}{1-r_1}\left[\dfrac{1-r_1^{m+1}}{1-r_1} - (m+1)r_1^{m+1}\right], & \text{if } r_1 = r_2 \neq 1 \end{cases}$$

Simplifying:

$$
\nu_{(0,0)} = \begin{cases} \dfrac{(1-r_1)(1-r_2)}{1 - \dfrac{1}{r_2 - r_1}\{r_2^{m+2}(1-r_1) - r_1^{m+2}(1-r_2)\}}, & r_1 \neq r_2, \\[2ex] \dfrac{(1-r_1)^2}{1-(m+2)r_1^{m+1} + (m+1)r_1^{m+2}}, & r_1 = r_2 \neq 1 \\[2ex] \dfrac{2}{(m+1)(m+2)}, & r_1 = r_2 = 1. \end{cases}
$$

The probability of overflow is given by:

$$
\begin{aligned}
P_{ov} &= \sum_{i+j=m} \nu_{(i,j)}(b_1 p_1 + b_2 p_2) \\
&= \nu_{(0,0)} \sum_{i=0}^{m} r_1^i r_2^{m-i}(b_1 p_1 + b_2 p_2) \\
&= \begin{cases} (b_1 p_1 + b_2 p_2)\nu_{(0,0)}\, r_2^m \dfrac{1-(r_1/r_2)^{m+1}}{1-(r_1/r_2)}, & r_1 \neq r_2, \\ (m+1)(b_1 p_1 + b_2 p_2)\nu_{(0,0)} r_1^m, & r_1 = r_2. \end{cases}
\end{aligned}
\tag{7.44}
$$

#

Example 7.15

We want to implement two stacks within $2k$ memory locations. The first solution is to divide the given area into two equal areas and preallocate the two areas to the two stacks. Assume $p_1 = p_2 = 1/2$, $b_1 = b_2 = b$, $d_1 = d_2 = d$, and hence $r_1 = r_2 = r = b/d$. Under the first scheme, the overflow in each stack occurs with probability [using equation (7.42)]:

$$
\frac{br^k(1-r)}{1-r^{k+1}}
$$

with the total overflow probability being twice as much.

Under the second scheme, where the two stacks grow toward each other, we have:

$$
\nu_{(0,0)} = \frac{(1-r)^2}{1-(2k+2)r^{2k+1} + (2k+1)r^{2k+2}}.
$$

and the overflow probability is given by [using equation (7.44)]:

$$
(2k+1)(b)\frac{(1-r)^2\, r^{2k}}{1-(2k+2)r^{2k+1} + (2k+1)r^{2k+2}}.
$$

Then the condition under which the second scheme is better than the first one is given by:

$$b(2k+1)\frac{(1-r)^2 r^{2k}}{1-(2k+2)r^{2k+1}+(2k+1)r^{2k+2}} \leqslant 2br^k \frac{1-r}{1-r^{k+1}}.$$

Assuming for simplicity that $r < 1$, the above condition is rewritten as:

$$\frac{(2k+1)r^k}{\dfrac{1-r^{2k+1}}{1-r} - (2k+1)r^{2k+1}} \leqslant \frac{2}{1-r^{k+1}}$$

and hence as:

$$(2k+1)r^k + (2k+1)r^{2k+1} \leqslant 2\,\frac{1-r^{2k+1}}{1-r}$$

$$= [2(\sum_{i=0}^{k-1} r^i) + r^k] + \{r^k + 2(\sum_{i=k+1}^{2k} r^i)\}. \quad (7.45)$$

In order to show that inequality (7.45) holds, observe that there are $(2k+1)$ terms in the expression within the square brackets and each term is greater than or equal to r^k (since $r < 1$). Similarly each of the $(2k+1)$ terms of the expression within braces is greater than or equal to r^{2k+1}.

Thus we conclude that the second scheme of sharing the $2k$ locations between the two stacks is superior to the first scheme of preallocating half the available area to each stack. #

7.8 FINITE MARKOV CHAINS WITH ABSORBING STATES: ANALYSIS OF PROGRAM EXECUTION TIME

In Chapter 5 we discussed the analysis of properly nested programs. Many programs, however, are not properly nested in that they contain **goto** statements. Program graphs associated with such a program can be treated as the state diagram of a discrete-parameter Markov chain, with appropriate assumptions [RAMA 1965, DEO 1974]. Since the program is eventually expected to terminate, it will contain certain "final" or "stopping" states. In the terminology of Markov chains, such states are called absorbing states. Also since the number of statements in the program will be finite, the corresponding chain will have a finite number of states.

Consider a program with its associated directed graph as shown in Figure 7.17. Each vertex s_j in the figure represents a group of statements with a single entry point and a single exit point. The last statement in the group is a multiway branch as typified by a **goto** statement using a label vector (e.g., PL/I). Vertex s_1 is the start vertex. Vertex s_5 has no outgoing edges and thus is a stop vertex. The weight p_{ij} of edge (s_i, s_j) is interpreted as the conditional probability that the program will next execute statement group s_j,

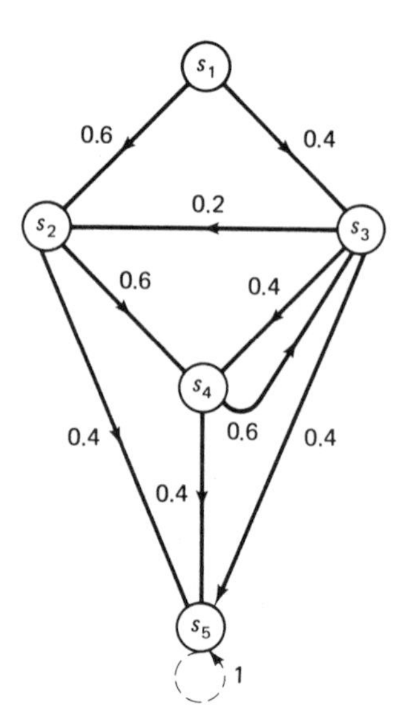

Figure 7.17 A program flow graph

given that it has just completed the execution of statement group s_i. We have assumed that this probability depends only on the current statement group and not on the previous history, of the program. Therefore, the corresponding Markov chain will be homogeneous. Herein lies a serious deficiency of the model. In actual programs such probabilities are not likely to be independent of previous history and a more accurate model involving the use of nonhomogeneous Markov chains is desirable. We will, however, continue with the simplified, albeit inaccurate, model using homogeneous Markov chains.

While interpreting the program flow graph of Figure 7.17 as the state diagram of a finite, discrete-time, homogeneous Markov chain, we encounter one difficulty. From the absorbing state s_5 there are no outgoing edges. Then $p_{5j} = 0$ for all j. But the assumption of Markov chain requires that $\sum_j p_{ij} = 1$ for each i. To avoid this difficulty, we imagine a "dummy" edge forming a self-loop on the absorbing state s_5. With this modification, the transition probability matrix of the Markov chain is given by:

$$P = \begin{array}{c} \\ s_1 \\ s_2 \\ s_3 \\ s_4 \\ s_5 \end{array} \begin{array}{c} \begin{array}{ccccc} s_1 & s_2 & s_3 & s_4 & s_5 \end{array} \\ \left[\begin{array}{ccccc} 0 & 0.6 & 0.4 & 0 & 0 \\ 0 & 0 & 0 & 0.6 & 0.4 \\ 0 & 0.2 & 0 & 0.4 & 0.4 \\ 0 & 0 & 0.6 & 0 & 0.4 \\ 0 & 0 & 0 & 0 & 1 \end{array}\right] \end{array}$$

Note that states s_1 through s_4 are transient states while state s_5 is an absorbing state.

In general, we consider a Markov chain with n states, $s_1, s_2, \ldots, s_n$. s_n will be the absorbing state, and the remaining states will be transient. The transition probability matrix of such a chain may be partitioned so that

$$P = \left[\begin{array}{c|c} Q & \mathbf{C} \\ \hline \mathbf{0} & 1 \end{array}\right],$$

where Q is an $(n-1)$ by $(n-1)$ substochastic matrix (with at least one row sum less than 1) describing the probabilities of transition only among the transient states. $\mathbf{C}$ is a column vector and $\mathbf{0}$ is a row vector of $(n-1)$ zeros.

Now the k-step transition probability matrix P^k has the form:

$$P^k = \left[\begin{array}{c|c} Q^k & \mathbf{C}' \\ \hline \mathbf{0} & 1 \end{array}\right],$$

where $\mathbf{C}'$ is a column vector whose elements will be of no further use and hence need not be computed. The (i, j) entry of matrix Q^k denotes the probability of arriving in (transient) state s_j after exactly k steps starting from (transient) state s_i. It can be shown that $\sum_{k=0}^{t} Q^k$ converges as t approaches infinity [PARZ 1962]. This implies that the inverse matrix $(I-Q)^{-1}$, called the **fundamental matrix,** M, exists and is given by:

$$M = (I-Q)^{-1} = I + Q + Q^2 + \cdots = \sum_{k=0}^{\infty} Q^k.$$

The fundamental matrix M is a rich source of information on the Markov chain, as seen below. Let X_{ij} $(1 \leqslant i, j < n)$ be the random variable denoting the number of times the program visits state s_j before entering the absorbing state, given that it started in state s_i. Let $\mu_{ij} = E[X_{ij}]$.

THEOREM 7.5 For $1 \leqslant i, j < n$, $E[X_{ij}] = m_{ij}$, the (i,j)th element of the fundamental matrix M.

Proof [BHAT 1972]:

Initially the process is in the transient state s_i. In one step it may enter the absorbing state s_n with probability p_{in}. The corresponding number of visits to state s_j is equal to zero unless $j = i$. Thus, $X_{ij} = \delta_{ij}$ with probability p_{in}, where δ_{ij} is the Kronecker δ function ($\delta_{ij} = 1$ if $i = j$ and 0 otherwise). Alternately, the process may go to transient state s_k at the first step (with probability p_{ik}). The subsequent number of visits to state s_j is given by X_{kj}. If $i = j$, the total number of visits, X_{ij}, to state s_j will be $X_{kj}+1$, otherwise it will be X_{kj}. Therefore:

$$X_{ij} = \begin{cases} \delta_{ij} & \text{with probability } p_{in}, \\ X_{kj} + \delta_{ij} & \text{with probability } p_{ik},\ 1 \leqslant k < n. \end{cases}$$

If the random variable Y denotes the state of the process at the second step (given that the initial state is i), we can summarize:

$$E[X_{ij}|Y = n] = \delta_{ij},$$
$$E[X_{ij}|Y = k] = E[X_{kj} + \delta_{ij}] = E[X_{kj}] + E[\delta_{ij}] = E[X_{kj}]+\delta_{ij}.$$

Now since the pmf of Y is easily derived as $P(Y = k) = p_{ik}$, $1 \leqslant k \leqslant n$, we can use the theorem of total expectation to obtain:

$$\begin{aligned}\mu_{ij} = E[X_{ij}] &= \sum_k E[X_{ij}|Y = k]\ P(Y = k) \\ &= p_{in}\delta_{ij} + \sum_{k=1}^{n-1} p_{ik}(E[X_{kj}] + \delta_{ij}) \\ &= \sum_{k=1}^{n} p_{ik}\delta_{ij} + \sum_{k=1}^{n-1} p_{ik}\ E[X_{kj}] \end{aligned}$$

$$\mu_{ij} = \delta_{ij} + \sum_{k=1}^{n-1} p_{ik}\mu_{kj}. \tag{7.46}$$

Forming the $(n - 1)$ by $(n - 1)$ matrix consisting of elements μ_{ij}, we have:

$$[\mu_{ij}] = I + Q[\mu_{ij}]$$

or

$$[\mu_{ij}] = (I - Q)^{-1} = M. \tag{7.47}$$

Let V_j denote the average number of times the statement group s_j is executed in a typical run of the program. Then, since s_1 is the starting state of the program, $V_j = m_{1j}$, the element in the first row and the jth column of the fundamental matrix, M. Now if t_j denotes the average execution time of statement group s_j (per visit), then the expected execution time $\bar{t}$ of the program is given by:

$$\bar{t} = (\sum_{j=1}^{n-1} V_j t_j) + t_n = (\sum_{j=1}^{n-1} m_{1j} t_j) + t_n,$$

since the number of visits, V_n, to the stop vertex s_n is one. An alternative form of (7.46) can be used to simplify computations of the visit counts as follows. From (7.47) we have:

$$M(I - Q) = I \quad \text{or} \quad M = I + MQ.$$

Therefore, the (i, j) element of matrix M can be computed using the formula:

$$m_{ij} = \delta_{ij} + \sum_{k=1}^{n-1} m_{ik}p_{kj}. \tag{7.48}$$

Recalling that $m_{1j} = V_j$, we have:

$$V_j = \delta_{1j} + \sum_{k=1}^{n-1} V_k p_{kj}, \qquad j = 1, 2, \ldots, n-1. \tag{7.49}$$

Thus, the visit counts are obtained by solving a system of $(n-1)$ linear equations (7.49).

Example 7.16

Returning to the program discussed earlier in this section, the matrix Q is given by:

$$Q = \begin{bmatrix} 0 & 0.6 & 0.4 & 0 \\ 0 & 0 & 0 & 0.6 \\ 0 & 0.2 & 0 & 0.4 \\ 0 & 0 & 0.6 & 0 \end{bmatrix}.$$

The fundamental matrix M is computed to be:

$$M = (I - Q)^{-1} = \begin{bmatrix} 1 & 0.7791 & 0.8953 & 0.8254 \\ 0 & 1.105 & 0.5236 & 0.8721 \\ 0 & 0.2907 & 1.453 & 0.7558 \\ 0 & 0.1744 & 0.8721 & 1.453 \end{bmatrix}$$

Thus the vertices s_1, s_2, s_3, and s_4 are respectively executed 1, 0.7791, 0.8953, 0.8254 times on the average. The average execution time of the program is equal to:

$$t_1 + 0.7791 \cdot t_2 + 0.8953 \cdot t_3 + 0.8254 \cdot t_4 + t_5 \text{ time units.}$$

#

Example 7.17

Often we are not interested in the details of computation states of a program but only wish to distinguish between the computation state and one of the m I/O states. Thus the program may appear as shown in Figure 7.18. Vertex (labeled 0) COMP is

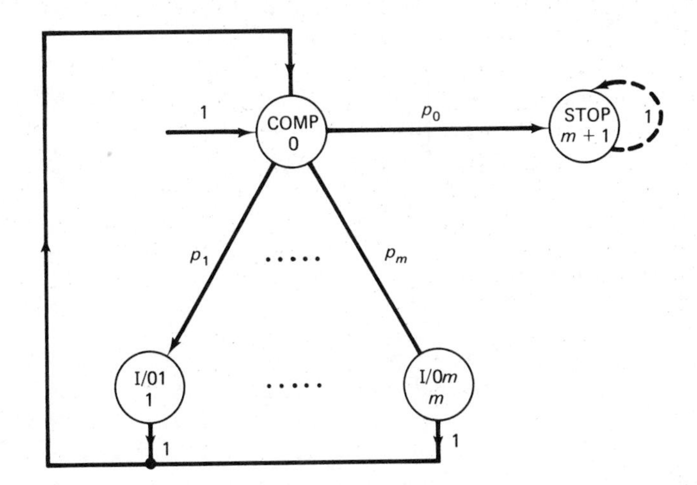

Figure 7.18 A program flow graph

the start vertex. The transition probability matrix P is given by:

$$
\begin{array}{c}
\\ \text{COMP} \\ \text{I/O1} \\ \text{I/O2} \\ \vdots \\ \text{I/O}i \\ \vdots \\ \text{I/O}m \\ \text{STOP}
\end{array}
\begin{array}{c}
\begin{array}{cccccccc}
\text{COMP} & \text{I/O1} & \cdots & \text{I/O}i & \cdots & \text{I/O}m & \text{STOP}
\end{array}\\
\left[\begin{array}{ccccccc}
0 & p_1 & \cdots & p_i & \cdots & p_m & p_0 \\
1 & 0 & & 0 & & 0 & 0 \\
1 & 0 & & 0 & & 0 & 0 \\
\vdots & \vdots & & \vdots & & \vdots & \vdots \\
1 & 0 & & 0 & & 0 & 0 \\
\vdots & \vdots & & \vdots & & \vdots & \vdots \\
1 & 0 & & 0 & & 0 & 0 \\
0 & 0 & & 0 & & 0 & 1
\end{array}\right]
\end{array}
$$

The portion of the transition probability matrix for the transitions among the transient states is given by:

$$
Q = \begin{bmatrix}
0 & p_1 & \cdots & p_m \\
1 & 0 & \cdots & 0 \\
\cdot & \cdot & \cdots & \cdot \\
1 & 0 & \cdots & 0 \\
1 & 0 & \cdots & 0
\end{bmatrix}
$$

Then:

$$
(I - Q) = \begin{bmatrix}
1 & -p_1 & \cdots & -p_m \\
-1 & 1 & 0 & 0 \\
\cdot & 0 & \cdot & \cdot \\
\cdot & \cdot & \cdot & \cdot \\
\cdot & \cdot & \cdot & \cdot \\
-1 & 0 & 0 & 1
\end{bmatrix}
$$

and the fundamental matrix:

$$
M = (I - Q)^{-1} = \begin{bmatrix}
1/p_0 & p_1/p_0 & p_2/p_0 & \cdots & p_m/p_0 \\
1/p_0 & 1 + p_1/p_0 & p_2/p_0 & \cdots & p_m/p_0 \\
\cdot & & & & \\
\cdot & & & & \\
\cdot & & & & \\
1/p_0 & p_1/p_0 & p_2/p_0 & \cdots & 1 + p_m/p_0
\end{bmatrix}
$$

Thus $V_0 = m_{00} = 1/p_0$ is the average number of times the COMP state is visited, and $V_j = m_{0j} = p_j/p_0$ is the average number of times the state I/Oj is visited.

This result can be derived directly using equation (7.49):

$$V_0 = \delta_{00} + \sum_{k=0}^{m} p_{k0} V_k, \qquad j = 0,$$

$$= 1 + \sum_{k=1}^{m} V_k$$

and

$$V_j = \delta_{0j} + \sum_{k=0}^{m} p_{kj} V_k, \qquad j = 1, 2, \ldots, m,$$

$$= p_j V_0.$$

Solving, we get:

$$V_j = \begin{cases} \dfrac{1}{p_0}, & j = 0, \\ \dfrac{p_j}{p_0}, & j = 1, 2, \ldots, m. \end{cases} \tag{7.50}$$

#

So far, we have assumed that there is a unique entry point (START state) to the program. The results above are easily generalized to the case with multiple entry points. Suppose that the program starts from vertex s_j with probability q_j $(1 \leqslant j \leqslant n)$ so that $\sum_{j=1}^{n} q_j = 1$. Since m_{ij} denotes the average number of times node j is visited given that the process started in node i, the average number, V_j, of times node j is visited without conditioning on the START state is given by the theorem of total expectation as:

$$V_j = \sum_{i=1}^{n} m_{ij} q_i. \tag{7.51}$$

Substituting expression (7.48) for m_{ij}, we have:

$$V_j = \sum_{i=1}^{n} q_i [\delta_{ij} + \sum_{k=1}^{n-1} m_{ik} p_{kj}]$$

$$= \sum_{i=1}^{n} q_i \delta_{ij} + \sum_{i=1}^{n} q_i \sum_{k=1}^{n-1} m_{ik} p_{kj}$$

$$= q_j + \sum_{k=1}^{n-1} p_{kj} \sum_{i=1}^{n} m_{ik} q_i,$$

interchanging the order of summation. Now, using (7.51), we have:

$$V_j = q_j + \sum_{k=1}^{n-1} p_{kj} V_k, \qquad j = 1, 2, \ldots, n - 1. \tag{7.52}$$

Clearly, for the STOP state s_n, the VISIT count still remains at 1. Equation (7.52) will be used again in Chapter 9.

Problems

1. [KNUT 1973] Given the stochastic program flow graph shown in Figure 7.P.2, compute the average number of times each vertex s_i is visited, and assuming that the execution time of s_i is given by $t_i = 2i + 1$ time units, find the average total execution time τ of the program.

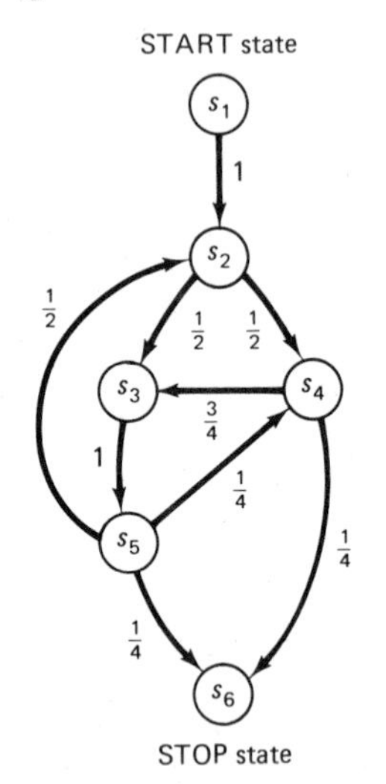

Figure 7.P.2 Another program flow graph

Review Problems

*1. [MCCA 1965] In Example 4.2 we stated that the average time to sequentially search a linear list is minimized if the keys are arranged in decreasing order of request probabilities. If the request probabilities are not known in advance, near-optimal behavior can still be achieved by using a self-organized list. One such technique, known as a "move to front" heuristic, moves the requested key, if located, to the front of the list.

   Show that the average number of key comparisons needed for a successful search is given by:

$$E[X] = 1 + \sum_{i=1}^{n} \sum_{\substack{j=1 \\ i \neq j}}^{n} \frac{\alpha_i \alpha_j}{\alpha_i + \alpha_j},$$

where n is the number of distinct keys in the list and $0 < \alpha_i < 1$ is the probability of accessing the key labeled i (which may be located in any one of the n positions). You may proceed to show this result first in case $n = 3$ and then attempt to generalize it.

*2. Reconsider Example 7.13 and assume that an overflow forces the program to abort, but an underflow is simply ignored. Compute the average number of operations until an abort occurs. Simplify the problem by first considering a data structure with three nodes and then attempt to generalize your result.

References

[ASH 1970] R. B. ASH, *Basic Probability Theory,* John Wiley & Sons, New York.

[BASK 1976] F. BASKETT and A. J. SMITH, "Interference in Multiprocessor Computer Systems with Interleaved Memory," *CACM,* 19:6, (June 1976), 327–34.

[BHAT 1972] U. N. BHAT, *Elements of Applied Stochastic Processes,* John Wiley & Sons, New York.

[CHAN 1977] D. CHANG, D. J. KUCK, and D. H. LAWRIE, "On the Effective Bandwidth of Parallel Memories," *IEEE Transactions on Computers,* C-26:5 (May 1977), 480–90.

[CLAR 1970] A. B. CLARKE and R. L. DISNEY, *Probability and Random Processes for Engineers and Scientists,* John Wiley & Sons, New York.

[COFF 1973] E. G. COFFMAN, Jr. and P. J. DENNING, *Operating System Theory,* Prentice-Hall, Englewood Cliffs, N.J.

[COUR 1977] P. J. COURTOIS, *Decomposability: Queuing and Computer System Applications,* Academic Press, New York.

[DEO 1974] N. DEO, *Graph Theory with Applications to Engineering and Computer Science,* Prentice-Hall, Englewood Cliffs, N.J.

[FELL 1968] W. FELLER, *An Introduction to Probability Theory and its Applications,* 2 Volumes, John Wiley & Sons, New York.

[FULL 1975] S. H. FULLER, "Performance Evaluation," in *Introduction to Computer Architecture,* H.S. Stone (ed.), Science Research Associates, Chicago.

[KNUT 1973] D. E. KNUTH, *The Art of Computer Programming,* Vol. 1, Addison-Wesley, Reading, Mass.

[MCCA 1965] J. MCCABE, "On Serial Files with Relocatable Records," *Operations Research,* Vol. 13, pp. 609–18.

[PARZ 1962] E. PARZEN, *Stochastic Processes,* Holden-Day, San Fransisco, California.

[RAMA 1965] C. V. RAMAMOORTHY, "Discrete Markov Analysis of Computer Programs," *Proceedings, ACM National Conference,* pp. 386–392.

[SHED 1972] G. S. SHEDLER and C. TUNG, "Locality in Page Reference String," *SIAM Journal on Computing,* 1:3 (September 1972), 218–41.

[SPIR 1977] J. R. SPIRN, *Program Behavior: Models and Measurements,* Elsevier, New York.

Chapter 8

Continuous-Parameter Markov Chains

8.1 INTRODUCTION

The analysis of continuous-parameter Markov chains is similar to that for the discrete-parameter case, except that the transitions from a given state to another state can take place at any instant of time. As in the last chapter, we confine our attention to **discrete-state** processes. This implies that, although the parameter t has a continuous range of values, the set of values of $X(t)$ is discrete. We let $I = \{0, 1, 2, \ldots\}$ denote the state space of the process and we let $T = [0, \infty)$ be its parameter space. As we recall from Chapter 6, a discrete-state continuous-parameter stochastic process $\{X(t),\ t \geqslant 0\}$ is called a Markov chain if for $t_0 < t_1 < t_2 \cdots < t_n < t$, with t and $t_r \geqslant 0$ $(r = 0, 1, \ldots, n)$, its conditional pmf satisfies the relation

$$\begin{aligned} P[X(t) = x | X(t_n) = x_n,\ X(t_{n-1}) = x_{n-1},\ \ldots,\ X(t_0) = x_0] \\ = P[X(t) = x |\ X(t_n) = x_n]. \end{aligned} \tag{8.1}$$

The behavior of the process is characterized by (1) the distribution of the initial state of the system given by the pmf of $X(t_0)$, $P[X(t_0) = k]$, $k = 0, 1, 2, \ldots$, and (2) the transition probabilities:

$$p_{ij}(v, t) = P[X(t) = j \mid X(v) = i] \tag{8.2}$$

for $0 \leqslant v \leqslant t$ and $i, j = 0, 1, 2, \ldots$, where we define:

$$p_{ij}(t, t) = \begin{cases} 1, & \text{if } i = j, \\ 0, & \text{otherwise.} \end{cases}$$

The Markov chain $\{X(t), t \geqslant 0\}$ is said to be **(time)-homogeneous** (or is said to have **stationary transition probabilities**) if $p_{ij}(v, t)$ depends only on the time difference $(t - v)$. In this case, we abbreviate the notation for the transition probabilities:

$$p_{ij}(t) = P[X(t + u) = j \mid X(u) = i] \qquad \text{for any } u \geqslant 0. \tag{8.2'}$$

Since (8.2) is a conditional pmf, it satisfies the relation:

$$\sum_j p_{ij}(v, t) = 1 \qquad \text{for all } i;\ 0 \leqslant v \leqslant t. \tag{8.3}$$

Let us denote the pmf of $X(t)$ (or the **state probabilities** at time t) by:

$$P_j(t) = P[X(t) = j], \qquad j = 0, 1, 2, \ldots, \quad t \geqslant 0. \tag{8.4}$$

It is clear that:

$$\sum_j P_j(t) = 1$$

for each $t \geqslant 0$, since at any given time the process must be in *some* state.

By using the theorem of total probability, for given $t > v$, we can express the pmf of $X(t)$ in terms of the transition probabilities $p_{ij}(v, t)$ and the pmf of $X(v)$:

$$\begin{aligned} P_j(t) &= P[X(t) = j] \\ &= \sum_i P[X(t) = j | X(v) = i] P[X(v) = i] \\ &= \sum_i p_{ij}(v, t)\, P_i(v). \end{aligned} \tag{8.5}$$

If we let $v = 0$ in (8.5), then:

$$P_j(t) = \sum_i p_{ij}(0, t) P_i(0). \tag{8.5'}$$

Hence, the probabilistic behavior of a Markov chain is completely determined once the transition probabilities $p_{ij}(v, t)$ and the initial probability vector $\mathbf{P}(0) = [P_0(0), P_1(0), \ldots]$ are specified.

The transition probabilities of a Markov chain $\{X(t), t \geqslant 0\}$ satisfy the **Chapman-Kolmogorov equation**: for all $i, j \in I$,

$$p_{ij}(v, t) = \sum_{k \in I} p_{ik}(v, u) p_{kj}(u, t) \qquad 0 \leqslant v < u < t. \tag{8.6}$$

To prove (8.6), we use the theorem of total probability:

$$\begin{aligned} P[X(t) = j \mid X(v) = i] = \sum_{k \in I} & P[X(t) = j \mid X(u) = k, X(v) = i] \\ & \cdot P[X(u) = k \mid X(v) = i]. \end{aligned}$$

The subsequent application of the Markov property (8.1) yields (8.6).

The direct use of (8.6) is difficult. Usually we obtain the transition probabilities by solving a system of differential equations that we derive next. For this purpose, under certain regularity conditions, we can show that for each j there is a nonnegative continuous function $q_j(t)$ defined by:

$$q_j(t) = -\frac{\partial}{\partial t} p_{jj}(v, t)\big|_{v=t}$$

$$= \lim_{h \to 0} \frac{p_{jj}(t, t) - p_{jj}(t, t+h)}{h} = \lim_{h \to 0} \frac{1 - p_{jj}(t, t+h)}{h}. \quad (8.7)$$

Similarly for each i and j ($\neq i$) there is a nonnegative continuous function $q_{ij}(t)$ defined by:

$$q_{ij}(t) = \frac{\partial}{\partial t} p_{ij}(v, t)\big|_{v=t}$$

$$= \lim_{h \to 0} \frac{p_{ij}(t, t+h) - p_{ij}(t, t)}{h} = \lim_{h \to 0} \frac{p_{ij}(t, t+h)}{h}. \quad (8.8)$$

Then the transition probabilities and the transition rates are related by:[1]

$$p_{ij}(t, t+h) = q_{ij}(t) \cdot h + o(h), \qquad i \neq j,$$

and

$$p_{jj}(t, t+h) = 1 - q_j(t) \cdot h + o(h), \qquad i = j.$$

Substituting $t + h$ for t in equation (8.6), we get:

$$p_{ij}(v, t+h) = \sum_k p_{ik}(v, u) p_{kj}(u, t+h)$$

which implies

$$p_{ij}(v, t+h) - p_{ij}(v, t) = \sum_k p_{ik}(v, u)[p_{kj}(u, t+h) - p_{kj}(u, t)].$$

Dividing both sides by h and taking the limit $h \to 0$ and $u \to t$, we get the differential equation known as **Kolmogorov's forward equation:** for $0 \leqslant v < t$ and $i, j \in I$:

$$\frac{\partial p_{ij}(v, t)}{\partial t} = [\sum_{k \neq j} p_{ik}(v, t) q_{kj}(t)] - p_{ij}(v, t) q_j(t). \quad (8.9)$$

In a similar fashion we can also derive **Kolmogorov's backward equation:**

$$\frac{\partial p_{ij}(v, t)}{\partial v} = [\sum_{k \neq i} p_{kj}(v, t) q_{ik}(v)] - p_{ij}(v, t) q_i(v).$$

[1] $o(h)$ is any function of h that approaches zero faster than h:

$$\lim_{h \to 0} \frac{o(h)}{h} = 0.$$

Using (8.5) and (8.9) we can also derive a differential equation for the unconditional probability $P_j(t)$ as:

$$\frac{dP_j}{dt} = [\sum_{i \neq j} P_i(t) q_{ij}(t)] - P_j(t) q_j(t). \tag{8.9'}$$

We use (8.9) when we want specifically to show the initial state, (8.9′) when the initial state (or initial distribution) is implied.

In many important applications the transition probabilities $p_{ij}(t, t + h)$ do not depend on the initial time t but only on the elapsed time h (that is, the resulting Markov chain is **time-homogeneous**). This implies that the transition rates $q_{ij}(t)$ and $q_j(t)$ are independent of t. Unless otherwise stated, we will be concerned only with time-homogeneous situations. In this case the transition rates are denoted by q_{ij} and the transition probabilities $p_{ij}(t, t + h)$ by $p_{ij}(h)$. Equations (8.9) and (8.9′) are rewritten as:

$$\frac{dp_{ij}(t)}{dt} = [\sum_{k \neq j} p_{ik}(t) q_{kj}] - p_{ij}(t) q_j, \tag{8.10}$$

$$\frac{dP_j}{dt} = \sum_{i \neq j} P_i(t) q_{ij} - P_j(t) q_j. \tag{8.10'}$$

Even in this simpler case of a time-homogeneous Markov chain, solution of equation (8.10) to obtain the time-dependent probabilities $P_j(t)$ is quite difficult. Nevertheless, in many interesting situations a further reduction is possible in that the probabilities $P_j(t)$ approach a limit p_j as t approaches infinity. We wish to explore the conditions under which such a limiting distribution exists.

A classification of states for a continuous-parameter Markov chain is similar to the discrete-parameter case. A state i is said to be an **absorbing state** provided that $q_{ij} = 0$ for all $j \neq i$, so that, once entered, the process is destined to remain in that state. For a Markov chain with two or more absorbing states, the limiting probabilities $\lim_{t \to \infty} p_{ij}(t)$ may well depend upon the initial state.

A state j is said to be **reachable** from state i if for some $t > 0$, $p_{ij}(t) > 0$. A continuous-parameter Markov chain is said to be **irreducible** if every state is reachable from every other state.

THEOREM 8.1.

For an irreducible continuous-parameter Markov chain, the limits:

$$p_j = \lim_{t \to \infty} p_{ij}(t) = \lim_{t \to \infty} P_j(t), \qquad i, j \in I, \tag{8.11}$$

always exist and are independent of the initial state i.

If the limiting probabilities p_j exist, then:

$$\lim_{t \to \infty} \frac{dP_j(t)}{dt} = 0, \tag{8.12}$$

and, substituting into equation (8.10′), we get the following system of linear homogeneous equations (one for each state j):

$$0 = \sum_{i \neq j} p_i q_{ij} - p_j q_j. \tag{8.13}$$

This is the continuous analog of equation (7.14).

For the homogeneous system of equations, one possible solution is that $p_j = 0$ for all j. If another solution exists, then an infinite number of solutions can be obtained by multiplying by scalars. To determine a nonzero unique solution, we use the condition:

$$\sum_j p_j = 1. \tag{8.14}$$

Irreducible Markov chains that yield positive limiting probabilities $\{p_j\}$ in this way are called **recurrent non-null** or **positive recurrent** and the probabilities $\{p_j\}$, satisfying (8.13) and (8.14) are also known as steady-state probabilities. It is clear that a finite irreducible Markov chain must be positive recurrent, hence we can obtain its unique limiting probabilities by solving the finite system of equations (8.13) under the condition (8.14).

We have seen in Chapter 6 that the distribution of times that a continuous-parameter homogeneous Markov chain spends in a given state must be memoryless. This implies that holding times in a state of a continuous-parameter Markov chain of the homogeneous type are exponentially distributed. In the next section we study the limiting distribution of a special type of Markov chain, called the birth-death process. In Section 8.4, we study limiting distributions of several non-birth-death processes.

The study of transient behavior $[P_j(t),\ t \geqslant 0]$ is quite complex for a general Markov chain. In Sections 8.3 and 8.5 we consider special cases where it is possible to obtain an explicit solution for $P_j(t)$.

Problems

*1. For a homogeneous Markov chain, define the **transition-rate matrix** Q so that its diagonal elements are given by $-q_j$ and the $(i,\ j)$ element is given by q_{ij} $(i \neq j)$. Define the matrix $P(t) = [p_{ij}(t)]$ and show that the forward and backward Kolmogorov equations can be rewritten as:

$$\frac{dP}{dt} = P(t)Q \quad \text{and} \quad \frac{dP}{dt} = QP(t)$$

with the initial condition $P(0) = I$, the identity matrix. Show that the solution to these matrix equations can be written as:

$$P(t) = e^{Qt} = I + \sum_{n=1}^{\infty} Q^n \frac{t^n}{n!},$$

assuming that matrix series converges. Generalize this result to the case of a nonhomogeneous Markov chain.

***2.** For a homogeneous Markov chain show that the Laplace transform of the transition probability matrix $P(t)$, denoted by $\bar{P}(s)$, is given by:

$$\bar{P}(s) = (sI - Q)^{-1}.$$

***3.** Show that the integral (convolution) form of the Kolmogorov forward equation is given by:

$$p_{ij}(v, t) = \delta_{ij}\, e^{-\int_v^t q_i(\tau)\,d\tau} + \int_v^t \sum_k p_{ik}(v, x)\, q_{kj}(x)\, e^{-\int_x^t q_j(\tau)\,d\tau}\, dx,$$

where δ_{ij} is the Kronecker delta function defined by $\delta_{ij} = 1$ if $i = j$ and 0 otherwise. Specialize this result to the case of a homogeneous Markov chain.

8.2 THE BIRTH AND DEATH PROCESS

A continuous-parameter homogeneous Markov chain $\{X(t), t \geqslant 0\}$ with the state space $\{0, 1, 2, \ldots\}$ is known as a **birth-death process** if there exist constants λ_i $(i = 0, 1, \ldots)$ and μ_i $(i = 1, 2, \ldots)$ such that the transition rates are given by:

$$q_{i,i+1} = \lambda_i,$$

$$q_{i,i-1} = \mu_i,$$

$$q_i = \lambda_i + \mu_i,$$

$$q_{ij} = 0 \qquad \text{for } |i - j| > 1.$$

The **birth rate** λ_i $(\geqslant 0)$ is the rate at which births occur in state i, and the **death rate** μ_i $(\geqslant 0)$ is the rate at which deaths occur in state i. These rates are assumed to depend only on state i and are independent of time. Note that only "nearest-neighbor" transitions are allowed. In a given state, births and deaths occur independently of each other. Such a process is a useful model of many situations in queuing theory and reliability theory.

The process will be in state k at time $t + h$ if one of the following mutually exclusive and collectively exhaustive events occurs:

1. The system is in state k at time t, and no changes of state occur in the interval $(t, t + h]$; the associated conditional probability is:

$$p_{k,k}(t, t + h) = 1 - q_k(t) \cdot h + o(h) = 1 - (\lambda_k + \mu_k) \cdot h + o(h).$$

2. The system is in state $k - 1$ at time t, and one birth occurs in the interval $(t, t + h]$; the associated conditional probability is:

$$p_{k-1,k}(t, t + h) = q_{k-1,k}(t) \cdot h + o(h) = \lambda_{k-1} \cdot h + o(h).$$

3. The system is in state $k + 1$ and one death occurs in the interval $(t,\ t + h]$; the associated conditional probability is

$$p_{k+1,k}(t,\ t + h) = q_{k+1,k}(t) \cdot h + o(h) = \mu_{k+1} \cdot h + o(h).$$

4. Two or more transitions occur in the interval $(t,\ t + h]$, resulting in $X(t + h) = k$, with associated conditional probability $o(h)$.

Then by the theorem of total probability we have:

$$\begin{aligned} P[X(t + h) = k] &= P_k(t + h) \\ &= P_k(t)p_{k,k}(t,\ t + h) + P_{k-1}(t)p_{k-1,k}(t,\ t + h) \\ &\quad + P_{k+1}(t)p_{k+1,k}(t,\ t + h) + o(h). \end{aligned}$$

After rearranging, dividing by h, and taking the limit as $h \to 0$, we get:

$$\frac{dP_k(t)}{dt} = -(\lambda_k + \mu_k)P_k(t) + \lambda_{k-1}P_{k-1}(t) + \mu_{k+1}P_{k+1}(t), \qquad k \geqslant 1, \tag{8.15}$$

$$\frac{dP_0(t)}{dt} = -\lambda_0 P_0(t) + \mu_1 P_1(t), \qquad k = 0,$$

where the special equation for $k = 0$ is required because the state space of the process is assumed to be $\{0, 1, 2, \ldots\}$. Equation (8.15) is a special case of equation (8.10′), with $q_{k-1,k} = \lambda_{k-1}$, $q_{k+1,k} = \mu_{k+1}$, and $q_k = (\lambda_k + \mu_k)$.

The solution of this system of differential-difference equations is a formidable task. However, if we are not interested in the transient behavior, then we can set the derivative $dP_k(t)/dt$ equal to zero, and the resulting set of difference equations provide the steady-state solution of the Markov chain. Let p_k denote the steady-state probability that the chain is in state k; that is, $p_k = \lim_{t \to \infty} P_k(t)$ (assuming it exists). Then the above differential-difference equations reduce to [a special case of equation (8.13)]:

$$0 = -(\lambda_k + \mu_k)p_k + \lambda_{k-1}p_{k-1} + \mu_{k+1}p_{k+1}, \qquad k \geqslant 1, \tag{8.16}$$

$$0 = -\lambda_0 p_0 + \mu_1 p_1. \tag{8.17}$$

These are known as the **balance equations** and we can obtain them directly from the state diagram, shown in Figure 8.1, by equating the rates of flow

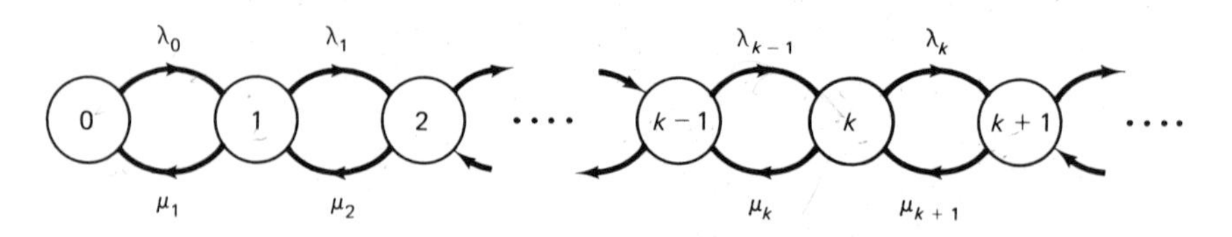

Figure 8.1 The state diagram of the birth-death process

into and out of each state. From the state diagram we have the rate of transition into state k as $\lambda_{k-1}p_{k-1} + \mu_{k+1}p_{k+1}$ and the rate of transition out of state k as $(\lambda_k + \mu_k)p_k$. In the steady state no build-up occurs in state k, hence these two rates must be equal.

We should note the difference between this state diagram (of a continuous-parameter Markov chain) and the state diagram of a discrete-parameter Markov chain (Chapter 7). In the latter, the arcs are labeled with conditional proabilities; in the former they are labeled with state transition rates (hence, the name transition-rate diagram is sometimes used).

By rearranging equation (8.16), we get:

$$\lambda_k p_k - \mu_{k+1}p_{k+1} = \lambda_{k-1}p_{k-1} - \mu_k p_k = \cdots = \lambda_0 p_0 - \mu_1 p_1.$$

But from (8.17) we have $\lambda_0 p_0 - \mu_1 p_1 = 0$. It follows that:

$$\lambda_{k-1}p_{k-1} - \mu_k p_k = 0$$

and hence

$$p_k = \frac{\lambda_{k-1}}{\mu_k} p_{k-1}, \qquad k \geqslant 1.$$

Therefore:

$$p_k = \frac{\lambda_0\lambda_1 \cdots \lambda_{k-1}}{\mu_1\mu_2 \cdots \mu_k} p_0 = p_0 \prod_{i=0}^{k-1} \left(\frac{\lambda_i}{\mu_{i+1}}\right), \qquad k \geqslant 1. \tag{8.18}$$

Since $\sum_{k \geqslant 0} p_k = 1$, we have:

$$p_0 = \frac{1}{1 + \sum_{k \geqslant 1} \prod_{i=0}^{k-1} \left(\frac{\lambda_i}{\mu_{i+1}}\right)}. \tag{8.19}$$

Thus, the limiting distribution $(p_0, p_1, \ldots .)$ is now completely determined. Note that the limiting probabilities are nonzero, provided that the series:

$$\sum_{k \geqslant 1} \prod_{i=0}^{k-1} \left(\frac{\lambda_i}{\mu_{i+1}}\right)$$

converges (in which case, all the states of the Markov chain are recurrent non-null).

Next we consider several special cases of the birth-death process.

8.2.1 The *M/M/1* Queue

We consider a single-server Markovian queue shown in Figure 8.2. Customer arrivals form a Poisson process with rate λ. Equivalently the customer interarrival times are exponentially distributed with mean $1/\lambda$. Service times of

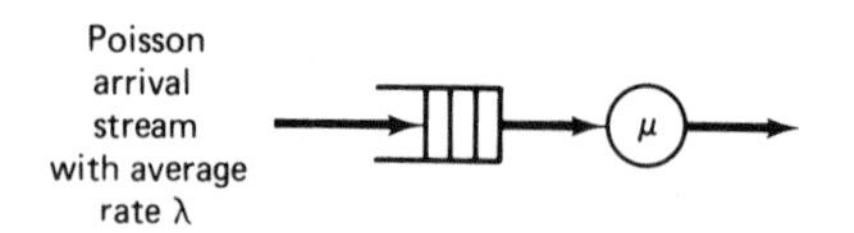

Figure 8.2 The $M/M/1$ queuing system

customers are independent identically distributed random variables, the common distribution being exponential with mean $1/\mu$. Assume that customers are served in their order of arrival (FCFS scheduling). If the "customer" denotes a job arriving into a computer system, then the server represents the computer system. [Since most computer systems consist of a set of interacting resources (and hence a network of queues), such a simple representation may be acceptable for "small" systems with little concurrency.] In another interpretation of the $M/M/1$ queue, the customer may represent a message and the server a communication channel.

Let $N(t)$ denote the number of customers in the system (those queued plus the one in service) at time t. [We change the notation from $X(t)$ to $N(t)$ to conform to standard practice.] Then $\{N(t),\ t \geqslant 0\}$ is a birth-death process with:

$$\lambda_k = \lambda, \quad k \geqslant 0; \qquad \mu_k = \mu, \quad k \geqslant 1.$$

The ratio $\rho = \lambda/\mu$ = mean service time /mean interarrival time, is an important parameter, called the **traffic intensity** of the system. Equations (8.18) and (8.19) in this case reduce to:

$$p_k = (\frac{\lambda}{\mu})^k p_0 = \rho^k p_0$$

and

$$p_0 = \frac{1}{\sum\limits_{k \geqslant 0} \rho^k} = 1 - \rho,$$

provided $\rho < 1$—that is, when the traffic intensity is less than unity. In the case that the arrival rate λ exceeds the service rate μ (i.e., $\rho \geqslant 1$), the geometric series in the denominator of the expression for p_0 diverges. In this case all the states of the Markov chain are either recurrent null or transient, hence the number of customers in the system tends to increase without bound. Such a system is called **unstable.** For a stable system ($\rho < 1$), the steady-state probabilities have a modified geometric distribution with parameter $1 - \rho$; that is,

$$p_k = (1 - \rho)\rho^k, \qquad k \geqslant 0. \tag{8.20}$$

The server utilization, $U_0 = 1 - p_0 = \rho$, is interpreted as the proportion of time the server is busy.

The mean and variance of the number of customers in the system are obtained using the properties of the modified geometric distribution as:

$$E[N] = \frac{\rho}{1-\rho} \tag{8.21}$$

and

$$\text{Var}\ [N] = \frac{\rho}{(1-\rho)^2}. \tag{8.22}$$

Let the random variable R denote the response time in the steady state. In order to compute the average response time $E[R]$ we use the well-known **Little's formula**, which states that the mean number of jobs in a queuing system in the steady-state is equal to the product of the arrival rate and the mean response time. When applied to the present case, Little's formula gives us:

$$E[N] = \lambda E[R].$$

Little's formula holds for a broad variety of queuing systems. For a proof see [STID 1974], and for its limitations see [BEUT 1980].

Using (8.21) and applying Little's formula to the present case, we have:

$$E[R] = \lambda^{-1}\frac{\rho}{1-\rho} = \frac{1/\mu}{1-\rho} = \frac{\text{average service time}}{\text{probability that the server is idle}}. \tag{8.23}$$

Note that the congestion in the system and hence the delay build rapidly as the traffic intensity increases (see Figures 8.3 and 8.4).

We may often employ a scheduling discipline other than FCFS. We distinguish between preemptive and nonpreemptive scheduling disciplines. A **nonpreemptive discipline** such as FCFS allows a job to complete execution once scheduled, whereas a **preemptive discipline** may interrupt the currently executing job in order to give preferential service to another job. A common example of a preemptive discipline is RR (round robin), which permits a job to remain in service for an interval of time referred to as its **quantum**. If the job does not finish execution within the quantum, it has to return to the end of the queue, awaiting further service. This gives preferential treatment to

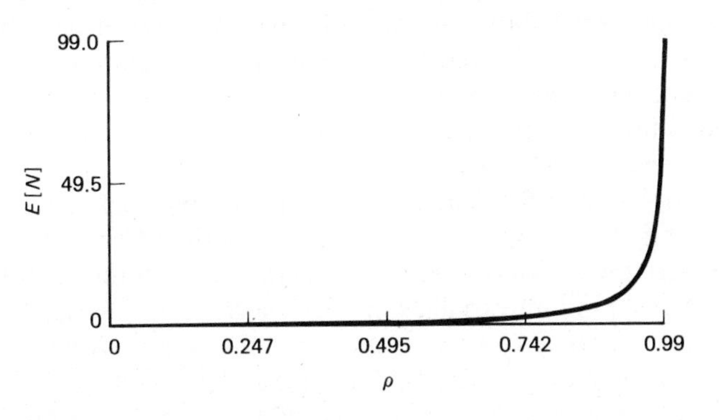

Figure 8.3 Expected number of jobs in system versus traffic intensity

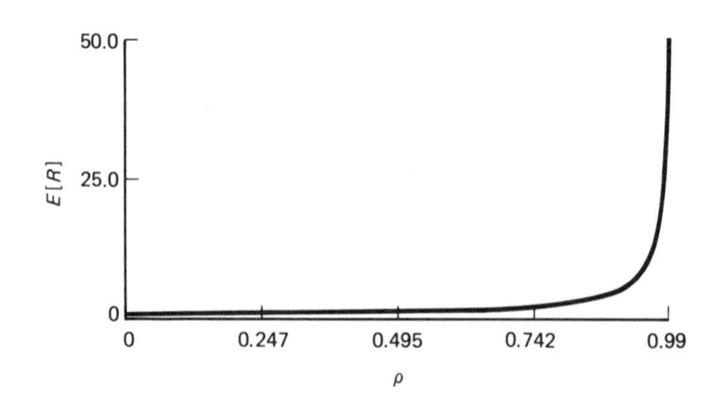

Figure 8.4 Average response time versus traffic intensity

short jobs at the expense of long jobs. When the time quantum approaches zero, the RR discipline is known as the PS (processor sharing) discipline.

Although we assumed FCFS scheduling discipline in deriving formula (8.20) for the queue-length distribution and formula (8.23) for the average response time, they hold for any scheduling discipline that satisfies the following conditions [COFF 1973, KOBA 1978]:

1. The server is not idle when there are jobs waiting for service.
2. The scheduler is not allowed to use any deterministic a priori information about job service times. Thus, for instance, if all job service times are known in advance, the use of a discipline known as SRPT (shortest remaining processing time first) is known to reduce $E[R]$ below that given by (8.23).
3. The service time distribution is not affected by the scheduling discipline.

Formulas (8.20) and (8.23) also apply for preemptive scheduling disciplines such as RR and PS, provided the overhead of preemption can be neglected (otherwise condition 3 above will be violated). We have also assumed in the above that a job is not allowed to leave the system before completion (see problems 3 and 4 below for exceptions).

Although the expression for the average response time (8.23) holds under a large class of scheduling disciplines, the distribution of the response time *does* depend upon the scheduling discipline. We shall derive the distribution function of the response time R assuming the FCFS scheduling discipline. If an arriving job finds n jobs in the system, then the response time is the sum of $n + 1$ random variables, $S + S'_1 + S_2 + \cdots + S_n$. Here S is the service time of the tagged job, S'_1 is the remaining service time of the job undergoing service, and $S_2, \ldots, S_n$ denote the service times of $(n - 1)$ jobs waiting in the queue. By our assumptions and the memoryless property of exponential distribution, these $(n + 1)$ random variables are independent and exponentially distributed with parameter μ. Thus, the conditional Laplace transform of R given $N = n$ is the convolution:

$$L_{R|N}(s|n) = \left(\frac{\mu}{s+\mu}\right)^{n+1}. \tag{8.24}$$

Kleinrock [KLEI 1975] shows that the distribution of the number of jobs in the system as seen by an arriving job is the same as that given by (8.20). Then, applying the theorem of total Laplace transform, we obtain:

$$\begin{aligned} L_R(s) &= \sum_{n=0}^{\infty} \left(\frac{\mu}{s+\mu}\right)^{n+1}(1-\rho)\rho^n \\ &= \frac{\mu(1-\rho)}{s+\mu} \frac{1}{1-\dfrac{\mu\rho}{s+\mu}} \\ &= \frac{\mu(1-\rho)}{s+\mu(1-\rho)}. \end{aligned} \tag{8.25}$$

It follows that the response time R is exponentially distributed with parameter $\mu(1-\rho)$.

Other measures of system performance are easily obtained. Let the random variable W denote the waiting time; that is, let:

$$W = R - S. \tag{8.26}$$

Then:

$$E[R] = E[S] + E[W] = \frac{1}{\mu} + E[W].$$

It follows that the average waiting time is given by:

$$E[W] = \frac{1}{\mu(1-\rho)} - \frac{1}{\mu} = \frac{\rho}{\mu(1-\rho)}. \tag{8.27}$$

If we now let the random variable Q denote the number of jobs waiting in the queue (excluding those if any, in service), then to determine the average number of jobs $E[Q]$ in the queue, we apply Little's formula to the queue excluding the server to obtain:

$$E[Q] = \lambda E[W] = \frac{\rho^2}{1-\rho}. \tag{8.28}$$

Note that the average number of jobs found in the server is:

$$E[N] - E[Q] = \rho. \tag{8.29}$$

Example 8.1

The capacity of a communication line is 2,000 bits per second. This line is used to transmit eight-bit characters, so the maximum rate is 250 characters per second. The application calls for traffic from many devices to be sent on the line with a total volume of 12,000 characters per minute. In this case:

$$\lambda = \frac{12,000}{60} = 200 \text{ cps}, \qquad \mu = 250 \text{ cps},$$

and line utilization $\rho = \lambda/\mu = 4/5 = 0.8$.

The average number of characters waiting to be transmitted is $E[Q] = 0.8 \cdot 0.8/(1-0.8) = 3.2$, and the average transmission (including queuing delay) time per character is $E[N]/\lambda = 4/200$ seconds $= 20$ ms. #

Example 8.2

We wish to determine the maximum call rate that can be supported by one telephone booth. Assume that the mean duration of a telephone conversation is three minutes, and that no more than a three-minute (average) wait for the phone may be tolerated; what is the largest amount of incoming traffic that can be supported?

1. $\mu = 1/3$ calls per minute; therefore, λ must be less than $1/3$ calls per minute, for the line to be stable.
2. The average waiting time $E[W]$ is given as three minutes; that is:

$$E[W] = \frac{\rho}{\mu(1-\rho)} = 3,$$

and since $\mu = 1/3$, we get:

$$1 - \rho = \rho$$

or

$$\rho = 1/2.$$

Therefore, the call arrival rate is given by

$$\lambda = 1/6 \text{ calls per minute.}$$

#

Problems

1. Consider an $M/M/1$ queue with an average arrival rate λ and the average service rate μ. We have derived the distribution function of the response time R. Now we are interested in deriving the distribution function of the waiting time W. The waiting time W is the response time minus the service time. To get started, first compute the conditional distribution of W conditioned upon the number of jobs in the system, and later compute the unconditional distribution function. Note that W is a mixed random variable since its distribution function has a jump equal to $P(W = 0)$ at the origin.

2. A group of telephone subscribers is observed continuously during a 80-minute busy-hour period. During this time they make 30 calls, with the total conversation time being 4,200 seconds. Compute the call arrival rate and the traffic intensity.

3. Consider an $M/M/1$ queuing system in which the total number of jobs is limited to n owing to a limitation on queue size.
   (a) Find the steady-state probability that an arriving request is rejected because the queue is full.

(b) Find the steady-state probability that the processor is idle.
(c) Given that a request has been accepted, find its average response time.

4. The arrival of large jobs at a computing center forms a Poisson process with rate two per hour. The service times of such jobs are exponentially distributed with mean 20 minutes. Only four large jobs can be accommodated in the system at a time. Assuming that the fraction of computing power utilized by smaller jobs is negligible, determine the probability that a large job will be turned away because of lack of storage space.

5. Let the random variable T_k denote the holding time in state k of the $M/M/1$ queue. Starting from the given assumptions on the interarrival time and the service time distributions show that the distribution of T_k is exponential (for each k).

6. Derive an expression for the frequency of entering state 0 (server idle) in an $M/M/1$ queue. This quantity is useful in estimating the overhead of scheduling. Plot this probability as a function of ρ for a fixed μ.

8.2.2 The *M/M/m* Queue

Consider a queuing system with arrival rate λ as before, but where $m \geqslant 1$ servers, with rate μ each, share a common queue (see Figure 8.5). This gives rise to a birth-death model with the rates:

$$\lambda_k = \lambda, \qquad k = 0, 1, 2, \ldots,$$

$$\mu_k = \begin{cases} k\mu, & 0 \leqslant k < m, \\ m\mu, & m \leqslant k. \end{cases}$$

The state diagram of this system is shown in Figure 8.6. The steady-state probabilities are given by [using equation (8.18)]:

$$\begin{aligned} p_k &= p_0 \prod_{i=0}^{k-1} \frac{\lambda}{(i+1)\mu} \\ &= p_0 \left(\frac{\lambda}{\mu}\right)^k \frac{1}{k!}, \qquad k < m, \\ p_k &= p_0 \prod_{i=0}^{m-1} \frac{\lambda}{(i+1)\mu} \prod_{j=m}^{k-1} \frac{\lambda}{m\mu} \\ &= p_0 \left(\frac{\lambda}{\mu}\right)^k \frac{1}{m!\, m^{k-m}}, \qquad k \geqslant m. \end{aligned} \tag{8.30}$$

Defining $\rho = \lambda/(m\mu)$, the condition for stability is given by $\rho < 1$. The expression for p_0 is obtained using (8.30) and the fact that $\sum_{k=0}^{\infty} p_k = 1$:

$$p_0 = \left[\sum_{k=0}^{m-1} \frac{(m\rho)^k}{k!} + \frac{(m\rho)^m}{m!} \frac{1}{1-\rho}\right]^{-1}. \tag{8.31}$$

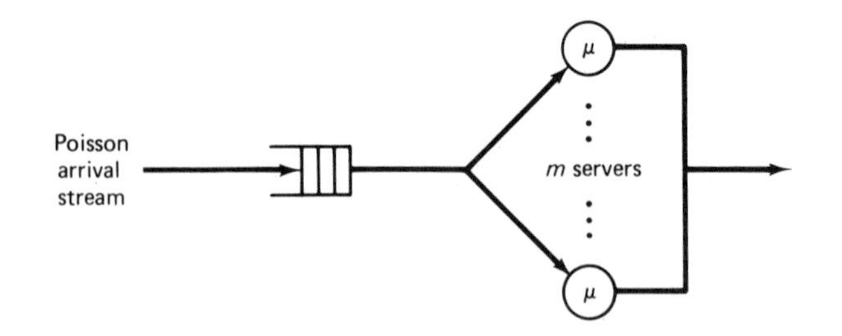

Figure 8.5 The $M/M/m$ queuing system

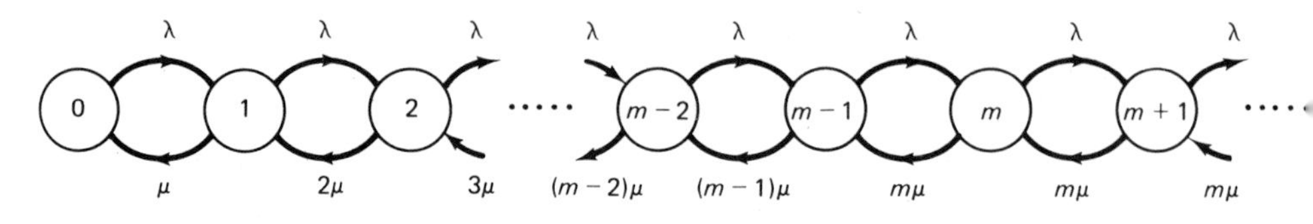

Figure 8.6 The state diagram of the $M/M/m$ queue

The expression for the average number of jobs in the system is (see problem 4 at the end of this section):

$$E[N] = \sum_{k \geqslant 0} kp_k = m\rho + \rho \frac{(m\rho)^m}{m!} \frac{p_0}{(1-\rho)^2}. \tag{8.32}$$

Let the random variable M denote the number of busy servers; then:

$$P(M = k) = \begin{cases} P(N = k) = p_k, & 0 \leqslant k \leqslant m-1, \\ P(N \geqslant m) = \displaystyle\sum_{k=m}^{\infty} p_k = \frac{p_m}{1-\rho}, & k = m. \end{cases}$$

The average number of busy servers is then:

$$E[M] = \sum_{k=0}^{m-1} kp_k + \frac{mp_m}{1-\rho},$$

which can be simplified (see problem 4 at the end of this section):

$$E[M] = m\rho = \frac{\lambda}{\mu}. \tag{8.33}$$

Thus, the utilization of any individual server is $\rho = \lambda/(m\mu)$, while the average number of busy servers is equal to the traffic intensity λ/μ.

The probability that an arriving customer is required to join the queue (or the probability of congestion) is derived as:

$$P[\text{queuing}] = \sum_{k=m}^{\infty} p_k = \frac{p_m}{1-\rho}$$

$$= \frac{(m\rho)^m}{m!} \cdot \frac{p_0}{1-\rho}, \tag{8.34}$$

where p_0 is given in (8.31). Formula (8.34) finds wide application in telephone traffic theory and gives the probability that no trunk is available for an arriving call in an exchange with m trunks. This formula is referred to as Erlang's C formula (or Erlang's delayed-call formula).

Example 8.3

While designing a multiprocessor operating system, we wish to compare two different queuing schemes shown in Figure 8.7. The criterion for comparison will be the average response times $E[R_s]$ and $E[R_c]$. It is clear that the first organization corresponds to two independent $M/M/1$ queues, with $\rho = \lambda/(2\mu)$. Therefore, using equation (8.23), we have:

$$E[R_s] = \frac{\frac{1}{\mu}}{1 - \frac{\lambda}{2\mu}} = \frac{2}{2\mu - \lambda}.$$

On the other hand, the common queue organization corresponds to an $M/M/2$ system. To obtain $E[R_c]$, we first obtain $E[N_c]$ [using equation (8.32)] as:

$$E[N_c] = 2\rho + \frac{\rho(2\rho)^2}{2!} \frac{p_0}{(1-\rho)^2} \quad \text{where} \quad \rho = \frac{\lambda}{2\mu},$$

and using equation (8.31), we have:

$$p_0 = [1 + 2\rho + \frac{(2\rho)^2}{2!} \frac{1}{1-\rho}]^{-1}$$

$$= \frac{1-\rho}{(1-\rho)(1+2\rho) + 2\rho^2} = \frac{1-\rho}{1+\rho}.$$

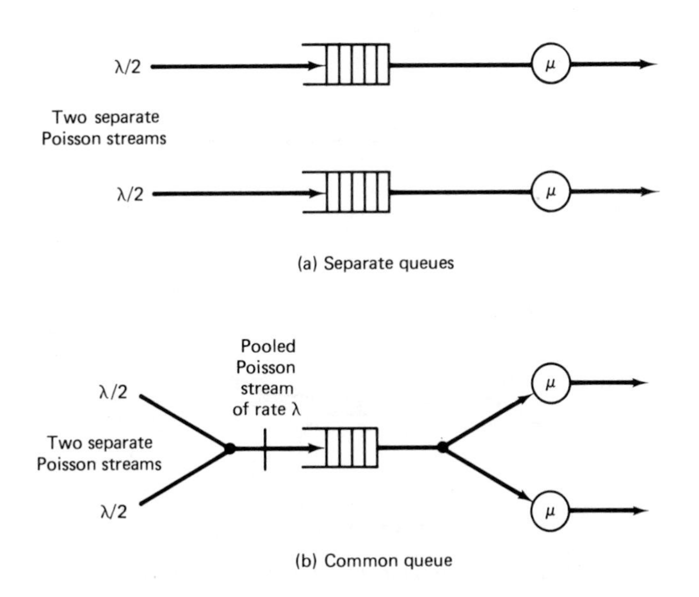

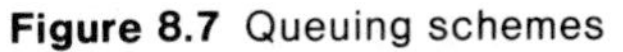

Figure 8.7 Queuing schemes

Thus:

$$E[N_c] = 2\rho + 2\rho^3 \frac{1-\rho}{(1+\rho)(1-\rho)^2} = \frac{2\rho(1-\rho^2+\rho^2)}{1-\rho^2} = \frac{2\rho}{1-\rho^2}$$

and, using Little's formula, we have:

$$E[R_c] = \frac{E[N_c]}{\lambda} = \frac{2\,\dfrac{1}{2\mu}}{1-(\dfrac{\lambda}{2\mu})^2}$$

$$= \frac{1}{\mu(1-\rho^2)} = \frac{4\mu}{4\mu^2-\lambda^2}. \tag{8.35}$$

Now:

$$E[R_s] = \frac{2}{2\mu-\lambda} = \frac{4\mu+2\lambda}{4\mu^2-\lambda^2} > E[R_c].$$

This implies that a common queue organization is better than a separate queue organization. This result generalizes to the case of m servers [KLEI 1976]. #

Example 8.4

Once again consider the problem of designing a system with two identical processors. We have two independent job streams with respective average arrival rates $\lambda_1 = 20$ and $\lambda_2 = 15$ per hour. The average service time for both job types is $1/\mu =$ 2 min $= 1/30$ hours. Should we dedicate a processor per job stream, or should we pool the job streams and processors together (see Figure 8.8)? Let $E[R_{s1}]$ and $E[R_{s2}]$ be

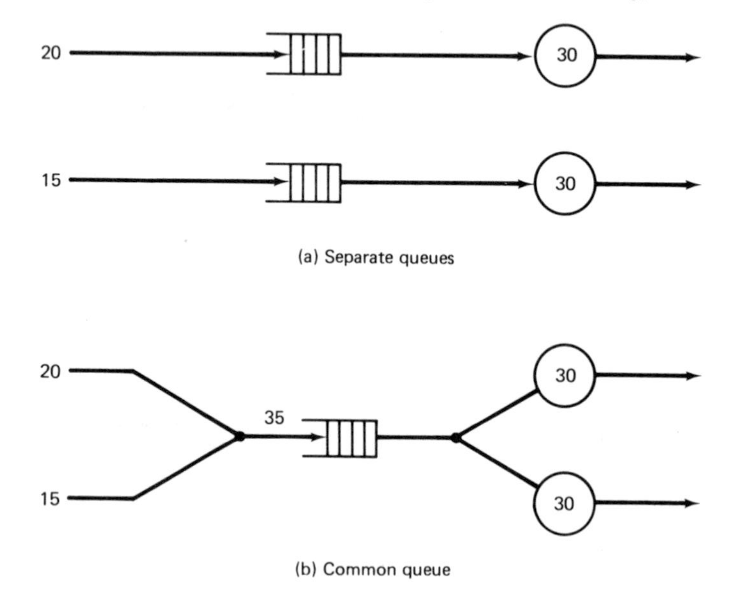

Figure 8.8 Queuing schemes for Example 8.4

the average response times of the two job streams in the separate queue organization and let $E[R_c]$ be the response time in the common-queue situation. Let $\rho_1 = \lambda_1/\mu = 20/30$, $\rho_2 = \lambda_2/\mu = 15/30$, $\rho = (\lambda_1 + \lambda_2)/(2\mu) = 35/60$. Then:

$$E[R_{s1}] = \frac{\frac{1}{\mu}}{1 - \rho_1} = \frac{\frac{1}{30}}{1 - \frac{20}{30}}, \qquad \text{using formula (8.23)}$$

$$= \frac{1}{30 - 20} = \frac{1}{10} \text{ hour} = 6 \text{ minutes.}$$

$$E[R_{s2}] = \frac{\frac{1}{\mu}}{1 - \rho_2} = \frac{\frac{1}{30}}{1 - \frac{15}{30}}, \qquad \text{using formula (8.23)}$$

$$= \frac{1}{15} \text{ hour} = 4 \text{ minutes.}$$

$$E[R_c] = \frac{\frac{1}{\mu}}{1 - \rho^2}, \qquad \text{using formula (8.35)}$$

$$= \frac{\frac{1}{30}}{1 - (\frac{35}{60})^2} = 3.03 \text{ minutes.}$$

Clearly, it is much better to form a common pool of jobs. #

Problems

1. Consider a telephone switching system consisting of n trunks with an infinite caller population. The arrival stream is Poisson with rate λ, and call holding times are exponentially distributed with average $1/\mu$. The traffic offered, A (in Erlangs), is defined to be the average number of call arrivals per holding time. Thus, $A = \lambda/\mu = \rho$. We assume that an arriving call is lost if all trunks are busy. This is known as BCC (blocked calls cleared) scheduling discipline. Draw the state diagram and derive an expression for p_i, the steady-state probability that i trunks are busy. Show that this distribution approaches the Poisson distribution in the limit $n \to \infty$ (i.e., ample-trunks case). Therefore, for finite n, the above distribution is known as the "truncated Poisson distribution." Define the **call congestion**, B, as the proportion of lost calls in the long run. Then show that:

$$B = \frac{\frac{\rho^n}{n!}}{\sum_{i=0}^{n} \frac{\rho^i}{i!}}.$$

This is known as Erlang's B formula. Define traffic carried, C (in Erlangs), to be the average number of calls completed in a time interval $1/\mu$. Then:

$$C = \sum_{i=0}^{n} ip_i.$$

Verify that:

$$B = 1 - \frac{C}{A}.$$

2. Derive the steady-state distribution of the waiting time W for an $M/M/2$ queuing system as follows:
   (a) First show that:

$$P(W = 0) = p_0 + p_1.$$

   (b) Now, conditioned on $n \geqslant 2$ jobs being present in the system at the time of arrival of the tagged job, argue that the distribution of W is $(n-1)$-stage Erlang with parameter 2μ.

   Compute the distribution function and hence compute the expected value of W.

3. [$M/M/\infty$ Queuing System] Suppose $P_n(t)$ is the probability that n telephone lines are busy at time t. Assume that infinitely many lines are available and that average call arrival rate is λ while average call duration is $1/\mu$. Derive the differential equation for $P_n(t)$. Solve the equation for $P_n(t)$ as $t \to \infty$. Let $E[N(t)]$ denote the average number of busy lines. Derive the differential equation for $E[N(t)]$. Obtain an expression for average length of queue $E[N]$ in the steady state. What is the average response time?

4. Show that the average number of busy servers for an $M/M/m$ queue in the steady state is given by:

$$E[M] = \frac{\lambda}{\mu}.$$

   Also verify formula (8.32) for the average number in the system.

8.2.3 Finite State Space

We consider a special case of the birth-death process having a finite state space $\{0, 1, \ldots, n\}$, with constant birth rates $\lambda_i = \lambda, 0 \leqslant i \leqslant n-1$, and constant death rates $\mu_i = \mu, 1 \leqslant i \leqslant n$. Also let $\rho = \lambda/\mu$, as before. The state diagram is given by Figure 8.9, and the steady-state probabilities are:

$$p_i = \rho^i p_0, \qquad 0 \leqslant i \leqslant n,$$

$$p_0 = \frac{1}{\sum_{i=0}^{n} \rho^i} = \begin{cases} \dfrac{1-\rho}{1-\rho^{n+1}}, & \rho \neq 1 \\ \dfrac{1}{n+1}, & \rho = 1. \end{cases} \tag{8.36}$$

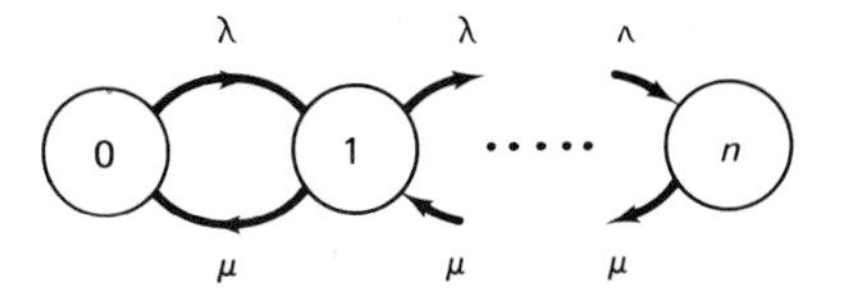

Figure 8.9 State diagram of a birth-death process with a finite state space

Note that such a system with a finite customer population will always be stable, no matter what value of ρ. Thus, (8.36) gives the steady-state probabilities for all finite values of ρ.

Example 8.5 (Machine Breakdown)

Consider a component with a constant failure rate λ. Upon a failure, it is repaired with an exponential repair-time distribution of parameter μ. Thus, the MTTF is $1/\lambda$ and the MTTR is $1/\mu$. This is an example of the Markov chain just discussed with $n = 1$ (a two-state system). Hence:

$$p_0 = \frac{1-\rho}{1-\rho^2} = \frac{1}{1+\rho}$$

and

$$p_1 = \frac{\rho}{1+\rho}.$$

The steady-state availability is the steady-state probability that the system is in state 0, the state with the system functioning properly. Thus:

$$\text{steady-state availability, } A = p_0 = \frac{1}{1+\rho} = \frac{1}{1+\dfrac{\lambda}{\mu}} = \frac{\dfrac{1}{\lambda}}{\dfrac{1}{\lambda}+\dfrac{1}{\mu}} = \frac{\text{MTTF}}{\text{MTTF}+\text{MTTR}}. \tag{8.37}$$

Note that a system with a low reliability will have a small MTTF, but if the repairs can be made fast enough (implying a low MTTR), the system may possess a high availability. #

Such availability models assume that all failures are recoverable. Consequently, the Markov chains of such systems are irreducible. If we assume that some failures are irrecoverable, then the system will have an absorbing state. In such cases, we study the distribution of time to reach the absorbing state (or failure state), and system reliability (see Section 8.5 for some examples).

Example 8.6 (Cyclic Queuing Model of a Multiprogramming System)

Consider the cyclic queuing model shown in Figure 8.10. Assume that the lengths of successive CPU execution bursts are independent exponentially distribut-

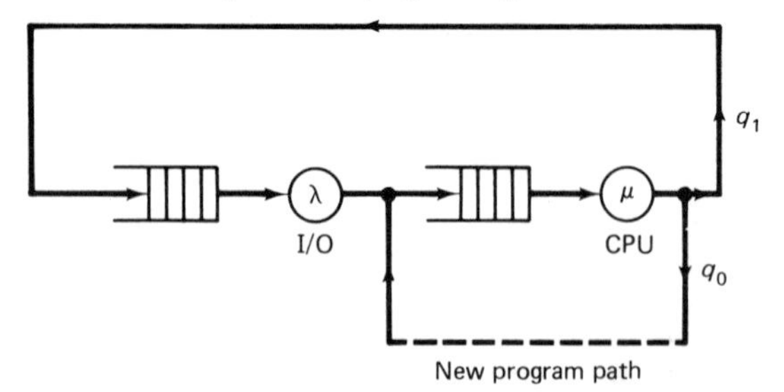

Figure 8.10 The cyclic queuing model of a multiprogramming system

ed random variables with mean $1/\mu$ and that successive I/O burst times are also independent exponentially distributed with mean $1/\lambda$. At the end of a CPU burst a program requests an I/O operation with probability $0 \leqslant q_1 \leqslant 1$, and it completes execution with probability q_0 ($q_1 + q_0 = 1$). At the end of a program completion another statistically identical program enters the system, leaving the number of programs in the system at a constant level n (known as the **degree of multiprogramming**).

Let the number of programs in the CPU queue including any being served at the CPU denote the state of the system, i, where $0 \leqslant i \leqslant n$. Then the state diagram is given by Figure 8.11. Denoting $\lambda/(\mu q_1)$ by ρ, we see that the steady-state probabilities are given by:

$$p_i = (\frac{\lambda}{\mu q_1})^i p_0 = \rho^i p_0, \quad \text{and} \quad p_0 = \frac{1}{\sum_{i=0}^{n} \rho^i},$$

so that:

$$p_0 = \begin{cases} \dfrac{1-\rho}{1-\rho^{n+1}}, & \rho \neq 1, \\ \dfrac{1}{n+1}, & \rho = 1. \end{cases}$$

The CPU utilization is given by:

$$U_0 = 1 - p_0 = \begin{cases} \dfrac{\rho - \rho^{n+1}}{1-\rho^{n+1}}, & \rho \neq 1, \\ \dfrac{n}{n+1}, & \rho = 1. \end{cases} \tag{8.38}$$

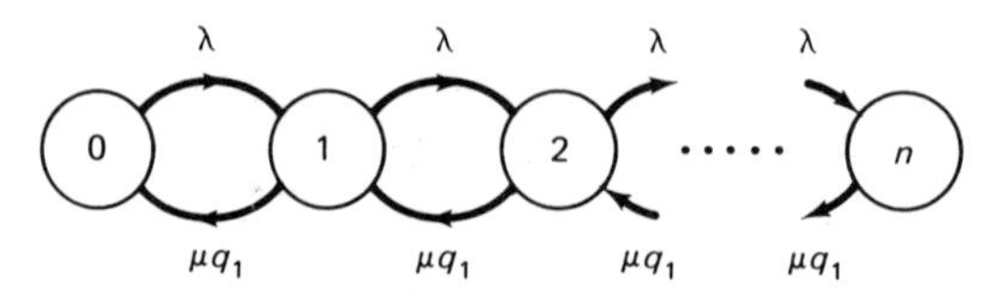

Figure 8.11 The state diagram for the cyclic queuing model

Let $C(t)$ denote the number of jobs completed by time t. Then the (time) average $C(t)/t$ converges, under appropriate conditions, to a limit as t approaches ∞ [ROSS 1970]. This limit is the average system throughput in the steady-state, and (with a slight abuse of notation) it is denoted here by $E[T]$. Whenever the CPU is busy, the rate at which CPU bursts are completed is μ, and a fraction q_0 of these will contribute to the throughput. Then:

$$E[T] = \mu q_0 U_0. \tag{8.39}$$

For fixed values of μ and q_0, $E[T]$ is proportional to the CPU utilization.

Let the random variable B_0 denote the total CPU time requirement of a tagged program. Then $B_0 \sim \text{EXP}(\mu q_0)$. This is true because B_0 is the random sum of K CPU service bursts, which are independent EXP (μ) random variables. Here the random variable K is the number of visits to the CPU per program and hence is geometrically distributed with parameter q_0. The required result is then obtained from our discussion on random sums in Chapter 5. Alternatively, the average number of visits V_0 to the CPU is $V_0 = 1/q_0$ (see Example 7.17), and thus $E[B_0] = V_0 E[S_0] = 1/(\mu q_0)$, where $E[S_0] = 1/\mu$ is the average CPU time per burst.

The average throughput can now be rewritten as:

$$E[T] = \frac{U_0}{E[B_0]}. \tag{8.40}$$

If B_1 represents the total I/O service time per program, then as in the case of CPU:

$$E[B_1] = \frac{q_1}{q_0}\frac{1}{\lambda} = V_1\, E[S_1],$$

where the average number of visits V_1 to the I/O device is given by $V_1 = q_1/q_0$ (by Example 7.17), and $E[S_1] = 1/\lambda$ is the average time per I/O operation. [Note that if U_1 denotes the utilization of the I/O device then, similar to (8.40), we have $E[T] = U_1/E[B_1]$.] Now the parameter ρ can be rewritten:

$$\rho = \frac{\lambda}{\mu q_1} = \frac{q_0 \lambda}{q_1} \cdot \frac{1}{\mu q_0} = \frac{E[B_0]}{E[B_1]}. \tag{8.41}$$

Thus ρ indicates the relative measure of the CPU versus I/O requirements of a program. If the CPU requirement $E[B_0]$ is less than the I/O requirement $E[B_1]$ (that is, $\rho < 1$), the program is said to be **I/O-bound**; if $\rho > 1$, then program is said to be **CPU-bound**; and otherwise it is called **balanced**.

In Figure 8.12 we have plotted U_0 as a function of the balance factor ρ and of the degree of multiprogramming n. When $\rho << 1$ or $\rho >> 1$, U_0 is insensitive to n. Thus, multiprogramming is capable of improving throughput only when the workload is nearly balanced (that is, ρ is close to 1). #

Example 8.7

Let us return to the availability model of Example 8.5 and augment the system with $(n - 1)$ identical copies of the component, which are to be used in a standby spare mode. Assume that an unpowered spare does not fail and that switching a spare is a fault-free process. Then the system is in state k provided that $n - k$ units are in

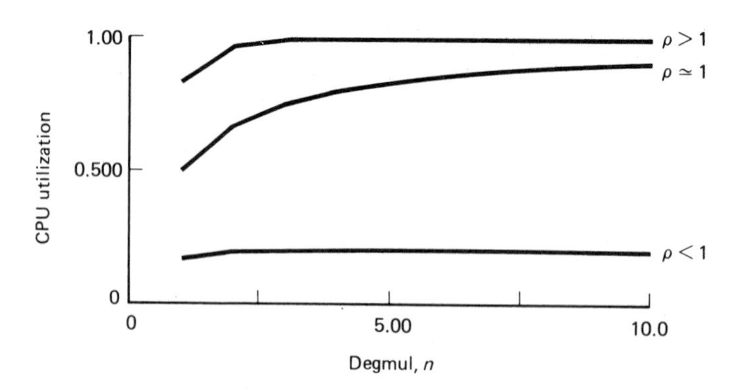

Figure 8.12 The CPU utilization as a function of the degree of multiprogramming

working order and k units are under repair. We picture this situation as the cyclic queue model of Figure 8.10, with $q_0 = 0$ and $q_1 = 1$. The total number of components, n, is the analogue of the degree of multiprogramming. The queue of available components is the analogue of the I/O queue; the queue of components under repair is represented by the CPU queue; MTTR $= 1/\mu$; and MTTF $= 1/\lambda$. The steady-state availability is given by:

$$\begin{aligned} A &= P(\text{``at least one copy is functioning properly''}) \\ &= p_0 + p_1 + \cdots + p_{n-1} = 1 - p_n \\ &= \begin{cases} 1 - \rho^n \cdot \dfrac{1-\rho}{1-\rho^{n+1}}, & \rho \neq 1, \\[2ex] \dfrac{n}{n+1}, & \rho = 1, \end{cases} \end{aligned}$$

or

$$A = \begin{cases} \dfrac{1-\rho^n}{1-\rho^{n+1}}, & \rho \neq 1, \\[2ex] \dfrac{n}{n+1}, & \rho = 1, \end{cases} \tag{8.42}$$

where ρ = MTTR/MTTF. For $n = 1$,

$$A = \frac{1}{1+\rho} = \frac{\text{MTTF}}{\text{MTTF} + \text{MTTR}}$$

as in Example 8.5, and as $n \rightarrow \infty$,

$$A \rightarrow \min\{1, \frac{1}{\rho}\} = \min\{1, \frac{\text{MTTF}}{\text{MTTR}}\}.$$

In the usual case, MTTR << MTTF, and steady-state availability will approach unity as the number of spares increases. In Table 8.1 we have shown how the system availability A increases with the number of spares for $\rho = 0.01$. #

Table 8.1 Availability of a Standby Redundant System

n	Availability	Number of spares
1	0.99009900	0
2	0.99990099	1
3	0.99999900	2
4	0.99999999	3

Problems

1. [FULL 1975] Consider a variation of the cyclic queuing network model of Example 8.6, in which the I/O device is a drum that uses the SLTF (shortest latency time first) scheduling discipline. The I/O service rate, λ_k, is a function of the number of requests in the drum queue and is given by:

$$\frac{1}{\lambda_k} = \frac{\tau}{k+1} + \frac{1}{r} = \frac{r\tau + k + 1}{r(k+1)},$$

where r^{-1} is the mean record transmission time and τ is the rotation time of the drum. Obtain an expression for the CPU utilization and the average system throughput, assuming $q_0 = 0.05$, $r\tau = 1/3$, and $\tau = 10$ milliseconds. Plot the average system throughput as a function of the CPU service rate μ (ranging from $0.1r$ to $10r$) for various values of the degree of multiprogramming $n = 1, 2, 5$, and 10. For $\mu = r$, compare the throughputs obtained by the SLTF scheduling and the FCFS scheduling algorithm (for the latter case, the average drum service time is constant at $1/\lambda_1$).

2. Consider an application of the cyclic queuing model (CPU, paging device) to demonstrate a phenomenon called "thrashing" that could occur in paged virtual memory systems. Assume a fixed number, M, of main-memory page frames, equally divided among n active programs. Increasing n implies a smaller page allotment per program, which in turn implies increased paging activity—that is, an increased value of μ. Often it is assumed that

$$\mu(n) = \frac{1}{a}\left(\frac{M}{n}\right)^{-b}.$$

Assuming $\lambda = 0.0001$, $a = 0.2$, $b = 2.00$, and $M = 100$, plot the average system throughput as a function of the degree of multiprogramming. [Note that q_0 and q_1 are functions of n and that when $n = 1$, a job is assumed to have all the memory it needs, requiring only initial paging, so that, $q_0(1) = 0.9$.] Unlike the model of nonpaged systems, the average throughput here will *not* increase monotonically, but after a critical value of n it will drop sharply. This is the phenomenon of thrashing [COFF 1973].

8.2.4 Machine Repairman Model

An interesting special case of the birth-death process occurs when the birth rate λ_j is of the form $(M - j)\lambda$, $j = 0, 1, \ldots, M$, and the death rate

$\mu_j = \mu$. Such a situation occurs in the modeling of interactive computer systems where a single terminal issues a request at the rate λ whenever it is in the "thinking state." If j out of the total of M terminals are currently waiting for a response to a pending request, the effective request rate is $(M - j)\lambda$. Here μ denotes the request completion rate. A similar situation arises when M machines share a repair facility. The failure rate of each machine is λ and the repair rate is μ.

The state diagram of such a finite-population system is given in Figure 8.13. The expressions for steady-state probabilities are obtained using equations (8.18) and (8.19) as:

$$p_k = p_0 \prod_{i=0}^{k-1} \frac{\lambda(M-i)}{\mu}, \quad 0 \leqslant k \leqslant M,$$

or

$$p_k = p_0 (\frac{\lambda}{\mu})^k \frac{M!}{(M-k)!} = p_0 \rho^k \frac{M!}{(M-k)!}.$$

Hence:

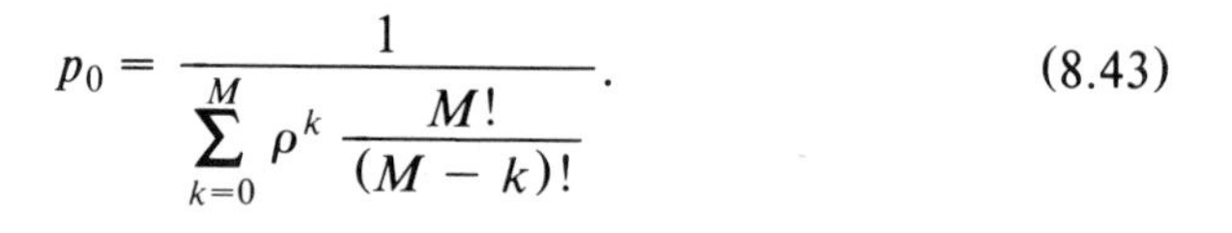

$$p_0 = \frac{1}{\sum_{k=0}^{M} \rho^k \frac{M!}{(M-k)!}}. \tag{8.43}$$

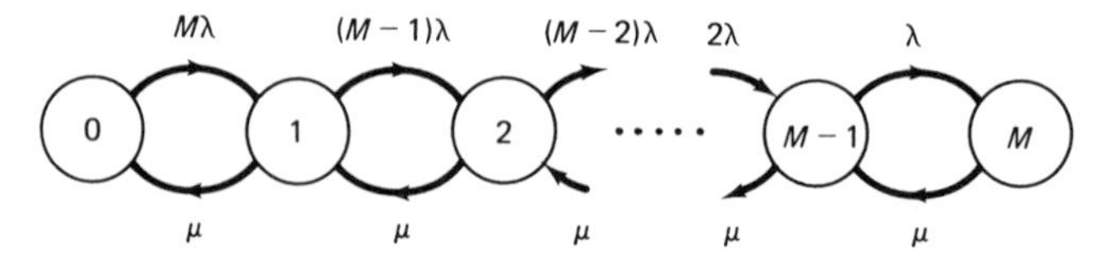

Figure 8.13 The state diagram of a finite-population queuing system

Example 8.8

Consider a parallel-redundant system with M components, each with a constant failure rate λ. The system is unavailable for use whenever all M components have failed and are waiting for repairs. We wish to compare the following designs of the repair facility.

1. Each component has its own repair facility with repair rate μ. Then the availability of an individual component is given by formula (8.37) as:

$$\frac{1}{1 + \frac{\lambda}{\mu}} = \frac{1}{1+\rho},$$

and the system availability is easily computed to be:

$$A_{\rm I} = 1 - (\frac{\rho}{1+\rho})^M.$$

Note that in this scheme, no machine has to wait for a repair facility to be available.

2. We want to economize on the repair facilities and share a single repair facility of rate μ among all M machines. Then equation (8.43) applies, and noting that the system is down only when all the components are undergoing repair, we compute the steady-state availability by:

$$A_{II} = 1 - p_M = 1 - \frac{\rho^M M!}{\sum_{k=0}^{M} \rho^k \frac{M!}{(M-k)!}}.$$

3. If we find that the availability A_{II} is low, we may speed up the rate of the repair facility to $M\mu$, while retaining a single repair facility. Then, using equation (8.43), we have:

$$A_{III} = 1 - \frac{(\frac{\lambda}{M\mu})^M M!}{\sum_{k=0}^{M} (\frac{\lambda}{M\mu})^k \frac{M!}{(M-k)!}}.$$

Table 8.2 shows the values of A_I, A_{II}, and A_{III} for various values of M, assuming that $\lambda/\mu = 0.1$. It is clear that:

$$A_{III} \geqslant A_I \geqslant A_{II}.$$

#

Table 8.2 *Availabilities for Parallel Redundant System*

M (number of components)	*Individual repair facility* A_I	*Single repair facility of rate* μ A_{II}	*Single repair facility of rate* $M\mu$ A_{III}
1	0.909091	0.909091	0.909091
2	0.991736	0.983607	0.995475
3	0.999249	0.995608	0.999799

Example 8.9 (Response Time in a Terminal-Oriented System)

Consider an interactive system with M terminals in which individual think times are exponentially distributed with mean $1/\lambda$ seconds (see Figure 8.14). Assume that the service time per request, B_0, is exponentially distributed with mean $E[B_0] = 1/\mu$ seconds. Then the steady-state probability that there are n requests executing or waiting on the CPU is given by:

$$p_n = p_0 \rho^n \frac{M!}{(M-n)!}, \quad n = 0, 1, 2, \ldots, M,$$

and the probability that the CPU is idle is:

$$p_0 = \frac{1}{\sum_{n=0}^{M} \rho^n \frac{M!}{(M-n)!}}$$

where $\rho = \lambda/\mu$.

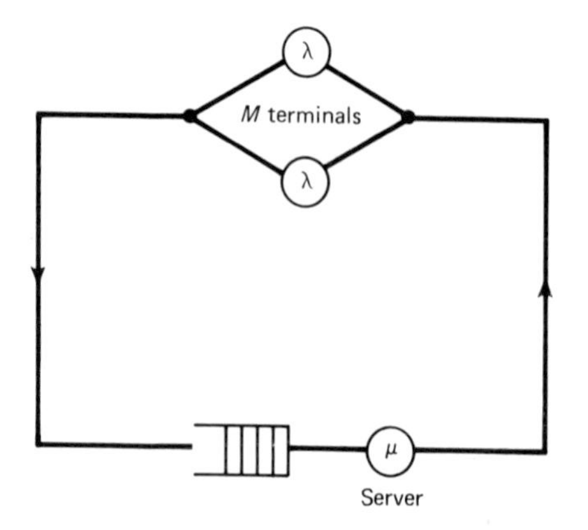

Figure 8.14 An interactive system

The CPU utilization U_0 is $1 - p_0$, and the average rate of request completion is $E[T] = \mu(1 - p_0) = U_0/E[B_0]$. If $E[R]$ denotes the average response time, then on the average a request is generated by a given terminal in $E[R] + (1/\lambda)$ seconds. Thus, the average request-generation rate of the terminal subsystem is $M/[E[R] + (1/\lambda)]$. In the steady-state, the request-generation and completion rates must be equal. Therefore, we have:

$$\frac{M}{E[R] + \dfrac{1}{\lambda}} = \mu(1 - p_0)$$

or

$$E[R] = \frac{M}{\mu(1 - p_0)} - \frac{1}{\lambda} = \frac{M \cdot E[B_0]}{U_0} - \frac{1}{\lambda} = \frac{M}{E[T]} - \frac{1}{\lambda}$$

$$= \frac{\text{number of terminals}}{\text{average throughput}} - \text{average think time.} \tag{8.44}$$

The last expression for average response time can also be derived using Little's formula, and as such it is known to hold under rather general conditions [DENN 1978]. In Figure 8.15, $E[R]$ is plotted as a function of the number of terminals, M, assuming $1/\lambda = 15$ seconds and $1/\mu = 1$ second.

When the number of terminals $M = 1$, there is no queuing and the response time $E[R]$ equals the average service time $E[B_0]$. As the number of terminals increases, there is increased congestion as the server utilization U_0 approaches unity. In the limit $M \to \infty$, $E[R]$ is a linear function $[ME[B_0] - (1/\lambda)]$ of M. In this limit, the installation of an additional terminal increases every other terminal user's response time by the new user's service time $E[B_0]$. This complete state of interference is to be generally avoided. The number of terminals, M^*, for which the heavy-load asymptote $E[R] = ME[B_0] - (1/\lambda)$ intersects with the light-load asymptote $E[R] = E[B_0]$ is therefore called the **saturation number** [KLEI 1976] and is given by:

$$M^* = \frac{E[B_0] + 1/\lambda}{E[B_0]} = 1 + \frac{\mu}{\lambda}. \tag{8.45}$$

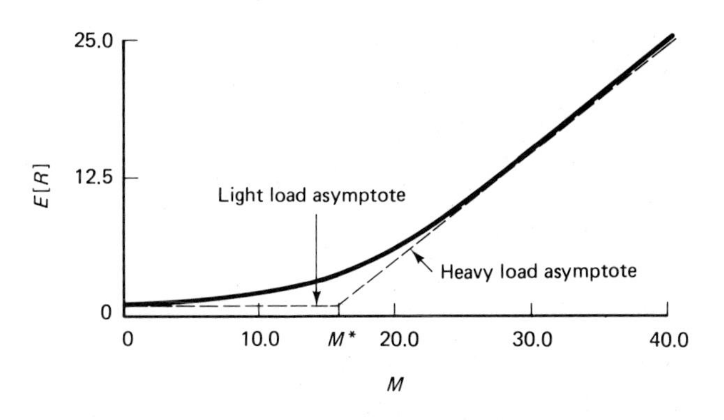

Figure 8.15 Average response time as a function of the number of terminals

For our example, the number of terminals beyond which we call the system saturated is given by $M^* = 16$. #

Problems

1. Assuming an average think time of 10 seconds, design an interactive system to support 101 terminals without saturation ($M \leqslant M^*$). We assume that on the average each request from any terminal requires the execution of 100,000 machine instructions. The result of the design process should be the minimum number of instructions that the machine should be able to execute per unit time.

2. Consider a model of a telephone switching system consisting of n trunks with a finite caller population of M callers. The average call rate of an idle caller (free source) is λ calls per unit time, and the average holding time of a call is $1/\mu$. If an arriving call finds all trunks busy, then it is lost (i.e., BCC scheduling discipline is used). Assuming that the call holding times and the intercall times of each caller are exponentially distributed, draw the state diagram and derive an expression for the steady-state probability $p_i = P(i \text{ trunks are busy})$, $i = 0, 1, \ldots, n$. The resulting distribution is known as the Engset distribution.

   Show that the expected total traffic offered (in Erlangs) by the M sources per holding time is given by

$$A = \sum_{i=0}^{n} p_i (M - i) \frac{\lambda}{\mu}$$

$$= \frac{M \sum_{i=0}^{n} \binom{M-1}{i} (\frac{\lambda}{\mu})^{i+1}}{\sum_{i=0}^{n} \binom{M}{i} (\frac{\lambda}{\mu})^{i}}.$$

   Let $\rho = \lambda/\mu$.

   Next obtain an expression for the traffic carried (in Erlangs) by the switching system per holding time:

$$C = \sum_{i=0}^{n} i p_i.$$

Now the probability, B, that a given call is lost is computed as:

$$B = 1 - \frac{C}{A} = \frac{\binom{M-1}{n}\rho^n}{\sum_{i=0}^{n}\binom{M-1}{i}\rho^i}.$$

The quantity B is also known as call congestion.

Additional Problems

1. You are given a hybrid NMR system in which N units are active and S units are in a standby status so that the failure rate of an active unit is λ and a unit in standby mode does not fail. There is a single repairman with repair rate μ. Give a queuing network that will model the behavior of this system. Draw the state diagram.

2. Components arrive at a repair facility with a constant rate λ, and the service time is exponentially distributed with mean $1/\mu$. The last step in the repair process is a quality-control inspection, and with probability p, the repair is considered inadequate, in which case the component will go back into the queue for repeated service. Determine the steady state distribution of the number of components at the repair facility.

3. In the availability model of a standby redundant system of Example 8.7, we made an assumption that the failure rate of an unpowered spare is zero. Extend the model so that the failure rates of powered and unpowered units are λ_1 and λ_2, respectively with $\lambda_1 \geqslant \lambda_2 \geqslant 0$. Obtain an expression for the steady-state availability. Verify that the expression derived yields the availability expression derived in Example 8.7 when $\lambda_2 = 0$. Similarly, verify that it gives the availability expression for a parallel redundant system when $\lambda_2 = \lambda_1 = \lambda$ as in Example 8.8 of Section 8.2.4.

8.3 OTHER SPECIAL CASES OF THE BIRTH-DEATH MODEL

We noted that the solution of the differential-difference equations (8.15) to obtain the probabilities $P_k(t)$ is a formidable task, in general. However, the calculation of the limiting probabilities $p_k = \lim_{t\to\infty} P_k(t)$ is relatively simple. In this section we consider several special cases of the birth-death model when the time-dependent probabilities $P_k(t)$ can be computed by simple techniques.

8.3.1 The Pure Birth Process

If the death rates $\mu_k = 0$ for all $k = 1, 2, \ldots$, we have a **pure birth process**. If, in addition, we impose the condition of constant birth rates—that is,

$\lambda_k = \lambda$ $(k = 0, 1, 2, \ldots)$—then we have the familiar Poisson process. The equations (8.15) now reduce to:

$$\frac{dP_0(t)}{dt} = -\lambda P_0(t) \qquad k = 0, \tag{8.46}$$

$$\frac{dP_k(t)}{dt} = \lambda P_{k-1}(t) - \lambda P_k(t), \quad k \geqslant 1,$$

where we have assumed that the initial state $N(0) = 0$, so that:

$$P_0(0) = 1, \quad P_k(0) = 0 \quad \text{for } k \geqslant 1. \tag{8.47}$$

One method of solving such differential equations is to use the Laplace transform, which simplifies the system of differential equations to a system of algebraic equations. The Laplace transform of $P_k(t)$, denoted by $\bar{P}_k(s)$, is defined in the usual way, namely:

$$\bar{P}_k(s) = \int_0^\infty e^{-st} P_k(t)\, dt,$$

and the Laplace transform of the derivative dP_k/dt is given by (see Appendix D):

$$s\bar{P}_k(s) - P_k(0).$$

Now, taking Laplace transforms on both sides of equations (8.46) above, we get:

$$s\bar{P}_0(s) - P_0(0) = -\lambda \bar{P}_0(s),$$

$$s\bar{P}_k(s) - P_k(0) = \lambda \bar{P}_{k-1}(s) - \lambda \bar{P}_k(s), \quad k \geqslant 1.$$

Using (8.47) and rearranging, we get:

$$\bar{P}_0(s) = \frac{1}{s + \lambda}$$

and

$$\bar{P}_k(s) = \frac{\lambda}{s + \lambda}\, \bar{P}_{k-1}(s),$$

from which we have

$$\bar{P}_k(s) = \frac{\lambda^k}{(s + \lambda)^{k+1}}, \quad k \geqslant 0.$$

(This expression can also be obtained using the result of problem 2 at the end of Section 8.1)

In order to invert this transform, we note that if Y is a $(k + 1)$-stage Erlang random variable with parameter λ, then:

$$L_Y(s) = \bar{f}_Y(s) = \frac{\lambda^{k+1}}{(s + \lambda)^{k+1}}.$$

It follows that:

$$P_k(t) = \frac{1}{\lambda} f_Y(t).$$

Therefore:

$$P_k(t) = P[X(t) = k] = \frac{(\lambda t)^k}{k!} e^{-\lambda t}, \quad k \geqslant 0;\ t \geqslant 0. \tag{8.48}$$

Thus, $X(t)$ is Poisson distributed with parameter λt.

The Poisson process can be generalized to the case where the birth rate λ is varying with time. Such a process is called a **nonhomogeneous Poisson Process**. The generalized version of equation (8.48) in this case is given by:

$$P_k(t) = e^{-\Lambda(t)} \frac{[\Lambda(t)]^k}{k!}, \quad k \geqslant 0, \tag{8.49}$$

where

$$\Lambda(t) = \int_0^t \lambda(x)\, dx.$$

The nonhomogeneous Poisson process finds its use in reliability computations when the constant-failure-rate assumptions cannot be tolerated. Thus, for instance, if $\lambda(t) = c\alpha t^{\alpha-1}$ $(\alpha > 0)$, then the time to failure of each component is Weibull distributed with parameters c and α, and the pmf of the number of failures $N(t)$ in the interval $(0, t]$ is:

$$P_k(t) = P[N(t)=k] = e^{-ct^\alpha} \frac{(ct^\alpha)^k}{k!}, \quad k \geqslant 0.$$

Problems

1. Set up the differential equation for $P_k(t)$ for the case of the nonhomogeneous Poisson process. Show that (8.49) is a solution to this equation.

*2. [PARZ 1962] Consider the nonhomogeneous Poisson process with $\lambda(t) = c\alpha t^{\alpha-1}$. Let T_k denote the occupancy time in state k. Note that T_k is the interevent time of the process. Show that T_0 has the Weibull distribution with parameters c and α. Next show that T_1 does not, in general, have the Weibull distribution by first showing that the conditional pdf of T_1 given T_0 is:

$$f_{T_1|T_0}(t|u) = e^{-\Lambda(t+u)+\Lambda(u)} \lambda(t+u),$$

and hence show that:

$$f_{T_1}(t) = \int_0^\infty \lambda(t+u)\lambda(u) e^{-\Lambda(t+u)}\, du.$$

3. [GOEL 1979] A nonhomogeneous Poisson model of software failures has been proposed with the failure rate given by $\lambda(t) = abe^{-bt}$. Determine the pmf of

$N(t)$, the total number of failures detected by time t. Show that the average number of failures to be eventually encountered is given by the parameter a.

8.3.2 Pure Death Processes

Another special case of a birth-death process occurs when the birth rates are all assumed to be zero; that is, $\lambda_k = 0$ for all k. The system starts in some state $n > 0$ at time $t = 0$ and eventually decays to state 0. Thus, state 0 is an absorbing state. We consider two special cases of interest.

8.3.2.1 Death Process with a Constant Rate. Besides $\lambda_i = 0$ for all i, we have $\mu_i = \mu$ for all i. This implies that the differential-difference equations (8.15) reduce to:

$$\begin{aligned}
\frac{dP_n(t)}{dt} &= -\mu P_n(t), && k = n,\\
\frac{dP_k(t)}{dt} &= -\mu P_k(t) + \mu P_{k+1}(t), && 1 \leqslant k \leqslant n-1,\\
\frac{dP_0(t)}{dt} &= \mu P_1(t), && k = 0,
\end{aligned}$$

where we have assumed that the initial state $N(0) = n$, so that:

$$P_n(0) = 1, \quad P_k(0) = 0, \quad 0 \leqslant k \leqslant n-1.$$

Taking Laplace transforms and rearranging, we reduce the above system of equations to:

$$\bar{P}_k(s) = \begin{cases} \dfrac{1}{s+\mu}, & k = n,\\[2ex] \dfrac{\mu}{s+\mu}\,\bar{P}_{k+1}(s), & 1 \leqslant k \leqslant n-1,\\[2ex] \dfrac{\mu}{s}\,\bar{P}_1(s), & k = 0, \end{cases}$$

so:

$$\bar{P}_k(s) = \frac{1}{\mu}\left(\frac{\mu}{s+\mu}\right)^{n-k+1}, \quad 1 \leqslant k \leqslant n.$$

If Y is an $(n-k+1)$-stage Erlang random variable with parameter μ, then the Laplace transform of its pdf is known to be $\bar{f}_Y(s) = [\mu/(s+\mu)]^{n+k-1}$. It follows that:

$$\begin{aligned}
P_k(t) &= \frac{1}{\mu} f_Y(t)\\
&= e^{-\mu t}\frac{(\mu t)^{n-k}}{(n-k)!}, \quad 1 \leqslant k \leqslant n.
\end{aligned}$$

Now, recalling that $\sum_{k=0}^{n} P_k(t) = 1$, we have:

$$\begin{aligned} P_0(t) &= 1 - \sum_{k=1}^{n} P_k(t) \\ &= 1 - \sum_{k=1}^{n} e^{-\mu t} \frac{(\mu t)^{n-k}}{(n-k)!} \\ &= 1 - \sum_{k=0}^{n-1} e^{-\mu t} \frac{(\mu t)^k}{k!}. \end{aligned}$$

$P_0(t)$ is easily recognized to be the CDF an n-stage Erlang random variable with mean n/μ.

Example 8.10

Consider a standby redundant system with n components, each with a constant failure rate μ. Then $P_0(t)$ above gives the distribution of the time to failure of such a system. This verifies our earlier result (see equation (3.64)). #

8.3.2.2 Death Process with a Linear Rate. In this case we assume that $\lambda_i = 0$ for all i and $\mu_i = i\mu$, $i = 1, 2, \ldots, n$. The state diagram is given in Figure 8.16. The differential-difference equations in this case are given by:

$$\begin{aligned} \frac{dP_n(t)}{dt} &= -n\mu P_n(t), & k &= n, \\ \frac{dP_k(t)}{dt} &= -k\mu P_k(t) + (k+1)\mu P_{k+1}(t), & 1 &\leqslant k \leqslant n-1, \\ \frac{dP_0(t)}{dt} &= \mu P_1(t), & k &= 0, \end{aligned}$$

where we have assumed that the initial state $N(0) = n$, so that:

$$P_n(0) = 1, \quad P_k(0) = 0, \quad 0 \leqslant k \leqslant n-1.$$

Using the method of Laplace transforms, we can obtain the solution to this system of equations as:

$$P_k(t) = \binom{n}{k} (e^{-\mu t})^k (1 - e^{-\mu t})^{n-k}, \quad 0 \leqslant k \leqslant n, \ t \geqslant 0,$$

which can be verified by differentiation. For a fixed t, this is recognized as a binomial pmf with parameters n and $p = e^{-\mu t}$.

Example 8.11

Consider a parallel-redundant system with n components, each having a constant failure rate μ. If we let k denote the number of components operating properly, then the death process with linear rate describes the behavior of this system. The distribu-

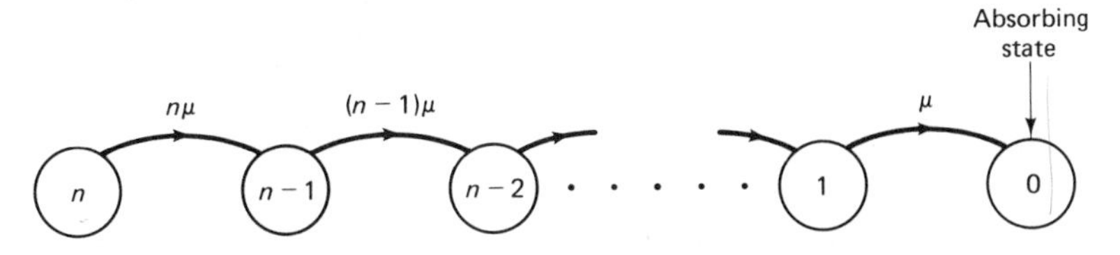

Figure 8.16 The state diagram of a pure death process with a linear death rate

tion of time to failure is then given by $P_0(t)$, and system reliability $R(t) = 1 - P_0(t) = 1 - (1-e^{-\mu t})^n$. This agrees with the expression derived in Chapter 3 [equation (3.56)]. The distinction between the Markov diagram (Figure 8.16) and a distribution represented as a network of exponential stages (e.g., Figure 3.27) should be noted. #

If we consider a more general death process with variable death rates and the initial state $N(0) = n$ then $P_0(t)$ can be shown to be the distribution function of a HYPO $(\mu_1, \mu_2, \ldots, \mu_n)$ random variable. Reliability of a hybrid-NMR system with perfect coverage can then be modeled by such a death process.

Problems

1. For the death process with linear death rate (see Figure 8.16), derive the formula for $P_k(t)$ given in the text starting from its differential equations and using the method of Laplace transforms. Also derive the formula for $P_k(t)$ using the convolution-integral approach developed in problem 3 at the end of Section 8.1.

8.4 NON-BIRTH-DEATH PROCESSES

So far, we have discussed special cases of the birth-death process. Not all Markov chains of interest satisfy the restriction of nearest-neighbor-only transitions. In this section we study several examples of non birth-death processes.

Example 8.12 ***M/M/2 Queue with Heterogeneous Servers [BHAT 1972]***

We consider a variant of the $M/M/2$ queue where the service rates of the two processors are not identical. This would be the case, for example, in a heterogeneous multiprocessor system. The queuing structure is shown in Figure 8.17. Assume without loss of generality that $\mu_1 > \mu_2$.

The state of the system is defined to be the tuple (n_1, n_2) where $n_1 \geqslant 0$ denotes the number of jobs in the queue including any at the faster server, and $n_2 \in \{0, 1\}$ denotes the number of jobs at the slower server. Jobs wait in line in the order of their arrival. When both servers are idle, the faster server is scheduled for service before the slower one. The state diagram of the system is given in Figure 8.18.

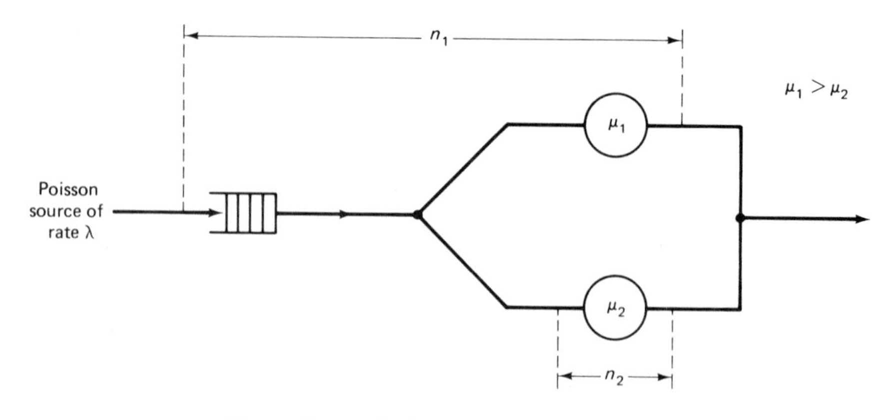

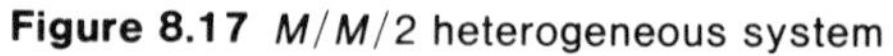

Figure 8.17 $M/M/2$ heterogeneous system

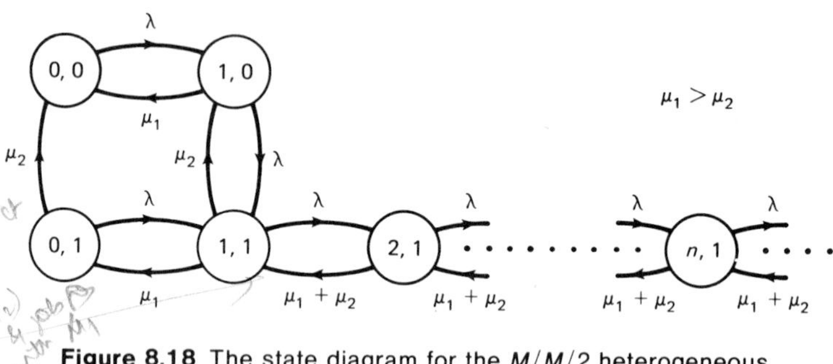

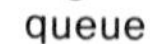

Figure 8.18 The state diagram for the $M/M/2$ heterogeneous queue

Balance equations, in the steady state, can be written by equating the rate of flow into a state to the rate of flow out of that state:

$$\lambda p(0, 0) = \mu_1 p(1, 0) + \mu_2 p(0, 1), \tag{8.50}$$

$$(\lambda + \mu_1) p(1, 0) = \mu_2 p(1, 1) + \lambda p(0, 0), \tag{8.51}$$

$$(\lambda + \mu_2) p(0, 1) = \mu_1 p(1, 1), \tag{8.52}$$

$$(\lambda + \mu_1 + \mu_2) p(1, 1) = (\mu_1 + \mu_2) p(2, 1) + \lambda p(0, 1) + \lambda p(1, 0), \tag{8.53}$$

$$(\lambda + \mu_1 + \mu_2) p(n, 1) = (\mu_1 + \mu_2) p(n + 1, 1) + \lambda p(n - 1, 1), \quad n > 1. \tag{8.54}$$

The traffic intensity for this system is:

$$\rho = \frac{\lambda}{\mu_1 + \mu_2}.$$

The form of equation (8.54) is similar to the balance equation of a birth-death process [equation (8.16)]. Therefore:

$$p(n, 1) = \frac{\lambda}{\mu_1 + \mu_2} p(n - 1, 1), \quad n > 1, \tag{8.55}$$

can easily be seen to satisfy equation (8.54). By repeated use of equation (8.55), we have:

$$p(n, 1) = \rho p(n-1, 1) = \rho^{n-1} p(1, 1), \quad n > 1. \tag{8.56}$$

From equations (8.50–8.52) we can obtain by elimination:

$$p(0, 1) = \frac{\rho}{1+2\rho} \frac{\lambda}{\mu_2} p(0, 0),$$

$$p(1, 0) = \frac{1+\rho}{1+2\rho} \frac{\lambda}{\mu_1} p(0, 0),$$

$$p(1, 1) = \frac{\rho}{1+2\rho} \frac{\lambda(\lambda+\mu_2)}{\mu_1\mu_2} p(0, 0).$$

Now, observing that:

$$\left[\sum_{n\geqslant 1} p(n, 1)\right] + p(0, 1) + p(1, 0) + p(0, 0) = 1,$$

we have:

$$(\sum_{n\geqslant 1} \rho^{n-1}) p(1, 1) + p(0, 0) \left[\frac{\rho}{1+2\rho} \frac{\lambda}{\mu_2} + \frac{1+\rho}{1+2\rho} \frac{\lambda}{\mu_1} + 1\right] = 1$$

or

$$\frac{1}{1-\rho} \frac{\rho}{1+2\rho} \frac{\lambda(\lambda+\mu_2)}{\mu_1\mu_2} p(0,0) + p(0,0)\left[\frac{\rho}{1+2\rho} \frac{\lambda}{\mu_2} + \frac{1+\rho}{1+2\rho} \frac{\lambda}{\mu_1} + 1\right] = 1,$$

from which we get:

$$p(0, 0) = \left[1 + \frac{\lambda(\lambda+\mu_2)}{\mu_1\mu_2(1+2\rho)(1-\rho)}\right]^{-1}. \tag{8.57}$$

The average number of jobs in the system may now be computed by observing that the number of customers in the system in state (n_1, n_2) is $n_1 + n_2$. Therefore, the average number of jobs is given by:

$$\begin{aligned} E[N] &= \sum_{k\geqslant 0} kp(k, 0) + \sum_{k\geqslant 0} (k+1)p(k, 1) \\ &= p(1, 0) + p(0, 1) + \sum_{k\geqslant 1} (k+1)p(k, 1) \\ &= p(1, 0) + p(0, 1) + \sum_{k\geqslant 1} p(k, 1) + \sum_{k\geqslant 1} kp(k, 1) \\ &= 1 - p(0, 0) + p(1, 1) \sum_{k=1}^{\infty} k\rho^{k-1} \\ &= 1 - p(0, 0) + \frac{p(1, 1)}{(1-\rho)^2}, \end{aligned}$$

so:

$$E[N] = \frac{1}{A(1-\rho)^2}, \tag{8.58}$$

where:

$$A = [\frac{\mu_1\mu_2(1+2\rho)}{\lambda(\lambda+\mu_2)} + \frac{1}{1-\rho}].$$

#

Example 8.13 ***[FULL 1975]***

A computing center currently has an IBM 360/50 computer system. The job stream can be modeled as a Poisson process with rate λ jobs/minute, and the service times are exponentially distributed with an average service rate μ_2 jobs/minute. Thus, $\rho_2 = \lambda/\mu_2$ and the average response time is given by equation (8.23) as $E[R_2] = (1/\mu_2)/(1-\rho_2)$.

Suppose this response time is considered intolerable by the users and an IBM 370/155 is purchased and added to the system. Let the service rate μ_1 of the 370/155 be equal to $\alpha\mu_2$ for some $\alpha > 1$. Assuming that a common-queue heterogeneous $M/M/2$ structure is used, we can compute the response time $E[R]$ as follows: Let

$$\rho = \frac{\lambda}{\mu_1+\mu_2} = \frac{\rho_2}{1+\alpha},$$

$$A = \frac{\alpha\mu_2^2(1+2\rho)}{\lambda(\lambda+\mu_2)} + \frac{1}{1-\rho}$$

$$= \frac{\alpha(1+2\rho)}{\rho_2(1+\rho_2)} + \frac{1}{1-\rho}.$$

Using equation (8.58), the average number of customers in the system is:

$$E[N] = \frac{1}{A(1-\rho)^2}$$

$$= \frac{\rho\,(1+\alpha)(1+\rho+\rho\alpha)}{(1-\rho)(\alpha+\rho+\rho^2+2\alpha\rho+\rho^2\alpha^2)}.$$

The average response time is then computed using Little's formula as:

$$E[R] = \frac{E[N]}{\lambda} = \frac{1+\rho+\rho\alpha}{(1-\rho)\mu_2(\alpha+\rho+\rho^2+2\alpha\rho+\rho^2\alpha^2)}. \tag{8.59}$$

For any $\alpha \geqslant 1$, $E[R] < E[R_2]$.

Suppose we want to consider the possibility of disconnecting the 360/50 and using the 370/155 by itself, thus reducing the response time. In this case we have $\rho_1 = \lambda/(\alpha\mu_2) = \rho_2/\alpha$, and the average response time is given by equation (8.23) as:

$$E[R_1] = \frac{1/\mu_1}{1-\rho_1} = \frac{1/\mu_2}{\alpha-\rho_2}.$$

The condition under which $E[R_1] \leqslant E[R]$ can be simplified to:

$$\rho^2(1+\alpha^2) - \rho(1+2\alpha^2) + (\alpha^2-\alpha-2) \geqslant 0,$$

or, in terms of ρ_2 and α:

$$\frac{\rho_2^2(1+\alpha^2)}{(1+\alpha)^2} - \rho_2\,\frac{1+2\alpha^2}{1+\alpha} + \alpha^2 - \alpha - 2 \geqslant 0.$$

Thus, for example, if $\lambda = 0.2$ and $\mu_2 = 0.25$ so that $\rho_2 = 0.8$, then if the 370/155 is more than three times faster than the 360/50, the above inequality is satisfied, and, surprisingly, it is better to disconnect the slower machine altogether. Of course, this conclusion holds only if we want to minimize response time. If we are interested in processing a larger throughput (λ), particularly if $\lambda \geqslant \alpha\mu_2$, then we are forced to use both machines.

Table 8.3 gives the response times (in minutes) of the three configurations for different values of α with $\lambda = 0.2$ and $\mu_2 = 0.25$.

Table 8.3

$\lambda = 0.2$	$\alpha = 1$	$\alpha = 2$	$\alpha = 3$	$\alpha = 4$	$\alpha = 5$
$E[R_1]$	20	3.33	1.818	1.25	0.95
$E[R_2]$	20	20	20	20	20
$E[R]$	4.7619	2.6616	1.875	1.459	1.20

Example 8.14

The two-state model of component failure-repair (Example 8.5) assumed that the failure and the repair-time distributions are both exponential. Assume now that the exponential failure law is reasonable but the repair process can be broken down into two phases: (1) fault detection and location, (2) actual repair. These two phases have exponential distributions with means $1/\mu_1$ and $1/\mu_2$, respectively. The overall repair-time distribution is then hypoexponential (see Section 3.8). Since the holding times of a Markov chain are exponentially distributed, the system being modeled is a semi-Markov process. However, by noting that the repair-time distribution is an instance of a Coxian phase-type distribution (see Figure 5.10), we can transform the given system into a Markov chain.

Define the following three states of the system:

0: the component is functioning properly
1: the component is in the detection-location phase
2: the component is in the final phase of repair.

The state diagram is given in Figure 8.19. Because of the transition from state 2 to state 0, this is not a birth-death process.

We may compute the steady-state probabilities by first writing down the balance equations:

$$\lambda p_0 = \mu_2 p_2, \qquad \mu_1 p_1 = \lambda p_0, \qquad \mu_2 p_2 = \mu_1 p_1,$$

which yield the relations:

$$p_2 = \frac{\mu_1}{\mu_2}\frac{\lambda}{\mu_1} p_0 = \frac{\lambda}{\mu_2} p_0, \qquad p_1 = \frac{\lambda}{\mu_1} p_0.$$

Now, since:

$$p_0 + p_1 + p_2 = 1,$$

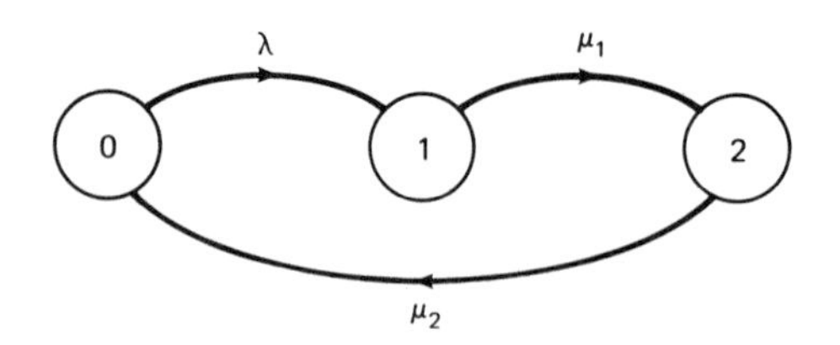

Figure 8.19 The state diagram for Example 8.14

we have:

$$p_0 = \frac{1}{1 + \frac{\lambda}{\mu_1} + \frac{\lambda}{\mu_2}}.$$

Thus, the steady-state availability A is given by:

$$A = p_0 = \frac{1}{1 + \lambda(\frac{1}{\mu_1} + \frac{1}{\mu_2})}.$$

This result can be extended to the case of a k-phase hypoexponential repair-time distribution with parameters $\mu_1, \mu_2, \ldots, \mu_k$ with the result:

$$A = \frac{1}{1 + \lambda(\frac{1}{\mu_1} + \frac{1}{\mu_2} + \cdots + \frac{1}{\mu_k})}.$$

If we denote the average total repair time by $1/\mu$, then from the formula of $E[X]$ for a hypoexponentially distributed X, we have:

$$\frac{1}{\mu} = \sum_{i=1}^{k} \frac{1}{\mu_i}.$$

With this value of μ, we can use formula (8.37) for availability derived from the two-state model (Example 8.5), even when the repair times are hypoexponentially distributed. #

Problems

1. Suppose we want to purchase a two-processor system, to be operated as an $M/M/2$ queue. Keeping the total amount of computing power constant, we want to investigate the trade-offs between a system with homogeneous processors (total power 2μ) and a heterogeneous system (total power $\mu_1 + \mu_2 = 2\mu$). Using the average response time as the criterion function, which system will be superior? To take a concrete case, let $\lambda/2\mu$ vary from 0.1 to 0.9 and compute the response times of the two systems, choosing different sets of values of μ_1 and μ_2 and assuming $\lambda = 1$. [It can be shown that the optimum values of μ_1 and μ_2 are given by

$$\lambda\left(\frac{1}{\rho} + 1 - \sqrt{1 + \frac{1}{\rho}}\right) \quad \text{and} \quad \lambda\left(\sqrt{1 + \frac{1}{\rho}} - 1\right)].$$

2. Write a simulation program to simulate the $M/M/2$ heterogeneous queuing system with $\mu_2 = 0.25$, $\lambda = 0.2$, and α varied as in Table 8.3. In order to estimate the steady-state response times $E[R_1]$, $E[R_2]$, and $E[R]$ as defined in Table 8.3, you have to execute three different simulations (two for an $M/M/1$ queue and one for the $M/M/2$ case), discard the statistics corresponding to initial transients, and then collect the steady-state values. The attainment of the steady state is determined by experimentation.

3. For the example of the $M/M/2$ heterogeneous queue let α_0 denote that value of α for which $E[R] = E[R_1]$. Study the variation of α_0 as a function of the job arrival rate λ. Graph this relationship, using the equation relating α and ρ_2 developed in Example 8.13.

4. Consider a variation of the two-state availability model so that the time to failure is a k-stage hypoexponentially distributed random variable with parameters $\lambda_1, \lambda_2, \ldots, \lambda_k$ and the repair times are exponentially distributed with parameter μ. Compute the steady-state availability. Recall that the time to failure of a hybrid-NMR system (which includes the class of parallel-redundant, standby-redundant, and TMR systems) is hypoexponentially distributed. The model of this example thus gives the steady-state availability for this class of systems, provided that the repair process cannot begin until the system breaks down. Show that the availability of such a system is obtained from the two-state model by substituting for λ, from the equation

$$\frac{1}{\lambda} = \sum_{i=1}^{k} \frac{1}{\lambda_i}.$$

5. Consider another variation of the two-state availability model where the time to failure of the unit is exponentially distributed with parameter λ, while the repair times are hyperexponentially distributed with phase-selection probabilities $\alpha_1, \alpha_2, \ldots, \alpha_k$, and individual phases have exponential distributions with parameters $\mu_1, \mu_2, \ldots, \mu_k$, respectively. Obtain an expression for the steady state availability.

6. In the example of maintenance models of Section 8.2, we assumed that a unit is available for repair as soon as it breaks down. However, in many systems it is not possible to service a failed unit until the complete system fails. This can happen if only the system's output is monitored rather than the status of individual units. Consider a two-unit parallel-redundant configuration in which repairs may not begin until both units break down. Assume a constant failure rate λ for each unit and an exponentially distributed repair time with mean $1/\mu$. Show that the steady-state availability is given by:

$$\frac{3\mu^2 + 2\lambda\mu}{3\mu^2 + 3\lambda\mu + \lambda^2}.$$

Compare the downtime of this maintenance policy with that of the maintenance policy that allows repairs as soon as the failure occurs. Use $\lambda = 0.001$ per hour, $\mu = 1.0$ per hour, and compute the expected downtime over a period of 10,000 hours for the two cases. In both cases, assume that each unit has its own repair facility.

*7. [TOWS 1978] Consider the following concurrent program with a cyclic structure:

```
repeat
  TCPU1;
  if B then TIO1;
    else
      cobegin
        TCPU2; TIO2
      coend
forever.
```

Assume that successive tests on condition B form a sequence of Bernoulli trials with probability of failure q. The execution times of the statement groups (or tasks) TCPU1 and TCPU2 are EXP (μ_1) and EXP (μ_2) random variables, respectively, while the execution times of TIO1 and TIO2 are both EXP (λ) random variables. Draw the Markov state diagram of this system and solve for the steady-state probabilties. Assuming that TCPU1 and TCPU2 are executed on a single CPU and that TIO1 and TIO2 are executed on a single I/O processor, compute steady-state utilizations of the two processors. Use $1/\mu_1 = 8$ milliseconds, $1/\mu_2 = 26.6$ milliseconds, $1/\lambda = 46.1$ milliseconds, and vary q from 0 to 1.

8.5 MARKOV CHAINS WITH ABSORBING STATES

With the exception of Section 8.3, our analysis has been concerned with irreducible Markov chains. We will introduce Markov chains with absorbing states through examples.

Example 8.15

Assume that we have a two-component parallel-redundant system with a single repair facility of rate μ. Assume that the failure rate of both components is λ. When both components have failed, the system is considered to have failed and no recovery is possible. Let the number of properly functioning components be the state of the system. The state space is $\{0, 1, 2\}$, where 0 is the absorbing state. The state diagram is given in Figure 8.20.

Assume that the initial state of the Markov chain is 2; that is, $P_2(0) = 1$, $P_k(0) = 0$ for $k = 0, 1$. Then $P_j(t) = p_{2j}(t)$, and the system of differential equations (8.10) becomes:

$$\begin{aligned}
\frac{dP_2(t)}{dt} &= -2\lambda P_2(t) + \mu P_1(t), \\
\frac{dP_1(t)}{dt} &= 2\lambda P_2(t) - (\lambda + \mu) P_1(t), \\
\frac{dP_0(t)}{dt} &= \lambda P_1(t).
\end{aligned} \tag{8.60}$$

Using the technique of Laplace transform, we can reduce the above system to:

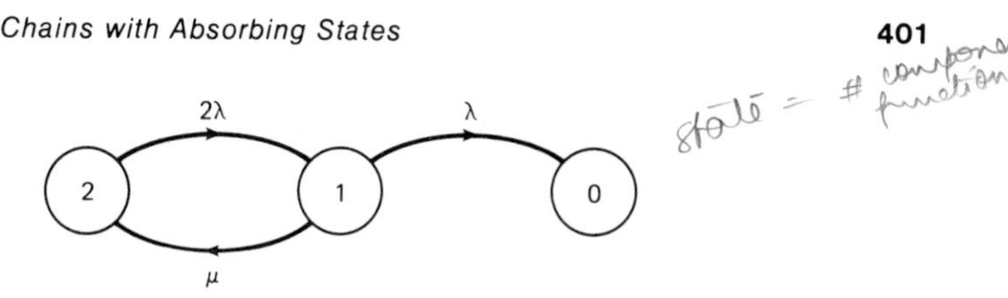

Figure 8.20 The state diagram for Example 8.15

$$\begin{aligned} s\bar{P}_2(s) - 1 &= -2\lambda \bar{P}_2(s) + \mu \bar{P}_1(s), \\ s\bar{P}_1(s) &= 2\lambda \bar{P}_2(s) - (\lambda + \mu)\bar{P}_1(s), \\ s\bar{P}_0(s) &= \lambda \bar{P}_1(s). \end{aligned} \tag{8.61}$$

Solving (8.61) for $\bar{P}_0(s)$, we get:

$$\bar{P}_0(s) = \frac{2\lambda^2}{s[s^2 + (3\lambda + \mu)s + 2\lambda^2]}.$$

After an inversion, we can obtain $P_0(t)$, the probability that no components are operating at time $t \geqslant 0$. Let Y be the time to failure of the system; then $P_0(t)$ is the probability that the system has failed at or before time t. Thus, the reliability of the system is:

$$R(t) = 1 - P_0(t).$$

The Laplace transform of the failure density:

$$f_Y(t) = -\frac{dR}{dt} = \frac{dP_0(t)}{dt},$$

is then given by:

$$L_Y(s) = \bar{f}_Y(s) = s\bar{P}_0(s) - P_0(0^-) = \frac{2\lambda^2}{s^2 + (3\lambda + \mu)s + 2\lambda^2}.$$

The denominator can be factored so that:

$$s^2 + (3\lambda + \mu)s + 2\lambda^2 = (s + \alpha_1)(s + \alpha_2),$$

and the above expression can be rearranged so that:

$$L_Y(s) = \frac{2\lambda^2}{\alpha_1 - \alpha_2}\left(\frac{1}{s + \alpha_2} - \frac{1}{s + \alpha_1}\right), \tag{8.62}$$

where:

$$\alpha_1, \alpha_2 = \frac{(3\lambda + \mu) \pm \sqrt{\lambda^2 + 6\lambda\mu + \mu^2}}{2}.$$

Inverting the transform in (8.62), we get:

$$f_Y(t) = \frac{2\lambda^2}{\alpha_1 - \alpha_2}\left(e^{-\alpha_2 t} - e^{-\alpha_1 t}\right).$$

Then the MTTF of the system is given by:

$$E[Y] = \int_0^\infty y f_Y(y)\,dy = \frac{2\lambda^2}{\alpha_1 - \alpha_2}\left[\int_0^\infty y e^{-\alpha_2 y} dy - \int_0^\infty y e^{-\alpha_1 y} dy\right].$$

Recalling that $\int_0^\infty ye^{-\alpha y}\,dy = 1/\alpha^2$, we get:

$$E[Y] = \frac{2\lambda^2}{\alpha_1 - \alpha_2}\left[\frac{1}{\alpha_2^2} - \frac{1}{\alpha_1^2}\right] = \frac{2\lambda^2(\alpha_1 + \alpha_2)}{\alpha_1^2\alpha_2^2}$$

$$= \frac{2\lambda^2(3\lambda + \mu)}{(2\lambda^2)^2} = \frac{3}{2\lambda} + \frac{\mu}{2\lambda^2}. \tag{8.63}$$

Note that the MTTF of the two-component parallel-redundant system, in the absence of a repair facility (i.e., $\mu = 0$), would have been equal to the first term, $3/(2\lambda)$, in the above expression. Therefore, the effect of a repair facility is to increase the mean life by $\mu/2\lambda^2$, or by a factor

$$\frac{\mu/2\lambda^2}{3/2\lambda} = \frac{\mu}{3\lambda}.$$

#

Example 8.16

Next consider a modification of the above example proposed by Arnold [ARNO 1973] as a model of duplex processors of an electronic switching system. We assume that not all faults are recoverable and that c is the coverage factor denoting the conditional probability that the system recovers, given that a fault has occurred. The state diagram is now given by Figure 8.21. Note that this chain is *not* an example of a birth-death process.

Assume that the initial state is 2, so that

$$P_2(0) = 1, \qquad P_0(0) = P_1(0) = 0.$$

Then $p_{2j}(t) = P_j(t)$ and the system of equation (8.10) yields:

$$\begin{aligned}
\frac{dP_2(t)}{dt} &= -2\lambda cP_2(t) - 2\lambda(1-c)P_2(t) + \mu P_1(t),\\
\frac{dP_1(t)}{dt} &= -(\lambda+\mu)P_1(t) + 2\lambda cP_2(t),\\
\frac{dP_0(t)}{dt} &= \lambda P_1(t) + 2\lambda(1-c)P_2(t).
\end{aligned} \tag{8.64}$$

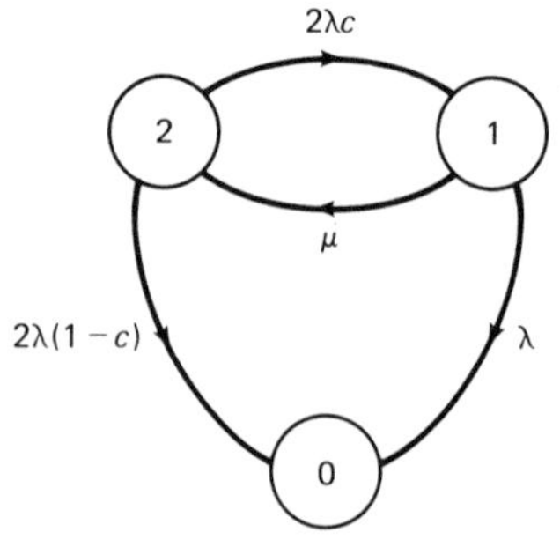

Figure 8.21 The state diagram of a duplex system with imperfect coverage

Using Laplace transforms as before, the above system reduces to:

$$\begin{aligned} s\bar{P}_2(s) - 1 &= -2\lambda \bar{P}_2(s) + \mu \bar{P}_1(s), \\ s\bar{P}_1(s) &= -(\lambda + \mu)\bar{P}_1(s) + 2\lambda c\bar{P}_2(s), \\ s\bar{P}_0(s) &= \lambda \bar{P}_1(s) + 2\lambda(1 - c)\bar{P}_2(s). \end{aligned} \tag{8.65}$$

This system of linear equations can be solved to yield:

$$\bar{P}_0(s) = \frac{2\lambda}{s} \frac{s + \lambda + \mu - c(s + \mu)}{(s + 2\lambda)(s + \lambda + \mu) - 2\lambda\mu c},$$

$$\bar{P}_1(s) = \frac{2\lambda c}{(s+2\lambda)(s + \lambda + \mu) - 2\lambda c\mu},$$

$$\bar{P}_2(s) = \frac{s + \lambda + \mu}{(s + 2\lambda)(s + \lambda + \mu) - 2\lambda c\mu}.$$

As before, if X is the time to system failure, then:

$$F_X(t) = P_0(t);$$

therefore:

$$\begin{aligned} \bar{f}_X(s) = L_X(s) &= sP_0(s) \\ &= \frac{2\lambda[(s + \lambda + \mu) - c(s + \mu)]}{(s + 2\lambda)(s + \lambda + \mu) - 2\lambda\mu c}. \end{aligned}$$

Let this be rewritten as:

$$L_X(s) = \frac{2\lambda U}{V},$$

where $U = s + \lambda + \mu - c(s + \mu)$ and $V = (s+2\lambda)(s + \lambda + \mu) - 2\lambda\mu c$. Instead of inverting this expression to obtain the distribution of X, we will be content with obtaining $E[X]$ using the moment generating property of Laplace transforms:

$$\begin{aligned} E[X] &= -\frac{dL_X}{ds}\Big|_{s=0} \\ &= \frac{2\lambda[U(2s + 3\lambda + \mu) - V(1 - c)]}{V^2}\Big|_{s=0} \\ &= \frac{2\lambda[(\lambda + \mu - \mu c)(3\lambda + \mu) - (2\lambda(\lambda + \mu) - 2\lambda\mu c)(1 - c)]}{[2\lambda(\lambda + \mu) - 2\lambda\mu c]^2}, \end{aligned}$$

which, when reduced, finally gives us the required expression for mean time to system failure:

$$E[X] = \frac{\lambda(1 + 2c) + \mu}{2\lambda\ [\lambda + \mu(1 - c)]}. \tag{8.66}$$

Note that as c approaches 1, this expression reduces to the MTTF given in equation (8.63) for Example 8.15. As the coverage factor c approaches 0, the above expression yields the value $1/(2\lambda)$, which corresponds to the MTTF of a series system consisting of the two components. It should be clear that the system MTTF is critically dependent on the coverage factor, as can be seen from Figure 8.22, in which the ratio $E[X]/E[Y]$ is plotted as a function of the coverage factor c. Recall that $E[Y]$ is the MTTF of a system with perfect coverage; it is obtained from equation (8.63). #

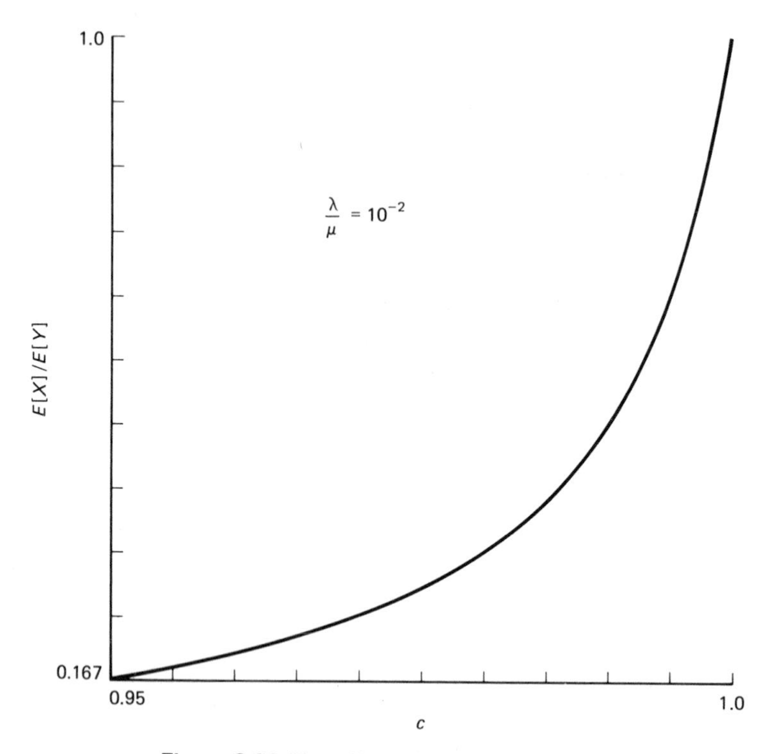

Figure 8.22 The effect of coverage on MTTF

Example 8.17

Consider a hybrid NMR system with three active units and one spare. The active configuration is operated in TMR mode. An active unit has a failure rate λ, while a standby spare unit has a failure rate μ. If i active units and j standby units are functioning properly, then the state of the system will be denoted by (i, j). The state F denotes the system-failure state. We assume that in state $(3, 1)$, failure of an active unit can be recovered with probability c ($\leqslant 1$). The state diagram of the Markov chain is given in Figure 8.23.

Differential equations for this Markov chain are written as follows:

$$\frac{dP_{3,1}}{dt} = -(3\lambda + \mu)P_{3,1}(t),$$

$$\frac{dP_{3,0}}{dt} = -3\lambda P_{3,0}(t) + (3\lambda c + \mu)P_{3,1}(t),$$

$$\frac{dP_{2,0}}{dt} = -2\lambda P_{2,0}(t) + 3\lambda P_{3,0}(t),$$

$$\frac{dP_F}{dt} = 3\lambda(1-c)P_{3,1}(t) + 2\lambda P_{2,0}(t),$$

where we have assumed that the initial state is (3, 1) so that $P_{3,1}(0) = 1$, and $P_{i,j}(0) = 0 = P_F(0)$ otherwise. Using Laplace transforms, we get:

$$s\bar{P}_{3,1}(s) - 1 = -(3\lambda + \mu)\bar{P}_{3,1}(s),$$

$$s\bar{P}_{3,0}(s) = -3\lambda\bar{P}_{3,0}(s) + (3\lambda c + \mu)\bar{P}_{3,1}(s),$$

$$s\bar{P}_{2,0}(s) = -2\lambda\bar{P}_{2,0}(s) + 3\lambda\bar{P}_{3,0}(s),$$

$$s\bar{P}_F(s) = 3\lambda(1-c)\bar{P}_{3,1}(s) + 2\lambda\bar{P}_{2,0}(s).$$

Solving this system of equations, we get:

$$\bar{P}_{3,1}(s) = \frac{1}{s + 3\lambda + \mu},$$

$$\bar{P}_{3,0}(s) = \frac{3\lambda c + \mu}{(s + 3\lambda + \mu)(s + 3\lambda)},$$

$$\bar{P}_{2,0}(s) = \frac{3\lambda(3\lambda c + \mu)}{(s + 3\lambda + \mu)(s + 3\lambda)(s + 2\lambda)},$$

and

$$s\bar{P}_F(s) = \frac{3\lambda(1-c)}{(s + 3\lambda + \mu)} + \frac{6\lambda^2(3\lambda c + \mu)}{(s + 2\lambda)(s + 3\lambda)(s + 3\lambda + \mu)}.$$

If X is the time to failure of the system, then $P_F(t)$ is the distribution function of X. It follows that:

$$\bar{f}_X(s) = s\bar{P}_F(s) - P_F(0) = s\bar{P}_F(s),$$

which can be rewritten as:

$$\bar{f}_X(s) = \frac{3\lambda + \mu}{(s + 3\lambda + \mu)}\left[\frac{3\lambda(1-c)}{(3\lambda + \mu)} + \frac{3\lambda c + \mu}{3\lambda + \mu}\left\{\frac{2\lambda}{s + 2\lambda} \cdot \frac{3\lambda}{s + 3\lambda}\right\}\right]. \tag{8.67}$$

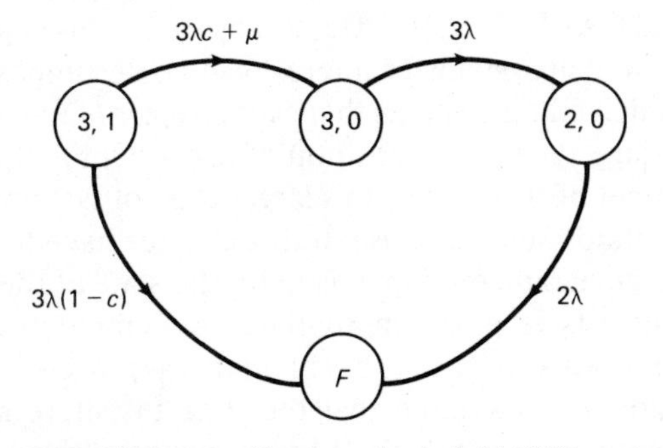

Figure 8.23 The state diagram of a hybrid-TMR system with imperfect coverage

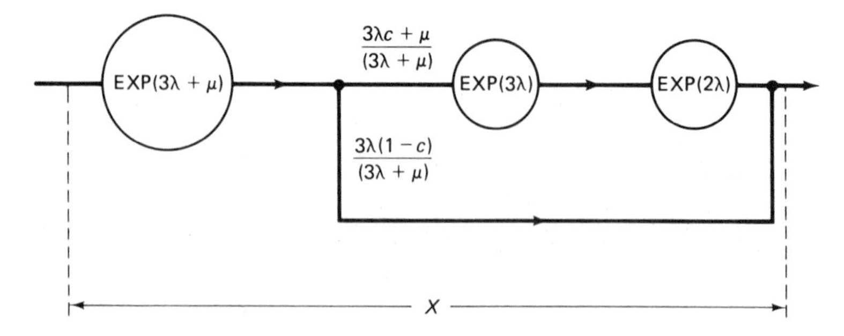

Figure 8.24 The stage-type lifetime distribution for the system of Example 8.17

The expression outside the square brackets is the Laplace transform of EXP $(3\lambda + \mu)$, while the expression within the braces is the Laplace transform of HYPO $(2\lambda, 3\lambda)$. Therefore, the system lifetime X has the stage-type distribution shown in Figure 8.24. It follows that system MTTF is:

$$\begin{aligned} E[X] &= \frac{1}{3\lambda + \mu} + \frac{3\lambda c + \mu}{3\lambda + \mu} [\frac{1}{2\lambda} + \frac{1}{3\lambda}] \\ &= \frac{1}{3\lambda + \mu} + \frac{3\lambda c + \mu}{3\lambda + \mu} \cdot \frac{5}{6\lambda}. \end{aligned}$$

This can be verified by computing $-d\bar{f}_X/ds\,|_{s=0}$, which is easy from (8.67). #

We have seen two different methods of reliability analysis. The first method, developed in Chapter 3 and further elaborated in Chapter 5, considered the distribution of time to failure as a network of a finite number of exponential phases. This culminated in the reliability expression of a hybrid NMR system with imperfect coverage, where spares are allowed to fail. Now if a given system is a series of several subsystems where each subsystem is of hybrid NMR type, then our method of Chapter 5 can be used to analyze the given system. Computer-based reliability analysis packages known as CARE [MATH 1972] and CARE II [STIF 1975] are based on this approach.

Nevertheless, not all systems of interest can be decomposed into a series of subsystems, and even for systems that can, if repair of units in a subsystem is allowed, then stage-type decomposition of subsystem lifetime could result in an infinite number of stages. For this larger class of systems, more general Markov chain methods such as those in this chapter have to be used. Similarly, if the coverage parameter depends upon the state of the system, then a Markov chain analysis is more appropriate. A computer-based reliability analysis package known as ARIES is based on this approach [NG 1976].

Both these approaches assume that the time to failure and the time to repair are exponentially distributed. If this assumption is violated, then we can model the time to failure as consisting of a Coxian network of exponential stages. The time to repair can be similarly modeled as a Coxian network. Quite general distributions can be accounted for by this method; however, the number of states in the Markov chain grows rapidly. Several approaches to deal with the problem of very large state spaces are currently under investigation [STIF 1979].

If the system is modeled so that it does not have a crash type of failure, then the corresponding Markov chain will not possess absorbing states. In this case we study the limiting distribution and the steady-state availability of the system, as in Examples 8.7 and 8.8. Transient analysis of such systems is usually quite difficult. In the analysis of chains with absorbing states, on the other hand, the steady-state analysis is trivial and uninteresting, while transient analysis is of interest.

Problems

1. Modify the structure of Example 8.15 so that it is a two-unit standby-redundant system rather than a parallel-redundant system. Assume that the failure rates of on-line and standby units are respectively given by λ_1 and λ_2, where $\lambda_1 \geqslant \lambda_2 \geqslant 0$. Repair times are exponentially distributed with mean $1/\mu$, and the system is considered to have failed upon the failure of the second unit before the first unit is repaired. Obtain expressions for system reliability and system MTTF, assuming that the detection and switching mechanism is fault-free.

*2. Consider a two-unit standby-redundant system where the spare failure rate is identical to the failure rate of the active unit. The system is modeled using the homogeneous Markov chain as shown in Figure 8.P.1. Here δ is the detection-reconfiguration rate and c is the coverage factor. Solve the system for its reliability $R(t)$, using the methods developed in this section. Next solve for $R(t)$ using the convolution-integral approach developed in problem 3 at the end of Section 8.1. Finally solve for $R(t)$ using the matrix series approach developed in problem 1, Section 8.1.

*3. Suppose that we wish to perform state aggregation on the state diagram of problem 2 above and reduce it to the state diagram shown in Figure 8.P.2. (Thus states 2 and 2_U of Figure 8.P.1 are aggregated into state 2′.) Derive expressions

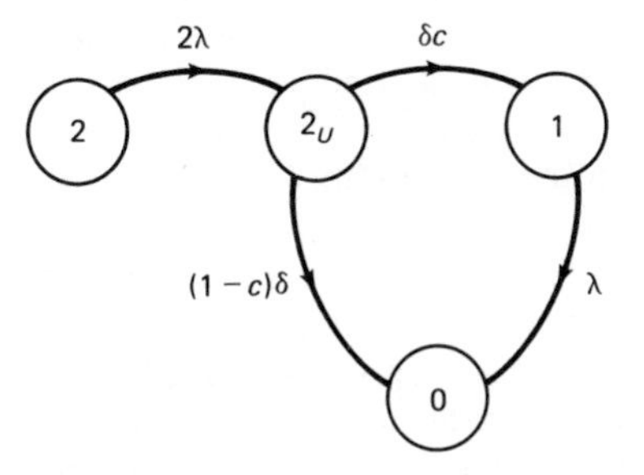

Figure 8.P.1 A two-unit system with non-zero detection latency

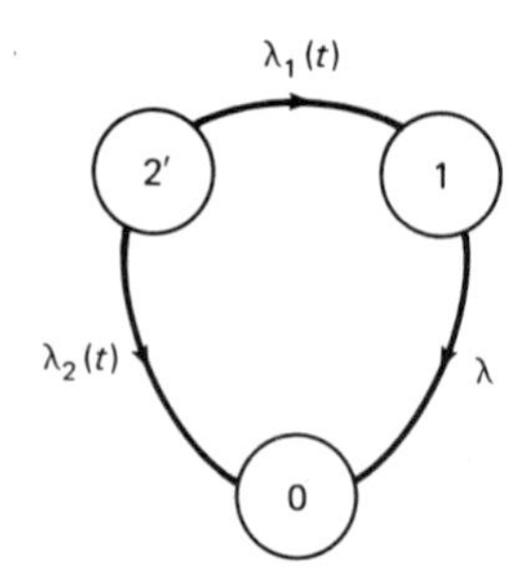

Figure 8.P.2 An aggregated version of Figure 8.P.1

for the transition parameters $\lambda_1(t)$ and $\lambda_2(t)$. Note that the reduced chain is a non-homogeneous Markov chain.

***4.** Consider a two-component parallel-redundant system with distinct failure rates λ_1 and λ_2, respectively. Upon failure of a component, a repair process is invoked with the respective repair rates μ_1 and μ_2. First assume that a failure of a component while another is being repaired causes a system failure. In this case set up the differential equations describing system behavior. Derive the reliability and the MTTF of the system.

5. Continuing with problem 4 above, assume that the failure of a component while another is being repaired is not catastrophic. Now compute the steady-state availability of the system.

6. Assuming $\lambda = 10^{-4}$, and $\mu = 1$, compare the reliability of a two-unit parallel-redundant system with repair (Example 8.15) with that of a two-unit parallel-redundant system without repair. Also plot the two expressions on the same graph paper.

7. Modify the reliability model of Example 8.17 to allow for a repair to occur from states (3, 0) and (2, 0) at a constant rate γ. State F is still assumed to be an absorbing state. Recompute the system MTTF.

8. Our assumption that the coverage probability is a given number is often unjustified in the modeling of fault-tolerant computers. In this problem we consider a "coverage model" of intermittent faults. (This is a simplified version of the model proposed by Stiffler [STIF 1980].) The model consists of five states as shown in Figure 8.P.3. In the active state A, the intermittent fault is capable of producing errors at the rate ρ and leading to the error state E. In the benign state B, the affected circuitry temporarily functions correctly. In state D the fault has been detected, and in the failure state F an undetected error has propagated so that we declare the system to have failed.

Set up the differential equations for the five state probabilities. If we assume that all transition rates are greater than 0, then states A, B, and E are transient while states D and F are absorbing states. Given that the process starts in state A, it will eventually end up in either state D or state F. In the former case the fault is covered; in the latter it is not. We can, therefore, obtain an expression for cov-

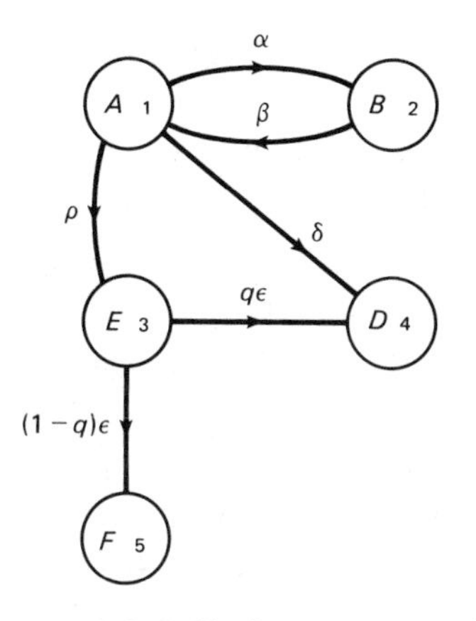

Figure 8.P.3 Stiffler's coverage model

erage probability, $c = \lim_{t \to \infty} P_D(t)$. Using the final value theorem of Laplace transforms (see Appendix D), show that:

$$c = \lim_{s \to 0} s\bar{P}_D(s) = \frac{\delta + \rho q}{\delta + \rho}.$$

Having obtained the value of c, we can then use an overall reliability model such as those in Examples 8.16 and 8.17. Such a decomposition is intuitively appealing, since the transition rates in the coverage model will be orders of magnitude larger than those in the overall reliability model. For a detailed study of this decomposition approach to reliability modeling see [STIF 1975, STIF 1979].

9. Modify the reliability model of Example 8.16 (Figure 8.21) to allow for repairs from state 0 at a constant rate μ_1. Now derive the expression for the steady-state availability of the system.

***10.** Return to the concurrent program analyzed in problem 7, Section 8.4. First, derive the Laplace transform for the execution time for one iteration of the **repeat** statement using the methods of Section 8.5. Now, assume that the **repeat** clause in the program is changed so that it terminates when a Boolean expression B' is true. Assuming that the testing of this condition forms a sequence of Bernoulli trials with probability of success p, compute the mean and variance of the program execution time.

References

[ARNO 1973] T. F. ARNOLD, "The Concept of Coverage and Its Effect on the Reliability Model of a Repairable System," *IEEE Trans. on Computers,* C-22 (March 1973), 251–54.

[BEUT 1980] F. J. BEUTLER, "Sojourn Times in Markov Queuing Networks: Little's Formula Revisited," Tech. Report, Computer Information and Control Engineering Program, University of Michigan, Ann Arbor.

[BHAT 1972] U. N. BHAT, *Elements of Applied Stochastic Processes,* John Wiley & Sons, New York.

[COFF 1973] E. G. COFFMAN, Jr., and P. J. DENNING, *Operating System Theory,* Prentice-Hall, Englewood Cliffs, N.J.

[DENN 1978] P. J. DENNING and J. P. BUZEN, "The Operational Analysis of Queuing Network Models," *ACM Computing Surveys,* Vol.10, pp. 225–41.

[FULL 1975] S. H. FULLER, "Performance Evaluation," in H. S. Stone, ed., *Introduction to Computer Architecture,* Science Research Associates, Chicago, pp. 474–545.

[GOEL 1979] A. L. GOEL and K. OKUMOTO, "A Time Dependent Error Detection Rate Model for Software Reliability and Other Performance Measures," *IEEE Trans. on Reliability* Vol. R-28, No.3 , August 1979, pp. 206–11.

[KLEI 1976] L. KLEINROCK, *Queuing Systems,* Vol. II, John Wiley & Sons, New York.

[KOBA 1978] H. KOBAYASHI, *Modeling and Analysis,* Addison-Wesley, Reading, Mass.

[MATH 1972] F. P. MATHUR, "Automation of Reliability Evaluation Procedures Through CARE—The Computer-Aided Reliability Estimation Program," *AFIPS Conference Proceedings,* Vol. 41, Fall Joint Computer Conference, pp. 65–82.

[NG 1976] Y-W. NG, "Reliability Modeling and Analysis for Fault-Tolerant Computers," Ph.D. Dissertation, Computer Science Department, University of California at Los Angeles.

[PARZ 1962] E. PARZEN, *Stochastic Processes,* Holden-Day, San Francisco, Calif.

[ROSS 1970] S. M. ROSS, *Applied Probability Models with Optimization Applications,* Holden-Day, San Francisco, Calif.

[STID 1974] S. STIDHAM, Jr., "A Last Word on $L = \lambda W$," *Operations Research,* 22 (1974), 417–21.

[STIF 1975] J. J. STIFFLER and others, "An Engineering Treatise of the CARE II Dual Mode and Coverage Models," Final Report, NASA Contract L-18084A.

[STIF 1979] J. J. STIFFLER and others, "CARE III Final Report, Phase I," NASA Contractor Report 159122, November 1979.

[STIF 1980] J. J. STIFFLER, "Robust Detection of Intermittent Faults," *Proc. of the 10th Int. Symp. on Fault-Tolerant Computing,* Kyoto, Japan, October 1980, pp. 216–18.

[TOWS 1978] D. F. TOWSLEY, J. C. BROWNE, and K. M. CHANDY, "Models for Parallel Processing Within Programs: Application to CPU:I/O and I/O:I/O Overlap," *CACM,* 21: 10 (October 1978), 821–31.

Chapter 9

Networks of Queues

9.1 INTRODUCTION

We have studied Markov processes of the birth death type in Chapter 8. Such processes are characterized by a simple product-form solution [equation (8.20)] and *a large number of applications.* When we remove the restriction of nearest neighbor transitions only, we may not have the convenient product-form solution. It is natural to ask whether there is a class of Markov processes that subsumes birth-death processes and that possesses a product-form solution. One important generalization of the birth-death process that we consider is a network of queues. Such networks can model problems of contention that arise when a set of resources are shared. Each resource is represented by a node or a service center. Thus, in a model for computer-system performance analysis, we may have a service center for the CPU(s), a service center for each I/O channel, and possibly others. A service center may have one or more servers associated with it. If a job requesting service finds all the servers at the service center busy, it will join the queue associated with the center, and at a later point in time, when one of the servers becomes idle, a job from the queue will be selected for service according to some scheduling discipline. After completion of service at one service center, the job may move to another service center for further service, reenter the same service center, or leave the system.

We shall consider two types of networks: open and closed. An **open queuing network** is characterized by one or more sources of job arrivals and correspondingly one or more sinks that absorb jobs departing from the network. In a **closed queuing network**, on the other hand, jobs neither enter nor depart from the network.

The behavior of jobs within the network is characterized by the probabilities of transitions between service centers and the distribution of job service times at each center. For each center the number of servers, the scheduling discipline, and the size of the queue must be specified. Unless stated otherwise, we assume that the scheduling is FCFS and that each server has a queue of unlimited capacity. For an open network, a characterization of job-arrival processes is needed, and for a closed network, the number of jobs in the network must be specified.

Consider the two-stage tandem network shown in Figure 9.1. The system consists of two nodes with respective service rates μ_0 and μ_1. The external arrival rate is λ. The output of the node labeled 0 is the input to the node labeled 1. The service-time distribution at both nodes is exponential and the arrival process to the node labeled 0 is Poisson.

This system can be modeled as a stochastic process whose states are specified by pairs (k_0, k_1), $k_0 \geqslant 0$, $k_1 \geqslant 0$, where k_i $(i = 0, 1)$ is the number of jobs at server i in the steady state. The changes of state occur upon a completion of service at one of the two servers or upon an external arrival. Since all interevent times are exponentially distributed (by our assumptions), it follows that the stochastic process is a Markov chain with the state diagram shown in Figure 9.2.

For k_0, $k_1 > 0$, the transitions into and out of that state are shown in Figure 9.3. Let $p(k_0, k_1)$ be the joint probability of k_0 jobs at node 0 and k_1 jobs at node 1 in the steady state. Equating the rates of flow into and out of the state, we obtain the following balance equations:

$$(\mu_0 + \mu_1 + \lambda)p(k_0, k_1) = \mu_0 p(k_0 + 1, k_1 - 1) + \mu_1 p(k_0, k_1 + 1) + \lambda p(k_0 - 1, k_1), \qquad k_0 > 0, \; k_1 > 0.$$

For the boundary states, we have:

$$(\mu_0 + \lambda)p(k_0, 0) = \mu_1 p(k_0, 1) + \lambda p(k_0 - 1, 0), \qquad k_0 > 0,$$

$$(\mu_1 + \lambda)p(0, k_1) = \mu_0 p(1, k_1 - 1) + \mu_1 p(0, k_1 + 1), \qquad k_1 > 0,$$

$$\lambda p(0, 0) = \mu_1 p(0, 1).$$

The normalization is provided by:

$$\sum_{k_0 \geqslant 0} \sum_{k_1 \geqslant 0} p(k_0, k_1) = 1.$$

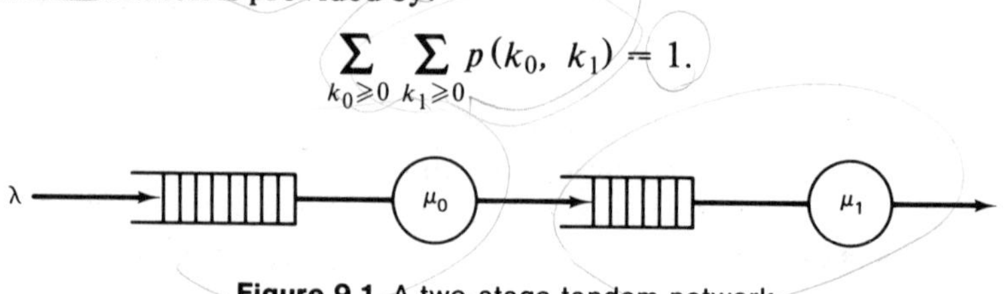

Figure 9.1 A two-stage tandem network

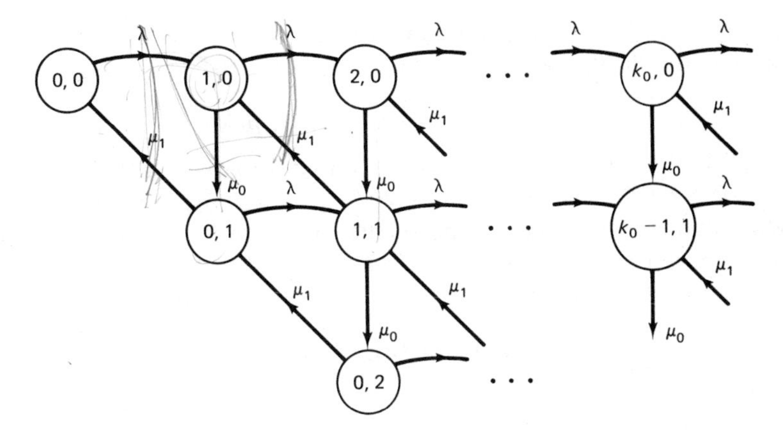

Figure 9.2 The state diagram for the two-stage tandem network

It is easily shown by direct substitution that the following equation is the solution to the above balance equations:

$$p(k_0, k_1) = (1 - \rho_0)\rho_0^{k_0}(1 - \rho_1)\rho_1^{k_1}, \tag{9.1}$$

where $\rho_0 = \lambda/\mu_0$ and $\rho_1 = \lambda/\mu_1$. The condition for stability of the system is that both ρ_0 and ρ_1 are less than unity.

Equation (9.1) is a product-form solution similar to that of an $M/M/1$ queue. Observe that the node 0 in Figure 9.1 has a Poisson arrival source of rate λ and exponentially distributed service time. Therefore, the node la-

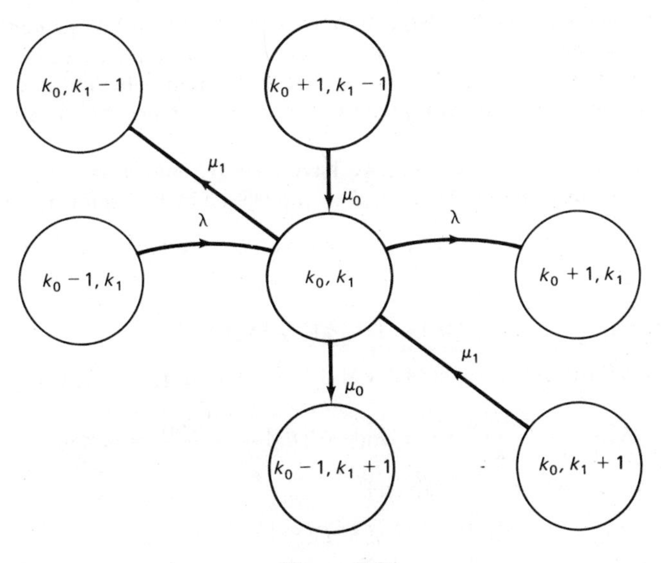

Figure 9.3

beled 0 is an $M/M/1$ queue. It follows that the pmf of the number of jobs N_0 at node 0 in the steady state is given by [formula (8.20)]:

$$P(N_0 = k_0) = p_0(k_0) = (1 - \rho_0)\rho_0^{k_0}.$$

Burke [BURK 1956] has shown that the output of an $M/M/1$ queue is also Poisson with rate λ (you are asked to verify this result in problem 2 at the end of this section). Thus, the second queue in Figure 9.1 is also an $M/M/1$ queue with server utilization $\rho_1 = \lambda/\mu_1$ (assumed to be < 1). Hence, the steady-state pmf of the number of jobs N_1 at node 1 is given by:

$$P(N_1 = k_1) = p_1(k_1) = (1 - \rho_1)\rho_1^{k_1}.$$

The joint probability of k_0 jobs at node 0 and k_1 jobs at node 1 is given by equation (9.1):

$$p(k_0, k_1) = (1 - \rho_0)\rho_0^{k_0}(1 - \rho_1)\rho_1^{k_1} = p_0(k_0)p_1(k_1).$$

Thus the joint probability $p(k_0, k_1)$ is the product of the marginal probabilities $p_0(k_0)$ $p_1(k_1)$; hence random variables N_0 and N_1 are independent in the steady state. Therefore, the two queues are independent $M/M/1$ queues. As the arrival rate λ increases, the node with the larger value of ρ will introduce instability. Hence the node with the largest value of ρ is called the "bottleneck" of the system.

The product form solution (9.1) can be generalized to an m-stage tandem queue.

Example 9.1

A repair facility shared by a large number of machines has two sequential stations with respective rates one per hour and two per hour. The cumulative failure rate of all the machines is 0.5 per hour. Assuming that the system behavior may be approximated by the two-stage tandem queue of Figure 9.1, determine the average repair time.

Given $\lambda = 0.5$, $\mu_0 = 1$, $\mu_1 = 2$, we have $\rho_0 = 0.5$ and $\rho_1 = 0.25$. The average length of the queue at station i ($i = 0, 1$) is then given by [using formula (8.21)]:

$$E[N_i] = \frac{\rho_i}{1 - \rho_i};$$

hence:

$$E[N_0] = 1 \quad \text{and} \quad E[N_1] = 1/3.$$

Using Little's formula, the repair delay at the two stations is respectively given by:

$$E[R_0] = \frac{E[N_0]}{\lambda} = 2 \quad \text{and} \quad E[R_1] = \frac{E[N_1]}{\lambda} = 2/3 \text{ hours}.$$

Hence the average repair time is given by:

$$E[R] = E[R_0] + E[R_1] = 8/3 \text{ hours}.$$

This can be decomposed into waiting time at station 0 (= 1 hour), the service time at station 0 ($= 1/\mu_0 =$ 1 hour), the waiting time at station 1 (= ⅙ hour), and the service time at station 1 ($1/\mu_1 =$ ½ hour). The probability that both service stations are idle is given by:

$$p(0, 0) = (1 - \rho_0)(1 - \rho_1) = 3/8.$$

Station 0 is the bottleneck of the repair facility. #

Problems

1. In Chapter 8 we derived the distribution function of the response time of an isolated $M/M/1$ FCFS queue [see equation (8.25)]. Using this result, derive the distribution function of the response time for the tandem network of Figure 9.1. From this obtain the variance of the response time.

*2. Consider an $M/G/1$ queue with FCFS scheduling. Let the random variables A, B, and D, respectively denote the interarrival time, the service time, and the interdeparture time. By conditioning on the number of jobs in the system and then using the theorem of total Laplace transforms show that in the steady state:

$$L_D(s) = \rho L_B(s) + (1 - \rho) L_A(s) L_B(s).$$

Point out why the assumption of Poisson arrival stream is needed to derive this result. Then, specializing to the case of $M/M/1$ queue, show that:

$$L_D(s) = L_A(s).$$

This verifies Burke's result that the output process of an $M/M/1$ FCFS queue is Poisson. Note that the independence of successive interdeparture times needs to be shown in order to complete the proof.

3. Using the result of problem 2, show that in the $M/G/1$ case, the coefficient of variation of the interdeparture time is given by:

$$C_D^2 = 1 + \rho^2 (C_B^2 - 1).$$

*4. Show that the interdeparture-time distribution of an $M/M/m$ FCFS queue is exponential. To simplify the problem, first consider an $M/M/2$ queue. Let D_i denote the interdeparture time conditioned on the number of jobs in the system $N = i$. Then show that $D_i \sim$ EXP (2μ) for $i \geq 2$. Next show that:

$$L_{D_1}(s) = \frac{\lambda \cdot 2 \cdot \mu}{(s + \lambda + \mu)(s + 2\mu)} + \frac{\mu}{s + \lambda + \mu}$$

and

$$L_{D_0}(s) = \frac{\lambda}{s + \lambda} L_{D_1}(s)$$

(a tree diagram may be helpful here). Then obtain the required result using the theorem of total Laplace transforms. The generalization to the $M/M/m$ case proceeds in a similar fashion.

9.2 OPEN QUEUING NETWORKS

The argument given in the previous section for the product-form solution of tandem queues can be generalized to any feed-forward network of exponential queues (in which a job may not return to previously visited nodes) that is fed from independent Poisson sources. Jackson [JACK 1957] showed that the product-form solution also applies to open networks of Markovian queues with feedback. Besides requiring that the distributions of job interarrival times and service times at all nodes be exponential, assume that the scheduling discipline is FCFS.

First we consider several examples illustrating Jackson's technique.

Example 9.2

Consider the simple model of a computer system shown in Figure 9.4a. Jackson's result is that two queues will behave like independent $M/M/1$ queues, and hence:

$$p(k_0, k_1) = (1 - \rho_0)\rho_0^{k_0}(1 - \rho_1)\rho_1^{k_1}, \tag{9.2}$$

where $\lambda_0/\mu_0 = \rho_0$ and $\lambda_1/\mu_1 = \rho_1$. To apply this result, we have to compute the average arrival rates λ_0 and λ_1 into the two nodes. Note that in the steady state the departure rates from the two nodes will also be λ_0 and λ_1 , respectively. Arrivals to the CPU node occur either from the outside world at the rate λ or from the I/O node at the rate λ_1. The total arrival rate to the CPU node is, therefore:

$$\lambda_0 = \lambda + \lambda_1.$$

Given that a job just completed a CPU burst, it will next request I/O service with probability p_1. Therefore, the average arrival rate to device 1 is given by:

$$\lambda_1 = \lambda_0 p_1.$$

Thus:

$$\lambda_0 = \frac{\lambda}{1 - p_1} = \frac{\lambda}{p_0} \quad \text{and} \quad \lambda_1 = \frac{p_1\lambda}{p_0}.$$

This implies that:

$$\rho_0 = \frac{\lambda}{p_0\mu_0} \quad \text{and} \quad \rho_1 = \frac{p_1\lambda}{p_0\mu_1}.$$

If we let B_0 denote the total CPU service requirement of a program, then $E[B_0] = 1/(p_0\,\mu_0)$. Similarly, $E[B_1] = p_1/(p_0\,\mu_1)$ denotes the expected value of the total service time required on the I/O device for a typical program. If $\rho_0 > \rho_1$ (that is, $E[B_0] > E[B_1]$), then the CPU is the bottleneck, in which case the system is said to be CPU-bound. Similarly, if $\rho_0 < \rho_1$, then the system is I/O-bound.

The average response time may be computed by summing the average number of jobs at the two nodes and then using Little's formula:

$$E[R] = \left(\frac{\rho_0}{1 - \rho_0} + \frac{\rho_1}{1 - \rho_1}\right)\frac{1}{\lambda}$$

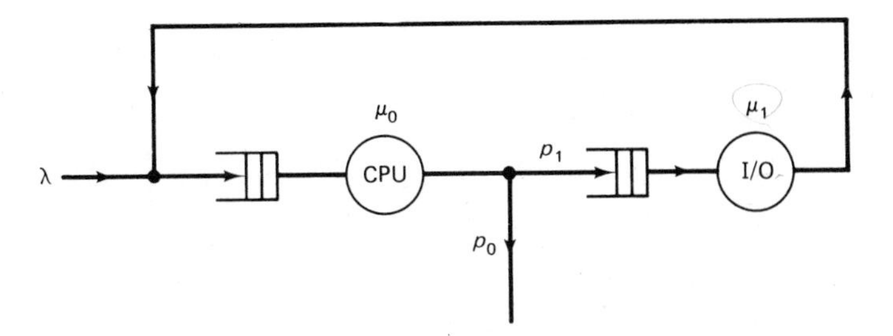

(a) An open network with feedback

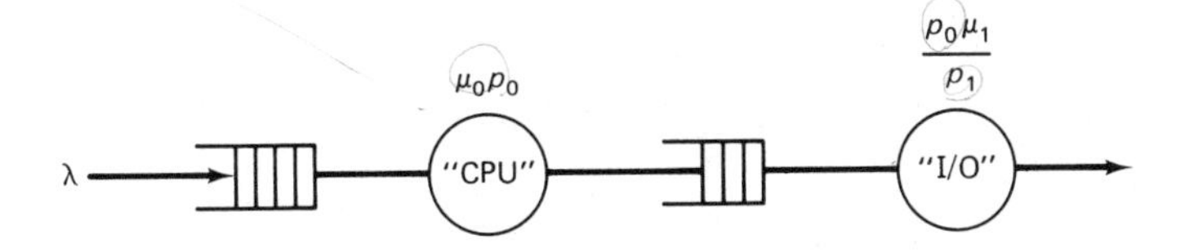

(b) An "equivalent" network without feedback

Figure 9.4

or

$$E[R] = \frac{1}{p_0\mu_0 - \lambda} + \frac{1}{\dfrac{p_0\mu_1}{p_1} - \lambda}$$

$$= \frac{E[B_0]}{1 - \lambda E[B_0]} + \frac{E[B_1]}{1 - \lambda E[B_1]}. \tag{9.3}$$

It is easily seen that this formula also gives the average turnaround time of the "unfolded" tandem network shown in Figure 9.4b. The service rate of the "equivalent" CPU in this system is $\mu_0 p_0$. Thus, a job requests an uninterrupted average CPU time equal to $E[B_0] = 1/(\mu_0 p_0)$ in the equivalent system, while in the original system a job requires $1/p_0$ CPU bursts of average time $1/\mu_0$ each. Therefore, to determine $E[R]$ and $E[N]$, it is sufficient to know only the aggregate resource requirements of a job; in particular, details of the pattern of resource usage are not important for computing these average values. We caution the reader that the equivalence between the networks of Figures 9.4a and 9.4b does not hold with respect to the distribution function $F_R(x)$ of the response time. Computation of response-time distribution is difficult even for Jacksonian networks without feedback [MELA 1980, SIMO 1979].

It is instructive to solve the network in Figure 9.4a by directly analyzing the stochastic process whose states are given by pairs (k_0, k_1), $k_0 \geqslant 0$, $k_1 \geqslant 0$. By the assumption of exponentially distributed interevent times, the process is a Markov process with the state diagram shown in Figure 9.5. For a state (k_0, k_1) with $k_0 > 0$, $k_1 > 0$, the steady-state balance equation is obtained by equating the rate of flow into the state with the rate of flow out of the state:

$$(\lambda + \mu_0 + \mu_1)\, p(k_0, k_1) = \lambda p(k_0 - 1, k_1) + \mu_0 p_1 p(k_0 + 1, k_1 - 1)$$
$$+ \mu_0 p_0 p(k_0 + 1, k_1) + \mu_1 p(k_0 - 1, k_1 + 1).$$

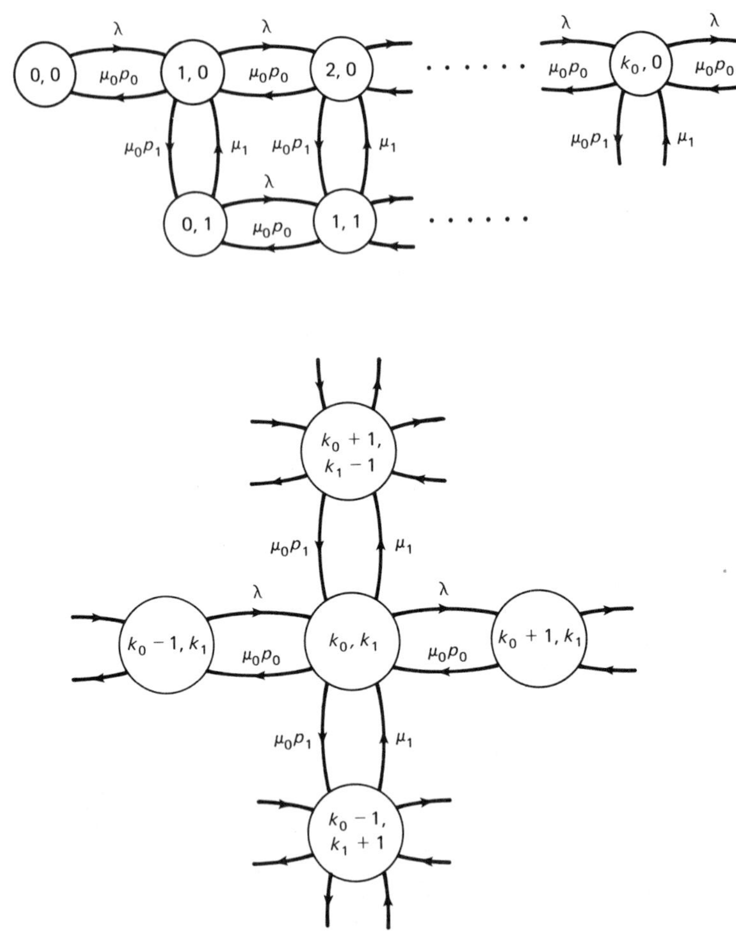

Figure 9.5

Similarly:

$$(\lambda + \mu_0)p(k_0, 0) = \lambda p(k_0 - 1,\ 0) + \mu_0 p_0 p(k_0 + 1,\ 0) + \mu_1 p(k_0 - 1,\ 1),\ k_0 > 0,$$

$$(\lambda + \mu_1)p(0,\ k_1) = \mu_0 p_0 p(1,\ k_1) + \mu_0 p_1 p(1,\ k_1 - 1),\quad k_1 > 0,$$

$$\lambda p(0,\ 0) = \mu_0 p_0 p(1,\ 0).$$

Also:

$$\sum_{k_0, k_1} p(k_0,\ k_1) = 1.$$

It may be verified by direct substitution that the solution (9.2) indeed satisfies these equations. #

Example 9.3

Consider the (open) central server queuing model of a computer system shown in Figure 9.6. Let us follow the path of a tagged program that just arrived. Temporarily

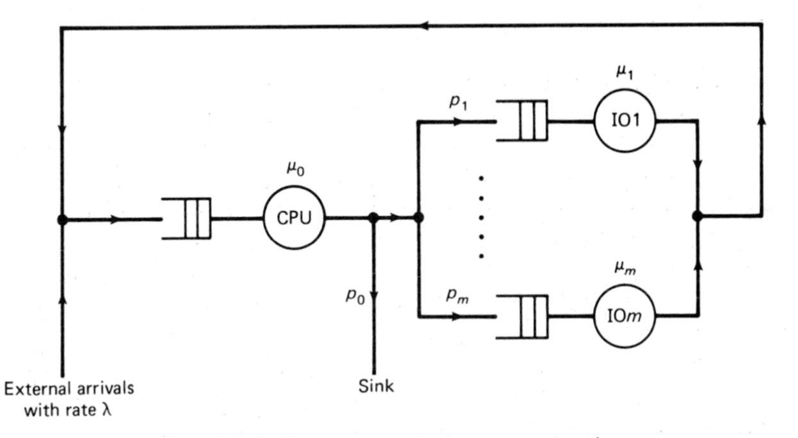

Figure 9.6 The open central server network

ignoring queuing delays, the program will be occupying one of $m + 1$ nodes at a time. With appropriate assumptions, we can model the behavior of a tagged program as a discrete-parameter Markov chain as in Example 7.17; the corresponding state diagram is given in Figure 7.18. [Note that we have added an absorbing state, labeled $m + 1$ (or STOP).] The transition probability matrix of this Markov chain is given by:

$$P = \left[\begin{array}{cccc|c} 0 & p_1 & \cdots & p_m & p_0 \\ 1 & 0 & & 0 & 0 \\ 1 & 0 & & 0 & 0 \\ 1 & \cdot & & \cdot & \cdot \\ 1 & \cdot & & \cdot & \cdot \\ 1 & \cdot & & \cdot & \cdot \\ 1 & 0 & & 0 & 0 \\ \hline 0 & 0 & \cdots & 0 & 1 \end{array}\right]$$

As discussed in Chapter 7 (Section 7.8), the boxed portion of this matrix, denoted here by X, is of interest:

$$X = \begin{bmatrix} 0 & p_1 & p_2 & \cdots & p_m \\ 1 & 0 & 0 & & 0 \\ 1 & \cdot & \cdot & & 0 \\ 1 & \cdot & \cdot & & 0 \\ 1 & \cdot & \cdot & & 0 \\ 1 & & & & 0 \\ 1 & 0 & 0 & & 0 \end{bmatrix}$$

X is known as the transition probability matrix of the queuing network of Figure 9.6. Analyzing the behavior of the tagged program, we are able to obtain the average number of visits (or visit counts) V_j as in Example 7.17 (we assume that $p_0 \neq 0$):

$$V_j = \begin{cases} 1/p_0, & j = 0, \\ p_j/p_0, & j = 1, 2, \ldots, m. \end{cases}$$

That is, the average number of visits made to node j by a typical program is p_j/p_0 $(j \neq 0)$ and $1/p_0$ $(j = 0)$. Since λ programs per unit time enter the network on the average, the overall rate of arrivals, λ_j, to node j is then given by:

$$\lambda_j = \begin{cases} \lambda/p_0, & j = 0, \\ \lambda p_j/p_0, & j = 1, 2, \ldots, m. \end{cases}$$

The utilization, ρ_j, of node j is given by $\rho_j = \lambda_j/\mu_j = \lambda\ V_j/\mu_j$ and we assume that $\rho_j < 1$ for all j. Jackson has shown that the joint probability of k_j customers at node j $(j = 0, 1, \ldots, m)$ is given by:

$$p(k_0, k_1, k_2, \ldots, k_m) = \prod_{j=0}^{m} p_j(k_j). \tag{9.4}$$

This formula implies that the queue lengths are mutually independent in the steady state, and the steady-state probability of k_j customers at node j is given by the $M/M/1$ formula:

$$p_j(k_j) = (1 - \rho_j)\rho_j^{k_j}.$$

The validity of this product-form solution can be established using the direct approach as in Examples 9.1 and 9.2. We leave this as an exercise.

The form of the joint probability (9.4) can mislead a reader to believe that the traffic along the arcs consists of Poisson processes. The reader is urged to solve problem 3 at the end of this section to realize that the input process to a service center in a network with feedback is not Poisson in general [BEUT 1978, BURK 1976]. It is for this reason that Jackson's result is remarkable.

The average queue length, $E[N_j]$, and the response time, $E[R_j]$, of node j are given by:

$$E[N_j] = \frac{\rho_j}{1 - \rho_j} \quad \text{and} \quad E[R_j] = \frac{1}{\lambda}\,\frac{\rho_j}{1 - \rho_j}.$$

From these, the average number of jobs in the system and the average response time are computed to be:

$$E[N] = \sum_{j=0}^{m} \frac{\rho_j}{1 - \rho_j}$$

and

$$\begin{aligned} E[R] &= \frac{1}{\lambda} \sum_{j=0}^{m} \frac{\rho_j}{1 - \rho_j} \\ &= \frac{1/(p_0\mu_0)}{1 - \lambda/(p_0\mu_0)} + \sum_{j=1}^{m} \frac{p_j/(p_0\mu_j)}{1 - \lambda p_j/(\mu_j\, p_0)} \\ &= \frac{1}{\mu_0 p_0 - \lambda} + \sum_{j=1}^{m} \frac{1}{\left(\dfrac{p_0\mu_j}{p_j} - \lambda\right)} \\ &= \frac{1}{\dfrac{\mu_0}{V_0} - \lambda} + \sum_{j=1}^{m} \frac{1}{\dfrac{\mu_j}{V_j} - \lambda} = \sum_{j=0}^{m} \frac{E[B_j]}{1 - \lambda\ E[B_j]}, \end{aligned} \tag{9.5}$$

where the total service requirement on device j, on the average, is given by $E[B_j] = V_j/\mu_j$. This last formula for the response time is a generalized version of formula (9.3). It also affords an "unfolded" interpretation of the queuing network of Figure 9.6, in the same way as Figure 9.4b is the "unfolded" version of the network in Figure 9.4a. #

Jackson's result applies in even greater generality. Consider an open queuing network with $(m + 1)$ nodes, where the ith node consists of c_i exponential servers each with mean service time of $1/\mu_i$ seconds. External Poisson sources contribute γ_i jobs/second to the average rate of arrival to the ith node so that the total arrival rate $\lambda = \sum_{i=0}^{m} \gamma_i$. If we let $q_i = \gamma_i/\lambda$, $i = 0, 1, \ldots, m$ so that $\sum_{i=0}^{m} q_i = 1$, then a job will first enter the network at node i, with probability q_i. Upon service completion at node i, a job next requires service at node j with probability x_{ij} or completes execution with probability $1 - \sum_{j=0}^{m} x_{ij}$.

First, we analyze the behavior of a tagged job through the network. This behavior can be modeled as a discrete-parameter Markov chain as in Example 7.17. Equation (7.52) is applicable here in computing V_i, the average number of visits made by the tagged program to node i. Therefore:

$$V_i = q_i + \sum_{k=0}^{m} x_{ki} V_k, \qquad i = 0, 1, \ldots, m.$$

Now the average job arrival rate to node i is obtained by multiplying V_i by λ, the average job arrival rate to the network. Thus:

$$\lambda_i = \lambda V_i = \lambda q_i + \sum_{k=0}^{m} x_{ki} \lambda V_k.$$

Noting that $\lambda q_i = \gamma_i$ and $\lambda V_k = \lambda_k$, the above expression simplifies to:

$$\lambda_i = \gamma_i + \sum_{k=0}^{m} \lambda_k x_{ki}, \qquad i = 0, 1, \ldots, m. \tag{9.6}$$

This system of equations has a unique solution if we assume that there is at least one node j such that $\gamma_j > 0$ and that the matrix power series:

$$\sum_{k=0}^{n} X^k$$

converges as n approaches infinity (see Section 7.8). This implies that after some number of visits to various service centers there is a positive probability that a job will depart from the system. Jackson's theorem states that each node behaves like an independent $M/M/c_i$ queue and, therefore, the steady-state probability of k_i customers at node i, $i = 0, 1, \ldots, m$ is given by the product form:

$$p(k_0, k_1, \ldots, k_m) = p_0(k_0) \cdots p_m(k_m), \tag{9.6a}$$

where $p_i(k_i)$ is the steady-state probability of finding k_i jobs in an $M/M/c_i$ queue with input λ_i and with average service time $1/\mu_i$ for each of the c_i servers. Thus, using equations (8.30) and (8.31), we have:

$$p_i(k_i) = \begin{cases} \dfrac{(\lambda_i/\mu_i)^{k_i}}{k_i!} p_i(0), & 1 \leqslant k_i < c_i \\ \dfrac{(\lambda_i/\mu_i)^{k_i} p_i(0)}{c_i! c_i^{k_i - c_i}}, & k_i \geqslant c_i, \end{cases}$$

where:

$$p_i(0) = \frac{1}{\displaystyle\sum_{k_i=0}^{c_i-1} \frac{(\lambda_i/\mu_i)^{k_i}}{k_i!} + \frac{(\lambda_i/\mu_i)^{c_i}}{c_i!} \frac{1}{1 - \dfrac{\lambda_i}{c_i\mu_i}}}.$$

Problems

1. Consider the open central server queuing model with two I/O channels with a common service rate of 1.2 second^{-1}. The CPU service rate is 2 second^{-1}, the arrival rate is $\frac{1}{7}$ job/second. The branching probabilities are given by $p_0 = 0.1$, $p_1 = 0.3$, and $p_2 = 0.6$. Determine steady-state probabilities, assuming all service times are independent exponentially distributed random variables. Determine the queue-length distributions at each node as well as the average response time from the source to the sink.

2. Consider a variation of the queuing model of Figure 9.4a, where the CPU node consists of two parallel processors with a service rate of μ_0 each. Draw a state diagram for this system and proceed to solve the balance equations. Obtain an expression for the average response time $E[R]$ as a function of μ_0, μ_1, p_0, and λ. Now compare your answer with that obtained using Jackson's result.

*3. [BURK 1976] Consider the $M/M/1$ FCFS queue with feedback as shown in Figure 9.P.1. By Jackson's theorem, it is easy to derive the steady-state pmf of the number of jobs N in the system:

$$P(N = i) = \left(1 - \frac{\lambda}{\mu p}\right) \left(\frac{\lambda}{\mu p}\right)^i, \qquad i = 0, 1, \ldots.$$

We wish to show that the actual input process (which is a merger of the external arrival process and the feedback process) is not Poisson, even though the exogenous arrival process and the departure process are both Poisson with rate λ.

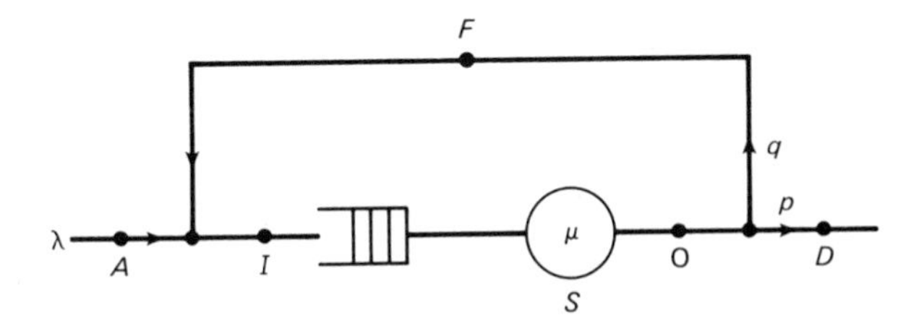

Figure 9.P.1 $M/M/1$ queue with Bernoulli feedback

Proceed by first observing that the complementary distribution function of the interinput times is given by:

$$R_I(t) = e^{-\lambda t} R_Y(t),$$

where Y is the time to the next feedback as measured from the time of the last input to the (queue, server) pair. Next derive the pdf of Y as:

$$f_Y(t) = \mu q e^{-(\mu - \lambda)t},$$

hence show that:

$$R_Y(t) = \frac{\mu q}{\mu - \lambda} e^{-(\mu - \lambda)t} + \frac{\mu p - \lambda}{\mu - \lambda}.$$

From this, conclude that I is hyperexponentially distributed.

In order to derive the Laplace transform of the interdeparture time D proceed by computing the conditional Laplace transform $L_{D|N_d = i}(s)$, where N_d is the number of jobs left in the system by a departing job. Using the result of problem 5, Section 7.6, show that:

$$P(N_d = i) = P(N = i)$$

and hence the unconditional Laplace transform of D is obtained as:

$$L_D(s) = \frac{\lambda}{s + \lambda}.$$

This verifies that the departure process is Poisson.

9.3 CLOSED QUEUING NETWORKS

One of the implicit assumptions behind the model of Example 9.3 is that immediately upon its arrival a job is scheduled into main memory and is able to compete for active resources such as the CPU and the I/O channels. In practice, the number of main-memory partitions will be limited, which implies the existence of an additional queue, called the **job-scheduler queue** (see Figure 9.7). However, such a network is said to involve multiple resource holding. This is because a job can simultaneously hold main memory and an active device. Such a network cannot be solved by product-form methods

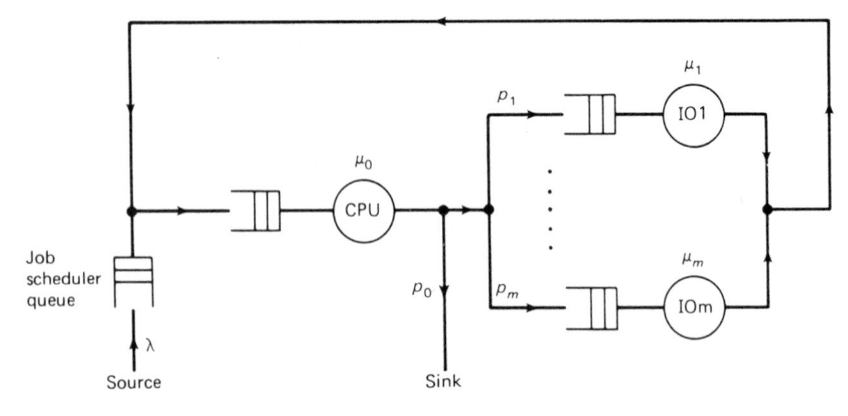

Figure 9.7 An open central server network with a blocking queue

[CHAN 1978b]. Nevertheless, an approximate solution to such a network is provided by the methods of the last section. If the external arrival rate λ is low, then the probability that a job has to wait in the scheduler queue will be low, and we expect the solution of Example 9.3 to be quite good. Thus, the model of Example 9.3 is a light-load approximation to the model of Figure 9.7. Let us now take the other extreme and assume a large value of λ, so that the probability that there is at least one customer in the job scheduler queue is very high.

We may then assume that the departure of a job from the active set immediately triggers the scheduling of an already waiting job into main memory. Thus, the closed network of Figure 9.8 will be a "good" approximation to the system of Figure 9.7, under heavy-load conditions. Each job circulating in this closed network is said to be an active job and must be allo-

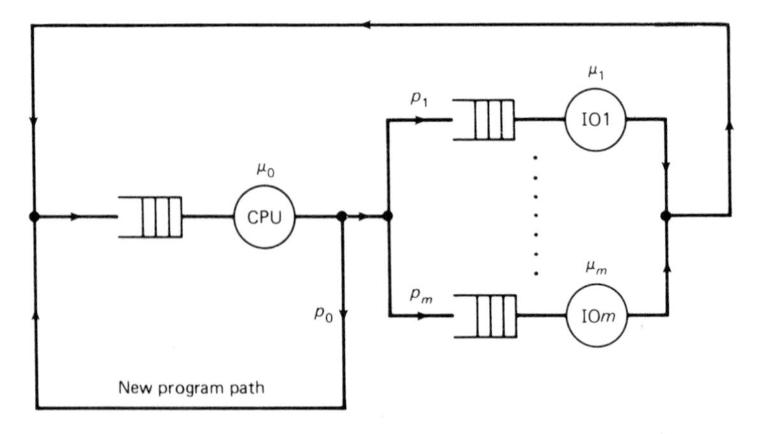

Figure 9.8 The (closed) central server model

cated a partition of main memory. The total number of active jobs is called **the degree** (or **level**) **of multiprogramming.**

Example 9.4

Let us return to the cyclic queuing model studied in the last chapter (Example 8.6), which is shown in Figure 9.9. Here, we choose to represent the state of the system by a pair, (k_0, k_1) where k_i denotes the number of jobs at node i ($i = 0, 1$). Recall that $k_0 + k_1 = n$, the degree of multiprogramming. Unlike the two-node open queuing network (Figure 9.4a), the state space in this case is finite. The dot pattern on the $(k_0 - k_1)$ plane of Figure 9.10 represents the infinite-state space of the open network (Figure 9.4a) while the dot pattern on the line $k_0 + k_1 = n$ is the finite-state space of the cyclic (closed) queuing network being studied here.

The state diagram for the cyclic queuing model is shown in Figure 9.11. The balance equations are given by:

$$(\mu_1 + \mu_0 p_1) p(k_0, k_1) = \mu_0 p_1 p(k_0 + 1, k_1 - 1) + \mu_1 p(k_0 - 1, k_1 + 1), \quad k_0, k_1 > 0,$$

$$\mu_1 p(0, n) = \mu_0 p_1 p(1, n - 1),$$

$$\mu_0 p_1 p(n, 0) = \mu_1 p(n - 1, 1).$$

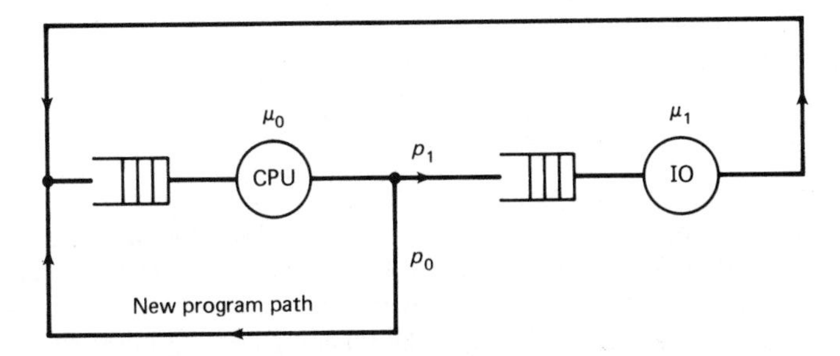

Figure 9.9 The closed cyclic queuing model

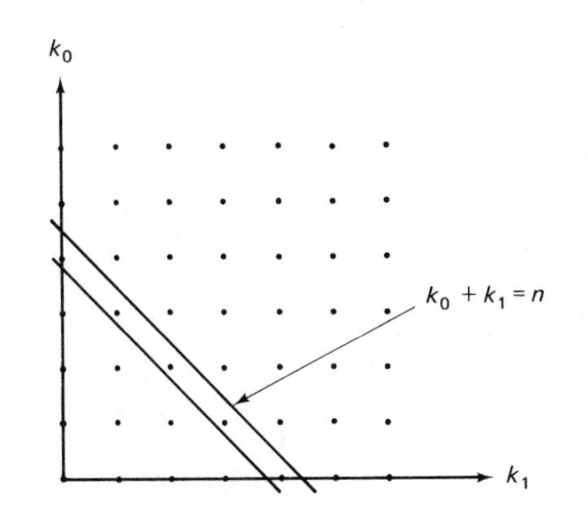

Figure 9.10 State spaces for two-stage networks

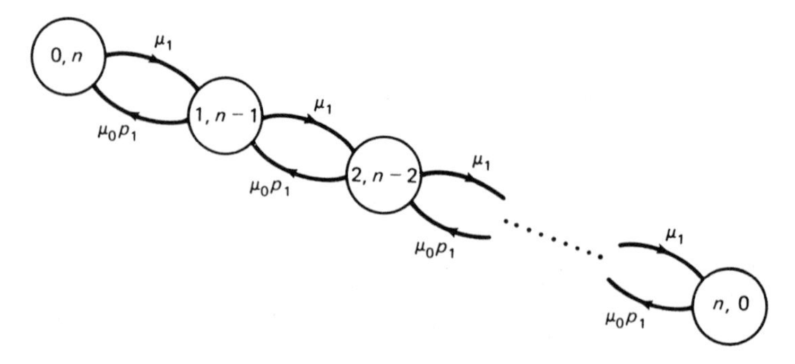

Figure 9.11 The state diagram for the closed cyclic queuing model

If we let $\rho_0 = a/\mu_0$ and $\rho_1 = ap_1/\mu_1$, where a is an arbitrary constant, then it can be verified by direct substitution that the steady-state probability $p(k_0, k_1)$ has the following product form:

$$p(k_0, k_1) = \frac{1}{C(n)} \rho_0^{k_0} \rho_1^{k_1}.$$

The normalizing constant $C(n)$ is chosen so that:

$$\sum_{\substack{k_0+k_1=n \\ k_0,k_1 \geq 0}} p(k_0, k_1) = 1.$$

The choice of the constant a is quite arbitrary in that the value of $p(k_0, k_1)$ will not change with a, although the intermediate values ρ_0, ρ_1, and $C(n)$ will depend upon a. If we define $\lambda_0 = a$ and $\lambda_1 = ap_1$, we may interpret the vector (λ_0, λ_1) as the relative throughputs of the corresponding nodes. Then $\rho_0 = (\lambda_0/\mu_0)$ and $\rho_1 = (\lambda_1/\mu_1)$ are interpreted as relative utilizations. Two popular choices of the constant a are $a = 1$ and $a = \mu_0$. Choosing $a = \mu_0$ we have $\rho_0 = 1$ and $\rho_1 = \mu_0 p_1/\mu_1$.
Also

$$p(k_0, k_1) = \frac{1}{C(n)} \rho_1^{k_1}.$$

Using the normalization condition, we get:

$$1 = \frac{1}{C(n)} \sum_{k_1=0}^{n} \rho_1^{k_1} = \frac{1}{C(n)} \frac{1-\rho_1^{n+1}}{1-\rho_1}$$

or

$$C(n) = \begin{cases} \dfrac{1-\rho_1^{n+1}}{1-\rho_1}, & \rho_1 \neq 1 \\ n+1, & \rho_1 = 1. \end{cases}$$

Now the CPU utilization U_0 may be expressed as:

$$U_0 = 1 - p(0,\ n) = 1 - \frac{\rho_1^n}{C(n)},$$

$$U_0 = \begin{cases} \dfrac{\rho_1 - \rho_1^{n+1}}{1 - \rho_1^{n+1}}, & \rho_1 \neq 1 \\[2ex] \dfrac{n}{n+1}, & \rho_1 = 1. \end{cases}$$

This agrees with the solution obtained in the last chapter [equation (8.38)]. The average throughput is given by:

$$E[T] = \mu_0 U_0 p_0.$$

#

Example 9.5

Consider the (closed) central server network shown in Figure 9.8. The state of the network is given by an $(m + 1)$-tuple, $(k_0,\ k_1,\ \ldots,\ k_m)$ where $k_i \geqslant 0$ is the number of jobs at server i (including any in service). Since the number of jobs in a closed network is fixed, we must further impose the constraint $\sum_{i=0}^{m} k_i = n$ on every state. Thus the state space of the network is finite, the number of states being equal to the number of partitions of n objects among $m + 1$ cells. You were asked to compute this number in problem 5, Section 1.12:

$$\binom{n+m}{m} = \frac{(n+m)!}{n!m!}. \tag{9.7}$$

If we assume that service times at all servers are exponentially distributed, the stochastic process modeling the behavior of the network is a finite-state continuous-parameter Markov chain, which can be shown to be irreducible and recurrent non-null (assuming that $0 < p_i < 1, i = 0,\ 1\ ,\ \ldots,\ m$). In principle, therefore, we can write down the steady-state balance equations and obtain the unique steady-state probabilities. However, the number of equations in this system will be equal to the number of states given by expression (9.7). This is a formidable number of states, even for relatively small values of n and m. Fortunately, Gordon and Newell [GORD 1967] have shown that such Markovian closed networks possess relatively simple product-form solutions.

In order to use their technique, we first analyze the behavior of a tagged program, ignoring all queues in the network. The movement of a tagged program through the network can be modeled by a discrete-parameter Markov chain with $m + 1$ states. The transition probability matrix X of this Markov chain is given by:

$$X = \begin{bmatrix} p_0 & p_1 & \cdot & \cdot & \cdot & p_m \\ 1 & 0 & \cdot & \cdot & \cdot & 0 \\ \cdot & \cdot & & & & \cdot \\ \cdot & \cdot & & & & \cdot \\ \cdot & \cdot & & & & \cdot \\ 1 & 0 & \cdot & \cdot & \cdot & 0 \end{bmatrix}.$$

The Markov chain is finite, and if we assume that $0 < p_i < 1$ for all i, then it can be shown to be irreducible and aperiodic. Then the unique steady-state probability vector $v = (v_0, v_1, \ldots, v_m)$ can be obtained by solving the system of equations:

$$\mathbf{v} = \mathbf{v}X \tag{9.8}$$

and

$$\sum_{i=0}^{m} v_i = 1. \tag{9.9}$$

If we observe the system for a real-time interval of duration τ, then $v_i\tau$ can be interpreted to be the average number of visits to node i in the interval. If we remove the normalization condition (9.9), then the v_i's cannot be interpreted as probabilities, but $av_i\tau$ will still yield the average number of visits to node i ($i = 0, 1, \ldots, m$) for some fixed constant a. In this sense, v_i can be thought of as the relative visit count for node i, and thus v_i is sometimes called the relative arrival rate or the relative throughput of node i.

For the central server network, equation (9.8) becomes:

$$v_0 = v_0 p_0 + \sum_{i=1}^{m} v_i$$

$$v_i = v_0 p_i, \qquad i = 1, 2, \ldots, m. \tag{9.8'}$$

Only m out of these $(m + 1)$ equations are independent, therefore [in absence of the normalization condition (9.9)], v_0 can be chosen as any real value that will aid us in our computations. The usual choices of v_0 are $1/p_0$, μ_0, and 1.

If we choose $v_0 = 1/p_0$, then from (9.8′) we have, $v_i = p_i/p_0$ ($i = 1, 2, \ldots, m$). Bearing in mind that the closed central server model is intended to be an approximation to the open model of Figure 9.7, it follows from our analysis of a tagged program for the open network of Figure 9.6 that with this choice of v_0, $v_i = V_i$, where V_i ($i = 0, 1, \ldots, m$) is the average number of visits a typical program makes to node i in order to complete its execution. Let the relative utilization of device i be given by $\rho_i = v_i/\mu_i$. If we let B_i be the total service requirement of a program on device i, then in this case ρ_i equals the expected value of the total service requirement $E[B_i]$ on device i. If we choose $v_0 = \mu_0$, then $\rho_0 = 1$, hence all device utilizations are scaled by the CPU utilization. This choice is often more convenient computationally.

Gordon and Newell [GORD 1967] have shown that the steady-state probability $p(k_0, k_1, \ldots, k_m)$ of finding k_i jobs at nodes i, $i = 0, 1, \ldots, m$, is given by:

$$\begin{aligned} p(k_0, k_1, \ldots, k_m) &= \frac{1}{C(n)} \prod_{i=0}^{m} \rho_i^{k_i} \\ &= \frac{1}{C(n)} \prod_{i=0}^{m} \left(\frac{v_i}{\mu_i}\right)^{k_i}. \end{aligned} \tag{9.10}$$

Here, the normalization constant $C(n)$ is evaluated using the condition that:

$$\sum_{\mathbf{k} \in I} p(k_0, k_1, \ldots, k_m) = 1,$$

where the state space:

$$I = \{(k_0, k_1, \ldots, k_m) \mid k_i \geqslant 0 \text{ for all } i \text{ and } \sum_{i=0}^{m} k_i = n\}$$

contains $\binom{n+m}{m}$ states. #

More generally, consider an arbitrary closed queuing network with exponential servers with respective rates μ_i $(i = 0, 1, \ldots, m)$ and the transition probability matrix $X = [x_{ij}]$. As in Example 9.5, we first remove all queues and model the behavior of a tagged program through the network. Since our only concern at this point is to count the average number of visits to device i, the behavior of the tagged program is then captured by a discrete-parameter, finite-state Markov chain with transition probability matrix X. We will assume that this chain is irreducible and aperiodic. Therefore, unlike the case of an open network, the transition probability matrix of a closed network is a stochastic matrix. As in Example 9.5, we can obtain the relative throughputs, v_i's, by solving the system of linear equations [analogous to equation (9.8)]:

$$v_i = \sum_{j=0}^{m} v_j x_{ji}, \qquad i = 0, 1, \ldots, m. \tag{9.11}$$

Again, since X is a stochastic matrix, only m out of the above $m + 1$ equations are independent, and the system of equations has a unique solution, up to a multiplying constant. Therefore, one of the components of $\mathbf{v}$ can be chosen arbitrarily. As before, common choices for v_0 are V_0, μ_0, and 1.

If v_0 is chosen to be equal to the average number of visits V_0 to node 0 per program, then the relative utilization ρ_i is equal to $V_i/\mu_i = E[B_i]$, the expected value of the total service requirement imposed by a typical program on the ith node. As we shall see below, U_i, the real utilization of node i, is a function only of the relative utilizations $\rho_0, \rho_1, \ldots, \rho_m$, and from the real utilization of device i, the average system throughput is computed as $U_i/E[B_i]$. Alternatively, we will see that in this case average system throughput equals the ratio $C(n-1)/C(n)$, which again depends only upon ρ_i, $i = 0, 1, \ldots, m$. Thus, measures of system performance such as device utilizations and average system throughput can be obtained from a specification of the $(m + 1)$ service requirements, $E[B_0], E[B_1], \ldots, E[B_m]$. In particular, the topology of the network, the branching probabilities x_{ij}, and the individual service rates μ_i's need not be specified. Also, equation (9.11) need not be solved for the relative throughputs v_i's. Such an interpretation is convenient since the quantities $E[B_i]$ are readily estimated from measured data [GIAM 1976, DENN 1978].

The reader is urged to verify in problems 1 through 3 at the end of this section that the choice of v_0 will not affect the performance measures of interest, although it will affect the values of intermediate quantities such as ρ_i and $C(n)$.

Continuing with our analysis of the general closed network, we see that since there are $n \geqslant 1$ programs circulating through the network, the state of the network at any time will be denoted by $(k_0, k_1, \ldots, k_m)$ where k_i is the number of jobs at node i. If $k_i = 0$, then device i is idle. If $k_i = 1$, a job is being processed by device i. If $k_i > 1$, a job is being processed by device i and $k_i - 1$ jobs are waiting to be served on device i. Since there are exactly n jobs in the system, we must further impose the restriction that $\sum_{i=0}^{m} k_i = n$. This implies, as before, that the state space, I, of the network contains $\binom{n+m}{m}$ states, specifically:

$$I = \{(k_0, k_1, \ldots, k_m) \mid k_i \geqslant 0, \sum_{i=0}^{m} k_i = n\}.$$

By assumption of exponentially distributed service times, all interevent times are exponentially distributed, and thus the network can be modeled by a continuous-parameter Markov chain. Since the chain is finite, if we assume that it is irreducible, and recurrent non-null, then a unique steady-state probability vector exists which can be obtained as a solution of steady-state balance equations:

For any state $s = (k_0, k_1, \ldots, k_m)$ the probability, p_s, of being in that state times the rate of transition from that state has to be equal to the sum over all states t of p_t times the rate of transition from t to s. Therefore, we have:

$$\sum_{j|k_j>0} \mu_j p(k_0, k_1, \ldots, k_m) \tag{9.12}$$

$$= \sum_{j|k_j>0} \sum_{i} x_{ij}\mu_i p(k_0, \ldots, k_i + 1, \ldots, k_j - 1, \ldots, k_m).$$

In other words, in steady state, the rate of flow out of a state must equal the rate of flow into that state.

As in Example 9.5, equation (9.12) has the following product-form solution [GORD 1967]:

$$p(k_0, k_1, \ldots, k_m) = \frac{1}{C(n)} \prod_{i=0}^{m} \rho_i^{k_i}. \tag{9.13}$$

The normalization constant $C(n)$ can be computed using the fact that the probabilities sum to unity; that is:

$$C(n) = \sum_{s \in I} \prod_{i=0}^{m} \rho_i^{k_i}, \tag{9.14}$$

where $s = (k_0, k_1, \ldots, k_m)$.

Since the number of states of the network grows exponentially with the number of customers and the number of service centers, it is not feasible to evaluate $C(n)$ by direct summation as in equation (9.14), because the computation would be too expensive and perhaps numerically unstable. Nevertheless, it is possible to derive stable and efficient computational algorithms to obtain the value of the normalization constant $C(n)$ [BUZE 1973, REIS 1978]. These algorithms also yield simple expressions for performance measures such as the average queue length, $E[N_i]$ and the utilization U_i of the ith server.

Consider the following polynomial in z [WILL 1976]:

$$\begin{aligned} G(z) &= \prod_{i=0}^{m} \frac{1}{1-\rho_i z} \\ &= (1+\rho_0 z+\rho_0^2 z^2+\cdots)(1+\rho_1 z+\rho_1^2 z^2+\cdots)\cdots \\ &\quad (1+\rho_m z+\rho_m^2 z^2+\cdots). \end{aligned} \tag{9.15}$$

It is clear that the coefficient of z^n in $G(z)$ is equal to the normalization constant $C(n)$, since the coefficient is just the sum of all the terms of the form $\rho_0^{k_0}\rho_1^{k_1}\cdots\rho_m^{k_m}$ with $\sum_{i=0}^{m} k_i = n$. In other words, $G(z)$ is the generating function of the sequence $C(1), C(2), \ldots$.

$$G(z) = \sum_{n=0}^{\infty} C(n) z^n, \tag{9.16}$$

where $C(0)$ is defined to be equal to unity. It should be noted that since $C(n)$ is not a probability, $G(z)$ is not a probability generating function and hence $G(1)$ is not necessarily equal to unity. In order to derive a recursive relation for computing $C(n)$, define:

$$G_i(z) = \prod_{k=0}^{i} \frac{1}{1-\rho_k z}, \qquad i = 0, 1, \ldots, m, \tag{9.17}$$

so that $G_m(z) = G(z)$. Also define $C_i(j)$ by:

$$G_i(z) = \sum_{j=0}^{\infty} C_i(j) z^j, \qquad i = 0, 1, \ldots, m,$$

so that $C_m(n) = C(n)$. Observe that:

$$G_0(z) = \frac{1}{1-\rho_0 z} \tag{9.18}$$

and

$$G_i(z) = G_{i-1}(z)\frac{1}{1-\rho_i z}, \qquad i = 1, 2, \ldots, m.$$

This last equation can be rewritten as:

$$G_i(z)[1 - \rho_i z] = G_{i-1}(z)$$

or

$$G_i(z) = \rho_i z G_i(z) + G_{i-1}(z)$$

or

$$\sum_{j=0}^{\infty} C_i(j) z^j = \sum_{j=0}^{\infty} \rho_i z C_i(j) z^j + \sum_{j=0}^{\infty} C_{i-1}(j) z^j.$$

Equating the coefficients of z^j on both sides, we have a recursive formula for the computation of the normalization constant:

$$C_i(j) = C_{i-1}(j) + \rho_i C_i(j-1), \qquad i = 1, 2, \ldots, m, \quad j = 1, 2, \ldots, n. \tag{9.19}$$

The initialization is obtained using (9.18) as:

$$C_0(j) = \rho_0^j, \qquad j = 0, 1, 2, \ldots, n.$$

Also from (9.17), we have that the coefficient of z^0 in $G_i(z)$ is unity, hence:

$$C_i(0) = 1, \qquad i = 0, 1, \ldots, m.$$

The computation of $C(n) = C_m(n)$ is illustrated in Table 9.1.

Table 9.1 *Computation of the Normalization Constant $C_i(j)$*

$C_i(j)$

j \ i	0	1	2	...	$i-1$	i	...	m
0	1	1	1	...	1	1	1	1
1	ρ_0	$\rho_0+\rho_1$	$\rho_0+\rho_1+\rho_2$	...		$\sum_{k=0}^{i} \rho_k$	...	$C_m(1)$
.	.					.		.
.	.					.		.
.	.					.		.
$j-1$	ρ_0^{j-1}	...				$C_i(j-1)$		
						$\downarrow {}^*\rho_i$		
j	ρ_0^j	...			$C_{i-1}(j) \rightarrow$	$C_i(j)$		
.	.							
.	.							
.	.							
n	ρ_0^n	...						$C_m(n)$

As we will see shortly [formula (9.20)], only the last column of the $C_i(j)$ matrix of Table 9.1 is needed for the computation of the device utilizations. It is possible to avoid the storage of the $(n + 1)$-by-$(m + 1)$ matrix suggested in the table. Because the matrix can be computed one column at at time, we need only store the column currently under computation. Assume a one dimensional array $C[0 \; .. \; n]$ initialized to contain all zeros, except for $C[0]$, which is initialized to 1, and representing the current column of the $C_i(j)$ matrix. Also let $\rho[0 \; .. \; m]$ denote the vector of relative utilizations. Then the set of all $C(j)$'s may be computed using the following program segment:

Program 9.1

```
{initialize} C[0]: = 1; for j:=1 to n do C[j]:=0;
for i:=0 to m do
      for j:=1 to n do
            C[j] := C[j] + ρ[i] * C[j − 1].
```

Next, let us derive an expression for $U_i(n)$, the utilization of the ith device. Consider a slight modification to the generating function $G(z)$, denoted by $H(z)$:

$$\begin{aligned} H(z) &= \left(\prod_{\substack{j=0 \\ j \neq i}}^{m} \frac{1}{1 - \rho_j z} \right) \left(\frac{1}{1 - \rho_i z} - 1 \right) \\ &= (1 + \rho_0 z + \rho_0^2 z^2 + \cdots) \cdots (1 + \rho_{i-1} z + \rho_{i-1}^2 z^2 + \cdots) \\ &\quad \cdot (\rho_i z + \rho_i^2 z^2 + \cdots)(1 + \rho_{i+1} z + \rho_{i+1}^2 z^2 + \cdots) \\ &\quad \cdots (1 + \rho_m z + \rho_m^2 z^2 + \cdots). \end{aligned}$$

The difference between $H(z)$ and $G(z)$ is that we have omitted the first term in the factor corresponding to the ith device. As a result, the coefficient of z^n in $H(z)$ will be the sum of all terms:

$$\rho_0^{k_0} \cdots \rho_i^{k_i} \cdots \rho_m^{k_m}$$

such that $k_i \geqslant 1$. From (9.13) we then see that the coefficient of z^n in $H(z)$ divided by the coefficient of z^n in $G(z)$ must yield the marginal probability $P(N_i \geqslant 1)$, which is exactly the utilization $U_i(n)$. Now:

$$H(z) = G(z) \frac{\dfrac{1}{1 - \rho_i z} - 1}{\dfrac{1}{1 - \rho_i z}} = G(z) \rho_i z.$$

Thus, the coefficient of z^n in $H(z)$ is simply ρ_i times the coefficient of z^{n-1} in $G(z)$. Therefore, we get:

$$U_i(n) = \frac{\rho_i C(n - 1)}{C(n)}. \tag{9.20}$$

From this formula we see that $U_i(n)/U_j(n) = \rho_i/\rho_j$ which explains the reason for calling ρ_i's "relative utilizations."

By a similar argument, we can obtain an expression for the probability that there are k or more jobs at node i:

$$P(N_i \geqslant k) = \frac{\rho_i^k C(n-k)}{C(n)}. \tag{9.21}$$

To get an expression for the average queue length at node i, as a function of the degree of multiprogramming n, observe that:

$$\begin{aligned}
E[N_i(n)] &= \sum_{k=1}^{n} kP(N_i = k) \\
&= \sum_{k=1}^{n} k[P(N_i \geqslant k) - P(N_i \geqslant k+1)] \\
&= \sum_{k=1}^{n} kP(N_i \geqslant k) - \sum_{k=1}^{n} kP(N_i \geqslant k+1) \\
&= \sum_{k=1}^{n} kP(N_i \geqslant k) - \sum_{j=2}^{n+1} (j-1)\, P(N_i \geqslant j) \\
&= \sum_{k=1}^{n} kP(N_i \geqslant k) - \sum_{j=2}^{n+1} jP(N_i \geqslant j) + \sum_{j=2}^{n+1} P(N_i \geqslant j) \\
&= P(N_i \geqslant 1) - (n+1)P(N_i \geqslant n+1) + \sum_{j=2}^{n+1} P(N_i \geqslant j) \\
&= \sum_{j=1}^{n} P(N_i \geqslant j)
\end{aligned}$$

since $P(N_i \geqslant n+1) = 0$. Now, using the expression (9.21), we have:

$$E[N_i(n)] = \frac{1}{C(n)} \sum_{j=1}^{n} \rho_i^j\, C(n-j). \tag{9.22}$$

Once again, in order to compute the average queue lengths only the last column of the $C_j(i)$ matrix is needed.

Formula (9.22) leads us to an alternative recursive formula for the computation of $C(n)$. Observe that:

$$\sum_{i=0}^{m} E[N_i(n)] = n$$

so from (9.22) we get:

$$n = \frac{1}{C(n)} \sum_{i=0}^{m} \sum_{j=1}^{n} \rho_i^j\, C(n-j),$$

so

$$C(n) = \frac{1}{n} \sum_{j=1}^{n} C(n-j) \left[\sum_{i=0}^{m} \rho_i^j \right] \tag{9.23}$$

with the initial condition $C(0) = 1$.

Formula (9.23) requires somewhat more arithmetic operations for its evaluation than does formula (9.19), but when many devices have equal ρ_i's and $m > n$, it will be more efficient to precompute the factors $\left[\sum_{i=0}^{m} \rho_i^j \right]$ above for each j and use formula (9.23).

Example 9.6

Consider a numerical instance of the central server model (Example 9.5) with the parameters as shown in Table 9.2.

Table 9.2 *Parameters of Example 9.6*

Parameter	*Symbol*	*Value*
Number of I/O channels	m	2
Degree of multiprogramming	n	From 1 to 10
Mean CPU time per burst	$1/\mu_0$	20 ms
Mean drum time per visit	$1/\mu_1$	30 ms
Mean disk time per visit	$1/\mu_2$	42.918 ms
Drum branching probability	p_1	0.667
Disc branching probability	p_2	0.233
Probability of job completion	p_0	0.1

We choose the relative throughput $v_0 = \mu_0 = 1/20$ (per millisecond). Then from (9.8′) we have, $v_i = \mu_0 \, p_i$, and hence $v_1 = 0.667/20$ and $v_2 = 0.233/20$. The relative utilizations are then computed to be $\rho_0 = 1$, $\rho_1 = 1$, and $\rho_2 = 0.5$. The values of $C(1), C(2), C(3), \ldots, C(10)$ are shown in Table 9.3. We compute the utilizations of the three nodes, using equation (9.20). Finally the average system throughput is computed by (in jobs per second):

$$E[T(n)] = \mu_0 \, p_0 \, U_0(n) = \frac{\mu_0 \, p_0 \, \rho_0 \, C(n-1)}{C(n)} = 5 \cdot \frac{C(n-1)}{C(n)}.$$

The average system throughput as a function of the degree of multiprogramming (degmul) is shown in Table 9.4 and is plotted in Figure 9.12.

Figure 9.12 shows that as the degree of multiprogramming increases, the average throughput also increases. An increase in the degree of multiprogramming generally implies that additional main memory must be purchased. In the case of systems employing paged virtual memory, the degree

Table 9.3 *Computation of the Normalization Constant*

$C_i(j)$

j \ i	0	1	2
1	1	2	2.5
2	1	3	4.25
3	1	4	6.125
4	1	5	8.062
5	1	6	10.031
6	1	7	12.016
7	1	8	14.008
8	1	9	16.004
9	1	10	18.002
10	1	11	20.001

Table 9.4 *Average System Throughput (jobs completed per second)*

n, degmul	*Average system throughput*
1	2.0
2	2.9412
3	3.4694
4	3.7985
5	4.0187
6	4.1743
7	4.2889
8	4.3764
9	4.4450
10	4.5003

of multiprogramming is not inherently limited by the size of the main memory, since a program is allowed to execute with only part of its address space in the main memory. For a fixed size of main memory, an increase in the degree of multiprogramming then implies a reduction in the page allotment per program, hence an increased frequency of page faults, which will tend to reduce average system throughput. On the other hand, an increased degree of multiprogramming implies that there is a greater chance of finding a job ready to run (on the CPU) whenever the currently executing job incurs a page fault. This will tend to increase the average throughput. The combined

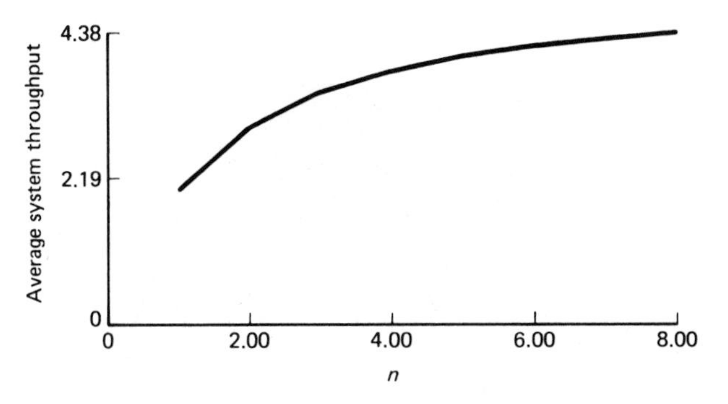

Figure 9.12 Average system throughput versus degree of multiprogramming

effect of these two conflicting factors on average system throughput is investigated in the next example.

Example 9.7

Consider a central server network with $m = 2$ I/O channels. The channel labeled 1 is a paging drum and the channel labeled 2 is a disk used for file I/O. Other parameters are specified in Table 9.5.

Table 9.5 Parameters for Example 9.7

Parameter	*Symbol*	*Value*
Number of I/O channels	m	2
Degree of multiprogramming	n	From 1 to 10
Mean total CPU time per job	$E[B_0]$	0.06667 sec
Mean drum service time	$1/\mu_1$	From 10 to 100 ms
Mean disc service time	$1/\mu_2$	50 ms
Average number of drum requests (page faults) per job	$V_1(n)$	$0.1e^{(0.415n)}$
Average number of disk requests per job	V_2	5

In this problem we set v_0, the relative throughput of the CPU, to be equal to the average number of visits V_0 to the CPU per job. Then ρ_0, the relative utilization of the CPU, equals $E[B_0]$, the average CPU service requirement per job. Note that V_0 is not even specified here. Also:

$$\rho_1(n) = E[B_1] = \frac{V_1(n)}{\mu_1}$$

is an increasing function of n and

$$\rho_2 = E[B_2] = \frac{V_2}{\mu_2} = 0.25\text{sec/job}.$$

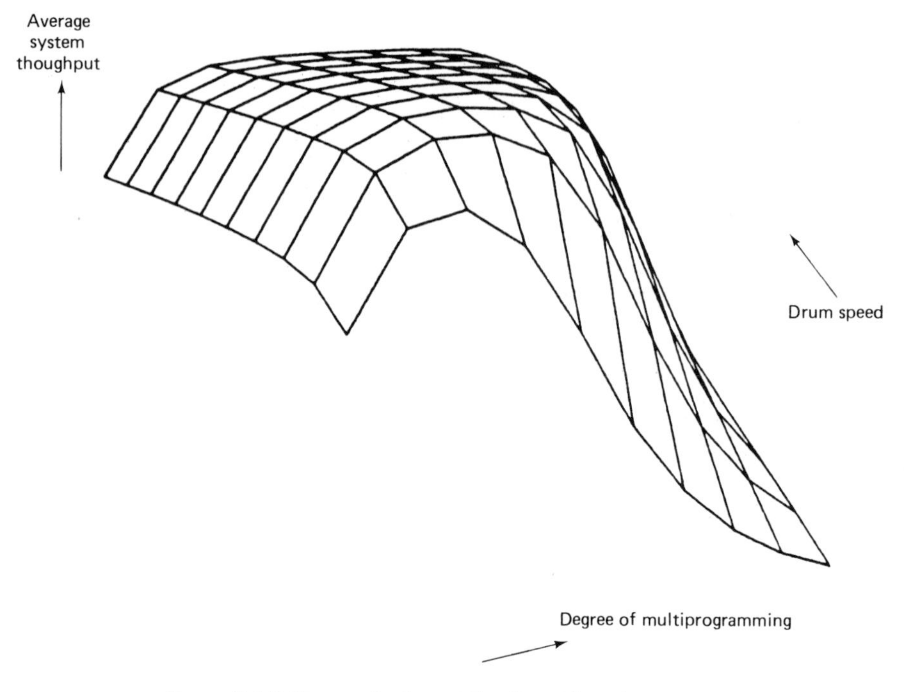

Figure 9.13 Demonstration of the thrashing phenomenon

Figure 9.13 is a three-dimensional plot showing the variation in average system throughput $E[T]$ with the degree of multiprogramming n and with the drum service rate μ_1. For a fixed value of μ_1, $E[T]$ first increases with n and, after reaching a maximum, starts to drop rather sharply. When n becomes rather large, the dramatic reduction in average system throughput, known as thrashing, occurs [COFF 1973]. The reduction in average system throughput can be compensated by an increase in the drum speed, which causes an increase in μ_1, as shown in Figure 9.13. Alternative methods to control thrashing are to purchase more main memory or to improve program locality so that programs will page fault less frequently at a given allotment of main memory. #

Example 9.8

The central server network discussed in the Example 9.5 may be interpreted from another viewpoint. Consider a system with n components, each with a powered failure rate μ_0. Standby redundancy is used so that at most one component is in powered status while the remaining $n - 1$ components are either in a powered-off standby status or waiting to be repaired. The failure rate of a powered-off spare is assumed to be zero. The failures are classified into $m + 1$ distinct classes with a conditional probability p_i for class i. For failures of classes $1, \ldots, m$, the average repair time is $1/\mu_i$, and failures of each class possess a dedicated repair facility. Failures of class 0 are transient, so the corresponding machine is returned immediately to the queue of good standby machines. The probability that at least one machine is available equals the steady-state availability and is given by formula (9.20):

$$A_0 = \rho_0 \frac{C(n-1)}{C(n)}. \qquad \#$$

The above treatment can be generalized to the case of a closed queuing network with c_i servers of rate μ_i at node i $(i = 0, 1, \ldots, m)$. If we compute v_i and ρ_i as before, then the joint probability of k_i jobs at node i $(i = 0, 1, \ldots, m)$ is given by [KLEI 1975]:

$$p(k_0, k_1, \ldots, k_m) = \frac{1}{C(n)} \prod_{i=0}^{m} \frac{\rho_i^{k_i}}{\beta_i(k_i)}, \tag{9.24}$$

where

$$\beta_i(k_i) = \begin{cases} k_i!, & k_i < c_i, \\ c_i! c_i^{k_i - c_i}, & k_i \geqslant c_i, \end{cases}$$

and

$$C(n) = \sum_{s \in I} \prod_{i=0}^{m} \frac{\rho_i^{k_i}}{\beta_i(k_i)},$$

where $I = \{(k_0, k_1, \ldots, k_m) \mid k_i \geqslant 0 \text{ and } \sum_{i=0}^{m} k_i = n\}$. The computation of $C(n) = C_m(n)$ may be performed using the following recursive scheme [WILL 1976]: For $i = 0, 1, \ldots, m$, let:

$$r_i(k) = \begin{cases} \dfrac{\rho_i^k}{\beta_i(k)}, & k \neq 0, \\ 1, & k = 0. \end{cases}$$

Then for $j = 1, 2, \ldots, n$:

$$C_i(j) = \begin{cases} r_0(j), & i = 0, \\ \sum_{k=0}^{j} C_{i-1}(j-k) r_i(k), & i \neq 0, \end{cases} \tag{9.25}$$

with the initialization, $C_i(0) = 1$ for all i. The computation of $C_j(n)$ is depicted in Table 9.6. A comparison of this table with Table 9.1 illustrates the greater complexity of the load-dependent case. Also note that, unlike the previous case, it is necessary to save all the values in column $j - 1$ while computing elements of column j. Thus two columns of the matrix need to be stored rather than just one.

***Table 9.6** Normalization Constant for Load-Dependent Servers*

$C_i(j)$

j \ i	0	1	2	3	...	$i-1$		i	....	m
0	1	1	1	1	...	$C_{i-1}(0) * r_i(n)$	$\rightarrow +$			
1	$r_0(1)$					$C_{i-1}(1) * r_i(n-1)$	$\rightarrow +$			
2	$r_0(2)$									
3							$\rightarrow +$			
.	.									
.	.									
.	.									
						$C_{i-1}(n-1) * r_i(1)$	$\rightarrow +$			
n	$r_0(n)$					$C_{i-1}(n)$	$\rightarrow$	$C_i(n)$	....	$C_m(n)$

A program to compute $C_i(j) = C(j)$, $j = 1, 2, \ldots, n$ can be easily written. Assume that a two dimensional array $C[0..n,\ 0..1]$ is declared and two binary variables PREV and CUR are also declared. The $r_i(k)$ values as specified above are assumed to be precomputed and stored in the two-dimensional array $r[0..m,\ 0..n]$. Program 9.2 below computes the desired value $C(n) = C[n, \text{CUR}]$.

Program 9.2 {normalization constant computation for a closed queuing network with a single job type and load-dependent servers}

```
begin
 {initialize}

  C[0,0]:=1; C[0,1]:=1;
  for j:=1 to n do
     C[j,0]:=0;
  PREV:=0; CUR:=1;
 {recursion}

  for i:=0 to m do
  begin
     for j:=1 to n do
     begin
        C[j,CUR]:=0;
        for k:=0 to j do
           C[j,CUR]:=C[j,CUR]+C[j-k,PREV]*r[i,k]
        end;
        PREV:=1-PREV;
        CUR :=1-CUR
  end
end.
```

The expression for the utilization of node i is a bit more complex in this general case, but for the node with a single server (load-independent service), the formula:

$$U_i(n) = \frac{\rho_i C(n-1)}{C(n)}$$

holds even though the nodes other than the ith node may give load-dependent service [BUZE 1971, WILL 1976].

Example 9.9

Consider a closed queuing network with two nodes. The CPU node is a multiple server node with the number of processors, c, varying from one to five. The total service requirement of a typical program on the CPU node is $E[B_0] = 10$ seconds. The I/O node is a single server node with the total service requirement $E[B_1] = 1$ second. The degree of multiprogramming $n = 5$.

Defining $\rho_0 = E[B_0]$ and $\rho_1 = E[B_1]$ we solve for average system throughput, $C(4)/C(5)$, as a function of the number of processors as shown in Table 9.7. Since programs are CPU-bound, increasing the number of processors improves average throughput substantially.

Table 9.7

Average throughput, $E[T(c)]$	*Number of processors,* c
0.0999991	1
0.1998464	2
0.2972297	3
0.3821669	4
0.4360478	5

Example 9.10 [TRIV 1978]

Consider the terminal-oriented distributed computing system shown in Figure 9.14. We use the following abbreviations:

T: the set of terminals
F: front-end interface processor
C: communication processor
D: database management processor
P: principal element processor

The average think time of a terminal user is assumed to be $1/\lambda$. The transition probability matrix X is given by:

$$X = \begin{array}{c} \\ T \\ F \\ C \\ D \\ P \end{array} \begin{array}{c} \begin{array}{ccccc} T & F & C & D & P \end{array} \\ \begin{bmatrix} 0 & 1 & 0 & 0 & 0 \\ p_0 & 0 & p_1 & 0 & 0 \\ 0 & p_2 & 0 & p_3 & p_4 \\ 0 & 0 & 1 & 0 & 0 \\ 0 & 0 & 1 & 0 & 0 \end{bmatrix} \end{array}$$

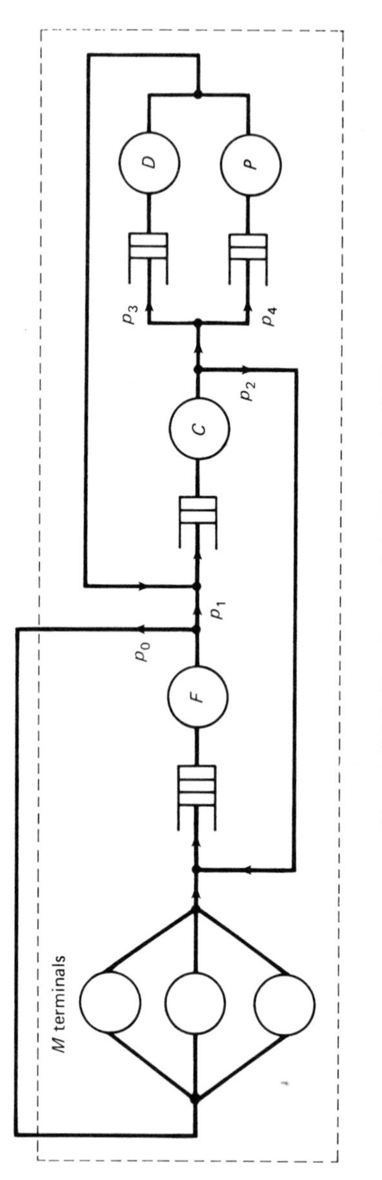

Figure 9.14 Queuing model of Example 9.10

where:

$$p_0 + p_1 = 1, \qquad p_2 + p_3 + p_4 = 1.$$

Solving for relative throughputs, we get:

$$\begin{aligned} v_T &= v_F p_0 \\ v_F &= v_T + v_C p_2 \\ v_C &= v_F p_1 + v_D + v_P \\ v_D &= v_C p_3 \\ v_P &= v_C p_4. \end{aligned}$$

Choose $v_T = 1$ and then: $v_F = \dfrac{1}{p_0}$, $v_C = \dfrac{1 - p_0}{p_0\, p_2}$, $v_D = \dfrac{(1 - p_0)\, p_3}{p_0\, p_2}$, and $v_P = \dfrac{(1 - p_0)\, p_4}{p_0\, p_2}$. Noting that the "service rate" of a terminal, μ_T, is given by λ, the device relative utilizations are given by:

$$\rho_T = \frac{1}{\lambda}, \qquad \rho_F = \frac{1}{p_0\, \mu_F}, \qquad \rho_C = \frac{1 - p_0}{p_0\, p_2\, \mu_C},$$

$$\rho_D = \frac{(1 - p_0) p_3}{p_0 p_2 \mu_D}, \qquad \rho_P = \frac{(1 - p_0) p_4}{p_0\, p_2\, \mu_P}.$$

(Note that with this choice of v_T, ρ_i equals the average service time $E[B_i]$ per terminal request on device i.) Then, using formula (9.24), we have:

$$p(k_T, k_F, k_C, k_D, k_P) = \frac{1}{C(M)} \frac{(\rho_T)^{k_T}}{k_T!} (\rho_F)^{k_F} (\rho_C)^{k_C} (\rho_D)^{k_D} (\rho_P)^{k_P}.$$

Note that the only node with load-dependent service is the terminal for which $\beta_T(k) = k!$, since necessarily $k \leqslant M$, the number of terminals. We can compute $C(i)$'s using formula (9.25). Now, since the F node has only one server, we have:

$$U_F = \rho_F \frac{C(M - 1)}{C(M)} = \frac{1}{p_0 \mu_F} \frac{C(M - 1)}{C(M)},$$

and the average throughput $E[T]$ or the rate of request completion is:

$$E[T] = \mu_F p_0 U_F = \frac{C(M - 1)}{C(M)}.$$

The average response time $E[R]$ can then be found from Little's formula, applied to the subsystem enclosed by dashed lines as $(E[R] + 1/\lambda)E[T] = M$, to be:

$$E[R] = \frac{M}{E[T]} - \frac{1}{\lambda} = \frac{M \cdot C(M)}{C(M - 1)} - \frac{1}{\lambda}.$$

As a numerical example, let $p_0 = 0.8$, $p_1 = 0.2$, $p_2 = 0.45833$, $p_3 = 0.33334$, and $p_4 = 0.20833$. Let $\mu_F = 1.5$, $\mu_C = 1.0$, $\mu_D = 0.2$, and $\mu_P = 0.2$. Let the average think time $1/\lambda = 15$ seconds (or $\lambda = 0.06667$). For this case, the average response time is plotted as a function of the number of terminals, M, in Figure 9.15. #

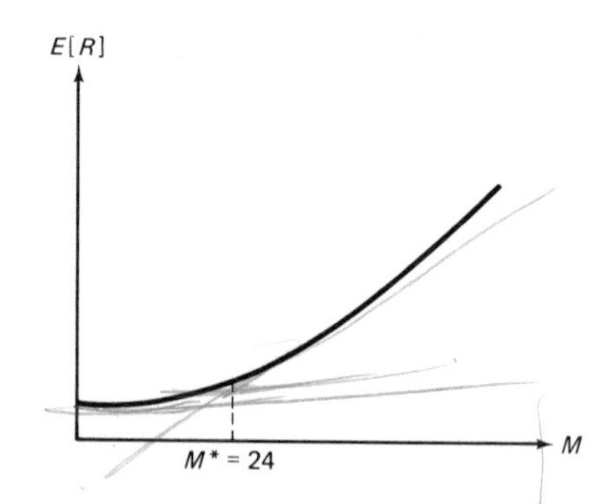

Figure 9.15 Average response time versus number of terminals

Problems

1. Solve Example 9.6, choosing the relative throughput $v_0 = 1/p_0 = 10$. Compare the values of $C(n)$, U_0, and $E[T]$ with those obtained in the text.

2. Let us write the normalization "constant" $C(n)$ as a function of the relative utilization vector $\boldsymbol{\rho} = (\rho_0, \rho_1, \ldots, \rho_m)$—that is, as $C_n(\boldsymbol{\rho})$. Show that:

$$C_n(a\boldsymbol{\rho}) = a^n C_n(\boldsymbol{\rho}).$$

3. Recall that while dealing with closed queuing networks, the relative throughputs (and hence, relative utilizations) can be computed within a multiplying constant. The relative throughput v_0 can be chosen arbitrarily. Of course, different choices of v_0 will result in different values of C_n, C_{n-1}, ... Assume that the first choice is $v_0 = \mu_0$, with the corresponding normalization constants C_n, C_{n-1}, ... Now let $v_0 = 1$ and denote the corresponding sequence of normalization constants by M_n, M_{n-1}, ... Show that although C_k may not be equal to M_k, the utilization U_i for each node i has the same final value no matter what choice of v_0 was made. Do the same for the expressions of average system throughput.

4. Derive a closed form expression for average system throughput for a closed queuing network under monoprogramming (i.e., $n = 1$).

5. Given a closed queuing network with a single server at each node and relative utilizations that are independent of n, show that in the limit as n approaches infinity, average system throughput is given by:

$$\lim_{n\to\infty} E[T(n)] = \frac{v_j}{V_j} \min_i \left\{\frac{1}{\rho_i}\right\} = \min_i \left\{\frac{1}{E[B_i]}\right\},$$

and it is for this reason that the network node with the largest value of $E[B_i]$ (or the largest relative utilization ρ_i) is called the "bottleneck" node. Show that in the limit as n approaches infinity, the real utilization for node i is given by:

$$\lim_{n\to\infty} U_i(n) = \frac{\rho_i}{\rho_b},$$

where b is the index of the bottleneck node.

6. Show that for a closed queuing network model of a terminal-oriented system with all other nodes in the subsystem having single servers, the heavy-load asymptote to the response time $E[R]$ is given by:

$$E[R] \text{ (heavy load)} = ME[B_b] - \frac{1}{\lambda},$$

where b is the index of the bottleneck node. Now show that the saturation number M^* is given by:

$$M^* = \frac{\sum E[B_i] + \frac{1}{\lambda}}{E[B_b]}.$$

7. Consider a central server network with two I/O channels and three CPU's. The average CPU time per program $E[B_0]$ is 500 ms, and the average time per program on the I/O devices is 175 ms and 100 ms, respectively. Compute the average system throughput for the degree of multiprogramming n varying from 1 to 5. First perform your computations manually, using the structure as in Table 9.6 and next using Program 9.2. Now vary the number of CPU's from one to three and study the effect on the average throughput.

8. For a closed network with a single server at each node and a total of n jobs, show that the average response time per visit to node i is given by:

$$E[R_i(n)] = \frac{1}{\mu_i}\{1 + E[N_i(n-1)]\}, \qquad \text{for all } i.$$

[*Hint*: Use equations (9.20) and (9.22) together with Little's formula.] Next, show that the average system throughput:

$$E[T(n)] = \frac{n}{\sum_i V_i E[R_i(n)]},$$

and the average number of jobs at node i is:

$$E[N_i(n)] = E[R_i(n)]\, V_i\, E[T(n)], \qquad \text{for all } i.$$

These three equations can be applied iteratively to compute $E[R_i(n)]$, $E[T(n)]$, and $E[N_i(n)]$ for any value of n starting at $n = 1$ and using the initial condition $E[N_i(0)] = 0$. This procedure for computing performance measures is known as **mean value analysis** [REIS 1979]. Since this procedure avoids the computation of the normalization constants $C(n)$, $C(n-1)$, . . . , in some cases it is numerically more stable than the procedure described in this section. Perform these calculations for the network of Example 9.6. Note that in the above analysis we have implied a choice of $v_0 = V_0$, the average number of visits per program to node 0.

The first of these equations is easily modified to $E[R_i(n)] = 1/\mu_i$ in case of a (terminal) node with no queuing. Recalculate the performance measures for the network of Example 9.10 using mean value analysis.

9. [SEVC 1980] An interactive system workload measurement showed that the average CPU time $E[B_0]$ per terminal request was 4.6 seconds and the average

disk time per request was 4.0 seconds. Since the response time was found to be very large, two alternative systems were considered for purchase. Compared to the existing system (denoted **ex**), one of the proposed systems (denoted **tr1**) had a CPU 0.9 times as fast and the disk was twice as fast. The alternative system (denoted **tr2**) had a CPU 1.5 times as fast as that in **ex** and a disk twice as fast. Decide whether a change from **ex** to **tr1** can be recommended (assuming first that **tr2** is intolerably expensive), first on the basis of asymptotic bounds analysis as in problem 6 and then by exactly computing the response times for the two cases as functions of the number of terminals. Next, assuming that **tr2** is affordable, show how much reduction in response time is possible as a function of the number of terminals.

9.4 NONEXPONENTIAL SERVICE-TIME DISTRIBUTIONS AND MULTIPLE JOB TYPES

More general queuing network models that overcome many of the restrictions of the models mentioned earlier have been formulated and have been found to have a product-form solution [CHAN 1977, CHAN 1978a, KLEI 1975, KOBA 1978]. For example, any differentiable service-time distribution can be allowed at a node, provided that the scheduling discipline at the node is PS (processor sharing) or LCFS-PR (last-come, first-served, preemptive resume). Any differentiable service-time distribution can also be allowed at a node with ample servers (so that no queuing is needed). Networks with multiple job types can also be analyzed. We will consider several examples of these more general models. For additional details, consult the references cited in this section.

Example 9.11

Consider an example of the cyclic queuing model of Figure 9.9 in which $p_0 = 0$. We assume that the I/O service times are exponentially distributed with parameter λ, but the CPU service times are hyperexponentially distributed with two phases so that its pdf is:

$$f(t) = \alpha_1\mu_1 e^{-\mu_1 t} + \alpha_2\mu_2 e^{-\mu_2 t}. \tag{9.26}$$

Let k, the number of jobs in the CPU node, denote the state of the system so that the state space is $\{0, 1, \ldots, n\}$. The corresponding stochastic process is not Markovian, since the future behavior of the process depends not only upon the current state but also on the time spent on the CPU by the job undergoing service (assuming that $k > 0$). If this time is denoted by τ, then we need to consider (k, τ) as the state of the system, and the corresponding state space is nondenumerable.

We can avoid the difficulty, however, by observing that the job at the CPU will be in one of two alternative phases, as shown in Figure 9.16 where a job scheduled for the CPU chooses phase i with probability α_i. We further simplify this discussion by considering only two jobs in the network, and we assume that the CPU scheduling discipline is processor sharing (PS). This discipline is a limiting case of the quantum-oriented RR discipline, where the quantum size is allowed to approach zero. As a result, the CPU is equally shared among all the jobs in the CPU queue. Thus, if there are k jobs in the queue, each job perceives the CPU to be slower by a factor k.

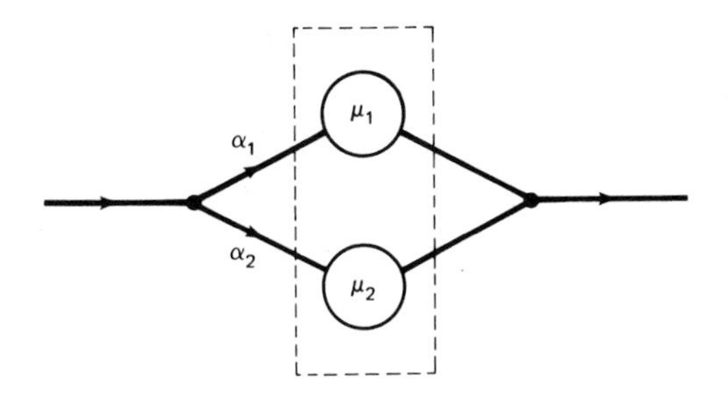

Figure 9.16 Two-phase hyperexponential CPU service-time distribution

If the state of the system is (i, j) then i jobs are in the first phase of the CPU execution and j jobs in the second. Clearly, $i \geqslant 0$, $j \geqslant 0$ and $i + j \leqslant 2$. The state space is thus $\{(0,0), (1,0), (0,1), (2,0), (0,2), (1,1)\}$. The state diagram is shown in Figure 9.17. Since all interevent times are now exponentially distributed, we have a Markov chain. When the job finishes execution on the I/O device, it selects one of the two CPU phases with respective probabilities α_1 and α_2. When the system is in state $(1,1)$, two jobs are in the CPU queue, each of which perceives a CPU of half the speed. Thus the job in phase 1 completes its CPU burst requirement with rate $\mu_1/2$, and similarly for the other job.

Job-flow balance equations in the steady state are written as:

$$\lambda p(0, 0) = \mu_1 p(1, 0) + \mu_2 p(0, 1),$$

$$(\mu_1 + \lambda)p(1, 0) = \mu_1 p(2, 0) + \lambda\alpha_1 p(0, 0) + \frac{\mu_2}{2} p(1, 1),$$

$$(\mu_2 + \lambda)p(0, 1) = \mu_2 p(0, 2) + \lambda\alpha_2 p(0, 0) + \frac{\mu_1}{2} p(1, 1),$$

$$\mu_1 p(2, 0) = \lambda\alpha_1 p(1, 0),$$

$$\mu_2 p(0, 2) = \lambda\alpha_2 p(0, 1),$$

$$\frac{\mu_1 + \mu_2}{2} p(1, 1) = \lambda\alpha_1 p(0, 1) + \lambda\alpha_2 p(1, 0).$$

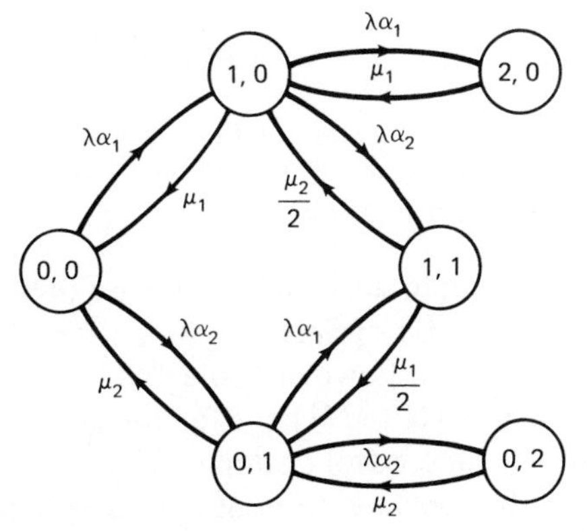

Figure 9.17 State diagram for the system of Example 9.11

The reader should verify by substitution that a parametric solution of these equations is:

$$
\begin{aligned}
p(1, 0) &= \frac{\lambda\alpha_1}{\mu_1} p(0, 0), \\
p(0, 1) &= \frac{\lambda\alpha_2}{\mu_2} p(0, 0), \\
p(0, 2) &= (\frac{\lambda\alpha_2}{\mu_2})^2 p(0, 0), \\
p(2, 0) &= (\frac{\lambda\alpha_1}{\mu_1})^2 p(0, 0), \\
p(1, 1) &= \frac{2(\lambda\alpha_1)(\lambda\alpha_2)}{\mu_1\mu_2} p(0, 0).
\end{aligned}
\tag{9.27}
$$

Now, using the normalization condition:

$$p(0, 0) + p(1, 0) + p(0, 1) + p(2, 0) + p(0, 2) + p(1, 1) = 1,$$

we can evaluate $p(0, 0)$.

Suppose that we want to pursue a reduced description of the system in which state 1 of the reduced version corresponds to the union of original states (1, 0) and (0, 1), and state 2 is the union of original states (2, 0), (0, 2) and (1, 1). Therefore:

$$p(0) = p(0, 0),$$

$$p(1) = p(1, 0) + p(0, 1),$$

and

$$p(2) = p(2, 0) + p(0, 2) + p(1, 1).$$

From equations (9.27) we get:

$$p(1) = \lambda\ (\frac{\alpha_1}{\mu_1} + \frac{\alpha_2}{\mu_2})\ p(0), \tag{9.28}$$

$$p(2) = \lambda^2\ (\frac{\alpha_1}{\mu_1} + \frac{\alpha_2}{\mu_2})^2\ p(0).$$

Let $1/\mu = \alpha_1/\mu_1 + \alpha_2/\mu_2$ be the average CPU service time per visit to the CPU. Then the above equations are the special case of birth-death recursions, and the system has the well known product-form solution. #

The argument presented in the example above can be generalized to show that the product-form solution of Sections 9.2 and 9.3 is valid for a queuing network under the conditions shown in Table 9.8.

Chandy and others [CHAN 1977] have further generalized these cases.

Now that we have considered an example of a network with nonexponential service-time distribution, we turn our attention to queuing network models that support multiple job classes.

Table 9.8 *Networks with Product-Form Solutions*

Scheduling discipline at node i	*Service time distribution at node i*
FCFS	Exponential
PS	Coxian phase type
LCFS-PR	Coxian phase type
IS (infinite server)	Coxian phase type

Example 9.12

Consider a central server network with three nodes (CPU node labeled 0 and two I/O nodes labeled 1 and 2). Let the number of jobs in the network be $n = 2$. These jobs are labeled 1 and 2. Job 1 does not access I/O node 2, and job 2 does not access I/O node 1. The mean service time of job 1 on CPU is $1/\mu_1$, and that of job 2 is $1/\mu_2$. The mean I/O service time of job 1 on device 1 is $1/\lambda_1$, and that of job 2 on device 2 is $1/\lambda_2$. For simplicity we assume that there is no new program path, so that a job completing a CPU burst enters its respective I/O node with probability 1. Assume that the CPU scheduling discipline is PS.

Define the state of the system as a triple (k_0, k_1, k_2) where for $i = 0, 1, 2$, k_i is the number of jobs at node i. The state diagram is shown in Figure 9.18. From the state diagram we obtain the following balance equations:

$$\begin{aligned}
\frac{\mu_1 + \mu_2}{2}\, p(2,0,0) &= \lambda_1 p(1,1,0) + \lambda_2 p(1,0,1), \\
\lambda_1 + \mu_2\, p(1,1,0) &= \lambda_2\, p(0,1,1) + \frac{\mu_1}{2}\, p(2,0,0), \\
(\lambda_2 + \mu_1)\, p(1,0,1) &= \lambda_1\, p(0,1,1) + \frac{\mu_2}{2}\, p(2,0,0), \\
(\lambda_1 + \lambda_2)\, p(0,1,1) &= \mu_1\, p(1,0,1) + \mu_2\, p(1,1,0).
\end{aligned} \tag{9.29}$$

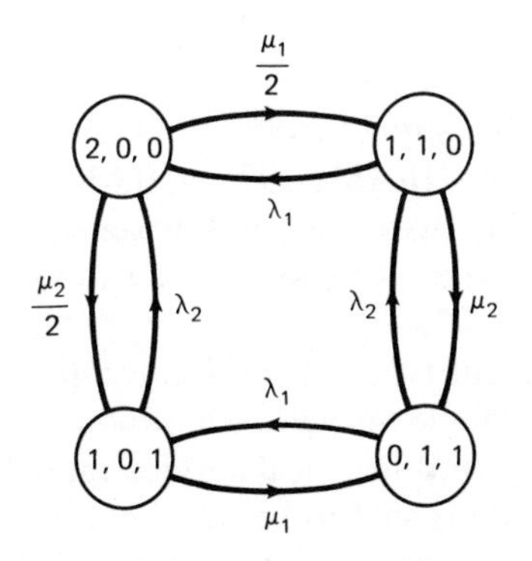

Figure 9.18 The state diagram for the central server network with two job types

The solution to this system of linear equations is easily seen to be:

$$\begin{aligned} p(2, 0, 0) &= \frac{1}{C}\frac{2}{\mu_1\mu_2}, \\ p(1, 1, 0) &= \frac{1}{C}\frac{1}{\lambda_1\mu_2}, \\ p(1, 0, 1) &= \frac{1}{C}\frac{1}{\lambda_2\mu_1}, \\ p(0, 1, 1) &= \frac{1}{C}\frac{1}{\lambda_1\lambda_2}, \end{aligned} \tag{9.30}$$

where the normalization constant C is evaluated by using the condition:

$$p(2, 0, 0) + p(1, 1, 0) + p(1, 0, 1) + p(0, 1, 1) = 1$$

as:

$$C = \frac{2}{\mu_1\mu_2} + \frac{1}{\lambda_1\mu_2} + \frac{1}{\mu_1\lambda_2} + \frac{1}{\lambda_1\lambda_2}. \tag{9.31}$$

The utilization of I/O device 1 is:

$$U_1 = p(1, 1, 0) + p(0, 1, 1) = \frac{1}{C}\frac{1}{\lambda_1}\left[\frac{1}{\mu_2} + \frac{1}{\lambda_2}\right]$$

and that of I/O device 2 is:

$$U_2 = p(1, 0, 1) + p(0, 1, 1) = \frac{1}{C}\frac{1}{\lambda_2}\left[\frac{1}{\mu_1} + \frac{1}{\lambda_1}\right].$$

The average throughput of type 1 jobs is therefore:

$$E[T_1] = U_1\lambda_1 = \frac{1}{C}\left[\frac{1}{\mu_2} + \frac{1}{\lambda_2}\right]$$

and that of type 2 jobs is:

$$E[T_2] = U_2\lambda_2 = \frac{1}{C}\left[\frac{1}{\mu_1} + \frac{1}{\lambda_1}\right]. \qquad \#$$

We can generalize this simple example by considering a closed network with r classes of customers [BASK 1975, CHAN 1977]. A class t customer has a transition probability matrix X_t, and its service rate at node i is denoted by μ_{it}. First we solve for the visit counts v_{it} by solving the eigenvector equation (9.11) for each $t = 1, 2, \ldots, r$.

We admit four different types of service at a node. Node i is said to be a type 1 node if it has a single server with exponentially distributed service times, FCFS scheduling, and identical service rates for all job types (that is $\mu_{it} = \mu_i$ for all t). A node is said to be of type 2 if it has a single server, PS scheduling, and a service-time distribution that is differentiable. Each job type may have a distinct service-time distribution. Node i is said to be a type 3 node if it has an ample number of servers so that no queue ever forms at the node. Any differentiable service-time distribution is allowed, and each job type may have a distinct service-time distribution. Finally, a node is said to be of type 4 if it has a single server with LCFS-PR scheduling. Any differentiable service-time distribution is allowed, and each job type may have a distinct service-time distribution.

Let k_{it} be the number of jobs of type t at node i. Assume that there are n_t jobs of type t in the network so that we have:

$$\sum_{i=0}^{m} k_{it} = n_t \qquad \text{for each } t.$$

Define vector $\mathbf{Y}_i$ by:

$$\mathbf{Y}_i = (k_{i1}, k_{i2}, \ldots, k_{ir}),$$

so that $(\mathbf{Y}_0, \mathbf{Y}_1, \ldots, \mathbf{Y}_m)$ is a state of the system. Let $k_i = \sum_{t=1}^{r} k_{it}$, be the total number of jobs at node i. The steady-state joint probability of such a state is given by [BASK 1975, CHAN 1977]:

$$p(\mathbf{Y}_0, \mathbf{Y}_1, \ldots, \mathbf{Y}_m) = \frac{1}{C(n_1, n_2, \ldots, n_r)} \prod_{i=0}^{m} g_i(\mathbf{Y}_i), \tag{9.32}$$

where:

$$g_i(\mathbf{Y}_i) = \begin{cases} k_i! \prod_{t=1}^{r} \frac{1}{(k_{it})!} (v_{it})^{k_{it}} (\frac{1}{\mu_i})^{k_i}, & \text{node } i \text{ is type 1,} \\ k_i! \prod_{t=1}^{r} \frac{1}{(k_{it})!} [\frac{v_{it}}{\mu_{it}}]^{k_{it}}, & \text{node } i \text{ is type 2 or 4,} \\ \prod_{t=1}^{r} \frac{1}{(k_{it})!} [\frac{v_{it}}{\mu_{it}}]^{k_{it}} & \text{node } i \text{ is type 3.} \end{cases}$$

Define the relative utilization of node i due to jobs of type t by:

$$\rho_{it} = \frac{v_{it}}{\mu_{it}}. \tag{9.33}$$

Then the real utilization of node i (of type 1, 2, or 4) due to jobs of type t is given by [WILL 1976]:

$$U_{it} = \frac{\rho_{it} C(n_1, n_2, \ldots, n_{t-1}, n_t - 1, n_{t+1}, \ldots, n_r)}{C(n_1, n_2, \ldots, n_{t-1}, n_t, n_{t+1}, \ldots, n_r)}, \tag{9.34}$$

and the utilization of node i is given by:

$$U_i = \sum_{t=1}^{r} U_{it}. \tag{9.35}$$

Assuming that all nodes are of type 1, 2 or 4, a recursive formula for the computation of the normalization constant C is derived in a way analogous to the single-job type case:

$$C(n_1, n_2, \ldots, n_r) = C_m(n_1, n_2, \ldots, n_r),$$

where for $i = 1, 2, \ldots, m$, and for $j_t = 1, 2, \ldots, n_t$:

$$C_i(j_1, j_2, \ldots, j_r) = C_{i-1}(j_1, j_2, \ldots, j_r) \tag{9.36}$$

$$+ \sum_{\substack{t=1 \\ j_t \neq 0}}^{r} \rho_{it} C_i\,(j_1, j_2, \ldots, j_{t-1}, j_t - 1. j_{t+1}, \ldots, j_r)$$

with the initial conditions:

$$C_0(j_1, j_2, \ldots, j_r) = \frac{(j_i + j_2 + \cdots + j_r)!}{j_1! j_2! \cdots j_r!} \prod_{t=1}^{r} \rho_{0t}^{j_t}$$

and

$$C_i(0, 0, 0, \ldots, 0) = 1.$$

A discussion of computational techniques for product-form networks is found in [REIS 1978, REIS 1979, SAUE 1981]. For further theoretical development see [LAM 1977, REIS 1975].

As with a single-job type, we can define the relative utilizations $\rho_{it} = E[B_{it}]$, the total service requirement of type t job on server i. In this case, the branching probability matrices X_t need not be specified and visit counts need not be computed.

Example 9.13

A computer center processes three types of jobs. Jobs of type 1 are I/O-bound: in order to complete execution, they need 1 second of CPU time (that is $E[B_{01}] = 1$), 10 seconds of I/O time, and 1 unit of main memory. Jobs of type 2 are balanced: they need 10 seconds each of CPU and I/O, and two units of main memory. Jobs of type 3 are CPU-bound: they consume 100 seconds of CPU time, 10 seconds of I/O time, and five units of main memory.

The total main memory available for user allocation is ten units. Therefore, we can admit either (one job of type 1, two jobs of type 2, and one job of type 3) or (three jobs of type 1, one job of type 2, and one job of type 3) in the active set. Evaluate the effects of the two choices.

We can let:

$$\rho_{01} = E[B_{01}] = 1, \qquad \rho_{11} = E[B_{11}] = 10,$$
$$\rho_{02} = E[B_{02}] = 10, \qquad \rho_{12} = E[B_{12}] = 10,$$
$$\rho_{03} = E[B_{03}] = 100, \qquad \rho_{13} = E[B_{13}] = 10.$$

For the first case with $n_1 = 1$, $n_2 = 2$, and $n_3 = 1$, we compute using formula (9.36):

$$C_1(1, 2, 1) = 1,410,000,$$
$$C_1(0, 2, 1) = 66,000,$$
$$C_1(1, 1, 1) = 56,400,$$
$$C_1(1, 2, 0) = 6,600.$$

Now the average throughputs by class in jobs per second are computed to be:

$$E[T_1] = \frac{C_1(0, 2, 1)}{C_1(1, 2, 1)} = 0.04681,$$
$$E[T_2] = \frac{C_1(1, 1, 1)}{C_1(1, 2, 1)} = 0.04,$$
$$E[T_3] = \frac{C_1(1, 2, 0)}{C_1(1, 2, 1)} = 0.004681.$$

For the second alternative, where $n_1 = 3$, $n_2 = 1$, and $n_3 = 1$, we have:

$$E[T_1] = \frac{C_1(2, 1, 1)}{C_1(3, 1, 1)} = 0.07758,$$
$$E[T_2] = \frac{C_1(3, 0, 1)}{C_1(3, 1, 1)} = 0.01667,$$
$$E[T_3] = \frac{C_1(3, 1, 0)}{C_1(3, 1, 1)} = 0.0055569.$$

We notice that with the second choice, the average throughput of class 1 jobs has gone up considerably at the expense of class 2 jobs. #

Problems

1. Consider a special case of an $M/G/1$ queue with processor-sharing scheduling discipline and an average arrival rate λ. The Laplace transform of the service time B is given by:

$$L_B(s) = b\left(\frac{\mu_1}{s + \mu_1}\right) + (1 - b)\left(\frac{\mu_1}{s + \mu_1}\right)\left(\frac{\mu_2}{s + \mu_2}\right)$$

with $0 \leqslant b \leqslant 1$. For instance, if $b = 0$ we have a two-stage hypoexponential distribution (Erlang, if $\mu_1 = \mu_2$), if $b = 1$ we have an exponential distribution, and if $b = p_1 + p_2\mu_2/\mu_1$ (where $p_1 + p_2 = 1$), we have a two-stage hyperexponential service time distribution.

Draw the state diagram of the system, write down the steady-state balance equations, and proceed to show that the steady-state probability of n jobs in the system has the product-form solution, identical to the $M/M/1$-FCFS solution.

2. Consider a mixed interactive-batch system with 30 terminals, each with an average think time of 15 seconds. The mean disk service time is 20 milliseconds, and the disk scheduling algorithm is FCFS. An interactive job makes an average of 30 disk requests, a batch job an average of 10 disk requests. The average CPU requirement ($E[B_{0,\text{batch}}]$) of a batch job is 5 seconds while that of a terminal user is 0.4 second. A round robin CPU scheduling algorithm with a small enough time quantum is employed so that the use of PS approximation is considered adequate. Determine the effect of the degree of multiprogramming of batch jobs on the average response time and batch throughput varying batch multiprogramming level from 0 to 10 (assuming that such a variation is permissible).

   The actual size of main memory will support a batch multiprogramming level of 1. In order to improve batch throughput, three alternatives are being considered:

   (a) Adding main memory so that batch multiprogramming level can be increased up to 5.
   (b) Purchasing a new CPU with a speed improvement factor of 1.4.
   (c) Adding another disk of the same type.

   For each alternative calculate the resulting improvement in batch throughput and the positive or adverse effect on the terminal response time.

9.5 NON-PRODUCT-FORM NETWORKS

Many useful queuing network models do not possess the properties required for a product-form solution. One such example is the network of Figure 9.7, where a job may hold one of the active resources (CPU or I/O device) simultaneously with a partition of main memory. One possible method of solution is to draw the state transition diagram of the Markov chain, write down the balance equations and proceed to solve them [STEW 1978]. For another instance, suppose that in Example 9.11 we require that CPU scheduling is FCFS rather than PS. Then it can be shown that the resulting network does not have a product-form solution, but the network can be represented by a Markov chain and its steady-state solution obtained by the methods discussed in Chapter 8. Nevertheless, this procedure is usually a formidable task, owing to the size of the state space.

Approximate solution techniques are applicable in many cases. A detailed exposition on approximation techniques may be found in [BRAN 1974, CHAN 1978b, COUR 1977, KOBA 1974, TRIP 1979]. We now illustrate approximation techniques by several examples.

Example 9.14

Returning to the queuing system shown in Figure 9.7, suppose we represent the CPU-I/O subsystem by one equivalent server as seen by the job scheduler (see Figure 9.19). The service rate of this equivalent server is obtained from a closed version of the central server model (Figure 9.8); specifically, the average throughput of the

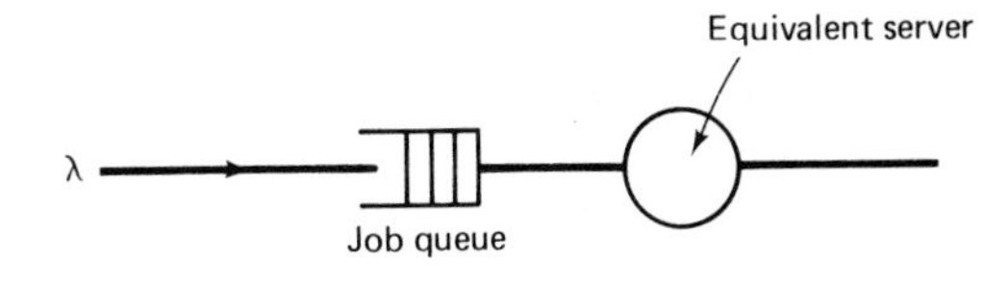

Figure 9.19

central server model determines the service rate of the equivalent server. Since the average throughput depends on the degree of multiprogramming, the equivalent server has load-dependent service rates given by:

$$\gamma_i = E[T(i)], \qquad 1 \leqslant i < n, \tag{9.37}$$
$$\gamma_i = E[T(n)], \qquad i \geqslant n,$$

where n is the upper bound on the degree of multiprogramming. Once the average throughput vector $(E[T(1)], E[T(2)], \ldots, E[T(n)])$ of the inner model is obtained, the outer model is recognized as a birth-death process with a constant birth rate λ and the death rates specified by equation (9.37).

Recall from our discussion of birth-death processes in Chapter 8 that if q_i is the steady-state probability that i jobs reside in the queuing system (of Figure 9.19), we have (assume that $\lambda E[T(n)] < 1$ for stability):

$$q_i = \frac{\lambda}{\gamma_i} q_{i-1}, \qquad i \geqslant 1,$$
$$= \frac{\lambda^i}{\prod_{j=1}^{i} \gamma_j} q_0$$

or

$$q_i = \begin{cases} \dfrac{\lambda^i}{\prod_{j=1}^{i} E[T(j)]} q_0, & 1 \leqslant i < n, \\[2ex] \dfrac{\lambda^i}{(E[T(n)])^{i-n} \prod_{j=1}^{n} E[T(j)]} q_0, & i \geqslant n. \end{cases}$$

From this, q_0 and $E[N] = \sum_{i=0}^{\infty} i q_i$ are determined as:

$$\frac{1}{q_0} = 1 + \sum_{i=1}^{n} \frac{\lambda^i}{\prod_{j=1}^{i} E[T(j)]} + \sum_{i=n+1}^{\infty} \frac{\lambda^i}{(E[T(n)])^{i-n} \prod_{j=1}^{n} E[T(j)]}$$
$$= 1 + \sum_{i=1}^{n} \frac{\lambda^i}{\prod_{j=1}^{i} E[T(j)]} + \frac{1}{\prod_{j=1}^{n} E[T(j)]} \frac{\lambda^{n+1}}{E[T(n)] - \lambda}$$

$$E[N] = q_0 \left[\{ \sum_{i=1}^{n} \frac{i\lambda^i}{\prod_{j=1}^{i} E[T(j)]} \} + \frac{1}{\prod_{i=1}^{n} E[T(i)]} \frac{\lambda^{n+1}}{E[T(n)] - \lambda} \right.$$

$$\left. \cdot \{ n + 1 + \frac{\lambda}{E[T(n)] - \lambda} \} \right].$$

Once we have computed $E[N]$, the average number of jobs in the system, we can determine the average response time $E[R]$ by using Little's formula:

$$E[R] = \frac{E[N]}{\lambda}.$$

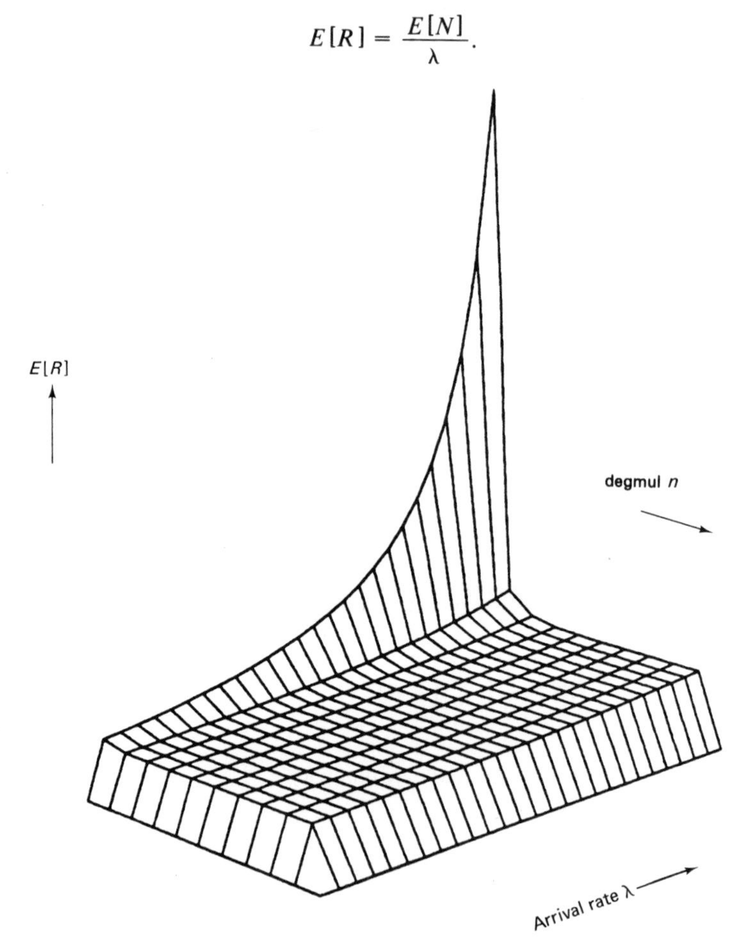

Figure 9.20 The behavior of the average response time as a function of the number of partitions and the arrival rate

In Figure 9.20 we have plotted the average response time, $E[R]$, as a function of the number of allowable partitions, n, and of the arrival rate λ. Various parameters of the system are specified in Table 9.9. #

Table 9.9 *Parameters for Example 9.14.*

Parameter	*Symbol*	*Value*
Number of I/O channels	m	2
Degree of multiprogramming	n	From 1 to 10
Branching probabilities	p_0	0.1
	p_1	0.7
	p_2	0.2
CPU service rate	μ_0	1
Drum service rate	μ_1	0.5
Disk service rate	μ_2	0.3
Average arrival rate	λ	From 0.01 to 0.03

The approximation technique illustrated in the example above was developed by Chandy, Herzog, and Woo and is generally referred to in the literature as Norton's theorem reduction (from the analogous Norton's theorem of electrical circuit theory). The primary justification of this method is that it can be shown to yield exact results when applied to product-form networks [SAUE 1981]. Errors incurred in applying this technique to non-product-form networks are discussed in [TRIP 1979], who also propose methods of adjusting Norton's theorem reduction to account for the errors. The following example illustrates the nature of some of these errors.

Example 9.15

Consider the queuing network of Example 9.14 with the restriction that the number of partitions $n = 1$. The approximation technique of Example 9.14 will then imply that the equivalent server has load-independent service time, hence the reduced network is an $M/M/1$ queue. However, the actual "overall" service time of a job is composed of many exponential parts and will have a general distribution. To perform an exact analysis of this network we can consider the system as an $M/G/1$ queue. In order to apply the P-K mean-value formula (7.38), we must compute the second moment of the service-time distribution.

Each individual job can be seen to execute the following program:

Program 9.3

```
begin
 COMP;
 while B do
 begin
  case i of
  1: I/O 1;
  2: I/O 2;
  .
  .
  .
  m: I/O m
  end;
  COMP
 end
end.
```

But this is precisely the program we analyzed in Chapter 5 (Example 5.18), where we derived an expression for the second moment $E[S^2]$ [equation (5.64)]. Then, using the $M/G/1$ formula (7.38), we obtain the average queue length $E[N]$, whence, using Little's result, we get the average response time $E[R]$. Note that with this analysis, service-time distribution at each individual server is allowed to be general. We consider the system whose parameters are specified in Table 9.9 and compute the average response time with $n = 1$ using the $M/G/1$ formula. A comparison of the $M/G/1$ results with the $M/M/1$ results (as in Example 9.14) is provided in Table 9.10. The reason that the two sets of results correspond fairly well is that $\sigma_S = (E[S^2] - (E[S])^2)^{1/2} = 32.26$ while $E[S] = 30.667$ implying that the coefficient of variation $C_S \simeq 1$ and thus the overall service time distribution is close to exponential in this case.

Table 9.10 $E[R]$

$\lambda =$	0.01	0.02	0.025	0.03
$M/G/1$	44.9556	81.9097	136.8145	402.2165
$M/M/1$	44.2315	79.3126	131.4347	383.3854

Example 9.16

Next consider a model of a terminal-oriented system shown in Figure 9.21. In this model, the number of active jobs concurrently sharing main memory is limited by n, the number of partitions. Whenever more than n terminal requests are pending, remaining jobs will have to queue up for memory. The resulting queuing network model does not belong to the product-form class. We resort to an approximation technique to solve this problem (as in Example 9.14).

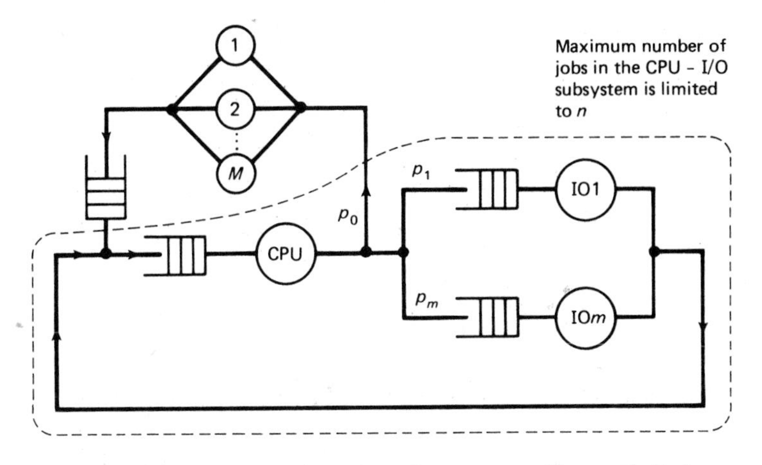

Figure 9.21 A terminal-oriented system with a limited number of memory partitions

First we replace the terminal subsystem with a short circuit and compute the average throughput of the resulting central server network as a function of the number of jobs in the network. Denote this average throughput vector by $(E[T(1)], E[T(2)], \ldots, E[T(n)])$. Now we replace the central server subnetwork by an equivalent server, as shown in Figure 9.22.

The stochastic process corresponding to the model of Figure 9.22 is a birth-death process with birth rates:

$$\lambda_i = (M - i)\lambda, \qquad i = 0, 1, \ldots, M$$

and death rates:

$$\gamma_i = \begin{cases} E[T(i)], & i = 1, 2, \ldots, n, \\ E[T(n)], & i > n. \end{cases}$$

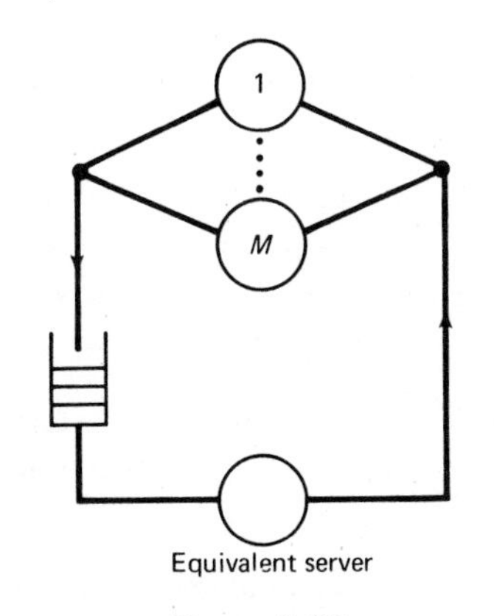

Figure 9.22

Here $1/\lambda$ is the average think time of the terminal users. It follows that the probability that the equivalent server is idle (denoted by q_0) can be obtained by equation (8.19) as:

$$q_0 = \frac{1}{1 + \sum_{k=1}^{M} \dfrac{\lambda^k M!}{\prod_{j=1}^{k} \gamma_j \, (M-k)!}};$$

also

$$q_i = \frac{\lambda^i}{(M-i)!} \, \frac{M!}{\prod_{j=1}^{i} \gamma_j} \, q_0, \qquad i = 1, 2, \ldots, M.$$

The expected throughput, $E[T]$, of the equivalent server is obtained from:

$$E[T] = \sum_{i=1}^{M} q_i \gamma_i = \sum_{i=1}^{n} q_i E[T(i)] + E[T(n)] \sum_{i=n+1}^{M} q_i.$$

Finally, the average response time $E[R]$ is computed from:

$$E[R] = \frac{M}{E[T]} - \frac{1}{\lambda}.$$

As a numerical example, let the average think time $1/\lambda = 15$ seconds, and let other system parameters be as shown in Table 9.11, where $m = 3$. Assuming n, the maximum number of programs allowed in the active set, is 4, we obtain the response time $E[R]$ as a function of the number of terminals M (see Table 9.12). $E[\hat{R}]$ denotes the value of average response time, assuming that the main memory is large enough so that no waiting in the job queue is required; that is, $n \geqslant M$. Sometimes $E[\hat{R}]$ is used as an approximation to $E[R]$, but this example indicates that this approximation can be quite poor.

Table 9.11

	CPU	IO/1	IO/2	IO/3
μ_i	89.3	44.6	26.8	13.4
p_i	0.05	0.5	0.3	0.15

Table 9.12

$M =$	10	20	30	40	50	60
$E[R]$	1.023	1.23	1.64	2.62	5.10	7.03
$E[\hat{R}]$	1.023	1.21	1.46	1.82	2.35	3.11

Figure 9.23 is a three-dimensional plot of the average response time as a function of the number of terminals M and the number of partitions n. We see that increasing n beyond 6 or 7 does not significantly reduce the response time. For instance, with $M = 40$ and $n = 8$, $E[R] = 1.86$ while $E[\hat{R}] = 1.82$. #

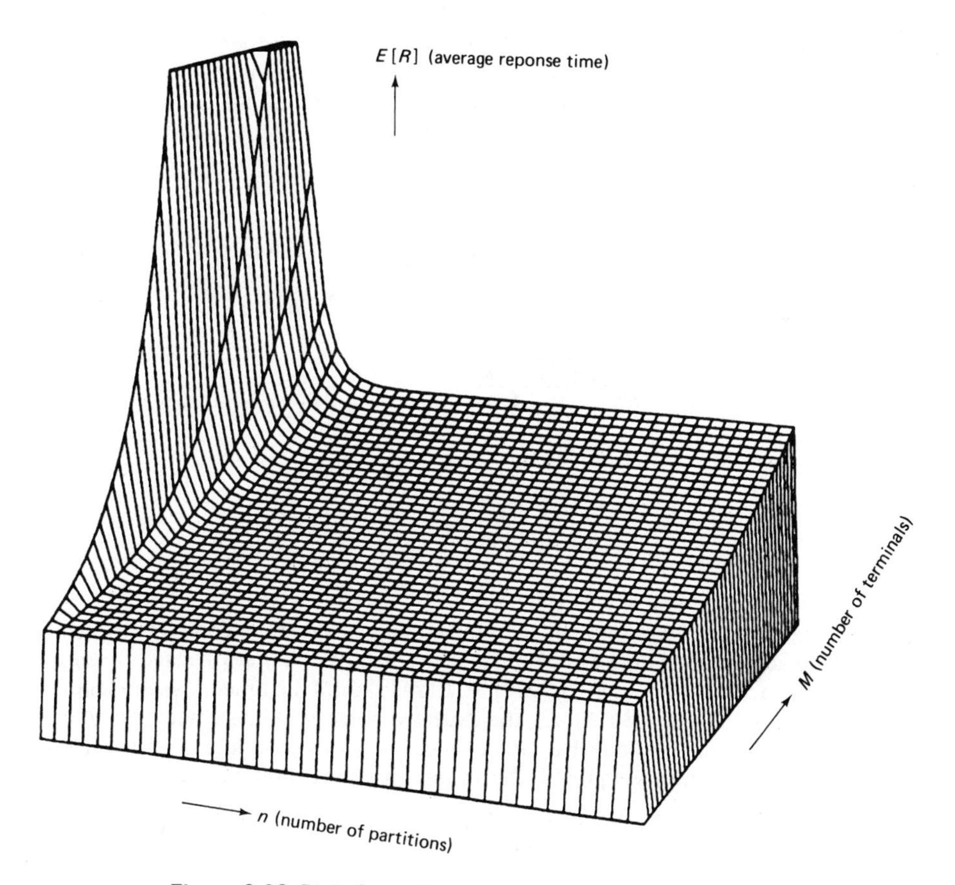

Figure 9.23 Plot of average response time for Example 9.16

Example 9.17

Let us return to the two-node system with multiple processors discussed in Example 9.9. Assume that the failure rate of the I/O device is practically zero while the failure rate of each CPU is $\lambda = 10^{-4}$ failures per hour. Initially the system begins operation with five processors, and it continues to operate in a degraded mode in the face of processor failures until all processors have failed.

The random time to failure X of the system is composed of five phases:

$$X = X_1 + X_2 + \dots + X_5,$$

where the end of each phase marks the failure of a processor. From our discussion of reliability models we know that:

$$X_k \sim \text{EXP}\ [(6 - k)\lambda],$$

Let Y_k denote the number of jobs completed in phase k. Then the total number of jobs completed before system failure is:

$$Y = \sum_{i=1}^{5} Y_k.$$

In order to accurately determine Y_k, we have to perform a transient analysis for the average throughput of the system with $(6 - k)$ processors. Noting that the frequency of processor-I/O interactions is several orders of magnitude higher than the frequency of failure events, we may assume that the system reaches steady state long before the next failure. In this case:

$$Y_k \simeq E[T(6 - k)] \cdot X_k.$$

From this we compute the mean number of jobs completed before system failure as:

$$\begin{aligned} E[Y] &= \sum_{k=1}^{5} E[Y_k] \simeq \sum_{k=1}^{5} E[T(6 - k)] \cdot E[X_k] \\ &= \sum_{k=1}^{5} \frac{E[T(6 - k)]}{(6 - k)\lambda} \\ &= \frac{0.4817419}{\lambda} \\ &= 17,342,708 \text{ jobs,} \end{aligned}$$

using the results of Example 9.9 (Table 9.7).

It is interesting to investigate an alternative mode of system operation where only one CPU is active at one time while remaining nonfaulty CPU's are in standby status. Now the average throughput is $0.099991 = E[T(1)]$ until the time to system failure. Since system MTTF is $5 \cdot 10^4$ hours, the expected number of jobs completed before system crash is actually somewhat larger, equal to 17,999,838 jobs! However, the first system organization will provide better response time until it is degraded to one CPU. #

Problems

*1. Reconsider Example 9.14. We wish to study the behavior of the average response time $E[R]$ as a function of the degree of multiprogramming n. Decompose $E[R]$ into three parts:

$$E[R] = E[R_{eq}] + E[R_{iq}] + E[R_p],$$

where $E[R_{eq}]$ is the average time in the job queue (external queuing time), $E[R_{iq}]$ is the average queuing time in the queues internal to the CPU-I/O subsystem, and $E[R_p]$ is the average processing time on the servers in the subsystem. Derive expressions for $E[R_{eq}]$, $E[R_{iq}]$, and $E[R_p]$. Note that $E[R_{iq}] + E[R_p]$ is the average time spent by a job in the subsystem. For the parameters specified in Table 9.9, plot $E[R_{eq}]$, $E[R_{iq}]$, $E[R_p]$, and $E[R]$ as functions of n on the same graph paper. Give intuitive explanations for the shapes of these curves.

2. Consider the interactive system shown in Figure 9.P.2 (note that the topology is slightly different from that in Figure 9.21).

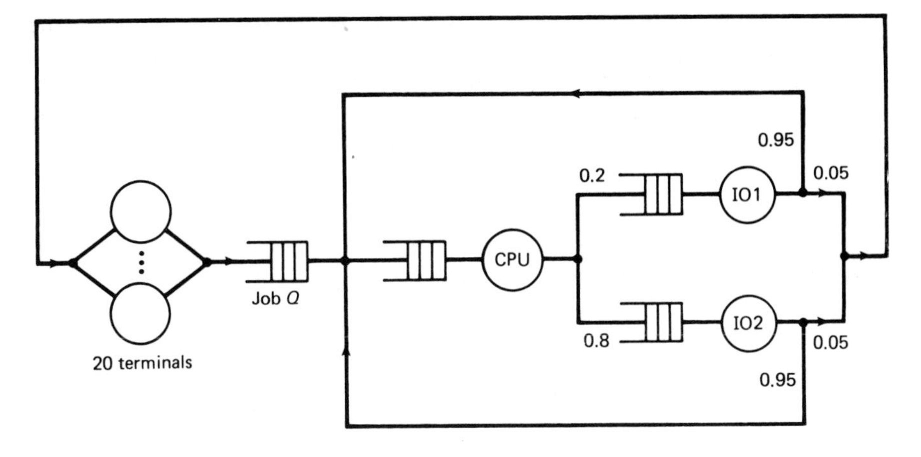

Figure 9.P.2 Another terminal-oriented system

CPU MIPS rate = 0.1.

Number of instructions between two successive I/O requests = 2,000.

Total number of instructions per program = 40,000.

$1/\mu_2 = 50$ msec, $1/\mu_1 = 8$ msec.

Average think time = 5 seconds.

Maximal degree of multiprogramming = 4.

Compute the average response time of the system.

3. [DOWD 1979] Consider a closed central server model with two I/O channels with respective rates 5 second^{-1}, and 3 second^{-1}. The CPU service rate is 7 second^{-1}, and the branching probabilities are $p_0 = 0.05$, $p_1 = 0.65$, and $p_2 = 0.3$. If the degree of multiprogramming is fixed then the given network has a product-form solution as specified in Section 9.3. In practice, however, the degree of multiprogramming varies. Let the degree of multiprogramming N be a random variable. Compute and plot the average system throughput as a function of the average degree of multiprogramming $E[N]$ (varying from 1 to 10), assuming:
   (a) N is a constant random variable with value $E[N]$.
   (b) N takes two different values: $N = E[N] - 1$ with probability 0.5 and $N = E[N] + 1$ with probability 0.5.
   (c) N is Poisson distributed with parameter $E[N]$.
   (d) N is binomially distributed with parameters $2E[N]$ and ½.
   (e) N has a discrete uniform distribution over $\{0, 1, 2, \ldots, 2E[N]\}$.
   (f) N takes two values: $N = 0$ with probability 0.5 and $N = 2E[N]$ with probability 0.5.

4. It can be shown that the procedure using the Norton's theorem reduction gives exact results when applied to a queuing network belonging to the product-form

class. Note that the interactive system model of Example 9.16 will be a product-form network if the number of partitions $n \geqslant M$. In this case, apply the technique [formula (9.24)] developed in Section 9.3 and compute the average response time $E[R]$ as a function of M. Now compute $E[R]$ using the procedure used in Example 9.16 and compare the results.

5. Modify the problem of Example 9.16 so that the I/O device labeled 1 is a paging drum. It is convenient to reparametrize the problem assuming $v_{\text{terminal-node}} = 1$. Then $\rho_0 = E[B_0] = 0.223964$ seconds, $V_2 = 6$, $V_3 = 3$, and $V_1(n)$ is computed assuming the following page-fault characteristics of programs:

Number of page faults, V_1	13	13	162	235	240	300	900
Page allotment	25	12	6	5	4	3	2

Compute average response time $E[R]$ as a function of the number of terminals M and as a function of the number of partitions n, assuming the total pageable memory is 50 page frames. Suppose now we purchase a new drum with average service time 10 milliseconds. Recompute $E[R]$ and compare with earlier results.

6. Write and run a simulation program to evaluate the performance of the system in Example 9.16. Compare the results thus obtained with those shown in Table 9.12.

7. Write and run a simulation program to evaluate the performance of the system in Example 9.14. Compare the results of the simulation with the results obtained analytically.

8. Consider another approximate solution technique for the system of Example 9.16. Specifically, the equivalent server in Figure 9.22 is assumed to have load-independent service rate given by $E[T(n)]$, the average throughput of the subsystem at the maximal degree of multiprogramming. The resulting system was studied in Example 8.9. Let the average response time thus obtained be denoted by $E[\tilde{R}]$. Compare the results obtained with those in Table 9.12. (Notice that the approximation $E[\tilde{R}]$ will be good under heavy-load conditions—that is, $M >> n$.)

9. Resolve Example 9.6 using Norton's theorem of queuing networks by first short-circuiting the CPU and determining the characteristics of the composite I/O server. Next solve a cyclic queuing network with a CPU and the composite I/O server (whose service rate is queue size dependent). Compare your answers with those obtained in the text.

9.6 SUMMARY

In this chapter we have introduced networks of queues, which are an important tool in computer system performance analysis and prediction.

Open queuing networks are useful in studying the behavior of computer-communication networks [KLEI 1976]. Closed networks are more useful in computer system performance evaluation [BASK 1975, CHAN 1977, FERR 1978, KLEI 1976, KOBA 1978]. Readers interested in case studies may see [BHAN 1974, BUZE 1978, GIAM 1976a, KEIN 1979, SAUE 1981, TRIV 1978].

When queuing networks (open, closed, or mixed) satisfy certain properties (a sufficient condition is known as **local-balance** [CHAN 1977]), we can often obtain a convenient product-form solution for the network. However, certain interesting queuing networks do not possess these properties and hence do not have convenient product-form solutions. In these cases we must make use of a number of approximation techniques to obtain a convenient solution. One such technique, Norton's theorem for queuing networks, was illustrated in this chapter by means of two examples. For a deeper study of the approximation techniques, see [CHAN 1978, COUR 1977, SAUE 1981, TRIP 1979]. Sevcik [SEVC 1977] discusses several approximation techniques for incorporating priority scheduling in multiclass closed queuing networks.

This chapter has only touched on the wide range of results and applications for networks of queues. For extensive treatments of queuing theory see [GROS 1974, KELL 1979, KLEI 1975] and for applications see [CHAN 1978a, COUR 1977, KOBA 1978, KLEI 1976]. The readers may also enjoy the more recent approach to queuing network analysis developed by Buzen and Denning [DENN 1978].

Review Problems

1. [HOGA 1975] *Two-Queue Blocking System.*

   Consider the closed cyclic queuing network of Figure 9.9 with the service rates of two servers equal to μ_0 and μ_1, respectively. Let the branching probabilities $p_0 = 0.1$ and $p_1 = 0.9$. Assume that there is a finite waiting room at node one so that the total number of jobs at the node (at its queue plus any at the server) is limited to three jobs. There is no such restriction at node 0. For the degree of multiprogramming $n = 4$, draw the state diagram for the system and proceed to derive an expression for the steady-state probabilities. Repeat for $n = 3, 5, 6$, and 10.

   Compute the average system throughput as a function of the degree of multiprogramming n using $\mu_0 = 1$ and $\mu_1 = 0.5$. Now remove the restriction on the queue size at node 1 and compute the average system throughput as a function of n (this then becomes a case of Example 9.4) and compare the results.

*2. Draw the state diagram and write down the steady-state balance equations for the queuing system $M/E_2/1$ with FCFS scheduling. Solve for the steady-state probabilities. Recall that E_2 implies that the service time distribution is two-stage Erlang.

*3. Recall that in problem 2 at the end of Section 9.4 we assumed a PS discipline for CPU scheduling. Now assume that a preemptive priority is given to interactive jobs over batch jobs. Then the resulting queuing network does not belong to the product-form class. Sevcik [SEVC 1977] has suggested the following approximation technique to solve such problems. We provide separate CPU's to both classes of jobs. The CPU for the batch jobs is known as the shadow CPU and it is slowed down by a factor equal to $(1 - U_{0a})$ where U_{0a} is the utilization of the CPU for interactive jobs. Since U_{0a} is not known initially, it may be estimated by solving a single-class queuing network with batch jobs deleted; then an iteration is performed using the above approximation until convergence is reached. Resolve problem 2 of Section 9.4 using this technique.

*4. Returning to the concurrent program of problem 7, Section 8.4, assume that two independent processes executing the same program are concurrently sharing the resources of a system with a single CPU and a single I/O processor. TCPUj tasks can be executed only on the CPU and TIOj tasks can be executed only on the I/O processor. Describe the state space of the continuous-parameter Markov chain for this system and draw its state diagram. Solve for the steady-state probabilities and hence obtain expressions for the utilizations of the two devices. Next, assuming that all tasks in a process are to be executed sequentially, solve for the utilizations of the two processors (the problem then becomes a special case of the cyclic queuing model). Using the parameters specified earlier, determine the percentage improvement in resource utilizations due to CPU/IO overlap.

References

[BASK 1975] F. BASKETT, K. M. CHANDY, R. R. MUNTZ, and F. G. PALACIOS, "Open, Closed and Mixed Networks of Queues with Different Classes of Customers," *JACM*, 22:2 (April 1975), 248–60.

[BEUT 1978] F. J. BEUTLER and B. MELAMED, "Decomposition and Customer Streams of Feedback Networks of Queues in Equilibrium," *Operations Research*, 26:6 (November–December 1978), 1059–72.

[BHAN 1974] R. BHANDIWAD and A. C. WILLIAMS, "Queuing Network Models of Computer Systems," *Third Texas Conference on Computing Systems*, November 1974.

[BRAN 1974] A. BRANDWAJN, "A Model of a Time-Sharing Virtual Memory System Solved Using Equivalence and Decomposition Methods," *Acta Informatica*, Vol. 4, pp. 11–47.

[BURK 1956] P. J. BURKE, "Output of a Queuing System," *Operations Research*, Vol. 4, pp. 699–704.

[BURK 1976] P. J. BURKE, "Proof of a Conjecture on the Interarrival Time Distribution in a $M/M/1$ Queue with Feedback," *IEEE Trans. on Comm.*, Vol. 24, pp. 175–78.

[BUZE 1973] J. P. BUZEN, "Computational Algorithms for Closed Queuing Networks with Exponential Servers," *CACM*, 16:9 (September 1973).

[BUZE 1978] J. P. BUZEN, "A Queuing Network Model of MVS," *ACM Computing Surveys*, September 1978.

[CHAN 1977] K. M. CHANDY, J. H. HOWARD, and D. F. TOWSLEY, "Product Form and Local Balance in Queuing Networks," *JACM*, 24:2, 250–263.

[CHAN 1978a] K. M. CHANDY and R. T. YEH, *Current Trends in Programming Methodology*, Vol. III: *Software Modeling*, Prentice-Hall, Englewood Cliffs, N. J.

[CHAN 1978b] K. M. CHANDY and C. H. SAUER, "Approximate Methods for Analyzing Queuing Network Models of Computer Systems," *ACM Computing Surveys*, 10 (September 1978), 281–317.

[COUR 1977] P. J. COURTOIS, *Decomposability: Queuing and Computer System Applications*, Academic Press, New York.

[DENN 1978] P. J. DENNING, and J. P. BUZEN, "The Operational Analysis of Queuing Network Models," *ACM Computing Surveys*, 10 (September 1978), 225–41.

[DOWD 1979] L. DOWDY and others, "On the Multiprogramming Level in Closed Queuing Networks," Technical Report, Department of Computer Science, University of Maryland, College Park, Md.

[FERR 1978] D. FERRARI, *Computer System Performance Evaluation*, Prentice-Hall, Englewood Cliffs, N.J.

[GIAM 1976] T. GIAMMO, "Extensions to Exponential Queuing Network Theory for Use in a Planning Environment," *Proc. IEEE Compcon*, September 1976.

[GIAM 1976a] T. GIAMMO, "Validation of a Computer Performance Model of the Exponential Queuing Network Family," *Acta Informatica*, 7, 137–52.

[GORD 1967] W. GORDON and G. NEWELL, "Closed Queuing Systems with Exponential Servers," *Operations Research*, 15, 254–265.

[GROS 1974] D. GROSS and C. M. HARRIS, *Fundamentals of Queuing Theory*, John Wiley & Sons, New York.

[HOGA 1975] J. HOGARTH, "Optimization and Analysis of Queuing Networks," Ph.D. Dissertation, Department of Computer Science, University of Texas at Austin.

[JACK 1957] J. R. JACKSON, "Networks of Waiting Lines," *Operations Research*, 5, 518–21.

[KEIN 1979] M. G. KEINZLE and K. C. SEVCIK, "A Systematic Approach to the Performance Modeling of Computer Systems," *Fourth International Symp. on Modeling and Performance Evaluation of Computer Systems*, Vienna, February 1979.

[KELL 1979] F. P. KELLY, *Reversibility and Stochastic Networks*, John Wiley & Sons, New York.

[KLEI 1975] L. KLEINROCK, *Queuing Systems*, Vol. I, *Theory*, John Wiley & Sons, New York.

[KLEI 1976] L. KLEINROCK, *Queuing Systems*, Vol. II, *Computer Applications*, John Wiley & Sons, New York.

[KOBA 1974] H. KOBAYASHI, "Applications of the Diffusion Approximation to Queuing Networks", (two parts), *JACM*, Vol. 21.

[KOBA 1978] H. KOBAYASHI, *Modeling and Analysis*, Addison-Wesley, Reading, Mass.

[LAM 1977] S. LAM, "Queuing Networks with Population Size Constraints," *IBM Journal of Research and Development*, 1977.

[MELA 1980] B. MELAMED, "Sojourn Times in Queuing Networks," Technical Report, Department of Industrial Engineering and Management Sciences, Northwestern University, Evanston, Ill.

[REIS 1975] M. REISER and H. KOBAYASHI, "Queuing Networks with Multiple Closed Chains: Theory and Computational Algorithms," *IBM Journal of Research and Development*, 1975.

[REIS 1978] M. REISER and C. H. SAUER, "Queuing Network Models: Methods of Solution and Their Program Implementations," in [CHAN 1978a].

[REIS 1979] M. REISER, "Mean Value Analysis—A New Look at an Old Problem," *Fourth International Symposium on Modeling and Performance Evaluation of Computer Systems*, Vienna, February 1979.

[SAUE 1981] C. H. SAUER and K. M. CHANDY, *Computer Systems Performance Modeling*, Prentice-Hall, Englewood Cliffs, N.J.

[SEVC 1977] K. C. SEVCIK, "Priority Scheduling Disciplines in Queuing Network Models of Computer Systems," *Proc. IFIP Congress*, IFIP Press, pp. 565–70.

[SEVC 1980] K. C. SEVCIK, G. S. GRAHAM, and J. ZAHORJAN, "Configuration and Capacity Planning in a Distributed Processing System," *Proc. Computer Performance Evaluation Users Group Meeting*, Orlando, Fla.

[SIMO 1979] B. SIMON and R. D. FOLEY, "Some Results on Sojourn Times in Acyclic Jackson Networks," *Management Science*, 25:10 (October 1979), 1027–34.

[STEW 1978] W. J. STEWART, "A Comparison of Numerical Techniques in Markov Modeling," *CACM*, February 1978.

[TRIP 1979] S. K. TRIPATHI, "On Approximate Solution Techniques for Queuing Network Models of Computer Systems," Computer Systems Research Group, Technical Report, University of Toronto, Canada.

[TRIV 1978] K. S. TRIVEDI and R. L. LEECH, "The Design and Analysis of a Functionally Distributed Computer System," *1978 International Conference on Parallel Processing*.

[WILL 1976] A. C. WILLIAMS and R. A. BHANDIWAD, "A Generating Function Approach to Queuing Network Analysis of Multiprogrammed Computers," *Networks*, 6, 1–22.

Chapter 10

Statistical Inference

10.1 INTRODUCTION

The probability distributions discussed in the preceding chapters will yield probabilities of the events of interest, provided that the form (or the type) of the distribution and the values of its parameters are known in advance. In practice, the form of the distribution and its associated parameters have to be estimated from data collected during the actual operation of the system under investigation.

In this chapter we investigate problems in which, from the knowledge of some characteristic of a suitably selected subset of a collection of elements, we draw inferences about the characteristics of the entire set. The collection of elements under investigation is known as the **population**, and its selected subset is called a **sample**. Methods of **statistical inference** help us in estimating the characteristics of the entire population based upon the data collected from (or the evidence produced by) a sample. Statistical techniques are useful in both the planning of the measurement activities and the interpretation of the collected data.

Two aspects of the sampling process seem quite intuitive. First, as the sample size increases, the estimate generally gets closer to the "true" value, with complete correspondence being reached when the sample embraces the entire population. Second, whatever the sample size, the sample should be 'representative' of the population. These two desirable aspects (not always satisfied) of the sampling process will lead us to definitions of the **consistency** and **unbiasedness** of an estimate.

When we say that the population has the distribution $F(x)$, we mean that we are interested in studying a characteristic X of the elements of this population and that this characteristic X is a random variable whose distribution function is $F(x)$.

The following issues will occupy us in this chapter:

1. Different samples from the same population will result, in general, in distinct estimates, and these estimates themselves will follow some form of statistical distribution, called a **sampling distribution**.
2. Assuming that the distributional form (for example, normal or exponential), of the parent population is known, unknown parameters of the population may be estimated. One also needs to define the confidence in such estimates. This will lead us to **interval estimates**, or **confidence intervals**.
3. In cases where the distributional form of the parent population is not known, we can perform a **goodness-of-fit** test against some specified family of distributions and thus determine whether the parent population can reasonably be declared to belong to this family.
4. Instead of estimating properties of the population distribution, we may be interested in **testing a hypothesis** regarding a relationship involving properties of the distribution function. Based on the collected data, we will perform statistical tests and either reject or fail to reject the hypothesis. We will also study the errors involved in such judgments.

Statistical techniques are extremely useful in algorithm evaluation, system performance evaluation, and reliability estimation. Suppose we want to experimentally evaluate the performance of some algorithm A with the input space S. Since the input space S is rather large, we execute the algorithm and observe its behavior for some randomly chosen subset of the input space. On the basis of this experiment we wish to estimate the properties of the random variable, T, denoting the execution time of algorithm A. It usually suffices to estimate some parameter of T such as its mean $E[T]$ or its variance $\text{Var}[T]$.

Assume that the interarrival times of jobs coming to a computing center are known to be exponentially distributed with parameter λ, but the value of λ is unknown. After observing the arrival process for some finite time, we could obtain an estimate $\hat{\lambda}$ of λ. Sometimes we may not be interested in estimating the value of λ but wish to test the hypothesis $\lambda = \lambda_0$ against an alternative hypothesis $\lambda > \lambda_0$. Thus if λ_0 represents some threshhold of job arrival rate beyond which the system becomes overloaded or unstable, then the acceptance of the above hypothesis will imply that no new equipment needs to be purchased. At other times we may wish to test the hypothesis that job interarrival times are exponentially distributed.

As another example of a statistical problem arising in system performance evaluation, assume we are interested in making a purchase decision between two interactive systems on the basis of their response times (respectively denoted by R_1 and R_2) to trivial requests (such as a simple editing command). Let the corresponding mean response times to trivial requests be denoted by θ_1 and θ_2. On the basis of measurement results, we would like to test the hypothesis $\theta_1 = \theta_2$ versus the alternative $\theta_1 < \theta_2$ [or the alternative $\theta_1 > \theta_2$]. We may also wish to test a hypothesis on variances: Var $[R_1] =$ Var $[R_2]$ versus the alternative Var $[R_1] <$ Var $[R_2]$ (or the alternative Var $[R_1] >$ Var $[R_2]$).

Now suppose that we are interested in tuning a time-sharing system by varying the parameters associated with resource schedulers. In this case we may want to investigate the functional dependence of the average response time on these parameters. We may want to further investigate the functional dependence of the average response time on various characteristics of the transaction such as its CPU time requirement, number of disk I/O requests, number of users logged on, and so on. In this case we will collect a set of measurements and perform a **regression analysis** to estimate and characterize the functional relationships.

A common method of reliability estimation is life testing. A random sample of n components is taken and the times to failure of these components are observed. Based on these observed values, the mean life of a component can be estimated or a hypothesis concerning the mean life may be tested.

First we discuss problems of estimating parameters of the distribution of a single random variable X. Next we discuss hypothesis testing. In the next chapter we discuss problems involving more than one random variable and the associated topic of regression analysis.

10.2 PARAMETER ESTIMATION

Suppose that the parent population is distributed in a form that is completely determinate except for the value of some parameter θ. The parameter θ being estimated could be the population (or true) mean $\mu = E[X]$ or the population (or true) variance $\sigma^2 =$ Var $[X]$. The estimation will be based on a collection of n experimental outcomes $x_1, x_2, \ldots, x_n$. Each experimental outcome x_i is a value of a random variable X_i. The set of random variables $X_1, X_2, \ldots, X_n$ is called a **sample** of size n from the population.

Definition 10.1 (Random Sample). The set of random variables $X_1, X_2, \ldots, X_n$ is said to constitute a **random sample** of size n from the population with the distribution function $F(x)$ provided that they are mutually independent and identically distributed with distribution function $F_{X_i}(x) = F(x)$ for all i and for all x.

Note that the above definition does not hold for sampling without replacement from a finite population (of size N), since the act of drawing an object changes the characteristics of the population. In this case the requirement of independence in the above definition is replaced by the requirement:

$$P(X_1 = x_1, X_2 = x_2, \ldots, X_n = x_n) = \frac{1}{N} \cdot \frac{1}{N-1} \cdot \ldots \cdot \frac{1}{N-n+1} = \frac{(N-n)!}{N!}.$$

Unless otherwise specified, we will assume that the population is very large or conceptually infinite, so that Definition 10.1 is applicable.

In general, we will want to obtain some desired piece of information about the population from a random sample. If we are lucky, the information may be obtained by direct examination or by pictorial methods. However, it is usually necessary to reduce the set of observations to a few meaningful quantities.

Definition 10.2 (Statistic). Any function $T(X_1,X_2,...,X_n)$ of the observations $X_1,X_2,\ldots,X_n$, is called a **statistic**.

Thus, a statistic is a function of n random variables, and assuming that it is also a random variable, its distribution function, called the *sampling distribution* of T, can be derived from the population distribution. The types of functions we will be interested in include the sample mean:

$$\overline{X} = \sum_{i=1}^{n} \frac{X_i}{n},$$

and the sample variance S^2 (to be defined later).

Definition 10.3 (Estimator). Any statistic $\hat{\Theta} = \hat{\Theta}(X_1, X_2, \ldots, X_n)$ used to estimate the value of a parameter θ of the population is called an **estimator** of θ. An observed value of the statistic $\hat{\theta} = \hat{\theta}(x_1, x_2, \ldots, x_n)$ is known as an **estimate** of θ.

A statistic $\hat{\Theta}$ cannot be guaranteed to give a close estimate of θ for every sample. We must design statistics that will give good results "on the average" or "in the long run."

Definition 10.4 (Unbiased). A statistic $\hat{\Theta} = \hat{\Theta}(X_1, X_2, \ldots, X_n)$ is said to constitute an **unbiased** estimator of parameter θ provided

$$E[\hat{\Theta}(X_1, X_2, \ldots, X_n)] = \theta.$$

In other words, on the average, the estimator is on target.

Example 10.1

The sample mean $\overline{X}$ is an unbiased estimator of the population mean μ whenever the latter exists:

$$E[\overline{X}] = E\left[\frac{\sum_{i=1}^{n} X_i}{n}\right]$$

$$= \frac{1}{n}\sum_{i=1}^{n} E[X_i]$$

$$= \frac{1}{n}\sum_{i=1}^{n} E[X]$$

$$= \frac{1}{n} nE[X]$$

$$= E[X]$$

$$= \mu.$$

We can also compute the variance of the sample mean (assuming that the population variance is finite) by noting the independence of $X_1, X_2, \ldots, X_n$ as:

$$\text{Var}[\bar{X}] = \sum_{i=1}^{n} \text{Var}[\frac{X_i}{n}]$$

$$= \frac{n\text{Var}[X_i]}{n^2}$$

$$= \frac{\text{Var}[X]}{n}$$

$$= \frac{\sigma^2}{n}.$$

This implies that the accuracy of the sample mean as an estimator of the population mean increases with the sample size n when the population variance is finite. #

If the population distribution is Cauchy, so that

$$f_X(x) = \frac{1}{\pi[1 + (x - \theta)^2]}, \qquad -\infty < x < \infty,$$

then by problem 1 at the end of this section it follows that the sample mean $\bar{X}$ is also Cauchy distributed. If $\bar{X}$ is used to estimate the parameter θ, then it does not increase in accuracy as n increases. Note that in this case neither the population mean nor the population variance exist.

If we take a sample of size n without replacement from a finite population of size N, then the sample mean $\bar{X}$ is still an unbiased estimator of the population mean but the Var $[\bar{X}]$ is no longer given by σ^2/n. In this case, it can be shown [KEND 1961] that:

$$\text{Var}[\bar{X}] = \frac{\sigma^2}{n}(1 - \frac{n}{N})(\frac{N}{N-1}).$$

Note that as the population size N approaches infinity, we get the formula σ^2/n for the sample variance.

Next consider the function

$$\hat{\Theta}(X_1, X_2, \ldots, X_n) = \frac{\sum_{i=1}^{n}(X_i - \bar{X})^2}{n}$$

as an estimator of the population variance. It can be seen that this function provides a biased estimator of the population variance. On the other hand, the function:

$$\frac{1}{n-1} \sum_{i=1}^{n} (X_i - \bar{X})^2$$

is an unbiased estimator of the population variance. These two functions differ little when the sample size is relatively large.

Example 10.2

The sample variance S^2, defined by:

$$S^2 = \frac{1}{n-1} \sum_{i=1}^{n} (X_i - \bar{X})^2$$

is an unbiased estimator of the population variance σ^2 whenever the latter exists. This can be shown as follows:

$$S^2 = \frac{1}{n-1} \sum_{i=1}^{n} (X_i^2 - 2X_i \bar{X} + \bar{X}^2)$$

$$= \frac{1}{n-1} (\sum_{i=1}^{n} X_i^2) - \frac{2n}{n-1} \left(\frac{\sum_{i=1}^{n} X_i}{n} \right) \bar{X} + \frac{n\bar{X}^2}{n-1}$$

$$= \frac{1}{n-1} \sum_{i=1}^{n} X_i^2 - \frac{n}{n-1} \bar{X}^2.$$

Therefore:

$$E[S^2] = \frac{1}{n-1} \sum_{i=1}^{n} E[X_i^2] - \frac{n}{n-1} E[\bar{X}^2].$$

But:

$$E[X_i^2] = \text{Var}\ [X_i] + (E[X_i])^2 = \sigma^2 + \mu^2$$

and

$$E[\bar{X}^2] = \text{Var}\ [\bar{X}] + (E[\bar{X}])^2 = \frac{\sigma^2}{n} + \mu^2.$$

Thus:

$$E[S^2] = \frac{1}{n-1} n(\sigma^2 + \mu^2) - \frac{n}{n-1} \left(\frac{\sigma^2}{n} + \mu^2 \right)$$

$$= \sigma^2.$$

Thus the sample variance S^2 is an unbiased estimator of the population variance σ^2.#

The above formula applies to the case of an infinite population. The unbiased estimator of the variance of a finite population of size N (assuming sampling without replacement) is given by:

$$S^2 = \frac{1 - \frac{1}{N}}{n - 1} \sum_{i=1}^{n} (X_i - \bar{X})^2.$$

Unbiasedness is one of the more desirable properties of an estimator, although not essential. This criterion by itself does not provide a unique estimator for a given estimation problem.

Example 10.3

$$\hat{\Theta} = \sum_{i=1}^{n} a_i X_i$$

is an unbiased estimator of the population mean μ (if it exists) for any set of real weights a_i such that $\sum a_i = 1$. #

Thus we need another criterion to choose the best among all unbiased estimators. For an estimator $\hat{\Theta}$ of parameter θ to be a good estimator, we would like the probability of the dispersion $P(|\hat{\Theta} - \theta| \geqslant \epsilon)$ to be small. Noting that for an unbiased estimator, $E[\hat{\Theta}] = \theta$, Chebyshev's inequality gives us:

$$P(|\hat{\Theta} - \theta| \geqslant \epsilon) \leqslant \frac{\text{Var}[\hat{\Theta}]}{\epsilon^2}, \qquad \text{for } \epsilon > 0.$$

Thus, one way of comparing two unbiased estimators is to compare their variances.

Definition 10.5 (Efficiency). An estimator $\hat{\Theta}_1$ is said to be a more efficient estimator of the parameter θ than the estimator $\hat{\Theta}_2$, provided:

1. $\hat{\Theta}_1$ and $\hat{\Theta}_2$ are both unbiased estimators of θ, and
2. $\text{Var}[\hat{\Theta}_1] < \text{Var}[\hat{\Theta}_2]$.

Example 10.4

The sample mean:

$$\bar{X} = \frac{\sum_{i=1}^{n} X_i}{n}$$

is the most efficient (minimum-variance) linear estimator of the population mean, whenever the latter exists. To show this, we first note that:

$$\begin{aligned}
\text{Var}[\hat{\Theta} = \sum a_i X_i] &= \sum_{i=1}^{n} a_i^2 \text{Var}[X_i] \\
&= \sum a_i^2 \text{Var}[X] \\
&= \sum a_i^2 \sigma^2,
\end{aligned}$$

since $X_1, X_2, \ldots, X_n$ are mutually independent and identically distributed. Thus we solve the following optimization problem:

$$\text{min:} \quad \text{Var}[\hat{\Theta}] = \sigma^2 \sum_{i=1}^{n} a_i^2 \qquad \text{s.t.:} \quad \sum_{i=1}^{n} a_i = 1, \quad i = 1, 2, \ldots n.$$

We can solve this problem using the method of Lagrange multipliers to obtain the result:

$$a_i = \frac{1}{n}, \qquad i = 1, 2, \ldots, n.$$

In other words, the estimator minimizing the variance is:

$$\sum a_i X_i = \bar{X}, \qquad \text{the sample mean.}$$

#

It can also be shown that under some mild conditions the sample variance, S^2, is a minimum-variance (quadratic) unbiased estimator of the population variance σ^2, whenever the latter exists. Thus, in most practical situations, the sample mean and the sample variance are acceptable estimators of μ and σ^2, respectively.

As in the case of the sample mean, the variance of the sampling distribution of an estimator generally decreases with increasing n. This leads us to another desirable property of an estimator:

Definition 10.6 (Consistency). An estimator $\hat{\Theta}$ of parameter θ is said to be consistent if $\hat{\Theta}$ **converges in probability** to θ; that is:

$$\lim_{n \to \infty} P(|\hat{\Theta} - \theta| \geqslant \epsilon) = 0.$$

As the sample size increases, a consistent estimator gets close to the true value. If we consider a population with finite mean and variance, then $\text{Var}[\bar{X}] = \sigma^2/n$, and using the Chebyshev inequality, we conclude that the sample mean is a consistent estimator of the population mean. In fact, any unbiased estimator $\hat{\Theta}$ of θ, with the property:

$$\lim_{n \to \infty} \text{Var}[\hat{\Theta}] = 0$$

is a consistent estimator of θ owing to the Chebyshev inequality.

The data collected in a sample may be summarized by the arithmetic methods (sample mean, sample variance, and so on) discussed so far. Alternative methods are pictorial in nature. For example, a bar chart or a histogram is often used. Yet another useful way of summarizing data is to construct an **empirical distribution function**, $\hat{F}(x)$: let k_x be the number of observed values x_i (out of a total of n values) which are less than or equal to x; then $\hat{F}(x) = k_x/n$.

The empirical distribution function is a consistent estimator of the true distribution function $F(x)$. To show this, suppose we perform n independent trials of the event, "Sample value observed is less than or equal to x." Each observation y_i is then a value of a Bernoulli random variable Y_i with the probability of success $p = F(x)$, so that:

$$\bar{Y} = \frac{\sum_{i=1}^{n} Y_i}{n} = \hat{F}(x)$$

and $E[\bar{Y}] = p$. Now the law of large numbers tells us that:

$$\lim_{n \to \infty} P(|\bar{Y} - E[\bar{Y}]| \geqslant \epsilon) = 0.$$

But here $\bar{Y} = \hat{F}(x)$ and $E[\bar{Y}] = E[Y] = p = F(x)$, so:

$$\lim_{n \to \infty} P[|\hat{F}(x) - F(x)| \geqslant \epsilon] = 0,$$

as desired.

Next we discuss two general methods of parameter estimation.

Problems

1. Given that the population X has the Cauchy distribution, show that the sample mean $\bar{X}$ has the same distribution.

2. Solve the optimization problem posed in Example 10.4 using the method of Lagrange multipliers. To obtain an algebraic proof, show that:

$$\sum_{i=1}^{n} a_i^2 \geqslant \sum_{i=1}^{n} \left(\frac{1}{n}\right)^2.$$

10.2.1 The Method of Moments

Suppose one or more parameters of the distribution of X are to be estimated based on a random sample of size n. Define the kth sample moment of the random variable X to be:

$$M'_k = \sum_{i=1}^{n} \frac{X_i^k}{n}, \qquad k = 1, 2, \ldots.$$

Of course the kth population moment is:

$$\mu'_k = E[X^k], \qquad k = 1, 2, \ldots,$$

and this moment will be a function of the unknown parameters.

The **method of moments** consists of equating the first few population moments with the corresponding sample moments to obtain as many equations as there are unknown parameters and then solving these equations simultaneously to obtain the required estimates. This method usually yields fairly simple estimators that are consistent. However, it can give estimators that are biased (see problem 1 at the end of this section) and inefficient.

Example 10.5

Let X denote the main memory requirement of a job as a fraction of the total user-allocatable main memory of a computing center. We suspect that the density function of X has the form:

$$f(x) = \begin{cases} (k+1)x^k, & 0 < x < 1, \ k > 0 \\ 0, & \text{otherwise.} \end{cases}$$

A large value of k implies a preponderance of large jobs. If $k = 0$, the distribution-of-memory requirement is uniform. We have a sample of size n from which we wish to estimate the value of k. Since one parameter is to be estimated, only the first moments need to be considered:

$$\mu'_1 = \int_0^1 (k+1)x^k x \, dx = \frac{k+1}{k+2}$$

and

$$M'_1 = \sum_{i=1}^{n} \frac{X_i^1}{n} = \text{sample mean, } \bar{X}.$$

Then the required estimate $\hat{k}$ is obtained by letting:

$$M'_1 = \mu'_1$$

and hence:

$$\hat{k} = \frac{2M'_1 - 1}{1 - M'_1} = \frac{2\bar{X} - 1}{1 - \bar{X}}.$$

As a numerical example, we are given the following sample of size 8:

0.25 0.45 0.55 0.75 0.85 0.85 0.95 0.90

The sample mean $M'_1 = 5.55/8 = 0.69375$. Thus $\hat{k} = 1.265306$. #

Example 10.6

Assume that the repair time of a computer system has a gamma distribution with parameters λ and α. This could be suggested, for instance, by the sequential (stage-type) nature of the repair process. After taking a random sample of n actual repair times, we compute the first two sample moments M'_1 and M'_2. Now the corresponding population moments for a gamma distribution are given by:

$$\mu'_1 = \frac{\alpha}{\lambda} \quad \text{and} \quad \mu'_2 = \frac{\alpha}{\lambda^2} + \mu'^2_1.$$

Then the estimates $\hat{\lambda}$ and $\hat{\alpha}$ can be obtained by solving:

$$\frac{\hat{\alpha}}{\hat{\lambda}} = M'_1 \quad \text{and} \quad \frac{\hat{\alpha}}{\hat{\lambda}^2} + \frac{\hat{\alpha}^2}{\hat{\lambda}^2} = M'_2.$$

Hence:

$$\hat{\alpha} = \frac{M_1'^2}{M_2' - M_1'^2} \quad \text{and} \quad \hat{\lambda} = \frac{M_1'}{M_2' - M_1'^2}. \qquad \#$$

Problems

1. Show that the method-of-moments estimators of the population mean and of the population variance are given by the sample mean, $\overline{X}$, and $(n-1)S^2/n$, respectively. Show that the method-of-moments estimator of the population variance is **biased**.

2. Consider the problem of deriving method-of-moments estimates for the three parameters α, λ_1, and λ_2 of a two-stage hyperexponential distribution. Clearly, three sample moments will be needed for this purpose. But if we have only the sample mean $\bar{x}$ and sample variance s^2 available, we can solve the problem by putting a restriction on the parameters. Assume that:

$$\frac{1}{\lambda_1 + \lambda_2} = \frac{\dfrac{\alpha}{\lambda_1} + \dfrac{1-\alpha}{\lambda_2}}{2}$$

is the chosen restriction. Show that the method-of-moments estimates of the parameters are given by (assuming that $s^2 \geqslant \bar{x}^2$):

$$\hat{\lambda}_1, \hat{\lambda}_2 = \frac{1}{\bar{x}} \pm \frac{1}{\bar{x}} \sqrt{\frac{s^2 - \bar{x}^2}{s^2 + \bar{x}^2}}$$

and

$$\hat{\alpha} = \frac{\hat{\lambda}_1(\hat{\lambda}_2 \bar{x} - 1)}{\hat{\lambda}_2 - \hat{\lambda}_1}.$$

3. [V. A. Abell and S. Rosen] The memory residence times of 13,171 jobs were measured and the sample mean was found to be 0.05 second and the sample variance 0.006724. Estimate the parameters α and λ using the method of moments, assuming that the memory residence time is gamma distributed. Using the result of problem 2, obtain the method-of-moments estimates for parameters α, λ_1, and λ_2, assuming that the memory residence time possesses a two-stage hyperexponential distribution.

4. Derive method of moments estimates for the parameters λ_1 and λ_2 of a two-stage hypoexponential distribution.

10.2.2 Maximum-Likelihood Estimation

The method of maximum likelihood produces estimators that are usually consistent and under certain regularity conditions can be shown to be most efficient in an asymptotic sense (that is, as the sample size n approaches infinity). However, the estimators may be biased for small sample sizes. The principle of this method is to select as an estimate of θ the value for which the observed sample is most "likely" to occur.

We introduce this method through an example. Suppose we want to estimate the probability, p, of a successful transmission of a message over a communication channel. We observe the transmission of n messages and observe that k have been transmitted without errors. From these data, we wish to obtain a maximum-likelihood estimate of the parameter p.

The transmission of a single message is modeled by a Bernoulli random variable X with the pmf:

$$p_X(x) = p^x(1-p)^{1-x}, \qquad x = 0,\ 1;\ 0 \leqslant p \leqslant 1.$$

A random sample $X_1, X_2, \ldots, X_n$ is taken, and the problem is to find an estimator $T(X_1, X_2, \ldots, X_n)$ such that $t(x_1, x_2, \ldots, x_n)$ is a good estimate of p, where $x_1, x_2, \ldots, x_n$ are the observed values of the random sample. The joint probability that $X_1, X_2, \ldots, X_n$ take these values is given by the compound pmf:

$$P(X_1 = x_1, X_2 = x_2, \ldots, X_n = x_n) = \prod_{i=1}^{n} p^{x_i}(1-p)^{1-x_i}$$

$$= p^{\sum x_i}(1-p)^{n-\sum x_i}.$$

If we fix n and the observed values $x_1, x_2, \ldots, x_n$ in the above pmf, then it can be considered a function of p, called a **likelihood function**:

$$L(p) = p^{\sum x_i}(1-p)^{n-\sum x_i}, \qquad 0 \leqslant p \leqslant 1.$$

The value of p, say $\hat{p}$, maximizing $L(p)$, is the maximum-likelihood estimate of p. Thus this method selects that value of the unknown parameter for which the probability of obtaining the measured data is maximum, and $\hat{p}$ is the "most likely" value of p.

Maximizing $L(p)$ is equivalent to maximizing the natural logarithm of $L(p)$; that is:

$$\ln L(p) = \left(\sum_{i=1}^{n} x_i\right) \ln p + \left(n - \sum_{i=1}^{n} x_i\right) \ln(1-p).$$

To find the maximum of the above function, when $0 < p < 1$, we take the first derivative and set it equal to zero:

$$\frac{d \ln L(p)}{dp} = \left(\sum_{i=1}^{n} x_i\right)\left(\frac{1}{p}\right) + \left(n - \sum_{i=1}^{n} x_i\right)\left(\frac{-1}{1-p}\right) = 0$$

and get:

$$p = \frac{\sum_{i=1}^{n} x_i}{n} = \bar{x}.$$

Verifying that the second derivative of ln $L(p)$ is negative, we conclude that $p = \bar{x}$ actually maximizes ln $L(p)$. Therefore, the statistic $\sum X_i/n = \bar{X}$, is known as the **maximum-likelihood estimator of** p:

$$\hat{P} = \frac{1}{n}\sum_{i=1}^{n} X_i = \bar{X}.$$

Since X_i is 0 for a garbled message and 1 for a successful transmission, the sum $\sum X_i$ is the number of successfully transmitted messages. Therefore, the above estimator of p is simply the one we would get by using a relative-frequency argument.

More generally for a discrete population, we define the likelihood function as the joint probability of the event, $[X_1 = x_1, X_2 = x_2, \ldots, X_n = x_n]$:

$$L(\boldsymbol{\theta}) = P(X_1 = x_1, X_2 = x_2, \ldots, X_n = x_n \,|\, \boldsymbol{\theta}) = \prod_{i=1}^{n} p_{X_i}(x_i \,|\, \boldsymbol{\theta})$$

where $\boldsymbol{\theta} = (\theta_1, \theta_2, \ldots, \theta_k)$ is the vector of parameters to be estimated. Analogously, in the case of a continuous population, the likelihood function is defined to be the product of the marginal densities:

$$L(\boldsymbol{\theta}) = \prod_{i=1}^{n} f_{X_i}(x_i \,|\, \boldsymbol{\theta}).$$

Thus, the likelihood function is the joint pmf or pdf of the random variables $X_1, X_2, \ldots, X_n$. Under certain regularity conditions, the maximum likelihood estimate of $\boldsymbol{\theta}$ is the solution of the simultaneous equations:

$$\frac{\partial L(\boldsymbol{\theta})}{\partial \theta_i} = 0, \qquad i = 1, 2, \ldots, k.$$

This method usually works quite well, but sometimes difficulties arise. There may not be a closed-form solution to the above system of equations. For example, if we wish to estimate the parameters α and λ of a gamma distribution, the method of maximum likelihood will produce two simultaneous equations that are impossible to solve in closed form. Alternatively, there may not be a unique solution to the above system of equations. In this case it is necessary to verify which solution, if any, maximizes the likelihood function. Another possibility is that the solution to the above system may not be in the parameter space, in which case a constrained maximization of the likelihood function becomes necessary.

Example 10.7

It is desired to estimate the average arrival rate of incoming calls to a telephone trunk, based on a random sample $X_1 = x_1, \ldots, X_n = x_n$ where X_i denotes the number of calls per hour in the ith observation period. Let the number of calls per hour, X, be Poisson distributed with parameter λ; that is:

$$p_X(x|\lambda) = e^{-\lambda}\frac{\lambda^x}{x!}.$$

The likelihood function is then:

$$L(\lambda) = \prod_{i=1}^{n} e^{-\lambda} \frac{\lambda^{x_i}}{x_i!} = \frac{1}{x_1!\, x_2! \cdots x_n!} e^{-n\lambda} \lambda^{\sum_{i=1}^{n} x_i}.$$

Taking logs, we have:

$$\ln L(\lambda) = -\ln(x_1! x_2! \cdots x_n!) - n\lambda + (\sum_{i=1}^{n} x_i) \ln(\lambda).$$

Taking the derivative with respect to λ and setting it equal to zero, we get:

$$-n + \frac{\sum_{i=1}^{n} x_i}{\lambda} = 0.$$

Thus the maximum likelihood estimator of the average arrival rate is the sample mean:

$$\hat{\Lambda} = \frac{\sum_{i=1}^{n} X_i}{n} = \bar{X}.$$

#

A common method of estimating parameters related to component (system) reliability is that of life testing. This consists of selecting a random sample of n components, testing them under specified environmental conditions, and observing the time to failure of each component.

Example 10.8

Assume that the time to failure, X, of a telephone switching system is exponentially distributed with a failure rate λ. We wish to estimate the failure rate λ from a random sample of n times to failure. Then:

$$L(\lambda) = \prod_{i=1}^{n} \lambda e^{-\lambda x_i} = \lambda^n e^{-\lambda \sum_{i=1}^{n} x_i},$$

$$\frac{dL}{d\lambda} = n\lambda^{n-1} e^{-\lambda \sum_{i=1}^{n} x_i} - \left(\sum_{i=1}^{n} x_i \right) \lambda^n e^{-\lambda \sum_{i=1}^{n} x_i} = 0,$$

from which we get the maximum-likelihood estimator of the failure rate to be the reciprocal of the sample mean:

$$\hat{\Lambda} = \frac{n}{\sum_{i=1}^{n} X_i} = \frac{1}{\bar{X}}.$$

The corresponding maximum-likelihood estimator of the mean life (MTTF) is equal to the sample mean $\bar{X}$.

#

Usually, the MTTF (mean time to failure) is so large as to forbid such exhaustive life tests, hence **truncated life tests** are common. Such a life test is terminated after the first r failures have occurred (sample-truncated) or after a specified time has elapsed (time-truncated). If a failed component is repaired or is replaced by a new one, then the life test is called a **replacement test**; otherwise it is a **non-replacement test**.

Example 10.9

Consider a sample truncated test of n components without replacement. Let $T_1, T_2, \ldots, T_r$ be the observed times to failure so that $T_1 \leqslant T_2 \leqslant \cdots \leqslant T_r$. Specific values of these random variables are denoted by $t_1, t_2, \ldots, t_r$. Let θ be the MTTF to be estimated and assume that components follow an exponential failure law.

Since $(n - r)$ components have not failed when the test is completed, the likelihood function is defined in the following way. Assume $T_{r+1}, \ldots, T_n$ are the times to failure of the remaining components, whose failures will not actually be observed. Then:

$$L(\theta) \prod_{i=1}^{r} h_i = P(t_i \leqslant T_i < t_i + h_i,\ i = 1, 2, \ldots, r;$$

$$T_i > t_r,\ i = r + 1, \ldots, n)$$

and, dividing by the product of h_i's and taking the limit as $h_i \to 0$, we get:

$$L(\theta) = \prod_{i=1}^{r} f(t_i|\theta) \prod_{j=r+1}^{n} R(t_r|\theta)$$

$$= \prod_{i=1}^{r} \frac{1}{\theta} e^{-t_i/\theta} \prod_{j=r+1}^{n} e^{-t_r/\theta}$$

$$= \frac{1}{\theta^r} \exp\left[-\frac{\sum_{i=1}^{r} t_i + (n - r)t_r}{\theta}\right].$$

Let:

$$s_{n:r} = \sum_{i=1}^{r} t_i + (n - r)t_r$$

be the accumulated life on test. Differentiating the likelihood function with respect to θ and setting it equal to zero, we get:

$$-\frac{r}{\theta^{r+1}} e^{-s_{n:r}/\theta} + \frac{1}{\theta^r} (-s_{n:r}) \left(-\frac{1}{\theta^2}\right) e^{-s_{n:r}/\theta} = 0.$$

Then the maximum-likelihood estimator of the mean life is given by:

$$\hat{\Theta} = \frac{S_{n:r}}{r} = \frac{\sum_{i=1}^{r} T_i + (n - r)T_r}{r}.$$

Thus the estimator of the mean life is given by the accumulated life on test, $S_{n:r}$, divided by the number of observed failures. #

Problems

1. Show that the maximum-likelihood estimator of the mean life θ with a replacement test until r failures is:

$$\hat{\Theta} = \frac{nT_r}{r},$$

where the random variable T_r denotes the time for the rth failure from the beginning of the experiment.

2. Suppose that the CPU service time X of a job is gamma distributed with parameters λ and α. Based on a random sample of n observed service times, $x_1, x_2, \ldots, x_n$, we wish to estimate parameters λ and α. Show that maximum-likelihood estimators of λ and α do not yield a closed-form solution. Recall that method-of-moments estimators of λ and α are simple closed-form expressions.

10.2.3 Confidence Intervals

The methods of parameter estimation discussed so far produce a **point estimate** of the desired parameter. Of course, it will be the rare case when the point estimate $\hat{\theta}$ coincides with the actual value of the parameter θ being estimated. It is, therefore, desirable to find an **interval estimate**. We construct an interval, called a **confidence interval**, in such a way that we have a certain confidence it contains the true value of the unknown parameter. If we can ascertain that the estimator $\hat{\Theta}$ satisfies the condition:

$$P(\hat{\Theta} - \epsilon_1 < \theta < \hat{\Theta} + \epsilon_2) = \gamma,$$

then we say that the random interval $A(\theta) = (\hat{\Theta} - \epsilon_1, \hat{\Theta} + \epsilon_2)$ is a $100 \times \gamma$ percent confidence interval for parameter θ. γ is called the **confidence coefficient**. Often we choose a symmetric confidence interval so that $\epsilon_1 = \epsilon_2 = \epsilon$.

The meaning of the above probability statement needs some clarification. For any specific set of observations, $x_1, x_2, \ldots, x_n$, the estimate $\hat{\theta}$ is a fixed value. The confidence interval $A(\theta)$ either will or will not contain the true value of θ, in which case the probabilities are one and zero, respectively. However, when $\hat{\Theta}$ is considered as a function of random variables $X_1, X_2, \ldots, X_n$, the end points of the interval $A(\theta)$ are then random variables, and it is possible to say that the probability is γ that the (random) interval $A(\theta)$ will contain the (fixed) true value of θ. The relative-frequency interpretation of the above probability implies that if this process of sampling is repeated many times, the fraction of the time in which the true value of θ will be contained in the confidence interval $A(\theta)$ will be γ.

One simple way to obtain a confidence interval (involving an unbiased estimator) is to apply Chebyshev's inequality:

$$P(\hat{\Theta} - \epsilon < \theta < \hat{\Theta} + \epsilon) \geqslant 1 - \frac{\text{Var}[\hat{\Theta}]}{\epsilon^2},$$

provided Var $[\hat{\Theta}]$ is known (or can be estimated).

Example 10.10

Let $\theta = \mu$ and $\hat{\Theta} = \bar{X}$, the sample mean. Assume that the population variance σ^2 is known. Then Var $[\bar{X}] = \sigma^2/n$ and the Chebyshev inequality yields:

$$P(\bar{X} - \epsilon < \mu < \bar{X} + \epsilon) \geqslant 1 - \frac{\sigma^2}{n\epsilon^2}.$$

Thus for a given $\epsilon > 0$, we may make the confidence coefficient arbitrarily close to 1 by choosing a sufficiently large value of n. #

Confidence intervals obtained by Chebyshev's inequality can usually be improved upon if the distribution of X is known. In general, the steps involved in obtaining a confidence interval for the parameter θ from a random sample, $X_1, X_2, \ldots, X_n$ are as follows:

1. Find a random variable that is a function of $X_1, X_2, \ldots, X_n$:

$$T = T(X_1, X_2, \ldots, X_n; \theta),$$

such that the distribution of T is known.

2. Find numbers a and b such that:

$$P(a < T < b) = \gamma.$$

3. After sampling the values x_i of X_i, determine the range of values that θ can take on while maintaining the condition:

$$a < t(\theta) < b$$

where $t(\theta) = T(x_1, x_2, \ldots, x_n; \theta)$.

This range of values is a $100\,\gamma$ percent confidence interval of θ.

It should be clear at the outset that the above procedure (step 1) depends upon the distribution of X. Therefore, our subsequent discussion is divided according to some common distributions of X.

10.2.3.1 Sampling from the Normal Distribution. Suppose a random sample of size n is taken from a normal population with unknown mean μ and a known variance σ^2; that is, $X \sim N(\mu, \sigma^2)$. Then it is easy to show that the sample mean is $N(\mu, \sigma^2/n)$, so that $Z = (\bar{X} - \mu)/(\sigma/\sqrt{n})$ is standard normal; that is, Z is $N(0, 1)$. Now if we want a 100γ percent confidence interval for the population mean μ, we find numbers a and b [from $N(0, 1)$ tables] such that:

$$P(a < Z < b) = \gamma.$$

Once the numbers a and b are determined, we obtain the required confidence interval as follows:

$$a < \frac{\bar{x} - \mu}{\sigma/\sqrt{n}} < b$$

or

$$\bar{x} - \frac{b\sigma}{\sqrt{n}} < \mu < \bar{x} - \frac{a\sigma}{\sqrt{n}}.$$

Therefore, $(\bar{x} - b\sigma/\sqrt{n},\ \bar{x} - a\sigma/\sqrt{n})$ is a 100γ percent confidence interval for μ.

Example 10.11

It is common to choose a symmetric confidence interval for μ so that we have $a = -b$. (If the estimator has a symmetrical pdf, as it does in this case, then the choice $a = -b$ is known to produce the confidence interval of minimum width.) Then:

$$P(-b < Z < b) = \gamma.$$

We let $\gamma = 1 - \alpha$ for convenience. Now from the symmetry of the pdf of Z;

$$P(Z < -b) = \frac{\alpha}{2} \quad \text{and} \quad P(Z > b) = \frac{\alpha}{2}.$$

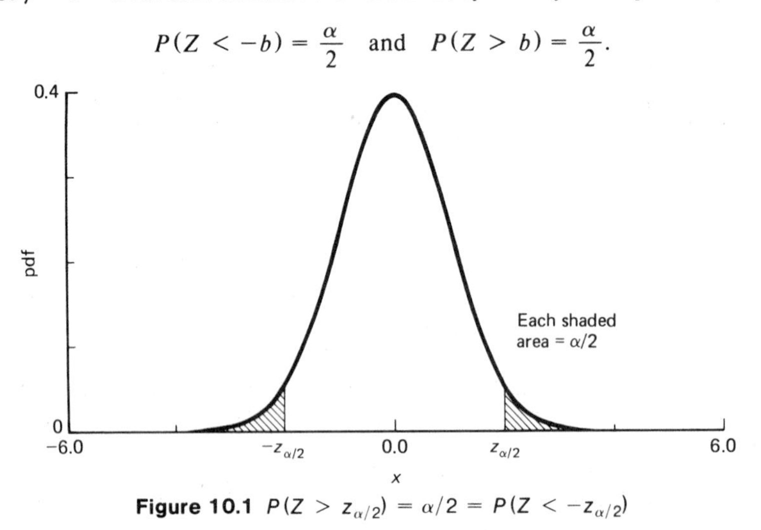

Figure 10.1 $P(Z > z_{\alpha/2}) = \alpha/2 = P(Z < -z_{\alpha/2})$

The above value of b is usually denoted by $z_{\alpha/2}$ (see Figure 10.1), and these values can be read from a table. Now:

$$\left(\bar{x} - z_{\alpha/2}\frac{\sigma}{\sqrt{n}},\ \bar{x} + z_{\alpha/2}\frac{\sigma}{\sqrt{n}}\right)$$

is a $(1 - \alpha)\,100$ percent confidence interval for the population mean μ. Usual values of $(1 - \alpha)$ and corresponding $z_{\alpha/2}$ are shown in Table 10.1.

Table 10.1

$1 - \alpha$	0.90	0.95	0.99
$z_{\alpha/2}$	1.645	1.96	2.576

We have $100(1-\alpha)$ percent confidence that the sample mean $\bar{X}$ deviates from the population mean μ by less than $E = z_{\alpha/2}\sigma/\sqrt{n}$. Then the sample size required in order to produce a symmetrical $100(1-\alpha)$ percent confidence interval of width $2E$ for the population mean is given by:

$$n = \lceil (\frac{z_{\alpha/2}\sigma}{E})^2 \rceil. \qquad \#$$

Example 10.12

The average working-set size X of a program is normally distributed with unknown mean μ and a known variance $\sigma^2 = 81$. The program was executed 36 times and the average working-set size for each run recorded. The sample mean was computed to be 100 page frames. Assuming successive runs of the program are independent, the 95 percent confidence interval for the mean average working-set size is given by:

$$(100 - 1.96 \cdot 9/6,\ 100 + 1.96 \cdot 9/6) = (97.06,\ 102.94). \qquad \#$$

Example 10.13

Suppose we wish to estimate the average CPU service time of a job and we wish to assert with a 99 percent confidence that the estimated value is within less than half a second of the true value. Suppose that past experience suggests that CPU service time is normally distributed with $\sigma^2 = 2.25$ sec^2. Then the required number of random samples is given by:

$$n = \lceil (\frac{2.576 \cdot 1.5}{0.5})^2 \rceil = \lceil 59.722 \rceil = 60. \qquad \#$$

Two difficulties with the interval estimation procedure discussed so far should be noted. First, the assumption that the population is normally distributed does not always hold. In the next few sections we will discuss interval estimation when the population is not normally distributed. Also note that, in practice, the assumption of normality does not pose a problem when the sample size is large, owing to the central limit theorem, which states that the statistic $(\bar{X} - \mu)/(\sigma/\sqrt{n})$ is asymptotically normal (under appropriate conditions).

The second difficulty with the formula for confidence interval given above is that it requires the knowledge of population variance σ^2. If σ^2 is unknown, we may replace it by its estimate s^2 to get an approximate confidence interval for μ:

$$\bar{x} - \frac{z_{\alpha/2}s}{\sqrt{n}} < \mu < \bar{x} + \frac{z_{\alpha/2}s}{\sqrt{n}},$$

which will be a good approximation for large values of n $(n > 30)$.

When the sample size is relatively small, the above approximation is poor. But in this case we can make use of the Student t distribution.

If $\bar{X}$ is the sample mean of a random sample of size n from a normal population having the mean μ and variance σ^2, then by Example 3.35 we have that the random variable:

$$T = \frac{\bar{X} - \mu}{S/\sqrt{n}}$$

has a Student t distribution with $n - 1$ degrees of freedom.

In Figure 10.2 a t distribution with three degrees of freedom is plotted, together with the standard normal distribution for comparison. Thus we obtain the $100(1 - \alpha)$ percent confidence interval of μ as:

$$\bar{x} - t_{n-1;\alpha/2} \frac{s}{\sqrt{n}} < \mu < \bar{x} + t_{n-1;\alpha/2} \frac{s}{\sqrt{n}},$$

where $t_{n-1;\alpha/2}$ is defined such that the area under the t pdf to its right is equal to $\alpha/2$ or $P(T > t_{n-1;\alpha/2}) = \alpha/2$ (Figure 10.3). This value can be read from a table (see Appendix C).

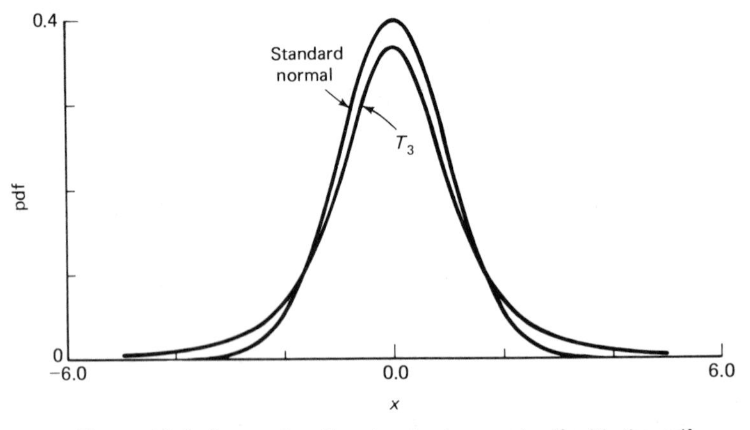

Figure 10.2 Comparing the standard normal pdf with the pdf of the t-distribution with three degrees of freedom

Example 10.14

We wish to estimate the average execution time of a program. The program was run six times with randomly chosen data sets, and the sample mean of the execution times was evaluated as $\bar{x} = 230$ ms and the sample standard deviation as $s = 14$ ms. To obtain a 98 percent confidence interval of the true mean execution time μ, we read $t_{5;0.01}$ from the table of t distribution with $n - 1 = 5$ degrees of freedom to be 3.365. Then the required confidence interval is:

$$230 - \frac{3.365 \cdot 14}{\sqrt{6}} < \mu < 230 + \frac{3.365 \cdot 14}{\sqrt{6}}$$

or

$$210.767 < \mu < 249.233 \qquad \text{(with 98 percent confidence).}$$

#

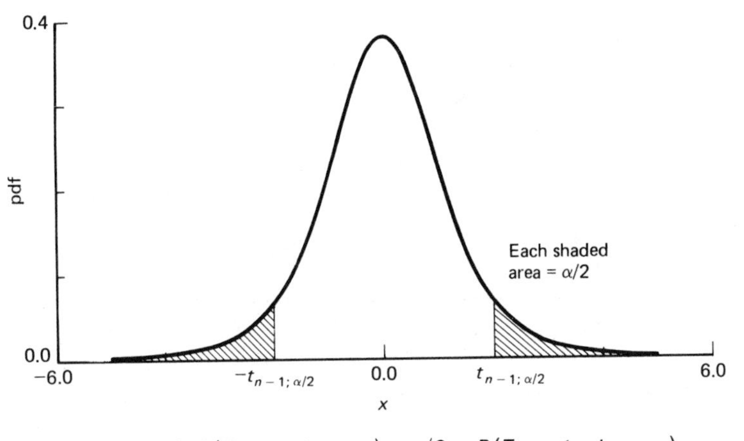

Figure 10.3 $P(T_{n-1} > t_{n-1;\alpha/2}) = \alpha/2 = P(T_{n-1} < -t_{n-1;\alpha/2})$

So far, we have considered confidence intervals for the population mean. Next we discuss confidence intervals for the population variance. If X is normally distributed, then we have shown in Example 3.33 that the random variable:

$$X_{n-1}^2 = \frac{(n-1)S^2}{\sigma^2}$$

possesses a chi-square distribution with $n - 1$ degrees of freedom.

To determine a $100(1-\alpha)$ percent confidence interval of σ^2, we find two numbers a and b such that:

$$P\left[a < \frac{(n-1)\ S^2}{\sigma^2} < b\right] = 1 - \alpha.$$

Since chi-square is a nonnegative random variable, its distribution is obviously not symmetrical about zero. Therefore, in choosing a and b, the requirement of "equal tails" is usually imposed (see Figure 10.4), so that:

$$P(X_{n-1}^2 > b) = \frac{\alpha}{2} \quad \text{and} \quad P(X_{n-1}^2 < a) = \frac{\alpha}{2}.$$

In this case, b and a are denoted by $\chi_{n-1;\alpha/2}^2$ and $\chi_{n-1;1-\alpha/2}^2$, respectively. The $100(1-\alpha)$ percent confidence interval of the population variance is then given by:

$$\frac{(n-1)s^2}{\chi_{n-1;\alpha/2}^2} < \sigma^2 < \frac{(n-1)s^2}{\chi_{n-1;1-\alpha/2}^2}.$$

Note that, like the confidence interval for the population mean μ found using the t distribution, this interval does not require knowledge of any parameters of the population distribution function.

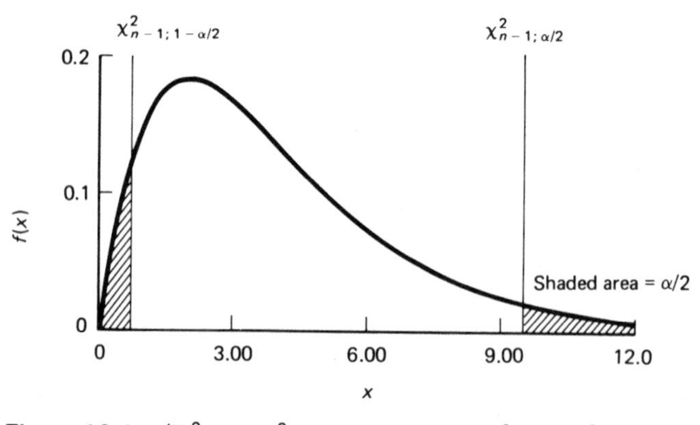

Figure 10.4 $P(X^2_{n-1} > \chi^2_{n-1;\alpha/2}) = \alpha/2 = P(X^2_{n-1} < \chi^2_{n-1;1-\alpha/2})$

Example 10.15

A usual complaint of interactive system users is the large variance of the response time. While contemplating the purchase of a new interactive computer system, we measure 30 random samples of response times, and the sample variance is computed to be 25 ms². Assuming the response times are approximately normally distributed, a 95 percent confidence interval for the population variance σ^2 is given by:

$$\frac{(n-1)s^2}{\chi^2_{n-1;\alpha/2}} < \sigma^2 < \frac{(n-1)s^2}{\chi^2_{n-1;1-\alpha/2}}, \qquad \text{where } \alpha = 0.05.$$

From a table of chi-square distribution with $n - 1 = 29$ degrees of freedom, $\chi^2_{29;.025} = 45.722$ and $\chi^2_{29;0.975} = 16.047$, so that the required confidence interval is obtained from the relation:

$$\frac{29 \cdot 25}{45.722} < \sigma^2 < \frac{29 \cdot 25}{16.047}$$

as (15.86, 45.18). #

The construction of confidence intervals discussed so far was based on the assumption that X is normally distributed. Empirical evidence suggests that the confidence interval for μ based on the normality of the statistic $(\bar{X} - \mu)\sqrt{n}/\sigma$ is highly reliable (in the sense of providing adequate coverage) even when the distribution of X is considerably different from normal. However, the confidence interval of σ^2 derived above can be quite poor when X has a distribution significantly different from normal.

Problems

1. In an exhaustive, nonreplacement life test of ten components, the observed times to failure (in hours) are: 1,200, 1,500, 1,625, 1,725, 1,750, 1,785, 1,800, 1,865, 1,900, and 1,950. Assuming that component lifetimes are normally distributed, compute an estimate of the mean life μ and the variance σ^2. Also compute a 90 percent confidence interval for the mean life.

2. Given the following twenty measurements of the mean length of a CPU queue, compute the best estimates of the mean queue length, variance of the queue length, and a 95 percent confidence interval of the mean queue length. Assume that queue-length distribution is approximately normal.

3.00	2.87	3.58	3.28	3.87
4.14	5.23	3.86	2.88	4.37
4.75	4.33	3.17	2.85	4.16
4.03	3.57	3.68	3.95	3.58

3. A program was tested using a random collection of 30 input data sets, and execution time was measured for each run. The sample mean and the sample variance of the execution time were found to be $\bar{x} = 65$ ms and $s^2 = 36$ ms^2, respectively. Derive a 95 percent confidence interval for the average execution time of the program assuming that the population is normal.

4. Execution times (in seconds) of 40 jobs processed by a computing center were measured and found to be:

10	19	90	40	15	11	32	17	4	152
23	13	36	101	2	14	2	23	34	15
27	1	57	17	3	30	50	4	62	48
9	11	20	13	38	54	46	12	5	26

Calculate the sample mean and the sample variance. Find the 90 percent confidence intervals for the mean and the variance of execution time of a job. Assume that the execution time is approximately normally distributed.

10.2.3.2 Sampling from the Exponential Distribution. Now we consider the special case when X is exponentially distributed. This case is of importance in queuing theory as well as in reliability theory (life testing).

We have noted that under exhaustive testing of n components, the maximum-likelihood estimator of mean life is given by the sample mean; that is:

$$\hat{\Theta} = \bar{X}.$$

Now if X is exponentially distributed with parameter λ, then $\sum_{i=1}^{n} X_i$ is an n-stage Erlang random variable with parameter λ, and $\bar{X} = (\sum_{i=1}^{n} X_i)/n$ is n-stage Erlang with parameter $n\lambda$. This implies that:

$$2n\lambda\bar{X} = \frac{2n\bar{X}}{\theta} = \frac{2n\hat{\Theta}}{\theta}$$

has an n-stage Erlang distribution with parameter 1/2, which is the chi-square distribution with $2n$ degrees of freedom. Then a $100(1 - \alpha)$ percent confidence interval of mean life θ is obtained from:

$$\chi^2_{2n;1-\alpha/2} < \frac{2n\hat{\theta}}{\theta} < \chi^2_{2n;\alpha/2}$$

as

$$\hat{\theta}\,\frac{2n}{\chi^2_{2n;\alpha/2}} < \theta < \hat{\theta}\,\frac{2n}{\chi^2_{2n;1-\alpha/2}}.$$

Next we consider a test (without replacement) terminated after $r \leqslant n$ failures have occurred. Consider the accumulated life on test, $S_{n;r}$:

$$S_{n;r} = \sum_{i=1}^{r} T_i + (n - r)T_r.$$

Let Y_i denote the time from the $(i - 1)$st failure to the ith failure so that:

$$T_i = \sum_{j=1}^{i} Y_j, \qquad i = 1, 2, \ldots, r.$$

Then:

$$\begin{aligned}
S_{n;r} &= \sum_{i=1}^{r} T_i + (n - r)T_r \\
&= \sum_{i=1}^{r}\sum_{j=1}^{i} Y_j + (n - r)\sum_{j=1}^{r} Y_j \\
&= (Y_1) + (Y_1 + Y_2) + (Y_1 + Y_2 + Y_3) + \cdots + (Y_1 + Y_2 + \cdots Y_r) \\
&\quad + (n - r)(Y_1 + Y_2 + \cdots + Y_r) \\
&= \sum_{i=1}^{r} (r - i + 1)Y_i + \sum_{i=1}^{r} (n - r)Y_i \\
&= \sum_{i=1}^{r} (n - i + 1)Y_i.
\end{aligned}$$

Now from our discussion in Example 3.22, Y_i is exponentially distributed with parameter $(n - i + 1)\lambda$, and therefore $(n - i + 1)Y_i$ is exponentially distributed with parameter λ. Therefore $S_{n;r}$ is r-stage Erlang with parameter λ, and hence $2\lambda S_{n;r} = 2S_{n;r}/\theta$ is r-stage Erlang with parameter 1/2,—that is, the χ^2_{2r} distribution. Thus a $100(1 - \alpha)$ percent confidence interval for θ is given by:

$$\frac{2s_{n;r}}{\chi^2_{2r;\alpha/2}} < \theta < \frac{2s_{n;r}}{\chi^2_{2r;1-\alpha/2}}.$$

Example 10.16

Assume that $n = 50$ chips are placed on a life test without replacement and the test is to be truncated after $r = 10$ failures have been observed. Observed failure times are $t_1 = 80$, $t_2 = 95$, $t_3 = 370$, $t_4 = 415$, $t_5 = 505$, $t_6 = 590$, $t_7 = 635$, $t_8 = 835$, $t_9 = 895$, $t_{10} = 960$ hours. Then:

$$s_{n;r} = (80 + 95 + 370 + 415 + 505 + 590 + 635 + 835 + 895 + 960) + (50 - 10)960 = 43,780 \text{ hours.}$$

The estimated mean life is $\hat{\theta} = 43,780/10 = 4{,}378$ hours, and the estimated failure rate is $\hat{\lambda} = 0.0002284$ failures per hour. Finally a 90 percent confidence interval for mean life is:

$$\frac{2(43780)}{31.410} < \theta < \frac{2(43780)}{10.851}$$

or

$$2,787 < \theta < 8,069 \text{ hours,}$$

where $\chi^2_{20;0.05} = 31.410$ and $\chi^2_{20;0.95} = 10.851$ values are obtained from a table of chi-square distribution (see Appendix C) with 20 degrees of freedom. #

Sometimes a one-sided confidence interval is sought in place of the two-sided interval given above. For example, an **upper one-sided** confidence interval of the mean life θ is denoted by (Θ_L, ∞) where Θ_L is known as the **lower confidence limit**. Since $2S_{n;r}/\theta$ is chi-square distributed with $2r$ degrees of freedom, we have:

$$P(\frac{2S_{n;r}}{\theta} < \chi^2_{2r;\alpha}) = 1 - \alpha.$$

It follows that:

$$\frac{2s_{n;r}}{\theta_L} = \chi^2_{2r;\alpha} \quad \text{or} \quad \theta_L = \frac{2s_{n;r}}{\chi^2_{2r;\alpha}}.$$

Similarly a **lower one-sided** confidence interval of the mean life is denoted by $(0, \Theta_U)$, where a value of the **upper confidence limit** Θ_U is given by:

$$\theta_U = \frac{2\ s_{n;r}}{\chi^2_{2r;1-\alpha}}.$$

Example 10.16 (continued).

Returning to Example 10.16, we note that for a chi-square distribution with $2r =$ 20 degrees of freedom, $\chi^2_{2r;\alpha} = \chi^2_{20;0.1} = 28.41$ and $\chi^2_{2r;1-\alpha} = \chi^2_{20;0.9} = 12.443$. It follows that the 90 percent lower confidence limit of the mean life is given by:

$$\theta_L = \frac{2\ s_{n;r}}{\chi^2_{20;0.1}} = \frac{87,560}{28.41} = 3,082 \text{ hours.}$$

Therefore, with 90 percent confidence we can assert that the true mean life is greater than 3,082 hours. The 90 percent upper confidence limit is:

$$\theta_U = \frac{2s_{n;r}}{\chi^2_{20;0.9}} = \frac{87,560}{12.443} = 7,036 \text{ hours.}$$

Therefore, with 90 percent confidence we can assert that the true mean life is less than 7,036 hours. #

For ultrahigh-reliability systems, the mean life may be much larger than the duration of a normal "mission." In this case we are more interested in obtaining a confidence interval for system reliability given a mission time t. We proceed to derive such a confidence interval starting from the $100(1-\alpha)$ percent upper one-sided confidence interval of the mean life θ. Thus:

$$1-\alpha = P(\theta \geqslant \Theta_L)$$
$$= P(e^{-t/\theta} \geqslant e^{-t/\Theta_L})$$

since the exponential is a monotonic function,

$$= P\left[R(t) \geqslant e^{-t/\Theta_L}\right].$$

That is:

$$R_L = e^{-t/\theta_L} = e^{\frac{-(t\chi^2_{2r;\alpha})}{2s_{n;r}}}$$

is a lower $100(1-\alpha)$ percent confidence limit for the reliability, given a mission time t. Note that the chi-square distribution here has $2r$ degrees of freedom, since we are discussing a test, without replacement, truncated after r failures.

In the running example of this section, we have 90 percent confidence that the chip reliability exceeds the threshold $R_L = e^{-t/3082}$. Thus, if we observe a large number of chips for 30.82 hours, we are 90 percent confident that at least $100 \cdot e^{-0.01} = 99$ percent of the chips will still be functioning properly.

Problems

1. Fifteen RAM chips are put into operation, and a truncated nonreplacement life test is conducted until three chips have failed. Corresponding failure times are noted as $t_1 = 850$ hours, $t_2 = 900$ hours, $t_3 = 1000$ hours. Assume that the devices follow an exponential failure law.
   (a) Obtain a point estimate of the mean life.
   (b) Obtain a 90 percent confidence interval for the mean life of a chip.

2. Show that the $100(1-\alpha)$ percent confidence interval for the mission time t_γ such that the reliability for this mission time satisfies:

$$R(t_\gamma) = P(X > t_\gamma) = \gamma$$

is given by:

$$\left[\frac{2s_{n;r}}{\chi^2_{2r;\alpha/2}} \ln\left(\frac{1}{\gamma}\right), \frac{2s_{n;r}}{\chi^2_{2r;1-\alpha/2}} \ln\left(\frac{1}{\gamma}\right)\right].$$

It is assumed that the lifetime X is exponentially distributed with the parameter λ (which is to be estimated from the data), and the remaining assumptions are the same as those made throughout the section.

3. Twenty items are placed on a life test. The first and the second failures occur at 3,001 and 7,030 hours, respectively, after which the test is terminated. Assuming that the lifetimes of items are exponentially distributed, compute:
   (a) The maximum-likelihood estimate of the MTTF.
   (b) The 95 percent confidence interval for the MTTF.
   (c) The 95 percent confidence interval for the length of the mission with the reliability 0.9.

10.2.3.3 Sampling from the Poisson Distribution. Recalling that the interevent times of a Poisson process are exponentially distributed, we can obtain a confidence interval for the average arrival rate. Assume that a Poisson process of rate λ is observed until a fixed number n of events have been counted. Let X_i denote the time between the $i-1$st and ith event. Then X_i is exponentially distributed with parameter λ, and the statistic:

$$S_n = \sum_{i=1}^{n} X_i$$

is n-stage Erlang with parameter λ. It follows that $2\lambda S_n$ is chi-square distributed with $2n$ degrees of freedom. Consequently:

$$\left(\frac{\chi^2_{2n;1-\alpha/2}}{2s_n}, \frac{\chi^2_{2n;\alpha/2}}{2s_n}\right)$$

is a confidence interval for λ, with confidence coefficient $(1 - \alpha)$.

Example 10.17

Arrival of jobs to a computing center was monitored and it was found that 50 jobs arrived within 100 minutes. Assuming a Poisson model, the maximum-likelihood estimate for the average job arrival rate is $\hat{\lambda} = 50/100 = 0.5$ jobs per minute. Noting that $\chi^2_{100;0.05} = 124.34$ and $\chi^2_{100;0.95} = 77.93$, we find the 90 percent confidence interval for λ to be $(0.39, 0.62)$. #

10.2.3.4 Sampling from the Bernoulli Distribution. In many situations we are interested in the percentage of certain components that will perform satisfactorily for a given period. At other times we may be interested in the proportion of terminal requests whose response times do not exceed a threshold. Each individual experimental observation, X_i, can then be treated as a Bernoulli random variable with an unknown parameter p. The statistic:

$$S_n = \sum_{i=1}^{n} X_i$$

is then binomially distributed; that is:

$$F_{S_n}(k) = B(k; n, p).$$

As we have seen earlier, the maximum-likelihood estimator of the *proportion* of successes is given by:

$$\hat{P} = \frac{S_n}{n} = \bar{X}.$$

Since $E[S_n] = np$, we have, $E[\hat{P}] = p$, and $\text{Var}[S_n] = np(1-p)$ implies that $\text{Var}[\hat{P}] = p(1-p)/n$. Thus, the sample proportion, $\hat{P}$, is a consistent and unbiased estimator of the "true" proportion, p.

Since the distribution of the statistic S_n is discrete, it may be impossible to obtain an interval with a degree of confidence exactly equal to $(1-\alpha)$. Therefore we derive a confidence interval for p with an approximate degree of confidence $(1-\alpha)$.

Let k_0 be the largest integer such that:

$$\sum_{k=0}^{k_0} b(k;\, n,\, p) = B(k_0;\, n,\, p) \leqslant \frac{\alpha}{2}$$

and let k_1 be the smallest integer such that:

$$\sum_{k=k_1}^{n} b(k;\, n,\, p) = 1 - B(k_1 - 1;\, n,\, p) \leqslant \frac{\alpha}{2}.$$

Note that k_0 and k_1 are functions of p. Then, since $P[k_0(p) < S_n < k_1(p)] \simeq 1 - \alpha$, an approximate $100(1-\alpha)$ percent confidence interval for p can be obtained by inverting:

$$k_0(p) < s_n < k_1(p).$$

Unfortunately, there are no closed-form expressions for k_0 and k_1 as functions of p. Therefore, a tabular technique is usually employed to obtain the desired confidence interval.

Example 10.18

From a large population of RAM chips, a sample of twenty is taken and a test carried out on each to see whether they perform correctly. In the test seven chips are found to perform correctly and the remaining thirteen do not perform to specifications. Therefore, a point estimate of the yield of these chips is $\hat{p} = 7/20 = 0.35$.

In order to determine a 95 percent confidence interval for p, we have to determine integers k_0 and k_1 such that k_0 is the largest integer satisfying:

$$B(k_0;\, 20,\, p) \leqslant 0.025,$$

while k_1 is the smallest integer satisfying:

$$1 - B(k_1 - 1;\, 20,\, p) \leqslant 0.025.$$

Using the binomial formula, and varying p from 0.1 to 0.9, we obtain the following table of values:

p	0.1	0.2	0.3	0.4	0.5	0.6	0.7	0.8	0.9.
k_0	—	0	1	3	5	7	9	11	14
k_1	6	9	11	13	15	17	19	20	—

Now since $s_n = k = 7$, the interval of p satisfying:

$$k_0(p) < 7 < k_1(p)$$

is (0.133, 0.600) from the above table (together with a little interpolation). This is the 95 percent confidence interval that was sought for p. #

If the sample size n is large or p is not suspected to be very close to either 0 or 1, we can use the normal approximation to the binomial distribution. As a rule of thumb, $np \geqslant 5$ and $nq \geqslant 5$ usually suffices. Thus S_n is approximately normal with $\mu = np$ and $\sigma^2 = np(1-p)$. Note that σ^2 contains the unknown parameter p, but we may approximate it by $\hat{\sigma}^2 = n\hat{p}(1-\hat{p})$. Then:

$$\frac{S_n - np}{\hat{\sigma}}$$

is approximately standard normal. It follows that an approximate $100(1-\alpha)$ percent confidence interval for p is obtained from:

$$-z_{\alpha/2} < \frac{S_n - np}{\hat{\sigma}} < z_{\alpha/2}$$

or from:

$$\frac{s_n}{n} - z_{\alpha/2}\sqrt{\frac{s_n\left(1 - \frac{s_n}{n}\right)}{n^2}} < p < \frac{s_n}{n} + z_{\alpha/2}\sqrt{\frac{s_n\left(1 - \frac{s_n}{n}\right)}{n^2}}.$$

For instance, using this approximation for the data of Example 10.18, and noting that $z_{0.025} = 1.96$, we get a 95 percent confidence interval of p to be:

$$0.35 - 1.96\sqrt{\frac{7(1-0.35)}{400}} < p < 0.35 + 1.96\sqrt{\frac{7(1-0.35)}{400}}.$$

Thus the required interval is (0.141, 0.559), which, considering the assumptions, is in fair agreement with the results obtained in Example 10.18.

In many practical situations of interest the Bernoulli parameter p is either very close to 0 or very close to 1. For example, while estimating the fault coverage of a fault-tolerant computer system we would expect the probability of successful recovery to be close to 1—say > 0.9. On the other hand, in a quality-control inspection plan $\hat{p}$ will represent a fraction of the total inspected items found to be defective. In this case we will expect the population parameter to be close to 0—say < 0.1. In such cases the normal approximation to the binomial distribution will be poor. However, for $p < 0.1$ we may use the Poisson approximation to the binomial, provided that the sample size is large enough. In the complementary case of $p > 0.9$, we obtain a confidence interval for $q = 1 - p$ using the same approach.

In the case $p < 0.1$, we will be interested in a one-sided confidence interval with an approximate confidence coefficient γ so that if k is the observed number of successes, we write:

$$\begin{aligned}\gamma &\leqslant P(S_n \leqslant k)\\ &= \sum_{i=0}^{k} \binom{n}{i} p^i (1-p)^{n-i}\\ &\simeq \sum_{i=0}^{k} e^{-np} \frac{(np)^i}{i!}.\end{aligned}$$

Now by comparison with the distribution function of $(k+1)$-stage Erlang random variable with parameter np, denoted by Y, we see that the right-hand side is equal to $1 - F_Y(1)$. But then $2npY$ is chi-square distributed with $2(k+1)$ degrees of freedom; hence we have:

$$P(X^2_{2(k+1)} > 2np) \geqslant \gamma.$$

It follows that the required confidence interval for p is given by:

$$p < \frac{1}{2n} \chi^2_{2(k+1);\gamma}.$$

For example, if a sample of 50 RAM chips selected at random from a large batch is found to contain nine defectives, an approximate 90 percent confidence interval for the fraction defective in the entire batch is given by:

$$p < \frac{1}{2 \cdot 50} \chi^2_{20;0.9} = 0.284.$$

The point estimate of p in this case is given by:

$$\hat{p} = \frac{9}{50} = 0.18.$$

Problems

1. Assume that CPU activity is being probed and let the ith observation $X_i = 0$ if the CPU is idle and 1 otherwise. Assume that the successive observations are sufficiently separated in time so as to be independent. Assume that X is a Bernoulli random variable with parameter p. Thus the expected utilization is p. We use the sample mean $\overline{X}$ as an estimator $\hat{P}$ of p. Determine an approximate $100(1-\alpha)$ percent confidence interval for p.

   Next assume that we wish to determine the sample size that is required to attain a measurement error at most equal to E with confidence coefficient $1-\alpha$. Then show that:

$$n = \lceil\, [p(1-p)(\frac{z_{\alpha/2}}{E})^2]\, \rceil$$

Since this formula requires a prior knowledge of p (or its estimate), we would like an approximation here. Since for values of p for which $0.3 < p < 0.7$, $p(1 - p)$ is close to 0.25, an approximate sample size for this range of p is obtained as:

$$n = \left\lceil \left[\frac{1}{4} \left(\frac{z_{\alpha/2}}{E} \right)^2 \right] \right\rceil.$$

(This expression is known to give conservative results.)

2. Returning to the text example of inspection of a lot of RAM chips, obtain a 95 percent confidence interval for p using the following two methods and compare with the one obtained in the text using the Poisson approximation:
   (a) Either by consulting an extensive table for the CDF of the binomial distribution or by writing a computer program, obtain a one-sided confidence interval for p using exact binomial probabilities.
   (b) Compute the required one-sided confidence interval using the normal approximation to the binomial.

3. Obtain a distribution-free confidence interval for population median $\pi_{0.5}$ of a continuous population by first ordering the random sample $X_1, X_2, \ldots, X_n$, with the resulting order statistics denoted by $Y_1, Y_2, \ldots, Y_n$. Now the observed values y_i and y_j for $i < j$ can be used to provide a confidence interval (y_i, y_j) of the median $\pi_{0.5}$. Show that the corresponding confidence coefficient is given by:

$$\gamma = \sum_{k=i}^{j-1} \binom{n}{k} \left(\frac{1}{2} \right)^k \left(\frac{1}{2} \right)^{n-k}$$

   Generalize to obtain a confidence interval for population percentile π_p, where π_p is defined by

$$P(X \leqslant \pi_p) = p.$$

4. [GAY 1978] In order to estimate the fault-detection coverage c of a fault-tolerant computer system, 200 random faults were inserted. The recovery mechanism successfully detected 178 of these faults. Determine 95 percent confidence intervals for the coverage c using exact binomial probabilities and using the normal and the Poisson approximations to the binomial.

10.2.4 Estimating Parameters of a Markov Chain

So far, we have restricted our attention to estimating parameters related to the probability distribution of a single random variable X. In this section we will study the estimation of parameters of a Markov chain.

10.2.4.1 Discrete Parameter Markov Chains. First consider a discrete parameter Markov chain with a finite number of states. Let the state space be $\{1, 2, \ldots, m\}$. Assume that the chain is observed for a total of n transitions so that N_{ij} is the number of transitions from state i to state j $(i, j = 1, 2, \ldots, m)$. Let $N_i = \sum_{j=1}^{m} N_{ij}$ be all transitions out of state i, and note that

$n = \sum_{i=1}^{m} N_i$ is a fixed constant while N_i and N_{ij} are random variables. Particular values of these random variables are denoted by n_i and n_{ij} respectively. From these observations, we wish to estimate the m^2 elements of the transition probability matrix $P = [p_{ij}]$. It can be shown that the maximum-likelihood estimator, $\hat{P}_{ij}$, of p_{ij}, is given by [BHAT 1974]:

$$\hat{P}_{ij} = \frac{N_{ij}}{N_i}.$$

Example 10.19

Consider the CPU of a computer system modeled as a three-state discrete-parameter Markov chain. The states are indexed 1, 2, and 3 and respectively denote the CPU in supervisor state, user state, and idle state. We record the states of the CPU at 21 successive time instants, and the recorded sequence is:

1 2 3 3 2 1 1 2 2 3 2

3 1 3 2 3 1 2 3 1 2

With 21 observations, the number of transitions, n, is 20. From the data, we derive the values of n_{ij} and n_i to be:

n_{ij}	1	2	3	
1	1	4	1	6
2	1	1	5	7
3	3	3	1	7
				20

Thus, the maximum-likelihood estimate of the transition probability matrix P for the CPU (modeled as a Markov chain) is given by:

$$\hat{P} = \begin{bmatrix} 1/6 & 2/3 & 1/6 \\ 1/7 & 1/7 & 5/7 \\ 3/7 & 3/7 & 1/7 \end{bmatrix}.$$

#

A confidence interval for p_{ij} may be obtained by assuming that a fixed number n_i of transitions out of state i have been observed, out of which a random number N_{ij} of transitions are to state j. Owing to the assumptions of a Markov chain, N_{ij} is then binomially distributed so that N_{ij} is $B(k;\, n_i,\, p_{ij})$. Now the methods of Section 10.2.3.4 can be used to derive the confidence interval for p_{ij}. Thus, if n_i is sufficiently large and if p_{ij} is not close to 0 or 1, an approximate $100(1-\alpha)$ percent confidence interval for p_{ij} is given by:

$$\left[\frac{N_{ij}}{n_i} - z_{\alpha/2}\frac{\sqrt{N_{ij}(1 - N_{ij}/n_i)}}{n_i},\ \frac{N_{ij}}{n_i} + z_{\alpha/2}\frac{\sqrt{N_{ij}(1 - N_{ij}/n_i)}}{n_i}\right].$$

10.2.4.2 Estimating Parameters of an *M/M/1* Queue. Next consider estimating parameters of a continuous-parameter Markov chain, such as a simple birth-death process with constant birth rate λ and constant death rate μ. This corresponds to an $M/M/1$ queue with an arrival rate λ and a service rate μ. To estimate the average arrival rate λ, we observe that the arrival process is Poisson and therefore the method of Section 10.2.3.3 is applicable. If S_n is the time required to observe n arrivals, then the maximum-likelihood estimator of λ is:

$$\hat{\Lambda} = \frac{n}{S_n}.$$

Also, since $2\lambda S_n$ has a chi-square distribution with $2n$ degrees of freedom, a $100(1-\alpha)$ percent confidence interval for λ is given by:

$$\left(\frac{\chi^2_{2n;1-\alpha/2}}{2s_n}, \frac{\chi^2_{2n;\alpha/2}}{2s_n}\right).$$

In order to estimate the average service rate μ, we note that, owing to our assumptions, the service times are independent exponentially distributed random variables, hence the method of Section 10.2.3.2 is applicable. Thus, if service times $X_1, X_2, \ldots, X_m$ have been observed, and if we let:

$$Y_m = \sum_{i=1}^{m} X_i,$$

the maximum-likelihood estimator of μ is given by:

$$\hat{M} = \frac{m}{Y_m}.$$

Y_m may also be interpreted as the total busy time of the server during the observation period. Noting that $2\mu Y_m$ has a χ^2_{2m} distribution, we get a $100(1-\alpha)$ percent confidence interval for μ as:

$$\left(\frac{\chi^2_{2m;1-\alpha/2}}{2y_m}, \frac{\chi^2_{2m;\alpha/2}}{2y_m}\right).$$

Server utilization ρ is now estimated by:

$$\hat{R} = \frac{\hat{\Lambda}}{\hat{M}} = \frac{n/S_n}{m/Y_m} = \frac{Y_m/m}{S_n/n}.$$

To obtain confidence intervals for ρ, we use the ratio:

$$\frac{\hat{R}}{\rho} = \frac{(Y_m/m)/(S_n/n)}{\lambda/\mu} = \frac{2\mu Y_m/2m}{2\lambda S_n/2n}.$$

Now since Y_m and S_n are independent, and $2\mu Y_m$ and $2\lambda S_n$ are chi-square distributed, it follows that $\hat{R}/\rho$ has an F distribution with $2m$ and $2n$ degrees of freedom (see Theorem 3.9). To obtain a $100(1-\alpha)$ percent confidence interval for ρ we write for some constants c and d:

$$1-\alpha = P(c < \mathrm{F} < d) = P\left(c < \frac{\hat{R}}{\rho} < d\right).$$

Select c and d so that $P(\mathrm{F} \leqslant c) = \alpha/2$ and $P(\mathrm{F} < d) = 1-\alpha/2$. Then by our usual notation, $c = f_{2m,2n;1-\alpha/2}$ and $d = f_{2m,2n;\alpha/2}$ so the required confidence interval for ρ is given by (ρ_L, ρ_U) where:

$$\rho_L = \frac{\hat{\rho}}{f_{2m,2n;\alpha/2}} \quad \text{and} \quad \rho_U = \frac{\hat{\rho}}{f_{2m,2n;1-\alpha/2}}.$$

From the confidence interval of ρ, we can obtain a confidence interval for any monotonically increasing function, $H(\rho)$, of ρ. Thus:

$$(H(\rho_L),\ H(\rho_U))$$

is the $100(1-\alpha)$ percent confidence interval for $H(\rho)$.

Example 10.20

Assume that a communication channel can be modeled as an $M/M/1$ queue. Suppose we observe the time until 30 message arrivals to be 59.46 minutes and these 30 messages keep the channel busy for a total of 29 minutes. Thus, $m = n = 30$, $s_n = 59.46$ minutes, and $y_m = 29$ minutes. Point estimates of the average arrival rate, the average service rate, and the average channel utilization are given by:

$$\hat{\lambda} = \frac{n}{s_n} = \frac{30}{59.46} = 0.505 \text{ messages per minute,}$$

$$\hat{\mu} = \frac{m}{y_m} = \frac{30}{29} = 1.03 \text{ messages per minute,}$$

$$\hat{\rho} = \frac{\hat{\lambda}}{\hat{\mu}} = 0.488.$$

To obtain 95 percent confidence intervals for λ and μ, we use a chi-square distribution with $2m,\ 2n = 60$ degrees of freedom. Noting that:

$$\chi^2_{60;0.025} = 83.3 \quad \text{and} \quad \chi^2_{60;0.975} = 40.48,$$

the required confidence interval for λ is $(0.34, 0.7)$ and that for μ is $(0.698, 1.436)$.

To obtain a confidence interval for ρ, we use an F distribution with (60, 60) degrees of freedom. Noting that:

$$f_{60,60;\alpha/2} = f_{60,60;0.025} = 1.67$$

and

$$f_{60,60;1-\alpha/2} = f_{60,60;0.975} = 0.5988,$$

we obtain:

$$\rho_L = \frac{\hat{\rho}}{f_{60,60;0.025}} = 0.292 \quad \text{and} \quad \rho_U = \frac{\hat{\rho}}{f_{60,60;0.975}} = 0.815.$$

Thus, the 95 percent confidence interval for ρ is (0.292, 0.815).

From the confidence interval for ρ we can obtain a confidence interval for the average number of messages queued or in service, $E[N] = \rho/(1-\rho)$. Thus, a 95 percent confidence interval for $E[N]$ is given by:

$$\left(\frac{0.292}{1-0.292}, \frac{0.815}{1-0.815}\right) \quad \text{or} \quad (0.412,\ 4.405).$$

Much more data must be collected if we want to narrow down this interval. #

Problems

1. For an $M/M/1$ queue, 955 arrivals were observed in a period of 1,000 time units and the server was found to be busy for 660 time units. Compute 95 percent confidence intervals for the following quantities:
   (a) The average arrival rate λ.
   (b) The average service time $1/\mu$.
   (c) The server utilization ρ.
   (d) The average queue length $E[N]$.
   (e) The average response time $E[R]$.

2. Give an argument for determining a $100(1-\alpha)$ percent confidence interval for the server utilization ρ as:

$$\rho_L \leqslant \rho \leqslant \rho_U$$

   in the following cases:
   (a) $M/E_k/1$ queuing system:

$$\rho_L = \frac{\hat{\rho}}{f_{2mk,\,2n;\alpha/2}} \quad \text{and} \quad \rho_U = \frac{\hat{\rho}}{f_{2mk,\,2n;1-\alpha/2}}.$$

   (b) $E_k/M/1$ queuing system:

$$\rho_L = \frac{\hat{\rho}}{f_{2m,\,2nk;\alpha/2}} \quad \text{and} \quad \rho_U = \frac{\hat{\rho}}{f_{2m,\,2nk;1-\alpha/2}}.$$

3. We have noted (in Chapter 8) that in an $M/M/1$ queue the response time is exponentially distributed. Given a sample of observations of response times for n successive jobs, can we use methods of Section 10.2.3.2 to obtain confidence intervals for the parameter δ $[=\mu(1-\rho)]$ of the response-time distribution?

10.2.5 Estimation with Dependent Samples

So far, we have assumed that the sample random variables $X_1, X_2, \ldots, X_n$ are mutually independent. Measurements obtained from real systems, however, often exhibit dependencies. For example, there is a high correlation between the response times of consecutive requests in an interactive computer system. Although observations taken from such a system do not satisfy the definition of a random sample, the behavior of the system can be modeled as a stochastic process, and the observations made are then a portion of one particular realization of the process. If the stochastic process is a Markov chain, then, by noting special properties of such a process, we can make use of the methods discussed in Section 10.2.4. We consider the more general case here.

Consider a discrete-parameter stochastic process (or a stochastic sequence) $\{X_i \mid i = 1, 2, \ldots\}$ (the treatment can also be generalized to the case of a continuous-parameter stochastic process). We observe the sequence for n time units to obtain the values $x_1, x_2, \ldots, x_n$. The observed quantities are values of dependent random variables $X_1, X_2, \ldots, X_n$. Assume that the process has an index-invariant mean, $\mu = E[X_i]$, and an index-invariant variance, $\sigma^2 = \text{Var}[X_i]$.

As before, the sample mean:

$$\bar{X} = \sum_{i=1}^{n} \frac{X_i}{n}$$

is a consistent unbiased point estimator of the population mean. However, derivation of confidence intervals for μ poses a problem, since the variance of $\bar{X}$ is not σ^2/n any longer. Assume that the sequence $\{X_i\}$ is wide-sense stationary, so that the autocovariance function:

$$K_{j-i} = E[(X_i - \mu)(X_j - \mu)] = \text{Cov}(X_i, X_j)$$

is finite and is a function only of $|i - j|$. Then the variance of the sample mean is given by:

$$\begin{aligned}\text{Var}[\bar{X}] &= \frac{1}{n^2}\{\sum_{i=1}^{n} \text{Var}[X_i] + \sum_{\substack{i,j=1\\ i\neq j}}^{n} \text{Cov}(X_i, X_j)\} \\ &= \frac{\sigma^2}{n} + \frac{2}{n}\sum_{j=1}^{n-1}\left(1 - \frac{j}{n}\right) K_j.\end{aligned}$$

As n approaches infinity:

$$\lim_{n\to\infty} n\text{Var}[\bar{X}] = \sigma^2 + 2\sum_{j=1}^{\infty} K_j = \sigma^2 a, \qquad \text{where } a = 1 + 2\sum_{j=1}^{\infty} \frac{K_j}{\sigma^2}.$$

It can be shown under rather general conditions that the statistic:

$$\frac{\bar{X} - \mu}{\sigma\sqrt{a/n}}$$

of the correlated data approaches the standard normal distribution as n approaches infinity. Therefore, an approximate $100(1 - \alpha)$ percent confidence interval for μ is given by:

$$\bar{X} \pm \sigma z_{\alpha/2}\sqrt{\frac{a}{n}}.$$

It is for this reason that the quantity n/a is called the **effective** size of the independent samples when the correlated sample size is n. The quantity $\sigma^2 a$ is an unknown, however, and it must be estimated from the observed data.

The need to estimate $\sigma^2 a$ can be avoided by using the method of **independent replications**. (For other methods and additional details see [FISH 1978, KLEI 1974].) We replicate the experiment m times, with each experiment containing n observations. If the initial state of the stochastic sequence is chosen randomly in each of the m experiments, then the results of the experiments will be independent although n observations in a single experiment are dependent.

Let the ith observation in the jth experiment be the value $x_i(j)$ of a random variable $X_i(j)$. Let the sample mean and the sample variance of the jth experiment be denoted by $\bar{X}(j)$ and $S^2(j)$, respectively, where:

$$\bar{X}(j) = \frac{1}{n} \sum_{i=1}^{n} X_i(j)$$

and

$$S^2(j) = \frac{1}{n-1} \left(\sum_{i=1}^{n} [X_i(j) - \bar{X}(j)]^2\right).$$

From the individual sample means, we obtain a point estimator of the population mean μ to be:

$$\bar{X} = \frac{1}{m} \sum_{j=1}^{m} \bar{X}(j) = \frac{1}{mn} \sum_{j=1}^{m} \sum_{i=1}^{n} X_i(j).$$

Note that $\bar{X}(1), \bar{X}(2), \ldots, \bar{X}(m)$ are m independent and identically distributed random variables (hence they define a random sample of size m). Let the common variance of $\bar{X}(j)$ be denoted by v^2. The variance v^2 can be estimated as:

$$V^2 = \frac{1}{m-1} \sum_{j=1}^{m} [\bar{X}(j) - \bar{X}]^2 = \frac{1}{m-1} \sum_{j=1}^{m} \bar{X}^2(j) - \frac{m}{m-1} \bar{X}^2.$$

Since an estimate of the variance is used, the statistic $(\bar{X} - \mu)\sqrt{m}/V$ is approximately t-distributed with $(m-1)$ degrees of freedom. Therefore, a $100(1-\alpha)$ percent confidence interval for μ is given by:

$$\bar{x} \pm \frac{t_{m-1;\alpha/2}\, v}{\sqrt{m}}.$$

Example 10.21

We are interested in estimating the average response time of a time-sharing system. For this purpose sixteen independent experiments are conducted, with each experiment measuring twenty successive response times. The following data are recorded.

j	$\bar{x}(j)$ (seconds)	$\bar{x}^2(j)$
1	0.52	0.2704
2	1.03	1.0609
3	0.41	0.1681
4	0.62	0.3844
5	0.55	0.3025
6	0.43	0.1849
7	0.92	0.8464
8	0.88	0.7744
9	0.67	0.4489
10	0.29	0.0841
11	0.87	0.7569
12	0.72	0.5184
13	0.61	0.3721
14	0.45	0.2025
15	0.98	0.9604
16	0.89	0.7921
		8.1274

The point estimate of the average response time is:

$$\bar{x} = \frac{1}{16} \sum_{j=1}^{16} \bar{x}(j) = \frac{10.84}{16} = 0.6775 \text{ second.}$$

Also:

$$\begin{aligned} v^2 &= \frac{1}{15} \sum_{j=1}^{16} \bar{x}^2(j) - \frac{16}{15}(0.6775)^2 \\ &= 0.5418 - 0.4896 \\ &= 0.052. \end{aligned}$$

Now for the t distribution with fifteen degrees of freedom we have:

$$t_{15;0.025} = 2.131.$$

Therefore:

$$\bar{x} \pm \frac{2.131v}{\sqrt{m}} = 0.6775 \pm 2.131 \cdot \sqrt{\frac{0.052}{16}}$$

or (0.556, 0.799) is a 95 percent confidence interval for the average response time. #

Problems

1. The sample mean and sample variance of response time for ten sets of 1,000 jobs were measured. For the first set, the total CPU busy time and total time of completion were also measured.

Sample number	Mean response time for 1,000 jobs (sec)	Sample variance of response time (sec^2)	Total time of completion (sec)	CPU busy time (sec)
1	1.8	1.5	1010	640
2	1.6	1.4		
3	1.83	2.18		
4	1.37	0.65		
5	1.67	1.52		
6	1.62	1.59		
7	1.84	2.10		
8	1.52	0.92		
9	1.59	1.01		
10	1.73	1.30		

Applying the method of Section 10.2.4.2, and assuming that the system is $M/M/1$, derive a 90 percent confidence interval for the average response time using the first sample. Next, using all ten samples and the method of this section, obtain a 90 percent confidence interval for the average response time.

10.3 HYPOTHESIS TESTING

Many practical problems require us to make decisions about populations on the basis of limited information contained in a sample. For instance, a computer center manager may have to decide whether or not to upgrade the capacity of his installation. The choice is binary in nature—either an upgrade takes place or it does not. In order to arrive at a decision, we often make an assumption or guess about the nature of the underlying population. Such an assertion, which may or may not be valid, is called a **statistical hypothesis**—a statement about one or more probability distributions associated with the population.

Procedures that enable us to decide whether to reject or accept hypotheses, based on the information contained in a sample, are called **statistical tests**. We typically form a **null hypothesis**, H_0, which is a claim (about a probability distribution) that we are interested in rejecting or refuting. The contradictory hypothesis is called the **alternative hypothesis**, H_1.

For example, based on experimental evidence, we may be interested in testing the hypothesis (H_0) that MTTF of a certain system exceeds a threshold value θ_0 hours. The alternate hypothesis may be MTTF $< \theta_0$. Similarly, we may be interested in testing the hypothesis that the job arrival rate λ for a certain computer system satisfies $\lambda = \lambda_0$.

The experimental evidence, upon which the test is based, will consist of a random sample $X_1, X_2, \ldots, X_n$, of size n as in the parameter estimation problem. Since X_i is a random variable for each i, the totality of all n-tuples will span the Euclidean n-space, $\mathbb{R}^n$. The hypothesis testing procedure consists of dividing the n-space of observations into two regions, $R(H_0)$ and $R(H_1)$. If the observed vector $(x_1, x_2, \ldots, x_n)$ lies in $R(H_1)$, we reject the null hypothesis H_0. On the other hand, if the observed n-tuple lies in $R(H_0)$, we fail to reject H_0. The region $R(H_0)$ is known as the **acceptance region** and the region $R(H_1)$ as the **critical** or the **rejection region**.

The possibility always exists that the null hypothesis is true but the sample lies in the rejection region, leading us to reject H_0. This is known as the **type I error**. The corresponding probability is denoted by α and is known as the **level of significance** of the test. Similarly, if the null hypothesis is false and the sample lies in the acceptance region, leading us to a rejection of H_1, a **type II error** is committed. The probability of type II error is denoted by β, and $1 - \beta$ is known as the **power** of the test. When we say that P(type I error) $= \alpha$ and P(type II error) $= \beta$, we mean that if a test is performed a large number of times, α proportion of the time we will reject H_0 when it is true, and β proportion of the time we will fail to reject H_0, when in fact it is false.

An error of type I or type II leads to a wrong decision, so we must attempt to minimize these errors. If we fix the sample size n, a decrease in one type of error leads to an increase in the other type. One can also associate a cost (or a penalty) with a wrong decision and minimize the total cost of the decision. The only way to simultaneously reduce both types of error is to increase the sample size n.

A hypothesis is said to be **simple** if all the parameters in the test are specified exactly. Thus, for example, a test of the form H_0: $\lambda = \lambda_0$ versus H_1: $\lambda = \lambda_1$ is a test concerning two simple hypotheses. A hypothesis such as H_0: $\lambda \in (\lambda_L, \lambda_U)$ is a **composite** hypothesis.

10.3.1 Tests on the Population Mean

Hypothesis testing is closely related to the procedure of interval estimation. Assume that we wish to test a simple hypothesis H_0: $\theta = \theta_0$ regarding a parameter θ of the population distribution based on a random sample $X_1, X_2, \ldots, X_n$. The following steps can be used for this purpose:

1. Find a random variable (called a test statistic) that is a function of $X_1, X_2, \ldots, X_n$:

$$T = T(X_1, X_2, \ldots, X_n; \theta),$$

such that the distribution of T is known.

2. Choose an interval (a, b) such that:

$$P[T \notin (a, b) | H_0 \text{ is true}] = \alpha.$$

Note that then:

$$P[a < T < b) | H_0 \text{ is true}] = 1 - \alpha:$$

that is,

$$P[a < T(X_1, X_2, \ldots, X_n; \theta_0) < b] = 1 - \alpha.$$

3. The actual test is then: take a sample $x_1, x_2, \ldots, x_n$ and compute $t = T(x_1, x_2, \ldots, x_{n;} \theta_0)$; if $t \notin (a, b)$, reject H_0 in favor of H_1; otherwise fail to reject H_0.

The implication is that if H_0 is true, we have $100(1 - \alpha)$ percent confidence that the observed value of the test statistic will lie in the interval (a, b). If the observed value lies outside this interval, we know that such an event could happen with probability α (given H_0 is true). In this case, we conclude that the observations differ **significantly** (at the level of significance α) from what would be expected if H_0 were true, and we are inclined to reject H_0.

If α is prespecified, then a and b can be determined so that the above relation is satisfied. Alternatively, if a and b are specified, then α can be determined from the above equation.

The similarity of the above procedure and the procedure described earlier for obtaining confidence intervals should be noted. Specifically, let θ be an unknown parameter of the population distribution. Let (a, b) be a confidence interval for parameter θ with confidence coefficient γ. Now, while testing the hypothesis H_0: $\theta = \theta_0$, if we accept H_0 whenever $\theta \in (a, b)$ and reject it otherwise, then the significance level α of this test is related to the above confidence coefficient by $\alpha = 1 - \gamma$.

Assume that we wish to test a hypothesis regarding the population mean μ based on a random sample of size n taken from a normal population with a known variance σ^2:

$$H_0: \mu = \mu_0.$$

A required statistic is easily obtained, for if $\bar{X}$ is the sample mean, then:

$$Z = \frac{\bar{X} - \mu}{\sigma/\sqrt{n}}$$

is known to be standard normal. Also let:

$$Z_0 = \frac{\bar{X} - \mu_0}{\sigma/\sqrt{n}}.$$

Assume that the alternative hypothesis is:

$$H_1: \mu \neq \mu_0.$$

Since the test statistic is symmetric about zero, we choose $a = -b$. The acceptance region in terms of the test statistic will then be $(-b, b)$. As a result the type I error-probability is specified by:

$$\alpha = 1 - P(-b < Z < b \mid \mu = \mu_0)$$

or

$$P(-b < Z_0 < b) = 1 - \alpha.$$

But this implies that $b = z_{\alpha/2}$. Thus, the acceptance region for a level of significance α is given by:

$$\{(X_1, X_2, \ldots, X_n) \in \mathbb{R}^n \mid -z_{\alpha/2} < Z_0 < z_{\alpha/2}\}.$$

which will be abbreviated as (see Figure 10.5):

$$-z_{\alpha/2} < Z_0 < z_{\alpha/2}$$

or in terms of the sample mean:

$$\mu_0 - \frac{z_{\alpha/2}\sigma}{\sqrt{n}} < \bar{X} < \mu_0 + \frac{z_{\alpha/2}\sigma}{\sqrt{n}}.$$

The corresponding rejection (or critical) region is:

$$\{(X_1, X_2, \ldots, X_n) \in \mathbb{R}^n \mid |\bar{X} - \mu_0| > \frac{z_{\alpha/2}\sigma}{\sqrt{n}}\},$$

abbreviated as:

$$|\bar{X} - \mu_0| > \frac{z_{\alpha/2}\sigma}{\sqrt{n}}.$$

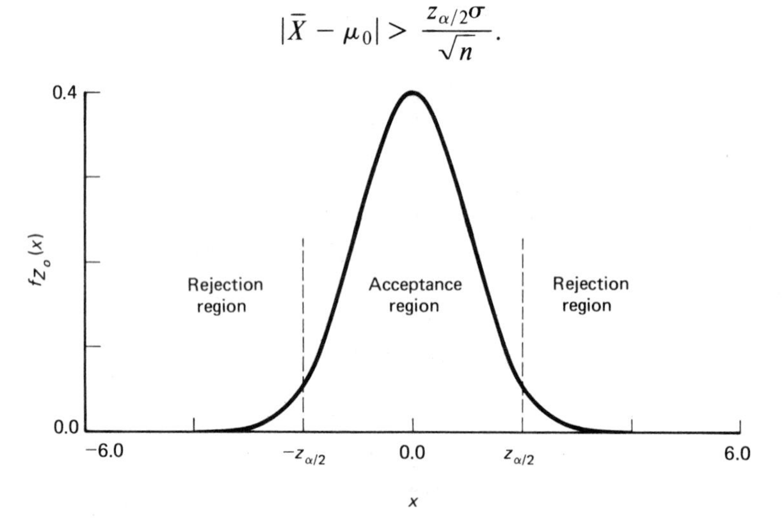

Figure 10.5 Acceptance and rejection regions for a two-sided test

If the alternative hypothesis is of the form:

$$H_1: \mu < \mu_0,$$

then we adopt an asymmetric acceptance region (such tests are known as **one-tailed** or **one-sided tests**) (see Figure 10.6):

$$Z_0 > -z_\alpha \quad \text{or} \quad \bar{X} > \mu_0 - \frac{z_\alpha \sigma}{\sqrt{n}},$$

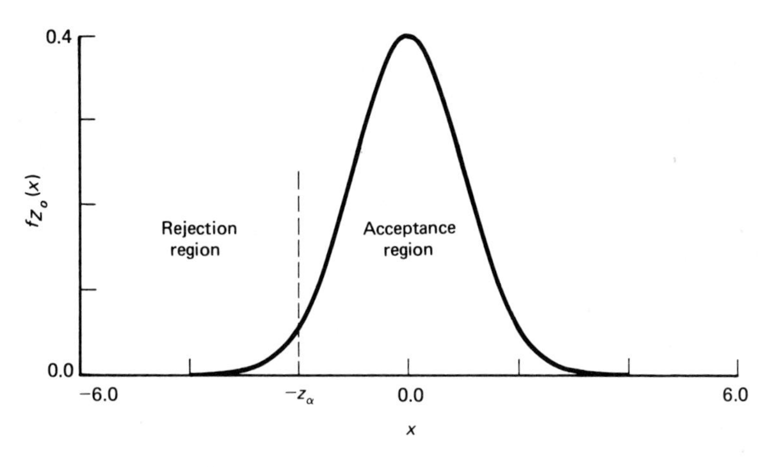

Figure 10.6 Acceptance and rejection regions for a one-sided test

and the rejection region is $\bar{X} < \mu_0 - z_\alpha \sigma/\sqrt{n}$. Similarly, if the alternative hypothesis is:

$$H_1\colon \mu > \mu_0,$$

then the rejection region is:

$$\bar{X} > \mu_0 + \frac{z_\alpha \sigma}{\sqrt{n}}.$$

Example 10.22

A program's average working-set size was found to be $\mu_0 = 50$ pages with a variance of $\sigma^2 = 900$ pages^2. A reorganization of the program's address space was suspected to have improved its locality and hence decreased its average working-set size. In order to judge the locality-improvement procedure, we test the hypothesis:

$$H_0\colon \mu = \mu_0 \quad \text{versus} \quad H_1\colon \mu < \mu_0.$$

Now since $z_\alpha = z_{0.05} = 1.645$, we have that at 5 percent level of significance, the critical region is:

$$\bar{X} < \mu_0 - \frac{\sigma}{\sqrt{n}} z_\alpha = 50 - \frac{30}{\sqrt{n}} 1.645$$

or

$$\bar{X} < 50 - \frac{49.35}{\sqrt{n}}.$$

Thus if 100 samples of the "improved" version of the program's working-set size were taken and the sample average found to be less than 45 pages, we would have reason to believe that the reorganization indeed improved program locality. #

Instead of fixing the significance level at a value α, we may be interested in computing the probability of getting a result as extreme as, or more extreme than, the observed result under the null hypothesis. Such a probability is known as the *descriptive level* (also called the *P*-value) of the test.

Definition (Descriptive Level). The **descriptive level** of a test H_0 is the smallest level of significance α at which the observed test result would be declared significant—that is, would be declared indicative of rejection of H_0.

For instance, in Example 10.22 above, the descriptive level δ for an observed sample mean $\bar{x}$ is given by:

$$\delta = P(\bar{X} \leqslant \bar{x} \mid H_0) = F_{\bar{X}}(\bar{x} \mid H_0) = F_{Z_0}\left[\frac{\bar{x} - \mu_0}{\sigma/\sqrt{n}}\right].$$

Hence for $n = 100$ and $\bar{x} = 45$:

$$\delta = F_{Z_0}\left[\frac{45 - 50}{30/10}\right] = 1 - F_{Z_0}\left[\frac{5}{3}\right] = 0.0475.$$

If the observed value of $\bar{x} = 40$, then:

$$\delta = F_{Z_0}\left[\frac{40 - 50}{3}\right] = 1 - F_{Z_0}(3.33) = 0.00045.$$

Thus if H_0 holds, observation $\bar{x} = 40$ is an extremely unlikely event, and we will be inclined to reject H_0. On the other hand, if the observed value of $\bar{x} = 47$, then:

$$\delta = F_{Z_0}\left[\frac{47 - 50}{3}\right] = F_{Z_0}(-1) = 1 - F_{Z_0}(1) = 0.1587.$$

The corresponding event can occur with about one chance in six under H_0 and in this case we are likely not to reject H_0.

The assumption that the population variance σ^2 is known is very unrealistic. We can use the sample variance S^2 in place of σ^2 and derive a critical region, using the fact that the statistic:

$$\frac{\bar{X} - \mu}{S/\sqrt{n}}$$

possesses the t distribution with $n - 1$ degrees of freedom. Thus, in Example 10.22, if the observed sample variance is $s^2 = 900$, then the critical region for the test H_0: $\mu = \mu_0$ versus H_1: $\mu < \mu_0$ is obtained as:

$$\bar{X} < \mu_0 - \frac{s}{\sqrt{n}}\, t_{n-1;\alpha} = 50 - \frac{30}{\sqrt{n}}\, t_{n-1;\alpha}.$$

Since from t tables, $t_{15;0.05} = 1.753$, the critical region for $n = 16$ and $\alpha = 0.05$ is:

$$\bar{X} < 50 - \frac{52.59}{\sqrt{n}},$$

while for $n = 30$ and $\alpha = 0.05$ the critical region is:

$$\bar{X} < 50 - \frac{30 \cdot 1.699}{\sqrt{n}} = 50 - \frac{49.97}{\sqrt{n}}.$$

For $n \geqslant 30$, the use of the t distribution will give nearly the same results as those obtained by using the standard normal distribution.

If X does not have a normal distribution, but the sample size is sufficiently large, the above procedure can be used to obtain an approximate critical region.

Example 10.23

Assume that we are interested in statistically testing the hypothesis that a given combinational circuit is functioning properly. Prior testing with a properly functioning circuit has shown that if its inputs are uniformly distributed over their range of values, then the probability of observing a 1 at the output is p_0. Thus we drive the given circuit with a sequence of n randomly chosen input sets and test the hypothesis H_0: $p = p_0$ versus H_1: $p \neq p_0$. The test statistic X is the number of observed 1's at the output in a sample of n clock ticks. Let the critical region for the test be $|X - np_0| > n\,\epsilon$. The quantity ϵ is known as the **test stringency** and the interval $(np_0 - n\epsilon,\ np_0 + n\epsilon)$ is the acceptance region. Assuming H_0 is true, X has a binomial distribution with parameters n and p_0. If n is large, and if p_0 is not close to either 0 or 1, we can use the normal approximation to the binomial distribution with mean $\mu_0 = np_0$ and variance $\sigma^2 = np_0(1 - p_0)$. Then:

$$Z_0 = \frac{X - np_0}{\sqrt{np_0(1 - p_0)}}$$

has a standard normal distribution. Thus an approximate acceptance region for a level of significance α is given by:

$$-z_{\alpha/2} < Z_0 < z_{\alpha/2}$$

or

$$np_0 - z_{\alpha/2}\sqrt{np_0(1 - p_0)} < X < n\,p_0 + z_{\alpha/2}\sqrt{np_0(1 - p_0)}.$$

Thus the test stringency is derived as:

$$\epsilon \simeq \frac{1}{n}\, z_{\alpha/2}\sqrt{np_0(1 - p_0)}.$$

Since a type I error in this circuit-testing situation implies that a properly functioning circuit is declared defective, this type of error is known as a **false alarm** in this connection. For a given test stringency ϵ, the type I error:

$$\alpha = P(X \leqslant np_0 - n\epsilon \text{ or } X \geqslant np_0 + n\epsilon \text{ given } H_0)$$

$$\simeq P\left(|Z_0| \geqslant \frac{n\epsilon}{\sqrt{np_0(1 - p_0)}}\right)$$

using the normal approximation,

$$= 2P\left(Z_0 \geqslant \frac{n\epsilon}{\sqrt{np_0(1 - p_0)}}\right)$$

by symmetry of the normal distribution. Now, since $p_0(1 - p_0) \leqslant 1/4$ for all $0 \leqslant p_0 \leqslant 1$, we have:

$$\alpha \leqslant 2P(Z_0 \geqslant 2\epsilon\sqrt{n}) = 2\left[1 - F_{Z_0}(2\epsilon\sqrt{n})\right].$$

This bound will be close to the actual value of α for $0.3 \leqslant p_0 \leqslant 0.7$. Thus the bound on the probability of declaring a fault-free circuit defective depends only upon the test stringency ϵ and the sample size n. It is clear that α decreases as the test stringency ϵ is increased and as the sample size is increased (Figure 10.7). However, as we will see later on, an increase in ϵ will imply an increase in the probability of type II error (also called the **probability of escape** in this case). #

The inverse relationship between the type I and the type II errors can be demonstrated by devising a simple test. Suppose we always accept the null hypothesis H_0: $\mu = \mu_0$, no matter what the outcome of sampling may be.

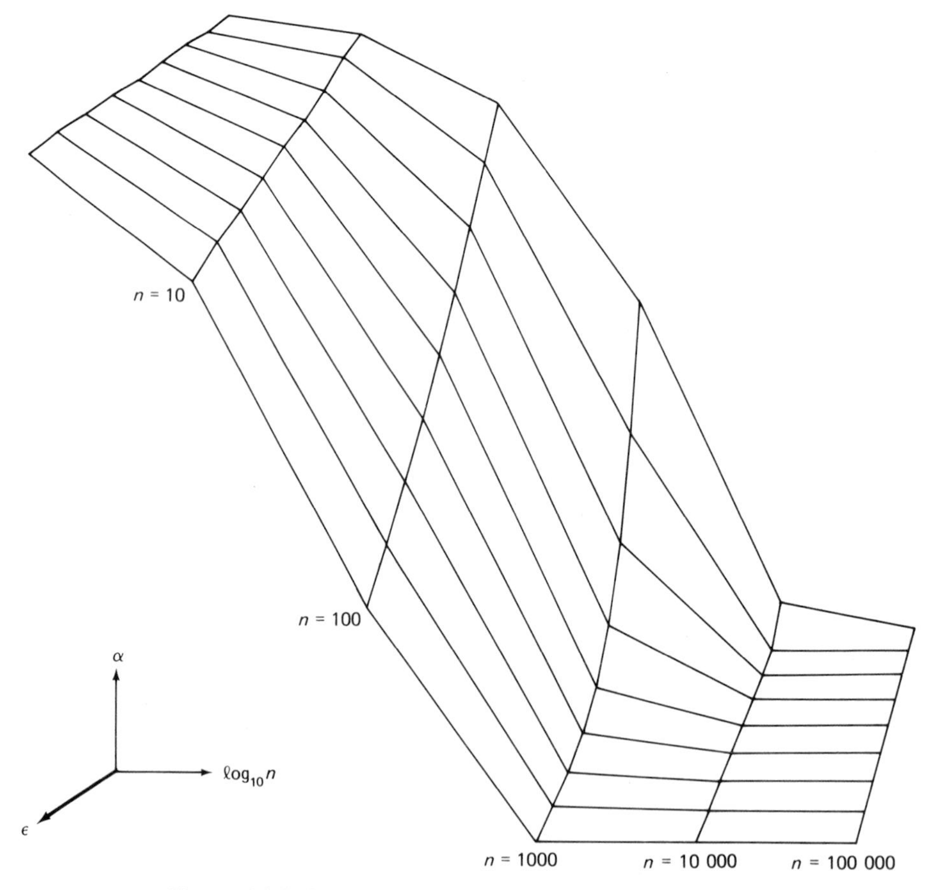

Figure 10.7 Probability of false alarm as a function of the test stringency ϵ and the sample size n

Then clearly, the probability of rejecting the null hypothesis when it is true is zero, hence $\alpha = 0$. Simultaneously, the probability of accepting H_0 when it is false is one—that is, $\beta = 1$. Similarly, if we always reject H_0, independent of the outcome of sampling, then $\beta = 0$ but $\alpha = 1$. In practice, we usually want to fix the probability of the type I error at some small value α, typically $\alpha = 0.01$ or 0.05, and then devise a test that has P(type II error) as small as possible.

To derive an expression for the type II error probability β, consider H_0: $\mu = \mu_0$ and H_1: $\mu = \mu_1 > \mu_0$ for fixed values of μ_0 and μ_1. Referring to Figure 10.8, suppose that the critical region for the test is $\bar{X} > C$. Now if the population $X \sim N(\mu, \sigma^2)$, then $\bar{X} \sim N(\mu, \sigma^2/n)$, hence:

$$\alpha = P(\bar{X} > C|H_0), \qquad \beta = P(\bar{X} < C|H_1),$$

where α is the area under the pdf of the normal distribution $N(\mu_0, \sigma^2)$ from C to ∞ and β is the area under the pdf of the normal distribution $N(\mu_1, \sigma^2)$ from $-\infty$ to C. For a given level of significance α, the dividing line of the criteria can be determined as:

$$C = \mu_0 + \frac{z_\alpha \sigma}{\sqrt{n}}.$$

If the allowable type II error probability β is also specified, then the minimum acceptable sample size can also be determined. We want:

$$P(\bar{X} < C|H_1) \leqslant \beta;$$

that is,

$$P(Z < \frac{C - \mu_1}{\sigma/\sqrt{n}}) \leqslant \beta, \quad \frac{C - \mu_1}{\sigma/\sqrt{n}} \leqslant -z_\beta,$$

so we want n large enough so that $(\mu_1 - C)/(\sigma/\sqrt{n}) \geqslant z_\beta$; that is, for fixed α:

$$\frac{\mu_1 - (\mu_0 + z_\alpha \sigma/\sqrt{n})}{\sigma/\sqrt{n}} \geqslant z_\beta;$$

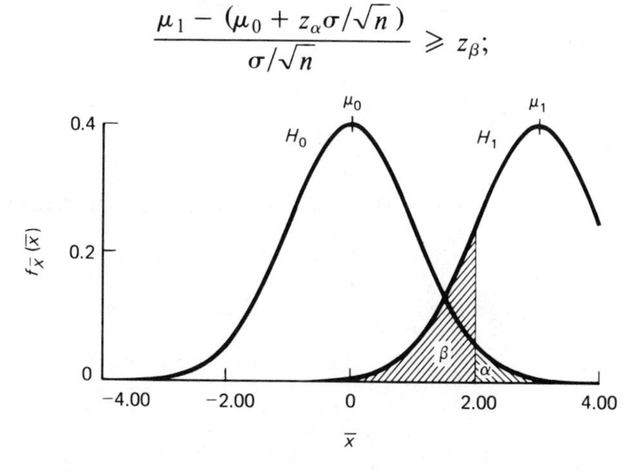

Figure 10.8 Computing the type I and type II errors

hence:

$$n \geqslant \frac{\sigma^2 (z_\alpha + z_\beta)^2}{(\mu_1 - \mu_0)^2}.$$

Example 10.24

We wish to test the hypothesis that the response time to a trivial request for a given time-sharing system is two seconds against an alternative hypothesis of three seconds. The probabilities of the two error types are specified to be $\alpha = 0.05$ and $\beta = 0.10$. Since $z_{0.05} = 1.645$ and $z_{0.10} = 1.28$ from Table 10.1, we determine the required number of response-time samples to be (assuming population variance $\sigma^2 = 5.8$ seconds^2):

$$n \geqslant \frac{5.8 \cdot (1.645 + 1.28)^2}{(3 - 2)^2}$$

$$= 50 \quad \text{rounded up to the nearest integer.}$$

The dividing line of the criteria is:

$$C = 2 + 1.645 \cdot \sqrt{\frac{5.8}{50}} = 2.556 \text{ seconds.}$$

Thus if the sample mean of the observed response times exceeds 2.556 seconds, then the hypothesis that the system provides a two second response should be rejected. #

In the above analysis, we assumed that the actual mean μ was suspected to be larger than μ_0 and therefore, we used a **one-tailed** test. If we did not have such knowledge, we would use a **two-tailed** test so that the acceptance region would be $C_1 < \bar{X} < C_2$. Hence we have:

$$\alpha = P(\bar{X} < C_1 \text{ or } \bar{X} > C_2 | H_0),$$

$$\beta = P(C_1 < \bar{X} < C_2 | H_1),$$

$$C_1 = \mu_0 - z_{\alpha/2} \cdot \frac{\sigma}{\sqrt{n}},$$

$$C_2 = \mu_0 + z_{\alpha/2} \cdot \frac{\sigma}{\sqrt{n}}.$$

Let $\text{n}(\mu, \sigma^2)$ denote the pdf of the normal random variable with mean μ and variance σ^2. Then:

$$\beta = \int_{C_1}^{C_2} \text{n}(\mu_1, \sigma^2)\, dx$$

$$= \int_{(C_1-\mu_1)/(\sigma/\sqrt{n})}^{(C_2-\mu_1)/(\sigma/\sqrt{n})} \text{n}(0, 1)\, dx$$

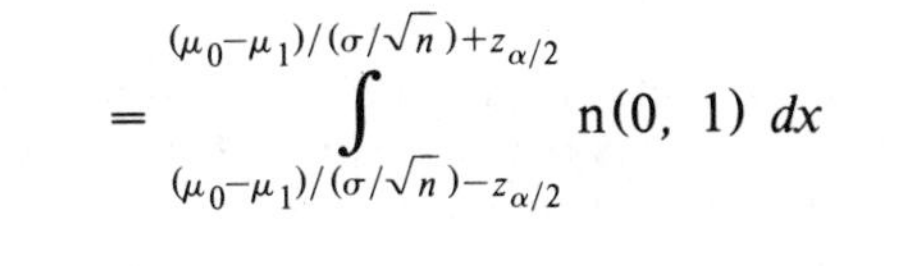

$$= \int_{(\mu_0-\mu_1)/(\sigma/\sqrt{n})-z_{\alpha/2}}^{(\mu_0-\mu_1)/(\sigma/\sqrt{n})+z_{\alpha/2}} \text{n}(0,\ 1)\ dx$$

$$= F_Z\left(\frac{\mu_0 - \mu_1}{\sigma/\sqrt{n}} + z_{\alpha/2}\right) - F_Z\left(\frac{\mu_0 - \mu_1}{\sigma/\sqrt{n}} - z_{\alpha/2}\right).$$

Often the alternative value μ_1 of the actual mean will not be specified. In this case we could compute β as a function of μ_1. A plot of $1 - \beta$ as a function of μ_1 is known as a **power curve**. A typical power curve for a two-tailed test is shown in Figure 10.9. Note that $1 - \beta(\mu_1)$ is the probability of rejecting H_0: $\mu = \mu_0$ when actually $\mu = \mu_1$, so that for $\mu_1 \neq \mu_0$, the power is the probability of a correct decision. Now if $\mu_1 = \mu_0$, then $1 - \beta(\mu_1)$ is the probability of rejecting H_0 when it should be accepted. Thus the value of the power curve at $\mu_1 = \mu_0$, $1 - \beta(\mu_0) = \alpha$. A typical power curve for a one-tailed test (H_0: $\mu = \mu_0$, H_1: $\mu > \mu_0$) is shown in Figure 10.10.

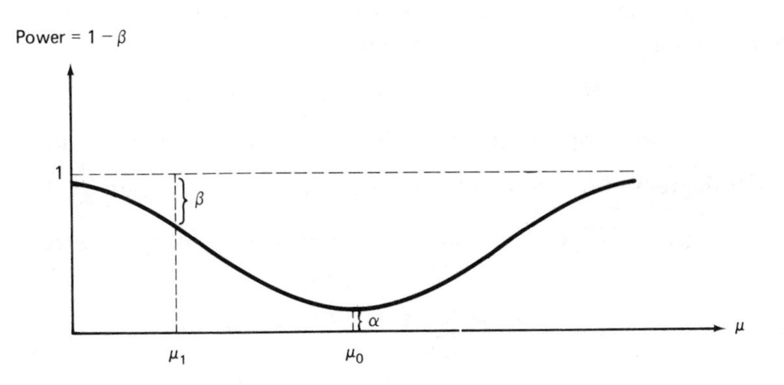

Figure 10.9 A typical power curve for a two-tailed test

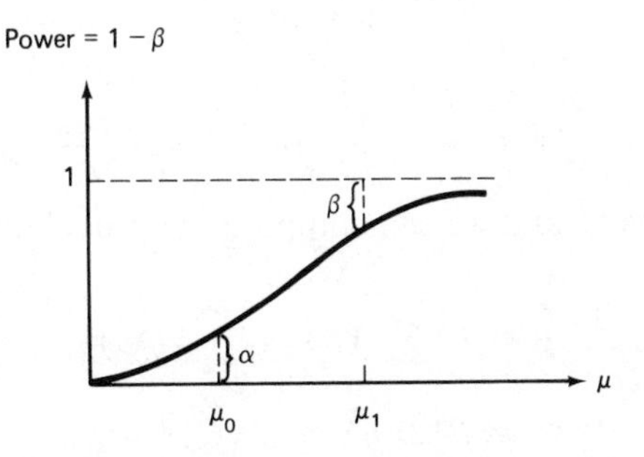

Figure 10.10 A typical power curve for a one-tailed test

Example 10.25

Continuing with our example of statistically monitoring a circuit, we are testing the hypothesis H_0: $p = p_0$ versus the alternative H_1: $p \neq p_0$. The null hypothesis H_0 corresponds to a fault-free circuit and the alternative corresponds to a faulty circuit. Different faults may give rise to a different value of the test statistic X, the number of observed 1's at the output.

If we assume that X is binomially distributed with mean np, then:

$$\begin{aligned}\beta &= P(np_0 - n\epsilon < X \leqslant np_0 + n\epsilon \mid H_1)\\ &= \sum_{np_0 - n\epsilon < k \leqslant np_0 + n\epsilon} \binom{n}{k} p^k (1-p)^{n-k}\\ &= F_X(np_0 + n\epsilon) - F_X(np_0 - n\epsilon)\end{aligned}$$

Now since X is binomially distributed with mean np and variance $np(1-p)$, it is approximately $N[np,\ np(1-p)]$; then:

$$\beta \simeq \left[F_Z\left(\frac{\mu_0 - np}{\sigma/\sqrt{n}} + \frac{\sigma_0}{\sigma} z_{\alpha/2}\right) - F_Z\left(\frac{\mu_0 - np}{\sigma/\sqrt{n}} - \frac{\sigma_0}{\sigma} z_{\alpha/2}\right)\right],$$

where $\sigma = \sqrt{np(1-p)}$, $\sigma_0 = \sqrt{np_0(1-p_0)}$, and $\mu_0 = np_0$. #

Now assume we are sampling from an exponential distribution with parameter λ and we wish to test the hypothesis:

$$H_0\colon \lambda = \lambda_0 \quad \text{versus} \quad H_1\colon \lambda = \lambda_1 < \lambda_0.$$

Recall that when population $X \sim \text{EXP}(\lambda)$, $2\lambda n\bar{X}$ has chi-square distribution with $2n$ degrees of freedom. Suppose we use the critical region $\sum_{i=1}^{n} X_i = n\bar{X} \geqslant C$ for some constant C. It follows that the probability of type I error is given by:

$$\begin{aligned}\alpha &= P\left(\sum_{i=1}^{n} X_i \geqslant C \mid \lambda = \lambda_0\right)\\ &= P(2\lambda n\bar{X} \geqslant 2\lambda C \mid \lambda = \lambda_0)\\ &= P(\chi^2_{2n} \geqslant 2\lambda_0 C)\end{aligned}$$

and so:

$$2\lambda_0 C = \chi^2_{2n;\alpha} \quad \text{or} \quad C = \frac{\chi^2_{2n;\alpha}}{2\lambda_0}.$$

Fixing α thus fixes C, and we can compute the probability of type II error by observing that:

$$\begin{aligned}\beta &= P\left(\sum_{i=1}^{n} X_i \leqslant C \mid \lambda = \lambda_1\right)\\ &= P(2\lambda n\bar{X} \leqslant 2\lambda C \mid \lambda = \lambda_1)\\ &= P(\chi^2_{2n} \leqslant 2\lambda_1 C).\end{aligned}$$

Example 10.26

Returning to Example 10.16, suppose we wish to test the hypothesis:

$$H_0: \lambda = .00025 \quad \text{versus} \quad H_1: \lambda < .00025.$$

Assume further that we wish to attain $\alpha = 0.05$. Then:

$$C = \frac{\chi^2_{20;0.05}}{2 \cdot 0.00025}$$

$$= 31.41/.0005 = 62,820 \text{ hours.}$$

We therefore reject the null hypothesis if the accumulated life on test, $S_{n;r}$, is greater than the critical value 62,820 and fail to reject H_0 otherwise. The type II error probability β can be computed for any given value of λ_1. For example, if $\lambda_1 = .0002$, then $2\lambda_1 C = 25.128$, and, since $\chi^2_{20;.2} = 25.04$, we have $\beta(0.0002) \simeq 0.8$. On the other hand, for $\lambda_1 = 0.0001$, we have $2\lambda_1 C = 12.564$, and, since $\chi^2_{20;.90} = 12.44$, we have $\beta(0.0001) \simeq .1$, which shows a dramatic reduction in the type II error!

Suppose now that we wish to reduce the value of β to 0.1 for $\lambda_1 = 0.0002$. This can be done either by increasing the sample size (in this case, the number of observed failures) or by allowing a larger value of α:

$$C = \frac{\chi^2_{2r;\alpha}}{2\lambda_0} = \frac{\chi^2_{2r;1-\beta}}{2\lambda_1}$$

so:

$$\chi^2_{2r;\alpha} = \frac{0.00025}{0.0002}\,\chi^2_{2r;0.9}.$$

We can now specify either α or the sample size r; should we choose $r = 40$, we get:

$$\chi^2_{2r;\alpha} = 1.25 \cdot 64.25 = 80.35.$$

Since $\chi^2_{80;0.5} = 79.33$, the value of α is approximately 0.5. Of course we could choose smaller α at the expense of further increase in the sample size. #

Problems

1. To test whether a given circuit is fault free, we drive it for a sequence of 100 inputs and observe 37 ones at the output (63 zeros). If the circuit is fault free, 50 ones are expected. At a significance level of 0.05 (probability of a false alarm), can we reject the hypothesis that the circuit is fault free? Compute the descriptive level of the test.

2. In statistical pattern recognition, one method used to distinguish the letter B from the numeral 8 is to compute the straightness ratio, defined to be a value of the random variable X, which is the ratio of the height of the symbol to its arc length (on the left-hand side), and perform a hypothesis test on it. Suppose that the conditional distribution of X given that the symbol is 8 is normal with mean 0.8 and variance 0.01, while the conditional distribution of X given B is normal with mean

0.96 and variance 0.01. The pattern-recognition problem is now cast as a hypothesis-testing problem:

$$H_0\colon E[X] = 0.8 \quad \text{versus} \quad H_1\colon E[X] = 0.96.$$

Suppose after measurement of the given symbol we reject H_0 if $x > 0.90$. Compute the error probabilities α and β.

3. Consider the combinational circuit in problem 1 of the Review Problems for Chapter 1. First compute the probability of a 1 at the output assuming that at each of the inputs the probability of a 1 is ½. Then test the hypothesis that the circuit is fault free versus the hypothesis that there is a stuck-at-0 type fault at input x_1. For this case compute the probability of false alarm (α) and the probability of escape (β), assuming the length of test $n = 400$ and test stringency $\epsilon = 0.005$. Repeat the calculation of β for each of the remaining thirteen fault types.

4. In selecting a computer service we are considering three alternative systems. The first criterion to be met is that the response time to a simple editing command should be less than three seconds at least 70 percent of the time. We would like the type I error probability to be less than 0.05. On $n = 64$ randomly chosen requests the number m of requests that met the criterion of less than three seconds response were found to be as shown in the following table:

Service number	*m*
1	52
2	47
3	32

First determine critical value c so that if $m < c$ for a service, that service will be rejected, and then determine which of the three services will be rejected from further consideration.

5. Consider the problem of acceptance sampling from a large batch of items (integrated-circuit chips). From a sample of size n, the number of defectives found, X, is noted. If $X \leqslant k$, the batch is accepted; otherwise the batch is rejected. Let p denote the actual probability of defective items in the batch. Using the Poisson approximation to the binomial, show that the probability of accepting the batch as a function of p is obtained by solving:

$$2np = \chi^2_{2(k+1);\beta(p)}.$$

The plot of $\beta(p)$ against p is known as the **operating characteristic**. Plot this curve for $n = 20$ and $k = 8$.

The producer of the items demands that if $p = p_0$, where p_0 is the **acceptable quality level**, then the probability α of the batch being rejected should be small. In this connection α is called the **producer's risk**. Note that $\beta(p_0) = 1 - \alpha$. The consumer demands that if the lot is relatively bad ($p \geqslant p_1$), the probability of its being acceptable should be small. The probability $\beta(p_1)$ is called the **consumer's risk**.

Show that for fixed values of α, β, p_0, and p_1, the value of the critical point k is determined by solving:

$$\frac{\chi^2_{2(k+1);\beta}}{\chi^2_{2(k+1);1-\alpha}} \leqslant \frac{p_1}{p_0}.$$

Given $p_0 = 0.05$, $p_1 = 0.10$, $\alpha = 0.05$, and $\beta\ (p_1) = 0.10$, determine the values of k and n. Plot the operating-characteristic curve for this case and mark the above values on the curve.

6. *The Sign Test.* For a continuous population distribution, develop the test for median (based on the random sample $X_1, X_2, \ldots, X_n$):

$$H_0\colon \pi_{0.5} = m_0 \quad \text{versus} \quad H_1\colon \pi_{0.5} = m_1 > m_0.$$

Let the random variable $Z_i = 0$ if $X_i - m_0 \leqslant 0$, and otherwise $Z_i = 1$. Let $Z = \sum Z_i$ and show that if H_0 is true, Z is binomially distributed with parameters n and 0.5. Derive an expression for the significance level α of the test based on the critical region $Z > k$. This test is known as the **sign test**, since the statistic Z is equal to the number of positive signs among:

$$X_1 - m_0,\ X_2 - m_0,\ \ldots,\ X_n - m_0.$$

10.3.2 Hypotheses Concerning Two Means

While making a purchasing decision, two vendors offer computing systems with nearly equal costs (and with all other important attributes except throughput differing insignificantly). After running benchmarks and measuring throughputs on the two systems, we are interested in making a comparison of the performances and finally selecting the better system. In such cases we wish to test the null hypothesis that the difference between the two population means, $\mu_X - \mu_Y$, equals some given value d_0. We shall discuss three separate cases.

Case 1. Suppose we wish to test the null hypothesis, $H_0\colon \mu_X - \mu_Y = d_0$, for a specified constant d_0 on the basis of independent random samples of size n_1 and n_2, assuming that the population variances σ_X^2 and σ_Y^2 are known. The test will be based on the difference of the sample means $\bar{X} - \bar{Y}$. If we assume that both populations are normal, then the statistic:

$$Z = \frac{\bar{X} - \bar{Y} - (\mu_X - \mu_Y)}{(\mathrm{Var}\,[\bar{X} - \bar{Y}])^{1/2}}$$

can be shown to have the standard normal distribution. Here:

$$\mathrm{Var}\,[\bar{X} - \bar{Y}] = \frac{\sigma_X^2}{n_1} + \frac{\sigma_Y^2}{n_2}.$$

Now for a significance level α, the critical regions for the test statistic Z can be specified:

$$Z < -z_\alpha, \qquad \text{if } H_1\text{: } \mu_X - \mu_Y < d_0,$$

$$Z > z_\alpha, \qquad \text{if } H_1\text{: } \mu_X - \mu_Y > d_0,$$

$$Z < -z_{\alpha/2} \quad \text{or} \quad Z > +z_{\alpha/2}, \qquad \text{if } H_1\text{: } \mu_X - \mu_Y \neq d_0.$$

Example 10.27

Two time-sharing systems are compared according to their response time to an editing command. The mean response time of 50 such requests submitted to system 1 was measured to be 682 milliseconds with a known standard deviation of 25 milliseconds. A similar measurement on system 2 resulted in a sample mean of 675 milliseconds with a standard deviation of 28 milliseconds. To test the hypothesis that system 2 provides better response than system 1, we form the hypotheses:

$$H_0\text{: } \mu_X = \mu_Y,$$

that is, there is no difference in response times, and

$$H_1\text{: } \mu_X > \mu_Y,$$

that is, system 2 is better than system 1. Then, the null hypothesis:

$$\mu_X - \mu_Y = 0,$$

$$\sigma_{\bar{X}-\bar{Y}} = \sqrt{\frac{\sigma_X^2}{n_1} + \frac{\sigma_Y^2}{n_2}} = \sqrt{\frac{(25)^2}{50} + \frac{(28)^2}{50}} = 5.3,$$

and the test statistic

$$Z = \frac{\bar{X} - \bar{Y}}{\sigma_{\bar{X}-\bar{Y}}} = \frac{\bar{X} - \bar{Y}}{5.3}$$

has the value:

$$z = \frac{682 - 675}{5.3} = 1.32.$$

Using a one-tailed test at a 5 percent level of significance, we fail to reject the hypothesis, since the observed value of z $(= 1.32)$ is less than the critical value $z_\alpha = z_{0.05} = 1.645$. Thus, based on the given data, we cannot support the claim that system 2 is more responsive than system 1.

Note that the null hypothesis can be rejected in this case at a 10 percent (rather than 5 percent) level of significance. This would mean that we are willing to take a 10 percent chance of being wrong in rejecting the null hypothesis. #

Case 2. The assumption that the population variances σ_X^2 and σ_Y^2 are known rarely holds in practice. If we use sample estimates in place of population variances, then we can use the t distribution in place of the normal distribution, provided we further assume that the two population variances are equal ($\sigma_X^2 = \sigma_Y^2 = \sigma^2$) and that the two populations are approximately normal.

Suppose we select two independent random samples, one from each population, of sizes n_1 and n_2, respectively. Using the two sample variances S_X^2 and S_Y^2, the common population variance σ^2 is estimated by S_p^2, where:

$$S_p^2 = \frac{(n_1 - 1)S_X^2 + (n_2 - 1)S_Y^2}{(n_1 + n_2 - 2)}.$$

Now if the null hypothesis H_0: $\mu_X - \mu_Y = d_0$ holds, then it can be shown that the statistic:

$$T = \frac{\bar{X} - \bar{Y} - d_0}{S_p\sqrt{\dfrac{1}{n_1} + \dfrac{1}{n_2}}}$$

has a t distribution with $n_1 + n_2 - 2$ degrees of freedom.

Small departures from the assumption of equal variances may be ignored if $n_1 \simeq n_2$. If σ_X^2 is much different from σ_Y^2, a modified t statistic can be used as described in [WALP 1968].

Example 10.28

Elapsed times for a synthetic job were measured on two different computer services. The sample sizes for the two cases were 15 each, and the sample means and sample variances were computed to be:

$$\bar{x} = 104 \text{ seconds}, \qquad \bar{y} = 114 \text{ seconds},$$

$$s_X^2 = 290 \text{ second}^2, \qquad s_Y^2 = 510 \text{ second}^2.$$

To test the hypothesis that the population means $\mu_X = \mu_Y$ against the alternative $\mu_X < \mu_Y$, we first calculate an estimate of the (assumed) common variance:

$$\begin{aligned} s_p^2 &= \frac{14 \cdot 290 + 14 \cdot 510}{28} \\ &= 400. \end{aligned}$$

Now, since:

$$\frac{\bar{X} - \bar{Y}}{S_p/\sqrt{7.5}}$$

is approximately t distributed with 28 degrees of freedom, we get the descriptive level of this test:

$$\begin{aligned} \delta &= P(T_{28} \leqslant \frac{104 - 114}{\sqrt{400/7.5}}) \\ &= P(T_{28} \leqslant -\frac{10}{7.30}) = P(T_{28} \geqslant 1.3693) \\ &\simeq 0.0972. \end{aligned}$$

Thus the observed results have a chance of about one in ten of occurring, hence we do not reject H_0. #

Case 3. The test procedures discussed in the two cases above are valid only if the populations are approximately normal. We now describe a test due to Wilcoxon, Mann, and Whitney that allows for arbitrary continuous distributions for X and Y and therefore is known as a **distribution-free** or a **nonparametric** test. Specifically, we consider the problem of testing: for all x,

$$H_0\colon f_X(x) = f_Y(x) \quad \text{versus} \quad H_1\colon f_X(x) = f_Y(x + c) \qquad (10.1)$$

where c is a positive constant (see Figure 10.11). This test is often posed as a test for the equality of the two population medians; in case of symmetric densities f_X and f_Y it is equivalent to a test of equality of the two population means (provided both exist).

Assume that two independent random samples of respective sizes n_1 and n_2 are collected from the two populations and denoted by $x_1, x_2, x_3, \ldots, x_{n_1}$ and $y_1, y_2, \ldots, y_{n_2}$. We now combine the two samples, arrange these values in order of increasing magnitude, and assign to the $(n_1 + n_2)$-ordered values the ranks $1, 2, 3, \ldots, n_1 + n_2$. In the case of ties, assign the average of the ranks associated with the tied values. Let $r(Y_i)$ denote the rank of Y_i in the combined ordered set and define the statistic:

$$W = \sum_{i=1}^{n_2} r(Y_i).$$

The statistic W is a sum of ranks, hence the test we describe is commonly known as the **rank-sum test**.

Under H_1, the density of Y is shifted to the right of the density of X and the values in the Y-sample would tend to be larger than the values in the X-sample. Thus under H_1, W would tend to be larger than expected under H_0. Therefore, the critical region for the test (10.1) will be of the form $W > w_0$.

We can derive the distribution of W under H_0 by noting that the combined ordered set represents a random sample of size $n_1 + n_2$ from the popu-

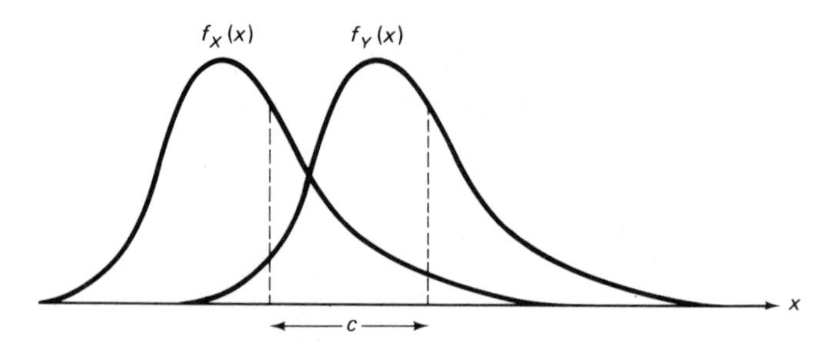

Figure 10.11 A nonparametric test

lation of X. Further, since the ranks depend upon the relative (and not the absolute) magnitudes of the sample values, it is sufficient to consider the positions of the Y-values in the combined set in order to evaluate $F_W(w)$. Let $\#(W = w)$ denote the set of all combinations of y-ranks that will sum to w. The total ways of picking all combinations of ranks given n_1 and n_2 is:

$$\binom{n_1 + n_2}{n_2},$$

and since each of these combinations is equally likely under H_0, we get (assuming no ties):

$$P(W = w \mid H_0) = \frac{\#(W = w)\,n_1!\,n_2!}{(n_1 + n_2)!}.$$

Then the significance level α is determined by:

$$P(W \geqslant w \mid H_0) \leqslant \alpha.$$

Since the distribution function $F_W(w)$ depends only upon relative ranks, it can be computed by combinatorial methods. For small values of n_1 and n_2, w_α-values have been precomputed and listed in Appendix C. For larger values of n_1 and n_2 we use a normal approximation to the distribution of W. It can be shown that if H_0 is true, and $n_1 \geqslant 10$ and $n_2 \geqslant 10$, W possesses an approximate normal distribution with:

$$E[W] = \frac{n_1(n_1 + n_2 + 1)}{2}$$

and

$$\text{Var}\,[W] = \frac{n_1 n_2(n_1 + n_2 + 1)}{12}.$$

Example 10.29

The times between two successive crashes are recorded for two competing computer systems as follows (time in weeks):

System x:	1.8	0.4	2.7	3.0	
System y:	2.0	5.4	1.3	4.5	0.8

In order to test the hypothesis that both systems have the same mean time between crashes against the alternative that system x has a shorter mean time between crashes, we first arrange the combined data in ascending order and assign ranks (y-ranks are underlined for easy identification):

Original data	0.4	0.8	1.3	1.8	2.0	2.7	3.0	4.5	5.4
Ranks	1	<u>2</u>	<u>3</u>	4	<u>5</u>	6	7	<u>8</u>	<u>9</u>

The rank sum $W = 2 + 3 + 5 + 8 + 9 = 27$. Looking up the table of rank-sum critical values with $n_1 = 4$ and $n_2 = 5$, we find that:

$$P(W \geqslant 27 \mid H_0) = 0.056.$$

Therefore, we reject the null hypothesis at 0.056 level of significance. #

Noether [NOET 1967] points out several reasons why in general the rank-sum test is preferable to the t test. Since the rank-sum test is distribution free, whatever the true population distribution, as long as both samples come from the same population (that is, $f_X = f_Y$), the significance level of the test is known. On the other hand, for nonnormal populations, the significance level of the t test may differ considerably from the calculated value. In contrast to the t test, the rank-sum test is not overly affected by large deviations (so called **outliers**). On the other hand, when there is sufficient justification for assuming that the population distribution is normal, it would be a mistake not to use that information [HOEL 1971].

Problems

1. Returning to the system-selection problem considered in problem 4, Section 10.3.1, the sample means of the response times for the first two systems are 2.28 and 2.52 seconds, respectively. The sample size in both cases is 64, and the variances can be assumed to be 0.6 second^2 and 0.8 second^2 respectively. Test the hypothesis that the mean response times of the two systems are the same against the alternative that system 1 has a smaller mean response time.

2. In this section we assumed that X and Y samples were chosen independently. In practice, the observations often occur in pairs, $(x_1, y_1), (x_2, y_2), \ldots, (x_n, y_n)$. Based on pairwise differences $d_i = x_i - y_i$ construct a t test using the statistic:

$$T = \frac{\bar{D} - d_0}{S_D/\sqrt{n}}$$

to test the hypothesis H_0: $\mu_X - \mu_Y = d_0$. Also show that in case of nonnormal populations, the sign test of problem 1, Section 10.3, can be adapted to test the null hypothesis H_0: $\mu_X = \mu_Y$. Apply these two tests to the claim that two computer systems are about equal in their processing speeds, based on the following data:

Benchmark program	Run time in seconds	
	System A	*System B*
Payroll	42	55
Linear programming (simplex)	201	195
Least squares	192	204
Queuing network solver	52	40
Puzzle	10	12
Simulation	305	290
Statistical test	10	13
Synthetic 1	1	1
Synthetic 2	350	320
Synthetic 3	59	65

10.3.3 Hypotheses Concerning Variances

First consider the problem of testing the null hypothesis that a population variance σ^2 equals some fixed value σ_0^2 against a suitable one-sided or two-sided alternative. For instance, many time-sharing users may not mind a relatively long average response time so long as the system is consistent. In other words, they wish the variance of the response to be small.

Assuming that we are sampling from a normal population $N(\mu, \sigma_0^2)$, we have shown in Chapter 3 that the statistic:

$$X_{n-1}^2 = \frac{(n-1)S^2}{\sigma_0^2}$$

is chi-square distributed with $n - 1$ degrees of freedom. From this, the critical regions for testing H_0: $\sigma^2 = \sigma_0^2$ are:

Reject H_0 if, for $X_{n-1}^2 = \frac{(n-1)S^2}{\sigma_0^2}$	*when H_1 is*
$X_{n-1}^2 < \chi_{n-1;1-\alpha}^2$	$\sigma^2 < \sigma_0^2$
$X_{n-1}^2 > \chi_{n-1;\alpha}^2$	$\sigma^2 > \sigma_0^2$
$\chi_{n-1;\alpha/2}^2 < X_{n-1}^2$ or $X_{n-1}^2 < \chi_{n-1;1-\alpha/2}^2$	$\sigma^2 \neq \sigma_0^2$

Example 10.30

In the past, the standard deviation of the response time to editing commands of a time-sharing system was 25 milliseconds and the mean response time was 400 milliseconds. A new version of the operating system was installed, and it was claimed to be biased against the time-sharing users. With the new system, a random sample of 21 editing commands experienced a standard deviation of response times of 32 milliseconds. Is this increase in variability significant at a 5 percent level of significance? Is it significant at a 1 percent level? We wish to test:

$$H_0\colon \sigma^2 = (25)^2 \quad \text{versus} \quad \sigma^2 > (25)^2.$$

The observed value of the chi-square statistic is:

$$\frac{(n-1)s^2}{\sigma_0^2} = \frac{20(32)^2}{(25)^2} = 32.8.$$

Since $\chi^2_{20;0.05} = 31.41$, we conclude at the 5 percent level of significance that the new version of the system is unfair to time-sharing users. On the other hand, since $\chi^2_{20;0.01} = 37.566$, we cannot reject H_0 at the 1 percent level of significance. #

We should caution the reader that the above test is known to give poor results if the population distribution deviates appreciably from the normal distribution. The reader is advised to use a suitable nonparametric test in such cases [NOET 1967].

If we wish to compare the variances of two normal populations, with variances σ_X^2 and σ_Y^2, respectively, then we test the hypothesis:

$$H_0\colon \sigma_X^2 = \sigma_Y^2.$$

In this case we use the fact that the ratio $S_X^2\,\sigma_Y^2/(S_Y^2\,\sigma_X^2)$ has an F distribution with $(n_1 - 1,\ n_2 - 1)$ degrees of freedom. This statistic is simply the ratio of sample variances if H_0 is true. Here the sample size for the first population is n_1, while the sample size for the second population is n_2. Therefore, the critical region for testing H_0 against the alternative $H_1\colon \sigma_X^2 > \sigma_Y^2$ is that the observed value of the F statistic satisfies $F_{n_1-1,n_2-1} > f_{n_1-1,n_2-1;\alpha}$, where $f_{n_1-1,n_2-1;\alpha}$ is determined by the F distribution to be that value such that $P(F_{n_1-1,n_2-1} > f_{n_1-1,n_2-1;\alpha}) = \alpha$. Similarly if $H_1\colon \sigma_X^2 < \sigma_Y^2$, we use the critical region $F_{n_1-1,n_2-1} < f_{n_1-1,n_2-1;1-\alpha}$.

Example 10.31

In comparing two time-sharing systems based on sample means, suppose that at the desired level of significance the hypothesis $\mu_X = \mu_Y$ cannot be rejected. In other words, the difference in the average response times is not statistically significant. The next level of comparison is then the difference in the variances of the response times. Recalling Example 10.27, we have $s_X^2 = (25)^2 = 625$ and $s_Y^2 = (28)^2 = 784$. The observed value of the F statistic under H_0 is $625/784 = 0.797$. If we wish to test H_0 against $H_1\colon \sigma_X^2 < \sigma_Y^2$, we need to obtain the critical value of $f_{49,49;1-\alpha}$, so we use a table of the F distribution with (49, 49) degrees of freedom. Then

$f_{49,49;\alpha} = f_{49,49;0.05} = 1.62$, which we invert to obtain $f_{49,49;1-\alpha} = 0.617$. Since the observed value is larger than the critical value, we fail to reject H_0. In other words, system 1 does not provide a statistically lower variability in response time at the 5 percent level of significance. #

Problems

1. Returning to the data in Example 10.15, test the hypothesis at significance level 0.05 that the variance of response time is 20 ms^2 against the alternative that the variance is greater than 20 ms^2.

10.3.4 Goodness-of-Fit Tests

Most of the methods in the preceding sections require that the type of the distribution function of X is known and either its parameters are being estimated or a hypothesis concerning its parameters is being tested. It is important to have some type of test that can establish the "goodness of fit" between the postulated distribution type of X and the evidence contained in the experimental observations. Such experimental data are likely to be in the same basic form as the data used to estimate parameters of the distribution. Graphical methods are often used to establish goodness of fit but we will use analytical methods.

First assume that X is a discrete random variable with true (but unknown) pmf given by $p_X(i) = p_i$. We wish to test the null hypothesis that X possesses a certain specific pmf given by $p_i = p_{i_0}$, $0 \leqslant i \leqslant k-1$. Our problem then is to test H_0 versus H_1, where:

$$H_0\colon\ p_i = p_{i_0}, \qquad i = 0, 1, \ldots, k-1,$$

$$H_1\colon \text{ not } H_0.$$

Assume we make n observations and let N_i be the observed number of times (out of n) that the measured value of X takes the value i. N_i is clearly a binomial random variable with parameters n and p_i so that $E[N_i] = np_i$ and $\text{Var}[N_i] = np_i(1-p_i)$. Wilks [WILK 1962] shows that the statistic:

$$Q = \sum_{i=0}^{k-1} \frac{(N_i - np_i)^2}{np_i} \tag{10.2}$$

is approximately chi-square distributed with $(k-1)$ degrees of freedom. One degree of freedom is lost because only $k-1$ of the N_i are independent owing to the relation:

$$n = \sum_{i=0}^{k-1} N_i = \sum_{i=0}^{k-1} np_i.$$

Under the assumption H_0: $p_i = p_{i_0}$, the statistic (10.2) is just:

$$\chi^2_{k-1} = \sum \frac{(\text{observed} - \text{expected})^2}{\text{expected}}.$$

Example 10.32

A computer system has six I/O channels and the system personnel are reasonably certain that the load on the channels is balanced. If X is the random variable denoting the index of the channel to which a given I/O operation is directed, then its pmf is assumed to be:

$$p_X(i) = p_i = 1/6, \qquad i = 0, 1, \ldots, 5.$$

Out of $n = 150$ I/O operations observed, the numbers of operations directed to various channels were:

$$n_0 = 22, \quad n_1 = 23, \quad n_2 = 29, \quad n_3 = 31, \quad n_4 = 26, \quad n_5 = 19.$$

We wish to test the hypothesis that the load on the channels is balanced; that is, H_0: $p_i = 1/6$, $i = 0, 1, \ldots, 5$. Using the chi-square statistic, we obtain:

$$\begin{aligned}\chi^2 &= \frac{(22-25)^2}{25} + \frac{(23-25)^2}{25} + \frac{(29-25)^2}{25} \\ &\quad + \frac{(31-25)^2}{25} + \frac{(26-25)^2}{25} + \frac{(19-25)^2}{25} \\ &= 4.08.\end{aligned}$$

For the chi-square distribution with five degrees of freedom, the 55 percent critical value is:

$$\chi^2_{5;0.55} \simeq 4.00.$$

In other words there is a high probability under H_0 of observing such a small deviation, hence we cannot reject the null hypothesis that the channels are load-balanced. #

In deriving the distribution of test statistic (10.2) above, the multivariate normal approximation to the multinomial distribution is employed [WILK 1962]. For the approximation to be accurate, each np_i value should be moderately large (as a rule of thumb, $np_i \geqslant 5$). When the random variable X takes a large (perhaps, infinite) number of values, the condition $np_i \geqslant 5$ for all i will be difficult to meet even with a large value of n. If the expected numbers in several categories are small, then these categories should be combined to form a single category. Note that this process of combination of categories implies a concomitant loss in power of the test [WILK 1962].

While performing a goodness-of-fit test, often a null hypothesis specifies only that the population distribution belongs to a family of distributions $F_X(x;\boldsymbol{\theta})$ where $\boldsymbol{\theta}$ is a vector of unknown parameters. For example, we may want to test whether or not X has a Poisson distribution. This amounts to a null hypothesis:

$$H_0\colon\ p_i = \frac{\lambda^i e^{-\lambda}}{i!}, \qquad i = 0, 1, \ldots,$$

which we cannot test without some specification of λ. In such situations, the unknown parameters of the population (such as λ in the Poisson example) must first be estimated from the collected sample of size n. We then use a test statistic:

$$\hat{Q} = \sum_{i=0}^{k-1} \frac{(N_i - n\hat{p}_i)^2}{n\hat{p}_i},$$

where $\hat{p}_i = p_X(i;\hat{\boldsymbol{\theta}})$ is obtained using the maximum-likelihood estimates of the parameters $\boldsymbol{\theta}$. It can be shown that the statistic $\hat{Q}$ is approximately chi-square distributed with $k - m - 1$ degrees of freedom. Thus, if m population parameters are to be estimated, the chi-square statistic loses m degrees of freedom.

Example 10.33

It is suspected that the number of errors discovered in a system program is Poisson distributed. The number of errors discovered in each one-week period is given in Table 10.2. The total number of errors observed in the 50 weeks was 95.

Table 10.2

Number of errors (i) in one-week period	Number of one-week periods with i errors	Poisson probabilities $\hat{p}_{i_0}$	Expected frequencies $n\hat{p}_{i_0}$
0	14	0.150	7.50
1	11	0.284	14.20
2	9	0.270	13.50
3	6	0.171	8.55
4	5	0.081	4.05
5+	5	0.044	2.20

To compute the Poisson probabilities above, we must have an estimate of the rate parameter λ, which is computed from the data:

$$\hat{\lambda} = \frac{\text{total number of observed errors}}{\text{total number of weeks observed}} = \frac{95}{50} = 1.9.$$

$$\begin{aligned}
\chi^2 &= \sum \frac{(N_i - n\hat{p}_i)^2}{n\hat{p}_i} \\
&= \frac{(14 - 7.50)^2}{7.50} + \frac{(11 - 14.20)^2}{14.20} + \frac{(9 - 13.5)^2}{13.5} + \frac{(6 - 8.55)^2}{8.55} \\
&\quad + \frac{(5 - 4.05)^2}{4.05} + \frac{(5 - 2.20)^2}{2.20} \\
&= 5.633 + 0.7211 + 1.5 + 0.7605 + 0.2228 + 3.564 \\
&= 12.401
\end{aligned}$$

is the value of a chi-square random variable with $k - 2 = 6 - 2 = 4$ degrees of freedom. Since $\chi^2_{4;.05} = 9.488$ is lower than the observed value of 12.401, we conclude that the null hypothesis of Poisson distribution should be rejected at a 5 percent level of significance. The descriptive level for this test is approximately 0.016, indicating that the observations would be highly unlikely to occur under the Poisson assumption. #

Now suppose X is a continuous random variable and we wish to test the hypothesis that the distribution function of X is a specified function; that is:

$$H_0\text{: for all } x,\ F_X(x) = F_0(x),$$

versus:

$$H_1\text{: there exist } x \text{ such that, } F_X(x) \neq F_0(x).$$

The chi-square test described above is applicable in this case, but we will be required to divide the image of X into a finite number of categories. The subsequent loss of information results in a loss of power of the test [WILK 1962]. The Kolmogorov-Smirnov test to be described is the preferred goodness-of-fit test in case of a continuous population distribution. Conversely, when applied to discrete population distributions, the Kolmogorov-Smirnov test is known to produce conservative results. That is, the actual probability of type I error will be at most equal to the chosen value α but this advantage is offset by a corresponding loss of power (or increase in the probability of type II error).

The given random sample is first arranged in order of magnitude so that the values are assumed to satisfy $x_1 \leqslant x_2 \leqslant \cdots \leqslant x_n$. Then the empirical distribution function $\hat{F}_n(x)$ is defined by:

$$\hat{F}_n(x) = \begin{cases} 0, & x < x_1, \\ i/n, & x_i \leqslant x < x_{i+1}, \\ 1, & x_n \leqslant x. \end{cases} \tag{10.3}$$

The alternative definition of $\hat{F}_n(x)$ is:

$$\hat{F}_n(x) = \frac{\text{number of values in the sample that are} \leqslant x}{n}.$$

A natural measure of deviation of the empirical distribution function from $F_0(x)$ is the absolute value of the difference:

$$d_n(x) = |\hat{F}_n(x) - F_0(x)|.$$

Since $F_0(x)$ is known, the deviation $d_n(x)$ can be computed for each value of x. The largest among these values as x varies over its full range is an indicator of how well $\hat{F}_n(x)$ approximates $F_0(x)$. Since $\hat{F}_n(x)$ is a step function (see Figure 10.12) with n steps and $F_0(n)$ is continuous and nondecreasing, it suffices to evaluate $d_n(x)$ at the left and right end points of the intervals

$[x_i, x_{i+1}]$. The maximum value of $d_n(x)$ is then the value of the Kolmogorov-Smirnov statistic defined by:

$$D_n = \sup_x |\hat{F}_n(x) - F_0(x)|. \tag{10.4}$$

[The reason for the use of supremum rather than maximum in the above definition is that by definition (10.3), $\hat{F}_n(x)$ is constant over the interval $[x_i, x_{i+1})$.] The definition (10.4) simply says that we evaluate $|\hat{F}_n(x) - F_0(x)|$ at the end points of each interval $[x_i, x_{i+1}]$, treating $\hat{F}_n(x)$ as having a constant value in that interval, and then choosing the largest of these values as the value of D_n.

The usefulness of the statistic (10.4) is that it is *distribution free*, hence its exact distribution can be derived. In other words, for a continuous $F_0(x)$, the sampling distribution of D_n depends only on n and not on $F_0(x)$. Thus the D_n statistic possesses the advantage that its exact distribution is known even for small n whereas the Q statistic is only approximately chi-square distributed, and a fairly large sample size is needed in order to justify the approximation. We shall not derive the distribution function of D_n, but we give a table of critical values $d_{n;\alpha}$ in Appendix C.

We reject the null hypothesis at a level of significance α if the observed value of the statistic D_n exceeds the critical value $d_{n;\alpha}$, otherwise we reject the alternative hypothesis H_1.

Example 10.34

Given the random sample of size $n = 10$:

0.3407 0.1440 0.6960 0.8675 0.5649 0.5793 0.1514
0.5044 0.9859 0.4658,

we wish to test the hypothesis that the population distribution function is the uniform distribution over (0, 1); that is:

$$F_0(x) = \begin{cases} 0, & x < 0, \\ x,, & 0 \leqslant x < 1, \\ 1, & x \geqslant 1. \end{cases}$$

$\hat{F}_{10}(x)$ and $F_0(x)$ are plotted in Figure 10.12. The observed value of the D_n statistic is 0.1658. Now, using the table of critical values in Appendix C, we find that at $\alpha = 0.05$, $d_{10;\alpha} = 0.41$, hence the rejection region is $\{D_{10} > 0.41\}$. We therefore accept the null hypothesis at the 5 percent level of significance. #

Now consider the problem of obtaining a confidence interval for the unknown function $F_0(x)$. For a fixed value of x, $n\hat{F}_n(x)$ is a binomial random variable with parameters n and $p = F_0(x)$. Hence we can use the procedure of Section 10.2.3.4. Here we are considering a confidence interval not just for $F_0(x)$ at a number of isolated points but for $F_0(x)$ as a whole. A **confidence**

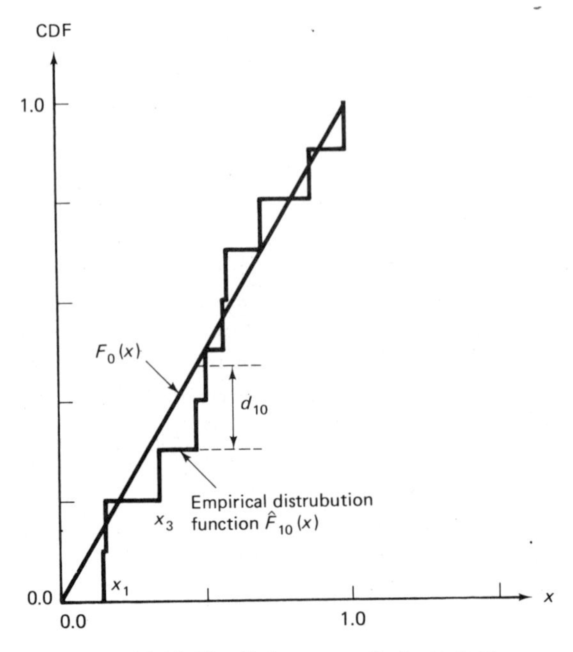

Figure 10.12 The Kolmogorov-Smirnov test

band with confidence coefficient γ for $F_0(x)$ is obtained using the D_n statistic by using:

$$\begin{aligned}\gamma &= P(D_n \leqslant d_{n;1-\gamma}) \\ &= P(\sup_x |\hat{F}_n(x) - F_0(x)| \leqslant d_{n;1-\gamma}) \\ &= P(|\hat{F}_n(x) - F_0(x)| \leqslant d_{n;1-\gamma} \text{ for all } x) \\ &= P(\hat{F}_n(x) - d_{n;1-\gamma} \leqslant F_0(x) \leqslant \hat{F}_n(x) + d_{n;1-\gamma} \text{ for all } x).\end{aligned}$$

Noting that $0 \leqslant F_0(x) \leqslant 1$, we have a confidence band for $F_0(x)$ with confidence coefficient γ as follows: for all x,

$$\max\{0,\ \hat{F}_n(x) - d_{n;1-\gamma}\} \leqslant F_0(x) \leqslant \min\{1,\ \hat{F}_n(x) + d_{n;1-\gamma}\} \quad (10.5)$$

Suppose that the null hypothesis does not specify the function $F_0(x)$ completely but specifies some parametric family of functions $F_0(x;\boldsymbol{\theta})$ where parameters $\boldsymbol{\theta}$ are to be estimated from the given sample. Analogous to the chi-square test, we will use the test statistic:

$$\hat{D}_n = \sup_x |\hat{F}_n(x) - F_0(x;\hat{\boldsymbol{\theta}})|,$$

where $\hat{\boldsymbol{\theta}}$ is an appropriate estimate of unknown vector of parameters $\boldsymbol{\theta}$. Unfortunately, there is no simple modification as in the case of the chi-square test. The sampling distribution of $\hat{D}_n$ must be separately studied for each family of population distribution functions. Lilliefors has studied the distribution of $\hat{D}_n$ in case $F_0(x;\theta)$ is the family of exponential distributions with the unknown mean θ [LILL 1969] and in case $F_0(x; \mu, \sigma^2)$ is the family of normal distributions with unknown mean μ and unknown variance σ^2 [LILL 1967]. In Appendix C we give tables of critical values for the statistic $\hat{D}_n$ in these two cases.

Problems

1. The number of busy senders in a panel-type switching machine of a telephone exchange was observed as follows:

Number busy	*Observed frequency*
0	0
1	5
2	14
3	24
4	57
5	111
6	197
7	278
8	378
9	418
10	461
11	433
12	413
13	358
14	219
15	145
16	109
17	57
18	43
19	16
20	7
21	8
22	3

Test whether the corresponding theoretical distribution is Poisson.

2. Perform a goodness-of-fit test at significance level 0.05 for the binomial model on the data of Example 2.4, assuming that the population parameter p is known to be 0.1.

3. [V.A. Abell and S. Rosen] Return to the question of memory residence times considered in problem 3, Section 10.2.1. The empirical distribution function for the memory residence times was measured and is shown below:

Time in milliseconds	*Count*	*Cumulative percent*
0 – 31	7,540	57.24
32 – 63	2,938	79.55
64 – 95	1,088	87.81
96 – 127	495	91.57
128 – 191	449	94.98
192 – 319	480	98.62
319 – 1,727	181	100.00

Graphically compare the empirical distribution with three theoretical distributions, normal, gamma, and hyperexponential (for the last two distributions, parameters were estimated in problem 3, Section 10.2.1, whereas for the normal distribution, parameters μ and σ^2 are readily estimated). Now perform chi-square goodness-of-fit tests against the three theoretical distributions at the 5 percent level of significance.

4. Since the Poisson model was found to be improper for the data in Example 10.33, try the (modified) geometric model:

$$H_0\colon\ p_i = p(1-p)^i, \qquad i = 0,\ 1,\ \ldots .$$

Note that the unknown parameter p must be estimated before the chi-square test can be performed.

5. Using formula (10.5), construct a 90 percent confidence band for the distribution function $F_0(x)$ based on the data in Example 10.34. Plot your results together with the distribution of F_0 under the null hypothesis.

6. Observed times between successive crashes of a computer system were noted for a six-month period as follows (time in hours):

1, 10, 20, 30, 40, 52, 63, 70, 80, 90, 100, 102, 130, 140, 190,

210, 266, 310, 530, 590, 640, 1340

Using the $\hat{D}_n$ statistic , test a goodness-of-fit against an exponential model and a normal model for the population distribution.

7. We wish to verify the analytical results of review problems 2 and 3 at the end of Chapter 3 by means of a simulation. Suppose the mantissas X and Y of two floating-point numbers are independent random variables with uniform (over $[1/\beta, 1))$ and reciprocal pdf $1/(y \ln \beta)$, respectively. Generate n random deviates of X and Y (for random deviate of Y use the formula derived in problem 6(a), Section 3.5) and compute the mantissas of the normalized product Z_N and the normalized quotient Q_N. From these values, obtain the empirical distributions of Z_N and Q_N. Now perform goodness-of-fit tests against the reciprocal distribution, using the Kolmogorov-Smirnov D_n statistic. Use $n = 10, 20, 30$, and $\beta = 2, 10$, and 16.

References

[FISH 1978] G. S. FISHMAN, *Principles of Discrete Event Digital Simulation*, John Wiley & Sons, New York.

[GAY 1978] F. A. GAY, "Evaluation of Maintenance Software in Real-Time Systems," *IEEE Trans. on Comput.*, June 1978.

[HOEL 1971] P. G. HOEL, S. C. PORT, and C. J. STONE, *Introduction to Statistical Theory*, Houghton Mifflin Company, Boston.

[KEND 1961] M. G. KENDALL and A. STUART, *The Advanced Theory of Statistics*, Vol.2: *Inference and Relationship*, Hafner Publishing Company, New York.

[KLEI 1974] J. P. C. KLEIJNEN, *Statistical Techniques in Simulation, Parts 1 and 2*, Marcel Dekker, New York.

[LILL 1967] H. W. LILLIEFORS, "On the Kolmogorov-Smirnov Test for Normality with Mean and Variance Unknown," *J. Amer. Stat. Assoc.*, 62, 399–402.

[LILL 1969] H. W. LILLIEFORS, "On the Kolmogorov-Smirnov Test for the Exponential Distribution with Mean Unknown," *J. Amer. Stat. Assoc.*, 64, 387–389.

[NOET 1967] G.E. NOETHER, *Elements of Nonparametric Statistics*, John Wiley & Sons, New York.

[WALP 1968] R. E. WALPOLE, *Introduction to Statistics*, The Macmillan Company, New York.

[WILK 1962] S. S. WILKS, *Mathematical Statistics*, John Wiley & Sons, New York.

Chapter 11

Regression, Correlation, and Analysis of Variance

11.1 INTRODUCTION

In this chapter, we study aspects of *statistical relationships* between two or more random variables. For example, in a computer system the throughput Y and the degree of multiprogramming X might well be related to each other. One indicator of the association (interdependence) between two random variables is their correlation coefficient $\rho(X, Y)$ and its estimator $\hat{\rho}(X, Y)$. Correlation analysis will be considered in Section 11.5.

A related problem is that of predicting a value of system throughput y at a given degree of multiprogramming x. In other words, we are interested here in studying the dependence of Y on X. The problem then is to find a *regression line* or a *regression curve* that describes the dependence of Y on X. Conversely, we may also study the inverse regression problem of dependence of X on Y. In the balance of this section we consider regression when the needed parameters of the population distribution are known exactly. Commonly, though, we are required to obtain a regression curve that best approximates the dependence based upon sampled information. This topic will be covered in Sections 11.3 and 11.4.

Another related problem is that of *least-squares curve fitting*. Suppose we have two variables (not necessarily random) and we hypothesize a relationship (e.g., linear) between the two variables. From a collection of n pairs of measurements of the two variables, we wish to determine the equation of a line (curve, in general) of closest fit to the data. Out of the many possible criteria in choosing the "best" fit, the criterion leading to simplest calculations is the least-squares criterion, which will be discussed in the next section.

Consider two random variables X and Y possessing a joint density $f(x, y)$. We would like to design a function $d(x)$ so that the random variable $d(X)$ will be as close as possible to Y in an appropriate sense. When $d(x)$ is used to predict a value of Y, a modeling error will usually result so that the actual value, $y = d(x) + \epsilon$.

Such errors are introduced, of course, because we are attempting to simplify the joint distribution of X and Y by postulating an elementary functional dependence d, of Y upon X. The extent of our error is thus the extent to which the random variable Y differs from the random variable $d(X)$, the most common measure of which is undoubtedly the expected value of the squared difference, $E[D^2]$, where $D = Y - d(X)$.

The function $d(x)$ for which $E[D^2]$ is at a minimum is commonly called the **least-squares regression curve** of Y on X. Now it is easy to show that this regression curve is necessarily given by $d(x) = E[Y \mid x]$; nevertheless, conditional distributions are, in practice, difficult to obtain, so most often we simply restrict our choices for $d(x)$ to a specific class of functions and minimize $E[D^2]$ over that class. In a similar fashion $g(y) = E[X \mid y]$ will give us the dependence of X on Y.

A common choice, of course, is the linear predictor function, $d(x) = a + bx$. In general, the appropriate functional class may be inferred from observed data. The n pairs of measurements of the variables X and Y may be plotted as points, $(x_1, y_1), (x_2, y_2), \ldots, (x_n, y_n)$, on the (x, y) plane. The resulting set of points is called a **scatter diagram**. If the observations are clustered about some curve in the (x, y) plane then we may infer that the curve describes the dependence of Y on X. Such a curve is called an **approximating curve**. The inferred relationship could be **linear** as in Figure 11.1, or **nonlinear** as in Figure 11.2. On the other hand, no particular functional dependence of Y on X can be inferred from the scatter diagram in Figure 11.3.

Assume that a linear model of the dependence of Y on X is acceptable; that is, we restrict d to the class of functions of the form $d(x) = a + bx$. The problem of regression (or optimal prediction) then reduces to the problem of choosing the parameters a and b to minimize:

$$G(a, b) = E[D^2] = E[(Y - d(X))^2] = E[(Y - a - bX)^2].$$

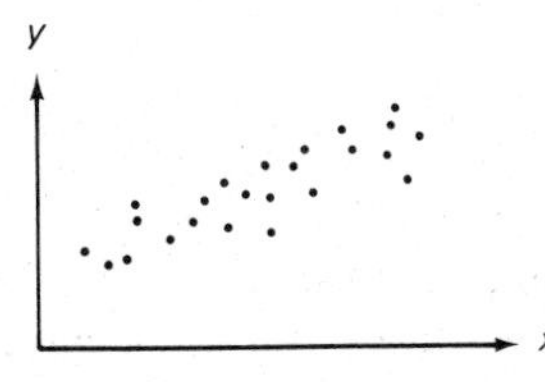

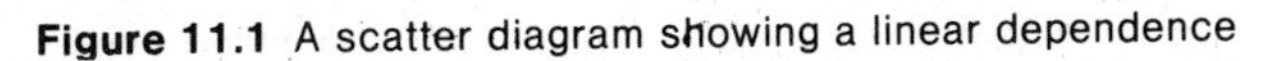
Figure 11.1 A scatter diagram showing a linear dependence

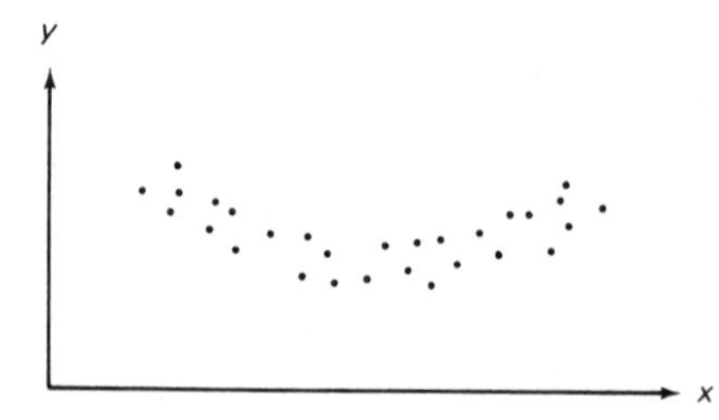

Figure 11.2 A scatter diagram showing a nonlinear dependence

Letting μ_X and μ_Y denote the respective expectations $E[X]$ and $E[Y]$, the mean squared error may be rewritten as:

$$\begin{aligned}
G(a, b) &= E[(Y - a - bX)^2] \\
&= E[((Y - \mu_Y) + (\mu_Y - a) - b(X - \mu_X) - b\mu_X)^2] \\
&= E[(Y - \mu_Y)^2 + (\mu_Y - a)^2 + b^2(X - \mu_X)^2 + b^2\mu^2{}_X \\
&\quad + 2(Y - \mu_Y)(\mu_Y - a) - 2b(Y - \mu_Y)(X - \mu_X) \\
&\quad - 2b(Y - \mu_Y)\,\mu_X \\
&\quad - 2b(\mu_Y - a)(X - \mu_X) \\
&\quad - 2b(\mu_Y - a)\mu_X + 2b^2(X - \mu_X)\mu_X]. \qquad (11.1)
\end{aligned}$$

Note that $E[Y - \mu_Y] = E[X - \mu_X] = 0$ and let:

$$\sigma_X^2 = \text{Var}\,[X], \quad \sigma_Y^2 = \text{Var}\,[Y], \quad \text{and } \rho = \frac{\text{Cov}\,(X, Y)}{\sigma_X \sigma_Y}.$$

Then:

$$\begin{aligned}
G(a, b) &= \sigma_Y^2 + b^2\sigma_X^2 + (\mu_Y - a)^2 + b^2\mu^2{}_X \\
&\quad - 2b\rho\sigma_X\sigma_Y - 2b\mu_X(\mu_Y - a) \\
&= \sigma_Y^2 + b^2\sigma_X^2 + (\mu_Y - a - b\mu_X)^2 - 2b\rho\sigma_X\sigma_Y.
\end{aligned}$$

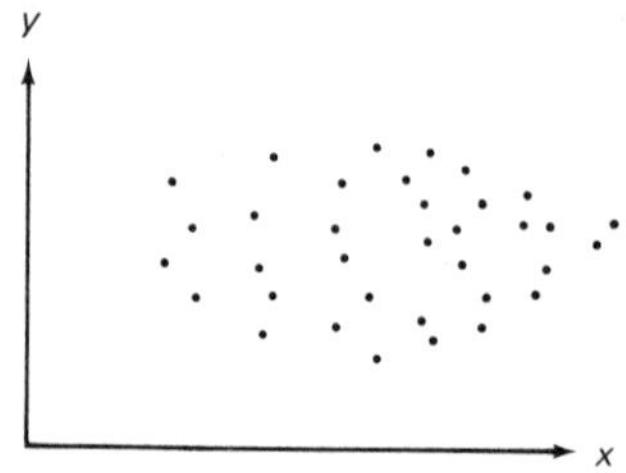

Figure 11.3 A scatter diagram showing no specific dependence

To minimize $G(a, b)$, we take its partial derivatives with respect to a and b and set them equal to zero:

$$\frac{\partial G}{\partial a} = -2(\mu_Y - a - b\mu_X) = 0$$

and

$$\frac{\partial G}{\partial b} = 2b\sigma_X^2 - 2\mu_X(\mu_Y - a - b\mu_X) - 2\rho\sigma_X\sigma_Y = 0.$$

Thus the optimal values of a and b are given by:

$$b = \rho\,\frac{\sigma_Y}{\sigma_X} \quad \text{and} \quad a = \mu_Y - b\mu_X = \mu_Y - \rho\,\frac{\sigma_Y}{\sigma_X}\,\mu_X$$

or

$$b = \frac{\text{Cov}\,(X, Y)}{\text{Var}\,[X]} \tag{11.2}$$

and

$$a = E[Y] - \frac{\text{Cov}\,(X, Y)}{\text{Var}\,[X]}\,E[X]. \tag{11.3}$$

The corresponding linear regression curve is:

$$y = E[Y] - \frac{\text{Cov}\,(X, Y)}{\text{Var}\,[X]}\,(E[X] - x), \tag{11.4}$$

which can be rewritten as:

$$\frac{y - \mu_Y}{\sigma_Y} = \rho\,\frac{x - \mu_X}{\sigma_X}. \tag{11.5a}$$

Similarly, the regression line of Y on X is given by:

$$\frac{x - \mu_X}{\sigma_X} = \rho\,\frac{y - \mu_Y}{\sigma_Y}. \tag{11.5b}$$

The square of the minimum prediction error is:

$$\begin{aligned} E[D^2] = G(a, b) &= \sigma_Y^2 + \rho^2\,\frac{\sigma_Y^2}{\sigma_X^2}\,\sigma_X^2 - 2\rho\,\frac{\sigma_Y}{\sigma_X}\,\rho\sigma_X\sigma_Y \\ &= \sigma_Y^2 + \rho^2\sigma_Y^2 - 2\rho^2\sigma_Y^2 \\ &= \sigma_Y^2 - \rho^2\sigma_Y^2 = (1 - \rho^2)\sigma_Y^2. \end{aligned} \tag{11.6}$$

Recall that $-1 \leqslant \rho \leqslant 1$. From (11.6), if $\rho = \pm 1$, $E[D^2] = 0$. In this case the regression line of Y on X (11.5a) and the regression line of X on Y

(11.5b) coincide with each other, hence X and Y are completely dependent and are said to have a **functional relationship** with each other. (A note of caution: a statistical relationship such as this, however strong, cannot logically imply a causal relationship.) In this case when $\rho^2 = 1$, the linear model is a perfect fit, and in terms of the joint distribution function of X and Y this means that the entire probability mass is concentrated on the regression line. Also note that the best linear model may well be the best model, even when the fit is not perfect (that is, $\rho^2 \neq 1$). For example, it can be shown that when X and Y have a joint normal distribution, the regression curve of Y on X is necessarily linear.

If $\rho = 0$, then X and Y are uncorrelated and the two regression lines [equations (11.5a) and (11.5b)] are at right angles to each other. If X and Y are independent, then $\rho = 0$, but the converse does not hold in general. In the special case of bivariate normal distribution it is true that $\rho = 0$ implies the independence of X and Y. For this reason ρ can be used as a measure of interdependence of X and Y only in cases of normal or near-normal variation. Otherwise ρ should be used as an *indicator* rather than a *measure* of interdependence. Furthermore, ρ is essentially a coefficient of *linear* interdependence, and more complex forms of interdependence lie outside its scope. If we suspect that $E[Y|x]$ is far from being linear in x, a linear predictor is of little value, and an appropriate nonlinear predictor should be sought.

In the case $\rho^2 \neq 1$, from (11.6) we conclude that there is a nonzero prediction error even when the parameters a and b are known exactly. Since, for the minimizing values of a and b:

$$E[D^2] = (1 - \rho^2)\sigma_Y^2,$$

we have

$$\sigma_Y^2 = \rho^2\sigma_Y^2 + E[D^2]. \tag{11.7}$$

The term $\rho^2\sigma_Y^2$ is interpreted as the variance of Y *attributable* to a linear dependence of Y on X and the term $E[D^2] = (1 - \rho^2)\sigma_Y^2$ is the **residual variance** that cannot be "explained" by the linear relationship. Additional errors will occur if the parameters are estimated from observed data.

The predictor discussed above is attractive, since it requires only the knowledge of the first two moments and the correlation coefficient of the random variables X and Y. If true (population) values of these quantities are unknown, then they have to be estimated based on a random sample of n pairs of observations of the two random variables.

More generally, suppose we are interested in modeling the input-output behavior of a stochastic system with m inputs (also called independent variables) $\mathbf{X} = (X_1, X_2, \ldots, X_m)$ and output (or dependent variable) Y. The appropriate predictor of Y, $d(\mathbf{x};\mathbf{a})$, is the conditional expectation:

$$E[Y|x_1, x_2, \ldots, x_m].$$

The predictor $d(\mathbf{x};\mathbf{a})$ is a parameterized family of functions with a vector of parameters **a**. For example, in a model of a moving-head disk, the response time to an I/O request may be the dependent variable, the record size and the request arrival rate the independent variables, and the device transfer rate and average seek time the parameters. If the internal system behavior is well understood (that is, the conditional distribution of Y is known), then the function $d(\mathbf{x};\mathbf{a})$ can be derived analytically. Otherwise, the form of the function must be inferred empirically from observed data.

Once we have determined an appropriate functional class for the predictor d, the parameters **a** have to be determined just as a and b were determined in our linear model $d(x) = a + bx$. Nonlinear models are, of course, more difficult to handle than linear models, because the equations:

$$\frac{\partial}{\partial a_i}(E[(Y - d(\mathbf{X};\mathbf{a}))^2]) = 0, \qquad \text{for all } i$$

are more difficult to solve. We note that even if the regression function d is nonlinear in the independent variable $\mathbf{X}$ it could still be a linear function of the parameters. The regression model will be linear in this case. Thus for instance the regression functions $d = a + b \log x$ or $d = a + b \sin x$ give rise to linear models.

11.2 LEAST-SQUARES CURVE FITTING

Suppose, after reviewing price lists of various manufacturers of high-speed random access memories, we have gathered the data plotted in Figure 11.4. Since price seems to be nearly a linear function of size, we might wish to present our findings in a compact form: the best line through these points. Of course "best" is open to interpretation, and the "eyeball" method has a

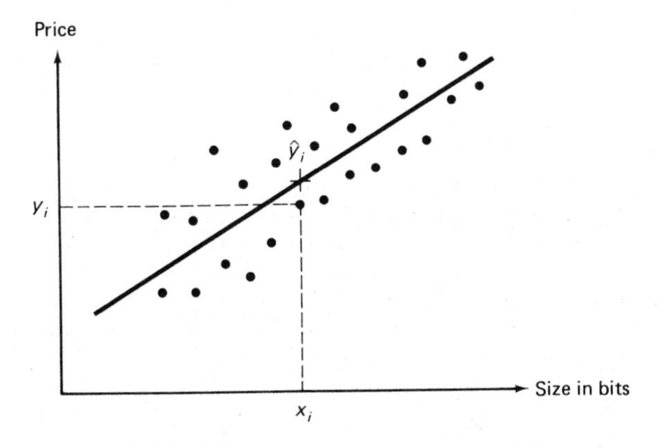

Figure 11.4 Least-squares linear fit

great following. Nevertheless, when a more precise analysis is required, "best" is usually defined to be that line which minimizes the sum of the squares of the y-coordinate deviations from it.

To be more specific, if our points are $\{(x_i, y_i) \mid i = 1, \ldots, n\}$ then we choose a and b so as to minimize

$$\sum_{i=1}^{n} [y_i - (a + b\, x_i)]^2. \tag{11.8}$$

Taking partial derivatives with respect to a and b and setting them equal to 0, we obtain:

$$b = \frac{\sum_{i=1}^{n} x_i y_i - n\bar{x}\bar{y}}{\sum_{i=1}^{n} x_i^2 - n(\bar{x})^2}$$

and (11.9)

$$a = \bar{y} - b\bar{x},$$

where:

$$\bar{x} = \frac{1}{n}\sum_{i=1}^{n} x_i \quad \text{and} \quad \bar{y} = \frac{1}{n}\sum_{i=1}^{n} y_i.$$

Such a technique does have a probabilistic counterpart. Consider two discrete random variables X and Y having joint pmf given by our plot of Figure 11.4, that is, $P(X = x_i, Y = y_i) = \frac{1}{n}$, $i = 1, 2, \ldots, n$. It is meaningful to ask: what is the regression line of Y on X?

From the definition of regression line, we want to minimize:

$$E[(Y - (a + bX))^2] = \sum_{i=1}^{n} \frac{1}{n} [y_i - (a + bx_i)]^2. \tag{11.10}$$

Obviously the values of a and b that minimize (11.10) also minimize (11.8) and thus are given by (11.9).

This empirical approach to regression is known as the **method of least squares** and equations (11.9) are known as the **normal equations of least squares**.

Example 11.1

The failure rate of a certain electronic device is suspected to increase linearly with its temperature. Fit a least-squares linear line through the data in Table 11.1 (two measurements were taken for each given temperature and hence we have twelve pairs of measurements).

Table 11.1 *The Failure Rate versus Temperature*

Temp. °F	55	65	75	85	95	105
Failure rate $\cdot 10^6$	1.90 1.94	1.93 1.95	1.97 1.97	2.00 2.02	2.01 2.02	2.01 2.04

The sample mean of the temperature is $\bar{x} = 80°\text{F}$ and the sample mean of failure rates is $\bar{y} = 1.98 \cdot 10^{-6}$ failures per hour. From these values we get $a = 1.80 \cdot 10^{-6}$ failures per hour and $b = 0.00226$ failures per hour per °F. Thus, $y = 1.80 + 0.00226x$ is the desired least-squares line. From this line we can obtain a predicted value of the failure rate for a specified temperature. For example, the predicted failure rate is $1.9582 \cdot 10^{-6}$ at 70°F. #

Suppose now that we were faced with fitting a summary curve to the data of Figure 11.5, rather than those of Figure 11.4. Though we could still perform a least-squares linear fit, the result would be of at best questionable value, and we would certainly prefer to fit a non-linear curve.

In general, as mentioned earlier in our discussion of regression, nonlinear fitting is much more difficult; nevertheless, in certain circumstances we can consider using our linear results to fit nonlinear curves. For example, suppose we want to fit an exponential curve, $y = ae^{bx}$, to the data of Figure 11.5. We might reason thus: if in each case $y \simeq ae^{bx}$ then $\ln y \simeq \ln a + bx$, so if we transform our data into pairs (x_i, z_i) where $z_i = \ln y_i$ and perform a least squares linear fit to obtain the line $z = a' + b'x$, then

$$y = e^z = e^{a'+b'x} = e^{a'}e^{b'x} = ae^{bx} \qquad \text{where } a = e^{a'} \text{ and } b = b'.$$

Example 11.2

Suppose we are interested in fitting the price of a CPU of an IBM 370 series as a function of its speed x [measured in millions of adds per second (MAPS)]. Grosch's law suggests that price, y, will be roughly proportional to the square root of the speed, x. We use a general power function:

$$y = a \cdot x^b,$$

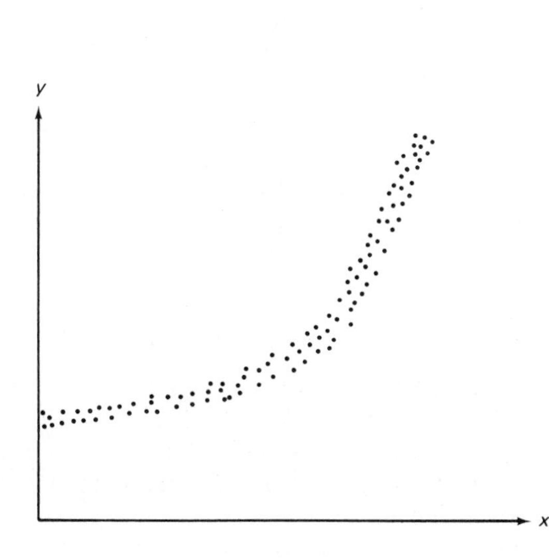

Figure 11.5 Nonlinear curve fitting

to model the relation between the price and the speed. Using the data from [HATF 1977] and using a transformation to $\ln y = \ln a + b \ln x$, Kinicki [KINI 1978] obtained the least-squares fit:

$$\text{estimated price, } \hat{y} = \$1,147,835 \cdot x^{0.55309}.$$

#

We must point out that the easy transformations illustrated here, though analytically precise, do not necessarily preserve the least-squares property; more specifically consider the $y = ae^{bx}$ model: the values of a and b that minimize:

$$\sum_{i=1}^{n} (y_i - ae^{bx_i})^2$$

are not necessarily the same as those we have chosen to use, which minimize:

$$\sum_{i=1}^{n} [\ln y_i - (\ln a + bx_i)]^2.$$

(For an easy example, consider the data $\{(-1, 1), (0, e), (1, 1)\}$). Nevertheless, the transformation fit may be preferable, owing to ease of application.

Problems

1. Consider an arithmetic unit of a computer system with a modulo-m on-line fault-detector. As the modulus m varies, the average detection latency y also varies. Given the following data, with two observations of y for each value of m:

m_i	y_i (μsec)	
3	1.45	1.5
5	1.30	1.26
7	1.20	1.23
11	1.10	1.08
13	1.05	1.03

Determine parameters a and b by performing a least-squares fit of the curve:

$$y = am^b$$

to the given data.

11.3 THE COEFFICIENT OF DETERMINATION

Having obtained a least-squares linear fit to data, $y = a + bx$, we next consider the *goodness* of fit between this line and the data. For a point x_i, the value predicted by the fitted line is

$$\hat{y}_i = a + b\, x_i.$$

The difference $|y_i - \hat{y}_i|$ between the observed and the predicted values should be low for a good fit. Observe that:

$$|y_i - \bar{y}| = |(y_i - \hat{y}_i) + (\hat{y}_i - \bar{y})|.$$

Squaring both sides and then summing, we get:

$$\sum_{i=1}^{n} (y_i - \bar{y})^2 = \sum_{i=1}^{n} (y_i - \hat{y}_i)^2 + \sum_{i=1}^{n} (\hat{y}_i - \bar{y})^2$$
$$+ 2 \sum_{i=1}^{n} (y_i - \hat{y}_i)(\hat{y}_i - \bar{y}).$$

The last sum can be shown to equal zero if we substitute the linear predictor for $\hat{y}_i$:

$$\begin{aligned}
&\sum_{i=1}^{n} (y_i - \hat{y}_i)(\hat{y}_i - \bar{y}) \\
&= \sum [(y_i - a - bx_i)(a + bx_i - \bar{y})] \\
&= a \sum (y_i - a - bx_i) + b \sum x_i (y_i - a - bx_i) \\
&\quad - \bar{y} \sum (y_i - a - b\,x_i) \\
&= 0,
\end{aligned}$$

since a and b are defined to have those values for which:

$$\frac{\partial}{\partial a} \sum_{i=1}^{n} (y_i - a - bx_i)^2 = 0$$

and

$$\frac{\partial}{\partial b} \sum_{i=1}^{n} (y_i - a - bx_i)^2 = 0.$$

It follows that:

$$\sum_{i=1}^{n} (y_i - \bar{y})^2 = \sum_{i=1}^{n} (y_i - \hat{y}_i)^2 + \sum_{i=1}^{n} (\hat{y}_i - \bar{y})^2. \qquad (11.11)$$

Note the similarity between this equation and equation (11.7). Here $y_i - \bar{y}$ is the deviation of the ith observed value of Y from its sample mean; therefore, the left hand side is the **sum of squares about the mean** (also called the **total variation**). $\hat{y}_i - \bar{y}$ is the difference between the predicted value and the sample mean. This quantity is "explained" by the least-squares line, since $\hat{y}_i - \bar{y} = b(x_i - \bar{x})$. Therefore, $\sum_{i=1}^{n} (\hat{y}_i - \bar{y})^2$ is called the **explained variation**. The quantity $\sum_{i=1}^{n} (y_i - \hat{y}_i)^2$ is the **residual sum of**

squares (also called the **unexplained variation**) and should be as small as possible. In fact, this sum would be zero if all the actual observations were to lie on the fitted line. This shows that the total variation can be partitioned into two components:

$$\text{total variation} = \text{unexplained variation} + \text{explained variation}.$$

The ratio of the explained variation to the total variation, called the **coefficient of determination**, is a good measure of how well the line $y = ax + b$ fits the given data.

$$\text{coefficient of determination} = \text{explained variation/total variation}$$

$$= \frac{\sum (\hat{y}_i - \bar{y})^2}{\sum (y_i - \bar{y})^2}. \tag{11.12}$$

The coefficient of determination must lie between 0 and 1, and the closer it is to 1, the better the fit.

Returning to Example 11.1, we compute the variation explained by regression to be 0.01788428 and the total variation to be 0.0206, hence the coefficient of determination is 0.8682, indicating a fairly good fit, since 86.82 percent of the variation in Y-values is explained by the fitted line.

Example 11.3

The failure rate (hazard rate) $h(t)$ of a system is thought to be a power function of its age. In other words a Weibull model seems appropriate, so that:

$$h(t) = ct^d, \qquad t \geqslant 0.$$

A large number of systems are put on test, and we divide the number of failures in each hourly interval by the number of surviving components to obtain an observed value, h_i, of the hazard rate in the interval. The data for this example are shown in Table 11.2.

Table 11.2 The Failure Rate Data for a Weibull Model

t, hours	$h_i \cdot 10^3$	$y_i = \log h_i$	$x_i = \log t_i$	$\hat{y}_i$
0.5–1.5	1.05	0.02119	0.0000	0.008946
1.5–2.5	3.95	0.59600	0.3010	0.607515
2.5–3.5	8.20	0.91381	0.4771	0.957707
3.5–4.5	17.60	1.24551	0.6021	1.206282
4.5–5.5	25.00	1.39794	0.6990	1.398977
5.5–6.5	38.00	1.57978	0.7782	1.556474
6.5–7.5	49.00	1.69020	0.8451	1.689512
7.5–8.5	59.00	1.77085	0.9031	1.804850
8.5–9.5	85.00	1.92942	0.9542	1.906468
9.5–10.5	97.50	1.98900	1.0000	1.997546
		13.1343	6.5598	

$$\sum y_i = 13.1343, \qquad \sum x_i = 6.5598,$$
$$\sum y_i^2 = 20.86416, \qquad \sum x_i^2 = 5.2152,$$
$$\sum x_i y_i = 10.42965.$$

Therefore:

$$a = (\log c) = 0.008930 \quad \text{or} \quad c \simeq 1.0208$$

and

$$b = d = 1.9886.$$

Also, $\sum (\hat{y}_i - \bar{y})^2 = 3.607$ and $\sum (y_i - \bar{y})^2 = 3.613$ which implies that the coefficient of determination is $3.607/3.613 = 0.9983$. Hence the quality of fit as indicated by the coefficient of determination is near perfect. The estimated hazard-rate function is $h(t) = 1.0208 \cdot t^{1.9886}$. #

Problems

1. Compute the coefficient of determination for the least-squares fit of the data in problem 1, Section 11.2.

11.4 CONFIDENCE INTERVALS IN LINEAR REGRESSION

We return to the problem of linear regression. Now we assume, not that required parameters of the population distributions are known, but rather that they are estimated based upon a random sample of n pairs of observations. Because the conditional expectation of Y given X minimizes the mean squared error, we usually consider X to be a controlled or nonrandom variable; that is, in sampling we can restrict ourselves to those points at which X takes on an a priori specified value. If we let Y_i denote the random variable Y restricted to those points at which $X = x_i$, then Y_i will be assumed to be normal with mean $\mu_i = a + bx_i$ and a common variance $\sigma^2 = \text{Var}[Y_i]$. In order to derive maximum-likelihood estimates of the two population parameters, a and b, we form the likelihood function of a and b:

$$L(a,b) = \prod_{i=1}^{n} f(y_i)$$
$$= \frac{\exp[-\frac{1}{2\sigma^2} \sum_{i=1}^{n} (y_i - a - bx_i)^2]}{(2\pi)^{n/2}\sigma^n}. \tag{11.13}$$

Taking logarithms and setting the derivatives with respect to a and b to zero, we get:

$$\sum_{i=1}^{n} (y_i - a - bx_i) = 0,$$
$$\sum_{i=1}^{n} (y_i - a - bx_i)x_i = 0.$$

The resulting maximum-likelihood estimates of a and b, denoted by $\hat{a}$ and $\hat{b}$, are then given by the normal equations (11.9). The corresponding estimators are:

$$\hat{A} = \overline{Y} - \hat{B}\overline{x} \tag{11.15}$$

and

$$\hat{B} = \frac{\sum_{i=1}^{n} (x_i - \overline{x})(Y_i - \overline{Y})}{\sum_{i=1}^{n} (x_i - \overline{x})^2}. \tag{11.16}$$

Now if Y_i is not normally distributed, we can still obtain the point estimates of the parameters a and b by the method of least squares. Similarly, the discussion about the coefficient of determination holds in the general case. However, in order to derive confidence intervals for these estimates we need to make assumptions similar to those we made while deriving maximum-likelihood estimates of a and b. Thus we continue to assume that X is a controlled variable. We further assume that $Y_1, Y_2, \ldots, Y_n$ are mutually independent, and that the conditional distribution of Y_i is normal with mean $E[Y_i] = a + bx_i$, and that variance of Y_i is equal to σ^2, that is $\text{Var}[Y_i] = \sigma_i^2 = \sigma^2$.

If we write:

$$Y_i = a + bx_i + \Delta_i, \tag{11.17}$$

then $\Delta_1, \Delta_2, \ldots, \Delta_n$ are mutually independent normal random variables with zero mean and a common variance σ^2, that is, $E[\Delta_i] = 0$ and $\text{Var}[\Delta_i] = \sigma^2$ for all i.

THEOREM 11.1

(a) $\hat{A}$ is an unbiased estimator of a; $E[\hat{A}] = a$,

(b) $\hat{B}$ is an unbiased estimator of b; $E[\hat{B}] = b$,

(c) $$\text{Var}[\hat{A}] = \frac{\sigma^2}{n}\left[1 + \frac{n\overline{x}^2}{\sum_{i=1}^{n} (x_i - \overline{x})^2}\right],$$

(d) $$\text{Var}[\hat{B}] = \frac{\sigma^2}{\sum_{i=1}^{n} (x_i - \overline{x})^2}.$$

Proof:

First we show part (b):

$$E[\hat{B}] = E\left[\frac{\sum (x_i - \overline{x})(Y_i - \overline{Y})}{\sum (x_i - \overline{x})^2}\right]$$

$$= \frac{\sum (x_i - \overline{x})\, E[Y_i - \overline{Y}]}{\sum (x_i - \overline{x})^2}.$$

Now since $E[Y_i] = a + b\, x_i$ and:

$$E[\overline{Y}] = \frac{1}{n} \sum_{i=1}^{n} E[Y_i] = a + b\, \overline{x},$$

we have:

$$E[\hat{B}] = \frac{\sum (x_i - \overline{x})(a + bx_i - a - b\overline{x})}{\sum (x_i - \overline{x})^2}$$

$$= b.$$

To show part (a):

$$\begin{aligned} E[\hat{A}] &= E[\overline{Y} - \hat{B}\overline{x}] \\ &= E[\overline{Y}] - E[\hat{B}]\overline{x} \\ &= a + b\overline{x} - b\overline{x} \\ &= a \end{aligned}$$

Next to show part (d), we rewrite:

$$\hat{B} = \frac{\sum (x_i - \overline{x})\, Y_i}{\sum (x_i - \overline{x})^2}.$$

Since the Y_i's are independent, we have:

$$\begin{aligned} \text{Var}\, [\hat{B}] &= \frac{1}{[\sum (x_i - \overline{x})^2]^2} \sum \text{Var}\, [(x_i - \overline{x})\, Y_i] \\ &= \sum \frac{(x_i - \overline{x})^2 \sigma^2}{\left[\sum (x_i - \overline{x})^2\right]^2} \\ &= \frac{\sigma^2}{\sum (x_i - \overline{x})^2}. \end{aligned}$$

The result of part (c) follows in a similar fashion.

It is also clear that $\hat{A}$ and $\hat{B}$ are both normally distributed. If the variance σ^2 of the error term Δ_i is known, we can use the Var $[\hat{A}]$ and Var $[\hat{B}]$ above to obtain confidence intervals for a and b. Usually, however, σ^2 will not be known in advance and it must be estimated by s^2:

$$\begin{aligned} s^2 &= \frac{\sum_{i=1}^{n} (y_i - \hat{a} - \hat{b}x_i)^2}{n - 2} \\ &= \frac{\sum (y_i - \hat{y}_i)^2}{n - 2} \qquad (11.18) \\ &= \frac{\text{unexplained variation}}{n - 2}. \end{aligned}$$

The denominator $(n - 2)$ reflects the fact that two degrees of freedom have been lost since $\hat{a}$ and $\hat{b}$ have been estimated from the given data.

If we use s^2 in place of σ^2 in estimating Var $[\hat{A}]$ and Var $[\hat{B}]$, we must use the Student t distribution with $(n - 2)$ degrees of freedom in place of a normal distribution. Now a $100(1 - \alpha)$ percent confidence interval for b is given by:

$$\hat{b} \pm t_{n-2;\alpha/2} \cdot s \cdot \left[\sum_{i=1}^{n} (x_i - \bar{x})^2\right]^{-1/2} \tag{11.19}$$

and for a by:

$$\hat{a} \pm t_{n-2;\alpha/2} \cdot s \cdot \left[\frac{1}{n} + \frac{\bar{x}^2}{\sum_{i=1}^{n} (x_i - \bar{x})^2}\right]^{1/2}. \tag{11.20}$$

Example 11.4

Returning to the Weibull model and the corresponding failure-rate data of Example 11.3, we compute the unexplained variation to be 0.006 and hence:

$$s^2 = 0.006/8 \quad \text{or} \quad s = 0.02738.$$

Also:

$$\sum (x_i - \bar{x})^2 = 0.91211 \quad \text{and} \quad \sqrt{\sum_i (x_i - \bar{x})^2} = 0.95504.$$

Thus, the required confidence intervals are obtained using the t distribution with $n - 2 = 8$ degrees of freedom. Since:

$$t_{8;0.05} = 1.860,$$

a 90 percent confidence interval for the exponent d is given by:

$$1.98864 \pm 0.05332.$$

Similarly, a 90 percent confidence interval for a ($=$ log c) is given by 0.008930 ± 0.03851. #

Problems

1. Complete the proof of Theorem 11.1 by showing part (c).
2. Compute 90 percent confidence intervals for the parameters a and b from the data in problem 1 of Section 11.2.

11.5 CORRELATION ANALYSIS

In the last section we assumed that X was a controlled or nonrandom variable. Consider again the case when X and Y are both random variables. A random sample will then consist of n pairs of observations $\{(x_1, y_1), (x_2, y_2), \ldots, (x_n, y_n)\}$ from a bivariate distribution.

Recall from Chapter 4 that the correlation coefficient $\rho(X, Y)$ gives an indication of the linearity of the relationship between X and Y. If X and Y are independent, then they are uncorrelated; that is, $\rho(X, Y) = 0$. The converse does not hold, in general. However, if X and Y are jointly normal, then $\rho(X, Y) = 0$ (X and Y are uncorrelated) implies that they are also independent. We will assume that $f(x, y)$ is a bivariate normal pdf, so that:

$$f(x, y) = \frac{1}{2\pi\, \sigma_X \sigma_Y \sqrt{1-\rho^2}} \cdot \exp\{-\frac{1}{2(1-\rho^2)}\,[(\frac{x-\mu_X}{\sigma_X})^2 - 2\rho\,(\frac{x-\mu_X}{\sigma_X})(\frac{y-\mu_Y}{\sigma_Y}) + (\frac{y-\mu_Y}{\sigma_Y})^2]\}, \tag{11.21}$$

where:

$$\mu_X = E[X], \quad \mu_Y = E[Y], \quad \sigma_X^2 = \text{Var}\,[X], \; \sigma_Y^2 = \text{Var}\,[Y]$$

and $\rho = \rho(X, Y)$.

We can form the likelihood function for a random sample of size n as:

$$L(\mu_X, \mu_Y, \sigma_X, \sigma_Y, \rho) = \prod_{i=1}^{n} f(x_i, y_i).$$

Taking natural log on both sides, we obtain:

$$\begin{aligned}\ln L &= \sum_{i=1}^{n} \ln f(x_i, y_i) \\ &= -n \ln\,(2\pi\sigma_X\,\sigma_Y\sqrt{1-\rho^2}) - \frac{1}{2(1-\rho^2)}\,\{\sum_{i=1}^{n} [(\frac{x_i-\mu_X}{\sigma_X})^2 \\ &\quad - 2\rho\,(\frac{x_i-\mu_X}{\sigma_X})(\frac{y_i-\mu_Y}{\sigma_Y}) + (\frac{y_i-\mu_Y}{\sigma_Y})^2]\}.\end{aligned}$$

Taking the partial derivatives with respect to the five parameters, and setting them equal to zero, we get the maximum-likelihood estimates:

$$\hat{\mu}_X = \frac{1}{n}\sum_{i=1}^{n} x_i = \bar{x}, \qquad \hat{\mu}_Y = \frac{1}{n}\sum_{i=1}^{n} y_i = \bar{y},$$

$$s_X = (\frac{1}{n}\sum_{i=1}^{n} (x_i - \hat{\mu}_X)^2)^{1/2}, \qquad s_Y = (\frac{1}{n}\sum_{i=1}^{n} (y_i - \hat{\mu}_Y)^2)^{1/2}, \tag{11.22}$$

and

$$\hat{\rho}(X, Y) = \frac{\sum_{i=1}^{n} (x_i - \bar{x})(y_i - \bar{y})}{(\sum_{i=1}^{n} (x_i - \bar{x})^2 \sum_{i=1}^{n} (y_i - \bar{y})^2)^{1/2}}. \tag{11.23}$$

The estimate $\hat{\rho}(X, Y)$ is also referred to as the sample correlation coefficient (let the corresponding estimator be denoted by $\hat{R}$). The distribution of $\hat{R}$ is difficult to obtain. When $\rho = \rho(X, Y) = 0$ (X and Y are independent), it can be shown [KEND 1961] that the statistic:

$$T = \hat{R}\left(\frac{n-2}{1-\hat{R}^2}\right)^{1/2} \tag{11.24}$$

has a Student t distribution with $(n - 2)$ degrees of freedom. The test for the hypothesis (of *independence* of normal random variables X and Y):

$$H_0\text{: } \rho(X, Y) = 0 \quad \text{versus the alternative} \quad H_1\text{: } \rho(X, Y) \neq 0$$

with level of significance α is to reject the hypothesis H_0 when $|T| > t_{n-2;\alpha/2}$, where $t_{n-2;\alpha/2}$ denotes the critical value of the t distribution with $(n - 2)$ degrees of freedom in the usual way. To test the hypothesis:

$$H_0\text{: } \rho(X, Y) = 0 \quad \text{versus the alternative} \quad H_1\text{: } \rho(X, Y) > 0$$

with a level of significance α, we reject H_0 if $T > t_{n-2;\alpha}$.

Example 11.5

It is believed that the number of breakdowns of a computing center is related to the number of jobs processed. Data was collected for two centers A and B as follows:

1. For center A, data on the number of breakdowns per month (y_i) and the number of jobs completed per month (x_i) were collected for ten months.
2. For center B, data of x_i and y_i were collected for twenty months.

The maximum-likelihood estimates of the correlation coefficients for the two cases were computed using formula (11.23) to be $\hat{\rho}_A = 0.49$ and $\hat{\rho}_B = 0.55$. The corresponding values of the t statistic were computed using formula (11.24): $t_A = 1.5899$ and $t_B = 2.794$. Recalling that a t distribution with eight degrees of freedom is applicable for center A, and that $t_{8;0.05} = 1.86$, we see that we cannot reject the hypothesis H_0: $\rho_A = 0$ in favor of H_1: $\rho_A \neq 0$ at the significance level 0.1.

On the other hand for center B we use a t distribution with eighteen degrees of freedom and $t_{18;0.01} = 2.552$ to reject H_0: $\rho_B = 0$ in favor of H_1: $\rho_B \neq 0$ at the 0.02 level of significance; that is, there is a significant correlation between breakdowns and workload. #

Problems

1. Associated with a job are two random variables: CPU time required (Y) and the number of disk I/O operations (X). Given the following data, compute the sample correlation coefficient.

i	Time (sec) y_i	Number x_i
1	40	398
2	38	390
3	42	410
4	50	502
5	60	590
6	30	305
7	20	210
8	25	252
9	40	398
10	39	392

Draw a scatter diagram from these data. Does a linear fit seem reasonable? Assuming we wish to predict the CPU time requirement given an I/O request count, perform a linear regression:

$$y = a + bx.$$

Compute point estimates of a and b as well as 90 percent confidence intervals.

Next suppose we want to predict a value of I/O request count, given a CPU time requirement. Thus perform a linear regression of X on Y. Calculate 90 percent confidence intervals for c and d with the regression line:

$$x = c + dy.$$

In both cases compute the coefficients of determination.

2. Since the method described for testing the hypothesis of no correlation in this section is based on a stringent assumption that the joint density of X and Y is bivariate normal, it is desirable to design a *nonparametric* alternative to this test in case the distributional assumption is not satisfied. First we arrange both the X and Y sample values separately in increasing order of magnitudes and assign ranks to these values. Now let δ_i be the difference between the ranks of the paired observations (x_i, y_i). The Spearman's **rank correlation coefficient** is defined to be [NOET 1976]:

$$r_S = 1 - \frac{6 \sum_{i=1}^{n} \delta_i^2}{n(n^2 - 1)}.$$

Compute r_S for the data in problem 1 above.

Note that r_S is a value of a random variable R_S. Suppose we want to perform a distribution-free test of the null hypothesis of no correlation. Then we can use the fact that under the null hypothesis, the distribution of R_S is approximately normal with:

$$E[R_S|H_0] = 0$$

$$\text{Var}[R_S|H_0] = \frac{1}{n-1}.$$

Carry out this test for the data in problem 1 above, and compute the descriptive level of the test.

11.6 SIMPLE NONLINEAR REGRESSION

Nonlinear regression presents special difficulties. The first and foremost of these, of course, is that the basic system of equations, given by:

$$\frac{\partial}{\partial a_i} E[(Y - d(X; a_1, a_2, \ldots, a_k))^2] = 0, \qquad i = 1, 2, \ldots, k,$$

is, in general, nonlinear in the unknown parameters. Nonlinear systems of equations are considerably more difficult to solve than linear systems. The difficulty is compounded when we realize that we are attempting to locate the lowest point (global minimum) of a multidimensional surface. If the surface has multiple valleys (that is, it is not unimodal), most numerical techniques can do no better than to find a local minimum (one of the valleys) rather than the global minimum.

Example 11.6

The price y of a semiconductor memory module is suspected to be a nonlinear function of its capacity x:

$$y = a + bx^c.$$

Even the logarithmic transformation that we used earlier does not help here. The empirical mean squared error is given by:

$$G(a, b, c) = \sum_{i=1}^{n} (y_i - a - bx_i^c)^2.$$

Either using a nonlinear (unconstrained) minimization routine directly, or using the derivative method and solving the resulting system of nonlinear equations is quite complex. We observe that if a is known, the problem can be transformed to a linear regression situation. Since a is not known, we can iterate on different values of a to find an optimum point.

Based on data collected from manufacturers (117 data points), Maieli [MAIE 1980] obtained the following least-squares fit for the above model of semiconductor memory: $b = 0.75608$ (indicating a price per bit of about ¾ cents), $c = 0.72739$ (indicating a significant economy of scale) and $a = -7488$ cents. The coefficient of determination was found to be 0.982, indicating a very good fit. #

Problems

1. Consider a computer system that is subject to periodic diagnosis and maintenance every 1,000 hours. The diagnosis and maintenance service is assumed not to be perfect, and the probability of its being able to correctly diagnose and correct the fault (if it exists) is c. The expected life y of the system is to be fitted as a power function of the coverage factor c:

$$y = a + be^c.$$

(Here e denotes the base of the natural logarithm.) Using the following data (adapted from [INGL 1976]):

c_i	y_i (hr)
0.2	11,960
0.4	15,950
0.6	23,920
0.8	47,830
0.9	95,670
0.92	120,000
0.94	159,500
0.96	239,200
0.98	478,340

Estimate the parameters a, b, and c and compute the coefficient of determination. If this is found not to be satisfactory, then plotting a scatter diagram may help you find a more appropriate function class.

2. Refit the data of problem 1, Section 11.2, to the curve:

$$y = a_1 m^{b_1} + c_1$$

Compare the quality of two fits by comparing the error sum of squares in the two cases.

11.7 HIGHER-DIMENSIONAL LEAST-SQUARES FIT

The treatment of least-squares linear fit to data can be extended in a simple way to cover problems involving several independent variables [DRAP 1966].

For example, suppose we wish to fit a three-dimensional plane:

$$y = a_0 + a_1x_1 + a_2x_2 + a_3x_3,$$

to a set of n points:

$$(y_1, x_{11}, x_{21}, x_{31}), \ldots, (y_n, x_{1n}, x_{2n}, x_{3n})$$

in four dimensions. The empirical mean squared error is given by:

$$G(a_0, a_1, a_2, a_3) = \sum_{i=1}^{n} (y_i - a_0 - a_1 x_{1i} - a_2 x_{2i} - a_3 x_{3i})^2. \quad (11.25)$$

Setting partial derivatives with respect to the four parameters equal to zero, we get the following solution for $\mathbf{a} = (a_0, a_1, a_2, a_3)$, given in matrix form:

$$\mathbf{a} = (X^T X)^{-1} X^T \mathbf{y} \quad (11.26)$$

where $\mathbf{y} = (y_1, y_2, y_3, \ldots, y_n)$,

$$X = \begin{bmatrix} 1 & x_{11} & x_{21} & x_{31} \\ 1 & x_{12} & x_{22} & x_{32} \\ \cdot & \cdot & \cdot & \cdot \\ \cdot & \cdot & \cdot & \cdot \\ \cdot & \cdot & \cdot & \cdot \\ 1 & x_{1n} & x_{2n} & x_{3n} \end{bmatrix}.$$

and superscript T denotes matrix transpose.

Example 11.7 (Overhead Regression)

In a given interval of time, the CPU usage may be divided into idle time, user time, and supervisory overhead time. To improve performance, we would like to investigate how to reduce the supervisory overhead. This overhead is caused by various calls to the supervisor. If x_i denotes the number of calls of type i and y denotes the overhead per unit of real time, then a linear model of overhead might be:

$$y = a_0 + \sum a_i x_i.$$

From the given measurement data, $(y_1, x_{11}, x_{21}, \ldots), (y_2, x_{12}, x_{22}, \ldots), \ldots, (y_n, x_{1n}, x_{2n}, \ldots)$, we can use our techniques [equation (11.26)] to estimate the parameters $a_0, a_1, \ldots$ A large value of a_i indicates that the ith type of supervisor call contributes heavily to the overhead. We could try to reduce the number of calls of this type or reduce the execution time of the corresponding service routine to reduce the value of a_i.

For example, this method was used by [BARD 1978] in a study of CP-67 system. Letting x_1 = number of virtual selector I/O, x_2 = number of of pages read, and x_3 = number of spool I/O, corresponding coefficients were obtained to be $a_1 = 9.7$, $a_2 = 6.1$, and $a_3 = 6.0$. A redesign of the supervisory routines led to a reduction in these numbers, $a_1 = 7.9$, $a_2 = 2.0$, and $a_3 = 4.6$. #

As a special case of the discussion in this section, consider polynomials of the form $y = \sum_{j=0}^{k} a_j x^j$, which are often useful in practice. These polynomials are linear in the unknown parameters and, hence, produce a linear system of equations for the unknowns. In fact, these equations can be obtained by substituting $(x_i)^j$ for x_{ji} in equation (11.26) where x_i is the ith observation of the independent variable.

Problems

1. [PUTN 1978] Let K denote the total man-years required for a software project over its life cycle. Let T_d be the development time until it is put into operation. The ratio $Y = K/T_d$ is to be fitted as a linear function of the number of report types (x_1) to be generated and the number of application subprograms (x_2) from the following data about nineteen software systems, where the y_i values are in man-years per year:

x_{1i}	x_{2i}	y_i
45	52	32.28
44	31	56.76
74	39	19.76
34	23	35.00
41	35	19.10
5	5	5.00
14	12	3.97
10	27	36.40
95	109	48.09
109	229	57.95
179	256	191.78
101	144	115.43
192	223	216.48
215	365	240.75
200	398	342.82
59	75	98.85
228	241	224.55
151	120	55.66
101	130	50.35

Obtain the parameters of the linear fit and compute the coefficient of determination.

11.8 ANALYSIS OF VARIANCE

In many practical situations we wish to determine which of the many independent variables affect the particular dependent variable. Regression analysis can be used to answer such questions, provided the independent variables take numeric values. Situations often arise where qualitative variables are encountered. For instance, we may wish to study the effect of disk scheduling algorithm on disk request throughput. The independent variable representing the scheduling algorithm may take FCFS, SCAN, SSTF, and so on as its values. (For a discussion of disk scheduling algorithms see [COFF 1973].) In such problems, regression is not directly applicable.

The technique known as **analysis of variance** is usually cast in the framework of **design and analysis of experiments**. The first step in planning a measurement experiment is to formulate a clear statement of the objectives. The second step is the choice of dependent variable (also called the **response variable**). The third step is to identify the set of all independent variables (called **factors**) that can potentially affect the value of the response variable. Some of these factors may be quantitative (e.g., the number of I/O operations performed by a job) while others may be qualitative (e.g., the disk scheduling algorithm). A particular value of a factor is called a **level**.

A factor is said to be **controllable** if its levels can be set by the experimenter, while the levels of an **uncontrollable** (or **observable**) factor cannot be set but only observed. For example, if we wish to test a logic circuit and if it is being tested on-line then its input variables are observable factors. On the other hand, if the circuit is being tested off-line, then it can be driven by a chosen set of inputs, and these factors are then controllable. Similarly, when a performance measurement is conducted on a computer system under its production workload, its workload factors are uncontrollable. In order to make the workload factors controllable, the experiment can be conducted using synthetic jobs. If there are m controllable factors, an m-tuple of assignments of a level to each of those factors is called a **treatment**.

The purpose in applying analysis of variance to an experimental situation is to compare the effect of several simultaneously applied factors on the response variable. This technique allows us to separate the effects of interest (those due to controllable factors) from the **uncontrolled** or **residual variation**. It allows us not only to gauge the effects of individual factors but also to study the interactions between the factors.

In order to study the effects of factors on the response variable, we must start with an empirical model. It is usual to assume that the effects of various factors and the interactions are **additive**.

First we deal with one-way analysis of variance, where we wish to study the effect of one controllable factor on the response variable. Assume that the factor can take c different levels. For the ith level we assume that the response variable Y_i takes the form:

$$Y_i = \mu_i + \Delta_i, \tag{11.27}$$

where Δ_i is a random variable that captures the effect of all uncontrollable factors. We will assume that $\Delta_1, \Delta_2, \ldots, \Delta_c$ are mutually independent, normally distributed random variables with zero means and a common variance σ^2; that is, $E[\Delta_i] = 0$ and Var $[\Delta_i] = \sigma^2$ for all i. Even though this last assumption may not hold in practice, it has been shown that the technique of analysis of variance is fairly robust, in that it is relatively insensitive to violations of the assumption of normality as well as the assumption of equal variances.

Suppose n observations are taken on each of the c levels for a total of $n_T = nc$ observations. Let y_{ij} be the jth observed value of the response variable at the ith level. The random variable Y_{ij} possesses the same distribution as Y_i:

$$Y_{ij} = \mu_i + \Delta_{ij}, \qquad i = 1, 2, \ldots, c, \; j = 1, 2, \ldots, n. \tag{11.28}$$

It is convenient to rewrite the above equation so that:

$$Y_{ij} = \mu + \alpha_i + \Delta_{ij}, \qquad i = 1, 2, \ldots, c, \; j = 1, 2, \ldots, n, \tag{11.29}$$

where: μ is the overall average $\dfrac{\sum_{i=1}^{c} \mu_i}{c}$,
α_i is the effect of the ith level,
Δ_{ij} is the random error term.

Note that $\mu_i = \mu + \alpha_i$ and that $\sum_{i=1}^{c} \alpha_i = 0$.

Such data are usually organized in the form of a table, as shown in Table 11.3. It is common to think of level i defining a separate population with corresponding population mean μ_i. Let the ith sample mean $\bar{Y}_{i.} = \dfrac{1}{n} \sum_{j=1}^{n} Y_{ij}$ where a dot in the subscript indicates the index being summed over. Then $\bar{Y}_{i.}$ is the minimum-variance unbiased estimator of the population mean μ_i.

Table 11.3 *Observations for One-Way Analysis of Variance*

	Observations	*Sample mean*	*Population mean*	*Sample variance*
Level 1	$y_{11}, y_{12}, \ldots, y_{1n}$	$\bar{y}_{1.}$	μ_1	s_1^2
Level 2	$y_{21}, y_{22}, \ldots, y_{2n}$	$\bar{y}_{2.}$	μ_2	s_1^2
.	.	.	.	.
.	.	.	.	.
.	.	.	.	.
Level c	$y_{c1}, y_{c2}, \ldots, y_{cn}$	$\bar{y}_{c.}$	μ_c	s_c^2

Our aim is to compare the observed sample means. If the observed means are close together, then the differences can be attributed to residual or chance variation. On the other hand, if the observed sample means are dispersed, then there is reason to believe that the effects of different treatments are significantly different. The problem can be formulated as a hypothesis test:

$$H_0\colon \mu_1 = \mu_2 = \cdots = \mu_c \quad \text{versus} \quad H_1\colon \mu_i \neq \mu_j \text{ for some } i \neq j. \tag{11.30}$$

If there are only two levels, then this is a special case of hypothesis test on two means as discussed in Chapter 10. In that case a t test can be used to compare $\bar{y}_{1.}$ and $\bar{y}_{2.}$ In the general case of c levels, we may be tempted to perform a series of t tests between each pair of sample means. As the number of levels c increases, however, this procedure not only encounters the problem of combinatorial growth but also suffers a loss of significance. For example, suppose there are 6 levels and hence $\binom{6}{2} = 15$ t tests need to be performed. Assume that each test is conducted at the 5 percent level of significance (that is, the probability of a type I error in each test is 0.05). Assuming that the composite hypothesis H_0 is true, the probability that each individual t test will lead us to accept the hypothesis $\mu_i = \mu_j$ is $1 - 0.05 = 0.95$. All fifteen individual hypotheses must be accepted in order for us to accept H_0. Assuming that these tests are mutually independent, the probability associated with this event is $(0.95)^{15}$, hence the probability of an overall type I error is $1 - (0.95)^{15} \simeq 0.537$. Clearly, this approach is not feasible for problems of reasonable size.

The method we develop for testing hypothesis (11.30) is based on comparing different estimates of the population variance σ^2. This variance can be estimated by any one of the sample variances:

$$S_i^2 = \frac{1}{n-1} \sum_{j=1}^{n} (Y_{ij} - \bar{Y}_{i.})^2 \tag{11.31}$$

and, hence, also by their mean:

$$S_W^2 = \sum_{i=1}^{c} \frac{S_i^2}{c} = \frac{\sum_{i=1}^{c} \sum_{j=1}^{n} (Y_{ij} - \bar{Y}_{i.})^2}{n_T - c}. \tag{11.32}$$

Note that $E[S_W^2] = \sigma^2$. The subscript W reminds us that this is a **within-sample variance**. By our assumptions, $Y_{ij} \sim N(\mu_i, \sigma^2)$ hence $\bar{Y}_i \sim N(\mu_i, \sigma^2/n)$. It follows that:

$$\frac{(n-1)S_i^2}{\sigma^2}$$

has a chi-square distribution with $(n - 1)$ degrees of freedom, and:

$$\frac{(n_T - c)S_W^2}{\sigma^2}$$

has a chi-square distribution with $(n_T - c)$ degrees of freedom.

An alternative method of estimating σ^2 is to obtain the variance of the c sample means:

$$S^2_{\bar{Y}_{..}} = \frac{\sum_{i=1}^{c} (\bar{Y}_{i.} - \bar{Y}_{..})^2}{c - 1}, \tag{11.33}$$

where the overall sample mean $\bar{Y}_{..}$ is defined by:

$$\bar{Y}_{..} = \frac{\sum_{i=1}^{c} \bar{Y}_{i.}}{c} = \frac{\sum_{i=1}^{c}\sum_{j=1}^{n} Y_{ij}}{n_T}. \tag{11.34}$$

Noting that:

$$E[\bar{Y}_{i.}] = \mu_i, \qquad E[\bar{Y}_{..}] = \mu,$$

$$\text{Var}\,[\bar{Y}_{i.}] = \sigma^2/n, \qquad \text{Var}\,[\bar{Y}_{..}] = \frac{\sigma^2}{n_T}$$

we have:

$$E[\bar{Y}^2_{i.}] = \text{Var}\,[\bar{Y}_{i.}] + \mu_i^2 = \frac{\sigma^2}{n} + \mu_i^2$$

and

$$E[\bar{Y}^2_{..}] = \text{Var}\,[\bar{Y}_{..}] + \mu^2 = \frac{\sigma^2}{n_T} + \mu^2.$$

Now, taking expectations on both sides of equation (11.33), we have:

$$\begin{aligned} E[S^2_{\bar{Y}_{..}}] &= E[\frac{\sum \bar{Y}^2_{i.}}{c - 1} - \frac{c}{c - 1}\bar{Y}^2_{..}] \\ &= \frac{1}{c - 1}\left(\frac{c\sigma^2}{n} + \sum_{i=1}^{c} \mu_i^2\right) - \frac{c}{c - 1}\left(\frac{\sigma^2}{nc} + \mu^2\right) \\ &= \frac{\sigma^2}{n} + \frac{\sum_{i=1}^{c} \mu_i^2 - c\,\mu^2}{c - 1}. \end{aligned} \tag{11.35}$$

Thus, if the null hypothesis H_0 is true (that is $\mu_i = \mu$ for all i), then the right-hand side above reduces to σ^2/n. Therefore, if the null hypothesis is true:

$$S^2_B = n\, S^2_{\bar{Y}_{..}} = \frac{n}{c - 1}\sum_{i=1}^{c} (\bar{Y}_{i.} - \bar{Y}_{..})^2 \tag{11.36}$$

is an unbiased estimator of σ^2. The subscript B refers to the fact that S_B^2 is a measure of the **between-sample variance**. Furthermore, under H_0:

$$\frac{(c-1)}{\sigma^2/n} S_{\bar{Y}_{..}}^2 = \frac{(c-1)\,S_B^2}{\sigma^2}$$

has a chi-square distribution with $(c-1)$ degrees of freedom.

If H_0 is true, S_W^2 and S_B^2 should have comparable values since they both estimate σ^2. It is reasonable, therefore, to base our test on their ratio. The statistic (often called the **variance ratio**):

$$\mathrm{F} = \frac{\dfrac{(c-1)S_B^2}{\sigma^2(c-1)}}{\dfrac{(n_T-c)S_W^2}{\sigma^2(n_T-c)}} = \frac{S_B^2}{S_W^2} \tag{11.37}$$

has an F distribution with $(c-1,\ n_T-c)$ degrees of freedom, by Theorem 3.9. Since between sample variance S_B^2 is expected to be larger than within-sample variance S_W^2 when H_0 is false, we reject H_0 at α percent level of significance if the observed variance ratio exceeds the critical value $f_{c-1,n_T-c;\alpha}$ of the F distribution.

To gain further insight into the problem, we observe that we can partition the total sum of squares (SST):

$$SST = \sum_{i=1}^{c} \sum_{j=1}^{n} (Y_{ij} - \bar{Y}_{..})^2 \tag{11.38}$$

as follows:

$$\begin{aligned} SST &= \sum \sum (Y_{ij} - \bar{Y}_{..})^2 \\ &= n \sum_{i=1}^{c} (\bar{Y}_{i.} - \bar{Y}_{..})^2 + \sum_{i=1}^{c} \sum_{j=1}^{n} (Y_{ij} - \bar{Y}_{i.})^2. \end{aligned} \tag{11.39}$$

The first term on the right-hand side is S_B^2 times its degrees of freedom, hence it is known as the **sum of squares between treatments** (or groups), $SS(\mathrm{Tr})$. The second term on the right-hand side is S_W^2 times its degrees of freedom, hence it is a measure of the chance error. This sum is referred to as the **residual variation**, or **error sum of squares**, SSE. Thus, an alternative form of equation (11.39) is:

$$SST = SS(Tr) + SSE \tag{11.40}$$

and the variance ratio is expressed by:

$$\mathrm{F} = \frac{\dfrac{SS(Tr)}{c-1}}{\dfrac{SSE}{n_T-c}}. \tag{11.41}$$

Numerical calculation of this F statistic is usually expressed in terms of a so-called analysis of variance (ANOVA) table as shown in Table 11.4.

Table 11.4 *One-Way ANOVA Table*

Source of variation	*Sum of squares*	*Degrees of freedom*	*Mean square*
Between treatments	$SS(\text{tr}) = n \sum_{i=1}^{c} (\bar{Y}_{i.} - \bar{Y}_{..})^2$	$c - 1$	S_B^2
Error (within groups)	$SSE = \sum_{i=1}^{c} \sum_{j=1}^{n} (Y_{ij} - \bar{Y}_{i.})^2$	$n_T - c$	S_W^2
Total variation	$SST = \sum_{i=1}^{c} \sum_{j=1}^{n} (Y_{ij} - \bar{Y}_{..})^2$	$n_T - 1$	

Example 11.8

Three different interactive systems are to be compared with respect to their response times to an editing request. Owing to chance fluctuations among the other transactions in process, it was decided to take ten sets of samples at randomly chosen times for each system and record the mean response time as follows:

	System response time (sec)		
Session	*A*	*B*	*C*
1	0.96	0.82	0.75
2	1.03	0.68	0.56
3	0.77	1.08	0.63
4	0.88	0.76	0.69
5	1.06	0.83	0.73
6	0.99	0.74	0.75
7	0.72	0.77	0.60
8	0.86	0.85	0.63
9	0.97	0.79	0.59
10	0.90	0.71	0.61

The ANOVA table for this problem can be formulated as follows:

Source of variation	*Sum of squares*	*Degrees of freedom*	*Mean square*	F
Between treatments	0.34041	2	0.1702033	17.4276
Error	0.26369	27	0.0097663	
Total	0.60410	29		

Since $f_{2,27;0.01}$ is 5.49, and the observed value is 17.4276, we reject the null hypothesis at the 1 percent level of significance. In other words there is a significant difference in the responsiveness of the three systems. #

Next we consider a two-way analysis of variance so that we have two controllable factors, and the response variable Y_{ij} is modeled as:

$$Y_{ij} = \mu + \alpha_i + \beta_j + (\alpha\beta)_{ij} + \Delta_{ij}, \qquad (11.42)$$
$$i = 1, 2, \ldots, c_1, \; j = 1, 2, \ldots, c_2.$$

There are c_1 levels of the first factor, c_2 levels of the second factor, and the total number of **cells** or treatments is $c_T = c_1c_2$. The term μ is the overall mean. The term α_i is the main effect of the first factor at level i, the term β_j accounts for the main effect of the second factor at level j, and $(\alpha\beta)_{ij}$ denotes the **interaction** (or joint) effect of the first factor at level i and the second factor at level j. As before, we will assume that Δ_{ij}'s ($i = 1, 2, \ldots, c_1$; $j = 1, 2, \ldots, c_2$) are mutually independent, normally distributed random variables with zero means and the common variance σ^2. We will further assume without loss of generality that:

$$\sum_{i=1}^{c_1} \alpha_i = \sum_{j=1}^{c_2} \beta_j = \sum_{i=1}^{c_1} (\alpha\beta)_{ij} = \sum_{j=1}^{c_2} (\alpha\beta)_{ij} = 0. \qquad (11.43)$$

These restrictions allow us to uniquely estimate the parameters μ, α_i, β_j, and $(\alpha\beta)_{ij}$.

As in the case of one-way analysis of variance, we assume that n independent observations are taken for each i and j so that:

$$Y_{ijk} = \mu + \alpha_i + \beta_j + (\alpha\beta)_{ij} + \Delta_{ijk}, \qquad (11.44)$$
$$i = 1, 2, \ldots, c_1, \; j = 1, 2, \ldots, c_2, \; k = 1, 2, \ldots, n,$$

denote the random variables corresponding to $n_T = c_1c_2n$ total observations of the response variable. These observations are usually organized as shown in Table 11.5.

The following quantities are calculated:

$$\overline{Y}_{ij.} = \frac{\sum_{k=1}^{n} Y_{ijk}}{n}, \qquad \text{the sample average in } (i, j)\text{th cell,}$$

$$\overline{Y}_{i..} = \frac{\sum_{j=1}^{c_2} \overline{Y}_{ij.}}{c_2} = \frac{\sum_{j=1}^{c_2}\sum_{k=1}^{n} Y_{ijk}}{nc_2}, \qquad \text{sample average of row } i,$$

Table 11.5 *Observations for Two-Way Analysis of Variance*

	Factor B				
Factor A	*Level 1*	*Level 2*	...	*Level c_2*	*Row avg.*
Level 1	$y_{111}, \ldots, y_{11n}$	$y_{121}, \ldots, y_{12n}$	...	$y_{1c_21}, \ldots, y_{1c_2n}$	$\bar{y}_{1..}$
Level 2	$y_{211}, \ldots, y_{21n}$	$y_{221}, \ldots, y_{22n}$		$y_{2c_21}, \ldots, y_{2c_2n}$	$\bar{y}_{2..}$
.	. .				
.	. .				
.	. .				
Level c_1	$y_{c_111}, \ldots, y_{c_11n}$	$y_{c_121}, \ldots, y_{c_12n}$		$y_{c_1c_21}, \ldots, y_{c_1c_2n}$	$\bar{y}_{c_1..}$
Column average	$\bar{y}_{.1.}$	$\bar{y}_{.2.}$	...	$\bar{y}_{.c_2.}$	$\bar{y}_{...}$

$$\bar{Y}_{.j.} = \frac{\sum_{i=1}^{c_1} \bar{Y}_{ij.}}{c_1} = \frac{\sum_{i=1}^{c_1} \sum_{k=1}^{n} Y_{ijk}}{nc_1}, \qquad \text{sample average of column } j,$$

$$\bar{Y}_{...} = \frac{\sum_{i=1}^{c_1} \bar{Y}_{i..}}{c_1} = \frac{\sum_{j=1}^{c_2} \bar{Y}_{.j.}}{c_2}$$

$$= \frac{\sum_{i=1}^{c_1} \sum_{j=1}^{c_2} \sum_{k=1}^{n} Y_{ijk}}{nc_1c_2}, \qquad \text{overall sample average.}$$

It should be clear that the above averages allow us to compute the best unbiased estimates of the model parameters:

$$\hat{\mu} = \bar{y}_{...}, \quad \hat{\alpha}_i = \bar{y}_{i..} - \bar{y}_{...}, \quad \hat{\beta}_j = \bar{y}_{.j.} - \bar{y}_{...},$$

$$\widehat{(\alpha\beta)}_{ij} = (\bar{y}_{ij.} - \bar{y}_{...}) - [(\bar{y}_{i..} - \bar{y}_{...}) + (\bar{y}_{.j.} - \bar{y}_{...})].$$

We are usually interested in testing one or more hypotheses regarding model parameters. For example, one hypothesis could be that the main effect of factor A is zero:

$$H_{01}\colon \alpha_i = 0 \qquad \text{for all } i.$$

Another hypothesis relates the main effect of factor B:

$$H_{02}\colon \beta_j = 0 \qquad \text{for all } j.$$

Finally, we may be interested in testing the hypothesis that the effects of the two factors are additive; that is, there is no interaction:

$$H_{03}: \ (\alpha\beta)_{ij} = 0 \qquad \text{for all } i \text{ and } j.$$

To proceed further, we break down the total sum of squares SST into four components: row sum of squares SSA, the main effect of factor A; column sum of squares SSB, the main effect of factor B; the interaction sum of squares $SSAB$; and the residual or error sum of squares SSE. Thus we have:

$$SST = SSA + SSB + SSAB + SSE, \tag{11.45}$$

where:

$$SST = \sum_i \sum_j \sum_k (Y_{ijk} - \bar{Y}_{...})^2,$$

$$SSA = nc_2 \sum_i (\bar{Y}_{i..} - \bar{Y}_{...})^2,$$

$$SSB = nc_1 \sum_j (\bar{Y}_{.j.} - \bar{Y}_{...})^2,$$

$$SSAB = n \sum_i \sum_j (\bar{Y}_{ij.} - \bar{Y}_{i..} - \bar{Y}_{.j.} + \bar{Y}_{...})^2,$$

$$SSE = \sum_i \sum_j \sum_k (Y_{ijk} - \bar{Y}_{ij.})^2.$$

These quantities are usually tabulated as shown in Table 11.6.

In order to test the hypothesis H_{01}, we compute the F statistic:

$$F_1 = \frac{\dfrac{SSA}{c_1 - 1}}{\dfrac{SSE}{n_T - c_T}}.$$

We reject H_{01} at the α percent level of significance, provided that the computed value of F_1 exceeds the critical value $f_{c_1-1,n_T-c_T;\alpha}$ of the F distribution. Similarly, we compute:

$$F_2 = \frac{\dfrac{SSB}{c_2 - 1}}{\dfrac{SSE}{n_T - c_T}} \quad \text{and} \quad F_3 = \frac{\dfrac{SSAB}{(c_1 - 1)(c_2 - 1)}}{\dfrac{SSE}{n_T - c_T}}$$

and respectively reject H_{02} and H_{03} at the α percent level of significance if the computed values exceed the critical values of F distributions with $(c_2 - 1,\ n_T - c_T)$ and $[(c_1 - 1)(c_2 - 1),\ n_T - c_T]$ degrees of freedom.

Clearly, the above procedure can be extended to the case of multifactor analysis of variance [NETE 1974].

Table 11.6 *Two-way ANOVA with Interaction*

Source	Sum of squares	Mean squared deviation	Degrees of freedom	Test statistic	E[mean squared deviation]
Main effect A (rows)	SSA	$S_A^2 = \frac{SSA}{c_1 - 1}$	$c_1 - 1$	$F_1 = \frac{S_A^2}{S_W^2}$	$\sigma^2 + nc_2 \sum \frac{\alpha_i^2}{c_1 - 1}$
Main effect B (columns)	SSB	$S_B^2 = \frac{SSB}{c_2 - 1}$	$c_2 - 1$	$F_2 = \frac{S_B^2}{S_W^2}$	$\sigma^2 + nc_1 \sum \frac{\beta_j^2}{c_2 - 1}$
Interaction	$SSAB$	$S_{AB}^2 = \frac{SSAB}{(c_1 - 1)(c_2 - 1)}$	$(c_1 - 1)(c_2 - 1)$	$F_3 = \frac{S_{AB}^2}{S_W^2}$	$\sigma^2 + n \sum \frac{(\alpha\beta)_{ij}^2}{(c_1 - 1)(c_2 - 1)}$
Residual	SSE	$S_W^2 = \frac{SSE}{n_T - c_T}$	$c_1 c_2 (n - 1) = n_T - c_T$		σ^2
Total	SST		$n_T - 1$		

The following example is a slightly modified version of the example in [LIU 1978].

Example 11.9

Suppose we wish to study the variation in the throughput of a paged multiprogramming system with the following five factors:

1. POL—The memory-partitioning policy at one of fourteen possible levels.
2. ML—Multiprogramming level at levels 3, 4, and 5.
3. MPP—Memory allotment per program at levels 20, 26, and 40 pages per program
4. DRM—Drum rotation time at three levels: 10, 35, and 70 milliseconds
5. WRK—four different workload types.

For each one of the $14 \cdot 3 \cdot 3 \cdot 3 \cdot 4 = 1512$ cells, a simulation is conducted with a sample size of $n = 25$. Selecting system throughput as the response variable, the ANOVA is shown in Table 11.7.

From this table we conclude that all main effects and two-way interactions are significant at the one percent level. An ordering of factors from the largest to the smallest effect is: drum rotation time, memory allotment, workload type, the degree of multiprogramming, and the memory-partitioning policy. The largest two-way interaction effect is the one due to the combination of workload levels and drum rotation times. The information gained from such an analysis can be used to improve system performance. #

Problems

1. Suppose we wish to study the effect of computer-aided instruction (CAI) on the performance of college students. We randomly divide the incoming class into three sections. Section A is taught in a conventional way, Section B is (nearly) completely automated, and in Section C a mixed approach is used. The following test scores are observed:

Section A	*Section B*	*Section C*
77	70	79
68	69	74
72	73	77
75	74	80
60	59	73
59	63	60
82	80	79

Perform an analysis of variance and determine whether the differences among the means obtained for the three sections are significant at $\alpha = 0.05$.

2. The **Kruskal-Wallis** H **test** is a nonparametric analog of test (11.30) in one-way analysis of variance [NOET 1976]. This test is a generalization of the Wilcoxon

Table 11.7: ANOVA Table for Example 11.9

Source of variation	Sum of squares	Degrees of freedom	Mean-squared deviation	Test statistic	$f_{.,.,0.01}$
Main effects:	21.238				
POL	0.401	13	0.03085	136.5	2.21
ML	0.253	2	0.12650	559.7	4.60
MPP	4.128	2	2.064	9132.7	4.60
DRM	15.053	2	7.5265	33303.1	4.60
WRK	1.403	3	0.4677	2069.5	3.78
Two-way interaction:	2.206				
POL-ML	insig.	26	—	—	—
POL-MPP	0.261	26	0.01004	44.425	1.76
POL-DRM	0.043	26	0.001654	7.319	1.76
POL-WRK	0.067	39	0.001718	7.602	1.60
ML-MPP	insig.	4	—	—	—
ML-DRM	0.022	4	0.0055	24.337	3.32
ML-WRK	0.013	6	0.002167	9.588	2.80
MPP-DRM	0.204	4	0.051	225.66	3.32
MPP-WRK	0.179	6	0.029834	132.01	2.80
DRM-WRK	1.417	6	0.236167	1045.00	2.80
Three-way interaction:	0.411				
MPP-DRM-WRK	0.230	12	0.019167	84.81	2.18
Other	0.181				
Residual	00.82	36,288	0.000226		
Total	24.775	37,799			

rank-sum test considered in Section 10.3.2. Suppose that n_i independent observations for treatment level i have been taken. We combine $k = \sum_{i=1}^{c} n_i$ observations and order them by increasing magnitude. Now we assign ranks to the combined sample values and let R_i be the sum of the ranks assigned to the n_i observations of treatment level i. The test is based on statistic H, defined by:

$$H = \frac{12}{k(k+1)} \sum_{i=1}^{c} \frac{R_i^2}{n_i} - 3(k+1).$$

If $n_i > 5$ for all i and if the null hypothesis that the k samples come from identical populations holds, then the distribution of the H statistic is approximately chi-square with $(c - 1)$ degrees of freedom. Apply the H test to the data in Example 11.8 and to the data in problem 1 above.

References

[BARD 1978] Y. BARD and M. SCHATZOFF, "Statistical Methods in Computer Performance Analysis," in Chandy and Yeh (eds.) *Current Trends in Programming Methodology*, Vol. III, Prentice-Hall, Englewood Cliffs, N.J.

[COFF 1973] E. G. COFFMAN, Jr. and P. J. DENNING, *Operating System Theory*, Prentice-Hall, Englewood Cliffs, N.J.

[DRAP 1966] N. R. DRAPER and H. SMITH, *Applied Regression Analysis*, John Wiley & Sons, New York.

[HATF 1977] M. A. HATFIELD and D. J. SLISKI (eds.) *Computer Review*, GML Corporation, Lexington, Mass.

[INGL 1976] A. D. INGLE and D. P. SIEWIOREK, "Reliability Models for Multiprocessor Systems with and without Periodic Maintenance," Technical Report, Department of Computer Science, Carnegie-Mellon University, Pittsburgh, Pa.

[KEND 1961] M. G. KENDALL and A. STUART, *The Advanced Theory of Statistics*, Vol. 2, *Inference and Relationship*, Hafner Publishing Co., New York.

[KINI 1978] R. E. KINICKI, "Queuing Models of Computer Configuration Planning," Ph.D. dissertation, Department of Computer Science, Duke University, Durham, N.C.

[LIU 1978] M. LIU, "A Simulation Study of Memory Partitioning Algorithms," in D. Ferrari. ed., *Performance of Computer Installations*, North-Holland, Amsterdam.

[MAIE 1980] M. V. MAIELI, "The Significant Parameters Affecting the Price of Matrix-Configured Random Access Computer Memory and Their Functional Relationship," A.M. dissertation, Department of Computer Science, Duke University, Durham, N.C.

[NETE 1974] J. NETER and W. WASSERMAN, *Applied Linear Statistical Models*, Richard D. Irwin, Homewood, Ill.

[NOET 1976] G. E. NOETHER, *Introduction to Statistics, A Nonparametric Approach*, Houghton Mifflin, Boston, MA.

[PUTN 1978] L. PUTNAM, "A General Empirical Solution to the Macro Software Sizing and Estimating Problem," *IEEE Trans. on Soft. Eng.*, July 1978.

Appendix A

Bibliography

I. THEORY

A. Probability Theory

1. R. B. Ash, *Basic Probability Theory*, John Wiley & Sons, New York, 1970 (Introductory).
2. A. B. Clarke and R. L. Disney, *Probability and Random Processes for Engineers and Scientists*, John Wiley & Sons, New York, 1970 (Introductory).
3. A. W. Drake, *Fundamentals of Applied Probability Theory*, McGraw-Hill, New York, 1967 (Introductory).
4. W. Feller, *An Introduction to Probability Theory and Its Applications*, 2 vols., John Wiley & Sons, New York, 1957 and 1966 (Introductory and Advanced).
5. P. G. Hoel, S. C. Port, and C. J. Stone, *Introduction to Probability Theory*, Houghton Mifflin, 1971 (Introductory).
6. E. Parzen, *Modern Probability Theory*, John Wiley & Sons, New York, 1960.

B. Stochastic Processes

1. U. N. Bhat, *Elements of Applied Stochastic Processes*, John Wiley & Sons, New York, 1972 (Intermediate).
2. E. Cinclar, *Introduction to Stochastic Processes*, Prentice-Hall, Englewood Cliffs, N.J., 1975 (Advanced).
3. D. R. Cox, *Renewal Theory*, Methuen, London, 1962 (Advanced).
4. D. R. Cox and H. D. Miller, *The Theory of Stochastic Processes*, Chapman and Hall, London, 1965.

5. S. KARLIN, *A First Course in Stochastic Processes*, Academic Press, New York, 1966 (Advanced).

6. S. KARLIN and H. M. TAYLOR, *A First Course in Stochastic Processes*, Academic Press, New York, 1975.

7. G. KEMENY and J. L. SNELL, *Finite Markov Chains*, D. Van Nostrand-Reinhold, New York, 1960.

8. E. PARZEN, *Stochastic Processes*, Holden-Day, San Francisco, 1962 (Intermediate).

9. N. U. PRABHU, *Stochastic Processes: Basic Theory and Its Applications*, The Macmillan Company, New York, 1965 (Advanced).

10. S. M. ROSS, *Applied Probability Models with Optimization Applications*, Holden-Day, San Francisco, 1970 (Advanced).

C. Queuing Theory

1. R. B. COOPER, *Introduction to Queuing Theory*, MacMillan, New York, 1972 (Introductory).

2. P. J. COURTOIS, *Decomposability: Queuing and Computer System Application*, Academic Press, New York, 1977 (Advanced).

3. D. R. COX and W. L. SMITH, *Queues*, Methuen, London, 1961.

4. F. P. KELLEY, *Reversibility and Stochastic Networks*, John Wiley & Sons, New York, 1979 (Advanced).

5. L. KLEINROCK, *Queuing Systems*, Vol I, *Theory*, John Wiley & Sons, New York, 1975 (Intermediate to Advanced).

6. L. TAKACS, *Introduction to the Theory of Queues*, Oxford University Press, New York, 1972.

7. D. GROSS and C. M. HARRIS, *Fundamentals of Queuing Theory*, John Wiley & Sons, New York, 1974.

D. Reliability Theory

1. B. L. AMSTADTER, *Reliability Mathematics*, McGraw-Hill, New York, 1971 (Introductory).

2. L. J. BAIN, *Statistical Analysis of Reliability and Life-Testing Models*, Marcel Dekker, New York, 1978 (Intermediate to Advanced).

3. R. E. BARLOW and F. PROSCHAN, *Mathematical Theory of Reliability*, John Wiley & Sons, New York (Advanced).

4. R. E. BARLOW and F. PROSCHAN, *Statistical Theory of Reliability and Life Testing: Probability Models*, Holt, Rinehart and Winston, 1975 (Advanced).

5. A. E. GREEN and A. J. BOURNE, *Reliability Technology*, Wiley Interscience, New York, 1972 (Introductory).

6. D. K. LLOYD and M. LIPOW, *Reliability: Management, Methods, and Mathematics*, Prentice-Hall, Englewood Cliffs, N.J., 1962.

7. G. H. SANDLER, *System Reliability Engineering*, Prentice-Hall, Englewood Cliffs, N.J., 1963 (Introductory).

8. M. L. SHOOMAN, *Probabilistic Reliability: An Engineering Approach*, McGraw-Hill, 1968 (Introductory).

9. M. ZELEN, *Statistical Theory of Reliability*, The University of Wisconsin Press, Madison, Wisc., 1963.

E. Statistics

1. C. CHATFIELD, *Statistics in Technology*, Chapman and Hall, London, 1970 (Introductory).

2. N. R. DRAPER and H. SMITH, *Applied Regression Analysis*, John Wiley & Sons, New York, 1967 (Intermediate).

3. P. G. HOEL, S. C. PORT, and C. J. STONE, *Introduction to Statistical Theory*, Houghton Mifflin, Boston, 1971 (Intermediate).

4. R. V. HOGG and E. A. TANIS, *Probability and Statistical Inference*. The Macmillan Company, New York, 1977 (Intermediate).

5. M. HOLLANDER and D. A. WOLFE, *Nonparametric Statistical Methods*, John Wiley & Sons, New York, 1973 (Intermediate).

6. M. G. KENDALL and A. STUART, *The Advanced Theory of Statistics*, Vol. 2, *Inference and Relationship*, Hafner Publishing Company, New York, 1961 (Advanced).

7. I. MILLER and J. E. FREUND, *Probability and Statistics for Engineers*, Prentice-Hall, Englewood Cliffs, N.J., 1977 (Introductory).

8. A. M. MOOD and F. A. GRAYBILL, *Introduction to the Theory of Statistics*, McGraw-Hill, New York, 1963 (Intermediate to Advanced).

9. M. R. SPIEGEL, *Theory and Problems of Statistics*, Schaum's Outline Series in Mathematics, McGraw-Hill Book Company, New York, 1961 (Introductory).

10. S. S. WILKS, *Mathematical Statistics*, John Wiley & Sons, New York, 1962 (Advanced).

II. APPLICATIONS

A. Computer Performance Evaluation

1. *ACM Computing Surveys*, Special Issue on Queuing Network Models, 10:3 (September 1978).

2. A. O. ALLEN, *Probability, Statistics, and Queuing Theory with Computer Science Applications*, Academic Press, New York, 1978 (Introductory).

3. S. C. BRUELL and G. BALBO, *Computational Algorithms for Closed Queuing Networks*, Elsevier, New York, 1980 (Introductory).

4. K. M. CHANDY and R. T. YEH, *Current Trends in Programming Methodology*, Vol. III, *Software Modeling*, Prentice-Hall, Englewood Cliffs, N.J., 1978 (Intermediate to Advanced).

5. E. G. COFFMAN, Jr. and P. J. DENNING, *Operating System Theory*, Prentice-Hall, Englewood Cliffs, N.J., 1973 (Intermediate).

6. P. J. COURTOIS, *Decomposability: Queuing and Computer System Applications*, Academic Press, New York, 1977 (Advanced).

7. W. EVERLING, *Exercises in Computer Systems Analysis*, Lecture Notes in Computer Science, Springer Verlag, New York, 1975 (Intermediate).

8. D. FERRARI, *Computer Systems Performance Evaluation*, Prentice-Hall, Englewood Cliffs, N.J., 1978 (Intermediate).

9. W. FREIBERGER, ed., *Statistical Computer Performance Evaluation*, Academic Press, New York, 1972 (Intermediate).

10. H. HELLERMAN and T. F. CONROY, *Computer System Performance*, McGraw-Hill, New York, 1975 (Introductory).

11. L. KLEINROCK, *Queuing Systems*, Vol. II, *Computer Applications*, John Wiley & Sons, New York, 1976 (Advanced).

12. H. KOBAYASHI, *Modeling and Analysis*, Addison-Wesley, Reading, Mass., 1978 (Intermediate).

13. C. H. SAUER and K. M. CHANDY, *Computer Systems Performance Modeling*, Prentice-Hall, Englewood Cliffs, N.J., 1981 (Introductory).

B. Communications

1. H. STARK and F. B. TUTEUR, *Modern Electrical Communications*, Prentice-Hall, Englewood Cliffs, N.J., 1979.

2. J. B. THOMAS, *An Introduction to Statistical Communication Theory*, John Wiley & Sons, New York, 1969.

3. R. E. ZIEMER and W. H. TRANTER, *Principles of Communications*, Houghton Mifflin, Boston, 1976 (Introductory).

C. Analysis of Algorithms

1. A. V. AHO, J. V. HOPCROFT, and J. D. ULLMAN, *The Design and Analysis of Algorithms*, Addison-Wesley, Reading, Mass., 1974 (Advanced).

2. B. BEIZER, *Micro-Analysis of Computer System Performance*, Van Nostrand-Reinhold, New York, 1978 (Introductory).

3. S. E. GOODMAN and S. T. HEDETNIEMI, *Introduction to the Design and Analysis of Algorithms*, McGraw-Hill, New York, 1977 (Introductory).

4. E. HOROWITZ and S. SAHNI, *Fundamentals of Computer Algorithms*, Computer Science Press, Potomac, Md., 1978 (Intermediate).

5. D. E. KNUTH, *The Art of Computer Programming*, 3 vols., Addison-Wesley, Reading, Mass., 1969, 1973 (Intermediate to Advanced).

D. Simulation

1. G. S. FISHMAN, *Concepts and Methods in Discrete Event Digital Simulation*, John Wiley & Sons, New York, 1973.

2. G. S. FISHMAN, *Principles of Discrete Event Digital Simulation*, John Wiley & Sons, New York, 1978.

3. G. GORDON, *System Simulation*, Prentice-Hall, Englewood Cliffs, N.J., 1978.

4. D. L. IGLEHART, "The Regenerative Method for Simulation Analysis," in Chandy and Yeh, eds., *Current Trends in Programming Methodology*, Vol. III, *Software Modeling*, Prentice-Hall, Englewood Cliffs, N.J., 1978.

5. J. P. C. KLEIJNEN, *Statistical Techniques in Simulation, Parts 1 and 2*, Marcel Dekker, New York, 1974.

6. H. KOBAYASHI, *Modeling and Analysis*, Addison-Wesley, Reading, Mass., 1978.

7. H. MAISEL and G. GNUGNOLI, *Simulation of Discrete Stochastic Systems*, Science Research Associates, Chicago, 1972.

8. T. H. NAYLOR, J. L. BALINTFY, D. S. BURDICK, and K. CHU, *Computer Simulation Techniques*, John-Wiley & Sons, New York, 1966.

E. Teletraffic Engineering

1. D. BEAR, *Principles of Telecommunication-Traffic Engineering*, Peter Peregrinus (on behalf of IEE), England, 1976 (Introductory).

2. R. SYSKI, *Introduction to Congestion Theory in Telephone Systems*, Oliver and Boyd, Edinburgh, 1960 (Advanced).

F. Computer-Communication Networks

1. N. ABRAMSON and F. F. KUO, eds., *Computer-Communication Networks*, Prentice-Hall, Englewood Cliffs, N.J., 1973.

2. L. KLEINROCK, *Queuing Systems*, Vol. II, *Computer Applications*, John Wiley & Sons, New York, 1976.

3. H. KOBAYASHI and A. G. KONHEIM, "Queuing Models for Computer Communications System Analysis," *IEEE Transactions on Communications*, 25:1 (1977), 2–29.

4. M. SCHWARZ, *Computer-Communication Network Design and Analysis*, Prentice-Hall, Englewood Cliffs, N.J., 1977.

G. Operations Research

1. D. P. GAVER and G. L. THOMPSON, *Programming and Probability Models in Operations Research*, Brooks/Cole Publishing Co., Monterey, Calif., 1973.

2. F. H. HILLIER and G. J. LIEBERMAN, *Operations Research*, Holden-Day, San Francisco, 1974.

3. J. MODER and S. E. ELMAGHRABY, *Handbook of Operations Research*, Vols. 1 and 2, Van Nostrand-Reinhold, New York, 1978.

4. B. D. SIVAZLIAN and L. E. STANFEL, *Analysis of Systems in Operations Research*, Prentice-Hall, Englewood Cliffs, N.J., 1975.

H. Fault-Tolerant Computing

1. J. E. ARSENAULT and J. A. ROBERTS, *Reliability and Maintainability of Electronic Systems*, Computer Science Press, Potomac, Md., 1980.

2. M. A. BREUER and A. D. FRIEDMAN, *Diagnosis & Reliable Design of Digital Systems*, Computer Science Press, Potomac, Md., 1976.

3. *IEEE Proceedings*, Special Issue on Fault-Tolerant Computing, 66:10 (October 1978).

4. *IEEE Transactions on Computers*, Special Issue on Fault-Tolerant Computing, June 1978.

5. S. OSAKI and T. NISHIO, "Reliability Evaluation of Some Fault-Tolerant Computer Architectures," Lecture Notes in Computer Science, Springer-Verlag, Berlin, 1980.

6. W. H. PIERCE, *Failure-Tolerant Computer Design*, Academic Press, New York, 1965.

7. B. RANDELL and H. ANDERSON, *Computing Systems Reliability*, Cambridge University Press, New York, 1979.

Appendix B

Properties of Distributions

Discrete Distributions

Distribution	Parameters	pmf, $p_X(i)$	PGF	Mean $E[X]$	Variance $Var[X]$
Bernoulli	p, $0 \leqslant p \leqslant 1$	$p_X(0) = p$, $p_X(1) = q = 1 - p$	$(1 - p) + pz$	p	$p(1 - p)$
Binomial	$n \geqslant 1$, p, $0 \leqslant p \leqslant 1$	$\binom{n}{i} p^i (1 - p)^{n-i}$, $i = 0, 1, \ldots, n$	$(1 - p + pz)^n$	np	$np(1 - p)$
Geometric	p, $0 < p \leqslant 1$	$p(1 - p)^{i-1}$, $i = 1, 2, \ldots$	$\frac{pz}{1 - (1 - p)z}$	$\frac{1}{p}$	$\frac{1 - p}{p^2}$
Modified geometric	p, $0 < p \leqslant 1$	$p(1 - p)^i$, $i = 0, 1, \ldots$	$\frac{p}{1 - (1 - p)z}$	$\frac{1 - p}{p}$	$\frac{1 - p}{p^2}$
Negative binomial	$0 < p \leqslant 1$, $r = 1, 2, \ldots$	$\binom{i - 1}{r - 1} p^r (1 - p)^{i-r}$, $i = r, r + 1, \ldots$	$[\frac{pz}{1 - (1 - p)z}]^r$	$\frac{r}{p}$	$\frac{r(1 - p)}{p^2}$
Poisson	α, $\alpha > 0$	$\frac{e^{-\alpha}\alpha^i}{i!}$, $i = 0, 1, 2, \ldots$	$e^{-\alpha(1 - z)}$	α	α
Uniform	n	$p_X(i) = 1/n$, $i = 1, 2, \ldots, n$	$\frac{1}{n} \sum_{i=1}^{n} z^i$	$\frac{n + 1}{2}$	$\frac{n^2 - 1}{12}$

Continuous Distributions

Distribution	Parameters	pdf, $f(x)$	Transform	Mean	Variance
Erlang, ERL (λ, r)	$r \geqslant 1, \lambda > 0$	$\frac{\lambda e^{-\lambda x}(\lambda x)^{r-1}}{(r-1)!}$	$L_X(s) = (\frac{\lambda}{\lambda + s})^r$	$\frac{r}{\lambda}$	$\frac{r}{\lambda^2}$
Exponential EXP (λ)	$\lambda > 0$	$\lambda e^{-\lambda x}, x > 0$	$L_X(s) = \frac{\lambda}{\lambda + s}$	$\frac{1}{\lambda}$	$\frac{1}{\lambda^2}$
Gamma, GAM (λ, α)	$\alpha > 0, \lambda > 0$	$\frac{\lambda e^{-\lambda x}(\lambda x)^{\alpha-1}}{\Gamma\alpha}, x > 0$	$L_X(s) = \frac{\lambda^\alpha}{(\lambda + s)^\alpha}$	$\frac{\alpha}{\lambda}$	$\frac{\alpha}{\lambda^2}$
Hyper-exponential	$\alpha_1, \alpha_2, \ldots, \alpha_r$ $\alpha_i \geqslant 0, \sum_{i=1}^{r} \alpha_i = 1,$ $\lambda_1, \lambda_2, \ldots, \lambda_r$ $\lambda_i > 0,$	$\sum_{i=1}^{r} \alpha_i \lambda_i e^{-\lambda_i x}$	$\sum_{i=1}^{r} \frac{\alpha_i \lambda_i}{\lambda_i + s}$	$\sum_{i=1}^{r} \frac{\alpha_i}{\lambda_i}$	$2 \sum_{i=1}^{r} \frac{\alpha_i}{\lambda_i^2} - (\sum_{i=1}^{r} \frac{\alpha_i}{\lambda_i})^2$
Hypo-exponential	$\lambda_1, \ldots, \lambda_r > 0$	$\sum_{i=1}^{r} a_i \lambda_i e^{-\lambda_i x},$	$\prod_{i=1}^{r} \frac{\lambda_i}{\lambda_i + s}$	$\sum_{i=1}^{r} \frac{1}{\lambda_i}$	$\sum_{i=1}^{r} \frac{1}{\lambda_i^2}$
HYPO $(\lambda_1, \lambda_2, \ldots, \lambda_r)$	$\lambda_i \neq \lambda_j, i \neq j$	$a_i = \prod_{\substack{j=1 \\ j \neq i}}^{r} \frac{\lambda_j}{(\lambda_j - \lambda_i)}$			

Distribution	Parameters	pdf, $f(x)$	Transform	Mean	Variance
Normal	μ, σ^2	$\frac{1}{\sigma\sqrt{2\pi}} e^{-\frac{(x-\mu)^2}{2\sigma^2}}$	$N_X(\tau) =$	μ	σ^2
$N(\mu, \sigma^2)$	$-\infty < \mu < \infty$, $\sigma^2 > 0$	$-\infty < x < \infty$	$e^{i\tau\mu - \frac{\tau^2\sigma^2}{2}}$		
Uniform	a, b $-\infty < a < b < \infty$	$\frac{1}{b-a}, a < x < b$	$\frac{e^{-as} - e^{-bs}}{s(b-a)}$	$\frac{a+b}{2}$	$\frac{(b-a)^2}{12}$
Weibull	$\lambda > 0$ $\alpha > 0$	$\lambda\alpha x^{\alpha-1}e^{-\lambda x^\alpha}$, $x > 0$		$(\frac{1}{\lambda})^{1/\alpha}\Gamma(1 + 1/\alpha)$	$[(\frac{1}{\lambda})^{2/\alpha}\Gamma(1 + 2/\alpha)] - (E[X])^2$

Appendix C

Statistical Tables

Table 1 *Binomial Distribution Function*

$$B(x; n, p) = \sum_{k=0}^{x} \binom{n}{k} p^k (1-p)^{n-k}$$

		p									
n	*x*	0.05	0.10	0.15	0.20	0.25	0.30	0.35	0.40	0.45	0.50
2	2	0.9025	0.8100	0.7225	0.6400	0.5625	0.4900	0.4225	0.3600	0.3025	0.2500
	1	0.9975	0.9900	0.9775	0.9600	0.9375	0.9100	0.8755	0.8400	0.7975	0.7500
3	0	0.8574	0.7290	0.6141	0.5120	0.4219	0.3430	0.2746	0.2160	0.1664	0.1250
	1	0.9928	0.9720	0.9392	0.8960	0.8438	0.7840	0.7182	0.6480	0.5748	0.5000
	2	0.9999	0.9990	0.9966	0.9920	0.9844	0.9730	0.9571	0.9360	0.9089	0.8750
4	0	0.8145	0.6561	0.5220	0.4096	0.3164	0.2401	0.1785	0.1296	0.0915	0.0625
	1	0.9860	0.9477	0.8905	0.8192	0.7383	0.6517	0.5630	0.4752	0.3910	0.3125
	2	0.9995	0.9963	0.9880	0.9728	0.9492	0.9163	0.8735	0.8208	0.7585	0.6875
	3	1.0000	0.9999	0.9995	0.9984	0.9961	0.9919	0.9850	0.9744	0.9590	0.9375
5	0	0.7738	0.5905	0.4437	0.3277	0.2373	0.1681	0.1160	0.0778	0.0503	0.0312
	1	0.9774	0.9185	0.8352	0.7373	0.6328	0.5282	0.4284	0.3370	0.2562	0.1875
	2	0.9988	0.9914	0.9734	0.9421	0.8965	0.8369	0.7648	0.6826	0.5931	0.5000
	3	1.0000	0.9995	0.0078	0.9933	0.9844	0.9692	0.9460	0.9130	0.8688	0.8125
	4	1.0000	1.0000	0.9999	0.9997	0.9990	0.9976	0.9947	0.9898	0.9815	0.9688
6	0	0.7351	0.5314	0.3771	0.2621	0.1780	0.1176	0.0754	0.0467	0.0277	0.0156
	1	0.9672	0.8857	0.7765	0.6554	0.5339	0.4202	0.3191	0.2333	0.1636	0.1094
	2	0.9978	0.9842	0.9527	0.9011	0.8306	0.7443	0.6471	0.5443	0.4415	0.3438
	3	0.9999	0.9987	0.9941	0.9830	0.9624	0.9295	0.8826	0.8208	0.7447	0.6562
	4	1.0000	0.9999	0.9996	0.9984	0.9954	0.9891	0.9777	0.9590	0.9308	0.8906
	5	1.0000	1.0000	1.0000	0.9999	0.9998	0.9993	0.9982	0.9959	0.9917	0.9844
7	0	0.6983	0.4783	0.3206	0.2097	0.1335	0.0824	0.0490	0.0280	0.0152	0.0078
	1	0.9556	0.8503	0.7166	0.5767	0.4449	0.3294	0.2338	0.1586	0.1024	0.0625
	2	0.9962	0.9743	0.9262	0.8520	0.7564	0.6471	0.5323	0.4199	0.3164	0.2266
	3	0.9998	0.9973	0.9879	0.9667	0.9294	0.8740	0.8002	0.7102	0.6083	0.5000
	4	1.0000	0.9998	0.9988	0.9953	0.9871	0.9712	0.9444	0.9037	0.8471	0.7734
	5	1.0000	1.0000	0.9999	0.9996	0.9987	0.9962	0.9910	0.9812	0.9643	0.9375
	6	1.0000	1.0000	1.0000	1.0000	0.9999	0.9998	0.9994	0.9984	0.9963	0.9922
8	0	0.6634	0.4305	0.2725	0.1678	0.1001	0.0576	0.0319	0.0168	0.0084	0.0039
	1	0.9428	0.8131	0.6572	0.5033	0.3671	0.2553	0.1691	0.1064	0.0632	0.0352
	2	0.9942	0.9619	0.8948	0.7969	0.6785	0.5518	0.4278	0.3154	0.2201	0.1445
	3	0.9996	0.9950	0.9786	0.9437	0.8862	0.8059	0.7064	0.5941	0.4770	0.3633
	4	1.0000	0.9996	0.9971	0.9896	0.9727	0.9420	0.8939	0.8263	0.7396	0.6367
	5	1.0000	1.0000	0.9998	0.9988	0.9958	0.9887	0.9747	0.9502	0.9115	0.8555
	6	1.0000	1.0000	1.0000	0.9999	0.9996	0.9987	0.9964	0.9915	0.9819	0.9648
	7	1.0000	1.0000	1.0000	1.0000	1.0000	0.9999	0.9998	0.9993	0.9983	0.9961
9	0	0.6302	0.3874	0.2316	0.1342	0.0751	0.0404	0.0207	0.0101	0.0046	0.0020
	1	0.9288	0.7748	0.5995	0.4362	0.3003	0.1960	0.1211	0.0705	0.0385	0.0195
	2	0.9916	0.9470	0.8591	0.7382	0.6007	0.4628	0.3373	0.2318	0.1495	0.0898
	3	0.9994	0.9917	0.9661	0.9144	0.8343	0.7297	0.6089	0.4826	0.3614	0.2539
	4	1.0000	0.9991	0.9944	0.9804	0.9511	0.9012	0.8283	0.7334	0.6214	0.5000
	5	1.0000	0.9999	0.9994	0.9969	0.9900	0.9747	0.9464	0.9006	0.8342	0.7461
	6	1.0000	1.0000	1.0000	0.9997	0.9987	0.9957	0.9888	0.9750	0.9502	0.9102
	7	1.0000	1.0000	1.0000	1.0000	0.9999	0.9996	0.9986	0.9962	0.9909	0.9805
	8	1.0000	1.0000	1.0000	1.0000	1.0000	1.0000	0.9999	0.9997	0.9992	0.9980

		p									
n	*x*	0.05	0.10	0.15	0.20	0.25	0.30	0.35	0.40	0.45	0.50
10	0	0.5987	0.3487	0.1969	0.1074	0.0563	0.0282	0.0135	0.0060	0.0025	0.0010
	1	0.9139	0.7361	0.5443	0.3758	0.2440	0.1493	0.0860	0.0464	0.0232	0.0107
	2	0.9885	0.9298	0.8202	0.6778	0.5256	0.3828	0.2616	0.1673	0.0996	0.0547
	3	0.9990	0.9872	0.9500	0.8791	0.7759	0.6496	0.5138	0.3823	0.2660	0.1719
	4	0.9999	0.9984	0.9901	0.9672	0.9219	0.8497	0.7515	0.6331	0.5044	0.3770
	5	1.0000	0.9999	0.9986	0.9936	0.9803	0.9527	0.9051	0.8338	0.7384	0.6230
	6	1.0000	1.0000	0.9999	0.9991	0.9965	0.9894	0.9740	0.9452	0.8980	0.8281
	7	1.0000	1.0000	1.0000	0.9999	0.9996	0.9984	0.9952	0.9877	0.9726	0.9453
	8	1.0000	1.0000	1.0000	1.0000	1.0000	0.9999	0.9995	0.9983	0.9955	0.9893
	9	1.0000	1.0000	1.0000	1.0000	1.0000	1.0000	1.0000	0.9999	0.9997	0.9990
11	0	0.5688	0.3138	0.1673	0.0859	0.0422	0.0198	0.0088	0.0036	0.0014	0.0005
	1	0.8981	0.6974	0.4922	0.3221	0.1971	0.1130	0.0606	0.0302	0.0139	0.0059
	2	0.9848	0.9104	0.7788	0.6174	0.4552	0.3127	0.2001	0.1189	0.0652	0.0327
	3	0.9984	0.9815	0.9306	0.8389	0.7133	0.5696	0.4256	0.2963	0.1911	0.1133
	4	0.9999	0.9972	0.9841	0.9496	0.8854	0.7897	0.6683	0.5328	0.3971	0.2744
	5	1.0000	0.9997	0.9973	0.9883	0.9657	0.9218	0.8513	0.7535	0.6331	0.5000
	6	1.0000	1.0000	0.9997	0.9980	0.9924	0.9784	0.9499	0.9006	0.8262	0.7256
	7	1.0000	1.0000	1.0000	0.9998	0.9988	0.9957	0.9878	0.9707	0.9390	0.8867
	8	1.0000	1.0000	1.0000	1.0000	0.9999	0.9994	0.9980	0.9941	0.9852	0.9673
	9	1.0000	1.0000	1.0000	1.0000	1.0000	1.0000	0.9998	0.9993	0.9978	0.9941
	10	1.0000	1.0000	1.0000	1.0000	1.0000	1.0000	1.0000	1.0000	0.9998	0.9995
12	0	0.5404	0.2824	0.1422	0.0687	0.0317	0.0138	0.0057	0.0022	0.0008	0.0002
	1	0.8816	0.6590	0.4435	0.2749	0.1584	0.0850	0.0424	0.0196	0.0083	0.0032
	2	0.9804	0.8891	0.7358	0.5583	0.3907	0.2528	0.1513	0.0834	0.0421	0.0193
	3	0.9978	0.9744	0.9078	0.7946	0.6488	0.4925	0.3467	0.2253	0.1345	0.0730
	4	0.9998	0.9957	0.9761	0.9274	0.8424	0.7237	0.5833	0.4382	0.3044	0.1938
	5	1.0000	0.9995	0.9954	0.9806	0.9456	0.8822	0.7873	0.6652	0.5269	0.3872
	6	1.0000	0.9999	0.9993	0.9961	0.9857	0.9614	0.9154	0.8418	0.7393	0.6128
	7	1.0000	1.0000	0.9999	0.9994	0.9972	0.9905	0.9745	0.9427	0.8883	0.8062
	8	1.0000	1.0000	1.0000	0.9999	0.9996	0.9983	0.9944	0.9847	0.9644	0.9270
	9	1.0000	1.0000	1.0000	1.0000	1.0000	0.9998	0.9992	0.9972	0.9921	0.9807
	10	1.0000	1.0000	1.0000	1.0000	1.0000	1.0000	0.9999	0.9997	0.9989	0.9968
	11	1.0000	1.0000	1.0000	1.0000	1.0000	1.0000	1.0000	1.0000	0.9999	0.9998
13	0	0.5133	0.2542	0.1209	0.0550	0.0238	0.0097	0.0037	0.0013	0.0004	0.0001
	1	0.8646	0.6213	0.3983	0.2336	0.1267	0.0637	0.0296	0.0126	0.0049	0.0017
	2	0.9755	0.8661	0.6920	0.5017	0.3326	0.2025	0.1132	0.0579	0.0269	0.0112
	3	0.9969	0.9658	0.8820	0.7437	0.5843	0.4206	0.2783	0.1686	0.0929	0.0461
	4	0.9997	0.9935	0.9658	0.9009	0.7940	0.6543	0.5005	0.3530	0.2279	0.1334
	5	1.0000	0.9991	0.9925	0.9700	0.9198	0.8346	0.7159	0.5744	0.4268	0.2905
	6	1.0000	0.9999	0.9987	0.9930	0.9757	0.9376	0.8705	0.7712	0.6437	0.5000
	7	1.0000	1.0000	0.9998	0.9988	0.9944	0.9818	0.9538	0.9023	0.8212	0.7095
	8	1.0000	1.0000	1.0000	0.9998	0.9990	0.9960	0.9874	0.9679	0.9302	0.8666
	9	1.0000	1.0000	1.0000	1.0000	0.9999	0.9993	0.9975	0.9922	0.9797	0.9539
	10	1.0000	1.0000	1.0000	1.0000	1.0000	0.9999	0.9997	0.9987	0.9959	0.9888
	11	1.0000	1.0000	1.0000	1.0000	1.0000	1.0000	1.0000	0.9999	0.9995	0.9983
	12	1.0000	1.0000	1.0000	1.0000	1.0000	1.0000	1.0000	1.0000	1.0000	0.9999
14	0	0.4877	0.2288	0.1028	0.0440	0.0178	0.0068	0.0024	0.0008	0.0002	0.0001
	1	0.8470	0.5846	0.3567	0.1979	0.1010	0.0475	0.0205	0.0081	0.0029	0.0009

n	x	p = 0.05	0.10	0.15	0.20	0.25	0.30	0.35	0.40	0.45	0.50
14	2	0.9699	0.8416	0.6479	0.4481	0.2811	0.1608	0.0839	0.0398	0.0170	0.0065
	3	0.9958	0.9559	0.8535	0.6982	0.5213	0.3552	0.2205	0.1243	0.0632	0.0287
	4	0.9996	0.9908	0.9533	0.8702	0.7415	0.5842	0.4227	0.2793	0.1672	0.0898
	5	1.0000	0.9985	0.9885	0.9561	0.8883	0.7805	0.6405	0.4859	0.3373	0.2120
	6	1.0000	0.9998	0.9978	0.9884	0.9617	0.9067	0.8164	0.6925	0.5461	0.3953
	7	1.0000	1.0000	0.9997	0.9976	0.9897	0.9685	0.9247	0.8499	0.7414	0.6074
	8	1.0000	1.0000	1.0000	0.9996	0.9978	0.9917	0.9757	0.9417	0.8811	0.7880
	9	1.0000	1.0000	1.0000	1.0000	0.9997	0.9983	0.9940	0.9825	0.9574	0.9102
	10	1.0000	1.0000	1.0000	1.0000	1.0000	0.9989	0.9989	0.9961	0.9886	0.9713
	11	1.0000	1.0000	1.0000	1.0000	1.0000	1.0000	0.9999	0.9994	0.9978	0.9935
	12	1.0000	1.0000	1.0000	1.0000	1.0000	1.0000	1.0000	0.9999	0.9997	0.9991
	13	1.0000	1.0000	1.0000	1.0000	1.0000	1.0000	1.0000	1.0000	1.0000	0.9999
15	0	0.4633	0.2059	0.0874	0.0352	0.0134	0.0047	0.0016	0.0005	0.0001	0.0000
	1	0.8290	0.5490	0.3186	0.1671	0.0802	0.0353	0.0142	0.0052	0.0017	0.0005
	2	0.9638	0.8159	0.6042	0.3980	0.2361	0.1268	0.0617	0.0271	0.0107	0.0037
	3	0.9945	0.9444	0.8227	0.6482	0.4613	0.2969	0.1727	0.0905	0.0424	0.0176
	4	0.9994	0.9873	0.9383	0.8358	0.6865	0.5155	0.3519	0.2173	0.1204	0.0592
	5	0.9999	0.9978	0.9832	0.9389	0.8516	0.7216	0.5643	0.4032	0.2608	0.1509
	6	1.0000	0.9997	0.9964	0.9819	0.9434	0.8689	0.7548	0.6098	0.4522	0.3036
	7	1.0000	1.0000	0.9996	0.9958	0.9827	0.9500	0.8868	0.7869	0.6535	0.5000
	8	1.0000	1.0000	0.9999	0.9992	0.9958	0.9848	0.9578	0.9050	0.8182	0.6964
	9	1.0000	1.0000	1.0000	0.9999	0.9992	0.9963	0.9876	0.9662	0.9231	0.8491
	10	1.0000	1.0000	1.0000	1.0000	0.9999	0.9993	0.9972	0.9907	0.9745	0.9408
	11	1.0000	1.0000	1.0000	1.0000	1.0000	0.9999	0.9995	0.9981	0.9937	0.9824
	12	1.0000	1.0000	1.0000	1.0000	1.0000	1.0000	0.9999	0.9997	0.9989	0.9963
	13	1.0000	1.0000	1.0000	1.0000	1.0000	1.0000	1.0000	1.0000	0.9999	0.9995
	14	1.0000	1.0000	1.0000	1.0000	1.0000	1.0000	1.0000	1.0000	1.0000	1.0000
16	0	0.4401	0.1853	0.0743	0.0281	0.0100	0.0033	0.0010	0.0003	0.0001	0.0000
	1	0.8108	0.5147	0.2839	0.1407	0.0635	0.0261	0.0098	0.0033	0.0010	0.0003
	2	0.9571	0.7892	0.5614	0.3518	0.1971	0.0994	0.0451	0.0183	0.0066	0.0021
	3	0.9930	0.9316	0.7899	0.5981	0.4050	0.2459	0.1339	0.0651	0.0281	0.0106
	4	0.9991	0.9830	0.9209	0.7982	0.6302	0.4499	0.2892	0.1666	0.0853	0.0384
	5	0.9999	0.9967	0.9765	0.9183	0.8103	0.6598	0.4900	0.3288	0.1976	0.1051
	6	1.0000	0.9995	0.9944	0.9733	0.9204	0.8247	0.6881	0.5272	0.3660	0.2272
	7	1.0000	0.9999	0.9989	0.9930	0.9729	0.9256	0.8406	0.7161	0.5629	0.4018
	8	1.0000	1.0000	0.9998	0.9985	0.9925	0.9743	0.9329	0.8577	0.7441	0.5982
	9	1.0000	1.0000	1.0000	0.9998	0.9984	0.9929	0.9771	0.9417	0.8759	0.7728
	10	1.0000	1.0000	1.0000	1.0000	0.9997	0.9984	0.9938	0.9809	0.9514	0.8949
	11	1.0000	1.0000	1.0000	1.0000	1.0000	0.9997	0.9987	0.9951	0.9851	0.9616
	12	1.0000	1.0000	1.0000	1.0000	1.0000	1.0000	0.9998	0.9991	0.9965	0.9894
	13	1.0000	1.0000	1.0000	1.0000	1.0000	1.0000	1.0000	0.9999	0.9994	0.9979
	14	1.0000	1.0000	1.0000	1.0000	1.0000	1.0000	1.0000	1.0000	1.0000	0.9997
	15	1.0000	1.0000	1.0000	1.0000	1.0000	1.0000	1.0000	1.0000	1.0000	1.0000
17	0	0.4181	0.1668	0.0631	0.0225	0.0075	0.0023	0.0007	0.0002	0.0000	0.0000
	1	0.7922	0.4818	0.2525	0.1182	0.0501	0.0193	0.0067	0.0021	0.0006	0.0001
	2	0.9497	0.7618	0.5198	0.3096	0.1637	0.0774	0.0327	0.0123	0.0041	0.0012
	3	0.9912	0.9174	0.7556	0.5489	0.3530	0.2019	0.1028	0.0464	0.0184	0.0063
	4	0.9988	0.9779	0.9013	0.7582	0.5739	0.3887	0.2348	0.1260	0.0596	0.0245

n	x	0.05	0.10	0.15	0.20	0.25	0.30	0.35	0.40	0.45	0.50
17	5	0.9999	0.9953	0.9681	0.8943	0.7653	0.5968	0.4197	0.2639	0.1471	0.0717
	6	1.0000	0.9992	0.9917	0.9623	0.8929	0.7752	0.6188	0.4478	0.2902	0.1662
	7	1.0000	0.9999	0.9983	0.9891	0.9598	0.8954	0.7872	0.6405	0.4743	0.3145
	8	1.0000	1.0000	0.9997	0.9974	0.9876	0.9597	0.9006	0.8011	0.6626	0.5000
	9	1.0000	1.0000	1.0000	0.9995	0.9969	0.9873	0.9617	0.9081	0.8166	0.6855
	10	1.0000	1.0000	1.0000	0.9999	0.9994	0.9968	0.9880	0.9652	0.9174	0.8338
	11	1.0000	1.0000	1.0000	1.0000	0.9999	0.9993	0.9970	0.9894	0.9699	0.9283
	12	1.0000	1.0000	1.0000	1.0000	1.0000	0.9999	0.9994	0.9975	0.9914	0.9755
	13	1.0000	1.0000	1.0000	1.0000	1.0000	1.0000	0.9999	0.9995	0.9981	0.9936
	14	1.0000	1.0000	1.0000	1.0000	1.0000	1.0000	1.0000	0.9999	0.9997	0.9988
	15	1.0000	1.0000	1.0000	1.0000	1.0000	1.0000	1.0000	1.0000	1.0000	0.9999
	16	1.0000	1.0000	1.0000	1.0000	1.0000	1.0000	1.0000	1.0000	1.0000	1.0000
18	0	0.3972	0.1501	0.0536	0.0180	0.0056	0.0016	0.0004	0.0001	0.0000	0.0000
	1	0.7735	0.4503	0.2241	0.0991	0.0395	0.0142	0.0046	0.0013	0.0003	0.0001
	2	0.9419	0.7338	0.4797	0.2713	0.1353	0.0600	0.0236	0.0082	0.0025	0.0007
	3	0.9891	0.9018	0.7202	0.5010	0.3057	0.1646	0.0783	0.0328	0.0120	0.0038
	4	0.9985	0.9718	0.8794	0.7164	0.5187	0.3327	0.1886	0.0942	0.0411	0.0154
	5	0.9998	0.9936	0.9581	0.8671	0.7175	0.5344	0.3550	0.2088	0.1077	0.0481
	6	1.0000	0.9988	0.9882	0.9487	0.8610	0.7217	0.5491	0.3743	0.2258	0.1189
	7	1.0000	0.9998	0.9973	0.9837	0.9431	0.8593	0.7283	0.5634	0.3915	0.2403
	8	1.0000	1.0000	0.9995	0.9957	0.9807	0.9404	0.8609	0.7368	0.5778	0.4073
	9	1.0000	1.0000	0.9999	0.9991	0.9946	0.9790	0.9403	0.8653	0.7473	0.5927
	10	1.0000	1.0000	1.0000	0.9998	0.9988	0.9939	0.9788	0.9424	0.8720	0.7597
	11	1.0000	1.0000	1.0000	1.0000	0.9998	0.9986	0.9938	0.9797	0.9463	0.8811
	12	1.0000	1.0000	1.0000	1.0000	1.0000	0.9997	0.9986	0.9942	0.9817	0.9519
	13	1.0000	1.0000	1.0000	1.0000	1.0000	1.0000	0.9997	0.9987	0.9951	0.9846
	14	1.0000	1.0000	1.0000	1.0000	1.0000	1.0000	1.0000	0.9998	0.9990	0.9962
	15	1.0000	1.0000	1.0000	1.0000	1.0000	1.0000	1.0000	1.0000	0.9999	0.9993
	16	1.0000	1.0000	1.0000	1.0000	1.0000	1.0000	1.0000	1.0000	1.0000	0.9999
19	0	0.3774	0.1351	0.0456	0.0144	0.0042	0.0011	0.0003	0.0001	0.0000	0.0000
	1	0.7547	0.4203	0.1985	0.0829	0.0310	0.0104	0.0031	0.0008	0.0002	0.0000
	2	0.9335	0.7054	0.4413	0.2369	0.1113	0.0462	0.0170	0.0055	0.0015	0.0004
	3	0.9868	0.8850	0.6841	0.4551	0.2630	0.1332	0.0591	0.0230	0.0077	0.0022
	4	0.9980	0.9648	0.8556	0.6733	0.4654	0.2822	0.1500	0.0696	0.0280	0.0096
	5	0.9998	0.9914	0.9463	0.8369	0.6678	0.4739	0.2968	0.1629	0.0777	0.0318
	6	1.0000	0.9983	0.9837	0.9324	0.8251	0.6655	0.4812	0.3081	0.1727	0.0835
	7	1.0000	0.9997	0.9959	0.9767	0.9225	0.8180	0.6656	0.4878	0.3169	0.1796
	8	1.0000	1.0000	0.9992	0.9933	0.9713	0.9161	0.8145	0.6675	0.4940	0.3238
	9	1.0000	1.0000	0.9999	0.9984	0.9911	0.9674	0.9125	0.8139	0.6710	0.5000
	10	1.0000	1.0000	1.0000	0.9997	0.9977	0.9895	0.9653	0.9115	0.8159	0.6762
	11	1.0000	1.0000	1.0000	1.0000	0.9995	0.9972	0.9886	0.9648	0.9129	0.8204
	12	1.0000	1.0000	1.0000	1.0000	0.9999	0.9994	0.9969	0.9884	0.9658	0.9165
	13	1.0000	1.0000	1.0000	1.0000	1.0000	0.9999	0.9993	0.9969	0.9891	0.9682
	14	1.0000	1.0000	1.0000	1.0000	1.0000	1.0000	0.9999	0.9994	0.9972	0.9904
	15	1.0000	1.0000	1.0000	1.0000	1.0000	1.0000	1.0000	0.9999	0.9995	0.9978
	16	1.0000	1.0000	1.0000	1.0000	1.0000	1.0000	1.0000	1.0000	0.9999	0.9996
	17	1.0000	1.0000	1.0000	1.0000	1.0000	1.0000	1.0000	1.0000	1.0000	1.0000

		p									
n	*x*	0.05	0.10	0.15	0.20	0.25	0.30	0.35	0.40	0.45	0.50
20	0	0.3585	0.1216	0.0388	0.0115	0.0032	0.0008	0.0002	0.0000	0.0000	0.0000
	1	0.7358	0.3917	0.1756	0.0692	0.0243	0.0076	0.0021	0.0005	0.0001	0.0000
	2	0.9245	0.6769	0.4049	0.2061	0.0913	0.0355	0.0121	0.0036	0.0009	0.0002
	3	0.9841	0.8670	0.6477	0.4114	0.2252	0.1071	0.0444	0.0160	0.0049	0.0013
	4	0.9974	0.9568	0.8298	0.6296	0.4148	0.2375	0.1182	0.0510	0.0189	0.0059
	5	0.9997	0.9887	0.9327	0.8042	0.6172	0.4164	0.2454	0.1256	0.0553	0.0207
	6	1.0000	0.9976	0.9781	0.9133	0.7858	0.6080	0.4166	0.2500	0.1299	0.0577
	7	1.0000	0.9996	0.9941	0.9679	0.8982	0.7723	0.6010	0.4159	0.2520	0.1316
	8	1.0000	0.9999	0.9987	0.9900	0.9591	0.8867	0.7624	0.5956	0.4143	0.2517
	9	1.0000	1.0000	0.9998	0.9974	0.9861	0.9520	0.8782	0.7553	0.5914	0.4119
	10	1.0000	1.0000	1.0000	0.9994	0.9961	0.9829	0.9468	0.8725	0.7507	0.5881
	11	1.0000	1.0000	1.0000	0.9999	0.9991	0.9949	0.9804	0.9435	0.8692	0.7483
	12	1.0000	1.0000	1.0000	1.0000	0.9998	0.9987	0.9940	0.9790	0.9420	0.8684
	13	1.0000	1.0000	1.0000	1.0000	1.0000	0.9997	0.9985	0.9935	0.9786	0.9423
	14	1.0000	1.0000	1.0000	1.0000	1.0000	1.0000	0.9997	0.9984	0.9936	0.9793
	15	1.0000	1.0000	1.0000	1.0000	1.0000	1.0000	1.0000	0.9997	0.9985	0.9941
	16	1.0000	1.0000	1.0000	1.0000	1.0000	1.0000	1.0000	1.0000	0.9997	0.9987
	17	1.0000	1.0000	1.0000	1.0000	1.0000	1.0000	1.0000	1.0000	1.0000	0.9998
	18	1.0000	1.0000	1.0000	1.0000	1.0000	1.0000	1.0000	1.0000	1.0000	1.0000

Table 2 *Poisson Distribution Function*

$$F(x;\alpha) = \sum_{k=0}^{x} e^{-\alpha} \frac{\alpha^k}{k!}$$

α \ x	0	1	2	3	4	5	6	7	8	9
0.02	0.980	1.000								
0.04	0.961	0.999	1.000							
0.06	0.942	0.998	1.000							
0.08	0.923	0.997	1.000							
0.10	0.905	0.995	1.000							
0.15	0.861	0.990	0.999	1.000						
0.20	0.819	0.982	0.999	1.000						
0.25	0.779	0.974	0.998	1.000						
0.30	0.741	0.963	0.996	1.000						
0.35	0.705	0.951	0.994	1.000						
0.40	0.670	0.938	0.992	0.999	1.000					
0.45	0.638	0.925	0.989	0.999	1.000					
0.50	0.607	0.910	0.986	0.998	1.000					
0.55	0.577	0.894	0.982	0.998	1.000					
0.60	0.549	0.878	0.977	0.997	1.000					
0.65	0.522	0.861	0.972	0.996	0.999	1.000				
0.70	0.497	0.844	0.966	0.994	0.999	1.000				
0.75	0.472	0.827	0.959	0.993	0.999	1.000				
0.80	0.449	0.809	0.953	0.991	0.999	1.000				
0.85	0.427	0.791	0.945	0.989	0.998	1.000				
0.90	0.407	0.772	0.937	0.987	0.998	1.000				
0.95	0.387	0.754	0.929	0.984	0.997	1.000				
1.00	0.368	0.736	0.920	0.981	0.996	0.999	1.000			
1.1	0.333	0.699	0.900	0.974	0.995	0.999	1.000			
1.2	0.301	0.663	0.879	0.966	0.992	0.998	1.000			
1.3	0.273	0.627	0.857	0.957	0.989	0.998	1.000			
1.4	0.247	0.592	0.833	0.946	0.986	0.997	0.999	1.000		
1.5	0.223	0.558	0.809	0.934	0.981	0.996	0.999	1.000		
1.6	0.202	0.525	0.783	0.921	0.976	0.994	0.999	1.000		
1.7	0.183	0.493	0.757	0.907	0.970	0.992	0.998	1.000		
1.8	0.165	0.463	0.731	0.891	0.964	0.990	0.997	0.999	1.000	
1.9	0.150	0.434	0.704	0.875	0.956	0.987	0.997	0.999	1.000	
2.0	0.135	0.406	0.677	0.857	0.947	0.983	0.995	0.999	1.000	

*Reprinted by kind permission from E. C. Molina, *Poisson's Exponential Binomial Limit,* D. Van Nostrand Company, Inc., Princeton, N.J., 1947.

α \ x	0	1	2	3	4	5	6	7	8	9
2.2	0.111	0.355	0.623	0.819	0.928	0.975	0.993	0.998	1.000	
2.4	0.091	0.308	0.570	0.779	0.904	0.964	0.988	0.997	0.999	1.000
2.6	0.074	0.267	0.518	0.736	0.877	0.951	0.983	0.995	0.999	1.000
2.8	0.061	0.231	0.469	0.692	0.848	0.935	0.976	0.992	0.998	0.999
3.0	0.050	0.199	0.423	0.647	0.815	0.916	0.966	0.988	0.996	0.999
3.2	0.041	0.171	0.380	0.603	0.781	0.895	0.955	0.983	0.994	0.998
3.4	0.033	0.147	0.340	0.558	0.744	0.871	0.942	0.977	0.992	0.997
3.6	0.027	0.126	0.303	0.515	0.706	0.844	0.927	0.969	0.988	0.996
3.8	0.022	0.107	0.269	0.473	0.668	0.816	0.909	0.960	0.984	0.994
4.0	0.018	0.092	0.238	0.433	0.629	0.785	0.889	0.949	0.979	0.992
4.2	0.015	0.078	0.210	0.395	0.590	0.753	0.867	0.936	0.972	0.989
4.4	0.012	0.066	0.185	0.359	0.551	0.720	0.844	0.921	0.964	0.985
4.6	0.010	0.056	0.163	0.326	0.513	0.686	0.818	0.905	0.955	0.980
4.8	0.008	0.048	0.143	0.294	0.476	0.651	0.791	0.887	0.944	0.975
5.0	0.007	0.040	0.125	0.265	0.440	0.616	0.762	0.867	0.932	0.968
5.2	0.006	0.034	0.109	0.238	0.406	0.581	0.732	0.845	0.918	0.960
5.4	0.005	0.029	0.095	0.213	0.373	0.546	0.702	0.822	0.903	0.951
5.6	0.004	0.024	0.082	0.191	0.342	0.512	0.670	0.797	0.886	0.941
5.8	0.003	0.021	0.072	0.170	0.313	0.478	0.638	0.771	0.867	0.929
6.0	0.002	0.017	0.062	0.151	0.285	0.446	0.606	0.744	0.847	0.916

α \ x	10	11	12	13	14	15	16
2.8	1.000						
3.0	1.000						
3.2	1.000						
3.4	0.999	1.000					
3.6	0.999	1.000					
3.8	0.998	0.999	1.000				
4.0	0.997	0.999	1.000				
4.2	0.996	0.999	1.000				
4.4	0.994	0.998	0.999	1.000			
4.6	0.992	0.997	0.999	1.000			
4.8	0.990	0.996	0.999	1.000			
5.0	0.986	0.995	0.998	0.999	1.000		
5.2	0.982	0.993	0.997	0.999	1.000		
5.4	0.977	0.990	0.996	0.999	1.000		
5.6	0.972	0.988	0.995	0.998	0.999	1.000	
5.8	0.965	0.984	0.993	0.997	0.999	1.000	
6.0	0.957	0.980	0.991	0.996	0.999	0.999	1.000

α \ x	0	1	2	3	4	5	6	7	8	9
6.2	0.002	0.015	0.054	0.134	0.259	0.414	0.574	0.716	0.826	0.902
6.4	0.002	0.012	0.046	0.119	0.235	0.384	0.542	0.687	0.803	0.886
6.6	0.001	0.010	0.040	0.105	0.213	0.355	0.511	0.658	0.780	0.869
6.8	0.001	0.009	0.034	0.093	0.192	0.327	0.480	0.628	0.755	0.850
7.0	0.001	0.007	0.030	0.082	0.173	0.301	0.450	0.599	0.729	0.830
7.2	0.001	0.006	0.025	0.072	0.156	0.276	0.420	0.569	0.703	0.810
7.4	0.001	0.005	0.022	0.063	0.140	0.253	0.392	0.539	0.676	0.788
7.6	0.001	0.004	0.019	0.055	0.125	0.231	0.365	0.510	0.648	0.765
7.8	0.000	0.004	0.016	0.048	0.112	0.210	0.338	0.481	0.620	0.741
8.0	0.000	0.003	0.014	0.042	0.100	0.191	0.313	0.453	0.593	0.717
8.5	0.000	0.002	0.009	0.030	0.074	0.150	0.256	0.386	0.523	0.653
9.0	0.000	0.001	0.006	0.021	0.055	0.116	0.207	0.324	0.456	0.587
9.5	0.000	0.001	0.004	0.015	0.040	0.089	0.165	0.269	0.392	0.522
10.0	0.000	0.000	0.003	0.010	0.029	0.067	0.130	0.220	0.333	0.458

	10	11	12	13	14	15	16	17	18	19
6.2	0.949	0.975	0.989	0.995	0.998	0.999	1.000			
6.4	0.939	0.969	0.986	0.994	0.997	0.999	1.000			
6.6	0.927	0.963	0.982	0.992	0.997	0.999	0.999	1.000		
6.8	0.915	0.955	0.978	0.990	0.996	0.998	0.999	1.000		
7.0	0.901	0.947	0.973	0.987	0.994	0.998	0.999	1.000		
7.2	0.887	0.937	0.967	0.984	0.993	0.997	0.999	0.999	1.000	
7.4	0.871	0.926	0.961	0.980	0.991	0.996	0.998	0.999	1.000	
7.6	0.854	0.915	0.954	0.976	0.989	0.995	0.998	0.999	1.000	
7.8	0.835	0.902	0.945	0.971	0.986	0.993	0.997	0.999	1.000	
8.0	0.816	0.888	0.936	0.966	0.983	0.992	0.996	0.998	0.999	1.000
8.5	0.763	0.849	0.909	0.949	0.973	0.986	0.993	0.997	0.999	0.999
9.0	0.706	0.803	0.876	0.926	0.959	0.978	0.989	0.995	0.998	0.999
9.5	0.645	0.752	0.836	0.898	0.940	0.967	0.982	0.991	0.996	0.998
10.0	0.583	0.697	0.792	0.864	0.917	0.951	0.973	0.986	0.993	0.997

	20	21	22
8.5	1.000		
9.0	1.000		
9.5	0.999	1.000	
10.0	0.998	0.999	1.000

Table 3 *Distribution Function of Standard Normal Random Variable*

z	0	1	2	3	4	5	6	7	8	9
.0	.5000	.5040	.5080	.5120	.5160	.5199	.5239	.5279	.5319	.5359
.1	.5398	.5438	.5478	.5517	.5557	.5596	.5363	.5675	.5714	.5753
.2	.5793	.5832	.5871	.5910	.5948	.5987	.6026	.6064	.6103	.6141
.3	.6179	.6217	.6255	.6293	.6331	.6368	.6406	.6443	.6480	.6517
.4	.6554	.6591	.6628	.6664	.6700	.6736	.6772	.6808	.6844	.6879
.5	.6915	.6950	.6985	.7019	.7054	.7088	.7123	.7157	.7190	.7224
.6	.7257	.7291	.7324	.7357	.7389	.7422	.7454	.7486	.7517	.7549
.7	.7580	.7611	.7642	.7673	.7703	.7734	.7764	.7974	.7823	.7852
.8	.7881	.7910	.7939	.7967	.7995	.8023	.8051	.8078	.8106	.8133
.9	.8159	.8186	.8212	.8238	.8264	.8289	.8315	.8340	.8365	.8389
1.0	.8413	.8438	.8461	.8485	.8508	.8531	.8554	.8577	.8599	.8621
1.1	.8643	.8665	.8686	.8708	.8729	.8749	.8770	.8790	.8810	.8830
1.2	.8849	.8869	.8888	.8907	.8925	.8944	.8962	.8980	.8997	.9015
1.3	.9032	.9049	.9066	.9082	.9099	.9115	.9131	.9147	.9162	.9177
1.4	.9192	.9207	.9222	.9236	.9251	.9265	.9278	.9292	.9306	.9319
1.5	.9332	.9345	.9357	.9370	.9382	.9394	.9406	.9418	.9430	.9441
1.6	.9452	.9463	.9474	.9484	.9495	.9505	.9515	.9525	.9535	.9545
1.7	.9554	.9564	.9573	.9582	.9591	.9599	.9608	.9616	.9625	.9633
1.8	.9641	.9648	.9656	.9664	.9671	.9678	.9686	.9693	.9700	.9706
1.9	.9713	.9719	.9726	.9732	.9738	.9744	.9750	.9756	.9762	.9767
2.0	.9772	.9778	.9783	.9788	.9793	.9798	.9803	.9808	.9812	.9817
2.1	.9821	.9826	.9830	.9834	.9838	.9842	.9846	.9850	.9854	.9857
2.2	.9861	.9864	.9868	.9871	.9874	.9878	.9881	.9884	.9887	.9890
2.3	.9893	.9896	.9898	.9901	.9904	.9906	.9909	.9911	.9913	.9916
2.4	.9918	.9920	.9922	.9925	.9927	.9929	.9931	.9932	.9934	.9936
2.5	.9938	.9940	.9941	.9943	.9945	.9946	.9948	.9949	.9951	.9952
2.6	.9953	.9955	.9956	.9957	.9959	.9960	.9961	.9962	.9963	.9964
2.7	.9965	.9966	.9967	.9968	.9969	.9970	.9971	.9972	.9973	.9974
2.8	.9974	.9975	.9976	.9977	.9977	.9978	.9979	.9979	.9980	.9981
2.9	.9981	.9982	.9982	.9983	.9984	.9984	.9985	.9985	.9986	.9986
3.	.9987	.9990	.9993	.9995	.9997	.9998	.9998	.9999	.9999	1.0000

Note 1: If a normal variable X is not "standard," its value must be standardized: $Z = (X - \mu)/\sigma$. Then $F_X(x) = F_Z(\frac{x - \mu}{\sigma})$.

Note 2: For $z \geqslant 4$, use $F_Z(z) = 1$ to four decimal places; for $z \leqslant -4$, $F_Z(z) = 0$ to four decimal places.

Note 3: The entries opposite $z = 3$ are for 3.0, 3.1, 3.2, etc.

Note 4: For $z < 0$ use $F_Z(z) = 1 - F_Z(-z)$.

Table 4 *Critical Values of the Students t-Distribution*

α	0.1	0.05	0.025	0.01	0.005	0.001
d.f.						
1	3.078	6.314	12.706	31.821	63.657	318.310
2	1.886	2.920	4.303	6.965	9.925	22.327
3	1.638	2.353	3.182	4.541	5.841	10.215
4	1.533	2.132	2.776	3.747	4.604	7.173
5	1.476	2.015	2.571	3.365	4.032	5.893
6	1.440	1.943	2.447	3.143	3.707	5.208
7	1.415	1.895	2.365	2.998	3.499	4.785
8	1.397	1.860	2.306	2.896	3.355	4.501
9	1.383	1.833	2.262	2.821	3.250	4.297
10	1.372	1.812	2.228	2.764	3.169	4.144
11	1.363	1.796	2.201	2.718	3.106	4.025
12	1.356	1.782	2.179	2.681	3.055	3.930
13	1.350	1.771	2.160	2.650	3.012	3.852
14	1.345	1.761	2.145	2.624	2.977	3.787
15	1.341	1.753	2.131	2.602	2.947	3.733
16	1.337	1.746	2.120	2.583	2.921	3.686
17	1.333	1.740	2.110	2.567	2.898	3.646
18	1.330	1.734	2.101	2.552	2.878	3.610
19	1.328	1.729	2.093	2.539	2.861	3.579
20	1.325	1.725	2.086	2.528	2.845	3.552
21	1.323	1.721	2.080	2.518	2.831	3.527
22	1.321	1.717	2.074	2.508	2.819	3.505
23	1.319	1.714	2.069	2.500	2.807	3.485
24	1.318	1.711	2.064	2.492	2.797	3.467
25	1.316	1.708	2.060	2.485	2.787	3.450
26	1.315	1.706	2.056	2.479	2.779	3.435
27	1.314	1.703	2.052	2.473	2.771	3.421
28	1.313	1.701	2.048	2.467	2.763	3.408
29	1.311	1.699	2.045	2.462	2.756	3.396
30	1.310	1.697	2.042	2.457	2.750	3.385
40	1.303	1.684	2.021	2.423	2.704	3.307
60	1.296	1.671	2.000	2.390	2.660	3.232
120	1.289	1.658	1.980	2.358	2.617	3.160
∞	1.282	1.645	1.960	2.326	2.576	3.090

Table 5 *Critical Values of the χ^2 Distribution*

The first column lists the number of degrees of freedom (ν). The headings of the other columns give probabilities (P) for χ^2 to exceed the entry value. For $\nu > 100$, treat $\sqrt{2\chi^2} - \sqrt{2\nu - 1}$ as a standard normal variable.

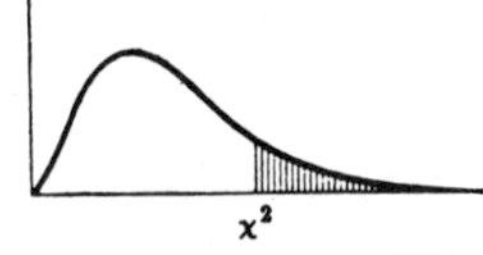

ν \ P	0.995	0.975	0.050	0.025	0.010	0.005
1	$0.0^{4}3927$	$0.0^{3}9821$	3.84146	5.02389	6.63490	7.87944
2	0.010025	0.050636	5.99147	7.37776	9.21034	10.5966
3	0.071721	0.215795	7.81473	9.34840	11.3449	12.8381
4	0.206990	0.484419	9.48773	11.1433	13.2767	14.8602
5	0.411740	0.831211	11.0705	12.8325	15.0863	16.7496
6	0.675727	1.237347	12.5916	14.4494	16.8119	18.5476
7	0.989265	1.68987	14.0671	16.0128	18.4753	20.2777
8	1.344419	2.17973	15.5073	17.5346	20.0902	21.9550
9	1.734926	2.70039	16.9190	19.0228	21.6660	23.5893
10	2.15585	3.24697	18.3070	20.4831	23.2093	25.1882
11	2.60321	3.81575	19.6751	21.9200	24.7250	26.7569
12	3.07382	4.40379	21.0261	23.3367	26.2170	28.2995
13	3.56503	5.00874	22.3621	24.7356	27.6883	29.8194
14	4.07468	5.62872	23.6848	26.1190	29.1413	31.3193
15	4.60094	6.26214	24.9958	27.4884	30.5779	32.8013
16	5.14224	6.90766	26.2962	28.8454	31.9999	34.2672
17	5.69724	7.56418	27.5871	30.1910	33.4087	35.7185
18	6.26481	8.23075	28.8693	31.5264	34.8053	37.1564
19	6.84398	8.90655	30.1435	32.8523	36.1908	38.5822
20	7.43386	9.59083	31.4104	34.1696	37.5662	39.9968
21	8.03366	10.28293	32.6705	35.4789	38.9321	41.4010
22	8.64272	10.9823	33.9244	36.7807	40.2894	42.7956
23	9.26042	11.6885	35.1725	38.0757	41.6384	44.1813
24	9.88623	12.4001	36.4151	39.3641	42.9798	45.5585
25	10.5197	13.1197	37.6525	40.6465	44.3141	46.9278
26	11.1603	13.8439	38.8852	41.9232	45.6417	48.2899
27	11.8076	14.5733	40.1133	43.1944	46.9630	49.6449
28	12.4613	15.3079	41.3372	44.4607	48.2782	50.9933
29	13.1211	16.0471	42.5569	45.7222	49.5879	52.3356
30	13.7867	16.7908	43.7729	46.9792	50.8922	53.6720
40	20.7065	24.4331	55.7585	59.3417	63.6907	66.7659
50	27.9907	32.3574	67.5048	71.4202	76.1539	79.4900
60	35.5346	40.4817	79.0819	83.2976	88.3794	91.9517
70	43.2752	48.7576	90.5312	95.0231	100.425	104.215
80	51.1720	57.1532	101.879	106.629	112.329	116.321
90	59.1963	65.6466	113.145	118.136	124.116	128.299
100	67.3276	74.2219	124.342	129.561	135.807	140.169

Table 6 *Critical Values of the F Distribution*

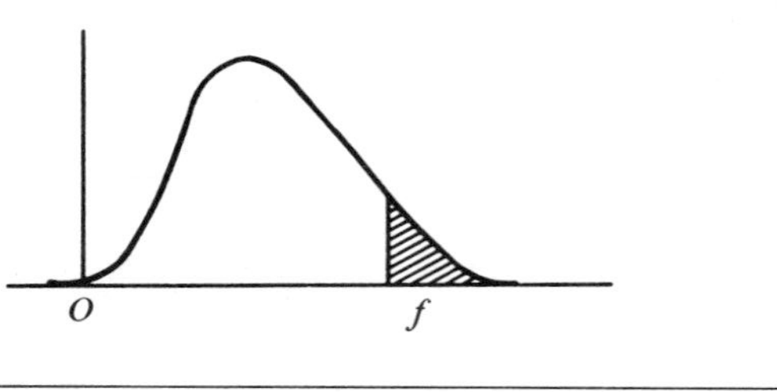

5% (Roman Type) and 1% (Boldface Type) Points for the Distribution of F

Degrees of freedom for denominator (ν_2)	Degrees of freedom for numerator (ν_1)																						
	1	2	3	4	5	6	7	8	9	10	11	12	14	16	20	24	30	40	50	75	100	200	500	∞
1	161	200	216	225	230	234	237	239	241	242	243	244	245	246	248	249	250	251	252	253	253	254	254	254
	4052	**4999**	**5403**	**5625**	**5764**	**5859**	**5928**	**5981**	**6022**	**6056**	**6082**	**6106**	**6142**	**6169**	**6208**	**6234**	**6258**	**6286**	**6302**	**6323**	**6334**	**6352**	**6361**	**6366**
2	18.51	19.00	19.16	19.25	19.30	19.33	19.36	19.37	19.38	19.39	19.40	19.41	19.42	19.43	19.44	19.45	19.46	19.47	19.47	19.48	19.49	19.49	19.50	19.50
	98.49	**99.01**	**99.17**	**99.25**	**99.30**	**99.33**	**99.34**	**99.36**	**99.38**	**99.40**	**99.41**	**99.42**	**99.43**	**99.44**	**99.45**	**99.46**	**99.47**	**99.48**	**99.48**	**99.49**	**99.49**	**99.49**	**99.50**	**99.50**
3	10.13	9.55	9.28	9.12	9.01	8.94	8.88	8.84	8.81	8.78	8.76	8.74	8.71	8.69	8.66	8.64	8.62	8.60	8.58	8.57	8.56	8.54	8.54	8.53
	34.12	**30.81**	**29.46**	**28.71**	**28.24**	**27.91**	**27.67**	**27.49**	**27.34**	**27.23**	**27.13**	**27.05**	**26.92**	**26.83**	**26.69**	**26.60**	**26.50**	**26.41**	**26.30**	**26.27**	**26.23**	**26.18**	**26.14**	**26.12**
4	7.71	6.94	6.59	6.39	6.26	6.16	6.09	6.04	6.00	5.96	5.93	5.91	5.87	5.84	5.80	5.77	5.74	5.71	5.70	5.68	5.66	5.65	5.64	5.63
	21.20	**18.00**	**16.69**	**15.98**	**15.52**	**15.21**	**14.98**	**14.80**	**14.66**	**14.54**	**14.45**	**14.37**	**14.24**	**14.15**	**14.02**	**13.93**	**13.83**	**13.74**	**13.69**	**13.61**	**13.57**	**13.52**	**13.48**	**13.46**
5	6.61	5.79	5.41	5.19	5.05	4.95	4.88	4.82	4.78	4.74	4.70	4.68	4.64	4.60	4.56	4.53	4.50	4.46	4.44	4.42	4.40	4.38	4.37	4.36
	16.26	**13.27**	**12.06**	**11.39**	**10.97**	**10.67**	**10.45**	**10.27**	**10.15**	**10.05**	**9.96**	**9.89**	**9.77**	**9.68**	**9.55**	**9.47**	**9.38**	**9.29**	**9.24**	**9.17**	**9.13**	**9.07**	**9.04**	**9.02**
6	5.99	5.14	4.76	4.53	4.39	4.28	4.21	4.15	4.10	4.06	4.03	4.00	3.96	3.92	3.87	3.84	3.81	3.77	3.75	3.72	3.71	3.69	3.68	3.67
	13.74	**10.92**	**9.78**	**9.15**	**8.75**	**8.47**	**8.26**	**8.10**	**7.98**	**7.87**	**7.79**	**7.72**	**7.60**	**7.52**	**7.39**	**7.31**	**7.23**	**7.14**	**7.09**	**7.02**	**6.99**	**6.94**	**6.90**	**6.88**
7	5.59	4.74	4.35	4.12	3.97	3.87	3.79	3.73	3.68	3.63	3.60	3.57	3.52	3.49	3.44	3.41	3.38	3.34	3.32	3.29	3.28	3.25	3.24	3.23
	12.25	**9.55**	**8.45**	**7.85**	**7.46**	**7.19**	**7.00**	**6.84**	**6.71**	**6.62**	**6.54**	**6.47**	**6.35**	**6.27**	**6.15**	**6.07**	**5.98**	**5.90**	**5.85**	**5.78**	**5.75**	**5.70**	**5.67**	**5.65**
8	5.32	4.46	4.07	3.84	3.69	3.58	3.50	3.44	3.39	3.34	3.31	3.28	3.23	3.20	3.15	3.12	3.08	3.05	3.03	3.00	2.98	2.96	2.94	2.93
	11.26	**8.65**	**7.59**	**7.01**	**6.63**	**6.37**	**6.19**	**6.03**	**5.91**	**5.82**	**5.74**	**5.67**	**5.56**	**5.48**	**5.36**	**5.28**	**5.20**	**5.11**	**5.06**	**5.00**	**4.96**	**4.91**	**4.88**	**4.86**

9	5.12	4.26	3.86	3.63	3.48	3.37	3.29	3.23	3.18	3.13	3.10	3.07	3.02	2.98	2.93	2.90	2.86	2.82	2.80	2.77	2.76	2.73	2.72	2.71
	10.56	**8.02**	**6.99**	**6.42**	**6.06**	**5.80**	**5.62**	**5.47**	**5.35**	**5.26**	**5.18**	**5.11**	**5.00**	**4.92**	**4.80**	**4.73**	**4.64**	**4.56**	**4.51**	**4.45**	**4.41**	**4.36**	**4.33**	**4.31**
10	4.96	4.10	3.71	3.48	3.33	3.22	3.14	3.07	3.02	2.97	2.94	2.91	2.86	2.82	2.77	2.74	2.70	2.67	2.64	2.61	2.59	2.56	2.55	2.54
	10.04	**7.56**	**6.55**	**5.99**	**5.64**	**5.39**	**5.21**	**5.06**	**4.95**	**4.85**	**4.78**	**4.71**	**4.60**	**4.52**	**4.41**	**4.33**	**4.25**	**4.17**	**4.12**	**4.05**	**4.01**	**3.96**	**3.93**	**3.91**
11	4.84	3.98	3.59	3.36	3.20	3.09	3.01	2.95	2.90	2.86	2.82	2.79	2.74	2.70	2.65	2.61	2.57	2.53	2.50	2.47	2.45	2.42	2.41	2.40
	9.65	**7.20**	**6.22**	**5.67**	**5.32**	**5.07**	**4.88**	**4.74**	**4.63**	**4.54**	**4.46**	**4.40**	**4.29**	**4.21**	**4.10**	**4.02**	**3.94**	**3.86**	**3.80**	**3.74**	**3.70**	**3.66**	**3.62**	**3.60**
12	4.75	3.88	3.49	3.26	3.11	3.00	2.92	2.85	2.80	2.76	2.72	2.69	2.64	2.60	2.54	2.50	2.46	2.42	2.40	2.36	2.35	2.32	2.31	2.30
	9.33	**6.93**	**5.95**	**5.41**	**5.06**	**4.82**	**4.65**	**4.50**	**4.39**	**4.30**	**4.22**	**4.16**	**4.05**	**3.98**	**3.86**	**3.78**	**3.70**	**3.61**	**3.56**	**3.49**	**3.46**	**3.41**	**3.38**	**3.36**
13	4.67	3.80	3.41	3.18	3.02	2.92	2.84	2.77	2.72	2.67	2.63	2.60	2.55	2.51	2.46	2.42	2.38	2.34	2.32	2.28	2.26	2.24	2.22	2.21
	9.07	**6.70**	**5.74**	**5.20**	**4.86**	**4.62**	**4.44**	**4.30**	**4.19**	**4.10**	**4.02**	**3.96**	**3.85**	**3.78**	**3.67**	**3.59**	**3.51**	**3.42**	**3.37**	**3.30**	**3.27**	**3.21**	**3.18**	**3.16**
14	4.60	3.74	3.34	3.11	2.96	2.85	2.77	2.70	2.65	2.60	2.56	2.53	2.48	2.44	2.39	2.35	2.31	2.27	2.24	2.21	2.19	2.16	2.14	2.13
	8.86	**6.51**	**5.56**	**5.03**	**4.69**	**4.46**	**4.28**	**4.14**	**4.03**	**3.94**	**3.86**	**3.80**	**3.70**	**3.62**	**3.51**	**3.43**	**3.34**	**3.26**	**3.21**	**3.14**	**3.11**	**3.06**	**3.02**	**3.00**
15	4.54	3.68	3.29	3.06	2.90	2.79	2.70	2.64	2.59	2.55	2.51	2.48	2.43	2.39	2.33	2.29	2.25	2.21	2.18	2.15	2.12	2.10	2.08	2.07
	8.68	**6.36**	**5.42**	**4.89**	**4.56**	**4.32**	**4.14**	**4.00**	**3.89**	**3.80**	**3.73**	**3.67**	**3.56**	**3.48**	**3.36**	**3.29**	**3.20**	**3.12**	**3.07**	**3.00**	**2.97**	**2.92**	**2.89**	**2.87**
16	4.49	3.63	3.24	3.01	2.85	2.74	2.66	2.59	2.54	2.49	2.45	2.42	2.37	2.33	2.28	2.24	2.20	2.16	2.13	2.09	2.07	2.04	2.02	2.01
	8.53	**6.23**	**5.29**	**4.77**	**4.44**	**4.20**	**4.03**	**3.89**	**3.78**	**3.69**	**3.61**	**3.55**	**3.45**	**3.37**	**3.25**	**3.18**	**3.10**	**3.01**	**2.96**	**2.89**	**2.86**	**2.80**	**2.77**	**2.75**
17	4.45	3.59	3.20	2.96	2.81	2.70	2.62	2.55	2.50	2.45	2.41	2.38	2.33	2.29	2.23	2.19	2.15	2.11	2.08	2.04	2.02	1.99	1.97	1.96
	8.40	**6.11**	**5.18**	**4.67**	**4.34**	**4.10**	**3.93**	**3.79**	**3.68**	**3.59**	**3.52**	**3.45**	**3.35**	**3.27**	**3.16**	**3.08**	**3.00**	**2.92**	**2.86**	**2.79**	**2.76**	**2.70**	**2.67**	**2.65**
18	4.41	3.55	3.16	2.93	2.77	2.66	2.58	2.51	2.46	2.41	2.37	2.34	2.29	2.25	2.19	2.15	2.11	2.07	2.04	2.00	1.98	1.95	1.93	1.92
	8.28	**6.01**	**5.09**	**4.58**	**4.25**	**4.01**	**3.85**	**3.71**	**3.60**	**3.51**	**3.44**	**3.37**	**3.27**	**3.19**	**3.07**	**3.00**	**2.91**	**2.83**	**2.78**	**2.71**	**2.68**	**2.62**	**2.59**	**2.57**
19	4.38	3.52	3.13	2.90	2.74	2.63	2.55	2.48	2.43	2.38	2.34	2.31	2.26	2.21	2.15	2.11	2.07	2.02	2.00	1.96	1.94	1.91	1.90	1.88
	8.18	**5.93**	**5.01**	**4.50**	**4.17**	**3.94**	**3.77**	**3.63**	**3.52**	**3.43**	**3.36**	**3.30**	**3.19**	**3.12**	**3.00**	**2.92**	**2.84**	**2.76**	**2.70**	**2.63**	**2.60**	**2.54**	**2.51**	**2.49**
20	4.35	3.49	3.10	2.87	2.71	2.60	2.52	2.45	2.40	2.35	2.31	2.28	2.23	2.18	2.12	2.08	2.04	1.99	1.96	1.92	1.90	1.87	1.85	1.84
	8.10	**5.85**	**4.94**	**4.43**	**4.10**	**3.87**	**3.71**	**3.56**	**3.45**	**3.37**	**3.30**	**3.23**	**3.13**	**3.05**	**2.94**	**2.86**	**2.77**	**2.69**	**2.63**	**2.56**	**2.53**	**2.47**	**2.44**	**2.42**
21	4.32	3.47	3.07	2.84	2.68	2.57	2.49	2.42	2.37	2.32	2.28	2.25	2.20	2.15	2.09	2.05	2.00	1.96	1.93	1.89	1.87	1.84	1.82	1.81
	8.02	**5.78**	**4.87**	**4.37**	**4.04**	**3.81**	**3.65**	**3.51**	**3.40**	**3.31**	**3.24**	**3.17**	**3.07**	**2.99**	**2.88**	**2.80**	**2.72**	**2.63**	**2.58**	**2.51**	**2.47**	**2.42**	**2.38**	**2.36**
22	4.30	3.44	3.05	2.82	2.66	2.55	2.47	2.40	2.35	2.30	2.26	2.23	2.18	2.13	2.07	2.03	1.98	1.93	1.91	1.87	1.84	1.81	1.80	1.78
	7.94	**5.72**	**4.82**	**4.31**	**3.99**	**3.76**	**3.59**	**3.45**	**3.35**	**3.26**	**3.18**	**3.12**	**3.02**	**2.94**	**2.83**	**2.75**	**2.67**	**2.58**	**2.53**	**2.46**	**2.42**	**2.37**	**2.33**	**2.31**
23	4.28	3.42	3.03	2.80	2.64	2.53	2.45	2.38	2.32	2.28	2.24	2.20	2.14	2.10	2.04	2.00	1.96	1.91	1.88	1.84	1.82	1.79	1.77	1.76
	7.88	**5.66**	**4.76**	**4.26**	**3.94**	**3.71**	**3.54**	**3.41**	**3.30**	**3.21**	**3.14**	**3.07**	**2.97**	**2.89**	**2.78**	**2.70**	**2.62**	**2.53**	**2.48**	**2.41**	**2.37**	**2.32**	**2.28**	**2.26**
24	4.26	3.40	3.01	2.78	2.62	2.51	2.43	2.36	2.30	2.26	2.22	2.18	2.13	2.09	2.02	1.98	1.94	1.89	1.86	1.82	1.80	1.76	1.74	1.73
	7.82	**5.61**	**4.72**	**4.22**	**3.90**	**3.67**	**3.50**	**3.36**	**3.25**	**3.17**	**3.09**	**3.03**	**2.93**	**2.85**	**2.74**	**2.66**	**2.58**	**2.49**	**2.44**	**2.36**	**2.33**	**2.27**	**2.23**	**2.21**
25	4.24	3.38	2.99	2.76	2.60	2.49	2.41	2.34	2.28	2.24	2.20	2.16	2.11	2.06	2.00	1.96	1.92	1.87	1.84	1.80	1.77	1.74	1.72	1.71
	7.77	**5.57**	**4.68**	**4.18**	**3.86**	**3.63**	**3.46**	**3.32**	**3.21**	**3.13**	**3.05**	**2.99**	**2.89**	**2.81**	**2.70**	**2.62**	**2.54**	**2.45**	**2.40**	**2.32**	**2.29**	**2.23**	**2.19**	**2.17**

Degrees of freedom for denominator (ν_2)	Degrees of freedom for numerator (ν_1)																							
	1	2	3	4	5	6	7	8	9	10	11	12	14	16	20	24	30	40	50	75	100	200	500	∞
26	4.22	3.37	2.89	2.74	2.59	2.47	2.39	2.32	2.27	2.22	2.18	2.15	2.10	2.05	1.99	1.95	1.90	1.85	1.82	1.78	1.76	1.72	1.70	1.69
	7.72	5.53	4.64	4.14	3.82	3.59	3.42	3.29	3.17	3.09	3.02	2.96	2.86	2.77	2.66	2.58	2.50	2.41	2.36	2.28	2.25	2.19	2.15	2.13
27	4.21	3.35	2.96	2.73	2.57	2.46	2.37	2.30	2.25	2.20	2.16	2.13	2.08	2.03	1.97	1.93	1.88	1.84	1.80	1.76	1.74	1.71	1.68	1.67
	7.68	5.49	4.60	4.11	3.79	3.56	3.39	3.26	3.14	3.06	2.98	2.93	2.83	2.74	2.63	2.55	2.47	2.38	2.33	2.25	2.21	2.16	2.12	2.10
28	4.20	3.34	2.95	2.71	2.56	2.44	2.36	2.29	3.24	2.19	2.15	2.12	2.06	2.02	1.96	1.91	1.87	1.81	1.78	1.75	1.72	1.69	1.67	1.65
	7.64	5.45	4.57	4.07	3.76	3.53	3.36	3.23	3.11	3.03	2.95	2.90	2.80	2.71	2.60	2.52	2.44	2.35	2.30	2.22	2.18	2.13	2.09	2.06
29	4.18	3.33	2.93	2.70	2.54	2.43	2.35	2.28	2.22	2.18	2.14	2.10	2.05	2.00	1.94	1.90	1.85	1.80	1.77	1.73	1.71	1.68	1.65	1.64
	7.60	5.52	4.54	4.04	3.73	3.50	3.33	3.20	3.08	3.00	2.92	2.87	2.77	2.68	2.57	2.49	2.41	2.32	2.27	2.19	2.15	2.10	2.06	2.03
30	4.17	3.32	2.92	2.69	2.53	2.42	2.34	2.27	2.21	2.16	2.12	2.09	2.04	1.99	1.93	1.89	1.84	1.79	1.76	1.72	1.69	1.66	1.64	1.62
	7.56	5.39	4.51	4.02	3.70	3.47	3.30	3.17	3.06	2.98	2.90	2.84	2.74	2.66	2.55	2.47	2.38	2.29	2.24	2.16	2.13	2.07	2.03	2.01
32	4.15	3.30	2.90	2.67	2.51	2.40	2.32	2.25	2.19	2.14	2.10	2.07	2.02	1.97	1.91	1.86	1.82	1.76	1.74	1.69	1.67	1.64	1.61	1.59
	7.50	5.34	4.46	3.97	3.66	3.42	3.25	3.12	3.01	2.94	2.86	2.80	2.70	2.62	2.51	2.42	2.34	2.25	2.20	2.12	2.08	2.02	1.98	1.96
34	4.13	3.28	2.88	2.65	2.49	2.38	2.30	2.23	2.17	2.12	2.08	2.05	2.00	1.95	1.89	1.84	1.80	1.74	1.71	1.67	1.64	1.61	1.59	1.57
	7.44	5.29	4.42	3.93	3.61	3.38	3.21	3.08	2.97	2.89	2.82	2.76	2.66	2.58	2.47	2.38	2.30	2.21	2.15	2.08	2.04	1.98	1.94	1.91
36	4.11	3.26	2.86	2.63	2.48	2.36	2.28	2.21	2.15	2.10	2.06	2.03	1.99	1.93	1.87	1.82	1.78	1.72	1.69	1.65	1.62	1.59	1.56	1.55
	7.39	5.25	4.38	3.89	3.58	3.35	3.18	3.04	2.94	2.86	2.78	2.72	2.62	2.54	2.43	2.35	2.26	2.17	2.12	2.04	2.00	1.94	1.90	1.87
38	4.10	3.25	2.85	2.62	2.46	2.35	2.26	2.19	2.14	2.09	2.05	2.02	1.96	1.92	1.85	1.80	1.76	1.71	1.67	1.63	1.60	1.57	1.54	1.53
	7.35	5.21	4.34	3.86	3.54	3.32	3.15	3.02	2.91	2.82	2.75	2.69	2.59	2.51	2.40	2.32	2.22	2.14	2.08	2.00	1.97	1.90	1.86	1.84
40	4.08	3.23	2.84	2.61	2.45	2.34	2.25	2.18	2.12	2.07	2.04	2.00	1.95	1.90	1.84	1.79	1.74	1.69	1.66	1.61	1.59	1.55	1.53	1.51
	7.31	5.18	4.31	3.83	3.51	3.29	3.12	2.99	2.88	2.80	2.73	2.66	2.56	2.49	2.37	2.29	2.20	2.11	2.05	1.97	1.94	1.88	1.84	1.81
42	4.07	3.22	2.83	2.59	2.44	2.32	2.24	2.17	2.11	2.06	2.02	1.99	1.94	1.89	1.82	1.78	1.73	1.68	1.64	1.60	1.57	1.54	1.51	1.49
	7.27	5.15	4.29	3.80	3.49	3.26	3.10	2.96	2.86	2.77	2.70	2.64	2.54	2.46	2.35	2.26	2.17	2.08	2.02	1.94	1.91	1.85	1.80	1.78
44	4.06	3.21	2.82	2.58	2.43	2.31	2.23	2.16	2.10	2.05	2.01	1.98	1.92	1.88	1.81	1.76	1.72	1.66	1.63	1.58	1.56	1.52	1.50	1.48
	7.24	5.12	4.26	3.78	3.46	3.24	3.07	2.94	2.84	2.75	2.68	2.62	2.52	2.44	2.32	2.24	2.15	2.06	2.00	1.92	1.88	1.82	1.78	1.75

46	4.05	3.20	2.81	2.57	2.42	2.30	2.22	2.14	2.09	2.04	2.00	1.97	1.91	1.87	1.80	1.75	1.71	1.65	1.62	1.57	1.54	1.51	1.48	1.46
	7.21	**5.10**	**4.24**	**3.76**	**3.44**	**3.22**	**3.05**	**2.92**	**2.82**	**2.73**	**2.66**	**2.60**	**2.50**	**2.42**	**2.30**	**2.22**	**2.13**	**2.04**	**1.98**	**1.90**	**1.86**	**1.80**	**1.76**	**1.72**
48	4.04	3.19	2.80	2.56	2.41	2.30	2.21	2.14	2.08	2.03	1.99	1.96	1.90	1.86	1.79	1.74	1.70	1.64	1.61	1.56	1.53	1.50	1.47	1.45
	7.19	**5.08**	**4.22**	**3.74**	**3.42**	**3.20**	**3.04**	**2.90**	**2.80**	**2.71**	**2.64**	**2.58**	**2.48**	**2.40**	**2.28**	**2.20**	**2.11**	**2.02**	**1.96**	**1.88**	**1.84**	**1.78**	**1.73**	**1.70**
50	4.03	3.18	2.79	2.56	2.40	2.29	2.20	2.13	2.07	2.02	1.98	1.95	1.90	1.85	1.78	1.74	1.69	1.63	1.60	1.55	1.52	1.48	1.46	1.44
	7.17	**5.06**	**4.20**	**3.72**	**3.41**	**3.18**	**3.02**	**2.88**	**2.78**	**2.70**	**2.62**	**2.56**	**2.46**	**2.39**	**2.26**	**2.18**	**2.10**	**2.00**	**1.94**	**1.86**	**1.82**	**1.76**	**1.71**	**1.68**
55	4.02	3.17	2.78	2.54	2.38	2.27	2.18	2.11	2.05	2.00	1.97	1.93	1.88	1.83	1.76	1.72	1.67	1.61	1.58	1.52	1.50	1.46	1.43	1.41
	7.12	**5.01**	**4.16**	**3.68**	**3.37**	**3.15**	**2.98**	**2.85**	**2.75**	**2.66**	**2.59**	**2.53**	**2.43**	**2.35**	**2.23**	**2.15**	**2.06**	**1.96**	**1.90**	**1.82**	**1.78**	**1.71**	**1.66**	**1.64**
60	4.00	3.15	2.76	2.52	2.37	2.25	2.17	2.10	2.04	1.99	1.95	1.92	1.86	1.81	1.75	1.70	1.65	1.59	1.56	1.50	1.48	1.44	1.41	1.39
	7.08	**4.98**	**4.13**	**3.65**	**3.34**	**3.12**	**2.95**	**2.82**	**2.72**	**2.63**	**2.56**	**2.50**	**2.40**	**2.32**	**2.20**	**2.12**	**2.03**	**1.93**	**1.87**	**1.79**	**1.74**	**1.68**	**1.63**	**1.60**
65	3.99	3.14	2.75	2.51	2.36	2.24	2.15	2.08	2.02	1.98	1.94	1.90	1.85	1.80	1.73	1.68	1.63	1.57	1.54	1.49	1.46	1.42	1.39	1.37
	7.04	**4.95**	**4.10**	**3.62**	**3.31**	**3.09**	**2.93**	**2.79**	**2.70**	**2.61**	**2.54**	**2.47**	**2.37**	**2.30**	**2.18**	**2.09**	**2.00**	**1.90**	**1.84**	**1.76**	**1.71**	**1.64**	**1.60**	**1.56**
70	3.98	3.13	2.74	2.50	2.35	2.32	2.14	2.07	2.01	1.97	1.93	1.89	1.84	1.79	1.72	1.67	1.62	1.56	1.53	1.47	1.45	1.40	1.37	1.35
	7.01	**4.92**	**4.08**	**3.60**	**3.29**	**3.07**	**2.91**	**2.77**	**2.67**	**2.59**	**2.51**	**2.45**	**2.35**	**2.28**	**2.15**	**2.07**	**1.98**	**1.88**	**1.82**	**1.74**	**1.69**	**1.62**	**1.56**	**1.53**
80	3.96	3.11	2.72	2.48	2.33	2.21	2.12	2.05	1.99	1.95	1.91	1.88	1.82	1.77	1.70	1.65	1.60	1.54	1.51	1.45	1.42	1.38	1.35	1.32
	6.96	**4.88**	**4.04**	**3.56**	**3.25**	**3.04**	**2.87**	**2.74**	**2.64**	**2.55**	**2.48**	**2.41**	**2.32**	**2.24**	**2.11**	**2.03**	**1.94**	**1.84**	**1.78**	**1.70**	**1.65**	**1.57**	**1.52**	**1.49**
100	3.94	3.09	2.70	2.46	2.30	2.19	2.10	2.03	1.97	1.92	1.88	1.85	1.79	1.75	1.68	1.63	1.57	1.51	1.48	1.42	1.39	1.34	1.30	1.28
	6.90	**4.82**	**3.98**	**3.51**	**3.20**	**2.99**	**2.82**	**2.69**	**2.59**	**2.51**	**2.43**	**2.36**	**2.26**	**2.19**	**2.06**	**1.98**	**1.89**	**1.79**	**1.73**	**1.64**	**1.59**	**1.51**	**1.46**	**1.43**
125	3.92	3.07	2.68	2.44	2.29	2.17	2.08	2.01	1.95	1.90	1.86	1.83	1.77	1.72	1.65	1.60	1.55	1.49	1.45	1.39	1.36	1.31	1.27	1.25
	6.84	**4.78**	**3.94**	**3.47**	**3.17**	**2.95**	**2.79**	**2.65**	**2.56**	**2.47**	**2.40**	**2.33**	**2.23**	**2.15**	**2.03**	**1.94**	**1.85**	**1.75**	**1.68**	**1.59**	**1.54**	**1.46**	**1.40**	**1.37**
150	3.91	3.06	2.67	2.43	2.27	2.16	2.07	2.00	1.94	1.89	1.85	1.82	1.76	1.71	1.64	1.59	1.54	1.47	1.44	1.37	1.34	1.29	1.25	1.22
	6.81	**4.75**	**3.91**	**3.44**	**3.13**	**2.92**	**2.76**	**2.62**	**2.53**	**2.44**	**2.37**	**2.30**	**2.20**	**2.12**	**2.00**	**1.91**	**1.83**	**1.72**	**1.66**	**1.56**	**1.51**	**1.43**	**1.37**	**1.33**
200	3.89	3.04	2.65	2.41	2.26	2.14	2.05	1.98	1.92	1.87	1.83	1.80	1.74	1.69	1.62	1.57	1.52	1.45	1.42	1.35	1.32	1.26	1.22	1.19
	6.76	**4.71**	**3.88**	**3.41**	**3.11**	**2.90**	**2.73**	**2.60**	**2.50**	**2.41**	**2.34**	**2.28**	**2.17**	**2.09**	**1.97**	**1.88**	**1.79**	**1.69**	**1.62**	**1.53**	**1.48**	**1.39**	**1.33**	**1.28**
400	3.86	3.02	2.62	2.39	2.23	2.12	2.03	1.96	1.90	1.85	1.81	1.78	1.72	1.67	1.60	1.54	1.49	1.42	1.38	1.32	1.28	1.22	1.16	1.13
	6.70	**4.66**	**3.83**	**3.36**	**3.06**	**2.85**	**2.69**	**2.55**	**2.46**	**2.37**	**2.29**	**2.23**	**2.12**	**2.04**	**1.92**	**1.84**	**1.74**	**1.64**	**1.57**	**1.47**	**1.42**	**1.32**	**1.24**	**1.19**
1000	3.85	3.00	2.61	2.38	2.22	2.10	2.02	1.95	1.89	1.84	1.80	1.76	1.70	1.65	1.58	1.53	1.47	1.41	1.36	1.30	1.26	1.19	1.13	1.08
	6.66	**4.62**	**3.80**	**3.34**	**3.04**	**2.82**	**2.66**	**2.53**	**2.43**	**2.34**	**2.26**	**2.20**	**2.09**	**2.01**	**1.89**	**1.81**	**1.71**	**1.61**	**1.54**	**1.44**	**1.38**	**1.28**	**1.19**	**1.11**
	3.84	2.99	2.60	2.37	2.21	2.09	2.01	1.94	1.88	1.83	1.79	1.75	1.69	1.64	1.57	1.52	1.46	1.40	1.35	1.28	1.24	1.17	1.11	1.00
	6.64	**4.60**	**3.78**	**3.32**	**3.02**	**2.80**	**2.64**	**2.51**	**2.41**	**2.32**	**2.24**	**2.18**	**2.07**	**1.99**	**1.87**	**1.79**	**1.69**	**1.59**	**1.52**	**1.41**	**1.36**	**1.25**	**1.15**	**1.00**

Table 7 *Rank-Sum Critical Values*

The sample sizes are shown in parentheses (n_1, n_2). The probability associated with a pair of critical values is the probability that $R \leq$ smaller value, or equally, it is the probability that $R \geq$ larger value. These probabilities are the closest ones to .025 and .05 that exist for integer values of R. The approximate .025 values should be used for a two-sided test with $\alpha = .05$, and the approximate .05 values for a one-sided test.

	(2, 4)	
3	11	.067
	(2, 5)	
3	13	.047
	(2, 6)	
3	15	.036
4	14	.071
	(2, 7)	
3	17	.028
4	16	.056
	(2, 8)	
3	19	.022
4	18	.044
	(2, 9)	
3	21	.018
4	20	.036
	(2, 10)	
4	22	.030
5	21	.061
	(3, 3)	
6	15	.050
	(3, 4)	
6	18	.028
7	17	.057
	(3, 5)	
6	21	.018
7	20	.036
	(3, 6)	
7	23	.024
8	22	.048
	(3, 7)	
8	25	.033
9	24	.058
	(3, 8)	
8	28	.024
9	27	.042
	(3, 9)	
9	30	.032
10	29	.050
	(3, 10)	
9	33	.024
11	31	.056

	(4, 4)	
11	25	.029
12	24	.057
	(4, 5)	
12	28	.032
13	27	.056
	(4, 6)	
12	32	.019
14	30	.057
	(4, 7)	
13	35	.021
15	33	.055
	(4, 8)	
14	38	.024
16	36	.055
	(4, 9)	
15	41	.025
17	39	.053
	(4, 10)	
16	44	.026
18	42	.053
	(5, 5)	
18	37	.028
19	36	.048
	(5, 6)	
19	41	.026
20	40	.041
	(5, 7)	
20	45	.024
22	43	.053
	(5, 8)	
21	49	.023
23	47	.047
	(5, 9)	
22	53	.021
25	50	.056
	(5, 10)	
24	56	.028
26	54	.050
	(6, 6)	
26	52	.021
28	50	.047

	(6, 7)	
28	56	.026
30	54	.051
	(6, 8)	
29	61	.021
32	58	.054
	(6, 9)	
31	65	.025
33	63	.044
	(6, 10)	
33	69	.028
35	67	.047
	(7, 7)	
37	68	.027
39	66	.049
	(7, 8)	
39	73	.027
41	71	.047
	(7, 9)	
41	78	.027
43	76	.045
	(7, 10)	
43	83	.028
46	80	.054
	(8, 8)	
49	87	.025
52	84	.052
	(8, 9)	
51	93	.023
54	90	.046
	(8, 10)	
54	98	.027
57	95	.051
	(9, 9)	
63	108	.025
66	105	.047
	(9, 10)	
66	114	.027
69	111	.047
	(10, 10)	
79	131	.026
83	127	.053

Table 8 *Critical Values of Kolmogorov-Smirnov D_n Statistic*

Sample size n	Level of a significance for $D_n = \sup\lvert\hat{F}_n(x) - F_0(x)\rvert$			
	$\alpha = 0.20$	$\alpha = 0.10$	$\alpha = 0.05$	$\alpha = 0.01$
1	0.9000	0.9500	0.9750	0.9950
2	0.6838	0.7764	0.8419	0.9293
3	0.5648	0.6360	0.7076	0.8290
4	0.4927	0.5652	0.6239	0.7342
5	0.4470	0.5095	0.5633	0.6685
6	0.4104	0.4680	0.5193	0.6166
7	0.3815	0.4361	0.4834	0.5758
8	0.3583	0.4096	0.4543	0.5418
9	0.3391	0.3875	0.4300	0.5133
10	0.3226	0.3687	0.4093	0.4889
11	0.3083	0.3524	0.3912	0.4677
12	0.2958	0.3382	0.3754	0.4491
13	0.2847	0.3255	0.3614	0.4325
14	0.2748	0.3142	0.3489	0.4176
15	0.2659	0.3040	0.3376	0.4042
16	0.2578	0.2947	0.3273	0.3920
17	0.2504	0.2863	0.3180	0.3809
18	0.2436	0.2785	0.3094	0.3706
19	0.2374	0.2714	0.3014	0.3612
20	0.2316	0.2647	0.2941	0.3524
25	0.2079	0.2377	0.2640	0.3166
30	0.1903	0.2176	0.2417	0.2899
35	0.1766	0.2019	0.2243	0.2690
Over 35	$1.07/\sqrt{n}$	$1.22/\sqrt{n}$	$1.36/\sqrt{n}$	$1.63/\sqrt{n}$

Adapted from Table 1 of L. H. Miller, Table of percentage points of Kolmogorov statistics, *J. Amer. Statist. Assoc.* **51** (1956), 113, and Table 1 of F. J. Massey, Jr., The Kolmogorov–Smirnov test for goodness-of-fit, *J. Amer. Statist. Assoc.* **46** (1951), 70, with permission of the authors and publisher.

Table 9 *Critical Values of Kolmogorov-Smirnov $\hat{D}_n$ Statistic for the Exponential Distribution with an Unknown Mean θ*

Sample size n	Level of significance for $\hat{D}_n = \sup\lvert F_n(x) - F_0(x, \theta)\rvert$				
	$\alpha = 0.20$	$\alpha = 0.15$	$\alpha = 0.10$	$\alpha = 0.05$	$\alpha = 0.01$
3	0.451	0.479	0.511	0.551	0.600
4	0.396	0.422	0.449	0.487	0.548
5	0.359	0.382	0.406	0.442	0.504
6	0.331	0.351	0.375	0.408	0.470
7	0.309	0.327	0.350	0.382	0.442
8	0.291	0.308	0.329	0.360	0.419
9	0.277	0.291	0.311	0.341	0.399
10	0.263	0.277	0.295	0.325	0.380
11	0.251	0.264	0.283	0.311	0.365
12	0.241	0.254	0.271	0.298	0.351
13	0.232	0.245	0.261	0.287	0.338
14	0.224	0.237	0.252	0.277	0.326
15	0.217	0.229	0.244	0.269	0.315
16	0.211	0.222	0.236	0.261	0.306
17	0.204	0.215	0.229	0.253	0.297
18	0.199	0.210	0.223	0.246	0.289
19	0.193	0.204	0.218	0.239	0.283
20	0.188	0.199	0.212	0.234	0.278
25	0.170	0.180	0.191	0.210	0.247
30	0.155	0.164	0.174	0.192	0.226
Over 30	$0.86/\sqrt{n}$	$0.91/\sqrt{n}$	$0.96/\sqrt{n}$	$1.06/\sqrt{n}$	$1.25/\sqrt{n}$

Adapted from Table 1 of H. W. Lilliefors, On the Kolmogorov–Smirnov test for the exponential with mean unknown, *J. Amer. Statist. Assoc.* **64** (1969), 388, with permission of the author and publisher.

Table 10 *Critical Values of Kolmogorov-Smirnov $\hat{D}_n$ Statistic for the Normal Distribution with an Unknown Mean μ and an Unknown Variance σ^2*

Sample size	Level of significance for $\hat{D}_n = \sup\lvert\hat{F}_n(x) - F_0(x, \mu, \sigma^2)\rvert$				
n	$\alpha = 0.20$	$\alpha = 0.15$	$\alpha = 0.10$	$\alpha = 0.05$	$\alpha = 0.01$
4	0.300	0.319	0.352	0.381	0.417
5	0.285	0.299	0.315	0.337	0.405
6	0.265	0.277	0.294	0.319	0.364
7	0.247	0.258	0.276	0.300	0.348
8	0.233	0.244	0.261	0.285	0.331
9	0.223	0.233	0.249	0.271	0.311
10	0.215	0.224	0.239	0.258	0.294
11	0.206	0.217	0.230	0.249	0.284
12	0.199	0.212	0.223	0.242	0.275
13	0.190	0.202	0.214	0.234	0.268
14	0.183	0.194	0.207	0.227	0.261
15	0.177	0.187	0.201	0.220	0.257
16	0.173	0.182	0.195	0.213	0.250
17	0.169	0.177	0.189	0.206	0.245
18	0.166	0.173	0.184	0.200	0.239
19	0.163	0.169	0.179	0.195	0.235
20	0.160	0.166	0.174	0.190	0.231
25	0.142	0.147	0.158	0.173	0.200
30	0.131	0.136	0.144	0.161	0.187
Over 30	$0.736/\sqrt{n}$	$0.768/\sqrt{n}$	$0.805/\sqrt{n}$	$0.886/\sqrt{n}$	$1.031/\sqrt{n}$

Adapted from Table 1 of H. W. Lilliefors, On the Kolmogorov–Smirnov test for normality with mean and variance unknown, *J. Amer. Statist. Assoc.* **62** (1967), 400, with permission of the author and publisher.

Appendix D

Laplace Transforms

We have had several occasions to use Laplace transforms in this text. This appendix examines useful properties of the Laplace transform and gives a table of Laplace transforms that are often used.

Suppose $f(t)$ is a piecewise-continuous function, defined at least for $t > 0$, and which is of *exponential* order α, meaning that it does not grow any faster than the exponential $e^{\alpha t}$; that is:

$$|f(t)| \leqslant Me^{\alpha t}, \qquad t > 0$$

for some constant M. Then the Laplace transform of $f(t)$, denoted by $\bar{f}(s)$, is defined by the integral:

$$\bar{f}(s) = \int_0^\infty e^{-st} f(t)\, dt \tag{D.1}$$

for any complex number s such that its real part Re $(s) > \alpha$.

Important $[f(t),\ \bar{f}(s)]$ pairs are given in Table D.1.

The Laplace transform is generally used to simplify calculations. However, the technique will not be of value unless it is possible to recover the function $f(t)$ from its transform $\bar{f}(s)$. In fact, we have the following theorem, which unfortunately does not have an elementary proof and so it is stated without proof.

THEOREM D.1 (CORRESPONDENCE OR UNIQUENESS THEOREM).

If $\bar{f}(s) = \bar{g}(s)$ for all s,

then $f(t) = g(t)$ for all t.

In other words, two functions that have the same transform are identical.

Table D.1 *Important Transform Pairs*

$f(t),\ t > 0$	$\bar{f}(s)$	
1. c a constant	$\dfrac{c}{s}$	
2. t	$\dfrac{1}{s^2}$	
3. t^a	$\dfrac{\Gamma(a+1)}{s^{a+1}}$,	$(a > -1)$
4. t^a	$\dfrac{a!}{s^{a+1}}$,	$(a = 0, 1, 2, \ldots)$
5. $t^{-1/2}$	$\sqrt{\dfrac{\pi}{s}}$,	
6. e^{-at}	$\dfrac{1}{s+a}$,	$[\text{Re}(s) > a]$
7. te^{-at}	$\dfrac{1}{(s+a)^2}$,	$[\text{Re}(s) > a]$
8. $t^b e^{-at}$	$\dfrac{b!}{(s+a)^{b+1}}$,	$[b = 0, 1, 2, \ldots, \text{Re}(s) > a]$

As a special case of Definition D.1., assume that $f(t)$ is the pdf of some nonnegative, absolutely continuous random variable X [in this case $f(t)$ is the short form for $f_X(t)$]; then $\bar{f}(s)$ is also called the Laplace transform $L_X(s)$ of the random variable X. In this case $\bar{f}(s)$ always exists for any positive α, since:

$$|\bar{f}(s)| \leqslant \int_0^\infty |e^{-st} f(t)|\, dt \leqslant \int_0^\infty f(t)\, dt = 1 \qquad \text{for } \text{Re}(s) > 0.$$

In this connection, the usefulness of the Laplace transform stems from the convolution property and the moment generating property (besides the uniqueness property):

THEOREM D.2 (THE CONVOLUTION THEOREM).

If $X_1, X_2, \ldots, X_n$ are independent random variables with respective Laplace transforms $L_{X_1}(s), \ldots, L_{X_n}(s)$, then the Laplace transform of the random variable:

$$Z = \sum_{i=1}^{n} X_i$$

is given by:

$$L_Z(s) = \prod_{i=1}^{n} L_{X_i}(s).$$

THEOREM D.3 (MOMENT GENERATING PROPERTY).
Let X be a random variable possessing a Laplace transform $L_X(s)$. Then the kth ($k = 1, 2, \ldots$) moment of X is given by:

$$E[X^k] = (-1)^n \frac{d^n L_X(s)}{ds^n} \Bigg|_{s=0}. \tag{D.2}$$

Thus if X denotes the time to failure of a system, then from a knowledge of the transform $L_X(s)$ we can quickly obtain the system MTTF, $E[X]$, while it may be considerably more difficult to obtain the density $f_X(t)$ and the reliability $R_X(t)$.

The Laplace transform is also used in solving differential equations, since it reduces an ordinary linear differential equation with constant coefficients into an algebraic equation in s. The solution in terms of s is then converted into a time function by an inversion that is unique by Theorem D.1.

The usefulness of the Laplace transform in solving differential equations is based on the fact that it is a *linear* operator and that the Laplace transform of any derivative of a function is easily computed from the transform of the function itself.

THEOREM D.4 (LINEARITY PROPERTY).
Define the function $g(t) = \sum_{i=1}^{n} C_i f_i(t)$ for some constants $C_1, C_2, \ldots, C_n$. Then:

$$\bar{g}(s) = \sum_{i=1}^{n} C_i \bar{f}_i(s). \tag{D.3}$$

THEOREM D.5 (INITIAL VALUE THEOREM).
Let f be a function such that f and its derivative f' are both of exponential order α. Then the Laplace transform of f' is given by:

$$\bar{f'}(s) = s\bar{f}(s) - f(0). \tag{D.4}$$

Proof: By Definition (D.1):

$$\bar{f'}(s) = \int_0^\infty \frac{df}{dt} e^{-st}\, dt.$$

Integrating by parts, we have:

$$\bar{f'}(s) = e^{-st} f(t) \Big|_0^\infty + s \int_0^\infty f(t) e^{-st}\, dt.$$

The second term on the right-hand side is simply $s\bar{f}(s)$. Consider the first term on the right-hand side. Since, by assumption, $f(t)$ grows more slowly than the exponential e^{+st} for sufficiently large s:

$$\lim_{t \to \infty} e^{-st} f(t) = 0.$$

At the lower limit we obtain:

$$\lim_{t \to 0} e^{-st} f(t) = f(0).$$

Hence the result follows. If a singularity occurs at $t = 0$, we must be careful to replace $f(0)$ by $f(0^-)$, which is the limit of $f(t)$ as t approaches 0 from the left hand side.

Important properties of the Laplace transform are summarized in Table D.2.

The procedure of solving a differential equation by means of Laplace transform is illustrated in Figure D.1.

We now discuss the inversion of a Laplace transform. Consider the solution to the differential equation of Figure D.1 in the s domain:

$$\bar{y}(s) = \frac{1}{(a + s)(b + s)}.$$

Since this function does not occur on the right-hand side of Table D.1, we cannot use table lookup and we must use alternate methods for its inversion. A procedure quite often used is **partial fraction expansion** (or **decomposition**). In this case we rewrite

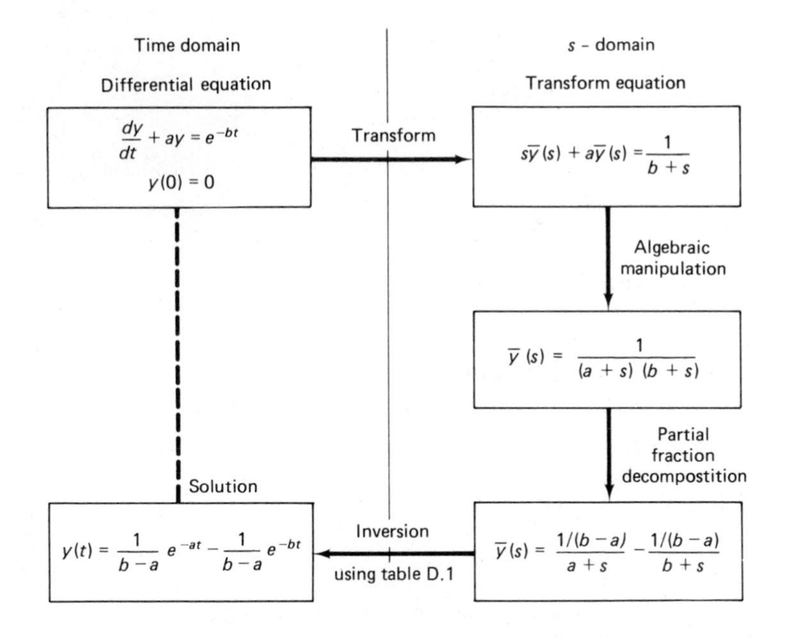

Figure D.1 Solution of a linear differential equation using Laplace transform procedure

$$\bar{y}(s) = \frac{C_1}{a+s} + \frac{C_2}{b+s}$$
$$= \frac{(C_1 b + C_2 a) + (C_1 + C_2)s}{(a+s)(b+s)}$$
$$= \frac{1}{(a+s)(b+s)}.$$

Table D.2 *Properties of Laplace Transforms*

Function	*Laplace transform*	
1. $f(t),\ t > 0$	$\bar{f}(s) = \int_0^\infty e^{-st} f(t)\, dt$	(definition)
2. $af(t) + bg(t)$	$a\bar{f}(s) + b\bar{g}(s)$	[linearity (superposition) property]
3. $f(at)$	$\frac{1}{a}\bar{f}\left(\frac{s}{a}\right)$	
4. $f(t-a)$	$e^{-as}\bar{f}(s)$	
5. $e^{-at}f(t)$	$\bar{f}(s+a)$	
6. $f'(t)$	$s\bar{f}(s) - f(0^-)$	
7. $f^{(n)}(t)$	$s^n\bar{f}(s) - s^{n-1}f(0^-) - s^{n-2}f'(0^-) - \ldots - f^{(n-1)}(0^-)$	
8. $\int_{-\infty}^{t} f(t')\, dt'$	$\frac{1}{s}\bar{f}(s)$	
9. $\int_{-\infty}^{t} f(\tau)g(t-\tau)\, d\tau$	$\bar{f}(s)\bar{g}(s)$	(convolution theorem)
10. $tf(t)$	$-\frac{d\bar{f}(s)}{ds}$	
11. $t^n f(t)$, n a positive integer	$(-1)^n \bar{f}^{(n)}(s)$	
12. $\frac{f(t)}{t}$	$\int_s^\infty \bar{f}(x)\, dx$	
13. $\int_0^\infty f(t)\, dt = \bar{f}(0)$		(integral property)
14. $\lim_{t\to 0} f(t) =$	$\lim_{s\to\infty} s\bar{f}(s)$	(initial-value theorem[a])
15. $\lim_{t\to\infty} f(t) =$	$\lim_{s\to 0} s\bar{f}(s)$	(final-value theorem[a])

[a]Initial-value and final-value theorems apply only when all the poles of $s\bar{f}(s)$ lie on the left half of the s plane; that is, if we write $s\bar{f}(s)$ as the ratio of two polynomials $N(s)/D(s)$, then all the roots of the equation $D(s) = 0$ are the poles of $s\bar{f}(s)$, and these must satisfy the condition $\mathrm{Re}\,(s) < 0$.

Hence:

$$C_1 b + C_2 a = 1 \quad \text{and} \quad C_1 + C_2 = 0$$

or

$$C_1 = \frac{1}{b-a} \quad \text{and} \quad C_2 = -\frac{1}{b-a}.$$

Thus:

$$\bar{y}(s) = \frac{\frac{1}{b-a}}{a+s} - \frac{\frac{1}{b-a}}{b+s}.$$

Now using the linearity property of Laplace transform and Table D.1 (entry 6), we conclude that:

$$y(t) = \frac{1}{b-a} e^{-at} - \frac{1}{b-a} e^{-bt}, \qquad t > 0.$$

More generally, suppose $\bar{f}(s)$ is a rational function of s:

$$\bar{f}(s) = \frac{N(s)}{D(s)} \tag{D.5}$$

$$= \frac{N(s)}{\prod_{i=1}^{d} (s + a_i)},$$

where we assume that the degree of polynomial $D(s)$ is at least one greater than the degree of $N(s)$ and that all the roots of $D(s) = 0$ are distinct (for a more general treatment see [WIDD 1946]). Then the partial fraction decomposition of $\bar{f}(s)$ is:

$$\bar{f}(s) = \sum_{i=1}^{d} \frac{C_i}{s + a_i}, \tag{D.6}$$

where:

$$C_i = \left(\frac{N(s)}{D(s)} (s + a_i) \right) \Bigg|_{s=-a_i}.$$

The inversion of (D.6) is easily obtained:

$$f(t) = \sum_{i=1}^{d} C_i e^{a_i t}.$$

References

[DOET 1961] G. DOETSCH, *Guide to the Applications of Laplace Transforms*, Van Nostrand-Reinhold, New York.

[WIDD 1946] D. V. WIDDER, *The Laplace Transform*, Princeton University Press, Princeton, N.J.

Appendix E

Program Analysis

Table E.1 *Analysis of Structured Control Statements*

Statement type	*Average execution time,* $E[T]$	*Variance of execution time,* $Var[T]$	*LaPlace transform,* $L_T(s)$
if B **then** S_1 **else** S_2	$pE[X_1] + (1-p)E[X_2]$	$p\, Var[X_1] + (1-p)\, Var[X_2] + p(1-p)(E[X_1] - E[X_2])^2$	$pL_{X_1}(s) + (1-p)L_{X_2}(s)$
case I **of** 1: S_1; 2: S_2; **end**	$\sum_i p_I(i)E[X_i]$	$\sum_i p_I(i)\left(E[X_i^2] - (E[T])^2\right)$	$\sum_i p_I(i)L_{X_i}(s)$
repeat S **until** B	$E[X]/p$	$\frac{Var[X]}{p} + \frac{1-p}{p^2}(E[X])^2$	$\sum_{n=1}^{\infty} p_N(n)[L_X(s)]^n$
while B **do** S	$\frac{pE[X]}{1-p}$	$\frac{p\, Var[X]}{1-p} + \frac{p(E[X])^2}{(1-p)^2}$	$\sum_{n=0}^{\infty} p_N(n)[L_X(s)]^n$
for i:=1 **to** n **do** S	$nE[X]$	$n\, Var[X]$	$[L_X(s)]^n$
begin S_1; S_2; . . . ;S_n **end**	$\sum_{i=1}^{n} E[X_i]$	$\sum_{i=1}^{n} Var[X_i]$	$\prod_{i=1}^{n} L_{X_i}(s)$

Notes: The execution time of statement group S_i is denoted by X_i. It is assumed that B is a Bernoulli random variable with parameter p. Independence assumptions are made wherever required. In the expressions for **while** and **repeat** statements, the roles of "success" and "failure" are reversed since a "success" of the loop test terminates execution of the **repeat** statement **while** a "failure" of the loop test terminates the execution of the **while** statement. Furthermore, the number of iterations of the **repeat** loop is geometrically distributed while the number of iterations of the **while** loop has a modified geometric distribution.

Author Index

Abell, V., 479, 536
Abramson, N., 577
Aho, A., 52, 576
Allen, A., 575
Amstadter, B., 574
Anderson, H., 578
Apostolakis, G., 299, 308
Arnold, T., 402, 409
Arsenault, J., 578
Ash, R., 28, 29, 52, 139, 179, 318, 324, 359, 573
Avizienis, A., 169, 180

Bain, L., 574
Balbo, G., 575
Balintfy, J., 577
Bard, Y., 558, 572
Barlow, R., 122, 179, 237, 267, 286, 291, 308, 574
Baskett, F., 329, 359, 450, 465, 466
Bear, D., 577
Beaudry, M., 160, 179
Beizer, B., 264, 267, 576
Beutler, F., 369, 409, 466
Bhandiwad, R., 465, 466, 468
Bhat, U., 118, 179, 297, 308, 314, 353, 359, 393, 410, 500, 573
Blake, I., 194, 231
Bouricius, W., 251, 267
Bourne, A., 574
Brandwajn, A., 454, 466
Breiman, L., 113, 179
Breuer, M., 578
Browne, J., 410
Bruell, S., 575
Burdick, D., 577
Burke, P., 414, 422, 466
Burks, A., 107
Buzen, J., 410, 431, 465, 466, 467

Carter, W., 267
Chandy, K., 410, 424, 446-68, 572, 575, 576
Chang, A., 335
Chang, D., 329, 359
Chatfield, C., 575
Chu, K., 577
Cinlar, E., 573
Clarke, A., 242, 267, 316, 359, 573
Clary, J., 150, 180
Coffman, E., 96, 107, 306, 308, 330, 341, 359, 370, 383, 410, 438, 560, 572, 575
Conroy, T., 576
Cooper, R., 574
Courtois, P., 330, 359, 454, 465, 467, 574, 576
Cox, D., 257, 267, 573, 574

Denning, P., 107, 308, 359, 386, 410, 429, 465, 467, 572, 575
Deo, N., 351, 359
Disney, R., 267, 359, 573
Doetsch, G., 607
Donovan, J., 308
Dowdy, L., 463, 467
Drake, A., 573
Draper, N., 557, 572, 575

Elmaghraby, S., 577
Everling, W., 576

Feller, W., 319, 359, 573
Ferrari, D., 465, 467, 572, 576
Fishman, G., 505, 537, 576, 577
Foley, R., 468
Freiberger, W., 576
Freund, J., 575
Friedman, A., 578
Fuller, S., 329, 359, 383, 396, 410
Furchtgott, D., 180

Gaver, D., 244, 267, 577
Gay, F., 499, 537
Giammo, T., 429, 465, 467
Gnugnoli, G., 577
Goel, A., 390, 410
Goldstine, H., 107
Goodman, S., 7, 18, 52, 576
Gordon, G., 577
Gordon, W., 427, 428, 430, 467
Graham, G., 468
Graybill, F., 575
Green, A., 574
Gross, D., 465, 467, 574

Hamming, R., 53, 114, 179, 180
Han, Y., 53
Harris, C., 467, 574
Hartley, H., 108
Hatfield, M., 546, 572
Hecht, H., 39, 53
Hedetniemi, S., 52, 576
Hellerman, H., 576
Herzog, U., 457
Hillier, F., 577
Hoel, P., 526, 537, 573. 575
Hogarth, J., 465, 467
Hogg, R., 575
Hollander, M., 575
Hopcroft, J., 52, 576
Horowitz, E., 576
Howard, J., 467
Hunter, D., 190, 231

Iglehart, D., 577
Ingle, A., 557, 572

Jackson, J., 416, 467
Jai, A., 180

Karlin, S., 574
Keinzle, M., 465, 467
Kellev. F., 466, 467, 574

Kemeny, G., 574
Kendall, M., 473, 537, 554, 572, 575
Khintchine, A., 286, 308
Kinicki, R., 546, 572
Kleijnen, J., 505, 537, 577
Kleinrock, L., 336, 371, 376, 386, 410, 439, 446, 464, 465, 467, 574, 576, 577
Knuth, D., 74, 84, 90, 107, 140, 180, 248, 249, 267, 358, 359, 576
Kobayashi, H., 202, 231, 370, 410, 446, 454, 465, 467, 468, 576, 577
Konheim, A., 577
Kuck, D., 154, 180, 359
Kuo, F., 577

Lam, S., 452, 468
Larson, H., 68, 107
Lawrie, D., 359
Leech, R., 468
Levy, A., 308
Lieberman, G., 577
Lilliefors, H., 535, 537
Linger, R., 264, 267
Lipow, M., 574
Liu, C., 16, 45, 53
Liu, M., 570, 572
Lloyd, D., 574

Madnick, S., 281, 308
Maieli, M., 556, 572
Maisel, H., 577
Mathur, F., 169, 180, 406, 410
McCabe, J., 358, 359
Melamed, B., 417, 466, 468
Mendelson, H., 239, 267
Meyer, J., 160, 180
Miller, H., 573
Miller, I., 575
Mills, H., 267
Moder, J., 577
Mood, A., 575
Muntz, R., 466

Naylor, T., 577
Neter, J., 568, 572
Newell, G., 427, 428, 467
Ng, Y., 169, 180, 259, 267, 406, 410
Nishio, T., 578
Noether, G., 526, 528, 537, 555, 570, 572

Okumoto, K., 410
Osaki, S., 578

Palacios, F., 466
Palm, C., 286, 308
Parzen, E., 318, 320, 353, 359, 390, 410, 573, 574
Pearson, E., 78, 108
Pierce, W., 578
Pliskin, J., 267
Port, S., 537, 573, 575
Prabhu, N., 574
Proschan, F., 179, 267, 308
Putnam, L., 559, 572

Ramamoorthy, C., 51, 53, 351, 359
Ramshaw, L., 266, 267
Randell, B., 39, 53, 250, 267, 578
Reiser, M., 431, 445, 452, 468
Roberts, J., 578
Romig, H., 68, 108
Rosen, S., 479, 536
Ross, S., 295, 308, 574
Rudin, W., 113, 180

Saeks, R., 180
Sahni, S., 576
Sandler, G., 574
Sauer, C., 452, 457, 465-68, 576
Schatzoff, M., 572
Schneider, P., 267
Schwarz, M., 577
Sedgewick, R., 205
Sevcik, K., 296, 305, 308, 445, 465, 467, 468
Shedler, G., 335, 359
Shooman, M., 575
Siewiorek, D., 180, 572
Simon, B., 417, 468
Sivazlian, B., 577
Sliski, D., 572
Smith, A., 359
Smith, H., 572, 575
Smith, W., 574
Snell, J., 574
Spiegel, M., 575
Spirn, J., 330, 332, 359
Stanfel, L., 577
Stark, H., 40, 53, 268, 308, 576
Sterbenz, P., 229, 231
Stewart, W., 454, 468
Stidham, S., 369, 410
Stiffler, J., 406, 407, 408, 409, 410
Stone, C., 537, 573, 575
Stone, H., 359, 410
Stuart, A., 537, 572, 575
Syski, R., 577

Takacs, L., 574
Tanis, E., 575
Taylor, H., 574
Thomas, J., 576
Thompson, G., 267, 577
Towsley, D., 400, 410, 467
Tranter, W., 576
Tripathi, S., 308, 454, 457, 465, 468
Trivedi, K., 441, 465, 468
Tsao, N., 230, 231
Tung, C., 335, 359
Tuteur, F., 53, 308, 576

Ullman, J., 52, 576

von Neumann, J., 107

Walpole, R., 523, 537
Wasserman, W., 572
Weide, B., 230, 231
Weikel, S., 180
Wetherell, C., 107, 108
Widder, D., 607
Wilks, S., 529, 530, 532, 537, 575
Williams, A., 431, 451, 466, 468
Wirth, N., 200, 231
Witt, B., 267
Wolfe, D., 575
Wolman, E., 187, 231
Woo, L., 457
Wu, L., 180

Yechiali, U., 267
Yeh, R., 467, 572, 575

Zahorjan, J., 308, 468
Zelen, M., 575
Ziemer, R., 576
Zipf, G., 183, 231

Subject Index

Algebra of events, 7
laws, 9, 10
Analysis of algorithms (programs), 4, 230, 245, 263, 351-58, 419, 421, 470, 576, 608
concurrent, 400, 409, 466
for statement, 229, 608
hashing, 248
if statement, 245, 608
MAX, 84, 199, 205, 282
repeat statement, 262, 608
searching, 182, 183, 215
sorting, 200
structured programs, 263, 264, 608
while statement, 266, 608
with two stacks, 106, 347
Analysis of variance (ANOVA), 538, 539
nonparametric (distribution-free), 570
two-way, 566-70
Arrival:
process, 412, 420
rate, 118, 184, 412, 421, 481, 507, 543
Autocorrelation function, 276, 278, 279
Autocovariance, 504
Availability analysis, 297-301, 379, 384, 388, 398, 399
Axioms of probability, 14, 15

Balance equation, 366, 412, 430
Bayes' rule, 33, 35
continuous analogue, 235
Bernoulli process, 280
nonhomogeneous, 282
Bernoulli theorem, 226
Bernoulli trials, 41, 56, 62, 70, 83, 206, 233
generalized, 45
Birth:
pure birth process, 388-91
rate, 344, 365, 383, 388
Birth-and-death process, 411
continuous-parameter, 365
finite, 378-83
discrete-parameter, 344-51
Bottleneck, 414, 444

Cauchy-Schwartz inequality, 193, 213
Central limit theorem, 132, 227, 228
Central server model, 418, 424, 445, 463
Channel diagram, 36, 40, 312
Chapman-Kolmogorov equation, 312, 361
Chebychev inequality, 223-29, 475, 476, 484
Codes:
error-correcting, 45, 154, 262
Hamming, 45, 52
repetition, 50
Coefficient:
confidence, 484, 534
of correlation, 194, 538, 542, 553
of determination, 546-49
Coefficient *(cont.)*:
of variation, 187, 205-15
Combinatorial problems, 20, 21, 22, 51
Communication channel, 33, 51, 84, 266, 480, 502, 576
binary, 36, 45, 50, 66, 311, 315
feedback, 83
probability of transmission error, 84, 138
trinary, 40
Computer system:
analysis, 1, 183, 222, 230, 271, 470, 478, 513, 514, 525, 530, 546, 554, 557
performance analysis, 242, 271, 291, 297, 322, 342, 380, 411-68, 470, 575, 576
reliability analysis (*see* Reliability)
Concentrated distribution, 186
Conditional:
distribution, 232-35, 550
expectation, 232, 245, 539, 542
pdf, 234, 304
pmf, 105, 232, 245, 309
probability, 23, 24, 105, 232
reliability, 122, 135
transforms, 246, 261
Confidence band, 533, 534, 536
Confidence interval, 470, 484-99, 504-7, 509
distribution-free, 499
in linear regression, 549-52
Convolution:
of densities, 156
discrete, 100
theorem of generating functions, 100, 101
theorem of Laplace transforms, 603
Correlation analysis, 538, 552-56
Covariance, 192, 276
Coxian stage-type distribution, 257, 258, 397, 406, 449
CPU (Central Processing Unit), 322, 545, 558
scheduling discipline, 369
quantum-oriented, 369
service-time distribution, 131, 215, 225, 228, 266, 554
utilization, 380, 427, 433
Cumulative distribution function (CDF) (*see* Distribution function)
Cyclic queuing model, 425
for availability analysis, 381
of multiprogramming system, 379

Data structure analysis, 345-51, 358
Death:
process, 391-93
rate, 344, 365
Degree of freedom, 128, 174, 176, 177, 531, 554, 562, 568
Dependent failures, 222, 237
Dependent (response) variable (in regression), 542, 560

Design of experimenis, 560
Disk, 143, 188, 194, 341, 437, 543
  average seek time, 189, 194, 543, 554
  scheduling, 559, 560
Distribution:
  Bernoulli, 206, 477, 480, 579
  binomial, 63, 206, 496-98, 533, 579, 583-87
    normal (Laplace) approximation, 68, 137, 497
    Poisson approximation, 76, 498
    reproductive property, 101
  bivariate normal, 148, 195, 542, 553
  chi-square, 128, 489-95, 501, 562, 593
    relation to Erlang and gamma distribution, 143, 171, 172
    relation to normal distribution, 140, 171, 172, 562
    reproductive property, 173
  concentrated, 186
  continuous uniform, 138, 209, 581
  diffuse, 186
  discrete uniform, 80, 182, 205, 579
  Engset, 387
  Erlang, 126, 288, 392, 495, 580
    relation to Poisson distribution, 160
  exponential, 114, 186, 199, 210, 283, 580, 600
    relation to Poisson distribution, 118, 124, 152, 283
  F (variance ratio), 175, 176, 501-3, 564, 568, 594-97
  gamma, 126, 172, 173, 210, 478, 479, 580
  Gaussian (*see* Distribution, normal)
  geometric, 70, 207, 579
  hyperexponential, 129, 212, 215, 240, 247, 479, 580
  hypergeometric, 80
  hypoexponential, 125, 202, 211, 479, 580
    generating a random deviate, 169
    reproductive property, 162
    as a series of exponentials, 161, 162, 202, 393
  log-normal, 142
  mixture, 130, 239, 241, 245
  modified geometric, 71, 243, 368, 579
  modified negative binomial, 75
  multinomial, 95, 287
  negative binomial, 75, 101, 105, 579
  normal, 132, 202, 204, 215, 485, 562, 581, 591, 601
    reproductive property, 170
    standard, 132, 139, 591
    truncated, 135
  Poisson, 79, 84, 208, 215, 233, 266, 285, 481, 530, 579, 588-90
    relation to exponential distribution, 118, 124, 152, 283
    reproductive property, 101
    trunated, 377
  Rayleigh, 178
  t (Student's), 177, 487, 488, 505, 552, 553, 592
  Weibull, 130, 154, 214, 222, 390, 548. 552, 581
Distribution function, 58, 61, 109, 181
  continuous, 110, 111
  discrete, 58, 59, 60
  joint (or compound), 144, 539
  marginal, 144
  mixed, 113
Distribution of products, 179
Distribution of quotients (ratios), 174-79
Distribution of sums, 154, 159, 162, 170, 173, 202, 204, 260
Drum:
  paging, 437, 464, 570
  scheduling, 383

Effect:
  interaction, 566
  main, 566
Empirical distribution, 476, 477, 532, 536
Erlang's *B* formula, 378
Erlang's *C* formula, 375
Error analysis, 229, 230
Error function, 134
Estimation, 471, 472, 535
  interval, 470
  with dependent samples, 503-7
Estimator, 173, 472, 538
  biased, 479
  consistent, 476
  efficient, 475
  maximum-likelihood, 479-84, 491, 495, 531, 549, 553
  method of moments, 477-79
  unbiased, 472, 550
Event(s), 6
  algebra, 7
  cardinality, 9
  collectively exhaustive, 10
  complement, 8
  elementary, 7
  impossible (null), 7
  independence, 26
  intersection, 8
  measurable, 17, 109
  mutually exclusive (disjoint), 9, 10
  union, 8
  universal, 7
Event space, 33, 43, 56
Expectation, 181, 184. 190, 215, 540
  of functions. 188-91, 246

Factor, 560, 567
  level, 560, 567
Failure, 187
  density (pdf), 118, 121
  rate, 118, 119, 121, 184, 222, 544, 548
    constant, 122, 184
    decreasing (DFR), 123
    increasing (IFR), 123, 135
Fault:
  coverage, 251, 402-9, 499, 557

Fault (*cont.*):
detection, 164, 397, 407-9, 499, 546, 557
recovery, 250
tolerance, 578
hardware (*see* Redundancy)
software, 39, 250
Final value theorem, 298, 299, 606
Finite population model, 383-87
Floating-point numbers, distributions related to, 230, 537
Functions of normal random variables, 170

Gamma function, 127
Goodness of fit test, 529-37
chi-square, 529-32

Hazard rate (*see* Failure, rate)
Hypothesis:
acceptance of, 508
alternative, 507
concerning two means, 521-27, 562
distribution-free (nonparametric), 524-26, 000
concerning variances, 527-29
critical region of, 508
null, 507
Hypothesis testing, 470, 507, 554, 562, 564, 568
descriptive level of, 512, 523, 556
power of, 508, 517
significance level of, 508, 554, 562, 564, 568
type I error, 508-15, 562
type II error, 508-15

Independence:
of events, 26, 27, 28
mutual, 28, 29, 63
pairwise, 29
of random variables, 98
mutual, 99, 146, 204, 471, 550, 560
pairwise, 99
Independent process, 275
Independent replications, 505-7
Inference, statistical, 469
Initial value theorem, 604, 606
Integration by parts, 184, 216
Interactive system, 215, 385, 441, 445, 454, 459, 462, 463, 490, 505, 522, 565
saturation, 386
Interarrival time, 118, 269, 280

Jackson's result, 416-22
Job scheduling, 423
Joint distribution, 144, 539
Joint probability density function, 145, 304, 481, 539
Joint probability mass function, 92, 93, 414, 420, 422, 428, 430, 439, 451, 481, 544

Kolmogorov's backward equation, 362
Kolmogorov's forward equation, 362
integral form, 365, 407
Kolmogorov-Smirnov test, 532-37, 599-601
Kronecker's delta function, 353, 365

Laplace transform, 196, 210-13, 241, 247, 252-59, 261, 298, 602-7
inverse, 202, 204, 605-7
use in solving differential equations, 389-93, 400-408, 605
Law of large numbers, 3, 227
Least-squares curve fitting, 538, 543-46
higher-dimensional, 557-59
Life testing, 482-84, 490-95, 518, 519
Lifetime, 114, 118, 147, 241, 254, 256, 258, 291, 406, 494
mean (*see* MTTF)
residual, 302
Likelihood function, 480-84, 549, 553
Linear dependence, 539, 553
Little's formula, 369, 386

Machine repairman model, 383-87
Marginal density, 145, 234, 481
Marginal distribution, 144
Markov chain, 275
absorbing state, 316-29, 351, 363, 400-408
aperodic, 319
continuous-parameter, 360, 454
discrete-parameter, 309, 499, 500
finite, 313, 351
fundamental matrix, 353
homogeneous, 275, 310, 325, 361
imbedded, 336, 343
irreducible, 319, 363
limiting probabilities of, 317, 363
n-step transition probabilities, 311
parameter estimation, 499-503
periodic, 316, 319
recurrent state, 318, 364
state aggregation, 407, 408
state classification, 317
stationary distribution of, 320
steady state probabilities, 320, 364
transition probability matrix, 310, 323
transition rate, 362, 364
Markov inequality, 223
Markov process, 275
Markov property, 312
Mathematical induction, principle, 15, 16, 317
$M/D/1$, 341
$M/E_k/1$, 272, 342, 503
Mean, 181, 549 (*see also* Expectation)
population, 472, 508-21, 561, 579, 580, 581 (*see also* Population)
sample, 190, 227, 472, 545, 561 (*see also* Sample mean)
time to failure (*see* MTTF)
Mean squared error, 540
Median, 181, 499, 521, 524

Memory:
main, 153, 187, 478, 543, 556
paged, 305, 383, 435, 570
Memoryless property:
of exponential distribution, 114, 116, 304
of geometric distribution, 72, 73, 325
$M/G/\infty$, 290
$M/G/1$, 336, 453, 457
with feedback, 342
imbedded Markov chain, 336, 343
output process, 415
P-K mean value formula, 341
P-K transform equation, 340
$M/H_k/1$, 342
$M/M/\infty$, 291, 378
$M/M/m$, 373-77, 422
$M/M/1$, 243, 367, 457
with feedback, 422
with finite queue size, 372
mean reponse time, 369, 420
output process, 414
parameter estimation, 501-3
response-time distribution, 371, 414
$M/M/2$, heterogeneous servers, 393-99
Mode, 181
Model:
binomial, 64, 65
parameters, 64, 173, 203, 538, 553, 567, 579, 580, 581
Poisson, 3, 530, 536
validation, 2
Moments, 185, 205
central, 185
Mortality curve, 123
MTTF (mean time to failure), 184, 215, 216, 220, 222, 262, 300, 507, 557
estimation, 482-84, 491-95
of hybrid-NMR system, 259
of NMR system, 221
of parallel system, 217, 218, 223, 401
of series system, 217, 222
of standby redundant system, 252, 256, 258
of TMR and TMR/simplex systems, 220
MTTR (mean time to repair), 299
Multinomial:
coefficient, 45
distribution, 95
expansion, 51
Multiplication rule, 25, 33, 234
continuous analogue, 234
generalized, 33, 310
Multiprocessor system, 153, 222, 326, 375
memory interference, 326, 327, 328, 329
Multiprogramming, degree (level), 380, 425, 463, 538, 570

Network of queues, 411 (*see also* **Queuing**)
Non-birth-death processes, 393-99
Normal equations of least-squares, 544
Norton's theorem, 454, 456-59, 463, 464

Operating system, 375, 379, 383, 437, 454, 558
Order statistics, 148
of exponential distribution, 166
Outlier, 526
Overhead, 558

Paging, 306, 330, 464, 558, 570
thrashing, 383, 438
Partial fraction expansion, 201, 605-7
Pascal program, 182
Performability, 160, 461, 462
Performance evaluation, 575, 576 (*see also* Computer system, performance analysis)
Poisson process, 276, 283, 285, 389
compound, 292
decomposition of, 287, 292
nonhomogeneous, 390
superposition of, 286, 287, 305
Pollaczek-Khinchine:
mean value formula, 341
transform equation, 340
Population, 469
distribution, 526, 530, 549
parameters, 549, 553
Probability, 573
assignment, 14, 18
axioms, 14, 15
conditional, 23
measure, 17
models, 2
tree (*see* Tree diagram)
Probability density function (pdf), 110, 580
Cauchy, 177, 228, 473
exponential, 114, 183, 580
joint (or compound), 145, 481, 539
marginal, 145, 234, 481
reciprocal, 114, 179
truncated normal, 135
Probability generating function (PGF), 87, 88, 90, 91, 92, 196, 206-9, 261, 579
convolution property, 100, 101
uniqueness property, 89
Probability mass function (pmf), 57, 58, 59, 62, 69, 70, 71, 73, 75, 77, 79, 80, 579
joint (or compound), 92, 93, 481, 544
marginal, 93, 95, 414
Program behavior, paging, 305, 330, 335
renewal model, 305

Qualitative variable, 559
Quality control, 520
Queuing, 411, 574
network(s), 411
closed, 411, 423-46
mean value analysis, 445
multiple job types, 449-54
nonexponential service time distributions, 446-49, 453
normalization constant, 428, 430-35, 439, 440, 444, 452

Queuing *(cont.)*:
relative throughput, 426-29
relative utilization, 426-29
utilization, 433, 441, 444, 451
joint probability, 414
marginal probabilities, 414
non-product-form, 454-64
open, 411, 416-23
product-form solution, 413, 414, 428, 430, 439, 449, 451, 465
notation, 270
theory, 114, 269, 411, 574

Random deviate (variate), 140, 141, 142, 245
Random experiment, 3, 4
Random incidence, 302-5
Random interval, 484
Random number, 140, 272
Random process (*see* Stochastic process)
Random sample, 171, 226, 471, 485, 508, 549 552
Random sums, 260
Random variable(s), 181
absolutely continuous, 110
continuous, 109, 110, 181
dependent, 539, 542
discrete, 54, 55, 57, 62, 70, 71, 73, 75, 79, 80, 81, 83, 181, 544
expectation, 181, 188, 246
functions, 139, 141, 170, 188
independent, 146, 204, 542, 553, 560
indicator, 82, 185
mixed, 112, 196, 245
orthogonal, 195
uncorrelated, 193, 195, 542, 553
Random vectors:
continuous, 144
discrete, 92
Random walk, 283
Rank correlation coefficient, 555
Recurrent process (*see* Renewal process)
Redundancy, 30
hybrid-NMR, 221, 399
NMR, 221
parallel, 30, 217, 218, 223, 399
standby, 158, 219, 399
subsystem level, 53
system level, 53
triple modular, 44, 220, 399
Regression, 471, 538
linear, 541, 549
nonlinear, 556, 557
Relative frequency, 3, 484
Reliability, 118, 119, 251, 291, 406, 482, 494, 574, 575
conditional, 122
of detector-redundant systems, 51
of diode cofigurations, 47, 48, 51
of hybrid-NMR system, 168, 259, 393, 404
of *m*-out-of-*n* (NMR) systems, 43, 51, 66, 149, 152, 166

Reliability *(cont.)*:
of non-series-parallel systems, 39, 40
of parallel systems, 30, 51, 149, 151, 392, 400, 407
product law, 30
of series-parallel systems, 31, 32
of series systems, 29, 51, 148, 149, 222, 236, 244, 287
software, 39, 250, 390
of standby redundant systems, 158, 252-54, 392, 407
of TMR system, 44, 51, 152, 165
of TMR/simplex system, 163
Renewal counting process, 276, 293
decomposition, 296
superposition, 305
Renewal process, 275, 292-96

Sample correlation coefficient, 553
Sample mean, 171, 190, 227, 472, 475, 491, 504, 545, 561
variance of, 227, 473, 504
Sample space, 3, 6, 54
continuous, 6
countably infinite, 6
discrete, 6, 17
finite, 5, 20
partition, 10
sequential, 12, 55
uncountable, 6, 18
Sampling:
from Bernoulli distribution, 405-99
from exponential distribution, 491-95, 518, 535
from finite population, 471, 473, 475
from normal distribution, 170-78, 485-91, 521, 535, 549
from Poisson distribution, 495
Sampling distribution, 170, 470, 472, 535
Scatter diagram, 539, 540, 555
Scheduling discipline, 341, 369, 449
BCC (blocked calls cleared), 377, 387
FCFS (first-come, first-served), 271, 336, 369, 383, 412, 415, 449
PS (processor sharing), 370, 449, 454
RR (round robin), 369
SLTF (shortest latency time first), 383
SRPT (shortest remaining processing time first), 370
Searching, analysis, 182, 183, 215
Semi-Markov process, 276
Service:
rate, 184
time-distribution, 412
Sigma field, 17, 18, 109
Sign test, 521
Simulation, 537, 576, 577
Standard deviation, 185
Statistic, 471, 508, 554, 569
chi-square, 527-32

Statistic *(cont.)*:
  F (variance ratio), 528
  Kolmogorov-Smirnov, 533, 549-601
  rank-sum, 524-26, 598
  t, 523, 526
Statistics, 170, 171, 469, 575
Stochastic process, 268, 504, 573, 574
  classification, 270, 274
  continuous-parameter, 269
  continuous state, 269
  discrete-parameter, 269, 504
  discrete state, 269
  distribution, 274
  sample function (realization), 268, 271-73, 279
  state space, 268
  stationary, strictly, 274
  stationary, wide-sense, 277, 504
Storage allocation, dynamic, 187
Sum of squares, 568
  about the mean, 547
  between treatments, 564
  residual (error), 548, 564

Telephone, 577
  exchange, 188, 215, 375, 377, 387, 402, 535
  trunks, 187, 481
Theorem:
  of total expectation, 246
  of total moments, 247
  of total probability, 34, 35

Theorem *(cont.)*:
    continuous analogue, 235
  of total transforms, 247, 253, 255
Throughput, 380, 427, 444, 460, 538, 570
Time-slicing, 42
Time-to-failure (*see* Lifetime)
Traffic intensity, 368
Transform(s), 196, 197, 205, 580, 581
  characteristic function, 203, 204
  convolution property, 100, 101, 197, 603
  correspondence (uniqueness) theorem, 89, 198, 602
  Fourier, 196
  Laplace, 196, 210-13, 252-59, 261, 389-93, 400-408, 602-7
  moment generating function, 196
  moment generating property, 198, 403, 604
  z (*see* Probability generating function)
Tree diagram, 12, 13, 34, 35, 255, 282, 347

Variance (population), 185, 191, 192, 227, 522, 527
  sample, 474, 476, 491, 523
Variation:
  explained, 547, 548
  residual, 542, 560, 564
  total, 547, 548, 565
  unexplained, 548, 551
Venn diagram, 11